D0938388

Marine Ecological Processes

Second Edition

Springer
New York
Berlin
Heidelberg
Barcelona
Budapest
Hong Kong
London
Milan
Paris
Tokyo

Photo by Jan Hahn, courtesy of Woods Hole Oceanographic Institution.

Ivan Valiela

Marine Ecological Processes

Second Edition

With 325 Illustrations

Springer

Ivan Valiela
Boston University Marine Program
Marine Biological Laboratory
Woods Hole, MA 02543, USA
and
Department of Biology
Boston University
Boston, MA 02215, USA

Library of Congress Cataloging-in-Publication Data
Valiela, Ivan.
Marine ecological processes / Ivan Valiela. — 2nd ed.
p. cm.
Includes bibliographical references (p.) and index.
ISBN 0-387-94321-8
1. Marine ecology. I. Title.
QH541.5.S3V34 1995
669'.94—dc20 94/41484

Printed on acid-free paper.

Production coordinated by Chernow Editorial Services, Inc. and managed by Bill Imbornoni; manufacturing supervised by Jacqui Ashri.
Typeset by Best-set Typesetter Ltd., Hong Kong.
Printed and bound by Edwards Brothers, Inc., Ann Arbor, MI.
Printed in the United States of America.

9 8 7 6 5 4 3 2 1

ISBN 0-387-94321-8 Springer-Verlag New York Berlin Heidelberg
ISBN 3-540-94321-8 Springer-Verlag Berlin Heidelberg New York

Preface to the Second Edition

Many areas in marine ecology have remained unchanged in the decade since the first edition appeared, but other areas have seen remarkable expansion in the last 10 years. The changes have sufficiently changed the field to suggest that a revision was timely.

The text of the first edition was already hirsute with references; in the intervening decade there has been an explosion of publications, with journal titles increasing exponentially, and most journals increasing the annual numbers of pages. The increasing pace of acquisition of information and publication has made comprehensive review of any field, let alone one as eclectic as marine ecology, more daunting and less feasible as we move into the next century. The overwhelming number of publications has forced selective use of references, and I have surely failed to include many meritorious papers. To be able to finish work on this new edition within reasonable time, I have also used a few more examples from my own work than in the first edition, simply because I had them readily available, rather than find other, perhaps better illustrations in the enormously expanded literature. The proliferation of published materials is a serious problem; our students' students will live in a different, certainly more specialized, perhaps paper-less world, but they will need more effective ways to seek and synthesize more and more information on narrower topics.

In this second edition I have updated expanding knowledge in topics covered in the first edition, and have added a few topics that have assumed greater importance since the first edition. Deletions and omissions have been more difficult, although necessary. The comprehensiveness of this second edition is less than I would have wished: many interesting issues and findings had to be left out to prevent the common elephantine growth of second editions.

The plan of this second edition is to first present information at the physiological and population levels, for both producers and consumers, in Chapters 1–7. Community ecology is introduced in Chapters 8–9 by discussion of notions of how producer and consumer populations relate to each other in food webs: further structural aspects of communities are addressed in Chapters 10–12. Integration at the level of marine ecosystems is discussed in Chapters 13 and 14, which focus on carbon and nutrient dynamics, based on much material from previous chapters. Then, having provided the essentials for understanding the workings of key processes at different levels of integration in marine ecosystems, Chapter 15 shows how these processes interact in determining annual seasonal patterns. Over the last 10 years it has become evident that whatever ecologists study over the coming decades will be increasingly modified by the effects of human activities. Ecologists no longer have the luxury of confining their work to pristine environments or to basic research. The major controls of ecological system function and structure will increasingly be altered by, or in fact, be, anthropogenically-determined. To encourage understanding of this theme, and to show how understanding of fundamentals interdigitates with applied aspects, I have increased coverage of how human activities interact with "natural" processes over the long term and at large spatial scales. Chapter 16 is devoted to these issues.

I have for the most part retained the focus on processes that occur at ecological rather than evolutionary and geological time scales, and I restricted coverage of evolutionary topics, choosing to emphasize proximate rather than ultimate causes. The actual space dedicated to material is probably a function of publishing activity in the community, as well as of importance: there is perhaps too much on large animals, less than would be desirable on microbial, geochemical, and physical aspects. In retrospect, though, I believe that the contents do convey how the field "looks" in the last decade of the 20th century.

I thank the many friends, collegues, and students that made suggestions as to topics that needed inclusion in this second edition, or provided information and critical comments: Merryl Alber, Randy Alberte, Karl Banse, Cheryl Ann Butman, Walter Boynton, David Caron, Edward Carpenter, Penny Chisholm, Cabell Davis, Paul Dayton, Carlos Duarte, John Field, Ken Foreman, Anne Giblin, Mark Hay, John Hobbie, Robert Howarth, Peter Jumars, Mimi Kohl, Jim Kremer, Michael Lamontagne, Michael Landry, Brian LaPointe, James McClelland, Bruce Menge, Michael Mullin, Mark Ohman, Candice Oviatt, Michael Pace, Robert Paine, Pete Peterson, Larry Pomeroy, Jennifer Purcell, Kenneth Sebens, Sybil Seitzinger and George Somero.

Once again, the supportive staff of the Marine Biological Laboratory-Woods Hole Oceanographic Institution Library devoted time and much effort to find materials and check references. New graphics were expertly prepared by Laurie Raymond and Robin MacDonald unstingtingly went over innumerable details during preparation of the revised manuscript. Lori Soucy was invaluable in tracing elusive references.

This book would have a much narrower scope and depth of experience if I had not had the support of the National Science Foundation, the Environmental Protection Agency, the National Oceanographic and Atmospheric Administration, the Woods Hole

Oceanographic Institution's Sea Grant, and other agencies and foundations, in a wide range of research activities during the last 25 years. It therefore belatedly thank Linda Duguay, Mary Alatalo, Phil Taylor, Tom Callahan, and Joan Mitchell at NSF, Bill Thomas, Michael Crosby, and Leon Cammen at NOAA, JoAnne Sulak, Rosemary Monahan, and Ron Manfredonia at EPA, and David Ross and Judy McDowell Capuzzo at WHOI Sea Grant, for their support.

Preface to the First Edition

This text is aimed principally at the beginning graduate or advanced undergraduate student, but was written also to serve as a review and, more ambitiously, as a synthesis of the field. To achieve these purposes, several objectives were imposed on the writing. The first was, since ecologists must be the master borrowers of biology, to give the flavor of the eclectic nature of the field by providing coverage of many of the interdisciplinary topics relevant to marine ecology. The second objective was to portray marine ecology as a discipline in the course of discovery, one in which there are very few settled issues. In many instances it is only possible to discuss diverse views and point out the need for further study. The lack of clear conclusions may be frustrating to the beginning student but nonetheless reflects the current—and necessarily exciting—state of the discipline. The third purpose is to guide the reader further into topics of specialized interest by providing sufficient recent references—especially reviews. The fourth objective is to present marine ecology for what it is: a branch of ecology. Many concepts, approaches, and methods of marine ecology are inspired or derived from terrestrial and limnological antecedents. There are, in addition, instructive comparisons to be made among results obtained from marine, freshwater, and terrestrial environments. I have therefore incorporated the intellectual antecedents of particular concepts and some non-marine comparisons into the text.

The plan of this book is to present information on specifics about physiological and populational levels of biological organization in Chapters 1–7. Notions of how populations relate to each other, and their environment, are documented (Chapters 8–9) and so community ecology is introduced. This is followed by Chapters 10–12, where major aspects of the chemistry of organic matter and nutrients in marine ecosystem are developed, based on much of the material of previous chapters. Then, having provided the essentials for understanding the working of various processes in marine ecosystems, the final chapters (Chapters 12–15) dwell on how the structure of marine communities and ecosystems may be maintained over space and time.

Although I am responsible for whatever errors remain, this book has been greatly improved by many people. I have to thank my colleages in Woods Hole, especially John Teal and John Hobbie, for many years of discussion and exchange of ideas. One or more chapters were criticized by Randy Alberte, Karl Banse, Judy Capuzzo, Hal Caswell, Jon Cole, Joseph Connell, Tim Cowles, Werner Deuser, Bruce Frost, Joel Goldman, Charles Greene, Marvin Grosslein, Loren Haury, John Hobbie, Robert Howarth, Michael Landry, Cindy Lee, Jane Lubchenco, Kenneth Mann, Roger Mann, Scott Nixon, Mark Ohman, Bruce Peterson, Donald Rhoads, Amy Schoener, Sybil Seitzinger, Charles Simenstad, and Wayne Sousa.

The graduate students associated with my laboratory during the writing of this book have served as a critical sounding board, and have substantially contributed in many ways. I therefore have to acknowledge the contributions of Gary Banta, Donald Bryant, Robert Buchsbaum, Nina Caraco, Charlotte Cogswell, Joseph Costa, Cabell Davis, William Dennison, Kenneth Foreman, Rod Fujita, Anne Giblin, Jean Hartman, Brian Howes, Alan Poole, Armando Tamse, Christine Werme, David White, and John Wilson. All of them have helped in some fashion with this text, especially Kenneth Foreman and Anne Giblin, who read and criticized most of the chapters. Virginia Valiela did much of the work on the index. Sarah Allen provided technical help throughout the writing of this book, and Jean Fruci was invaluable in helping put together the final manuscript. Lastly, I want especially to thank Virginia, Luisa, Cybele, and Julia Valiela for putting up with me while I was writing this book and my parents for providing a learning environment long ago.

Most of this text was written at the Boston University Marine Program, Marine Biological Laboratory, in Woods Hole. Arthur Humes and Richard Whittaker, Directors of BUMP, were always helpful and provided the time and academic environment in which to put this book together. Dorothy Hahn, Mark Murray-Brown, and Dale Leavitt patiently converted my endless sheets of illegible scribbles into neat piles of readable word processor output. I owe thanks also to Jane Fessenden and her staff, especially Lenora Joseph and Judy Ashmore, at the MBL Library for ever-ready help. The drafting skill of Laurie Raymond is obvious in the illustrations, and her sharp eye for errors was invaluable.

A necessary and stimulating stint of writing took place during a leave of absence at the Department of Oceanography, University of Washington. Karl Banse, Bruce Frost, Mike Landry, Amy Schoener, and the oceanography graduate students were hospitable and provided stimulation for my writing.

Contents

Part IV Functioning of Marine Ecosystems

Part I
Primary Producers in Marine Environments

A coccolithophorid, *Syracosphaera syrausa*. Photo courtesy of Susumo Honjo, Words Hole Oceanographic Institution.

All that takes place biologically in marine ecosystems—75% of the earth's surface—is based on the activity of organisms that use light or chemical energy, plus carbon and other essential elements, to produce organic matter. These organisms are the primary producers, the main subject of Part I. The major kinds of producers, the processes by which production takes place, the methods by which production may be measured, and comparisons of rates of production in various environments are included in Chapter 1. The rates of primary production varies over time and space; the major factors that directly influence rates of primary production are the subject of Chapter 2. This second chapter concludes with an examination of how the physically determined movements of water mediate the effects of limiting factors and effectively establish the regional patterns of primary production.

Chapter 1
Producers and Processes Involved in Primary Production

1.1 The Kinds and Amounts of Primary Producers in the Sea

There are many kinds of marine organisms that fix inorganic carbon into organic compounds using external sources of energy. There are many thousands of producer species with widely different phyletic origins (Taylor, 1978; Raven and Richardson, 1986), including single-celled phytoplankton and bottom-dwelling benthic algae, large many-celled macroalgae or seaweeds, symbiotic producers such as corals, and higher plants such as seagrasses.

The many species of phytoplankton may contribute 95% of marine primary production (Steeman Nielsen, 1975). This importance of phytoplankton is largely due to the very large area of the earth that is covered by the open sea. In shallow water near coasts, attached single-celled algae, larger multicellular algae, and vascular plants make considerably more important contributions to marine primary production.

1.1.1 Phytoplankton

1.1.1.1 Principal Taxonomic Groups

The principal taxa of microscopically visible planktonic producers that are found over most of the world's oceans are diatoms, dinoflagellates, coccolithophorids, silicoflagellates, and blue-green and other bacteria.

Diatoms (Fig. 1-1, top row) have cell walls of silica and pectin, often with complex sculptured surfaces and range in size from 0.01 to 0.2 mm. They occur floating in the water column or attached to surfaces, singly or as chains of cells. Diatoms are major contributors to production, especially, as discussed below, in coastal waters. Diatoms occur everywhere in the sea, but are most abundant in colder, nutrient-rich waters. Division occurs by fission, and each fission is accompanied by a reduction in the size of the cell and silicious cell wall; after several divisions the cells reach some lower limit in size. At that point the

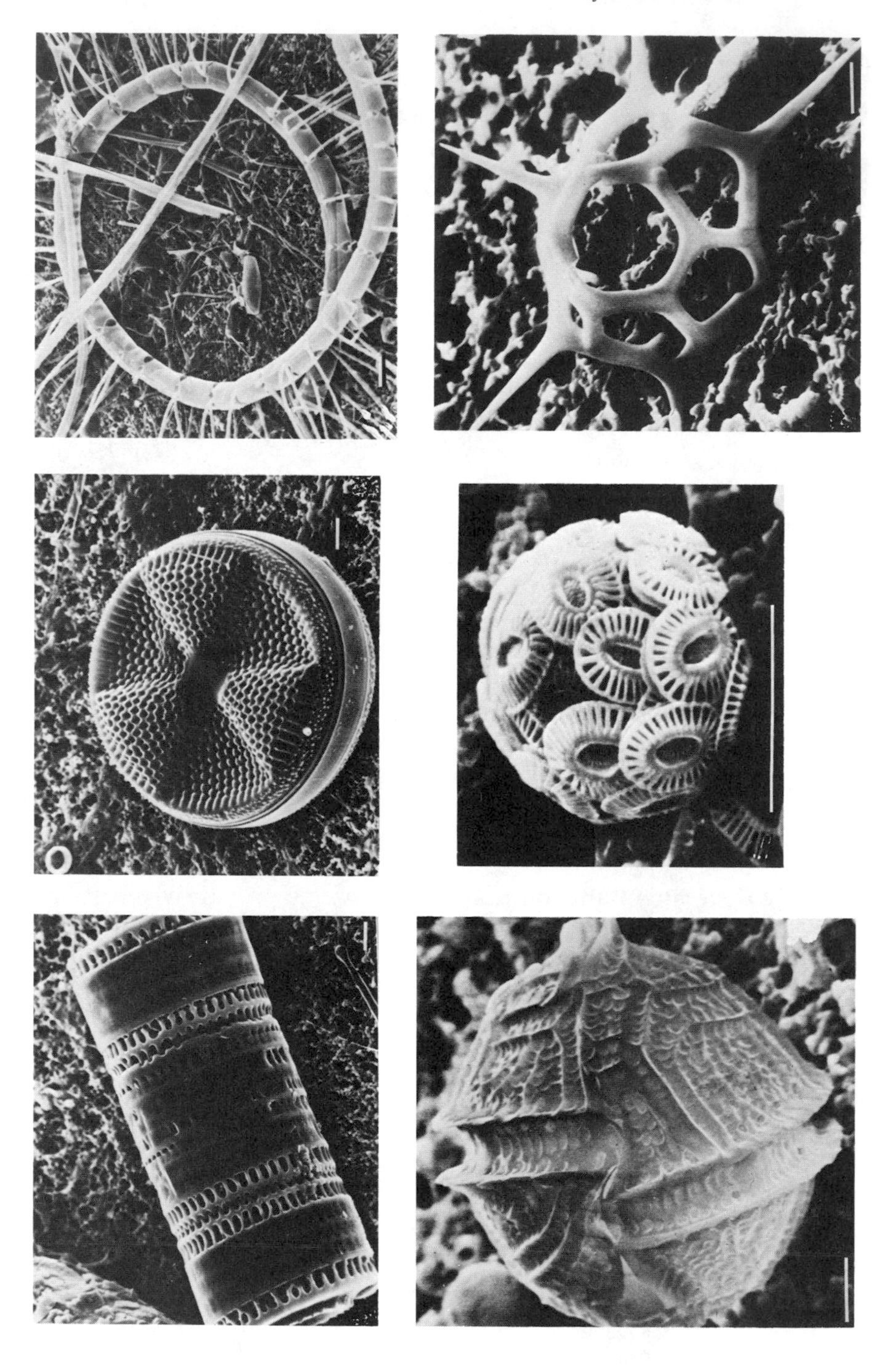

diatoms leave their silicious cell wall and the bare cells grow in size before forming a new cell wall.

Dinoflagellates (Fig. 1-1, left bottom) occur as single cells, either as naked cells or within cellulose cell walls, and range in size from 0.001 to 0.1 mm. They are planktonic, drifting with the water, although many species use flagella to move. Dinoflagellates are only second to diatoms in contributing to primary production, and are widespread over the sea. Metabolically they are very versatile: they may engage in photosynthesis, be parasitic or symbiotic, absorb dissolved organic matter, or ingest particles of organic matter. Reproduction in dinoflagellates is by division, and daughter cells grow to the size of the parent before dividing.

Coccolithophorids (Fig. 1-1, bottom center) occur as single cells and are protected by ornate calcareous plates (coccoliths) embedded in a gelatinous sheath that surrounds the cell. They are most abundant in warm, open-ocean waters, although sometimes abundant nearer shore. Coccolithophorids photosynthesize and may also absorb organic matter. Individuals may form cysts, from which spores develop to produce new individuals. Many species are flagellated.

Silicoflagellates (Fig. 1-1, right bottom) occur as single small (0.06-mm), flagellated cells. The silicoflagellates typically secrete a silicious outer skeleton. These organisms are photosynthetic, but some may consume organic matter. They are common in cold, nutrient-rich water, and reproduce by simple division.

Blue-green bacteria are photosynthetic procaryotes with cell walls made of chitin; they occur as single, very small coccoid cells or as longer filaments (Fig. 1-2, bottom). The latter may occur in bundles of many cells. Other groups of bacteria may be photosynthetic but are often restricted to waters with low oxygen content. The various types of bacteria may exist as single freefloating cells, as single cells attached to surfaces, or as long filaments composed of many cells. Most marine bacteria are motile and gram-negative, and divide by fission.

Prochlorophytes (*Prochlorococcus marinus,* described by Chisholm et al., 1992) are a recently discovered group of extremely abundant producers, barely visible by traditional microscopical means (Chisholm et al., 1988). These organisms are smaller than the coccoid bluegreens, are most abundant at the lower layers of the illuminated region of the water column, and can reach densities of over $10^5\,ml^{-1}$.

←

FIGURE 1-1. Scanning electron micrographs of planktonic unicellular producers. Diatoms—top left: *Paralia sulcata;* top center: *Actinoptychus sinarilus;* top right: *Chaetocerus carvisetus.* Dinoflagellate—bottom left: *Goniaulax polygramma.* Coccolithoporid—bottom center: *Emiliana huxleyi.* Silicoflagellate—bottom right: skeleton of *Distephanus speculum.* Scale bars = 5 μm. Photos from Honjo and Emery (1976).

FIGURE 1-2. Top: Water surface of the Sargasso Sea. The light-colored rafts are made up of the brown alga *Sargassum* and associated organisms. Bottom: Tufts of filaments of the blue-green *Oscillatoria*, common in the Sargasso Sea. The width of the filaments is about 10 μ. Photo by Natalie Pascoe.

1.1.1.2 Sizes of Producers in the Plankton

The size of cells of bacteria and algae is an important feature of marine plankton, as will be seen below, and diameters of cells vary over several orders of magnitude. Dussart (1965) and Sieburth et al. (1978) devised convenient subdivisions within the range in sizes of phytoplankton. Producers smaller than 2 μm—mainly bacteria, prochlorophytes and blue-greens—are picoplankton. Small phytoplankton between 2 and 20 μm are nanoplankton—including diatoms, coccolithophores, and silicoflagellates. Cells 20–200 μm in diameter are microplankton; diatoms and dinoflagellates are common in this size class.

1.1.1.3 Abundance of Various Groups of Phytoplankton

The taxonomic composition of the phytoplankton of a given body of seawater varies over time and space. There are many instances of localized short-term blooms, where one taxon may temporarily be very abundant, such as in the so-called red tides of estuaries and coastal waters, caused by rapid growth of red-pigmented dinoflagellates (*Gonyaulax*, *Gymnodinium*, and others).

Biomass of smaller producers in marine plankton is often greater than biomass of larger species. In the tropical Pacific, for example, the picoplankton makes up 39–63% of total chlorophyll, 27–42% is in the nanoplankton, and only 9–16% is in the microplankton fraction (Peña et al., 1990). Application of flow cytometry techniques (Olson et al., 1991) have permitted assessment of abundances of the smaller plankton (Fig. 1-3). Picoplanktonic bluegreens are often the numerically most dominant forms.

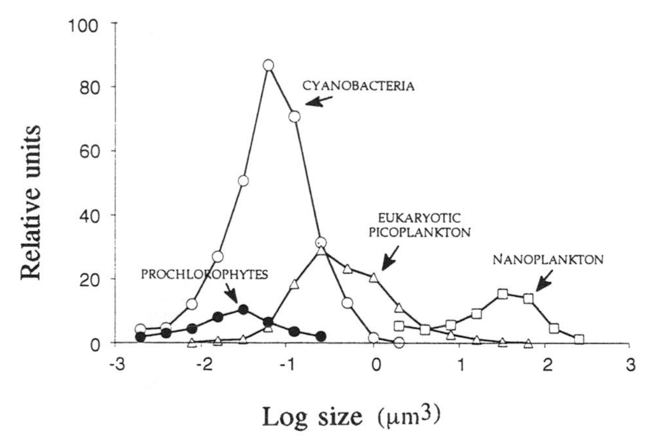

FIGURE 1-3. Relative abundance of planktonic organisms of different sizes in open waters of the Mediterranean. Data obtained by F. Jiménez-Gómez, in Rodríguez (1994).

There are, however, some notable and regular regional patterns. Phytoplankton abundance decreases along coastal to oceanic gradients. Densities of microplankton reaching several thousand algal cells per milliter are common near shore and on the continental shelf, while in the ocean it is more typical for algal microplankton to average a few to 100 cells ml^{-1} (Figs. 1-4 and 1-5).

The fertility of seawater also affects taxonomic composition. Diatoms, for example, may dominate nutrient-rich near-shore stations, while the relative abundance of coccolithophorids and dinoflagellates increases in nutrient-poor oceanic waters offshore (Fig. 1-5). The relative dominance of one or another taxon in the water column has many important consequences, since, for instance, phytoplankton dominated by coccolithophores provides only small particles for grazing animals or may produce a calcareous ooze in the sediments on the seafloor below. A diatom-dominated phytoplankton, in contrast, may furnish larger particles and a silicious ooze below. These are just two of many possible consequences.

In oceanic waters, colonial or at least aggregated phytoplankton cells can be found. Visible clumps of cells of the blue green bacterium *Oscillatoria* (also called *Trichodesmium*) are common in surface waters of the Sargasso Sea, Caribbean, and other nutrient-poor areas (Fig. 1-2, bottom). Mats of the diatom *Rhizosolenia* are frequent in the tropical Atlantic (Carpenter et al., 1977). Larger multicellular algae—referred to as macroalgae—may be locally abundant in the plankton. Species of the brown alga *Sargassum* are common in surface waters of the tropical Atlantic (Fig. 1-2, top). In certain estuaries, free-floating fronds of the brown macroalga *Ascophyllum nodosum* may be dense (Chock and Mathieson, 1976). Rafts of a red alga (*Antithamnion*) are found off the coast of Australia (Womersley and Norris, 1959).

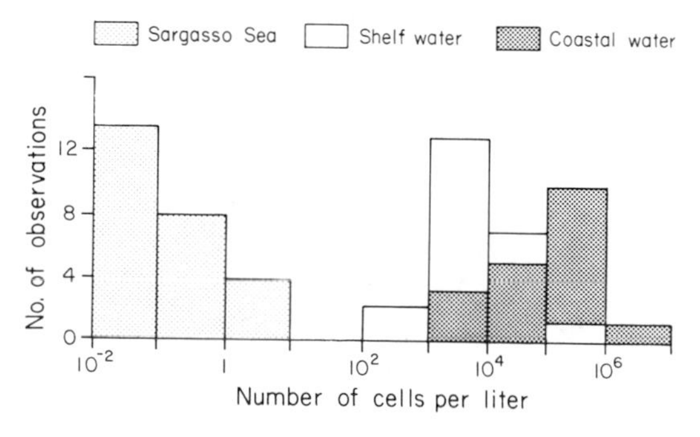

FIGURE 1-4. Abundance of algal microplankton cells in near-surface waters of three areas of the Atlantic [from data of Hulburt (1962)]. The Sargasso Sea represents truly oceanic waters, while the other two sites show the situations over the continental shelf and near-shore coastal waters. The bars are not plotted cumulatively. The Sargasso Sea is located within the northern semitropical gyre of the Atlantic Ocean.

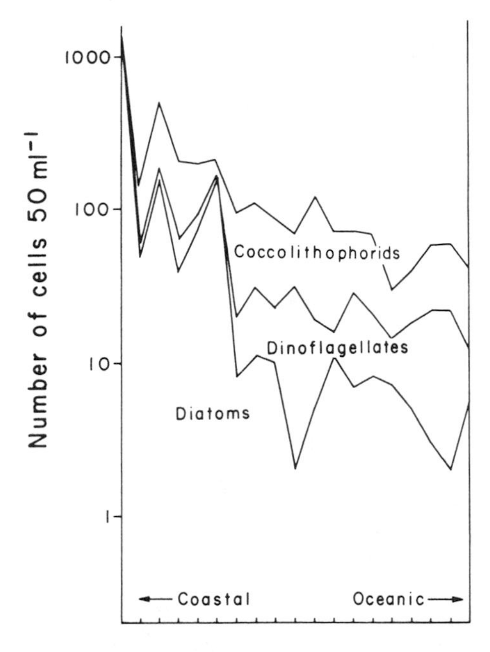

FIGURE 1-5. Change in composition of microphytoplankton in a series of stations (shown as tics) in a transect from the coast of Venezuela out to the Caribbean Sea. Note that abundance is expressed logarithmically and that the data are graphed cumulatively, so that abundance of each of the three groups is given between the lines. From data of Hulburt (1962).

1.1.2 Benthic Producers

1.1.2.1 Sand and Mud Bottoms

Mixtures of algae and vascular plants can often be found growing in or anchored to sand and mud bottoms in shallow waters. Unicellular or filamentous algae may be very abundant on the surface of sediments or rocks and on fronds and leaves of other producers. Diatoms are often dominant in algal mats common in shallow water environments (up to 40×10^6 diatoms cm^{-2}) (Grontved, 1960) and may occur as single cells or in colonial arrangements (Fig. 1-6). Green, red, and brown macroalgae, flagellates, and blue-green and photosynthetic bacteria can also be very abundant.

Marine vascular plants, with only about 60 species, are less diverse taxonomically than phytoplanktonic macroalgae taxa, but are widespread in coastal bottoms. In cold and temperate waters, stands of *Zostera* (eelgrass), *Phyllospadix* (surfgrass), *Ruppia* (widegeon grass), and other species can cover extensive subtidal and intertidal areas (Fig. 1-7). *Thalassia* (turtlegrass), *Posidonia,* and a variety of other genera occur in soft sediments of warmer waters. The taxonomy and natural history of sea grasses are reviewed in Den Hartog (1970).

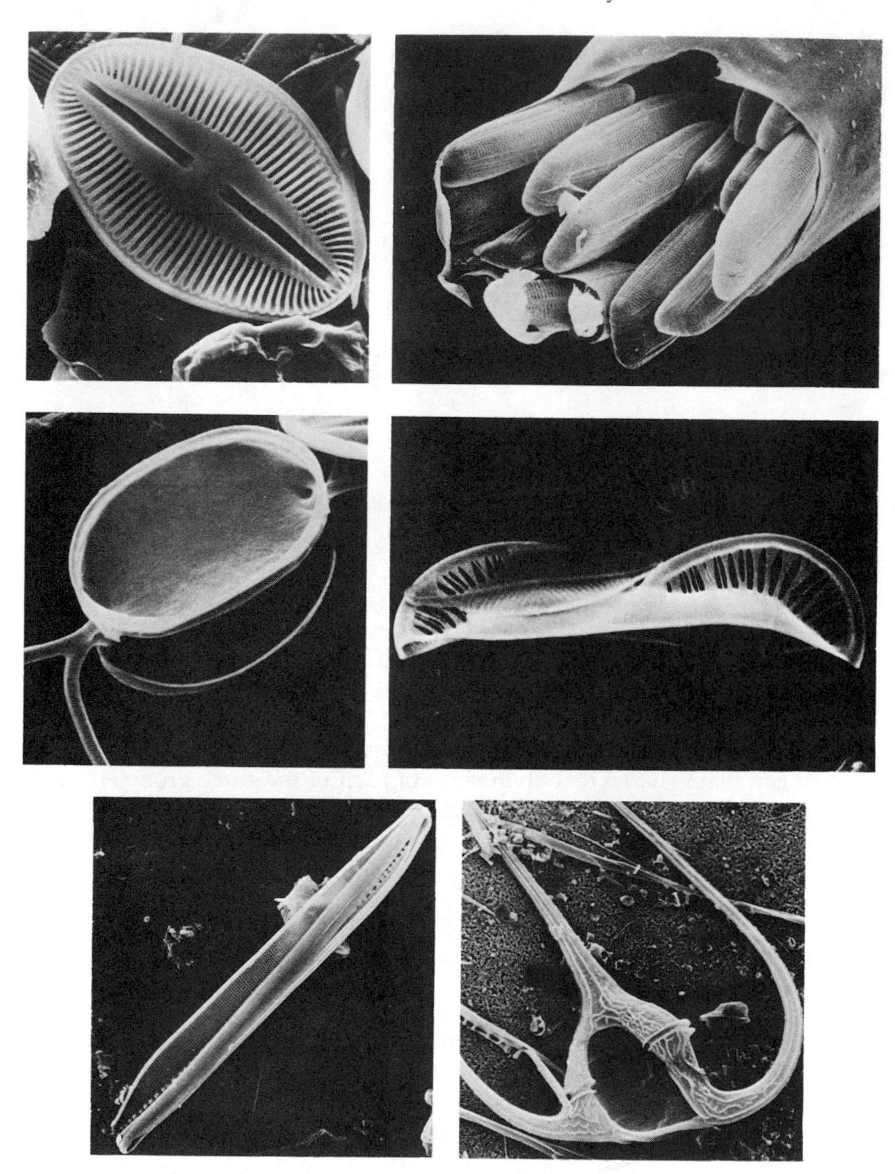

FIGURE 1-6. Scanning electron microphotographs of unicellular producers associated with the benthos: Diatoms (top left, *Navicula heufleri*, ×4300; top right, *Amphipleura elegans*, each cell about 5 μm wide; center left, *Chaetoceras*; center right, *Amphiprora*, ×5100; bottom left, *Nitzschia*, ×4300) and dinoflagellate (bottom right, *Ceratium* skeleton). Such algae are not obligate benthic dwellers and may be found in the water column; *Ceratium*, for example, may be more commonly planktonic, and skeleton or spores may sink to the benthos. *A. elegans* is a common colonial diatom found in salt marshes and lives within tubes large enough to be visible. Photos by Stjepko Golubic and his students, except that for *A. elegans* by Rod Fujita.

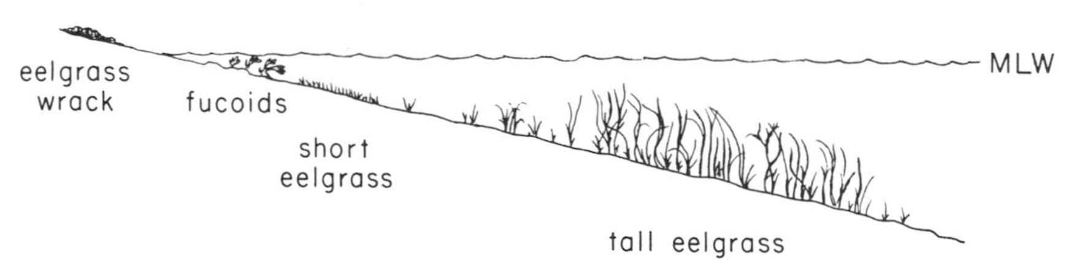

FIGURE 1-7. Top: Intertidal seagrass (*Zostera marina*) flat in Izembek Lagoon, Alaska. Photo by William Dennison. Middle: Underwater view of *Z. marina* stand. Photo by R. Phillips. Bottom: Profile of subtidal *Zostera* bed in Nova Scotia, Canada. Adapted from Harrison and Mann (1975).

Salt marshes fringe the intertidal zone of the muddy or sandy coasts of estuaries and protected shores in temperate and cold latitudes. The primary producers in salt marshes vary regionally; for example, in the coasts of the northeastern North America (Fig. 1-8), the major taxa are vascular plants in the genera *Spartina*, *Distichlis*, *Puccinellia*, *Carex*, *Juncus*, and *Salicornia*. The lowest part of the intertidal belt in these marshes is characteristically populated by attached or loose fronds of the brown algae *Ascophyllum* or *Fucus*. Tall (up to 3 m) plants of *Spartina*

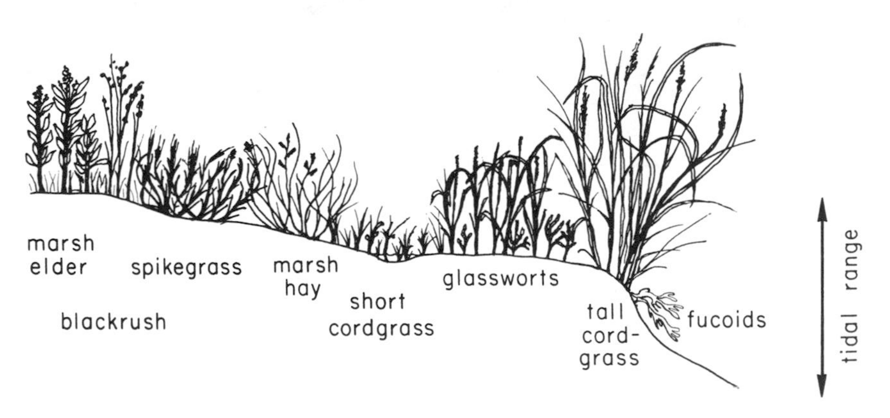

FIGURE 1-8. Salt marshes of New England, United States. Top: Tidal creek through an area dominated by the cordgrass (*Spartina alterniflora*). Bottom: Diagrammatic section through a New England salt marsh, corresponding to a transect from the upland on the left top of the photo to the tidal creek. The species included are (in the same sequence as in the diagram, left to right): *Iva frutescens*, *Juncus gerardi*, *Distichlis spicata*, *Spartina patens*, *Salicornia europaea*, *Spartina alterniflora*. The fucoids include *Fucus* and *Ascophyllum*. Drawing by Margery Taylor.

alterniflora are found slightly higher in the intertidal range. Shorter plants of *S. alterniflora* are present farther up the tidal range, while higher zones are dominated by other grasses (*Spartina patens* and *Distichlis spicata*). Elsewhere in the world, the species present and the vertical zonation of plants vary markedly; Chapman (1977) and Frey and Basan (1978) review some of these aspects of salt marsh vegetation.

Mangrove swamps (Fig. 1-9) replace salt marshes in protected coastlines with soft sediments in tropical latitudes (Chapman, 1977). In the American continent, the mangroves are mainly *Rhizophora* (red mangrove) and *Avicennia* (black mangrove). There are often well-defined horizontal belts of these species, with *Rhizophora* usually found lower in the intertidal range than *Avicennia* or other mangrove species. Mangrove swamps are best developed in Indomalasia, where canopies 10-20 m high are not unusual.

1.1.2.2 Rocky Bottoms and Reefs

On rocky coastlines we have the greatest variety of macroalgae found in the sea as well as many kinds of unicellular benthic algae. As in the case of mangrove swamps and salt marshes, there is usually a distinct zonation or producers in the subtidal and intertidal distribution of species. This zonation is extremetly variable from one geographical location to another, but within a given site the zones are clearly demarcated (Fig. 1-10). Brown algae are prominent in the various belts of vegetation, notably kelp such as *Macrocystis, Nereocystis, Laminaria,* and *Eklonia,* and fucoid rockweeds such as *Fucus* and *Ascophyllum.* There are also many red (*Chondrus, Gracilaria,* among many others) and green algae (*Enteromorpha, Ulua,* and others). Fucoids dominate the intertidal zone. The subtidal vegetation is usually dominated by kelp attached by holdfasts to the rocky bottom; the stipes and fronds of many kelp extend upward, often reaching the sea surface. These kelp may be found from the low tide mark to a depth of about 30 m. In gently sloping shores, this results in broad bands of kelp parallel to the coastline. These bands are referred to as kelp beds or forests (Fig. 1-11) and occur on rocky coasts* where seawater is relatively cold (Mann, 1973).

Coral reefs are found in shallow depths in tropical regions (Fig. 1-12). Reef-building corals are primary producers, with photosynthesis carried out by symbiotic dinoflagellates called zooxanthellae living within the tissues of the coral (cf. Chapter 5). In coral reefs and elsewhere, there are also heavily calcified red coralline algae (Dawson, 1966) that at

---

*The rocky coasts of Antarctica are uniquely devoid of kelp (Laminariales); they are replaced there by kelp-like macroalgae of the Desmarestiales, which form thickets rather than forests (Moe and Silva, 1977).

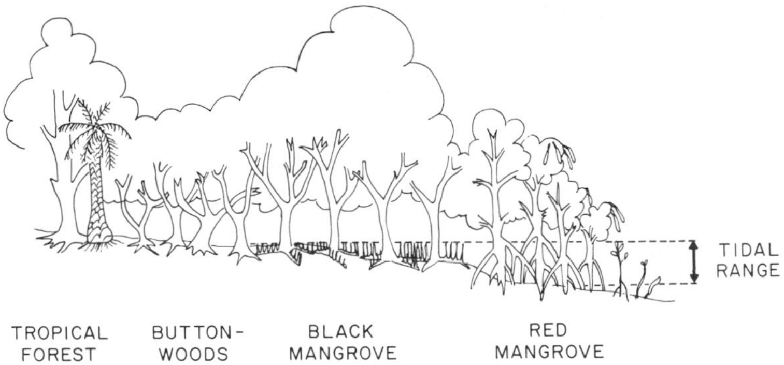

FIGURE 1-9. Mangrove swamps on the coast of Florida, United States. Top: Seaward edge of the mangrove, showing prop roots of the red mangrove (*Rhizophora mangle*). Bottom: Diagrammatic profile of a mangrove swamp, including, in sequence, *Conocarpus*, *Avicennia* (with respiratory organs, the pneumatophores, emerging vertically from the roots), and *Rhizophora*. The

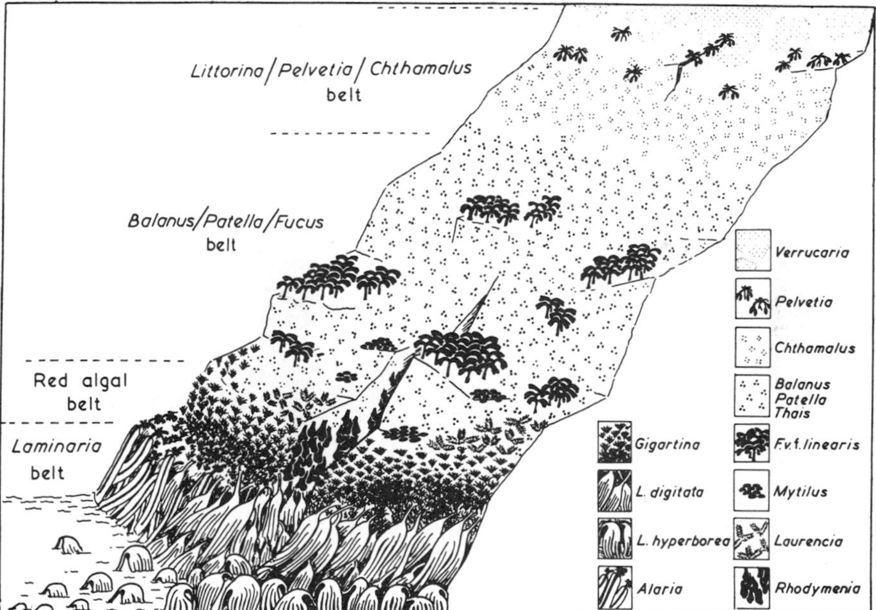

FIGURE 1-10. Top: Rocky shore on Sound of Jura, Scotland. This is a moderately exposed shore. Bottom: Diagram of rocky shore common in Scotland and Northern Ireland. Both producers and some animals are included in the diagram. (F.v.f.: *Fucus vesiculosus* form *linearis*). The tidal range is about 3–4 m. Adapted from Lewis (1964).

viviparous seedlings of the red mangrove have a long root and are shown hanging from the parent trees in the lower figure. The seedlings fall to the water and usually float off to establish new stands. The diagram has been extensively modified from Davis (1940).

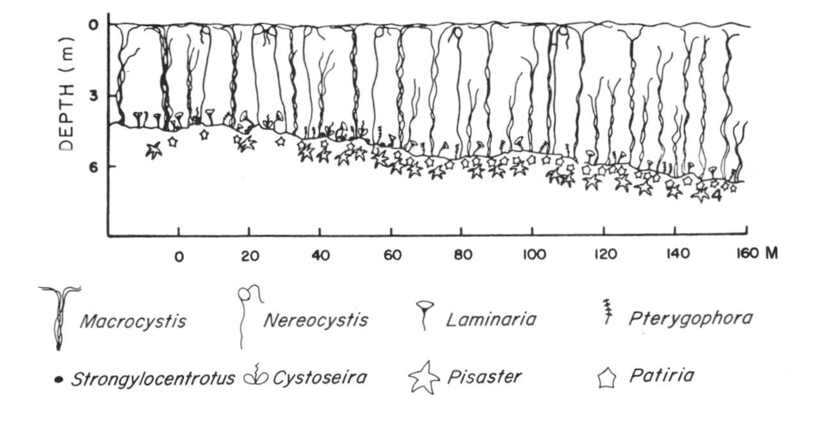
DEPTH (m)
0
3
6
0
20
40
60
80
100
120
140
160 M
Macrocystis
Nereocystis
Laminaria
Pterygophora
Strongylocentrotus
Cystoseira
Pisaster
Patiria

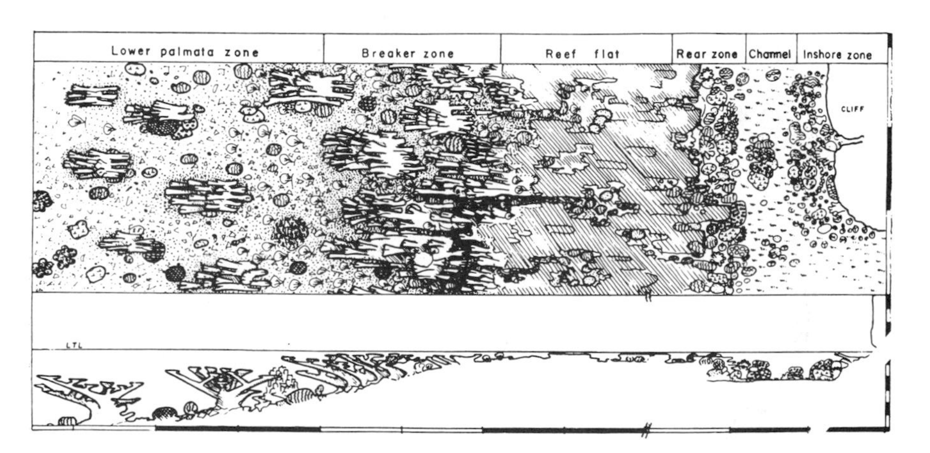

FIGURE 1-12. Coral reefs in the Caribbean. Top left: The coral *Acropora palmata* on the reef slope where waves approach from the left. Note the high degree of relief and percentage of the area that is covered by the coral. Photo by James W. Porter, Discovery Bay, Jamaica. Top right: The branching coral *A. palmata* grows over the massive head coral *Montastrea annularis*. The shading results in death of the head coral. Photo by James W. Porter, Guadeloupe. Bottom: Diagrammatic plan and lateral view of coral reef, Ocho Rios, Jamaica. Size of corals not to scale. Horizontal distances are marked in 10 m intervals; vertical distance in 1 m intervals. The different symbols refer to different species. From Gorean (1959).

←

FIGURE 1-11. Kelp forests. Top: Surface view of a *Macrocystis* stand near Oudekraal, South Africa. Photo by John Field. Middle: View of a kelp forest off Catalina Island, California, United States. The canopy is formed by *Macrocystis pyrifera* with an understory of *Eisenia arborea*. Photo by M.S. Foster. Bottom: Diagrammatic section through kelp forest off Point Santa Cruz, California. Adapted from Yellin et al. (1977).

times are responsible for a major portion of the reef-building activity. Coral reefs also show zonation, and the common occurrence of such stratification of species of coastal communities suggests that there are very powerful and ubiquitous ecological processes—to be discussed later—that determine where certain species may exist.

1.2 Production: The Formation of Organic Matter

Autotrophic organisms, including algae, some bacteria, and plants, produce energy-rich organic compounds from water and carbon dioxide. Autotrophs that use light as the energy source with which to fix carbon are termed photosynthetic organisms, while organisms that use energy stored in inorganic compounds, such as H_2S, methane, ammonia, nitrite, sulfur, hydrogen gas, or ferrous iron, are termed chemosynthetic.

1.2.1 Photosynthesis

Carbohydrates and other organic compounds are synthesized by photosynthetic autotrophs through the activity of the photosynthetic apparatus and associated biochemical pathways. The basic photosynthetic equation can be expressed as

$$CO_2 + 2H_2A \xrightarrow[\text{pigments}]{\text{light}} CH_2O + 2A + H_2O. \tag{1-1}$$

The actual details of photosynthesis are far more complicated than those that appear in Eq. (1-1) (see review by Govindjee, 1976). Photosynthesis is a multistep process comprised of two independent series of reactions. The "light" reactions take place only when light is available and depend on the capture of photons by the photosynthetic pigments. In this process an electron donor, H_2A, is split, liberating two electrons. In the case of oxygenic photosynthesis, the electron donor is water, and two H_2O are split to form an O_2 molecule and four protons. The energy in the excited electrons released in this reaction is transferred by a series of oxidation-reduction reactions involving various electron "carriers," to produce ATP and a strong reductant, $NADPH_2$.

Photosynthetic bacteria carry out anoxygenic photosynthesis in the anoxic environments in which they live. These organisms are relics of the time when the ancient atmosphere of the earth lacked oxygen; light is the source of energy but the electron donors are reduced inorganic compounds such as hydrogen sulfide or thiosulfate (in purple and green sulfur bacteria) and organic compounds (in the purple nonsulfur bacteria) (Table 1-1) (Fenchel and Blackburn, 1979).

The ATP and $NADPH_2$ provided by the light reactions is used to reduce CO_2 into complex organic molecules. These reactions are not light dependent and hence are called the "dark" reactions.

TABLE 1-1. Major Electron Donors, Acceptors, and End Products for the Three Major Types of Primary Production.[a]

	Electron donor (reductants)	Electron acceptor (oxidants)	Oxidized end products
Photosynthesis			
Oxygenic	H_2O	CO_2[b]	O_2
Anoxygenic	H_2S, H_2	CP_2[b]	S^0, SO_4^{2-}
Chemosynthesis			
Nitrifying bacteria	NO_2^-, NH_4^+, NH_2OH	O_2, NO_3^-	NO_3^-, NO_2^-
Sulfur bacteria[c]	H_2S, S^0, $S_2O_3^-$	O_2	S^0, SO_4^{2-}
Hydrogen bacteria[c]	H_2	O, SO_4	H_2O
Methane bacteria[c]	CH_4	O_2	CO_2
Iron bacteria[c]	Fe^{2+}	O_2	Fe^{3+}
Carbon monoxide bacteria[c]	CO	H_2	CH_4

[a] From Fenchel and Blackburn (1979) and Parsons et al. (1977). There are many other possible chemosynthetic reactions and end products (see Tables 13-7, 13-8).
[b] Takes place if light furnishes the large amounts of energy needed to reduce the CO_2.
[c] These groups may also live heterotrophically, using a variety of organic compounds manufactured by other organisms as sources of energy (or electron donors), and with CO_2, H_2O, or more oxidized organic compounds as the end products.

There are two major pathways of carbon fixation during the dark reactions. In the "C_3" pathway the first product of carbon fixation is a 3-carbon compound called 3-phosphoglyceric acid. Primary producers that manufacture this 3-carbon compound as an end product of carbon fixation are referred to as C_3 organisms. In the "C_4" pathway a 4-carbon compound (malic or aspartic acid) is the first end product of fixation. The malate is decarboxylated, and the CO_2 thus released is fixed through the C_3 pathway (Hatch and Black, 1970). There are important physiological differences associated with C_3 and C_4 metabolism, many of which have ecological significance (cf. Section 6.3.2). C_4 metabolism has been demonstrated in vascular plants but not in phytoplankton.

The organic compounds produced by photosynthesis may be stored or used immediately. The energy contained in the organic compounds is made available by a series of oxidative reactions. This oxidation is called dark respiration, and is the process that provides energy to sustain metabolic needs. The complete oxidation of glucose to carbon dioxide and water, for example, yields 36 ATP:*

$$C_6H_{12}O_6 + 6O_2 \rightarrow 6CO_2 + 6H_2O + 36ATP. \qquad (1\text{-}2)$$

* Respiration of course occurs in all organisms. In animals or many microbes, ingested or absorbed carbon compounds serve as the principal substrate for respiration (cf. Sections 7.3 and 13.2.2). Organisms whose metabolism is based on organic compounds fixed by autotrophs are called heterotrophs. This is in contrast to autotrophs, defined as organisms able to use CO_2 by reductive assimilation to supply carbon requirements.

Such oxidation-reduction reactions are not light dependent and in primary producers provide the sole source of energy during periods when photosynthesis is not active. Glucose is initially broken down by the process of glycolysis to produce NADH and ATP. The electrons in these reduced carriers and those produced by the Krebs cycle are donated to the electron transport chain where they are coupled to phosphorilation of ribulose.

The amount of dark respiration by producers relative to photosynthesis varies and is difficult to measure in the sea because rates are low and also because bacteria and small zooplankton also respire in the seawater where the producers are suspended. Dark respiration varies widely as a percent of gross production, but most values fall around 10% for benthic microalgae (Fig. 1-13) and for phytoplankton (Gerard, 1986). Only a small proportion of photosynthetically-fixed carbon is therefore lost via dark respiration (Kelly, 1989).

In C_3 plants there is an additional light-dependent respiration activity—referred to as photorespiration—that consumes oxygen and produces carbon dioxide (Tolbert, 1974; Harris, 1980). Under high partial pressures of oxygen, the apparent rate of photosynthesis by plant cells declines. When the intracellular concentrations of O_2 are high, one of the enzymes responsible for CO_2 fixation in the C_3 dark reactions—ribulose-bisphosphate carboxylase—can also catalyze the oxidation of a Calvin cycle intermediate, initiating a series of reactions

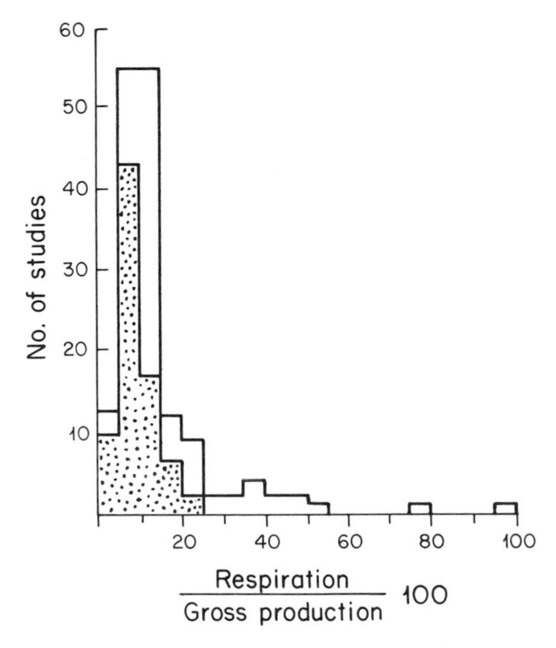

FIGURE 1-13. Frequency distribution of measurements of respiration by benthic microalgae, expressed as percent of gross production. Stippled data obtained by ^{14}C method, rest of data by oxygen method. Adapted from Foreman (1985).

that produce CO_2 and glycine as end products. Hence, when the splitting of water during the light reaction furnishes oxygen, photorespiration takes place. The function of photorespiration is not known. It has also proven difficult to ascertain the degree to which these reactions occur, and there is much contradictory information on the relative rates of photorespiration and photosynthesis by marine algae and angiosperms (Banse, 1980). Recent results showing that O_2 concentrations do not affect photosynthesis rate suggest that photorespiration is of modest importance in marine producers (Kelly, 1989).

The generalized equation for photosynthesis [Eq. (1-1)] is incomplete, because producers require a variety of inorganic nutrients to provide the building blocks for the synthesis of the many compounds present in cells. A more inclusive equation would be

$$\begin{aligned}&1{,}300\text{ kcal light energy} + 106\text{ mole } CO_2 + 90\text{ mole } H_2O\\&+ 16\text{ mole } NO_3 + 1\text{ mole } PO_4 + \text{small amounts of mineral}\\&\text{elements} = 3.3\text{ kg biomass} + 150\text{ mole } O_2 + 1{,}287\text{ kcal heat.}\end{aligned} \tag{1-3}$$

The biomass yield of Eq. (1-3) would contain on the average 13-kcal of energy, 106 g C, 180 g H, 46 g O, 16 g N, 1 g P, and 825 g of mineral ash. These values are based on the average contents of phytoplankton cells (Odum, 1971), but there is considerable variation in these elemental ratios. Although more comprehensive than Eq. (1-1), Eq. (1-3) still gives a very simplified picture of biomass production by producers. For example, phytoplankton may use ammonium rather than nitrate to satisfy their nitrogen requirement, and in the case of diatoms, substantial amounts of silica are required for growth since this element is a major component of their cell wall. Despite the intricacy of photosynthetic processes, the efficiency in capture of light energy is low. Only 1.5–2.4% of incident light energy is converted into energy stored in cells (Kelly, 1989).

1.2.2 Chemosynthesis

Chemosynthesis is carried out by bacteria that obtain energy from simple inorganic compounds (Table 1-1). The biochemical pathways of different kinds of chemosynthetic bacteria are very diverse (Fenchel and Blackburn, 1979; Parsons et al., 1977). The basic reaction of chemosynthesis is

$$nH_2A + nH_2O \rightarrow nAO + 4n[H^+ + e^-], \tag{1-4}$$

where H_2A represents a relatively reduced inorganic compound. Dehydrogenase converts this reduced compound to the oxidized end product, AO, and the reducing power gained is represented as $[H^+ + e^-]$. Some of the reducing power gained through Eq. (1-4) is devoted to energy production such as synthesis of ATP via transfer through the cytochrome system to O_2. Most chemosynthetic bacteria thus require free oxygen to function as the electron acceptor. Obligate anaerobic or facultative bacteria can use oxygen bound in nitrate or sulfate for this purpose. The rest of the reducing power obtained through Eq. (1-4) may be used to reduce NAD to $NADH_2$. The ATP and $NADH_2$ can then be used to assimilate CO_2 and make organic matter:

$$12NADH_2 + 18ATP + 6CO_2 \rightarrow C_6H_{12}O_6 + 6H_2O + 18ADP + 18P + 12NAD. \quad (1\text{-}5)$$

Chemosynthetic bacteria thus obtain both energy and reducing power from inorganic compounds (Table 1-1). Since the inorganic compounds used as electron donors are not assimilated by the bacteria, these reactions have been called dissimilative reactions, in contrast to assimilative reactions whose purpose is the uptake of elements or compounds.

The organic matter resulting from the carbon fixed as CO_2 by the chemosynthetic bacteria can be considered as truly primary production. The energy obtained from inorganic compounds and elements by chemosynthesizers was, however, fixed originally by photosynthesis in other organisms. The process of chemosynthesis therefore may not yield completely new primary production.

In this book, photosynthetic processes have been given the spotlight. This is justified in view of the extent to which photosynthesis today produces organic matter and so determines many of the dynamics of contemporary food webs. The book might have been organized differently, giving precedence to bacterial chemosynthesis. This latter approach would be justified because of the pervasive biogeochemical influence of chemosynthesis, as will be seen in Chapters 13 and 14, and from an evolutionary standpoint. Chemosynthetic transformations took place very early in the history of the earth. Photosynthesis, in fact, may be thought of as a derived process carried out by organisms that have access to light to provide the large amounts of energy needed to reduce CO_2, and have belatedly had O_2 to take up electrons to complete the process of fixing CO_2 and energy.

1.3 Measurement of Producer Biomass and Primary Production

1.3.1 *Phytoplankton*

1.3.1.1 Measurement of Algal and Microbial Biomass

The most frequently used method to estimate the standing crop of phytoplankton is to extract and measure the amount of chlorophyll *a* in seawater samples containing algae (Strickland and Parsons, 1968; Holm-Hansen and Rieman, 1978). The chlorophyll concentration can be used directly or, if desired, the algal standing crop can be obtained using a calculated ratio between weights of biomass and chlorophyll. For phytoplankton the ratio of biomass to chlorophyll averages about 62 and varies between 22 and 154. For benthic microalgae the ratio of carbon to chlorophyll in 42, with bounds of 22 and 79 (Foreman, 1985). The species composition of the sample, extent and variability of light adaptation, age, and nutritional state of the cells all affect these ratios.

More recently the content of ATP of phytoplankton has been used to estimate standing crop (Holm-Hansen and Booth, 1966). The ATP values are extrapolated to total carbon by multiplying by a derived factor of 250, although the ATP content of cells is variable due to the influence of several environmental and growth factors (Holm-Hansen, 1970; Banse, 1977; Banse and Mosher, 1980).

Hobbie et al. (1972) compared several methods of estimating microbial and algal biomass in seawater obtained from nutrient-poor and nutrient-rich stations in the Atlantic from the surface to a depth of 700 m. There were relatively high correlations among measurements of phytoplankton and organic carbon, volume of particulates, concentrations of ATP, DNA, and chlorophyll *a*. Comparisons among these correlated variables, however, are not straightforward, since, for example, not only algal but also fungal and bacterial biomass are included in the organic carbon, particulate volume, ATP, and DNA values. Organic carbon, particulate volume, and DNA may include nonliving materials. Knowledge of the ratio of algal abundance to that of other microbes or detritus would be needed to calculate phytoplankton biomass using these other variables. Hobbie et al. (1972) made an effort to obtain samples of water whose nutrient content ranged from rich to very poor. Their values of standing crops thus varied over a broad range, and they saw high correlations over that range. In any more local study, the range of concentration of organisms would likely be smaller, variation of the samples would be proportionally greater, and the correlation between methods could be considerably reduced.

All indirect methods to estimate phytoplankton biomass need to be used cautiously, but if the limitations are kept in mind and if they are applied with discrimination, these procedures can be useful.

1.3.1.2 Measurements of Production *in Situ*

The rate of primary production might be measured by determining the change over time of any of the components of Eq. (1-3), for example, the disappearance of phosphate or nitrate from the water column over a time interval. One major problem with this approach is that no natural water body is static and inhabited solely by producers. It is therefore difficult to ensure that the observed changes in nutrient concentrations are not due to movement of water in or out of the study area, or to uptake and release by bacteria or animals. Further, nutrients may be quickly released by algal cells after uptake; phosphate, for example, may be retained only a few minutes in phytoplankton cells (Pomeroy, 1960). This rapid turnover makes it difficult to determine exactly what the actual rate of disappearance of the specific nutrient actually was.

The changes in oxygen dissolved in water due to photosynthesis and respiration over night and day can be used to obtain net production* rates after correction for the diffusion of oxygen at the sea surface. This method has been used in places where there are well-defined water masses such as in fjords (Gilmartin, 1964), certain places in the Mediterranean (Minas, 1970), in coral reefs (Marsh and Smith, 1978), and elsewhere (Johnson et al., 1981).

Net production for whole communities or ecosystems can be inferred from hydrographic and chemical data (Broenkow, 1965). This method calculates what the resulting concentrations of oxygen and nitrate would have been if these substances were merely passively mixed during mixing of salty and fresher water masses. Comparison to field data provide a quantification of actual changes in oxygen and nitrate, which are then interpreted as net production using stoichiometric equations (Eq. 1-3 or those in Table 3-8). These calculations require much data on conditions in the field, and such data sets have been rare. This approach has been applied to estimate net ecosystem production in upwelling areas (Minas et al., 1986), shelf waters (Falkowski et al., 1993), in coastal estuaries (D'Avanzo et al., 1995), and estuaries and the oceans in general (Smith, 1991). The results of this approach furnish estimates of net ecosystem production (also referred to, less accurately, as net community production in some papers). Net ecosystem production includes gross production by *all producers* in the system, minus respiration by *all organisms* in the system. Net ecosystem production is therefore not the same as net production, since the latter is the sum of production by all producers in the system under study minus the respiration by the producers themselves. Implications of the two concepts are discussed below.

Carbon dioxide dissolved in water is the most frequent carbon source for marine photosynthesis and diffuses much more slowly than oxygen in seawater. The lowered concentrations due to consumption by photosynthesis thus tend to be measurable over periods of hours, so that CO_2 concentrations provide a relatively stable tracer of photosynthetic activity. Production of CO_2 by respiration of microorganisms and animals is unavoidably included in results obtained by this method (Johnson et al., 1981).

The rate of appearance of new algal biomass over time is a direct estimate of net primary production. To apply this approach to the sea,

---

*Gross production minus respiration is referred to as net production.

the losses of cells through grazing and sinking need to be measured. These are difficult to do, so this method has not been widely used to measure production.

Yet another *in situ* method involves estimating the amount of chlorophyll *a* and relating this to net production using a ratio of production rate to chlorophyll content. This ratio, called the coefficient of assimilation or the assimilation number, has been taken to be 3–4, but, averages 6.4 and ranges 0.2–20 mg C mg chlorophyll a^{-1} h^{-1} for microphytoplankton (Kelly, 1989). The assimilation numbers for picoplanktonic producers have a similar range, 0.03–15 mg C mg chlorophyll a^{-1} h^{-1} (Stockner and Antia, 1986), even though the cyanobacteria that dominate the picoplankton possess substantial amounts of other light-harvesting pigments besides chlorophyll *a*. Estimates of rates of production can be calculated using equations that consider the effect of light attenuation in the water column (Ryther and Yentsch, 1957). Although subject to all the problems of conversions using ratios, this procedure is attractive because of the simplicity of the chlorophyll *a* determination (Strickland and Parsons, 1968). The development of ideas based on these equations has resulted in the development of mathematical models to predict primary production rates under a variety of conditions (Hall and Moll, 1975; Kiefer, 1980).

1.3.1.3 Methods Using Containers

1.3.1.3.1 Light and Dark Bottle Oxygen Technique. The methods that have received the most attention involve placing seawater in small containers, some of which are transparent and some dark, and measuring the amount of oxygen dissolved in the water before and after some suitable time period has elapsed (Gaarden and Gran, 1927). Sensitive methods to measure oxygen make it now possible to measure production rates of oceanic phytoplankton using changes in oxygen (Tijssen, 1979; Friederich et al., 1991).

The increase in dissolved oxygen in the light bottle is a measure of photosynthetic activity (P_a) by the algae minus the respiration by algae (R_a), bacteria (R_b), and whatever zooplankton (R_z) were present in the sample; it is not net photosynthesis ($P_n = P_a - R_a$), since it includes R_b and R_z. The decrease in oxygen concentration in the dark bottle is a measure of respiration by all the aerobic organisms present:

$$\text{change in } O_2 \text{ in light bottle} = P_a - R_a - R_b - R_z, \tag{1-6}$$

$$\text{change in } O_2 \text{ in dark bottle} = R_a + R_b + R_z. \tag{1-7}$$

By adding the change in O_2 in the light bottle and the change in O_2 in the dark bottle we get P_g, the gross photosynthesis by the algae over the period of incubation.

The rates obtained from such light and dark bottle measurements can, if desired, be converted to amounts of CO_2 assimilated (Strickland and Austin, 1960). The photosynthetic ratios, i.e., the relation between O_2 released and CO_2 assimilated, vary. For example, where the CO_2 is devoted for synthesis of carbohydrates the ratio 1, while for lipid synthesis it is 1.4. The ratio is 1.05 for protein synthesis if the source of nitrogen is ammonium, and 1.6 or higher if nitrate is the source of nitrogen used to make protein (Williams et al., 1979). This difference reflects the fact that the nitrate has to be reduced intracellularly to ammonium prior to incorporation into amino acids. For fieldwork a value of about 1.2–1.4 is usually applied, but should be higher if phytoplankton are using nitrate as the nitrogen source.

1.3.1.3.2 Carbon-14 Method. The most common way to measure algal production is to measure the rate of uptake of ^{14}C, a radioactive isotope of carbon. A known, small amount of ^{14}C, usually in the form of dissolved $NaH^{14}CO_3$, is added to seawater samples contained in transparent and opaque 100- to 300-ml bottles. The seawater is incubated for a period that may vary from 0.5 to 24 hr, after which the sample is filtered, cells are killed, and the quantity of ^{14}C assimilated is determined by using a liquid scintillation counter to measure the beta-radiation emitted by the ^{14}C incorporated into the cells. The carbon uptake can be calculated as carbon uptake = ^{14}C incorporated into organic form/total ^{14}C added $\times$ available inorganic carbon $\times$ 1.05.

The available inorganic carbon has to be measured by chemical means and the value of 1.05 is a factor to account for the fact that $^{14}CO_2$ is taken up at a somewhat slower rate than the lighter $^{12}CO_2$. The rates of carbon uptake are subtracted from those of the light bottle if uptake in the light is the desired measurement. Strickland and Parsons (1968) provide further details of the method and Peterson (1980) and Carpenter and Lively (1980) evaluate the present status of the method.

The rate at which the ^{14}C is taken up by cells should be a measure of gross production. Comparisons of oxygen and ^{14}C methods yield relatively similar results (Williams et al., 1983; Bender et al., 1987; Grande et al., 1991), and suggest that both estimate gross primary production. To account for respiration of ^{14}C by cells during measurements, and hence estimate net production, uptake in the opaque bottles could be subtracted from the rates obtained in the light bottles. There is, however, still uncertainty about what is measured by the opaque bottles (Banse, 1993). In the opaque bottles there is respiration of ^{14}C by bacteria and animals, respiration by bacteria and phytoplankton in the dark may not be the same as in the light (Mortain-Bertrand et al., 1988), and there can be dark ^{14}C fixation by phytoplankton. Dark fixation of carbon is probably relatively more important where photosynthesis is limited by low light (Li et al., 1993) or where production is less than $1 \mu g C^{-1} h^{-1}$ (Prakash et al., 1991). It is unclear how these various factors balance out; it may be that they are not a problem overall.

The ^{14}C method has been the most widely used of all the available procedures, and knowledge of the rates of production by phytoplankton over the oceans is basically derived from ^{14}C measurements. At present a critical reassessment of the ^{14}C method is taking place, due to various independent observations that suggest that this method underestimates primary production.* In nutrient-poor oceanic waters, it might be that ^{14}C somewhat underestimates production, but it is not yet possible to say positively whether the reduction is significant.

1.3.2 Benthic Microalgae

The production of algal cells growing attached to sediments is best measured by adapting the O_2 or ^{14}C method. There are various procedures available (Gargas, 1970; Hargrave, 1969; Pomeroy, 1959; Van Raalte et al., 1974; Leach, 1970; Skauen et al., 1971).

1.3.3 Production by Macroalgae and Vascular Plants

Various methods exist to estimate the production of macroalgae and vascular plants (Wiegert and Evans, 1964; Williams and Murdoch, 1972; Hopkinson et al., 1980). The most used approach involves repeated harvest of standing crop and calculation of the growth increment during each time period. This procedure has often been used in salt marshes, but care must be taken to include the parts of the plant that die during the growth period (Valiela et al., 1975). Such methods provide estimates of net above-ground production; Hopkinson et al. (1980) compare various alternative methods applied to salt marsh production.

Leaf or frond marking methods have also been used. Growth and production in kelp have been measured by punching small holes at the

---

*These observations include (a) measurements of changes in O_2, CO_2, and dissolved organic carbon that suggest that CO_2 consumed or O_2 released exceeds primary production as measured by ^{14}C [Shulenberger and Reid (1981) and studies summarized in Johnson et al. (1981)]; (b) measurements of formation of particulate carbon exceeding rates of ^{14}C primary production (Postma and Rommets, 1979); (c) estimates of rates of O_2 formation in water masses whose age has been established by tritium and ^{3}He dating (Jenkins, 1977), where estimates of O_2 formed exceed rates obtained by ^{14}C: (d) measurement of high primary production (11–56% of total water column production) associated with fragile aggregates that are seldom sampled by usual water sampling methods (GA. Knauer, personal communication). All of these observations need confirmation and evaluation.

base of a frond and measuring the rate at which the hole moves away from the stipe as the frond grows (Mann, 1973; Chapman and Craigie, 1977). Production in seagrasses (Sand-Jensen, 1975) and mangroves (Onuf et al., 1977) has also been measured using leaf-marking techniques. Rates of elongation or of addition of tissues are then converted into production by using approximate conversion factors.

In marshes and seagrass beds, much of the plant production is translocated into sediments. Estimation of such below-ground "production" by harvest techniques is difficult and laborious (Valiela et al., 1976), since the living tissues are difficult to separate from the sediments.

Measurements of gas exchanges by a macroalgae or vascular plant contained in a bell jar or gas-proof tent is an accurate method to obtain total production (Kanwisher, 1966). In water, the changes in O_2 can be used to measure production. In air or water, a gas chromatograph or an infrared CO_2 analyzer can be used to estimate consumption of CO_2. Measurements under light and dark conditions can allow the calculation of gross and net production (Teal and Kanwisher, 1961; Lugo et al., 1975). Estimates of net production obtained using CO_2 exchange measurements agree well with estimates obtained by harvest at short intervals, at least in salt marshes (Valiela et al., 1976).

Estimates of potential production have been obtained based on the amount and interception of light and pigments in mangrove forests (Bunt et al., 1979) and in intertidal macroalgae (Brinkhuis, 1977). Such an approach may provide a fairly easy way to obtain production estimates. These methods are useful where light rather than nutrients is the dominant limiting factor. Careful measurements of local conditions are needed to make these methods reliable. Comparisons with other better established procedures should be carried out before extensive use of light-based calculations is made.

In certain situations where water traverses a stand of subtidal vegetation in a well-defined direction, as in some kelp forests, it is possible to estimate production by macroalgae by measuring the changes in O_2 and CO_2 in the seawater entering the stand compared to that leaving the stand (McFarland and Prescott, 1959). Such measurements must be interpreted carefully since microbial and animal respiration may be important.

1.3.4 Production by Chemosynthetic Bacteria

Chemosynthetic activity can be roughly measured by ^{14}C uptake measurements done in the dark. Uptake of CO_2 by photosynthetic organisms may take place in the dark, so these ^{14}C uptake values are not only due to chemosynthesis. Dark uptake of CO_2 by phytoplankton may reach at most 5% of the light uptake, so the error in measurement of chemosynthetic activity may be small.

1.4 Production Rates by Marine Primary Producers

1.4.1 *Rates of Production in Photosynthetic Producers*

Many measurements of net primary production in various marine environments have been made using the various methods sketched above. At any one site and at any one time the rates of production may vary greatly; in some environments there may be short-lived bursts of intense production. To make some sort of comparison among various marine producers, it is convenient therefore to express production in terms of amounts of carbon fixed per unit area per year.

The ranges of reported values of annual production by marine producers (Fig. 1-14, top) show that attached macroalgae, corals, and vascular plants are generally more productive than single-celled algae. Although attached microalgae can be very productive, phytoplanktonic

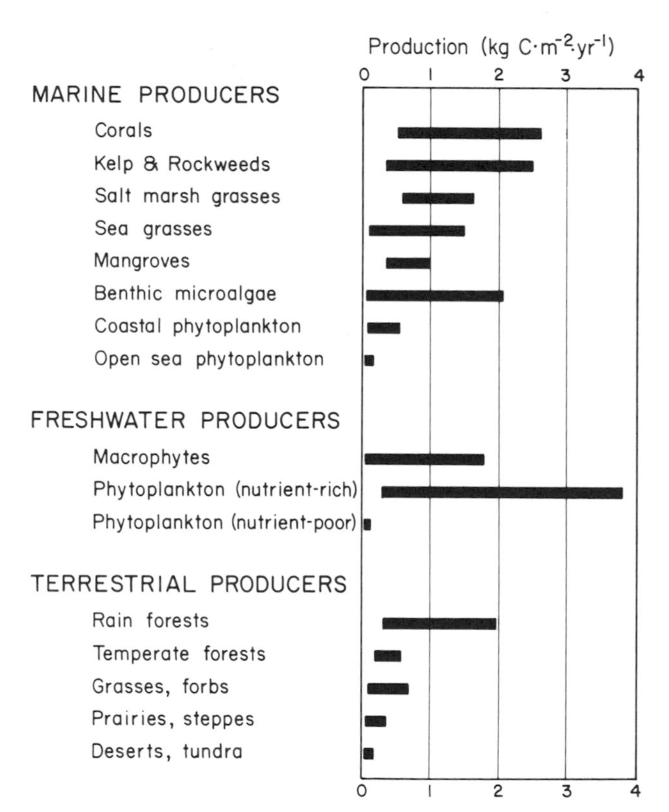

FIGURE 1-14. Annual net production rates in a variety of producers within marine, freshwater, and terrestrial environments. Modified from many sources and compilations by Margalef (1974), Bunt (1975), Wetzel (1975), Van Raalte (1975), Sournia (1977), and De Vooys (1979).

production is considerably lower than production by macroalgae. Production by oceanic phytoplankton is particularly low.

Primary production in water columns is carried out by taxa of different sizes, and often the smaller forms are most important. Production by picoplankton is quite variable, but can be significant: picoplankton carry out up to 90% and 60% of primary production in marine and freshwaters, respectively (Stockner and Antia, 1986). Picoplankton are responsible for 30–80% of total primary production in several different environments (Iturriaga and Mitchell, 1986). In the Celtic Sea, pico-, nano-, and microplankton carry out 23, 42, and 37% of annual production, respectively (Joint et al., 1986).

Reported values of production by diverse marine organisms lie within the same ranges measured in freshwater (Fig. 1-14, middle) and terrestrial (Fig. 1-14, bottom) environments. Production in rainforests is comparable to that of corals and macroalgae, while production in deserts is similar to oceanic production.

Average values of production by whole assemblages of producers in marine and terrestrial habitats are summarized in Table 1-2. The average annual production rates of estuaries, algal beds, and coral reefs are about four times those of shelf or upwelled water and more than an order of magnitude greater than those of the open ocean (Table 1-2). The open sea, however, comprises over 90% of the area of the world oceans, so that even though oceanic production rates are much lower than those of other marine environments, the area of open ocean is so large compared to coastal waters (Table 1-2) that over 60% of global marine production takes place in oceanic waters.

TABLE 1-2. Average Primary Production and Biomass, Turnover Time, and Chlorophyll in Major Environments.[a]

	Area (10^6 km^2)	Net production (g m^{-2} yr^{-1})	Biomass (kg m^{-2})	Turnover time (P/B, yr^{-1})	Chlorophyll (g m^{-2})
Open ocean	332	125	0.003	42	0.03
Upwellings	0.4	500	0.02	25	0.3
Continental shelf	27	300	0.001	300	0.2
Algal beds and reef	0.6	2,500	2	1.3	2
Estuaries (excl. march)	1.4	1,500	1	1.5	1
Total marine	361	155	0.01		0.05
Terrestrial environments	145	737	12	0.061	1.54
Swamp and marsh	2	3,000	15	0.2	3
Lakes and streams	2	400	0.02	20	0.2
Total continental	149	782	12.2	0.064	1.5

[a] The more productive habitats can have considerable more biomass and production than the average shown in this table (cf. Table 15-1 in Whittaker and Likens, 1975). Production and biomass expressed as dry matter. Terrestrial environments include all land habitats. Adapted from Whittaker and Likens (1975).

There is, however, more chlorophyll per m^2 on land communities, but marine chlorophyll is more active. On a per area basis, oceanic net production is about 20% that found on land (roughly 60 vs. 350 g C m^{-1} y^{-1}). On a per unit chlorophyll basis, marine production is six times that on land (0.3 vs. 0.05 mg C mg chlorophyll a^{-1} h^{-1}, assuming that dry organic matter is 45% C, and that there are 4380 h daylight y^{-1} (Kelly, 1989). These differences exist even though on a weight basis there are roughly the same amounts of chlorophyll in terrestrial plants (0.8 and 1.3 for shade- and sun-adapted land plants, assuming dry weight to be 1/6 of fresh weight) and marine algae (average 1%, range 0.4–3.7%) (Kelly, 1989).

The differences in standing crops of producers between terrestrial and ocean environments are far larger than the differences in production rates. The average producer biomass in the open ocean (Table 1-2) is several orders of magnitude lower than the average biomass per unit area on land and one order of magnitude smaller than those of freshwater environments. Even the much more productive coastal shelf waters have less biomass than most freshwater systems. Producer biomass in estuaries and marine algal beds is higher but does not match those of freshwater swamps and marshes. This contrast means that standing crop of marine plants turns over faster than that of terrestrial environments (Table 1-2). Terrestrial plants, because of the amount of nongrowing tissues needed for structural support, show very slow turnover times, as do plants in freshwater swamps and marshes, and to some extent estuarine producers and macroalgal beds in the sea. The most striking comparison, though, is that between phytoplankton and multicellular producers. Whether in freshwater or seawater, phytoplankton turn over at least 20–40 times a year, probably faster in oligotrophic waters. Rates of turnover vary seasonally, but there are times of year in which phytoplankton biomass turns over daily. This is in sharp contrast to the 1-2 times a year that most vascular plants achieve.

1.4.2 Production by Bacteria

1.4.2.1 Anoxygenic Photosynthesis

Anoxygenic photosynthesis demands anaerobic conditions and light. In the sea this combination is rare; for example, in the Black Sea the anaerobic layers lie below the depth to which light penetrates (Fig. 1-15, right), so bacterial photosynthesis is not likely to be high.

In seasonally stratified lakes with an anaerobic lower layer, production by anoxygenic photosynthetic bacteria may be 20–85% of the total annual production (Takahashi and Ichimura, 1968; Cohen et al., 1977a,b). In Solar Lake, a stratified coastal pond with very high salinities,

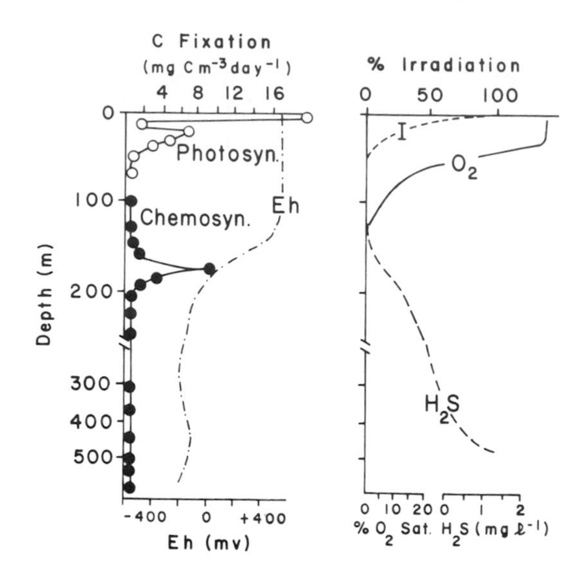

FIGURE 1-15. Profile of producer activity and certain physicochemical properties of the Black Sea. Eh is the redox potential; I is the % of incident radiation. Adapted from Sorokin (1964).

anoxygenic photosynthesis is 91% of total production (Cohen et al., 1977b).

In shallow coastal mud and sandflats, light penetrates the thin oxic surface layers of sediment, and some photosynthesis by anaerobic bacteria takes place. A thin plate of photosynthetic purple sulfur bacteria is often present just below the anoxic boundary. In such situations high rates of chemosynthetic production may take place.

Given the right circumstances, anoxygenic photosynthesis can therefore be locally important, and can take place at high rates. In general, it is not important for aerobic water columns.

1.4.2.2 Chemosynthesis

Chemosynthesis occurs in environments where anaerobic and aerobic conditions exist side by side, since oxidized reactants as well as anaerobic conditions are needed (cf. Table 1-1). Where such conditions are available, chemosynthetic activity can be substantial: in a small bay in Japan chemosynthetic rates exceed photosynthetic rates during spring, and amount to about 20% of annual carbon fixation (Seki, 1968).

The boundary between aerobic and anaerobic conditions is the most favorable site for chemosynthetic bacteria, and there is often a layer or plate of these organisms in such boundaries. The presence of an aerobic-anaerobic boundary depends on the balance of conditions that supply or deplete oxygen in the adjoining habitats.

The presence of oxygen is a function of renewal of water-borne O_2, and of consumption of O_2 by organisms. Oxygen-poor water may occur where the circulation of water is impaired. In the Black Sea and in fjords, for example, fairly shallow sills at the exits of these bodies of water prevent mixing below the depth of the sills. Any newly diffused or advected* oxygen that reaches the lower depths is quickly consumed by organisms. In the water column of the central Black Sea, aerobic photosynthesis due to algae takes place near the surface (Fig. 1-15, left). Below about 150 m, reducing conditions are found (note the lowered Eh at 150–200 m), dissolved O_2 disappears, and H_2S is present (Fig. 1-15, right) (cf. Section 14.3). At or below the interphase between the oxidized and the reduced zones of the water column, chemosynthetic bacteria show considerable carbon-fixing activity (Fig. 1-15, left).

Chemosynthetic activity is also important in anaerobic sediments. Where primary production is high, large amounts of organic matter are released into and onto sediments. The decay of these organic substrates consumes oxygen in excess of renewal by water exchange (cf. Section 13.2) and leads to anoxic conditions. Such a situation is common in sediments below many coastal waters, including especially regions of upwelling, mudflats, salt marshes, and mangrove swamps. The delivery of organic matter to deeper sediments is usually considerably less than that to coastal sediments (cf. Section 13.2.4.1), so that deeper sediments are seldom anoxic; chemosynthesis is therefore more active in shallower sediments.

A situation where chemosynthetic bacteria are responsible for all the production—and carry out truly primary production—is in the vicinity of hydrothermal vents in the deep sea, where the seafloor is spreading (Fig. 1-16). Geothermally reduced sulfur compounds are emitted from such vents, and chemosynthetic bacteria use the energy in these reduced sulfur compounds for the reduction of carbon dioxide to organic matter (Jannasch and Wirsen, 1979; Karl et al., 1980). The high chemosynthetic bacterial production near the vents serves as the food base for a complex community of heretofore unknown organisms (Lonsdale, 1977).

* Advection and turbulent (or eddy) diffusion can both be agents of transport of substances. Advection refers to the mass movement of a parcel of water. Diffusion in seawater tends to occur primarily due to small turbulent eddies rather than to molecular diffusion. In sediments molecular diffusion is more important.

FIGURE 1-16. Views of the communities of organisms—many of them new to science—associated with vents on the Galapagos rift zones. Photos taken from the research vessel *Alvin*. Top: View of water flowing through some of the vents. Some of the prominent animals are the large bivalves (up to 30 cm in length), *Calyptogena magnifica*, a new genus; the pogonophoran worms (left bottom), *Riftia pachypttila*, a new family; and the white crab *Dythograea thermydron*, a member of a new family. Photo by H. Sanders, courtesy Woods Hole Oceanographic Institution. Bottom: View of area at 2,800 m populated by *R. pachypttila*, whose tubes can reach up to 1.6 m in length, and limpets, crabs, worms, and an unidentified fish. Photo by John Edmond, courtesy Woods Hole Oceanographic Institution.

It should be emphasized again that the oxygen content in water of most marine environments is adequate for aerobic producers and therefore algal photosynthesis is the principal means by which carbon dioxide is converted into organic matter in marine environments. It is only in certain situations where the physicochemical conditions are such that bacterial photosynthesis or chemosynthesis may make more than trivial contributions to the production of organic matter.

Chapter 2
Factors Affecting Primary Production

The rate of primary production of a parcel of a marine environment depends on light and on the chemical conditions provided by the physics of water masses. There is thus a complex coupling of physics, chemistry, and biology in marine environments. In this chapter we start in reductionist fashion by examining how light, nutrients, and temperature affect primary producers. We focus on two major variables—light and nutrients—and their role in determining primary production. Grazing and sinking, the other major factors affecting producers, are discussed in Chapters 5 and 13. The interactions of controlling factors that determine seasonal cycles are discussed in Chapter 15. We end this chapter with some examples of how the motion of water masses affects production in the ocean through its effect on availability of light, nutrients, and temperature.

2.1 Light

2.1.1 Amount of Light

2.1.1.1 Availability of Light

The amount of solar radiation reaching the earth's surface is strongly influenced by absorption by water vapor, carbon dioxide, oxygen, and ozone present in the atmosphere (Fig. 2-1). The absorption by ozone of ultraviolet (UV) light is especially important to biological systems because UV light is harmful to organisms, particularly because of the absorption of these wavelengths by nucleic acids such as DNA. In fact, it was not until the evolution of oxygen-yielding photosynthetic cyanobacteria that the reducing atmosphere of the primitive earth was provided with oxygen and hence ozone. The ozone intercepted much of the UV radiation and its presence facilitated the subsequent evolution of aerobic organisms.

Passage through the atmosphere reduces the intensity of solar radiation, but absorption by the various compounds of the atmosphere over the broad spectrum of energy is very uneven (Fig. 2-1), as shown by the very irregular curve of irradiance that arrives at the surface of the sea (Fig. 2-1). Ozone, oxygen, water, and carbon dioxide are principal

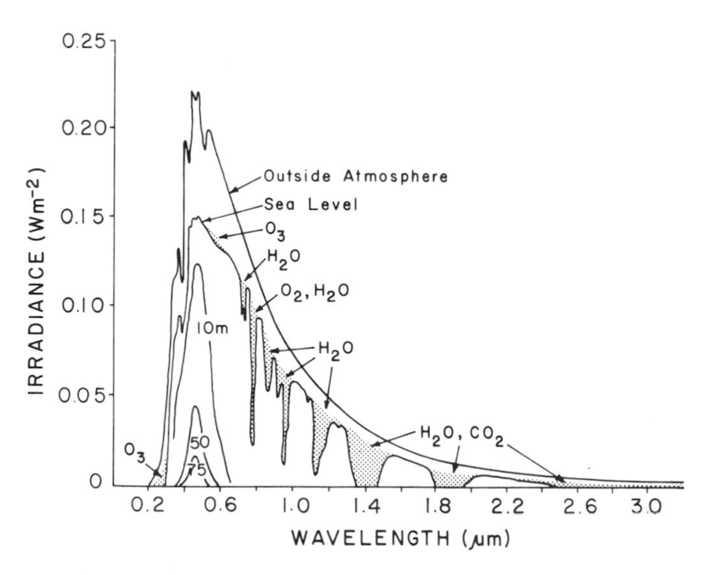

FIGURE 2-1. Absorption of solar radiation in the atmosphere and at depths of 10, 50, and 75 m in the sea. Adapted from Valley (1965) and Jerlov (1951). Note that little radiation in the ultraviolet radiation (below about 0.3 μm) reaches the surface of the sea and that long wave radiation (beyond 0.6 μm) is very effectively absorbed in the very upper layers of the water column. Marine organisms are left with a rather narrow window of wavelengths to use.

absorbers in the nonvisible ranges. Clouds may reduce radiation reaching the sea, and some solar energy is lost by scattering and reflection at the sea surface. When the sun is low in the sky or during rough seas, over 30% of radiation may be lost; on calm, clear days when the sun is high in the sky loss via these mechanisms is only a few percent of incident radiation (Von Arx, 1962).

There is further absorption in the water itself so that only small amounts of light penetrate the deeper layers of the sea (Fig. 2-1). Light is absorbed by water, suspended particles, and dissolved organic matter (Fig. 2-2). Absorption by algae peaks below the surface, as can be expected from the vertical distribution of chlorophyll or primary production (cf. Fig. 2-4). Deeper in the water column, water itself is responsible for most of the absorption. The amount of available light used by algae is directly proportional to the amount of photosynthetically active pigments in the photic zone. In a representative measurement in coastal water the photic zone was 10 m deep, the algae were dense and absorbed 56% of the light energy reaching the water surface. In an oceanic situation algal densities were low, the photic zone extended to 100 m, and only 1% of the energy available was absorbed by algae (Lorenzen, 1976).

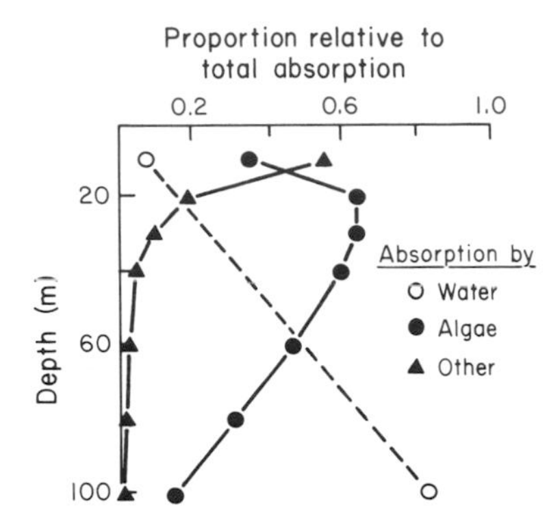

FIGURE 2-2. Relative absorption of downwelling light by water, algae, and organic materials in the water column. The "other" category refers to suspended particulate and dissolved matter. Adapted from Lorenzen (1976).

The intensity of light, or irradiance (photons arriving $m^{-2} sec^{-1}$), varies with time of day, season, and weather. The daily and seasonal effects occur because of the angle at which photons arrive at the water's surface. The greater the angle, the lower the irradiance. Moreover, the lower the sun is on the horizon, the greater the proportion of photons that are reflected at the surface and do not enter the water. Between solar altitudes of 90–45° reflective losses are only 2–3%, while surface reflection rises from 13 to 100% when the sun is lower than 20° (Kirk, 1992). Clouds reduce daily insolation (the sum of photons accumulated over a day). Wind increases the surface roughness of the sea, and increases reflection of photons at the sea surface, an effect that is specially important at low solar altitudes (Campbell and Aarup, 1989, Kirk, 1992).

The instantaneous rate of irradiance may not be the only aspect of light availability that matters. The total number of hours per day with irradiance above a threshold intensity, for example, determines whether eelgrass exists at a given depth (Dennison and Alberte, 1985). Another irradiance-related property is the photoperiod (number of hours of daylight), which often triggers seasonal physiological responses.

The total amount of light entering the water column from the surface and penetrating to any depth z can be described by Beer's Law:

$$I_z = I_0 e^{-kz}, \tag{2-1}$$

where I_z is the intensity of light at z, the depth of interest, I_0 is the intensity at the surface, and k is the extinction coefficient of water. The extinction coefficient varies from one place to another and is wavelength specific. When converted to logarithms and plotted, Eq. (2-1)

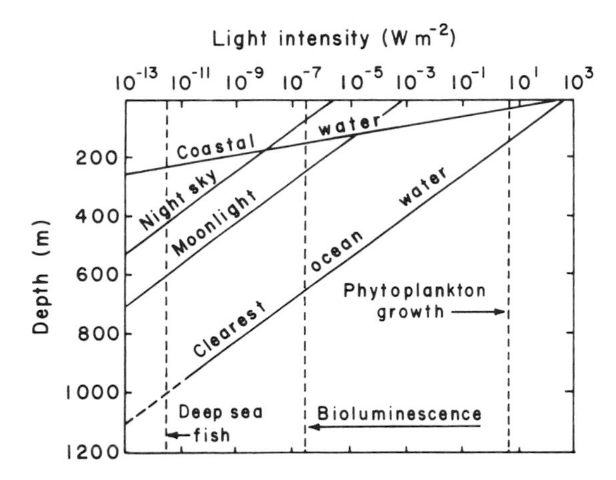

FIGURE 2-3. Logarithmic plot of $I_z = I_0 e^{-kz}$ plotted versus depth in the water column (diagonal lines) for daylight in coastal water, and for daylight, moonlight, and night sky for oceanic water. The approximate limits at which certain processes are activated (algal growth, vision in deep-sea fish) and the intensity of ambient light below which luminescence can be seen are also indicated as dashed vertical lines. Adapted from Clarke and Denton (1962).

becomes linear (Fig. 2-3). The vertical dashed lines indicate thresholds for various processes. One important feature of Fig. 2-3 is that the lower limit for photosynthesis (and hence for growth of phytoplankton) is very near the surface. This lower limit is the compensation depth, that depth at which light-supported photosynthesis just balances the metabolic losses due to respiration. Usually the compensation depth is taken to be about 1% of the surface radiation. The column of water above the compensation point is the photic zone.

The depth of the photic zone varies greatly, depending on the amounts of suspended particles and dissolved substances in the water. Mixing of nutrient-rich waters in spring often reduces light penetration to just a few meters because of resuspended bottom particles, presence of large amounts of dissolved organic matter, and blooms of phytoplankton. Of course, light intensity varies during the day and the depth of photic zone changes accordingly. Even in the clearest oceans, this depth seldom reaches 200 m, so that all the primary production in the water column of the sea takes place in a very thin upper layer.

2.1.1.2 Light Intensity and the Rate of Photosynthesis

Vertical profiles of photosynthetic rates typically have a subsurface peak (Fig. 2-4, top). Although there are exceptions (cf. Fig. 1-11), this pattern is the most commonly found in nature and is primarily due to

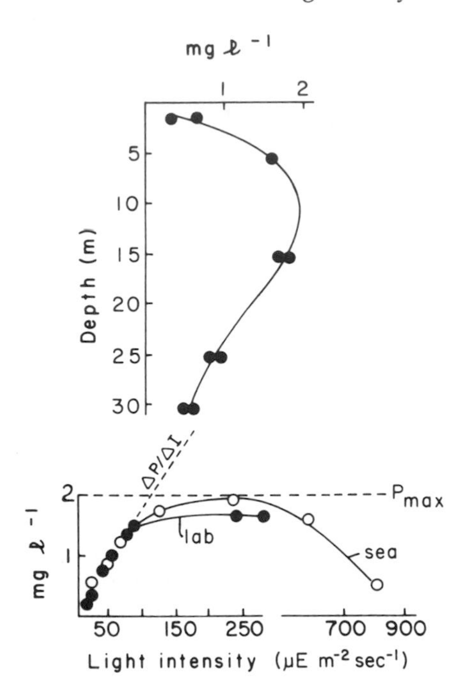

FIGURE 2-4. Top: Vertical profile of primary production in coastal water in milligrams of O_2 produced per liter. The experiment lasted three hours. Bottom: Comparisons of laboratory and field results of photosynthetic rate under varying light intensity. Adapted from Talling (1960).

the effects of light intensity* on the photosynthetic process. Laboratory experiments where light intensity is experimentally manipulated have borne this out (Fig. 2-4, bottom).

The rate of the light reaction of photosynthesis is strictly dependent on light intensity. Increases in light intensity lead to greater photosynthetic rates until some maximum is reached, defined as P_{max}. At this point the producers cannot use any more light; the enzymes involved in photosynthesis cannot act fast enough to process light quanta any faster, so rate of photosynthesis reaches an asymptote. The ratio of the change in photosynthesis with respect to the change in light ($\Delta P/\Delta I$) is a measure of how efficiently cells use changes in light. The ratio may be steeper for greens and diatoms than for dinoflagellates (Fig. 2-5), indicating that photosynthesis is saturated at lower light intensities for the former than the latter. Consequently greens and diatoms may be better suited for depths or latitudes where only dim light is available much of the year. Generalizations about the $\Delta P/\Delta I$

---

*The measurement of light intensity is reported in widely varying units. Phostosynthesis is driven by photons and so the preferred light unit quantifying photons is the einstein (E), where 1 E = 1 mole of photons. Full sunlight at the sea surface roughly equals $2{,}000\,\mu E\,m^{-2}\,s^{-1}$. Rough conversion among other light units used in the literature are: $1\,\mu E\,m^{-2}\,s^{-1} = 0.0187$ ly (langley) hr^{-1} $= 0.217\,W$ (watt) $m^{-2} = 51.2$ lux $= 4.78$ ft candles, where 1 ly $= 1\,g\,cal\,cm^{-2}$, $1\,W = 10^7\,ergs\,s^{-1}$; 1 lux = 1 candle on $1\,m^2$; 1 ft candle = a candle at 1 ft distance; all conversions refer to light in 400–700 nm, the range of wavelength that spans photosynthetically active radiation (PAR).

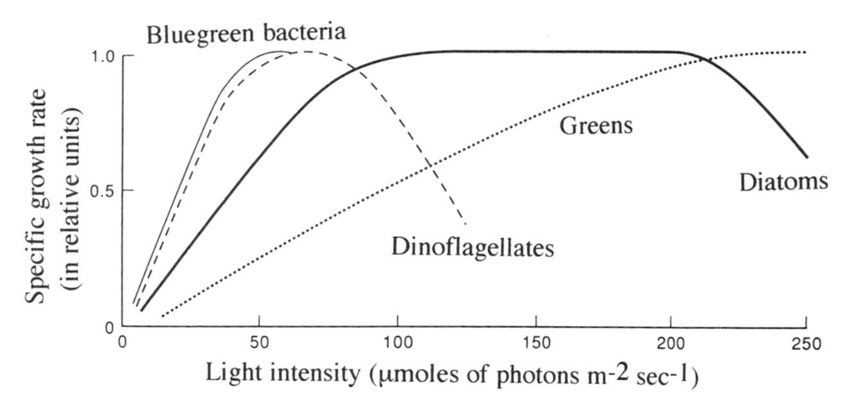

FIGURE 2-5. Specific growth rates in relation to light intensity in four major groups of marine producers. Adapted from Raven and Richardson (1986).

relationship are difficult, however, since the ratio varies substantially even within one species of algae and may be changed by nutrient deficiencies (Ketchum et al., 1958) and by cellular adaptations to low light in which the photosynthesis per cell may increase (Perry et al., 1981).

The actual light intensity at which P_{max} is reached depends on the algal taxon (Fig. 2-5). Bluegreens and dinoflagellates saturate at low intensities, and dinoflagellates are inhibited at relatively low light intensity. Diatoms show a broad plateau before being inhibited, while greens seem to not be saturated until the highest light intensities. It is not certain that these curves are expressed in the field as correspondingly different depth distributions of the various taxa.

Full sunlight usually inhibits photosynthesis in marine algae, as evidenced in the reduced photosynthetic rates at high light intensities (Figs. 2-4, 2-5). The reason for the inhibition of photosynthesis at high light intensities is not completely understood, but radiation in the UV range (Fig. 2-9) is probably implicated (Strickland, 1965) since in the laboratory under no UV radiation the $\Delta P/\Delta I$ curves do not show inhibition at intensities beyond the saturation point (McAllister et al., 1964; Harris, 1980). Ultraviolet radiation decreases photosynthetic rates (Lorenzen, 1979; Smith et al., 1980; Helbling et al., 1992).*

---

*The recent demonstrations of loss of stratospheric ozone over both the Antarctic (Farman et al., 1985; Newman et al., 1991) and Arctic (Stolarski et al., 1991) have prompted concern that the consequent increase in ultraviolet radiation (Lubin and Frederick, 1991) may have deleterious effects on organisms, especially those found at high latitudes (El-Sayed, 1988; Voytek, 1989). Screening off all ambient UV radiation below 378 nm (cf. Fig. 2-9) increased photosynthetic rates of Antarctic phytoplankton up to 200–300% (Helbling et al., 1992). Similar experiments with phytoplankton from tropical surface waters showed only 10–20% inhibition by ozone, but samples from greater depth were more sensitive. Although not everyone agrees, the decrease in ozone could have significant effects on certain marine food webs. This topic is discussed further in Chapter 16.

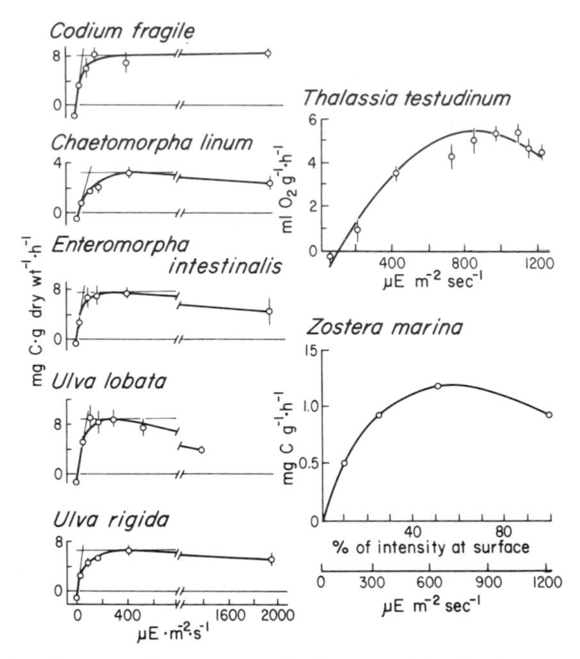

FIGURE 2-6. Net photosynthesis versus light intensities for five green macroalgae and two seagrasses. The $\Delta P/\Delta I$ and P_{max} are shown in the macroalgal graphs as straight lines. Points are mean and standard deviations, shown as vertical lines. Adapted from Arnold and Murray (1980), Buesa (1975), and McRoy (1974).

Another explanation for photoinhibition is that at high light intensities there is an excess of photosynthetically-generated energy, which has to be disposed of. Oversaturation of electron transport mechanisms may cause a chain of events that damages the reaction center of photosystem II (Kelly, 1989).

The relation of photosynthetic rate to light intensity in macroalgae and seagrasses (Fig. 2-6) is similar to that found in single-celled algae. The species with thinnest fronds generally show higher values for the $\Delta P/\Delta I$ ratio. Inhibition by high light intensities is less marked than for single-celled algae, probably because of self-shading within frond or leaf tissues. This is especially evident in the green alga *Codium fragile,* which has thick, optically dense fronds (Ramus, 1978).

The photosynthesis to light intensity relation is not constant over time. Producers may adapt physiologically to ambient light intensity, primarily by varying the amount of accessory pigments involved in harvesting photons. Photosynthesis in cells is carried out by photosynthetic units composed of a light-collecting antenna of accessory pigments and a reaction center made up of chlorophyll *a*. Under low light conditions, the size of the antenna increases. This shade

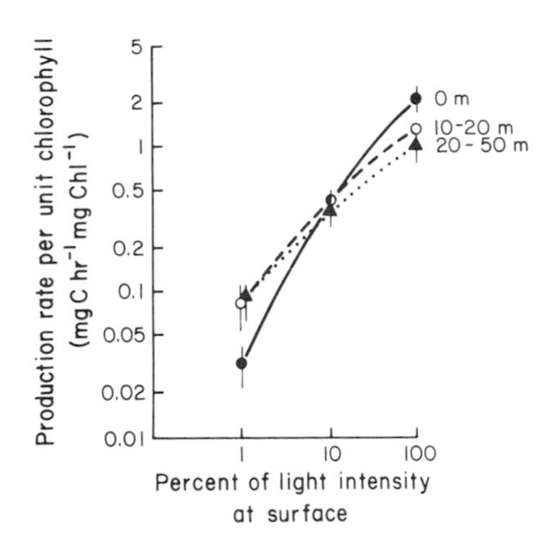

FIGURE 2-7. Light and shade adaptation in phytoplankton from the upwelling off northwest Africa. Samples were taken at the surface and at two deeper layers (10–20 m and 20–50 m) and incubated at 1, 10, and 100% of light at the surface of the sea. Adapted from Estrada (1974).

adaptation improves photosynthetic efficiency or ability to collect light (Prezelin and Sweeney, 1979; Perry et al., 1981), and results in higher photosynthesis rates per unit chlorophyll (Fig. 2-7).

The induction of shade adaptation in the photosynthetic unit may take 10–12 hr. If the water column is sufficiently stratified, phytoplankton can photoadapt, and photoinhibition is lessened. If the water column is mixed, phytoplankton will show photoinhibition (Fig. 2-7).

Phytoplankton may be repeatedly advected to depths where light is low during the course of a day. In such circumstances, a 10–12-hr period of adaptation is too long, and it may be advantageous to maintain photosynthetic units furnished with large antennae, so as to be able to photosynthesize effectively at least during the repeated periods of low light. Diatoms, a major component of phytoplankton, have such large photosynthetic units (Perry et al., 1981). Photosynthesis by picoplankton such as *Synechococcus* is most active at relatively low light intensities. These producers are most abundant toward the bottom of the photic zone (Stockner and Antia, 1986). Adaptation to low light is accomplished by increases in number of photosynthetic units per cell and by decreasing unit size (Barlow and Alberte, 1985).

The shape of photosynthesis versus irradiance curves depends on growth conditions and physiological adaptation; P_{max} and $\Delta P/\Delta I$ of *Synechococcus* spp. grown at high light intensities are much lower than for cells grown at low intensities (Fig. 2-8). In addition, some clones of *Synechococcus* spp. grown at low light intensity show inhibition (Fig. 2-8, left), while others do not (Fig. 2-8, right), depending on presence of specific photosynthetic pigments (Barlow and Alberte, 1985).

Macrophytes also show dark adaptation mechanisms. The estuarine vascular plant *Potamogeton perfoliatus* adjusts to lower light intensities by increases in $\Delta P/\Delta I$, increased leaf chlorophyll, and morphological responses, including elongation of stems, and thinning of lower leaves

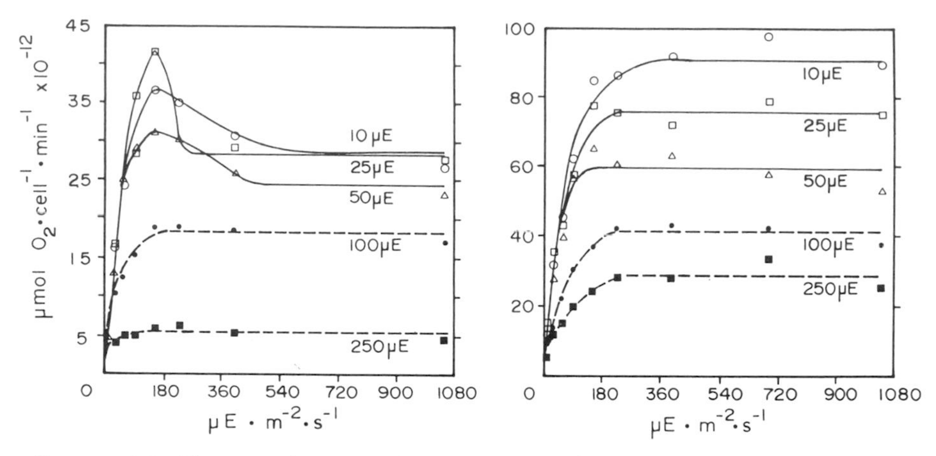

FIGURE 2-8. Photosynthesis-irradiance curves for two different clones of *Synechococcus* that contain different types of phycoerythrin. Cultures of each clone were grown at different light intensity regimes (10–250 μE m^{-2} s^{-1}). From Barlow and Alberte (1985).

(Goldsborough and Kemp, 1988). Such responses improve photosynthetic performance and growth, within limits. Irradiance of 11% of ambient was below the light intensity needed for survival; above that, there is a tradeoff, since the morphological adaptations (longer, thinner shapes) also made the plants more susceptible to damage by water turbulence.

Photosynthesis versus irradiance curves can be affected by nutrient supply. For phytoplankton (Cullen et al., 1992), macrophytes (Valiela et al., 1990), and in *Synechococcus* (Wyman et al., 1985), nutrient-richer conditions increase photosynthetic performance at any one irradiance, probably because nitrogen containing photosynthetic pigments increase when nitrogen supply increases. Nitrogen-containing photosynthetic pigments serve both as protein reserves as well as collectors of quanta for photosynthesis.

2.1.2 Quality of Light

2.1.2.1 Available Wavelengths in Water and Sediments

As light penetrates the water column, water molecules, dissolved substances, and suspended particles* not only attenuate the amount of

---

*Phytoplankton absorb light primarily at wavelengths about 400 and 700 nm. Water absorbs mainly near 700 nm, while the dissolved organic matter absorbs at wavelengths nearer 400 nm (Yentsch, 1980).

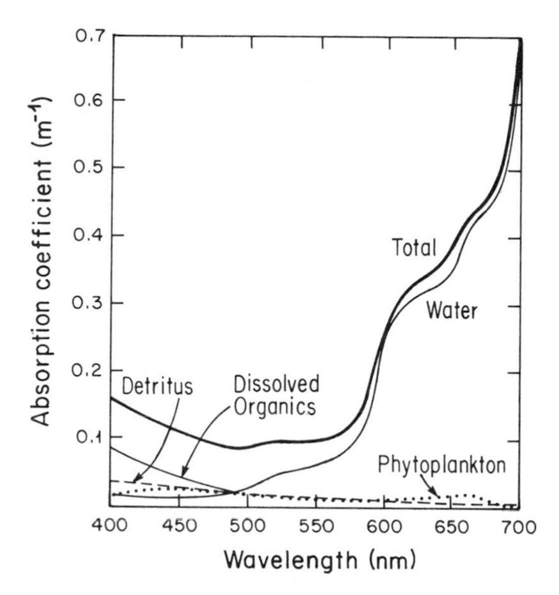

FIGURE 2-9. Diagrammatic spectrum of light absorption by water, dissolved organic matter, detritus, and phytoplankton in productive ocean water (1 mg chlorophyll *a* m^{-3}). Adapted from Kirk (1992).

light but also alter the spectral quality of light (Fig. 2-10, left and center). Phytoplankton and detrital particles, dissolved organic matter, and water have different effects on different parts of the light spectrum (Fig. 2-9). Phytoplankton absorb light at two peaks corresponding to the action of chlorophyll. Detrital particles and dissolved organic matter absorb most actively at shorter wavelengths. Water is the principal agent of absorption over the spectrum, but particularly at longer wavelengths (Fig. 2-9).

Absorption in different parts of the light wavelength changes the vertical distribution of the light spectrum. As light enters the water column, the longer (red and infrared) waves are absorbed quite near the surface (principally by absorption by water) and only blue light (short wavelengths) penetrates to some depth. Photosynthetic organisms in the water column therefore have to deal not only with a diminution of light quantity but also with a greatly altered spectral quality of light. These absorptional changes in quality occur much nearer the surface in coastal water than in oceanic water. Kirk (1992) reviews this topic in more detail.

In sediments where the photic zone is at most a few millimeters in depth (Fig. 2-10, right), photosynthetic organisms experience a similar effect of depth on the light regime. In contrast to what happens in the water column, in sediments blue light is absorbed first, and the longer

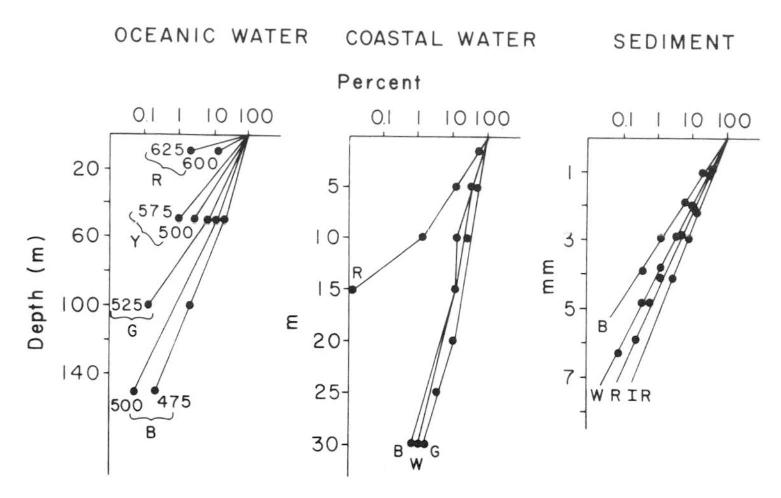

FIGURE 2-10. Percentage of incident light of differing wavelength penetrating to different depths in seawater and in a sediment. Note that depth is in meters for water and millimeters for sediment. Y, yellow; R, red; G, green; B, blue; and IR, infrared. W, white light is the sum of all wavelengths. Numbers on left-most graph are wavelengths in nanometers; adapted from Jerlov (1951, 1968) and Fenchel and Straarup (1971).

red and infrared wavelengths penetrate the farthest. This is because light penetration in sand is governed by refraction and scattering of light, and the refractive index of quartz, a typical component of sand, increases with shorter wavelengths (Fenchel and Straarup, 1971).

2.1.2.2 Absorption Spectrum of Marine Producers

The energy that drives photosynthesis is absorbed from sunlight by pigments found within internal membranes in chloroplasts in algae and within membranes located in the cytoplasm of bacterial cells. Each photosynthetic pigment has its own peculiar absorption pattern in the spectrum (Fig. 2-11). Absorption by ozone and water limits available radiation on the short (violet) and long (red) wavelengths of the visible spectrum (Figs. 2-1, 2-11), but the absorption peaks of the pigments commonly found in photosynthetic organisms cover virtually the entire remaining window of the energy spectrum of sunlight (Fig. 2-11). The range of photosynthetically active radiation for most algae and plants spans wavelengths of 400–700 nm, a range clearly adapted to available radiation.

Different groups of producers have somewhat different arrays of photosynthetic pigments, and show characteristic absorption spectra (Fig. 2-12). The green algae absorb light principally in the red and blue parts of the spectrum, due to the absorption spectrum of chlorophylls

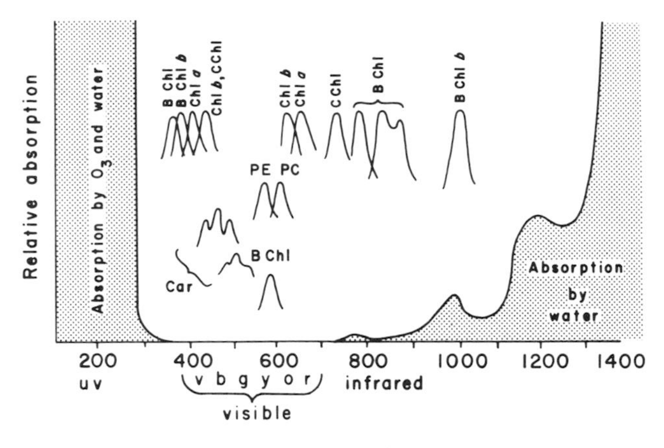

FIGURE 2-11. Absorption of radiation by photosynthetic pigments. For clarity only the principal absorption peaks are shown. Chl, chlorophyll; B Chl, purple photosynthetic bacterial chlorophyll; C Chl, green photosynthetic bacterial chlorophyll; Car, carotenes; PE, phycoerythrin; PC, phycocyanin. Adapted from Clayton (1971), Yentsch (1967), and Morel (1974).

a and *b* (Fig. 2-11). Both green algae and higher plants have these pigments. Other marine producers do not have chlorophyll *b*.

Brown algae contain chlorophylls *a* and *c* and use the green and yellow wavelengths more efficiently than the green algae. The shoulder in the absorption spectrum of browns around 500–540 mm (Fig. 2-12) is due to the activity of fucoxanthin, a photosynthetically active carotenoid pigment. Peridinin, a similar carotenoid, is responsible for absorption of green-yellow light in dinoflagellates.

Red algae and cyanobacteria contain only chlorophyll *a* and water-soluble pigments that allow absorption of light in the blue and green wavelengths (Fig. 2-12). They also contain phycobilins, of which phycoerythrin—red in color—is often the most common. Red algae may be found at deeper sites than other algal taxa, and show remarkable absorption of precisely those wavelengths available in the deeper water (Fig. 2-12), but their pigment compositions are largely adaptations to low light intensity rather than to light quality (Dring, 1981).

The cyanobacteria contain only chlorophyll *a*, and their absorption spectrum resembles that of green algae, except for the marked absorption around 600–640 nm due to phycocyanin, a phycobiliprotein that is blue in color (Fig. 2-11). Some cyanobacteria show a much enhanced absorption in the green part of the spectrum due to phycoerythrin; *Oscillatoria thiebautii*, for example, absorbs weakly at 600–640 nm but has a phycoerythrin peak at 495 nm (McCarthy and Carpenter, 1979).

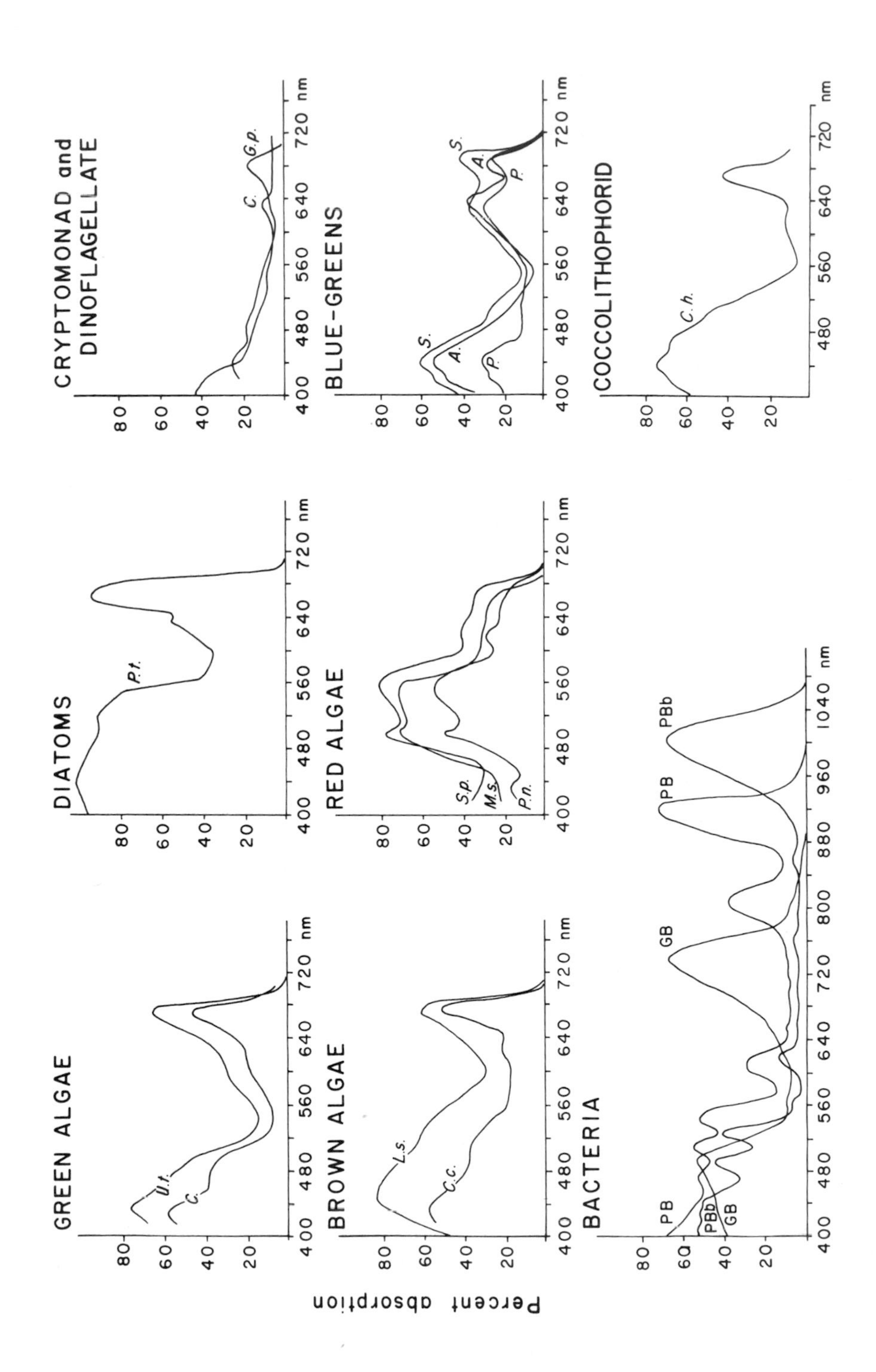
GREEN ALGAE
U.t.
C.
BROWN ALGAE
L.s.
C.c.
BACTERIA
PB
PBb
GB
DIATOMS
P.t.
RED ALGAE
S.p.
M.s.
P.n.
CRYPTOMONAD and DINOFLAGELLATE
C.
G.p.
BLUE-GREENS
S.
A.
P.
COCCOLITHOPHORID
C.h.
Percent absorption
400 480 560 640 720 nm
400 480 560 640 720 800 880 960 1040 nm
20 40 60 80

Development of new analytic methods reveal many new chlorophyll and carotenoid pigments in phytoplankton and bacteria, and new phycoerythrins in coccoid cyanobacteria (Letelier et al., 1993).

The large absorption peak in the red wavelengths might better suit phytoplankton for oceanic conditions, where red light penetrates somewhat deeper in the water column. Even in clear oceanic water, however, the red peak of chlorophyll absorption is not likely to absorb much light below 10 m (cf. Fig. 2-10, left) while the absorption of the blue band of chlorophyll will be functional deeper in the water column. There is also more energy in the blue range, since energy is negatively related to wavelength. Below 50 m absorption is mainly due to carotenoids that absorb near the blue wavelengths (Fig. 2-4). The presence of such accessory pigments is an important physiological adaptation, since turbulent mixing and sinking can readily transport phytoplankton cells to depths below 50 m.

Carotenoids are present in substantial amounts in virtually all major taxa of marine producers. In addition to absorbing light, they may also be involved in other functions. One such role may be to protect the photosynthetic apparatus from excess light. While in some green and brown algae three or four of the major carotenoids are photosynthetically active, it is not clear how photosynthetically active many of the remaining 60 identifiable carotenoids are, especially in the red algae (Halldal, 1974). Details of algal pigments and their properties are reviewed by Jeffrey (1980).

The photosynthetic bacteria (Fig. 2-2) absorb energy far to the right of the algal absorption spectra; bacteriochlorophylls absorb in the red and infrared (Fig. 2-9). As we have seen in Section 1.2.1, photosynthetic bacteria require reduced compounds for their photosynthesis, so that they exist only where oxygen is absent, such as in sediments. They are often found with a layer of algae above them, and the algae absorb in the visible range of wavelengths. The bacteria have adapted to use the long-waved parts of the spectrum that penetrate the deepest into the sediment (Fig. 2-10) and that are not absorbed by the algae on the surface.

FIGURE 2-12. Absorption spectra of marine photosynthetic organisms. The actual taxa involved are U.T., *Ulva taeniata*; C, *Chlorella pyrencidosa*; P.t., *Phaeodactylum tricornutum*; C, cryptomonad; G.p., *Gonyaulax polyedra*; L.s., *Laminaria saccharina*; C.c., *Coilodesme californica*; S.p., *Schizymenia pacifica*; M.s., *Myriogramme spectabilis*; P.n., *Porphyra nereocystis*; S., *Synechococcus* sp.; A., *Anacystis rubulans*; P., *Phormidium luridum*; PB, purple bacteria with bacteriochlorophyll; PBb, purple bacteria with bacteriochlorophyll b; GB, green bacteria; C.h., *Coccolithus huxleyi*. Adapted from Haxo and Blinks (1950), Clayton (1971), Govindjee and Mohanty (1972), Halldal (1974), and Paasche (1966).

2.2 The Uptake and Availability of Nutrients

Aquatic producers acquire nutrients by active uptake from adjacent water, so active producers are surrounded by a nutrient-depleted boundary layer of water. Further nutrient uptake is limited by the rate at which new externally-supplied nutrients can diffuse into the depleted boundary layer, but diffusion is not a fast process. To understand how organisms respond to the problem of diffusion limitation, we need to briefly examine the dynamics of boundary layers, specifically the workings of inertial and viscous forces at different size scales. Expanded discussion of these topics are available in Vogel (1981) and Mann and Lazier (1991).

The rate of transport of nutrients through the layer of water surrounding producers is determined by diffusion of molecules and of turbulent eddies.* Molecular diffusion dominates transport within the viscous layer next to the producer surface. Water molecules in contact with a solid, such as the surface of phytoplankton or plants, stick to that surface. Water viscosity inhibits movement of water molecules at distances of a few mm from the surface, so diffusion is the only process that can transport solutes through the viscous layer to the surface. Within the thin viscous layer water movement, and hence solute exchange, increase only slightly and linearly at greater distances away from the surface (Fig. 2-12), because diffusion is relatively slow.

Nondirectional diffusion of turbulent eddies is the major mechanism creating water motion within a few mm to m away from surfaces. The energy contained in turbulent eddies dissipates as it transmitted to the smaller viscous scales, so that velocity of water and solute exchanges decrease logarithmically nearer surfaces (Fig. 2-13). Molecular diffusion applies to smaller distances, is slower, and depends on the specific property (temperature, nutrient concentration, etc.), because it is the properties that are transported across gradients of concentration. In contrast, eddy diffusion applies to larger distances, is faster, and rates are the same for all properties, because it is the fluid that is moved.

The relative importance of viscous vs. inertial forces applied to any body, fluid or solid, is expressed by the dimensionless Reynolds number, $Re = ud/v$, where u is water velocity, d is a spatial dimension of the body or distance being considered, and v is the kinematic viscosity of water. Re increases logarithmically with spatial scale, for example with organism size (Fig. 2-14). A whale swimming at $10\,m\,s^{-1}$

*Turbulence is generated by winds, waves, currents, and tides. Turbulence is manifested as multiple eddies that have no well-established direction of translation. Turbulent eddies tend to disturb stratified waters and result in mixing of water masses and organisms.

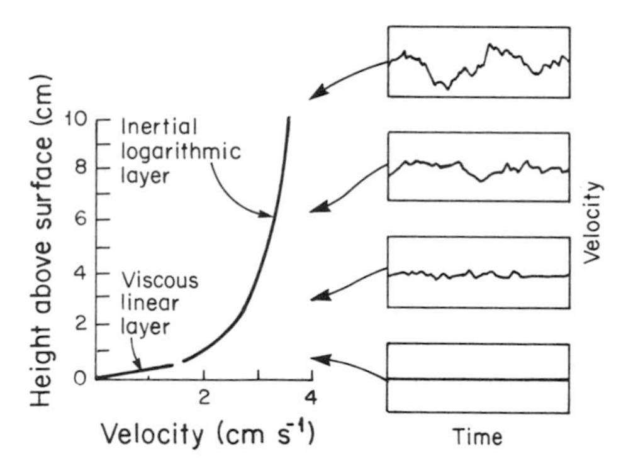

FIGURE 2-13. Idealized vertical profile of mean water velocity through viscous- and inertially-dominated sublayers of a boundary layer. Panels on right show time series of velocity at different heights above the surface, demonstrating the damping of turbulent fluctuations nearer the surface. Adapted from Mann and Lazier (1991).

has a *Re* of 3×10^9, while a sea urchin sperm advancing the species at $0.2\,\mathrm{mm\,s^{-1}}$ has an Re of 3×10^{-2} (Vogel, 1981).

Plankton overcome diffusion limitation by actively swimming or by sinking. Swimming organisms such a flagellates or ciliates drag their boundary layer, but the outer layers of water may be sheared away, to be replaced with water perhaps containing higher nutrient concentrations. Diffusion into the viscous layer would then be faster, since diffusion rates are concentration-dependant. Swimming allows microplankton to increase nutrient uptake by 50–200% (Sommer, 1988).

Larger phytoplankton such as diatoms and coccolithophorids cannot swim, but rather sink because their mineral content increases specific gravity. Sinking is less effective than swimming as a strategy to increase nutrient uptake because of the limited speeds generated for most except the largest cells. Sinking might increase nutirent uptake by 30–40% for a $50\,\mu$m cell, but have no effect for a $10\,\mu$m cell (Mann and Lazier, 1991; Kiørboe, 1993). There is a further disadvantage: sinking cells remain in illuminated upper layers of the sea only where there is upward movement of water. This tends to restrict these larger algae to times or places where water columns are vertically mixed.

Smaller organisms do not sink, as long as they remain unattached to particles. Sinking only applies to particles large enough to show $Re = 2$ or 3 (Fig. 2-14). Picoplanktonic organisms do, however, swim, but at slow speeds. To obtain the relation between size of organism (d) and swimming speed (u), Mann and Lazier (1991) calculated $Re \cong 1.4 \times 10^6 \times d^{1.86}$ from Fig. 2-14; and substituted $v = 10^{-6}\,\mathrm{m^2\,s^{-1}}$. Swimming speed (in m $\mathrm{s^{-1}}$) is then $u = 1.4d^{0.86}$. This result makes evident that swimming speed strongly depends on size of organism. Mann and

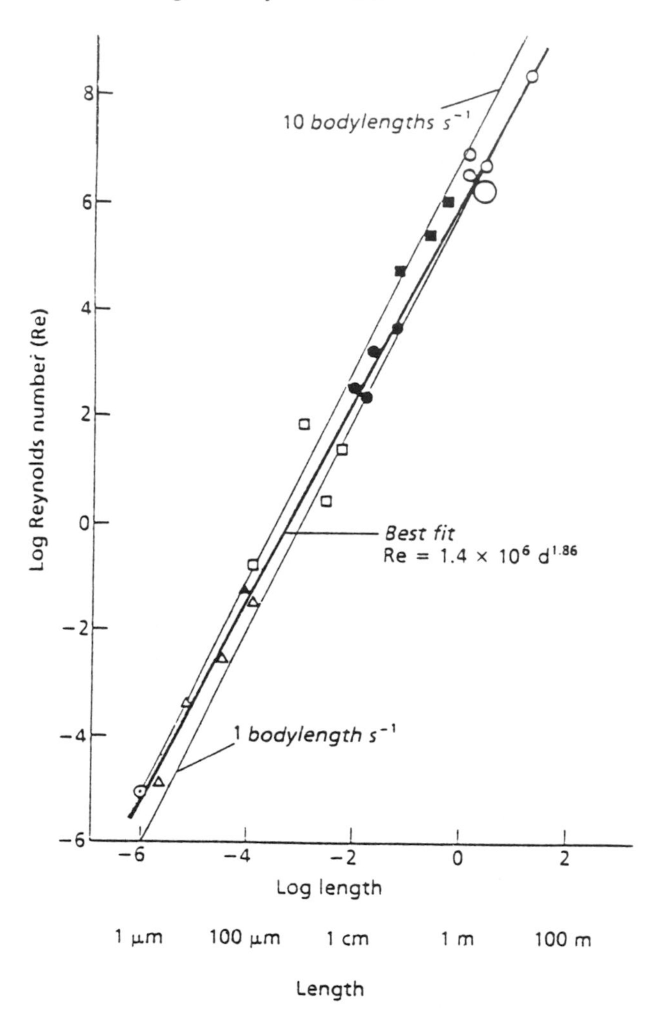

FIGURE 2-14. Relationship between Reynolds number and size of organism. The heavy line is best fit to the data. Thin lines show the relationship for swimming at 10 and 1 body lengths per sec. Small open circles are mammals, black squares are fish, black circles are amphipods, open squares are zooplankton, black triangles are protozoa, open triangles are phytoplankton, the dotted circle is for bacteria, and the large open circle are humans. Compiled by A. Okubo, adapted by Mann and Lazier (1991).

Lazier also added lines that show the presumed relationship if organisms moved at 1 and 10 body lengths s^{-1} (Fig. 2-14). These lines suggest that small organisms move at about 10 body lengths per second, while larger organisms move at perhaps 1 body length per second. Bacteria and other picoplankton are so small that their swim velocities ($<100\,\mu m\,sec^{-1}$) are about 100 times slower than rates of molecular diffusion of nutrients (Berg and Purcell, 1977). These organisms therefore depend exclusively on molecular diffusion to provide nutrients.

Most benthic plants, macroalgae, and corals are attached to the bottom, so water flows around the producer. Within certain limits, increased water flow leads to faster nutrient uptake (Wheeler, 1980; Gerard, 1982). Kelp growing in different current regimes develop fronds with different morphology (for example, corrugated fronds or frilly edges) to adjust the degree turbulent eddy diffusion (Gerard and Mann, 1979; Koehl and Alberte, 1988). Presumably such frond morphology results in higher nutrient uptake rates and favors higher photosynthesis rates. The tradeoff for attached seaweeds is that anatomical changes that enhance turbulence in small scales also make for increased drag, so enhance the danger of breakage or removal by episodic fast currents (Koehl, 1986).

In corals an experimentally increased Reynolds' number led to increased respiration and photosynthetic rates (Patterson et al., 1991). The net result is that primary production of coral heads increase with increasing weater motion. Water movement also increases the heterotrophic part of metabolism of some corals; flow increases particle capture in branching, but not in flat corals (Sebens and Johnson, 1991).

2.2.1 Kinetics of Nutrient Uptake

If algae are placed in a nutrient medium, concentration of nutrients decrease over time (Fig. 2-15) in the solution as the nutrients are incorporated into the plant cells. The velocity at which algal uptake removes nutrients depends on the nutrient concentrations in the medium (Fig. 2-16). Such measurements are usually obtained using

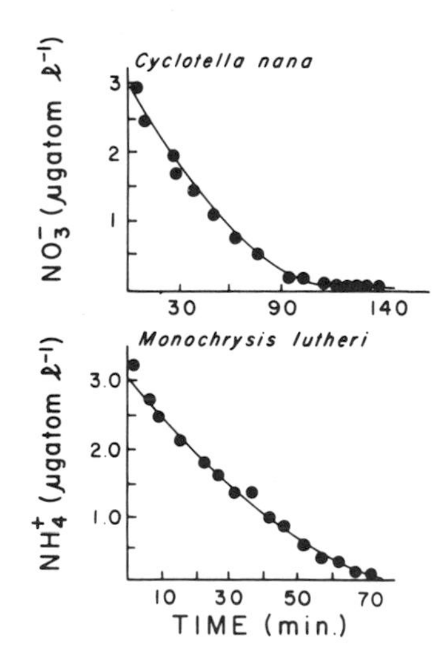

FIGURE 2-15. Consumption of nitrate and ammonium by two species of algae, a diatom *Cyclotella nana* and a flagellate *Monochrysis lutheri*. Adapted from Caperon and Meyer (1972).

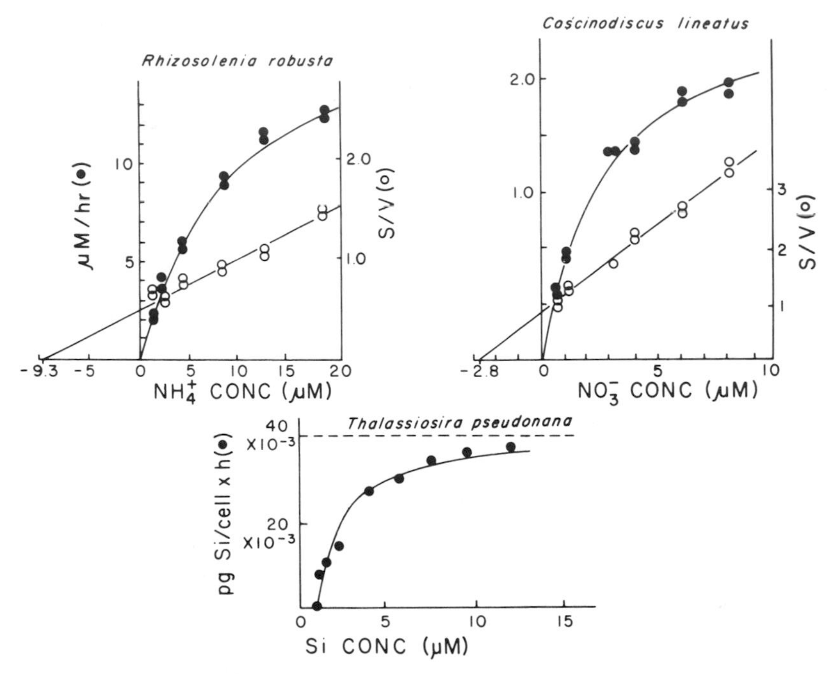

FIGURE 2-16. Michaelis-Menten curves (filled circles) and Woolf plots (open circles) fitted to data on uptake (μmole/hr) of ammonium, nitrate, and silica by three diatoms at different nutrient concentrations (μmole/liter). S is the concentration of nutrient being taken up, V is the uptake velocity. The x-intercepts of the top two graphs provide the estimate of K_s. Adapted from Eppley et al. (1969) and Paasche (1973).

chemostats, devices by which specific nutrient concentrations can be provided continuously to populations of algae. The uptake rates measured in chemostat cultures increase as the concentrations of the limiting nutrient increase, until an asymptote is reached where the specific nutrient is no longer limiting. This pattern is arithmetically similar to that observed in the saturation of inducible enzymes and has been described by the well-known Michaelis–Menten equation

$$V = V_{max}S/(K_s + S). \quad (2\text{-}2)$$

This equation relates the velocity of nutrient uptake (V) to the maximum rate of uptake (V_{max}), the concentration of limiting nutrient (S), and a constant (K_s) that represents the substrate concentration at which $V = V_{max}/2$.

The estimates of V_{max} and K_s can be obtained by fitting a hyperbola to the data using a least-squares method, or more simply, by plotting the reciprocal of the hyperbola ($1/V$ vs $1/S$) to obtain a straight line. The y intercept in this Lineweaver-Burk plot gives $1/V_{max}$ and the x intercept is $1/K_s$. Another procedure is the Woolf plot (Dowd and Riggs, 1964) (Fig. 2-16), where S/V is plotted versus S and the values of V_{max} and V_s are obtained from $S/V = K_s/V_{max} + S/V_{max}$.

The Michaelis–Menten equation describes fairly well the dynamics of uptake of a variety of nutrients by many species of producers (Fig. 2-16). In certain diatoms, uptake of silica apparently does not start until a threshold concentration is reached, so a modified Michaelis-Menten equation is needed to describe these data (Paasche, 1973). The Michaelis–Menten equation can be used to estimate K_s and S when it is known that the nutrient in question is in fact limiting. If the specific nutrient studied is not the limiting nutrient, spurious values of V_{max} may be obtained, determined by the low supply of the unidentified limiting nutrient. V_{max} value may also be affected by the presence of a competing substrate. If nitrate uptake rates are measured in the presence of ammonium, the uptake of nitrate may be severely underestimated because of the preference for ammonium by many algae (McCarthy, 1980). In view of such difficulties, it is not surprising to find instances where uptake of nutrients does not follow the simple Michaelis–Menten equation (Brown et al., 1978; De-Manche et al., 1979).

Uptake of nutrients is coupled to photosynthetic rate, and follows the same pattern as photosynthesis at different light intensities. Nutrients, especially ammonium (Dugdale and Goering, 1967), can be taken up at slow rates even in the dark. Increased temperatures may somewhat increase nitrate uptake (Eppley et al., 1969; Thomas and Dodson, 1974).

In most situations, however, the major factor affecting uptake rate is the concentration of the limiting nutrient in the environment (Fig. 2-16). The half-saturation constant (K_s) varies, depending on the nutrient concentrations present where specific phytoplankton species were collected* (Table 2-1). Since an average of about 0.5 μgat-liter NO_3-N liter^{-1} or less is present in oceanic waters, with lower or comparable amounts of ammonium, it is not surprising to find half-saturation coefficients of oceanic phytoplankton around 0.5–1 μgat liter^{-1}. Diatoms found in inshore (neritic) water have much larger K_s values, consonant with the greater nutrient concentrations in such waters. Flagellates generally have a wider range of K_s values than other kinds of phytoplankton. K_s values for nitrate in phytoplankton from oceanic areas

*There is some doubt about the validity of estimates of K_s obtained by incubation periods as long as hours, such as those in Table 2-1. Ammonium uptake can be very rapid during the first few minutes of exposure to nutrients (Glibert and Goldman, 1981), and slows later. Estimates obtained in exposures longer than 1 hr or so may therefore underestimate K_s (Glibert et al., 1982).

TABLE 2-1. Half-Saturation Constants (K_s) for Uptake of Nitrate and Ammonium by Cultured Marine Phytoplankton at 18°C.[a]

	K_s (μg-atoms liter)		Cell diameter (μm)
	Nitrate	Ammonium	
Oceanic coccolithophores and diatoms	0.1–0.7	0.1–0.5	5
Neritic diatoms	0.4–5.1	0.5–9.3	8–210
Neritic or littoral flagellates	0.1–10.3	0.1–5.7	5–47
Oligotrophic, Tropical Pacific	0.1–0.21	0.1–0.62	—
Eutrophic, Tropical Pacific	0.98	—	—
Eutrophic, Subarctic Pacific	4.21	1.3	—

[a] From MacIsaac and Dugdale (1969) and Eppley et al. (1969). The Pacific data from natural mixed phytoplankton.

with very different nutrient concentrations show the same trend as the coastal-oceanic comparison: K_s values are high in more eutrophic (nutrient-rich) water and low in oligotrophic (nutrient-poor) water.

The importance of ambient nutrient concentration is demonstrated by the occurrence of physiological races within the same species of algae or clones adapted to local nutrient regimes. Each of the bays in Table 2-2 are rich coastal lagoons; the Sargasso Sea on the other hand is one of the most oligotrophic regions of the ocean. In each of the three species of algae the K_s values for the strain from eutrophic waters are

TABLE 2-2. Half-Saturation Constants for Nitrate of Three Species of Algae Obtained from Coastal and Oceanic Environments in the Atlantic.[a]

Species	Source	K_s (Mean ± 95% confidence interval)
Cyclotella nana	Moriches Bay	1.87 ± 0.48
	Edge of shelf	1.19 ± 0.44
	Sargasso Sea	0.38 ± 0.17
Fragilaria pinnata	Oyster Bay	1.64 ± 0.59
	Sargasso Sea	0.62 ± 0.17
Bellerochia spp.	Great South	6.87 ± 1.38
	Bay	0.12 ± 0.08
	Off Surinam	0.25 ± 0.18
	Sargasso Sea	

[a] From Carpenter and Guillard (1971).

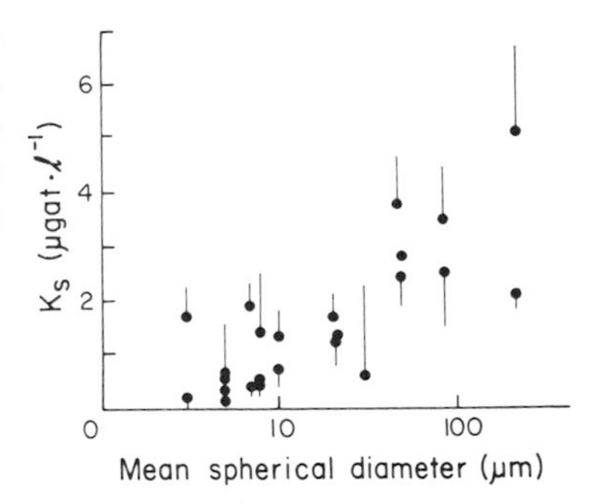

FIGURE 2-17. Half-saturation (K_s) values for nitrate uptake by phytoplankton of different size. The bars show the 95% confidence limits for the highest and lowest mean K_s reported. Adapted from Malone (1980).

higher than those of the oceanic strain. These strains maintain their physiological distinction even after being cultured for many generations in identical media in the laboratory.

Another factor that affects uptake rates is size of the phytoplankton cell (Fig. 2-17): larger cells have higher values of K_s. The higher surface-to-volume ratio of smaller cells may provide an advantage in nutrient uptake in oligotrophic water, since uptake rates are limited by the number of uptake sites on the surface of the cells. Thus, smaller cells may be favored in oligotrophic and nonturbulent waters. This fits in with the observation that oceanic phytoplankton tend, on the average, to be smaller than coastal algae, although there is great variability in the data [Table 2-1 and review by Malone (1980)]. The different average size may be a result of competitive exclusion of species of large algae with small surface-to-volume ratios by small algae with a high surface-to-volume ratio in stratified or oligotrophic waters. This may be one basis for the differences in species composition between coastal and oceanic floras.

2.2.2 *Internal Nutrient Pools and Growth*

The Michaelis–Menten equation can describe the growth rate (μ) of algal populations in chemostats

$$\mu = \mu_{max}S/(K_s + S), \tag{2-3}$$

with μ_{max}, the maximum rate of growth, and the other terms being the same as in Eq. (2-2). By using Eq. (2-3), we assume that growth rate is constant regardless of the ambient nutrient concentration or the state of nutrition of the cells. This is not always so, since nutrient-starved algae placed in a fresh culture medium quickly deplete nutrients in the medium at rates exceeding the growth rate of the algae. This "luxury consumption" results in the buildup of an internal pool of nutrients that may be later shared by successive generations of cells. These internal pools of compounds such as polyphosphates or nitrogen

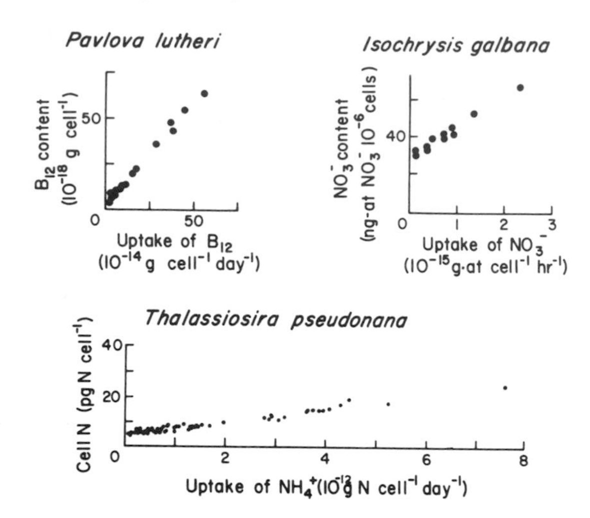

FIGURE 2-18. Relation of the internal pools and uptake of vitamin B_{12}, nitrate, and ammonium in three species of phytoplankton. Adapted from Droop (1973) and Goldman and McCarthy (1978).

storage compounds are responsible for continued growth sometimes observed after depletion of nitrate.

The relation between ambient nutrient concentration and growth rate of algae can thus, in theory at least, be expressed in three steps. First, the rate of nutrient uptake increases hyperbolically as nutrients in the medium increase, as discussed in the previous section. Second, the amount of nutrient stored in the plant cells is linearly related to the rate of nutrient uptake. This may apply to nutrients such as nitrate, silicate, and phosphate, as well as to elements and compounds required only in trace amounts such as vitamin B_{12} (Fig. 2-18). Third, the rate of growth is hyperbolically related to the internal pool of nutrients (Droop, 1968), with the growth rate decreasing as the internal pool becomes saturated (Fig. 2-19). The Droop equation is best applied to describe growth where the limiting element is required in relatively small amounts (Goldman and McCarthy, 1978).

The description of the relations of available nutrients, uptake, internal pools, and growth, and other theoretical advances such as inclusion of the effects of internal pools on uptake rates (DeManche et al., 1979) suggests some of the couplings between nutrients in the medium and the actual growth rate of producers. Such advances have been eagerly received because expressions such as Eqs. (2-2) and (2-3) are needed for building predictive mathematical models that can be used as tools to gain further insight into the dynamics of marine ecosystems. There is as yet no consensus that the theoretical models discussed here are completely applicable to natural situations. The use

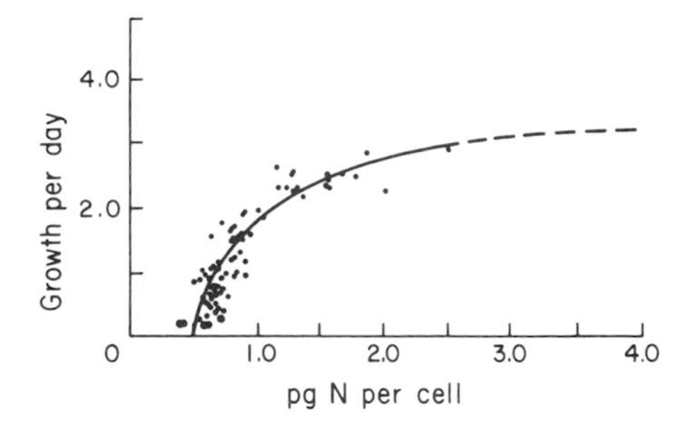

FIGURE 2-19. Specific growth rate and nitrogen cell content of the diatom *Thalassiosira pseudonana*. A picogram is 10^{-12} g. Adapted from Goldman and McCarthy (1978).

of the equations requires many suppositions, including steady state growth, and growth limitation by a single identified nutrient. Moreover, the complexities of uptake in the presence of other organisms that may release or conserve nutrients have not been dealt with, nor have physical processes of fluid motion been considered. There have been some advances in introducing additional factors into the equations, for example, temperature (Goldman and Carpenter, 1974). Further work in this area may lead to useful, but necessarily complex predictive models.

2.2.3 Availability of Nutrients

2.2.3.1 Availability of Nutrients in Sea and Freshwater

While seawater from different parts of the oceans may have variable concentrations* of different nutrients (cf. Chapter 2-14), representative vertical profiles for oceanic waters (Fig. 2-20) frequently have a marked decrease in nutrients near the surface. This reduction is due to uptake by phytoplankton in the photic zone. The profiles of Fig. 2-20 only show conditions for most of the year; in many regions of the sea there are times of year when physical processes result in vertical mixing of nutrient-rich water, so new nutrients reach the surface (cf. Chapter 15). Once in the photic zone, these nutrients are incorporated into particles and these particles in one fashion or another eventually sink from the photic zone into deeper waters where decay of the organic matter leads to the release of the combined nutrients (cf. Chapter 14).

---

*Oceanographers express concentrations in gram-atoms or moles of an element per liter, since this makes it clear that, for example, a reported value refers to the nitrogen in nitrate (NO_3^-) and does not involve the oxygen. The actual expression is usually shown as μgat NO_3-N liter^{-1}. Concentrations are also often stated in molar units M = moles l^{-1}. This is just as convenient a system of units, and μM NO_3^{-1} is equivalent to μgat or moles of NO_3-N liter^{-1} in the case of compounds such as NO_3^- or NH_4^+, where one atom of the element in question is present.

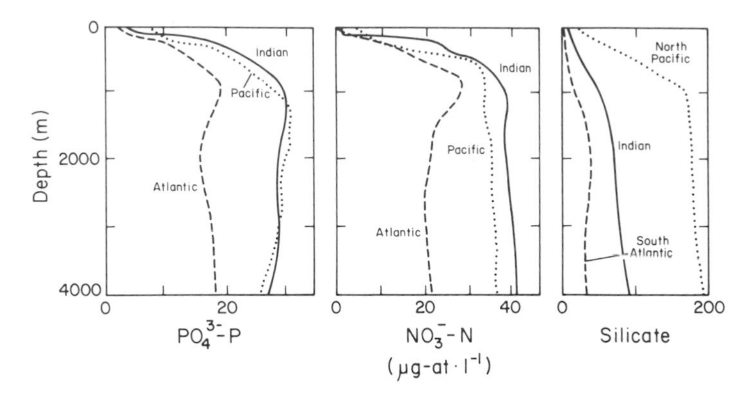

FIGURE 2-20. Vertical distribution of phosphate, nitrate, and silicate in the three major oceans. Adapted from Richards (1968).

Nutrients in coastal waters and estuaries are more abundant than those in the open ocean, as we will see below, but even in coastal water, nutrients are usually depleted in the photic zone. Thus, phytoplankton exert a major influence on the availability of nutrients and, by reducing the supply of their essential resources, limit their own growth.

It is not sufficient to speak merely of nutrients in general, since nutrient limitation of growth is invariably due to a specific deficit of an essential substance. We will therefore examine the availability of nitrate and phosphate. These two nutrients have received the most attention because analytic methods have long been available and, as will be seen, nitrogen and phosphorus are thought to be most often limiting to phytoplankton production. In addition, there is a sharp contrast in the role of nitrogen and phosphorus in regard to limitation of primary production in fresh and seawater, a comparison we will also examine.

Nitrate is generally far more abundant in freshwater than in seawater (Fig. 2-21, top), where nitrate concentrations are around 1 μgat NO_3-N liter^{-1} or less and rarely exceed 25 μgat liter^{-1}. The concentrations of phosphate in fresh and seawater cover approximately the same range, from 0 to about 3 μgat liter^{-1} (Fig. 2-20, bottom). Possible geochemical reasons for the differences in concentrations of these nutrients are discussed in Chapter 14.

Phosphate seldom achieves concentrations comparable to those of nitrate because it is strongly adsorbed to particles and forms insoluble salts (cf. Section 14.1), while nitrate is highly soluble (cf. Section 14.2). The concentration of other forms of nitrogen (reviewed by McCarthy, 1980) used by producers—ammonium, urea, amino acids—are as variable and generally as low or lower than nitrate values. Different forms of nitrogen may be transported from surrounding watersheds into freshwater bodies. These inputs may have N:P ratios of 40:1, so

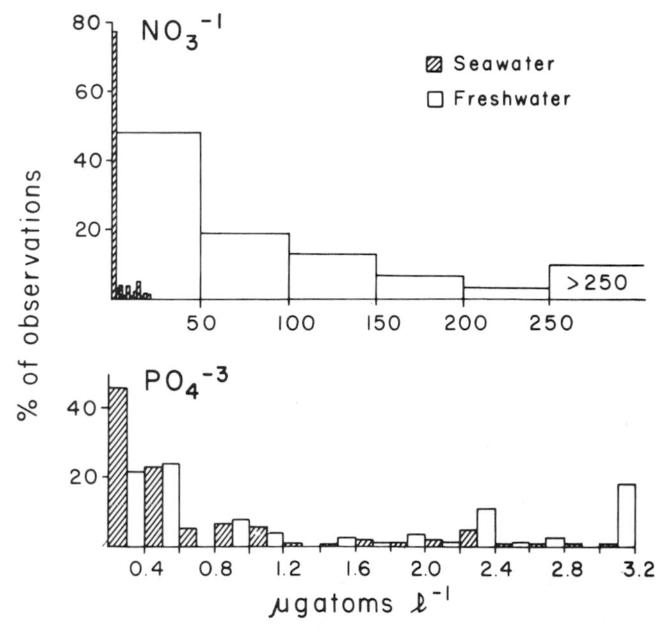

FIGURE 2-21. Concentrations of nitrate and phosphate in many samples of fresh and seawater. Values for seawater are for surface waters. Observations for this compilation were obtained by using values reported in *Limnology and Oceanography* since 1971 and reviews by Zenkevich (1963), Zeitzschel (1973), and Livingstone (1963). When many analyses of a specific site or station were available, representative values were chosen for each site.

that in freshwater nitrogen is considerably in excess of the 16:1 Redfield ratio* used by algae and plants during growth. Further, nitrogen-fixing cyanobacteria are more plentiful in freshwater and their activity may also increase N:P ratios.

2.2.3.2 Nutrient Contents of Producers in Seawater and Freshwater Environments

Marine phytoplankton contain less nitrogen than freshwater phytoplankton relative to phosphorus (Fig. 2-22). The ratios of N to P span a very large range, indicative of varied supply and physiological state of

* Early studies in the chemical composition of plankton (Redfield, 1934) showed that particulate matter had carbon:nitrogen:phosphorus of 106:16:1, a relationship now referred to as the Redfield ratio. This was extended to claim that uptake of N and P from seawater would also follow a 16:1 ratio (Redfield et al., 1963; Corner and Davies, 1971). These ratios are convenient averages but may not always be matched under field conditions (Banse, 1979).

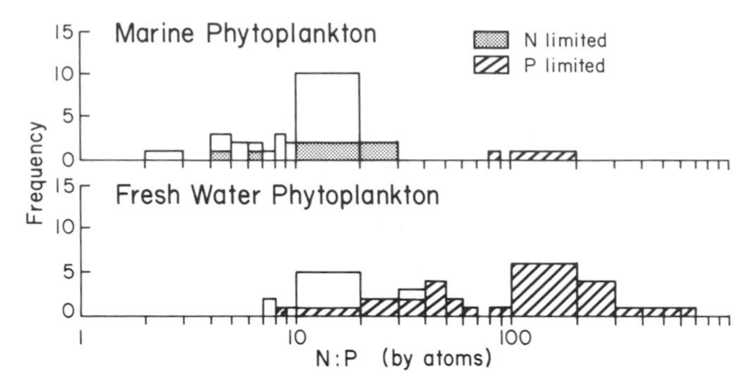

FIGURE 2-22. Nitrogen to phosphorus ratio in marine and freshwater phytoplankton. Where the ratio was obtained from cells growing in nitrogen or phosphorus limited cultures or come from low phosphorus environment (less than 0.1 mg P $liter^{-1}$), it is so indicated. Data from many sources, including Hutchinson (1975) and Jorgensen (1979).

cells, but marine phytoplankton most often have a N:P of about 10 or 20:1, comparable to the Redfield value of 16:1.

Marine macroalgae and vascular plants have ranges of nitrogen and phophorus content similar to those of freshwater species (Fig. 2-23). Note, however, that the modal nitrogen content of marine species is lower than that of freshwater species (Fig. 2-23, top). Marine species have a larger modal phosphorus content than do freshwater species (Fig. 2-23, bottom).*

The differences in nutrient contents of cells or tissues observed in Figs. 2-22 and 2-23 are likely to be due to differences in availability of nitrogen and phosphorus between fresh- and seawater (Fig. 2-21); this implies that in most marine waters nitrogen may be limiting to producers while in freshwater phosphorus is limiting. This contrast is corroborated by data in Fig. 2-22, where the dotted and dashed bars refer to N to P ratios measured in situations where the phytoplankton were clearly identified as being nitrogen or phosphorus limited. These data lead to the hypothesis that phosphorus limitation occurs more often in freshwater phytoplankton, while nitrogen-limited growth is more common in marine phytoplankton. The relationships described above are only descriptive correlations; more critical tests of the hypo-

---

*The anomalous high frequency of very low (most are 0 or trace) phosphorus contents in Fig. 2-23 is due to coralline algal species whose cells are encrusted with calcareous deposits that increase the total weight. The data on these species are thus difficult to compare to other algal taxa. The very high values of phosphorus for freshwater plants are from sedges (*Cyperus*) in environments with high phosphorus. Other *Cyperus* species growing in low phosphorus environments lie in the class with lowest P content.

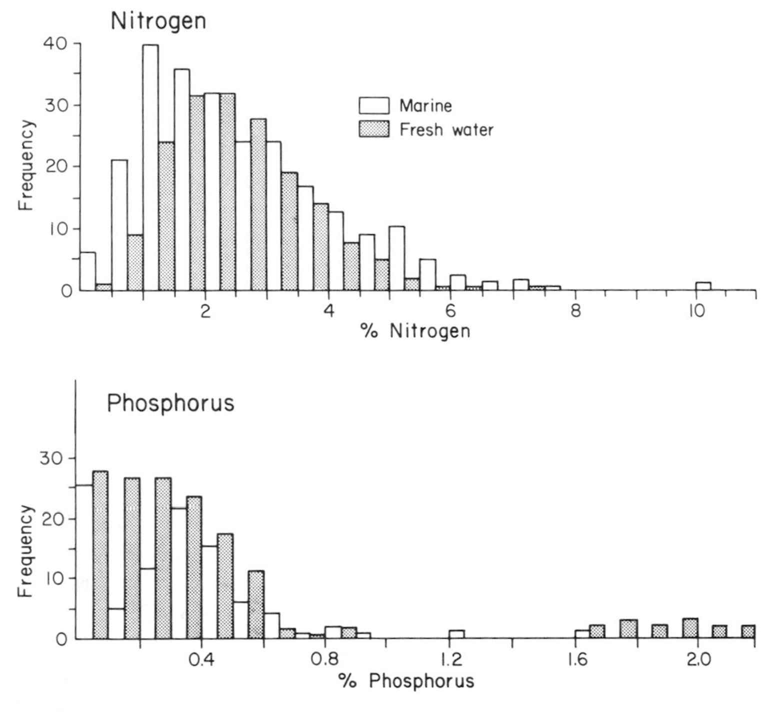

FIGURE 2-23. Nitrogen and phosphorus contents of marine and freshwater macroalgae and vascular plants. Compiled from many sources. The marine data from Vinogradov (1953) and articles in *Limnology and Oceanography, Journal of Phycology, Aquatic Botany, Botanica Marina, Marine Biology,* from 1973 to 1983.

thesis have been obtained by the experimental work done in freshwater, coastal, and oceanic waters that is described in the next section.

2.2.3.3 Evidence for Nutrient Limitation in Phytoplankton

The concept of nutrient limitation of producer growth and production has been applied in different ways. Before we review evidence for nutrient limitation, it is useful to examine different applications of the concept (Howarth, 1988; Codispoti, 1989).

First, nutrients could be said to limit growth of those producer populations that are currently present in the water body. Goldman et al. (1979) implicitly applied this use of nutrient limitation. The implications of their conclusions are discussed in Section 2.2.3.3.3.

Second, it has been argued that total production by all components within ecosystems could be limited by available nutrients (Smith, 1984, 1992). The effect of nutrient supply, use, and transformations on net ecosystem production (the difference between gross primary production and respiration by all organisms in the ecosystem) is examined by stoichiometric calculations based on accepted reactions. This version of the concept is discussed in Chapters 13, 14 and 15, where the topics include dynamics and stoichiometry of organic matter and nutrients.

Third, the post common use of the concept of nutrient limitation is that net primary production of the assemblage of producer species increases after the rate of supply of the limiting nutrient increases. The response of the system may be through increases in activity of estant species or by changes in the assemblage. As Howarth (1988) points out, this version of the concept is the one most concerned with the effects of eutrophication, where nutrient supply rates to the system increase. Here our concern is primarily with this usage of nutrient limitation.

2.2.3.3.1 Evidence from Freshwater Studies. For many years there was controversy among researchers working on freshwater systems about which element limited growth of freshwater producers. This became an especially important subject with the advent of phosphate-containing detergents, since the resulting phosphate-laden waste water entered water bodies and created eutrophic lakes and ponds, and obnoxious algal blooms. A series of laboratory and field experiments, especially the results of whole-lake fertilization experiments in Canadian lakes (Schindler, 1974, 1977), provided conclusive evidence that phosphorus is the limiting nutrient in freshwater bodies.

A typical response of phytoplankton in a freshwater-dominated coastal pond is seen in Fig. 2-24, bottom, where a growth response to phosphorus addition* is clear. There is a further stimulation of growth when both nitrogen and phosphorus are provided.

A nutrient is limiting only when in short supply; if concentrations of nitrogen, for example, are high, other elements become limiting to growth. Chemostat experiments where the ratio of N to P is varied show that there is a threshold below which algal cells are nitrogen limited and above which the culture is phosphorus limited (Fig. 2-25, top and middle). When one nutrient is limiting, the nutrient in excess accumulates in the cells (Fig. 2-25, bottom), as can also be seen in Fig. 2-22.

Different species of phytoplankton grow best at different concentrations and ratios of essential nutrients. Depending on the ambient ratios, therefore, competitive exclusion, or replacement of one species

---

*This kind of laboratory experiment is artificial in that sedimentation and grazing losses from the water are curtailed. Further, the regeneration of nutrients provided by grazers is also absent. Nonetheless, such experiments are a convenient, simple description of the situation.

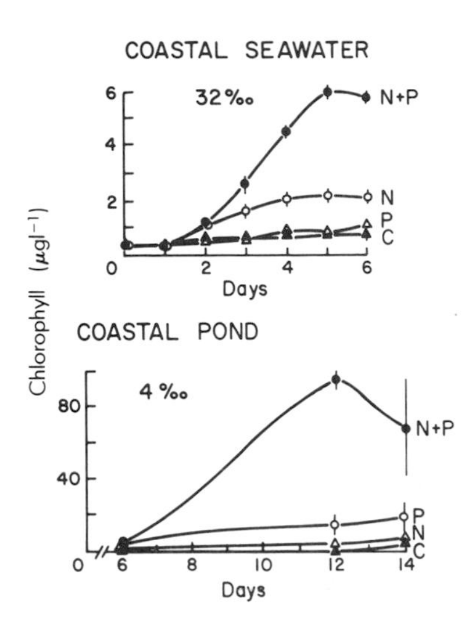

FIGURE 2-24. Enrichment experiments with coastal seawater of Vineyard Sound (salinity 32‰), Massachusetts, and a freshwater dominated coastal pond (salinity 4‰) in Falmouth, Massachusetts. N+P, addition of nitrogen and phosphorus; P, addition of phosphorus; N, addition of nitrogen; C, control, no nutrient addition. Adapted from Vince and Valiela (1973) and Caraco et al. (1982). Values are mean ± standard error of several replicates.

by another, can occur. Experiments where various concentrations of different kinds of nutrients are provided to natural phytoplankton do not, however, always provide easily interpretable results (Frey and Small, 1980). So far, it has not been possible to explain the species composition of phytoplankton only on the basis of the suite and concentration of nutrients available to the algae.

It seems odd that assemblages of multiple species of algae are the rule in the water column, an environment that appears to be very

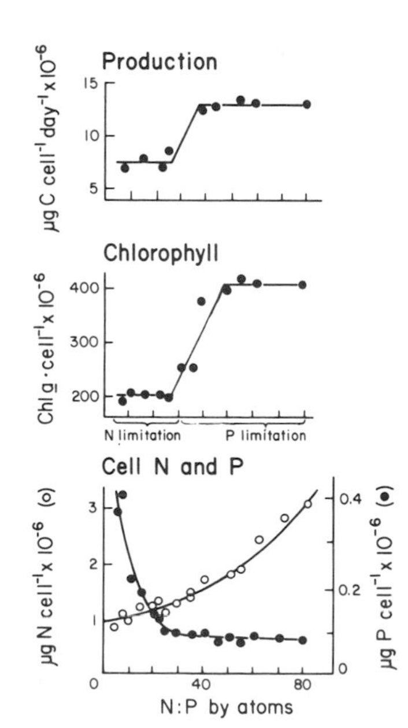

FIGURE 2-25. Production, chlorophyll, nitrogen, and phosphorus contents of *Scenedesmus* sp. in chemostats with varying N:P ratios. The N:P applies to both the cells and water in the system. Adapted from Rhee (1978).

homogeneous. This is Hutchinson's (1961) "paradox of the plankton." In a homogeneous environment competition should have long ago led to dominance of the competitively superior species. Richerson et al. (1970) argue that, in fact, conditions in local, small-scale parcels of water change frequently enough so that the water column is quite patchy. Such local heterogeneities may consist of differences in nutrient content, or presence or absence of grazers. This local heterogeneity can potentially prevent one species from continually having a competitive advantage in and therefore monopolizing all patches. This "contemporaneous disequilibrium" may therefore be behind the presence of multispecies phytoplankton floras.

2.2.3.3.2 Evidence from Studies in Coastal Waters. In coastal marine systems, nitrogen is the principal limiting element, as demonstrated by field observations and experiments. One of the early demonstrations of the role of nitrogen in coastal waters is the study by Ryther and Dunstan (1971) off New York Harbor, where organic particles and nutrients are high and decrease with distance from land (Fig. 2-26, top and middle). River and other freshwater inputs are often the major source of nutrients for some coastal waters. The decrease of dissolved inorganic nitrogen away from land is especially marked and suggests that this is the limiting nutrient. Stronger evidence is provided by experiments where seawater from a series of stations in the New York Bight area received additions of ammonium and phosphate (Fig. 2-26, bottom). Enrichment with nitrogen increased growth of the algae in 14 out of 15 stations, while additions of phosphorus did not increase growth. The stations nearest the nutrient source (a combination of eutrophic river water and urban sewage) were so nutrient rich that no response to further nutrient addition was seen. The phytoplankton from deeper waters responded to nitrogen additions, but the blooms were much smaller than those of harbor or shelf phytoplankton. This may be a secondary limitation due to a lack of trace nutrients or chelators of toxic substances, a topic discussed below.

The results of Ryther and Dunstan (1971) corroborated earlier conclusions by G. Riley and J. Conover based on interpretation of seasonal cycles of nutrients and plant growth, and also agreed with the results of the large bag experiments (Antia et al., 1963). Many other subsequent enrichment experiments (Fig. 2-24, top and Laws and Redalje, 1979, for example) in coastal water also support the conclusion that nitrogen is the primary limiting nutrient for coastal phytoplankton. This basic finding has fundamental applied importance: the nitrogen content of any effluent entering coastal waters is of considerable importance to management of the coastal zone, since it is closely linked to the level of primary production. The key effect of nitrogen loading rates (rather than concentrations) on entire estuarine systems has been

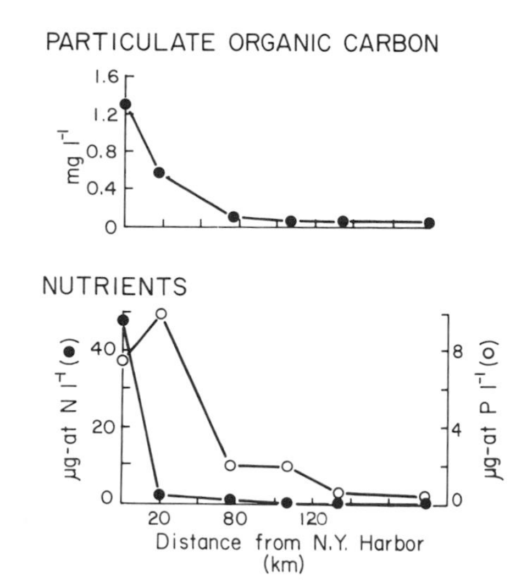

ENRICHMENT EXPERIMENTS

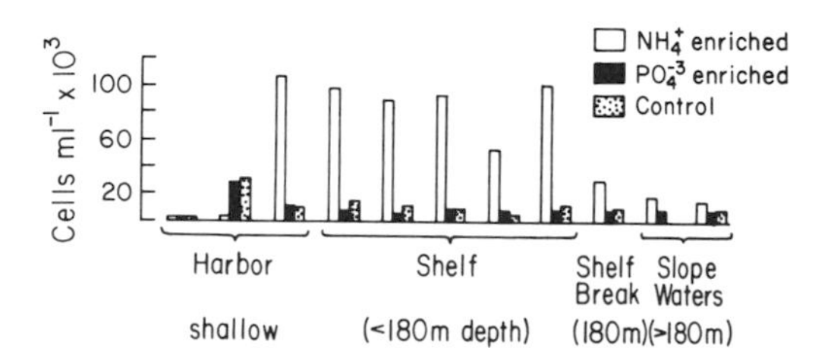

FIGURE 2-26. Top and middle: Concentrations of particulate organic carbon and nutrients in surface water in a transect from New York Harbor to offshore. Bottom: Growth of *Skeletonema costatum* in water samples that were enriched with ammonium or phosphate and in unenriched samples. The sequence of stations is in relation to their distance from the source of nutrients in New York Harbor. The inoculum with which the experiments were started was of the same size as the left-most station in the graph. Adapted from Ryther and Dunstan (1971).

documented in enrichment studies done in large tanks (5 m deep, 13 m^3 volume). Tanks subject to larger nitrogen inputs supported larger phytoplankton production (Fig. 2-27, top). Correlational evidence from many different marine systems that shows increased phytoplankton production in waters that receive larger nitrogen loads (Fig. 2-27, bottom). Similar comparisons with phosphorus loads do not show the same clear relationship.

Other nutrients, however, can be secondarily limiting. Once a nitrogen source is provided, there is a secondary limitation due to phosphorus (Fig. 2-24), so that addition of both N and P results in greater growth (Rudek et al., 1990; Fisher et al., 1992). Thus, if nitrogen is in excess of Redfield ratios in wastewaters entering a body of water, the added phosphorus may also be important. The supply of N and P may depend on the rates at which fresh and coastal water deliver nutrients to estuaries. During spring, when Chesapeake Bay systems are most influenced by river flow, P seems to restrict phytoplankton growth, while when waters are saltier, N resumes its limiting role (D'Elia et al.,

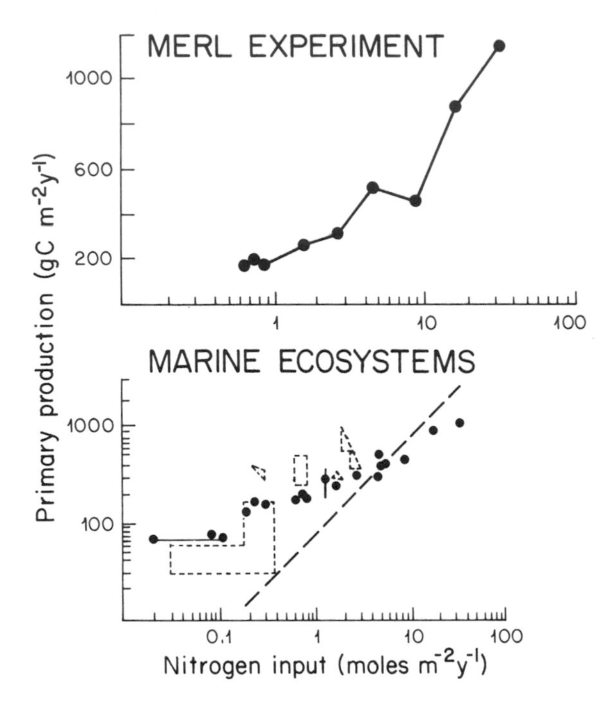

FIGURE 2-27. Relation of input of dissolved inorganic nitrogen and primary production by phytoplankton in MERL (Mesocosm Ecosytems Research Laboratory) experimental tanks (top), and in a variety of marine ecosystems (bottom). The dashed line represents the Redfield ratio of C to N (6.625) to show roughly how much C could be fixed at any given N input without any recycling of N. The polygons encompass many points from a system. Adapted from Nixon (1992).

1986; Fisher et al., 1992). In shallow coastal bays, the switch between N and P limitation seems to take place when salinites are about 10‰ (Caraco et al., 1991). In a few places phosphorus supply may limit phytoplankton. In the eastern Mediterranean, for example, phosphate in the water may absorb to iron oxide-rich dust particles, and be lost from surface waters as the particles sink below the photic zone (Krom et al., 1991). In many tropical coastal water bodies carbonate-rich sediments absorb phosphorus, and production by macroalgae is probably limited by phosphorus supply; in such environments ratios of N to P in producer tissues range 22–124, compared to 8–26 for temperate waters (Lapointe et al., 1992).

Other essential or toxic elements may also be important. In recently upwelled coastal waters rich in nitrate the availability of dissolved metals may limit phytoplankton production. Iron and manganese, essential components in many biochemical reactions, may be limiting. Cupric ion at certain concentrations may reduce phytoplankton growth.

On the other hand, appropriate concentrations of organic chelators may reverse the toxic effect of cupric ions (Sunda et al., 1981). It is not at all clear whether the role of such elements is widespread, or if they are only of local importance.

2.2.3.3.3 Evidence from Oceanic Regions. Nitrogen may also be the primary limiting element in nutrient-poor oceanic waters (Thomas, 1969). Nitrate and ammonium may be undetectable near the sea surface (McCarthy, 1980), undoubtedly because of uptake by phytoplankton. Sharp et al. (1980) find that phytoplankton grow slowly in the oligotrophic waters of the North Pacific, with doubling times of 5–9 days. The nitrogen-to-carbon ratios in the particulate matter are considerably lower than the Redfield ratios, suggesting a lack of nitrogen, a conclusion corroborated by Perry and Eppley (1981) based on uptake of radioactive isotopes.

Goldman et al. (1979) claim that despite low concentrations of nutrients in oceanic water, populations of oceanic phytoplankton can grow at high rates. In the oceanic waters of the Sargasso Sea, specific growth rates of phytoplankton cells range between 0.05 to about 1 doubling per day. Goldman et al. (1979) also argued that continuous laboratory cultures of single species of marine phytoplankton achieve Redfield ratios of carbon, nitrogen, and phosphorus when growing at maximal rates. Since oceanic phytoplankton often show elemental Redfield ratios, Goldman et al. (1979) suggested that in fact, growth of oceanic phytoplankton was likely to take place at similarly maximal rates, and by implication were not nutrient-limited, even in oligotrophic waters.

Furnas (1991) moreover, reports that doubling rates of marine phytoplankton vary widely, from less than 1 to >5 doublings d^{-1}, so that phytoplankton growth rates can be considerable higher than those in the Sargasso Sea. Nonetheless, some ideas have been proposed to reconcile the co-occurrence of relatively fast growth and low ambient nutrients. The first idea is that oceanic phytoplankton may saturate uptake rates and growth rates at very low nutrient concentrations, as in examples given by Goldman and McCarthy (1978) and McCarthy (1980). It is not known how common this is or if artifacts of measurement of nutrient uptake at low S are important. Oceanic assemblages may be made up of relatively rare phytoplankton species that are physiologically adapted to efficient and fast uptake of nutrients at low concentrations.

The second idea is based on the notion that the average bulk concentration of a nutrient in a typical sample of several liters may not be relevant to an algal a few to tens of micrometers in diameter. An algal cell extracts nutrients from a few microliters of surrounding seawater, not from larger volumes of seawater. It may be that small parcels of water exist where the concentration of nutrients is considerably higher than in the rest of the water volume. McCarthy and Goldman (1979), Turpin and Harrison (1979), and Lehman and Scavia (1982) hypothesized that excretion by zooplankton could produce tiny, nutrient-rich plumes that may be used by oceanic algae.

There are at least five difficulties with this second hypothesis. First, with current techniques it is difficult to measure nitrogen compounds in samples of a few microliters; new methods need to be developed to test directly the existence of such tiny parcels of high nutrient water. Second, recall that nonlimiting nutrients accumulate in internal pools (cf. Fig. 2-24). If some other essential element rather than N or P is limiting, the concentrations of N and P, and the N:P ratio could be high simply because the cells accumulate nitrogen and phosphorus while their growth or division are arrested by lack of something else, perhaps a trace element. Third, diffusion of nitrogen from tiny plumes may be faster than the rate of nitrogen uptake by cells, so that the plumes may be too short-lived to be useful to phytoplankton (Jackson, 1980). Fourth, the high-nutrient parcels provided by zooplankton at reasonable densities are too sparse to supply sufficient nutrients to support phytoplankton (Currie, 1964). Fifth, because C, N, and P are not the only elements that are involved in setting growth rates of phytoplankton, ratios of C, N, and P do not provide unequivocal indications of relative growth rate (Greene et al., 1991).

Although the issue of the use of small patches by algae is not settled, excretion from animals must be a major pathway by which dissolved nutrients are regenerated in the oceanic water column (McCarthy, 1980; Eppley, 1980). It may be, however, that smaller animals are involved in most of regeneration (Chapters 9 and 13).

A third explanation of high growth rates in nutrient-poor waters may be that there are high concentrations of nutrients associated with amorphous aggregates of organic matter ("marine snow") (Riley, 1963b; Shanks and Trent, 1979; Alldredge and Silver, 1988). Relatively high rates of production and respiration are associated with the layer of seawater surrounding these fragile particles. Presumably, phytoplankton cells in and near such particles could have access to relatively high nutrient supplies.

It has to be true that those species that are present at any one time in water should be well-adapted to grow in ambient conditions; it may also be true that additions of nutrients may not increase growth of these extant species. We would suppose, however, that in the face of an added resource, even if the extant species do not increase in abundance, the species assemblage would change, because species capable of exploiting the higher rates of supply of the limiting resource would dominate, and growth of the newly developed assemblage could increase. In sum, nutrient ratios of extant species provides some information about the physiological state of those species, rather than about behavior of the productive system under increased nutrient regimes.

The hypotheses offered by Goldman et al. (1979) and McCarthy and Goldman (1979) stimulated new questions. Critical examination and testing of such insights in the lab and field will further understanding of the processes controlling primary production in the sea. These hypotheses deal with specific growth rates or the doubling time of cells. We still need to make the jump from this physiological measure of growth of a specific population to a measure of production by the assemblage of producers present. Consider that oceanic phytoplankton occurs at very low cell densities. The production of phytoplankton in oceanic water, even with a division rate of four to five times a day, and average density of, say, 5 cells ml^{-1}, will not compare favorably with that of coastal water with a cell density of, say, 10,000 cells ml^{-1}, even if the coastal phytoplankton divide only once every 2 days. We need to understand not only the control of cell doubling times in individual species but also the control of abundances and seasonal cycles in of species assemblages oceanic water. This is discussed further in Chapters 8 and 15.

In oceanic waters it is possible that trace elements play more important roles limiting production rates that in coastal waters. There are some oceanic areas that are relatively nutrient-rich (more than 2–6 μM NO_3-N). Such areas include the Equatorial and Subarctic Pacific, and the Southern Ocean surrounding Antarctica. In the Equatorial Pacific, experimental addition of nutrients to the water column did not result in enhanced growth, much like some of the results obtained in enrichment experiments off New York. In these specific situations phytoplankton do not seem to be limited by nitrogen or phosphorus. We have so far emphasized nutrients—N and P—used in large amounts by producers,* but in oligotrophic waters limitation of growth by trace elements may be important. In very oligotrophic lake water, the short supply of molybdenum restricts algal growth (Goldman, 1972). In oligotrophic marine situations there have been a series of observations implicating iron as the primary controlling factor in the Sargasso Sea (Menzel et al., 1963) and in the Indian Ocean (Tranter and Newell, 1963). Early experiments with iron additions, however, did not consistently stimulate growth in experiments in the Central Pacific (Thomas, 1969).†

---

*There are low but detectable concentrations of Si over much of the oceans. In upwelling regions intense bloom of diatoms may deplete silicon, but there are few instances of direct evidence of Si limitations for the sea (Paasche, 1980).

†F. Morel (personal communication) proposes that this is because most iron exists as Fe^{3+} in seawater. Phytoplankton take up Fe^{2+} more readily than Fe^{3+} and have to rely on photoreduction of Fe^{3+} to Fe^{2+} to be able to take up iron. The iron added in the experiments may have been—or may have been quickly converted to—oxidized iron.

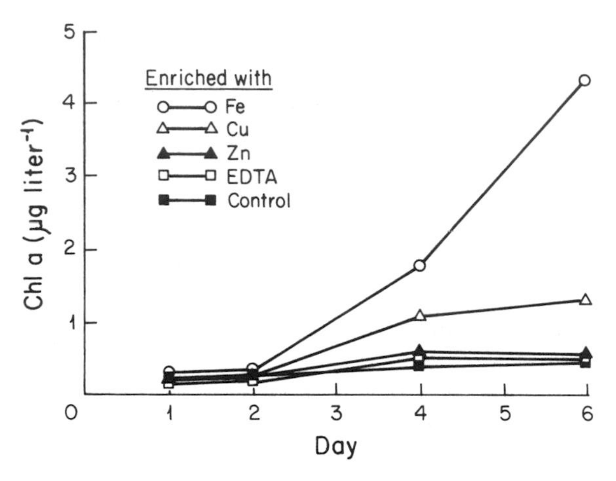

FIGURE 2-28. Time course of increase in chlorophyll a ($>0.7\,\mu$m) for different enrichment treatments and a control. EDTA is an organic metal chelator, which would bind metals and so acts as a further control. Adapted from Coale (1991).

Martin et al. (1991) concluded that in nutrient-rich oceanic waters far from continents (Antarctic Ocean, Gulf of Alaska, and the Equatorial Pacific), iron was the principal limiting factor for phytoplankton production. Shipboard iron enrichments using ultraclean methods increased phytoplankton division rates by about two-fold over unenriched controls. Chlorophyll concentrations increased in response to iron enrichments of Subarctic Pacific (Fig. 2-28) and pelagic Antarctic waters (Helbling et al., 1991). Further experiments showed that, at least in the equatorial Pacific, diatoms were limited by both iron and macronutrients, but prochlorophytes were not (DiTullio et al., 1993). Although there are differences of opinion regarding interpretation (Banse, 1991), limitation of phytoplankton growth by iron supply may be important in significantly large areas of the oceans.

The low chlorophyll concentrations found even in high-nutrient equatorial waters, however, are not just an expression of iron deficiency. Enrichments do increase phytoplankton growth, but doubling rates of phytoplankton in unenriched water from such areas is about once per day. At such doubling, rates increases in chlorophyll should be rapid. Losses of phytoplankton by sinking and vertical advection are at most a few percent per day, so only grazing remains as the brake that could maintain the low chlorophyll concentrations in these areas of the ocean (Banse and English, 1994). Grazing pressure has to be high, particularly on prochlorophytes (Cullen et al., 1992; DiTullio et al., 1993), and it is probably the proximate control on phytoplankton abundance, while iron availability may ultimately control doubling rates.

The issue of possible iron limitation (Chisholm and Morel, 1991) has received much public attention because it has raised the possibility that managed iron fertilization of large areas of the Southern Ocean, for example, could be a way to prompt phytoplankton to capture at least some of the carbon dioxide that is now accumulating in the atmosphere. The captured carbon would, it is claimed, then sink. Thus the carbon would be sequestered in the deep ocean. Not everyone agrees that the iron fertilization would work as proposed (Peng and Broecker, 1991; Fuhrman and Capone, 1991). Much more work is needed in this field, especially in regard to nutrient-metal interactions.

Trace substances may not only limit growth but also inhibit growth (Steemann Nielsen and Wium-Anderson, 1970). Cupric ion activity as low as $4 \times 10^{-11} M$ can reduce cell mobility in red tide organisms (Anderson and Morel, 1978). Moreover, there is evidence that synergistic and antagonistic effects among Cu, Mn, Fe, and Zn may influence productivity in oceanic areas away from land (Bruland et al., 1992). Other substances, including some organic compounds, can also inhibit growth of microalgae and bacteria (Harrison and Chan, 1980; Conover and Sieburth, 1968; Sieburth and Conover, 1965). Trace metals can be chelated by natural organic matter. Humic substances are products of decomposition of plant materials that bind copper and other toxic ions so as to reduce the activity of metals in solution (Sunda and Guillard, 1976; Sunda and Lewis, 1978; Prakash et al., 1973; Barber et al., 1971). Such organic compounds are often transported by river water into the sea, where they enhance growth of diatoms and dinoflagellates (Fig.

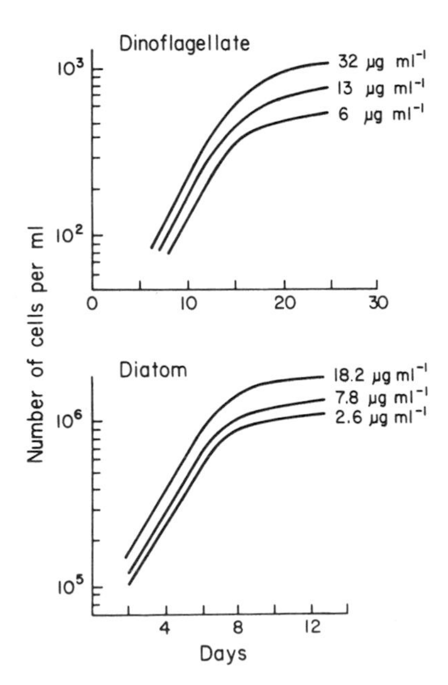

FIGURE 2-29. Top: Growth of the dinoflagellate *Gonyaulax tamarensis* in synthetic marine medium enriched with different amounts of humic acid extrated from marine sediments. Bottom: Growth of *Skeletonema costatum* in enriched seawater medium under additions of different amounts of humic acid extracted from decomposed seaweed (*Laminaria*). Adapted from Prakash et al. (1975).

2-29). Perhaps this is why red tide blooms are often recorded after periods of high discharge by rivers.

2.2.3.4 Evidence for Nutrient Limitation in Macroalgae and Vascular Plants

The relationships between ambient nutrients, uptake, storage, and growth by macrophytes are similar to those for one-celled algae. Growth of marine macroalgae and vascular plants is primarily limited by nitrogen supply. Experimental additions of nitrogen to salt marsh vegetation (Fig. 2-30) or to macroalgae (Peckol et al. 1994, and Fig. 2-31, top) result in increases in growth rate. As ambient nitrogen increases, growth rate increases up to a point, as in the case of phytoplankton. Beyond such a threshold, nitrogen is stored with no further increase in growth rate (Fig. 2-31, bottom). The lack of further increase is due to a secondary limitation by some other factor, again as in phytoplankton. In salt marsh plants, phosphate is the secondarily limiting factor (Fig. 2-30). As in the case of phytoplankton, growth of some brackish water benthic macrophytes, and of epiphytes growing on the macrophytes, can be N- or P-limited, depending on the time of year (Neundorfer and Kemp, 1993). At least one seagrass, *Zostera marina*, may not be nutrient-limited, judging by experiments in which sediments were enriched (Dennison et al., 1987; Short, 1987) and by model simulations of light and nutrient effects (Zimmerman et al., 1987). Other sediment enrichment studies, however, show some effects of nutrient additions on eelgrass growth (Williams and Ruckelshaus,

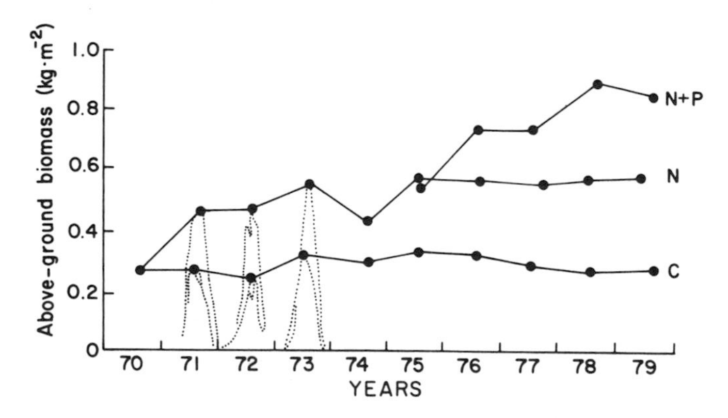

FIGURE 2-30. Response of above-ground standing crop of the salt marsh cordgrass (*Spartina alterniflora*) to experimental enrichment with nitrogen (1970–1979) and with a mixed nitrogen plus phosphorus fertilizer (1975–1979). Addition of phosphorus alone did not change the amount of biomass and is not shown. The dotted lines in 1971–1973 shows the seasonal pattern of growth of the plant. Adapted from Valiela et al. (1982).

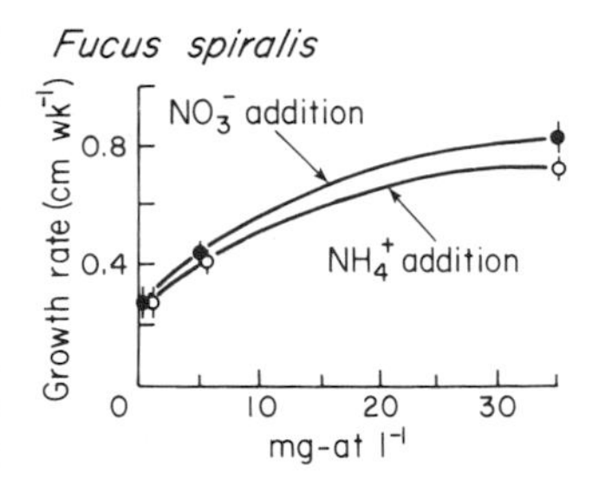

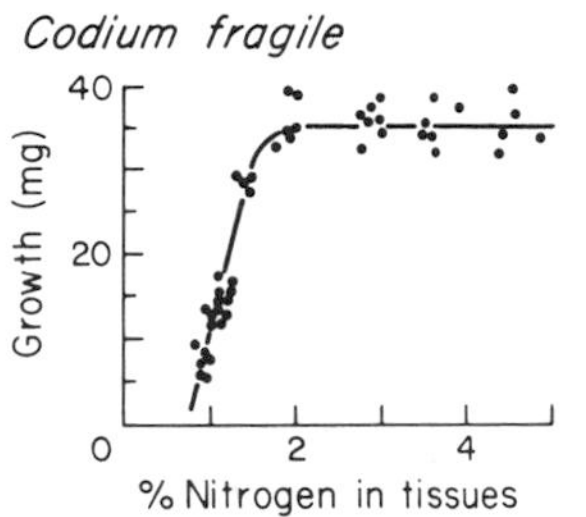

FIGURE 2-31. Top: Growth rate (elongation of fronds) of the brown alga *Fucus spiralis* in ambient concentrations (1.2 μg atom NO_3^- liter^{-1} and 1.7 μg atom NH_4^- liter^{-1} and in cultures where additional NH_4^- and NO_3^- were furnished. Adapted from Topinka and Robbins (1976). Bottom: Growth in weight of the green alga *Codium fragile* in relation to the percentage nitrogen in the tissues. Adapted from Hanisak (1979).

1993), so the issue is not settled. If the major nutrients occur at high concentrations, trace metals may limit growth in coastal macrophytes (Prince, 1974).

For rooted coastal plants where the sediment is anoxic, nutrient uptake may be inhibited by the reduced environment. Although nutrient concentrations may be high in interstitial water (Koch et al., 1990), anoxia inhibits the physiological mechanisms by which plants actively take up nutrients, so that in such plants nutrient limitation may be mediated by anaerobic conditions (cf. Section 15.3.3).

Nutrient and light supply may interact indirectly to determine growth rate, and even survival, of macrophytes. One of the first clues that a coastal water body is receiving increased nutrients, usually nitrogen, is the growth of epiphytic organisms on the surfaces of submerged macrophytes in response to increased nutrients in the water. The smaller, faster growing epiphytic species intercept light sufficiently to reduce growth of the macrophytes on which they grow (Orth, 1977; Tomasko and Lapointe, 1991; Neundorfer and Kemp, 1993). This indirect, nutrient-determined light limitation has been thought to be the cause of disappearance of many seagrass beds subject to nitrogen enrichment.

2.3 Temperature and Interactions with Other Factors

We will see in Chapter 15 that temperature does have some influence in seasonal production cycles, but there are few observations that demonstrate important effects of temperature on rates of primary

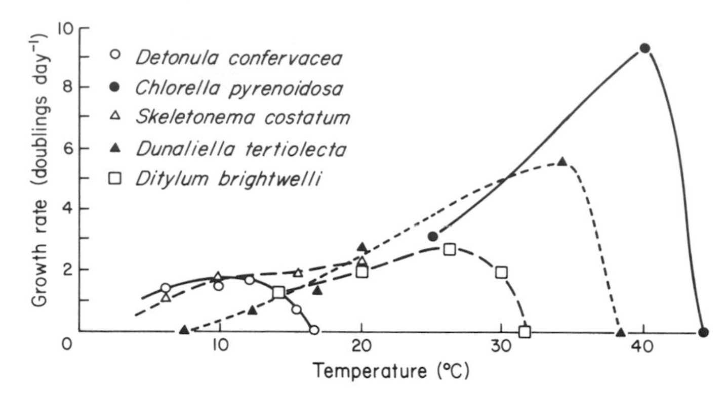

FIGURE 2-32. Growth rate of several phytoplankton species under different temperatures. Adapted from Eppley (1972).

production in the sea. In certain situations under clear ice in the Antarctic Ocean, the rate of photosynthesis and growth may be limited by low temperatures (Bunt and Lee, 1970). Phytoplankton photosynthesis was suppressed below 1°C, but potential production was as high at 2°C as at 8°C in Newfoundland Water (Pomeroy and Diebel, 1986). The effect of cold is not general, since in most temperate waters the spring bloom occurs at about the coldest part of the year.

There are measurable effects of temperature on growth rates of phytoplankton in laboratory cultures (Fig. 2-32), and local populations of many species of algae have peak growth rates in certain specific ranges of temperature (Hulburt and Guillard, 1968; Goldman and Carpenter, 1974). The doubling rates measured under optimal temperatures in the laboratory (up to 9 doublings day^{-1}) are in the range reported for the sea (Furnas, 1991).

The modest effect of temperature is further evidenced by the frequent occurrence of a poor correlation between ambient temperatures of the water where a particular species is found and its physiological temperature response. For example, *Thalassiosira nordenskioldii*, described as an Arctic species on the basis of where it occurs (Durbin, 1974), grows very well at temperatures warmer than 10°C. Such observations suggest that temperature does not normally determine the abundance and growth of algae in marine environments.

Temperature may be more important as a covariate with other factors than as an independent factor. For example, cells at low temperatures maintain greater concentrations of photosynthetic pigments, enzymes, and carbon (Steemann Nielsen and Jorgensen, 1968). This may result in more efficient use of light at low temperatures and higher photosynthetic rates of acclimated cells than would be predicted on the basis of temperature responses of unacclimated cells. Photosynthetic rates and growth rates of algae can be altered by light, as seen in Section 2.1.1.2, but increases in the ambient temperature can also increase maximal photosynthetic rates at any level of light intensity (Fig. 2-33, top left). In certain species, the day length may also significantly change rates of doublings, but does so only at high temperatures (Fig. 2-33, right). Higher temperatures also facilitate the uptake of nutrients (Fig. 2-33, top left), so that at any one concentration of phosphate in the medium, the maximum photosynthetic rate increases at warmer temperatures. In the case of algae under sea ice in the Antarctic Ocean, cited above, light was available, although low, while nutrients were high. In Tokyo Bay (Fig. 2-33, bottom left) nutrients were high, but the turbidity typical of shallow coastal waters created low light. From the data available it is difficult to say whether light or temperature was most important.

Though temperature may not be a primary limiting factor in the sea it may have other consequences. There are increasing numbers of nuclear power plants over many of the world's coasts; these plants use

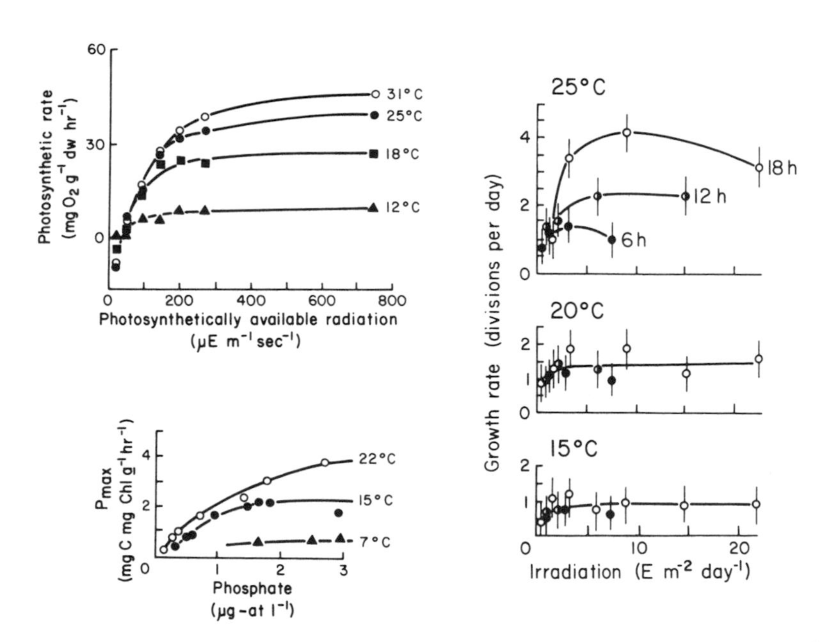

FIGURE 2-33. Interaction of temperature with light and nutrients. Top left: Photosynthetic rate of *Cladophora albida* under different levels of light intensity and temperatures in estuarine water. Adapted from Gordon et al. (1980). Right: Mean (±standard deviation) division rates during exponential phase of growth in *Talassiosira fluviatilis* at three temperatures and daylengths (18, 21, and 6 hr). Adapted from Hobson (1974). Bottom left: Maximum photosynthetic rate (P_{max}) of natural phytoplankton of Tokyo Bay under varying phosphate concentrations and temperatures. Adapted from Ichimura (1967).

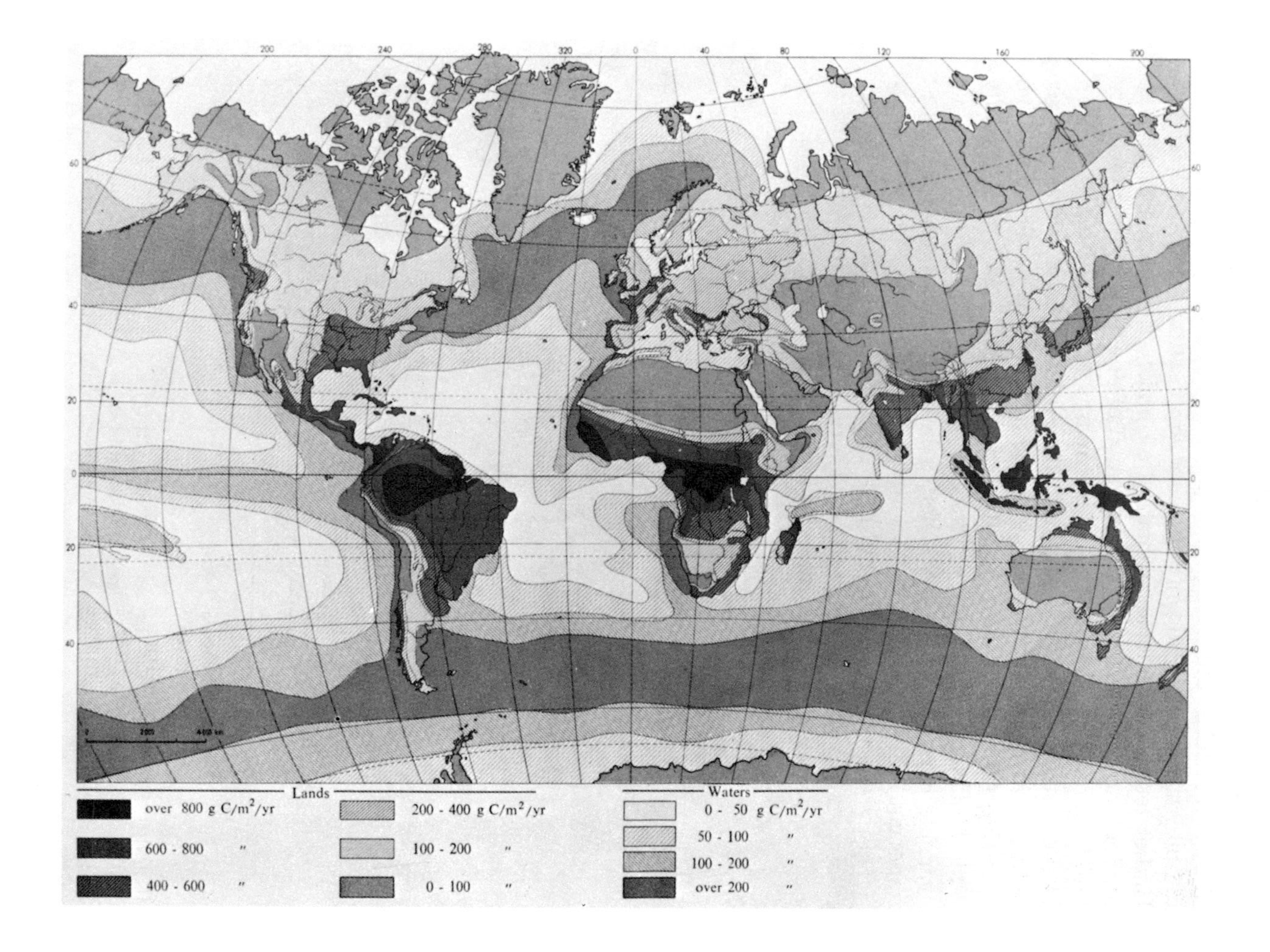
Lands
over 800 g C/m²/yr
600 - 800 "
400 - 600 "
200 - 400 g C/m²/yr
100 - 200 "
0 - 100 "
Waters
0 - 50 g C/m²/yr
50 - 100 "
100 - 200 "
over 200 "

seawater for cooling purposes and release heated water into the coastal zone. When the discharge results in increases in temperatures of the receiving water, changes in species composition may result. In freshwater, there is a transition from diatoms to greens to cyanobacteria in the immediate vicinity of the discharge canals. The abundance of certain dinoflagellates may increase in heated estuarine waters (Goldman and Carpenter, 1974). Disposal of heated water may therefore lead at least to changes in the species making up the phytoplankton community within the affected area. The species favored by higher temperatures, cyanobacteria and flagellates, moreover, may be less desirable as food for consumers than other kinds of phytoplankton.

2.4 Distribution of Phytoplankton Production Over the World Ocean

Knowledge of the concentration of nutrients, amount of light, and temperature are not sufficient to explain the distribution of primary production over the world's oceans (Fig. 2-34; also see cover illustration). A reductionist explanation including only these variables fails to explain the observed complexity. These variables certainly provide the basis for potential production, but actual production is the result of the movement of water masses coupled to nutrient and light conditions. Hydrography largely determines what the nutrient supply is within the photic zone, and how deep phytoplankton are carried by vertical mixing of water and hence how much light is available.

There are texts that describe water mass movements in the oceans (Pickard, 1964; Dietrich, 1968; Gross, 1977). Here we will only briefly discuss a few examples, enough to convey the idea of how major the effects of hydrography can be.

The most dramatic evidence of how hydrography controls production comes from regions of upwelling. Where upwelling of deeper water or other sorts of advection delivers nutrients in substantial amounts, primary production may be high. There are several types of hydrographic features where vertical advection of nutrients takes place, including coastal upwellings and divergences.

Coastal upwellings are common off Peru, northwest Africa, eastern India, southwest Africa, and the western coast of the United States. Their occurrence is marked by areas of high production located mainly along the western margins of continents (Fig. 2-34). Prevailing winds

←

FIGURE 2-34. Estimated world distribution of annual production. Notice that land patterns are markedly different from oceanic distribution of productivity. From Lieth (1975).

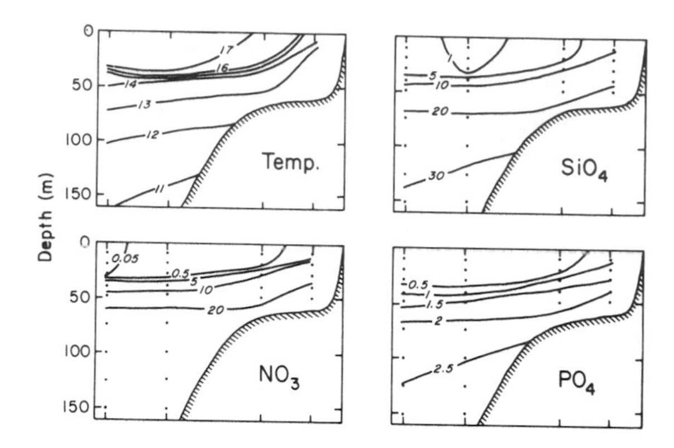

FIGURE 2-35. Distribution of temperature gradients, silicate, nitrate, and phosphate in the upwelling zone off Punta San Hipolito, Baja California. Temperature is in °C and nutrients in microgram-atoms per liter. The dots show the stations where water samples were taken. The farthest station was about 40 km from shore. Note the upward trend of the isoclines as the water approaches the coast. Adapted from Walsh et al. (1974).

blowing somewhat parallel to a coastline move water in the direction of the wind; forces resulting from the rotation of the earth deflect the actual path of the water. In the western margin of continents, this Coriolis effect results in surface water being moved offshore. Colder, nutrient-laden deeper water moves up to the surface to replace displaced surface water (Fig. 2-35). This is far too simple a description of the dynamics of upwelling; the actual movement of water and nutrients in coastal upwellings is very variable and the spatial and temporal patterns are complex (Boje and Tomczak, 1978) (cf. Fig. 15-9).

The effects of the rotation of the earth on wind-driven currents are reversed in the two hemispheres, so in the western margin of continents northerly winds result in upwellings in the northern hemisphere while southerly winds lead to upwellings in the southern hemisphere. The Coriolis effect does not lead to high productivity in eastern margins of continents because the deflection moves nutrient-poor oceanic water toward the coast. This may also result from northerly winds in the southern hemisphere and southerly winds in the northern hemisphere. An example of the latter is the "downwelling" in the Washington-Oregon coast during winter (Duxbury et al., 1966).

Upwelling of nutrient-rich water may occur through a variety of hydrological mechanisms and is not limited to the coastal zone. In the open ocean divergences bring nutrient-rich water to the surface. A prominent feature of the Pacific is the fast subsurface Equatorial Undercurrent, leading from west to east in a fairly narrow band around the

Equator. Near the Equator surface currents are slow. The contact of the moving equatorial surface water and the fast Equatorial Undercurrent results in shear stress that leads to a large amount of turbulent mixing and a renewal of nutrients in the photic zone. This produces a band of relatively high production (Table 2-3) in the mid-Pacific near the Equator (Fig. 2-34; also see cover illustration).

Another example of offshore upwelling of nutrient-rich water is the zone of high production found in the Southern Ocean (Fig. 2-34)—the waters surrounding Antarctica associated with the Antarctic Divergence (Fig. 2-36). We should note parenthetically that the high values of production reported in our map may reflect the concentration of studies in Antarctic waters during the spring and summer, times when phytoplankton production rates reach their highest levels (Glibert et al., 1982). The water near the surface is exposed to westerlies that drive the eastward-moving Antarctic Circumpolar Current. A northward component of flow of this current moves deeper water to the surface. This newly upwelled water, now called the Antarctic Surface Water (Fig. 2-36) provides a very rich supply of nutrients that supports the very productive assemblage of phytoplankton, krill, penguins and other sea birds, seals, and whales of the Southern Ocean (cf. Section 9.4.3). The upwelled water extends far from Antarctica; the northern limit of this zone of high production is marked by the Polar Front or Antarctic Convergence, near 40–50°S, where the still-rich Antarctic surface water sinks below the warmer subantarctic surface water (Fig. 2-36). After

TABLE 2-3. Estimated Nutrient Input, Primany Production and Other Properties of Three Different Subtropical Planktonic Ecosystems.[a]

	Coastal upwelling	Equatorial divergence	Central gyre
Vertical velocity[b] (cm sec^{-1})	10^2–10^3	10^{-3}–10^{-4}	10^{-4}–10^{-5}
NO_3-N (μg atom l^{-1}) at 100 m	20–30	20–30	0–5
Primary production (g m^{-2} day^{-1})	1–10	1–5	0.1–05
Phytoplankton standing crop (mg C mg $chl.^{-1}$ hr^{-1})	60–180	45–90	15–30
Size of cells	Large (variable)		Small
Hervivore mean weight (μg C $animal^{-1}$)	40–100	4–40	4–40
Herbivore mobility	High	Lower	Low
Temporal variability of herbivore populations	High	Lower	Low
Steps in the food chain	1–2	3–4	5–6
Yield at top of food chain	High	Intermediate	Low

[a] Adapted from Walsh (1976).

[b] These are estimates of flow when the upwelling is active. At other times, there may be no vertical movement of water and nutrients.

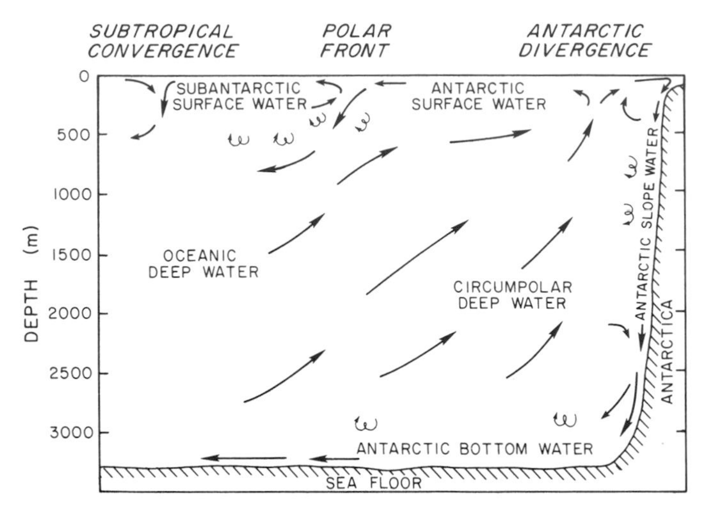

FIGURE 2-36. Schematic representation of flow of water masses in Southern Ocean, simplified from Gordon (1971). The arrows show direction of flow while the curlycues show areas of mixing.

sinking, the Antarctic surface water is renamed the Antarctic Intermediate Water and goes on to influence most of the oceans of the southern hemisphere. The northward-flowing Antarctic water masses are eventually warmed and their oxygen is consumed by biological activity deep in the North Atlantic. Eventually the deep water flows southward again. The conversion of deep water to surface water and back again with the consequent reacquisition of oxygen has been termed the "breathing" of the ocean (Gordon, 1971) and a key role in this process is played by the Southern Ocean.

Another prominent band of high annual production is at intermediate latitudes in the northern hemisphere. Its latitudinal location is more variable than that of its counterpart in the southern hemisphere due to the influence of land masses on circulation. Its geography is associated with the large-scale, clockwise gyres that circulate in the North Atlantic and North Pacific, and the western boundary currents (the Gulf Stream and the Kuroshio). Such major hydrographic phenomena lead to active turbulent mixing and larger eddies that result in renewal of nutrients in surface waters and hence in high production.

Other major areas of high production are the coastal zones (Fig. 2-34), where water- and wind-borne nutrients and particles originating from land are introduced into the sea. The shallowness of coastal water often allows surface disturbances to mix the entire water column vertically, a phenomenon that enhances transport of nutrients regenerated from the sediments into the photic zone. This frequent mixing, as we will see in Chapter 14, is one reason for high coastal primary production.

The low annual production rate evident in the Arctic Ocean is related to reduced amounts of light due to atmospheric filtering and the persistence of ice cover most of the year. Everywhere else in the ocean it is very hard to see broad regional effects of light limitation. For example, in the large subtropical gyres on either side of the Equator in the Atlantic and Pacific, where light is plentiful year-round, we find some of the lowest rates of primary production (Table 2-3). In subtropical gyres such as the Sargasso Sea and in other open ocean environments, stable thermoclines prevent or reduce vertical mixing. The consequence of this is that nutrients, especially nitrogen, are very dilute or undetectable; production is thus characteristically low in much of the world ocean.

There are many other features of Fig. 2-34 that merit attention, but the few examples mentioned emphasize the major points of this section. We have seen some evidence suggesting that low light may affect annual production; we will see more of this in Chapter 15, but if light were the dominant variable, one would not expect to find the distribution of production of Fig. 2-32. The same comments apply to temperature. In contrast, even a schematic representation of areas or relatively high physical mixing, which leads to delivery of nitrate to the photic zone, shows remarkable parallels to the geographical distribution of primary production (Harrison and Cota, 1991). In general, nutrients are the major factor limiting primary production in the world's oceans. The delivery and availability of nutrients are mediated in a major way by the hydrography of the oceans, and so hydrography and nutrient supply combine to establish the broad zones of primary production.

The same mix of effects of nutrients, light, and hydrography controls production by macrophytes in the margins of the sea, but local patterns of light and nutrients may be relatively more important to macroalgae and vascular plants than the regional patterns so obvious in phytoplankton. For example, salt marsh plants, sea grasses, or brown algae growing only meters apart can show differences in biomass and production as large as between their distributional extremes.

Note added in proof: Observations of low phytoplankton production in certain oceanic areas where nitrogen concentrations were relatively high, and small scale experiments that showed growth limitation by iron (Fig. 2-28) prompted an enrichment experiment in which iron supply was enhanced within an area of 64 Km^2 in the equatorial Pacific (Martin et al., 1994; Kolber et al., 1994). The iron enrichment doubled phytoplankton biomass and quadrupled production, thus showing that iron was indeed limiting producers. The growth, however, was insufficient to deplete the supply of inorganic nitrogen. Similar conclusions were reported for the Southern Ocean by de Baar et al. (1995). Other factors, probably vertical mixing and grazing by zooplankton, must be responsible for preventing depletion of nutrients by phytoplankton.

Part II
Consumers in Marine Environments

Comb jellies (*Mnemiopsis*), photo courtesy of Gray Museum, Marine Biological Laboratory, Woods Hole, Mass. U.S.A.

Organisms that depend on organic matter produced by photosynthetic organisms range from bacteria to whales and are astonishingly diverse. Such consumers are heterotrophic and engage in secondary production.

Marine bacteria are hard to identify and many species have probably never been cultured. Some of the kinds found in the sea are discussed in texts such as Kriss (1963) or Fenchel and Blackburn (1979).

The small invertebrates are better known but still need work. Macroinvertebrates have been studied more thoroughly (see reviews by Barnes, 1980; Kaestner, 1970; Andersen, 1969). There are, for example, about 9,000 species of cnidarians, 20,000 molluscs (mainly bivalves), 8,000 annelids and perhaps 30,000 crustacean species. These are underestimates since, for example, even though about 1,300 species of copepods parasitic on fish and easily twice that many on invertebrates have been described, every new expedition to tropical reefs or deep sea collection finds many new species (A. Humes, personal communication).

Knowledge of natural history and systematics of marine organisms is a prerequisite for understanding their role in nature. The diversity of types of consumers precludes our presentation here; rather we focus on the ecological relations of representative populations. For many ecological purposes it is not necessary for ecologists to have an exhaustive taxonomic knowledge of the environment under study but the present trend of few young scientists specializing in systematics portends future problems. When the currently active generation of systematic specialists retires, we will have increasing difficulties in having specimens reliably identified.

In Chapter 3 we discuss the growth potential of consumer populations and the resulting changes in abundance over time, how such changes can be measured, and then introduce some theories on the reproductive strategies of consumers. Having established that populations can outgrow their resources, we show in Chapter 4 that competition for resources exists and that it has significant consequences for populations of marine consumers and for secondary production. In Chapter 5 we discuss consequences of changes in abundance of food, and outline the mechanisms by which consumers respond to changed resource abundance. Since food is usually in low abundance or is of suboptimal quality, selection of food of high quality must be important, so in Chapter 6 we examine the process of discrimination among food items. In Chapter 7 we describe how food consumed is used by consumers in metabolism, secondary production, and excretion, and discuss some important consequences of the various fates of energy. Chapter 7 concludes with an examination of energetics of populations, building on the information of previous chapters.

Chapter 3
Dynamics of Populations of Consumers

3.1 Elements of the Mathematical Description of Growth of Populations

The growth or decline of a population over a period of time can be expressed in terms of the changes in biomass (ΔB) or of numbers (ΔN). The changes that occur are the result of the sum of the individuals born (R) minus those dying (M) during the interval, the growth (G) of the individuals in the population, and of the net difference between emigration (E) and immigration (I). Thus we can write for biomass

$$\Delta B = R - M + G + I - E, \tag{3-1}$$

all expressed in units of biomass. The change in the number of individuals (ΔN) is expressed by

$$\Delta N = R - M + I - E. \tag{3-2}$$

In this case the growth of each individual is disregarded. The two equations are interchangeable by simply accounting for the biomass of the individuals in the population. We will return to biomass expressions in Chapter 7, where we deal with secondary production. In this chapter we concentrate on numbers of individuals, since most of the work on growth of populations is done on this basis.

The simplest case of population growth is where reproduction is by fission. After 0, 1, 2, 3 . . . n generations there will be x, $2x$, $4x$, $8x$, $16x$, . . . , nx individuals. This can be represented by the series 2^0x, 2^1x, 2^2x, 2^3x, . . . , 2^nx, so that after n generations the number of individuals (N_n) will be

$$N_n = 2^n N_0. \tag{3-3}$$

Under circumstances with ample resources, the rate of increase of number of individuals in the population accelerates exponentially while the time to double the population (t_d) is constant.

If we wish to measure the size (N_t) of the population after an interval of time (t) during which the population grows, we calculate

$$N_t = N_0 2^{t/t_d}. \tag{3-4}$$

When the time elapsed equals the doubling time of the population, $t = t_d$, $N_t = 2N_0$. Equation (3-4) is usually linearized by taking logs of both sides of the equation.

$$\ln N_t = \ln N_0 + \ln 2 \cdot t/t_d. \tag{3-5}$$

A plot of $\ln N_t$ versus t [or versus the number of generations in Eq. (3-3)] shows the slope or growth rate of the population. This growth rate is equal to $\ln 2/t_d$, with units of growth rate per unit time.

A more general description of population growth, applicable to other forms of reproduction, is the differential form of the terms of Eq. (3-2) which describes instantaneous rates of population growth

$$\lim_{t \to 0} \frac{\Delta N}{\Delta t} = \frac{dN}{dt} \tag{3-6}$$

One difficulty here is that individuals are discrete, but if populations are large the steps in abundance due to additions of discrete individuals will be small and will approximate a continuous variable.

If we standardize Eqs. (3-2) and (3-6) by dividing by the number of individuals, we obtain an instantaneous, specific growth rate

$$\frac{1}{N}\frac{dN}{dt} = b - d + i - e, \tag{3-7}$$

where b, d, i, e are the instantaneous rates of birth, death, immigration, and emigration. If we assume for now that emigration and immigration are approximately equal, and solve for N_t by integrating, we get

$$N_t = N_0 e^{(b-d)t}. \tag{3-8}$$

This is the basic equation for exponential increase (e is the base of natural logs) where N_t and N_0 are the population size at the start and end of the time interval. Taking logs and shuffling terms, we get

$$r = b - d = \frac{\ln N_t - \ln N_0}{t}, \tag{3-9}$$

where r is the instantaneous rate of growth of the population.

3.2 Survival Life Tables

3.2.1 Life Table Definitions and Calculations

The rate of growth of a population [Eq. (3-9)] is a function of the number of births and deaths, so to measure r we need to assess reproductive performance and survival rates of individuals throughout

TABLE 3-1. Survivorship Table for the Barnacle *Balanus glandula* Settled at the Upper Shore on Pile Point, San Juan Island, Washington, U.S.[a]

Age in years (x)	Observed no. barnacles alive each year (n_x)	Proportion surviving at start of age interval x (l_x)	No. dying within age interval to $x + l_n$ (d_x)	Mortality rate (q_x)	Age structure (L_x)	Mean expectation of further life for animals alive at start of age x (e_x)
0	142	1.000	80	0.563	102	1.58
1	62	0.437	28	0.453	48	1.97
2	34	0.239	14	0.412	27	2.18
3	20	0.141	(4.5)	0.225	17.75	2.35
4	(15.5)	0.109	(4.5)	0.290	13.25	1.89
5	11	0.077	(4.5)	0.409	8.75	1.45
6	(6.5)	0.046	(4.5)	0.692	4.25	1.12
7	2	0.014	0	0.000	2	1.50
8	2	0.014	2	1.000	1	0.50
9	0	—	—	—	0	—

[a] Data obtained 1–2 months after the 1959-year class settled and each year until 1968, by which time all of the 1959-year class had died. Entries in parentheses were estimated rather than counted directly. Data of Connell (1970), table adapted from Krebs (1978).

their life span. These properties of populations can be quantified by using schedules of births and deaths called life tables.

Survivorship tables consist of several columns. Table 3-1 illustrates a survivorship table for a barnacle population.* The first column is x, the age classes into which we have divided the barnacle population. The next column, n_x, records the number of barnacles that were alive at the start of each age interval. The values of n_x are converted to a new column, l_x, by calculating the proportion of individuals initially present that survive to the start of x, $l_x = N_x/n_0$. In our example, for barnacles of 1 year of age, $l_x = 62/142 = 0.437$.

During any age interval deaths may occur, and an account of these is kept in the next column, d_x, the number of deaths during interval x to $x + 1$. In the barnacle data, 80 individuals died between x_0 and x_1, leaving 62 individuals at the start of x_1. The age specific mortality rate during any interval x to $x + 1$ is the next column, $q_x = d_x/n_x$. During x_0 to x_1, $q_x = 80/142 = 0.563$.

One other important life table statistic is e_x, shown in the right-most column in Table 3-1. This is the mean expectation of future life for

*Life tables can be applied to plants and seaweeds if there can be differentiated as individuals. Chapman (1986) reviews the demography of seaweeds.

individuals that are alive at the start of x. The entries of the e_x column are calculated from $e_x = T_x/n_x$. L_x is the survivorship table age structure (the average number of live individuals during each age interval), and $L_x = (n_x + n_{x+1})/2$. T_x is the time units (years in our example) still to be lived by individuals alive at the start of x, $T_x = \Sigma_x^n L_x$, for n age intervals. For the 5-year-old barnacles, for example, $T_5 = L_5 + L_6 + L_7 + L_8 + L_9 = 16$. Then we can calculate e_x by dividing T_x, the years still to be lived, bu the number of individuals (n_x) left alive at the start of x, and for the 5-year-old barnacles, $e_5 = T_5/n_5 = 16/11 = 1.45$ years. This is the average expectation of life for a 5-year-old barnacle.

3.2.2 Methods of Obtaining Numbers for Life Table Data

There are two different approaches that can be used in obtaining data for life tables. These two alternatives can be distinguished by considering the hypothetical population in Fig. 3-1. The occurrence of a series of births and deaths over time are indicated across the bottom of Fig. 3-1. This population has a more or less seasonal pattern of breeding, and individuals live for a maximum of 8 years. The "life line" of each individual is marked as a diagonal line that follows an individual from birth (top left) to death (bottom right) over time (indicated along the x axis), and through the various age classes (indicated along the y axis). We can follow the survival of a cohort of individuals born more or less synchronously, and from these data plot

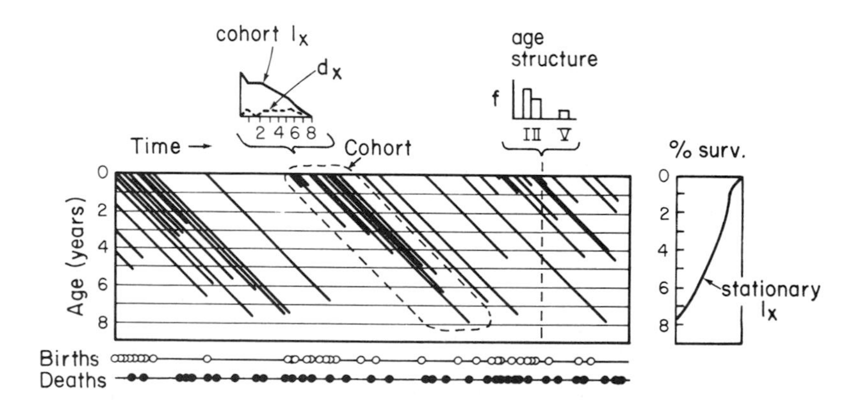

FIGURE 3-1. Diagram of the life and death of individuals of a hypothetical population over age classes and over time. The history of each individual is given as the diagonal lines. In the small graph of the top left of the figure, the vertical axis is the numbers of individuals, the horizontal axis is age in years. In the graph to the right, the vertical axis is age in years. Adapted from Margalef (1978).

the survival curve of the cohort, shown in the small graph on the top left of the diagram. This approach to obtain survivorship data is often referred to as the cohort, dynamic, horizontal, or direct method.

Another way to obtain a survivorship curve is to record the age of individuals that die. In Fig. 3-1 the age of carcasses was recorded for a collection of carcasses accumulated during the expanse of time comprised by the x axis. The survivorship curve obtained by this approach—in this case plotted as percentage survival—is shown on the small graph at the right of Fig. 3-1.* This second way to collect survivorship data is usually called the stationary, static, vertical, or indirect method (Caughley, 1966).

Specifically, the two approaches can be carried out by collecting data in several ways:

Direct methods. The direct following of a cohort over its lifetime can be done by either (1) recording the number of deaths occurring at each age, which provides entries for the d_x column, or (2) recording numbers of animals in a cohort still alive at various times, which provides data for the n_x column.

Indirect methods. In the more common situation where it is not possible to follow individuals in a cohort, there are four ways to obtain life table data: (1) record age at death of individuals marked at birth but not born at the same time, obtaining data for d_x; (2) age carcasses dead over a period of time, obtaining d_x data; (3) age carcasses from a population killed unbiasedly by a catastrophe, a method that provides n_x data; (4) census a living population, thus obtaining n_x data.

It does not really matter that data for different columns are obtained by different methods, since the columns of life tables are related:

$$n_{x+1} = n_x - d_x, \; q_x = d_x/n_x, \; l_x = n_x/n_0.$$

The results of indirect methods will resemble those direct methods for populations whose age distribution is stationary. Caughley (1966) discusses ways to test whether stationary age distributions are involved. In such stationary age distributions neither the abundance of the population nor the age structure changes over the period over which the censuses were taken. Since environmental conditions are seldom constant, there are usually changes in populations over time, so there will be differences among the two kinds of life tables. If there are large changes in the population during the period of study, the

* A collection of all the individuals present at any one time, indicated by the vertical dotted line, would include all the live individuals present. This is L_x the survivorship table age structure, and is shown in the histogram at top right of Fig. 3-1.

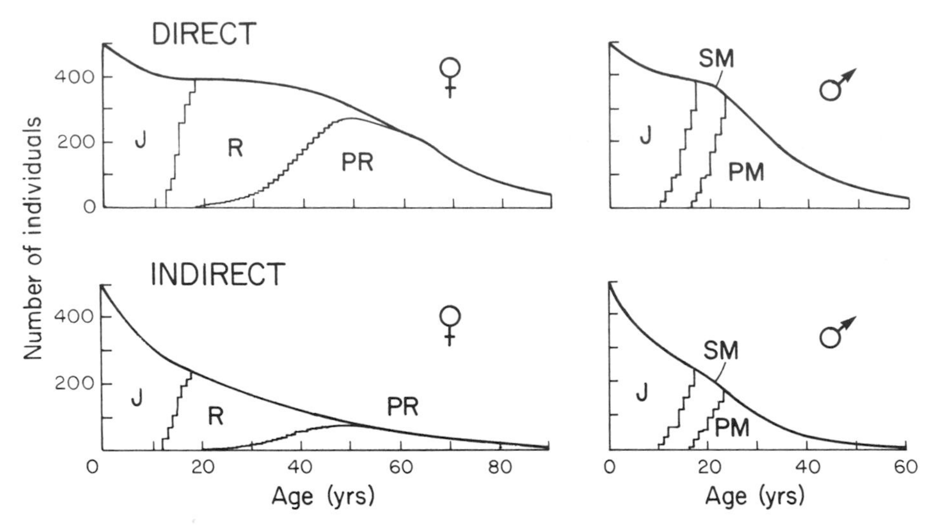

FIGURE 3-2. Age structure in female (left) and male (right) killer whales (*Orcinus orca*) in waters off British Colombia and Washington, as estimated by direct (following cohorts as they age) (top panels), and by indirect (considering all individuals found at one time) (bottom panels) methods. Sexual stages are shown as juvenile, reproductive, or post-reproductive in females, or as juvenile, submature, or post-mature in males. Adapted from Olesiuk et al. (1990).

indirect methods will provide data in which some age classes will be over- or underrepresented in the sampling.* Differences between data obtained directly and indirectly can be substantial. A comparison of the two approaches is available from data on pods of killer whales that were followed from 1974 to 1987 (Fig. 3-2). Individuals could be recognized by their markings. This particular killer whale population was growing at 2.9% y^{-1}. The age structure that is inferred by the two methods differ, and so are the derived life table statistics (Olesiuk et al., 1990). Although study by direct methods is not possible in many cases, the bias of indirect measurements needs to be kept in mind.

In most life table studies, age is not directly known, so that some way of estimating age is often required. The best technique is to mark individuals and follow their survival, if the marking method does not affect survival. Many other indirect criteria have been used as ways to estimate age. These include size of the organism, growth rings on shells of molluscs, genital plates of echinoderms, growth rings on

*This is the same problem presented by life insurance rates calculated on the basis of outdated human survivorship data. Since current human survival rates are nearly always higher than earlier rates, the fairer (and cheaper) rates will be those calculated for the very latest survival data. Incidentally, most of the development of life tables for ecological work was done based on actuarial techniques.

scales or otoliths of fish, teeth in marine mammals, and many others. Since young animals are the most fragile, have few hard parts, and have the highest growth and mortality rates, the young stages are almost invariably the weakest part of life table data.

3.3 Fecundity Life Tables

The survival life tables discussed above summarize the mortality schedule of a population. To determine the rate of population growth or decline we need to add reproduction schedules. Fecundity life tables (Table 3-2) contain an additional column, m_x, the number of female offspring per female of x age per unit time. The age-specific fecundity (m_x) is measured by examining the production of eggs or young for the various age classes of females. Histological examination of gonads of clams or fish, for example, can provide a measure of production of eggs.

To calculate the rate at which the population grows we have to estimate the number of young produced by each age class. This can be calculated by multiplying survivorship (l_x) by reproductive output (m_x) of each age class up to n, which gives the age-specific reproductive rate, R_x. The sum of these products is the multiplication rate per generation (R_0),

$$R_0 = \sum_{x=0}^{n} l_x m_x. \tag{3-10}$$

TABLE 3-2. Survivorship and Fecundity Table for the Sardine *Sardinops caerulea* Off California.[a]

x (years)	l_x	m_x	$l_x m_x$	u_x
0	1.000000000	0	0	1
2	0.000014084	36,543	0.5147	139,380
3	0.000009410	96,687	0.9098	215,814
4	0.000006328	119,414	0.7557	248,384
5	0.000004242	133,593	0.5667	269,760
6	0.000002843	143,824	0.4090	284,877
7	0.000001905	151,641	0.2890	296,288
8	0.000001277	158,743	0.2028	302,555
9	0.000000856	161,070	0.1379	300,818
10	0.000000574	161,070	0.0926	292,212
11	0.000000384	161,070	0.0619	274,862
12	0.000000257	161,070	0.0415	238,398
13	0.000000173	161,070	0.0279	161,070

[a] From Murphy (1967).

In Table 3-2 we have survival and fecundity data for sardines. The n_x data were obtained and are expressed as l_x, and the initial value of l_{x_0} is set at 1 for convenience in calculations. The right-most column depends on age specific survival and fecundity; R_0 is the sum of the terms in the $l_x m_x$ column.

The rate R_0 can also be called the net reproductive rate, since it equals the ratio between the numbers of individuals in a filial generation (N_T) relative to the number of parent females (N_0):

$$R_0 = N_T/N_0, \tag{3-11}$$

where T is the time interval between generations. From Eqs. (3-8) and (3-9)

$$N_T/N_0 = e^{rT} = R_0. \tag{3-12}$$

Taking logs and rearranging the terms, we get

$$r = \ln R_0/T. \tag{3-13}$$

This is an important result, because it allows us to calculate the instantaneous rate of population increase at any one time, if we know T. T can be estimated by calculating the average age at which a female reproduces. If at various age intervals x there are are m_x young born to a female,

$$T = \frac{\Sigma\, x l_x m_x}{\Sigma\, l_x m_x} = \frac{\Sigma\, x l_x m_x}{R_0}. \tag{3-14}$$

In the case of the California sardine (Table 3-2), $T = 4.111$ years, and $R_0 = 4.0125$. This sardine population is quadrupling its numbers per generation, a rather rapid rate of increase. The sardine is a species that often suffers wide variations in density. The rapid net reproductive rate calculated here reveals that sardines are an opportunistic species able to increase quickly under suitable conditions (Murphy, 1968).

T is only approximately estimated by Eq. (3-14), since offspring in most populations are born over a period of time rather than all at once and not all females give birth simultaneously. Equation (3-14) may be a good estimate of mean generation length in situations where a cohort colonizes a new habitat or in populations with strongly seasonal breeding patterns. Leslie (1966) discusses other ways to obtain T based on the average age of reproductive and nonreproductive females.

In populations where generations do overlap, Eq. (3-13) is not an accurate way to calculate r. In such cases r is best calculated by substituting values of r in Eq. (3-15) until we find the value such that

$$\sum_{x=0}^{\infty} e^{-rx} l_x m_x = 1, \tag{3-15}$$

or a reasonable approximation to 1. In the case of the sardine (Table 3-2), $r = 0.338$. The above equations are mainly the work of A.J. Lotka and associates and are derived in Lotka (1956) and Mertz (1970), where their applications are discussed.

A convenient way to carry out the calculations of population growth in the case of complex life histories is by means of matrix algebra. The techniques are reviewed in Poole (1974) and Pielou (1977), among other books. Most matrix calculations are done on the basis of age structures, but Caswell (1984) pioneered life stage-based matrix calculations that require fewer data and produce comparable results (Brault and Caswell, 1993).

3.4 Some Properties of Life Table Variables

3.4.1 Survival and Mortality Curves

Most marine animals, especially invertebrates and fish, suffer very large mortalities during the early stages of their life history (Fig. 3-3).

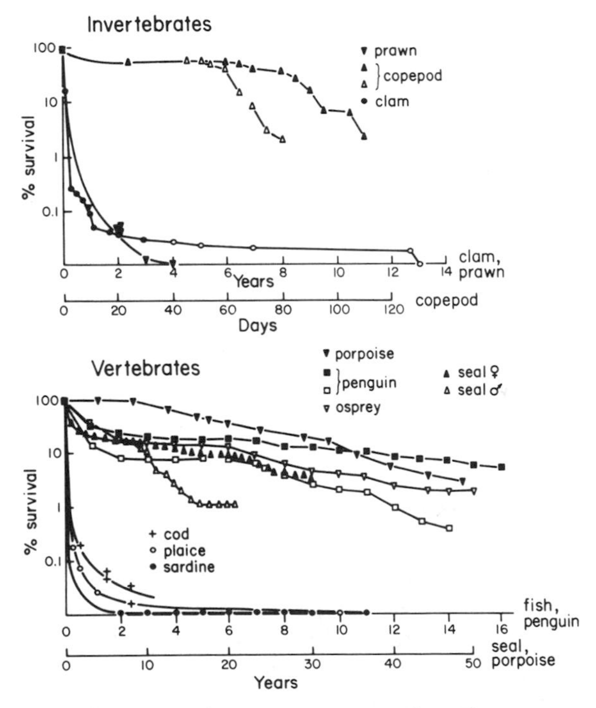

FIGURE 3-3. Survival curves for various animals. The copepods are *Tisbe reluctans* and *T. persimilis*; adapted from Volkmann-Rocco and Fava (1969). The prawn is *Leander squilla*; adapted from Kurten (1953). The clam is *Mya arenaria*; adapted from Brousseau (1978a). The porpoise is *Stenella attenuata*; adapted from Kasuya (1972). The osprey is *Pandion haliaetus*; data from Henny and Wight (1969), graphed on same scale as fish and penguin. The seal is *Halichoerus grypus*. Note that males and females of the same species can have very different survival curves. Adapted from Hewer (1964).

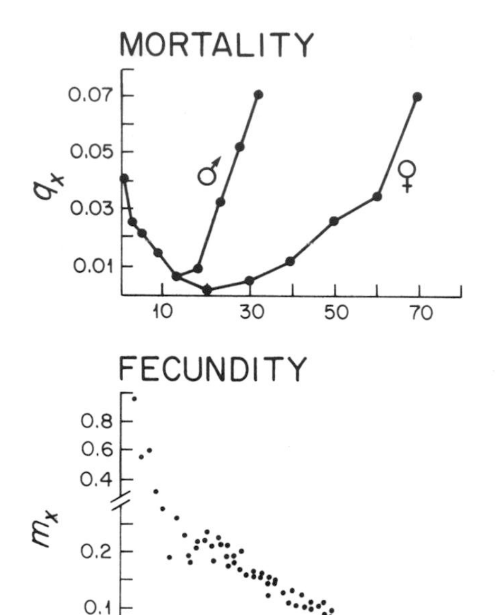

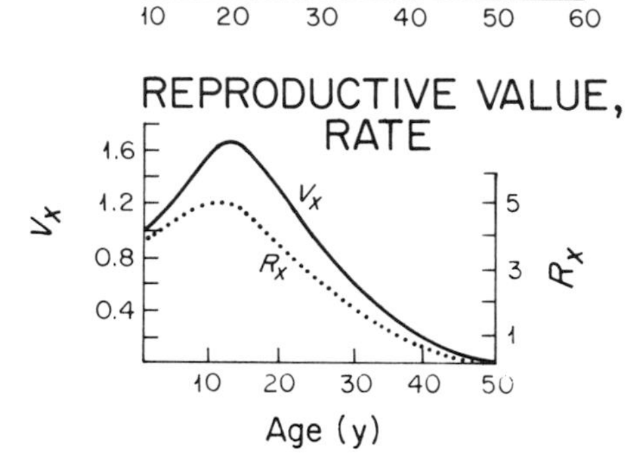

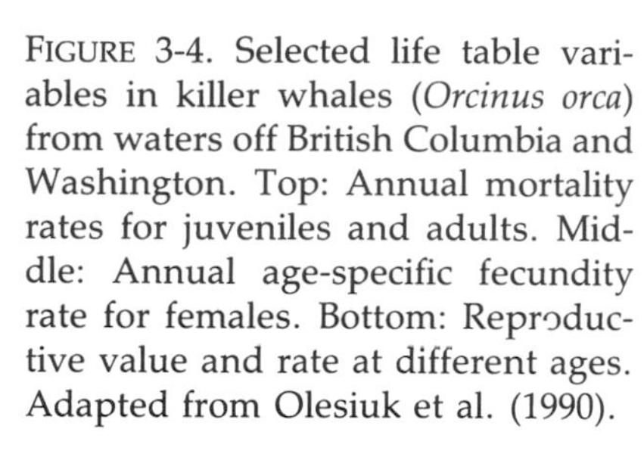
FIGURE 3-4. Selected life table variables in killer whales (*Orcinus orca*) from waters off British Columbia and Washington. Top: Annual mortality rates for juveniles and adults. Middle: Annual age-specific fecundity rate for females. Bottom: Reproductive value and rate at different ages. Adapted from Olesiuk et al. (1990).

Such survival curves show a very steep initial slope, with a two to three order of magnitude change in percentage survival. The intense mortality of young is not due to intrinsically low survival, since in a protected laboratory environment many species show very low mortalities during the earlier stages of life (cf., for example, the two copepods on Fig. 3-3, top, whose survival was measured in the laboratory). Mortality rates taper off during later stages and life expectancies are high once the hazardous early stages are past. The mortalities of young plaice in their first few months of life can be 80% per month; by the time plaice are 5–15 years old the mortality may be down to 10% per year (Cushing, 1975).

In contrast, there are marine organisms, mainly mammals and birds, whose survival in the field is relatively high during the early stages of their life history (Fig. 3-3, bottom). Marine mammals have especially low mortalities, clearly related to parental care and the birth of relatively large young in very small clutch sizes. Nonetheless, even in whales, mortality typically is higher early and late in the life history (Fig. 3-4, top). Whale life also seems more hazardous, and abbreviated, for moles.

3.4.2 Age or Size Structure of Populations

The age structure of a population is ofter the easiest demographic information to obtain. If age or size structures are followed through time (Fig. 3-5), the effects of mortality (reduction of cohort size) and growth (shift to larger size) of the individuals making up the cohorts is evident. Quite often there is a broad overlap among cohorts but there are methods to separate the cohorts (Cassie, 1954; MacDonald and Pitcher, 1979; Sardá et al., 1995a,b).

Sometimes it is not practical to follow populations through time, nor to identify cohorts. In these cases the decrease in the abundance of age classes to the right of the mode the entire population may be used as an estimate of mortality rates. Ebert (1977) and others review such calculations, all of which assume a stationary age distribution, constant mortality over all sizes, and exponential growth of individuals (Bertalanffy, 1957). These assumptions do not always hold (Frank, 1965; Yamaguchi, 1975; Van Sickle, 1977), but such calculations can

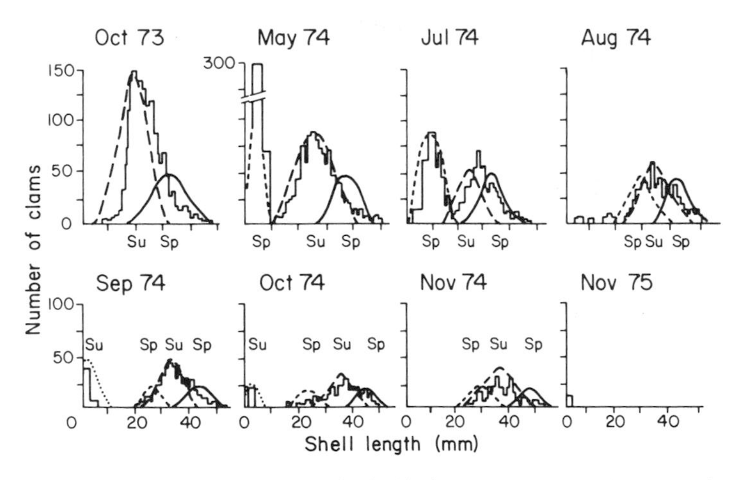

FIGURE 3-5. Age structure of the soft-shell clam, *Mya arenaria, over time*. The actual data are shown as histograms, while the curves are the cohorts separated by the use of probability paper (Cassie 1954). The pelagic larvae of this species settle onto sediments in spring (Sp) and summer (Su) cohorts. The pattern of growth and mortality can be seen by following, for example, the summer 1973 cohort (large dashes) through 1974. Adapted from Brousseau (1978).

give reasonable approximations (Brousseau, 1979). If data on size specific growth and size frequency are available, the changes in numbers (u_j) of a population of size class j can be calculated (Van Sickle, 1977) as

$$u_j = [g_j N_j - g_{j+1} N_{j+1}]/N_j, \quad (3\text{-}20)$$

where g_j and g_{j+1} are the growth rates of two adjoining size classes, N_j and N_{j+1} are the numbers of individuals in adjoining size classes. When frequency of size classes declines the growth rate will be negative, hence Eq. (3-20) also can be used to calculate mortality rates. Estimates of mortality rates for bivalves computed by Eq. (3-15) for size classes to the right of the mode can approximate estimates obtained by more complicated demographic means (Brousseau, 1979). In many cases there are few other feasible alternatives; for instance, most of the available estimates of mortality for whales use this general approach. It should be clear that there are limitations to this calculation, an obvious one being that only mortalities of older individuals are estimated.

A population subject to a constant schedule of births and deaths will increase geometrically [Eq. (3-8)] and will gradually approach a stable age distribution, in which the relative proportion of different age classes is fixed. This stable age distribution can be calculated as described by Mertz (1970). Seasonally breeding populations may not achieve a stable age distribution. We have earlier mentioned the stationary age distribution. The latter describes the age composition of a population exposed to given mortality rates (q_x), if birth rate were equal to death rate. Neither stable nor stationary age distributions are likely to exist for very long in nature, since populations seldom increase at the same rate for any length of time, and because the numbers of individuals in a population seldom remain constant.

The distribution of age classes can sometimes provide information on status of the population. Populations that are increasing tend to have more individuals in the youngest age classes; as each age class becomes older, the total numbers of the populations expand. Stable populations characteristically have about the same proportion of the population in all size categories; populations that are declining may have fewer individuals in younger age classes.

3.4.3 Fecundity and Reproductive Value

Fecundity, even if intrinsically high, does not guarantee reproductive success. A population must have sufficient individuals in the various age classes that contribute significantly to r. Age-specific fecundity can peak soon after the first reproduction, as in some copepods (Volkmann-Rocco and Fava, 1969) and whales (Fig. 3-4, middle), or later in reproductive life, as in clams (Brousseau, 1978) and sardines (Table 3-2).

The impact of individuals in different age classes on the reproductive rate of a population differs. Such a reproductive value is a result of the survival and reproductive performance of individuals in each age class. In an exponentially growing population the potential importance of each age class (x) for population growth is given by the reproductive value of that age class (v_x) relative to the reproductive value at birth (v_0)

$$\frac{v_x}{v_0} = \frac{e^{rx}}{l_x} \sum_{y=x}^{\infty} e^{-ry} l_y m_y. \tag{3-21}$$

What is summed in Eq. (3-21) is the number of offspring that will be produced by a female for age x as she passes through all the remaining age classes (y). Usually v_0 is made equal to 1 so v_x is expressed as multiples of v_0. The expression was first derived in its integral form by R.A. Fisher and its derivation is explained in Mertz (1970) and in Wilson and Bossert (1971).* In the case of killer whales, reproductive rate and value peak at an age of 11.5 (Fig. 3-4, bottom). Most of the changes in reproductive performance of killer whales are due to changes in fecundity (Fig. 3-4, middle), not to changes in survival. Female whales of reproductive age suffer little mortality, so that v_x reflects the reproductive rate.

In the example of the Pacific sardine (Table 3-2), the reproductive value at recruitment (v_2) is much greater than at v_0, due to the very high mortality of larval sardines. The reproductive value increases in subsequent age classes when mortalities are less marked and m_x reaches its peak. Ther is a slight decrease in v_x in the last few years due principally to mortality but in this species v_x is significant for most age classes.

In a rapidly growing sardine population several age classes contribute significantly to reproductive rate (Table 3-2). During the 1940s fishing pressure on the larger classes was heavy enough so that only two age classes contributed to reproduction, compared to eight or so age classes before that. Perhaps this situation was implicated in the collapse of certain stocks in 1949–1950, after two consecutive spawning failures. Fishing had removed the mechanism—reproduction by several age classes—that the population had evolved to compensate for reproductive failures due to fluctuations in environmental conditions (Murphy, 1968).

---

*Equation (3-21) is not completely correct, even though it appears very frequently in the ecological literature. See Goodman (1982) for a discussion of the correct version, $v_x = (e^{rx+1}/l_x)\Sigma_{y=x}^{\infty} e^{-ry} l_y m_y$. The values calculated using Eq. (3-21) are in error by e^r. Since we are generally interested in the patterns of the v_x over x, the error is not critical, since the pattern will still be evident.

The reproductive value can be useful to predict consequences of harvest of specific age classes, as for example, in prediction of consequences of size-selective fishing. Reproductive value could thus be an important tool in management of exploited stocks since, for example, it would be desirable to carry out harvests of stocks with the least effect on reproduction. Harvest of age or size classes with low v_x may be the most "prudent" fishing or predation strategy. In actuality, the optimization of harvest strategy is seldom so simple, and careful study of specific harvestable populations is required.

Knowledge of reproductive value may also be useful in studies of populations that colonize new habitats. The v_x of the propagule that has arrived at the new site may vary: if the colonizers are very young, reproduction and further expansion into the new habitat may be delayed until maturity; if very old, few offspring may be produced. Since colonization is often carried out by a few individuals, the reproductive ability of these few assumes a great importance (MacArthur and Wilson, 1967).

3.5 Reproductive Tactics

The various demographic characteristics of a population are interrelated traits, but some features affect population growth more significantly than others. For example, Cole (1954) found in simulation studies that age at first reproduction has a great effect on r (Fig. 3-6), so that delays in sexual maturity markedly lowered the attainable r.* Delays in age at first reproduction have the greatest impact on r in the left side of the graph both in species that reproduce once during their lifetime (semelparity) or in cases when reproduction takes place at repeated intervals (iteroparity). The effect of age at first reproduction is lessened at low clutch sizes (compare clutch size $b = 10$ to $b = 1$ or 2). The importance of early reproduction is such that if reproduction took place at the earliest possible age (Fig. 3-6) there were minimal differences in r between iteroparous and semelparous species, and only a small difference in later reproduction.

---

*In human populations there has been too much attention given to total family size and too little to age at first reproduction. Cole (1965) shows that human populations that started reproducing at 12 years of age could attain over twice the growth rate of a population whose age at first birth was 30 years of age. The effect of total number of children was small compared to the effect of age at first reproduction. The countries where growth of population has been most successfully curtailed are those where economic and social restrictions have led to delayed reproduction until the parents were in their 30s, as well as to reductions in the number of children per family.

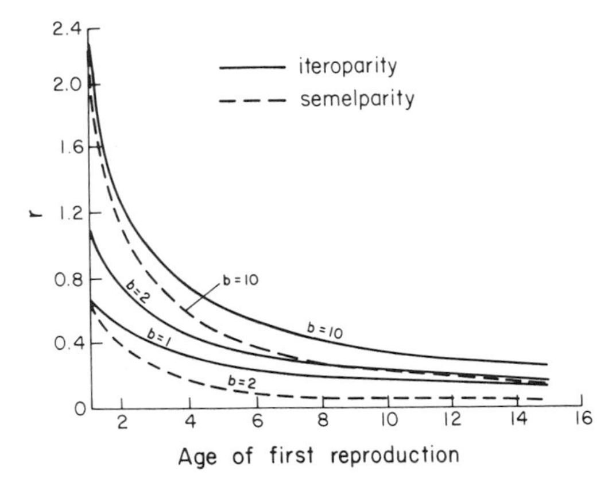

FIGURE 3-6. Simulation study of the effect of age at first reproduction and clutch size (*b*) on the rate of population increase (*r*). Age at first reproduction is shown for intervals representing the 1, 2, 3 . . . intervals at which reproduction is possible for iteroparous species and for reproduction at the 1, 2, 3 . . . interval of age for semelparous species. Adapted from Cole (1954).

The rich variety of demographic possibilities provided by semelparity, iteroparity, reproduction at any of a variety of ages, and variable clutch sizes allows species with different demographic repertoires to adapt to environments with different properties. Several theoretical constructs have been forwarded to codify the kinds of demographic features that would be most appropriate for environments of different properties. These include "the r–K dichotomy," "bet-hedging," and the "abundance/intermittency" idea.

3.5.1 r–K Dichotomy

Very different strategies of reproduction would be best suited for life in either sparsely inhabited but resource-rich environments where rapid and opportunistic population growth would be favored, or in resource-poor environments already saturated by organisms, where very slow population growth would be more adaptive. MacArthur and Wilson (1967) coined the terms "*r* selection" and "*K* selection" for these two alternatives.* Stearns (1976) reviews these theoretical constructs in some detail and summarizes the traits that are hypothetically characteristic of each: (1) *r* selection: early age of first reproduction,

* "*r*" refers to the intrinsic rate of increase of a population, and "*K*" to the carrying capacity of the environment, as used in the logistic equation of population growth (Chapter 4).

large clutch size, semelparity, no parental care, large reproductive effort, short generation time, small and numerous offspring, and low assimilation efficiency; (2) K selection: delayed reproduction, iteroparity, small clutches, parental care, smaller reproductive effort, longer generation time, a few large offspring, and high assimilation efficiency. Assimilation efficiency is taken up in Chapter 7. All the other characteristics are consistent with either prodigal or prudent reproductive tactics, to use Hutchinson's (1978) picturesque metaphor.

We should add a note about the "few large" and "many small" features, because we will come back to this in Chapter 9. Invertebrates and bony fish produce very numerous clutches, where each individual may have a 10^{-5}–10^{-6} chance of survival, as evident in the l_x curves of Fig. 3-2. Necessarily, the young are very small in relation to the size of the adult, and there is little or no parental care. Birds and mammals, on the other hand, produce far fewer young of a larger size in relation to the parents, and the chance of survival of each young can be 10^{-1}–10^{-2}. Birds and mammals carry out considerable parental care of the young.

Although the r–K dichotomy is an appealing idea and has stimulated much research, it is too much of a simplification. Not all species in an environment will be r or K selected; at the same time, any one environment may be r selective for a copepod but K selective for a whale. Further, there are many examples where a particular species shows combinations of traits that the r–K typology puts in the two separate categories. Finally, there are many species that are intermediate in demographic properties, so that the concept is probably best referred to as the r–K continuum. There has been, in general, insufficient independent verification of the r–K idea in part because we use many of the same properties to define environments that r or K select as we do to define whether a species is r or K selected. This circularity has made it difficult to test the idea.

3.5.2 Bet-hedging

Another theoretical scheme regarding adaptive reproductive strategies (Murphy, 1968; Schaeffer, 1974), is called "bet-hedging" by Stearns (1976), and is based on the idea that iteroparity may be a response to a changeable environment. The bet-hedging concept is derived from observations on reproduction in fish. In herring-like fish the species that suffer the most erratic breeding success are also the ones with the longest reproductive spans of life. Iteroparous breeding over a lengthy period of time may thus be a bet-hedging tactic that compensates for poor years in a variable environment. Results from terrestrial studies suggest that in variable environments, age at first reproduction is lower, size at sexual maturity smaller, reproductive effort higher, clutch size larger, and size of young smaller than in constant environments (Stearns, 1976). "Variable" and "constant" environments may not correspond exactly to *r*- and *K*-selective environments, so the predictions of the bet-hedging ad *r*- and *K*-selection hypotheses may not be mutually exclusive. This makes it hard to devise hypotheses with which to assess the two schemes.

3.5.3 Abundance/Intermittency Scheme

Whittaker and Goodman (1979) and Grime (1977) propose a third hypothetical scheme. The scheme identifies three kinds of species, considering both the abundance and the intermittent pattern of availability of resources:

Adversity-selected species. Some environments are depauperate and can sustain only a small number of organisms: species living there are selected for survival under adverse conditions. Perhaps reproduction can only occur in the very minimal periods of more favorable conditions, and the reduced resources allow only a small amount of reproduction. The total abundance over the habitat seldom increases very much.

Exploitatively selected species. In environments where conditions for life unpredictably change and may suddenly but repeatedly provide bountiful resources, species capable of rapid exploitative activity will be present, and these species will have opportunistic characteristics. There will be strong selection for sharp increases in birth rate (b) on the frequent occasions during which resources increase; the rest of the time, selection will favor those able to reduce death rates (d). For example, a species can evolve some mechanism to survive long periods of bad conditions, if there is a frequent and reliable return to favorable conditions. Thus, to explain reproductive strategy in exploitation-selected species we need to consider both b and d, rather than just the rate of increase, $r = b - d$. These populations will be generalists in use of resources (taking whatever resources are available) and flexible in demography (both b and d can change).

Saturation-selected species. In environments capable of continually sustaining an array of species, and where there are few intermittent lapses in available resources, there will be selection for low rates of birth and death, demographies that lead to near-complete use of the available resources, and the species can be thought of as "saturation-selected." In such situations it is likely that, due to the high density, competition rather that extrinsic vagaries of the environment determines the actual birth and death rates of the populations.

This third scheme of adaptive strategies is as hypothetical as the two prior ones. Its three kinds of environments are not exactly parallel to those of the $r-K$ or the bet-hedging alternatives, so comparisons are difficult. There is plenty of theory in this area; there is also a need for critical experiments to discriminate among competing ideas. Unfortunately, it is very difficult to make sure of just what is being tested. Reproductive strategies are an intrinsically interesting but perhaps intractable subject, one in which there is enough tautology to hinder the testing of hypotheses.

The reproductive strategy of specific species can, however, in some cases be clear, with important consequences for human exploitation of natural populations. There are many examples of this (Ricker, 1975), but we can consider a representative example: the relative ability of bony and or cartilaginous fish to sustain fishing pressure and its relation to reproductive strategy.

Most marine bony fish may be "exploitatively selected" species that use what can be called an "*r* strategy," and produce very numerous small eggs, most about 1 mm or so in diameter; sometimes, as in salmon and halibut, eggs may be several millimeters in diameter. As many as 10^4-10^7 eggs may be released at each spawning. The small larvae (3–10 mm in length) hatch and then undergo a very hazardous period of drifting in the water column. Larval mortality rates can be very high, reaching 10% per day in haddock (Jones, 1973), 80% per month in plaice (Cushing, 1975) (cf. Fig. 4-10, right). A female Arctic cod may release about 4 million eggs, and on the average only 2 of those eggs may survive to reproduce.

Many cartilaginous fish (elasmobranchs such as sharks, skates, and rays) may be "saturation-selected" species, in contrast to bony fish. Cartilaginous fish may use a "*K* strategy," in which the young are well developed when born and resemble the adult of the species. The number of young per clutch is far smaller than other fish. For example, the ray *Raja erinacea* only produces about 33 eggs per year (Richards et al., 1963), while *R. cavata* releases about 100 eggs, and *R. brachyura* and *R. montaghi* 50 eggs per year (Holden, 1978). The weight of the young decreases as litter size increases, and the smaller young may be more susceptible to predators.

A consequence of the low rate of reproduction and growth of most elasmobranch populations is that they generally do not sustain exploitation very well. A number of fisheries, including those exploiting the California fin shark (*Galeothinus zyopterus*), Australian school shark (*G. australis*), the spiny dogfish (*Squalus acanthias*), plus skates and rays (Holden, 1978) have collapsed or produced reduced catches. Young are produced in low numbers and the recruitment is probably a function of parent density (and probably parent size). When fishing lowers density of the adult population, recruitment is affected.

In contrast, in most species of pelagic marine bony fish, recruitment can take place even with a minimal production of eggs (cf. Section 4.4). Thus, reduction in parent stocks is less likely to affect recruitment, since the removal of a substantial proportion of the reproductive population does not necessarily reduce the rate of population increase.

Chapter 4
Competition for Resources Among Consumers

4.1 Population Growth in Environments with Finite Resources

In the previous chapter we examined growth of populations whose reproductive rate was constant. In nature, the rate of population growth very often varies. If r varies randomly, a population may become extinct or, alternatively, suffer enormous increases in abundance, since the reproductive potential of any species is capable of producing very large numbers of new individuals. Since few environments are overwhelmed by living organisms, there must be mechanisms that inhibit increase in abundance as density increases. This, in brief, is the theory behind the logistic equation,* first suggested by P.F. Verhulst in 1838 and later derived independently by R. Pearl and L.J. Reed in 1920 (Hutchinson, 1978). The logistic equation is

$$dN/dt = rN(K - N)/K, \tag{4-1}$$

which is the exponential growth equation modified by a factor $(K - N)/K$ that reduces dN/dt as the number of individuals (N) increases and approaches the carrying capacity of the particular environment (K). This results in a sigmoid pattern of population growth, in which N has a smooth, decelerating approach to K.

The integrated form of the logistic equation can be fitted to actual data by first rearranging the integrated form of the equation ($N = K/[1 + e^{a-rt}]$) into $\ln[(K - N)/N] = a - rt$, where a is a constant. Plotting $\ln[(K - N)/N]$ on the y axis, and time in the x axis, the slope of the line is r, and a is the y-intercept.

---

*The logistic is just one form of sigmoid growth. There are many other ways to describe such growth, including applications of Michaelis–Menten curves (Slater, 1979), Gompertz curves ($dN/dt = rN\log(K/N)$) (Margalef, 1974) or exponential curves (Gallopin, 1971). Further alternative expressions are discussed by Hutchinson (1978).

The logistic model is one of the cornerstones of much ecological theory due to its broad generality. Precision and realism are additional desirable attributes of models (Levins, 1966), but it is not possible to achieve all three attributes in one model: the logistic has broad generality, since it can apply to any population, but has only limited precision in predicting the abundance of a particular species (Krebs, 1978), since populations often overshoot K and subsequently crash, and seldom fit the logistic sigmoid carve. The logistic has limited realism, since it assumes that all individuals contribute identically to r, that K is a constant, and that there are no time lags. None of these assumptions is very realistic, since there are very significant differences among individuals of different sexes and ages, most environments and resources vary markedly over time, and consumer populations and their food resources seldom if ever react to changes instantly.

Models such as the logistic may have heuristic value, but actual data fail to support predictions of such models (Hall, 1988). Moreover, the heuristic fuction is impaired by the tautological nature of the models: if the model prediction fits actual data, the model is taken to apply; if the fit is poor, we conclude that some unspecified modification to the model is needed. Such testing of the models by comparison of model output to data may be useful in relatively simple models, where there are only a few alternative ways to modify the model once a lack of fit is discovered (Hall, 1988). In larger, more complex models testing by matching to data may be less useful, since model output depends on many variables. Lack of fit of model output to actual data therefore may not be an unambigous way to disprove specific hypotheses embodied in the model, a feature that is the hallmark of empirical science (Platt, 1964). Nonetheless, the logistic model of population growth is heuristically attractive because it pinpoints, among other things, one very important concept: competition for resources must take place.

4.2 The Nature of Competition

Competition is difficult to define nontautologically. It is usually taken to occur where resources are in short supply, and can take two forms (Nicholson, 1954; Elton and Miller, 1954). First, there is interference or contest competition, where access to a resource is denied to competitors by the dominant individual or species. Examples of interference competition are the release of antibiotics by microorganisms, territorial behavior, and social hierarchies. Interference competition results in unequal access to resources, and subordinate individuals may not be able to participate in reproduction. Interference competition tends to couple numbers to available resources and often reduces population fluctuations. Many proximate mechanisms involving behavior and physiology, such as responses to photoperiod, dormancy, and behavioral displays, are commonly involved in this kind of competition.

Most of the biological world, however, is more involved in exploitation or scramble competition, that is, the direct use of a resource, reducing its availability to a competing individual or species simply because of consumption. Exploitation competition tends, at least in theory, to involve large fluctuations in density since populations may be built up based on a temporarily abundant resource, and crash down when the resource is exhausted. Use of resources is not particularly efficient where allocation of resources is exploitative, since many organisms that will not reach maturity still consume resources.

Both interference and exploitation competition may occur simultaneously. *Patella cochlear*, for example, is territorial as are other limpets (Stimson, 1970; Branch, 1976) and actively prevents the encroachment of other animals in its feeding territory. *P. cochlear* preferentially eats the alga *Ralfsia* within its feeding territory, leaving only encrusting algae within it. *P. cochlear*, however, can continue to use the feeding territory by feeding on the encrusting algae, but other limpet species cannot eat encrusting algae. Thus, this is a case of both interference and exploitation competition mechanisms (Branch, 1976).

Competition pervades many dimensions of the life of any species, and what may appear as separate resources may in fact be inextricably linked. There is often no distinction, for example, between competition for food and space in rocky shore species. Experimental removal of the film of filamentous blue-green bacteria from territories of the owl limpet (*Lottia gigantea*) leads to a 25% increase in size of the territory 2 weeks later, while untreated territories changed by only −0.06% (Stimson, 1973). Another example involves encrusting bryozoans, organisms where overgrowth is a very common phenomenon. In shallow waters of the Pacific coast of Panama there are two bryozoans, *Onychosella alula* and *Antropora tincta*, that frequently live in close proximity. Buss (1979) fed radioactively labeled food pellets to colonies of these two bryozoans growing on separate surfaces and growing on the same surface. The feeding rate, indicated by the radioactive counts, did not change from the margin of a colony inward in monospecific colonies (Fig. 4-1, left). When the two species grow together on one substrate (Fig. 4-1, right) there was a notable decrease in feeding rate of *A. tincta* toward the margin of the colony in contact with *O. alula*. If growth rate depends on feeding rate, individuals on the edge of the *A. tincta* colony will grow slower, and the *O. alula* colony would overgrow the colony of *A. tincta*. This occurs despite the fact that growth of *A. tincta* in isolation is faster than that of *O. alula*. The specific proximate mechanism for this is that *O. alula* bears feeding tentacles that are considerably longer than those of *A. tincta*, and presumably can divert

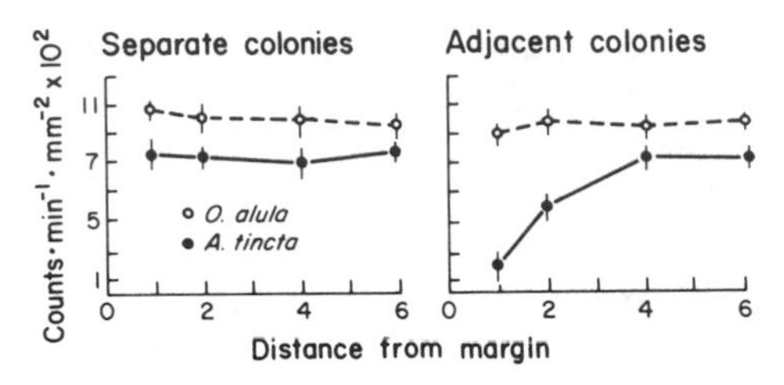

FIGURE 4-1. Feeding rates of two species of encrusting bryozoans (*Onychosella alula* and *Antropora tincta*) in colonies growing separately and with the two species growing adjacent to each other. Feeding rates are expressed as radioactivity due to ingestion of labeled food particles and were measured at various distances (in centimeters) away from the margin of the colonies. In the case of the adjacent colonies, the measurement was from the margin where the two species were in contact. Adapted from Buss (1979).

feeding currents away from *A. tincta*. Since in the thin layer of seawater above the rock surfaces where bryozoans live there are no other forces that move water effectively, *A. tincta* suffers a considerable reduction in food supply. Evidently, space and food are not independent resources.

Competition reflects the relative abundance of resources. The major determinant of relative abundance—since exploitation competition is so common—is the number of individuals in a population. The nature of the limitation of these numbers was heatedly debated in the 1950s and early 1960s. The crux of the argument was whether the rate of increase or decrease of a population could be controlled by factors that depended on the density of the population. These arguments are described in ecology textbooks (Krebs, 1978; Ricklefs, 1979) and elsewhere (Orians, 1962; Clark et al., 1967; Caswell, 1978) and need not be repeated here. Our interest below is to see if demographic properties are affected by changes in the density of a given population of marine organisms—intraspecific competition—and by changes in the density of other competing species—interspecific competition.

4.2.1 Intraspecific Competition

A very clear example of the effect of density within a species is provided by the studies of Branch (1975) on rocky shores of South Africa, where densities of the limpet *Patella cochlear* vary from 90 to over 1,700 limpets m^{-2}. This range of densities has a significant effect on survival and on fecundity.

High densities increase mortality of young limpets (Fig. 4-2, top), but once limpets grow older than about 20% of maximum age, survival is not affected by density. The increased mortality of the younger individuals at higher densities is due to the reduction of space available for settlement of small limpets. Limpets are territorial and aggressively defend their feeding territory so that at high densities higher proportions of larvae settle on the shell of older limpets (Fig. 4-2, bottom), presumably because these are the only surfaces that are not defended. Once the young limpets reach a certain age the shell surface of older individuals is not large enough to provide a sufficient grazing area, and the small limpets leave the shells of the older individuals and seek territories elsewhere. This is a hazardous stage in the life history that leads to the density-dependent early mortality seen in Fig. 4-2. Survival rate in the critical young stages is therefore density dependent.

As density increases, biomass of individual adult limpets increases, but the total biomass of the population eventually approaches an asymptote (Fig. 4-3, top). The more limpets per unit area, the lower the growth rate (Fig. 4-3, middle). Over time, reduced growth rates result in considerably smaller limpets at higher densities (Fig. 4-3, top and bottom). Smaller limpets release fewer larvae (Fig. 4-4, top). As density increases, gonad output per unit area of rock first increases due to the larger numbers of limpets present, but once competition becomes severe, there is a decrease in reproductive output (Fig. 4-4, bottom). Gonad output per individual decreases rapidly as limpet density increases. Fecundity is therefore density dependent.

In limpets space and food supply are one variable and determine growth of individuals and fecundity. The feeding territories containing the algae that limpets use as food are smaller at high limpet densities and do not sustain high growth rates. Consequently the animals are

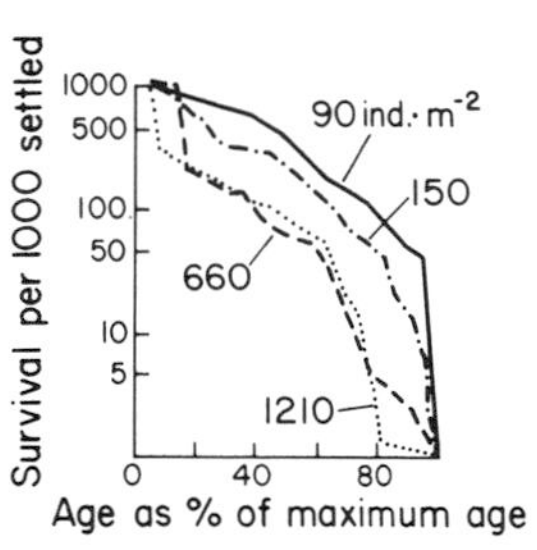

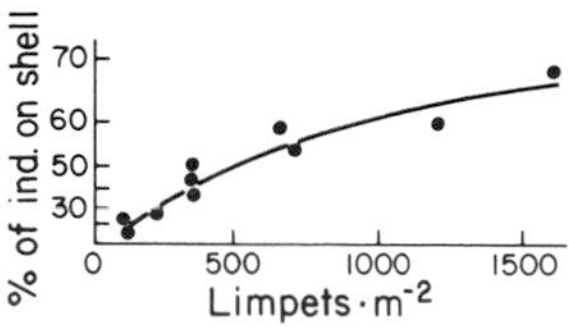

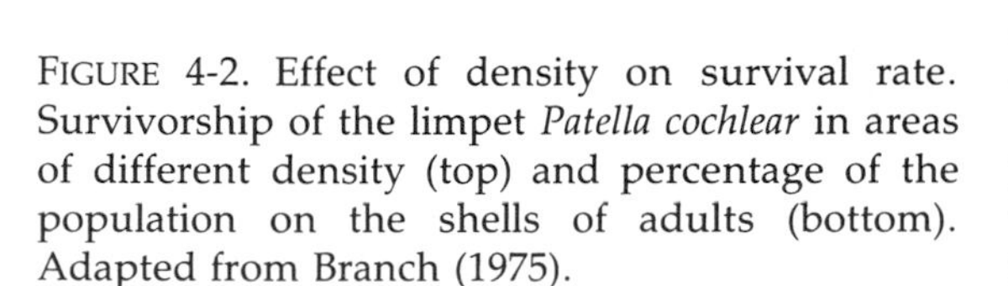
FIGURE 4-2. Effect of density on survival rate. Survivorship of the limpet *Patella cochlear* in areas of different density (top) and percentage of the population on the shells of adults (bottom). Adapted from Branch (1975).

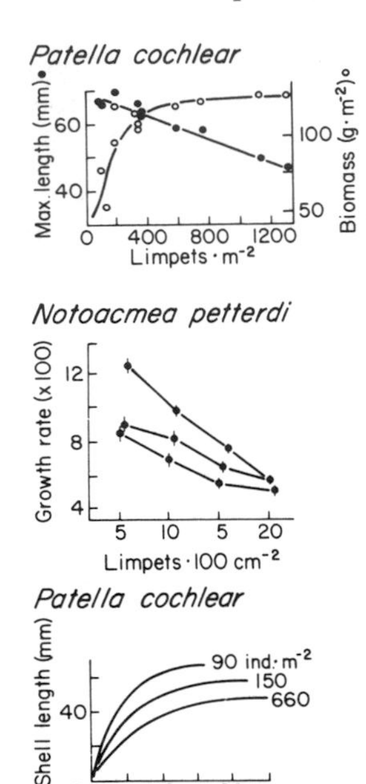

FIGURE 4-3. The effect of density on growth. Top: Length and biomass of *Patella cochlear* in areas of different density [adapted from Branch (1975)]. Middle: Growth rate ± standard error (= $[\ln(L_t - L_0)]/t$, L_t = final and L_0 = initial length of shell, and t = 4 months) of the limpet *Notoacmea petterdi* in relation to density. Measurements from Barrenjoey, Australia; the three sets of data come from areas at three elevations in the intertidal range, top line being the lowest location and bottom line being the highest [adapted from Creese (1980)]. Bottom: Growth over time of *P. cochlear* in areas of different density. Adapted from Branch (1975).

smaller and low fecundities ensue. These relations of food supply, size, and fecundity are common, judging by the many cases where there are clear correlations between size of individual and fecundity (Fig. 4-5), for both invertebrates and vertebrates.

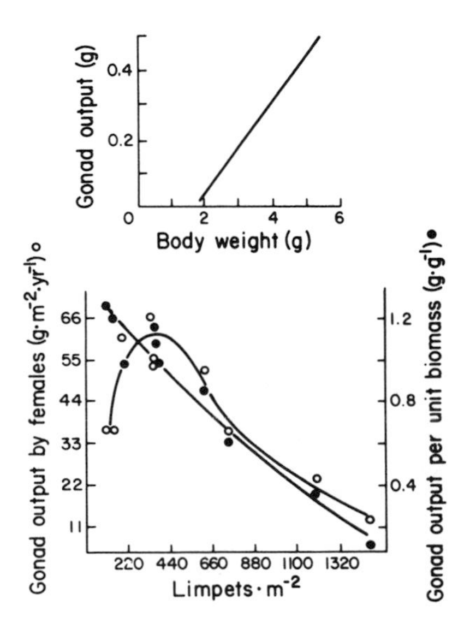

FIGURE 4-4. Effect of density and size on reproductive output. Release of reproductive materials by *P. cochlear* versus body weight (top) and versus density (bottom). Adapted from Branch (1976).

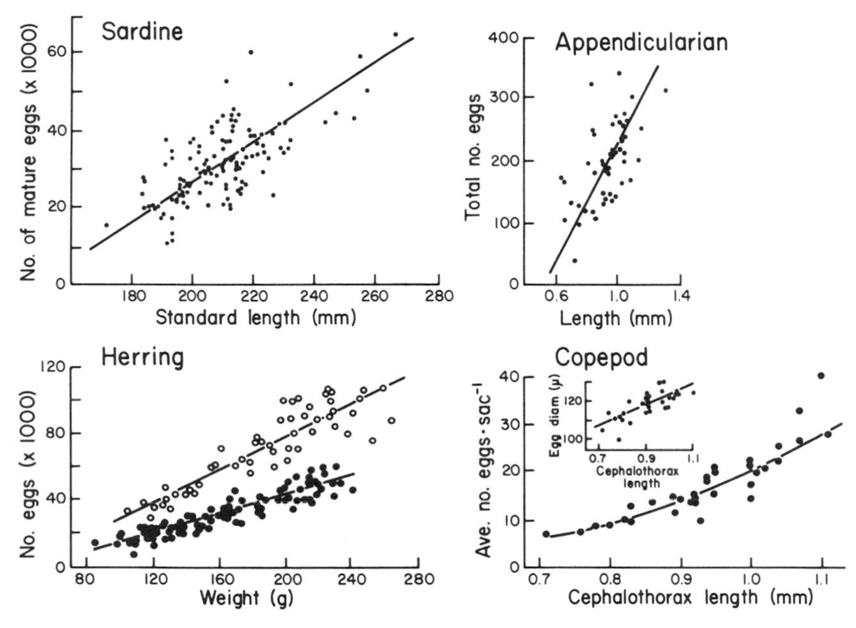

FIGURE 4-5. Relation of fecundity to body size. Sardine (*Sardinops caerulea*) data from McGregor (1957); appendicularian (*Oikopleura dioica*) data from Wyatt (1973); two populations of herring (*Clupea harengus*), data from Baxter (1959); copepod (*Pseudocalanus minutus*) data from Corkett and McLaren (1969).

4.2.2 Interspecific Competition

4.2.2.1 Models of Interspecific Competition

The logistic model [Eq. (4-1)] describes intraspecific competition. Lotka (1925) and Volterra (1926) modified the logistic to incorporate the effects of interspecific competition between two species.

The Lotka Volterra equations have been used to explore the possible consequences of competitive interactions. The Lotka–Volterra model can identify situations where, depending on the abundance of N_1 and N_2, and the values of a and K, the outcome of interspecific competition varies. Species 1 will outcompete and exclude species 2 if the effect of interspecific competition of 1 on 2 is greater than the effect of intraspecific competition of 2 is on itself. If both species are stronger interspecific than intraspecific competitors, the outcome is indeterminate and depends on the initial densities of the two species. Where there is clear partitioning of resources—that is, when intraspecific competition is more important than interspecific competition—it is possible to have stable coexistence of the two species. The model cannot predict exactly how much partitioning is needed to achieve coexistence. The Lotka–Volterra models predict that interspecific competition must lead to the exclusion of one of the competing species. In fact, the competitive exclusion principle (cf. Section 4.5) was inspired by these conclusions. The Lotka Volterra equations have inspired much thinking and theory. They are not easily subject to disproof, and the terms of the equations are difficult to define operationally (Peters, 1991). To do so we need to know that competition actually takes place, what the presumed competitors are competing for, and how the presence of a competitor quantitatively alters the supposed gains. These demands are not easily met even in simplified laboratory conditions. In the field, it is seldom the case that there are only two competitors present, and the surrounding conditions are rarely constant, so models based on Lotka Volterra equations have limitations. To get a more realistic view of competition, we need to examine actual evidence.

4.2.2.2 Evidence for Interspecific Competition

Evidence supporting the idea that different species may compete for a resource has been collected, such as many well-documented cases of partitioning of resources or of displacement in morphological characters in sympatric species. These phenomena are suggestive but only circumstantial evidence that competition may have taken place in the past, and are discussed in Section 4.5 below.

Manipulative experiments have shown explicitly that removal of one presumptive competitor indeed results in increased density of the other presumed competitor (Table 4-1).

The actual occurrence of interspecific competition in the field, at least for animals and its demographic consequences, was first demonstrated in a study of two barnacles, *Chthamalus stellatus* and *Balanus balanoides*, on a Scottish rocky shore by Connell (1961). Adults of the two species coexist in the upper reaches of the tidal range (Fig. 4-6, left), where

TABLE 4-1. Density of Black (*Embioteca jacksoni*) and Striped (*E. lateralis*) Surfperch in Experimental Parcels of Seafloor off the California Coast, in Which Abundance of the Presumptive Competitor Species was Reduced by 90% Over 4 Years. Density Reported Before and 4 Years After the Removals.[a]

Species	Treatment	No. fish ≥ 1 y old/240 m^2	
		Before	After
Black surfperch	Competitor removed	27.2 ± 0.8	35.8 ± 0.2
	Control	30.4 ± 1.8	29.8 ± 29.8
Striped surfperch	Competitor removed	10.9 ± 0.6	17.2 ± 3.7
	Control	19.9 ± 0.6	18.3 ± 0.9

[a] From Schmitt and Holbrook (1990).

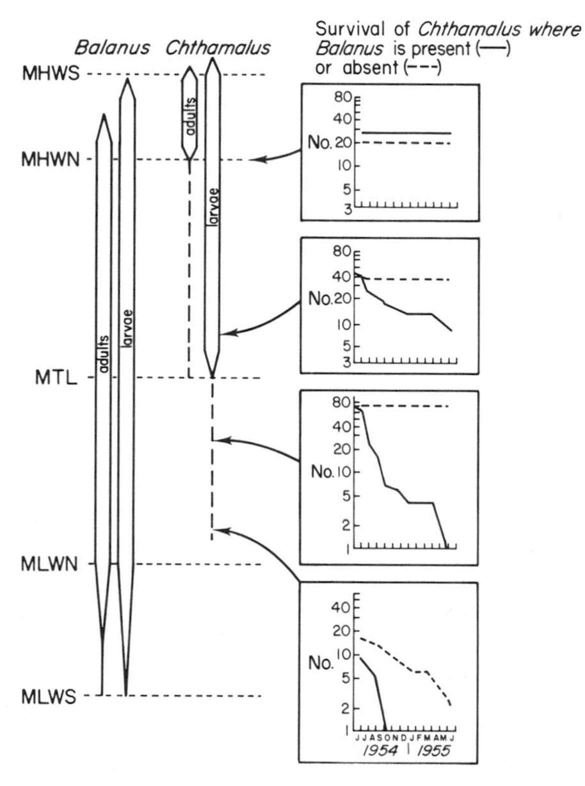

FIGURE 4-6. Vertical distribution of the barnacles *Balanus balanoides* and *Chthamalus stellatus* at Isle of Cumbrae, Scotland (left) and survival of *Chthamalus* in control situations and where all *Balanus* were experimentally removed (panels on right). The experiments on top two panels were done on the rocks and heights where *Chthamalus* had settled. The bottom two panels contain data from stones that were transplanted to the appropriate depths. The level of mean high and low water during spring and neap tides are indicated as MHWS, MLWS, MHWN, MLWS, MHWN, MLWN, respectively. Adapted from Connell (1961).

Balanus does not grow very vigorously or densely. Larvae of both species settle throughout a considerable portion of the tidal range, so a marked mortality of *Chthamalus* takes place throughout the lower to middle intertidal range. Connell set up pairs of experimental quadrats where he removed *Balanus* from one of the quadrats and left the other untouched. Then he followed the survival of the *Chthamalus* that settled on the quadrats. In the upper intertidal, above the level of mean high water during neap tides (MHWN), there was no effect of the removal of *Balanus* on survival of *Chthamalus*. At that elevation, these two populations do not interact, and the upper extent of the distribution of these barnacles is probably determined by desiccation. *Chthamalus* seems more tolerant of drying than *Balanus*. On the other hand, below

TABLE 4-2. Size and Reproduction in the Barnacle *Chthamalus stellatus* in Experimental Situations Where Its Competitor, *Balanus balanoides*, Is Present or Absent.[a]

Elevation in intertidal range above MLW (ft)	Average diameter (mm)		Percentage of individuals with larvae in mantle cavity	
	Balanus absent	*Balanus* present	*Balanus* absent	*Balanus* present
2.2	4.1	3.5	65	61
1.4	3.7	2.3	100	81
1.4	4	3.3	100	70
1.0	4.1	2.8	100	100
0.7	4.3	3.5	81	70

[a] From Connell (1961).

MHWN, there is significantly better survivorship of *Chthamalus* in the absence of *Balanus*. Mortality rates of *Chthamalus* increase where rates of growth of *Balanus* are highest. *Balanus* grows faster and outcompetes *Chthamalus* for space by covering, lifting, or crushing *Chthamalus*. As in the case of intraspecific competition, interspecific competition reduces the growth of individuals and reproductive output (Table 4-2).

Intraspecific and interspecific competition are seldom separable in real situations. The growth of *Patella granularis* is negatively correlated to density of its own species and to the density of barnacles on the rock surfaces (Branch, 1976). Where barnacles are scarce, biomass and reproductive output of limpets are directly related to limpet density and climbs steeply (Fig. 4-7) to about 100 limpets m^{-2}, beyond which

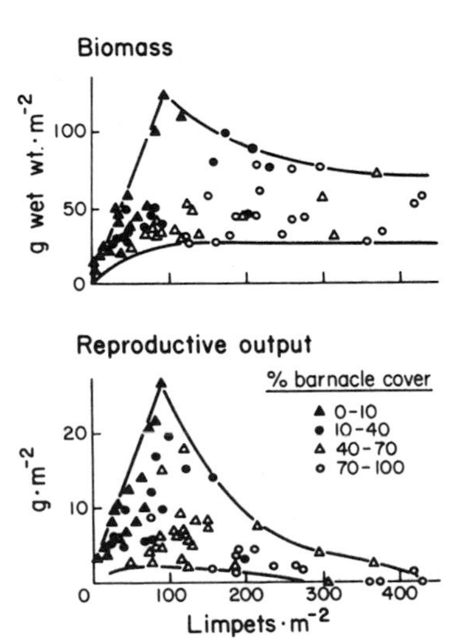

FIGURE 4-7. Effects of intraspecific and interspecific competition in *Patella granularis*. Adapted from Branch (1976).

biomass and reproductive output are reduced. This is the result of intraspecific competition. At any one density of limpets, however, the more barnacles that are present, the lower the limpet biomass and reproductive performance, so that interspecific competition is also important.

4.3 Density-Dependent Control of Abundance

Other examples of active competition besides those provided by Branch and Connell are now available. The collective results of these studies suggest that even though density-independent factors often greatly *affect* the numbers of organisms present, abundance is ultimately *controlled* by density-dependent mechanisms such as the ones reviewed in the previous section. *Control* implies that recruitment may increase at low density while it decreases at high densities.

Density-dependent mechanisms, such as proposed by Branch (1976) for the limpet *P. cochlear*, can hypothetically provide for control of abundance. High limpet densities lead to low growth rates and hence small size. At high densities more of the juveniles settle on adult shells, and this causes increases in mortality. Both small size and higher mortality depress reproduction, so the number of larvae released into the water is lower. Low densities of *P. cochlear* lead to low mortalities of juveniles, since most larvae are on the rock surfaces, and rapid growth, because large enough feeding territories are available, so large sizes are achieved. All this produces a large release of larvae into the water. If the density-dependent effects extend over a large enough area, and vagaries of weather and hydrography during pelagic existence do not uncouple the abundance of larvae from the parent density, the number of larvae settling and surviving on rock surfaces in that area may be density dependent. Sections 8.3 and 12.2 offer more on this issue.

The coupling of mortality, growth, and fecundity to density and food supply has also been demonstrated in other situations. Kneib (1981) manipulated densities of the salt marsh killifish *Fundulus heteroclitus* within enclosed parcels of natural salt marsh. Mortality (Kneib, 1981), growth rate (Fig. 4-8, top), and the reproductive output of the fish population (Fig. 4-8, bottom) depend on the density of the extant population. The reduction in growth rate associated with greater densities is mediated by food supply, since much higher growth rates were observed where extra food was added to the experimental marsh parcels. In addition, fed fish showed no mortality and fecundity was high. In these fish, then, there is density-dependent control of abundance.

In our discussion so far we have considered that high densities lead to reduced growth of populations. This is probably true for many

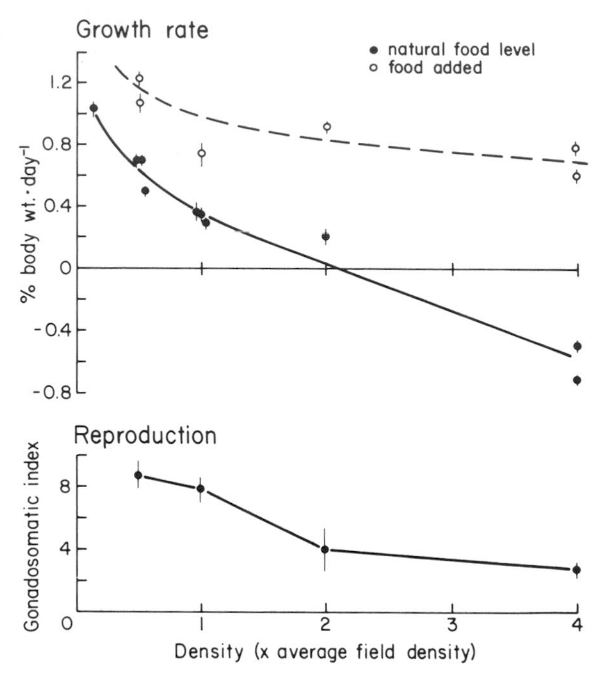

FIGURE 4-8. Changes in growth rate and reproductive potential in the killifish *Fundulus heteroclitus* in response to changes in density. Each point is the mean ±95% confidence interval. The average field density of fish was 180 fish per pen. Each pen consisted of an area of salt marsh artificially diked in a Delaware salt marsh. Gonadosomatic index = weight of the gonad/weight of fish × 100. Food additions were done within replicated diked areas by distributing fish flesh every 2–4 days into tidal pools within the areas. From Weisberg and Lotrich (1986)

populations, but there must be some low densities where population growth increases as density increases. Low densities might, for example, lower reproduction due to difficulty of males and females finding each other; as density increases, reproduction rates might increase. Some other benefits of high numbers might also be important; in the guillemot (*Uria aalge*), a sea bird that nests colonially on rocky promontories, breeding success increases with density. When colonies are numerous and dense, gulls are less able to prey on eggs and chicks (Birkhead, 1977). A complete description of the effect of population size on population increase thus needs to include both a positively density-dependent range and a negatively dependent range. The humped curve that results has been known as the "Allee (1931) effect." One important consequence of positively density-dependent reproduction is that (assuming death rates remain constant) a species can decline to extinction if its density has become lower than a critical threshold, even though there are actively reproducing individuals still alive.

4.4 Density-Dependent Versus Density-Independent Effects on Abundance

There are examples of marine animals where there is evidence that density-dependent factors are not important. Experiments involving removal of a dominant species of damsel fish (*Pomacentrus wardi*) in the Great Barrier Reef of Australia show that rates of recruitment and survival of new recruits are independent of density of adult fish on the reef (Doherty, 1983). Further, the adult fish do not occupy all potential territories in the reef. The density of populations of reef fishes may be reduced by meterological events such as storms, cold temperatures, predators, or lack of enough larvae that can settle in the reef (Williams, 1980; Victor, 1983).

The lack of enough recruits—even though the reef damsel fishes are among the most fecund of vertebrates (Sale, 1978)—is especially likely in case of fishes that have pelagic larvae, since these larvae are exposed to many mortality factors during the period of larval drift in the water column and mortality would be very high. Manipulation experiments by Doherty (1983) show that the densities of damsel fishes do not saturate the possible habitats of the reef. Perhaps these fishes could saturate the reef after several periods of unusually successful settlement, but this seems to be unusual. Here we have a situation where populations appear not to exhaust resources; competition among similar species may thus be rare, since the obvious resources seem not to be in short supply.

It is not clear if such a density-independent situation is common among other marine fish populations. All the previous examples involve coastal, shallow water fish. It has not been feasible to carry out experiments in which density or food is manipulated in the open sea, but the economic importance of fish and concern with fishery yields have stimulated many analyses of density-dependent features of the population dynamics of pelagic marine fish.

The abundance of pelagic fish populations varies markedly with time, sometimes due to natural causes, sometimes due to changes in fishing pressure. Such changes in fish abundance offer the best glimpses we have of competitive interactions as possible controls of fish populations in the open sea. If the food supply of pelagic fish is limited, we would, as discussed in Chapter 3, expect that survival, growth, and recruitment depend on the abundance of adult fish. In the following sections we examine the available evidence for such density dependence in fish stocks, focusing primarily on intraspecific competition.

4.4.1 Effect of Abundance on Survival

The larger the density of larval and young fish, the lower their survival rates (Fig. 4-9). The survival is therefore uniformly negatively density dependent, in the species included in Fig. 4-9 and in others (Table 4-3).

The density dependence of survival has been specifically examined in salmonid fish. Salmonids on the northwest coast of North America hatch from eggs in freshwater and migrate as young fish through estuaries to the sea, where they grow into maturity. Cohorts from a particular river basin tend to migrate together to the same area of the sea when mature. The salmon eventually return to their native stream or lake to spawn and then die.* Salmon are regularly censused as they depart and as they return to various estuaries in North America, and these data enable comparison of the original numbers of young fish that left the estuary to the number of adults returning (Peterman, 1980). Although the data are variable, out of 18 data sets from various breeding grounds, 12 show that the more young fish that leave an estuary, the greater the mortality. There is of course no assurance that density in the sea is predicted by the abundance of stock leaving the estuaries. Members of specific stocks migrate over very large areas of the Pacific and mix with other stocks. Thus, the salmon data are only suggestive. The correlations of survival and abundance of young salmon have a great deal of variability, but suggest that young salmon may be food limited in the sea. This conclusion has consequences for management and stocking efforts (Simenstad et al., 1982) in the freshwater and estuarine breeding areas: the stocking of individual estuaries by adding artificially reared young salmon may not result in a significant increase in the abundance of the stock, since there may not be enough food at sea for them. When a particular stock is almost depleted, of course, the stocking is useful.

The density-dependent survival of larval and young pelagic fish contrasts with the complete lack of density dependence of survival of adult fish (Table 4-3). This general lack of density dependence of

---

*Such a life cycle where fish return to freshwater to breed is referred to as anadromous. Fishes such as eels that live mainly in freshwater and return to the ocean to breed are referred to as catadromous. In the case of eels, breeding takes place in deep waters of the Sargasso Sea.

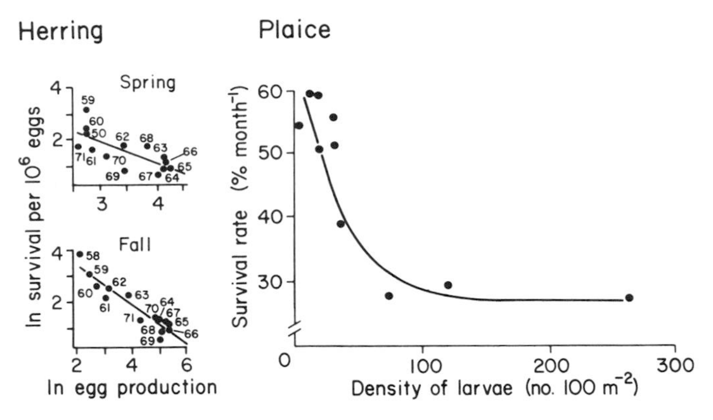

FIGURE 4-9. Survival of fish during the first year of life (age 0) at different densities. Data for spring and fall herring cohorts from Winters (1976). Plaice data from Lockwood (1978).

survival in adult fish could mean that adult fish, at least the ones in Table 4-2, are not abundant enough to tax their food resources.

4.4.2 Effect of Abundance on Growth

Rate of growth of young marine fish is lower when density of the cohort is higher (Table 4-3). This is not true for populations of adult fish (Table 4-3). If growth rate depends on food supply, larvae and young fish must exist in food-limited conditions, while food may not be in short supply for adults.

Additional evidence on the relatively small effect of abundance on growth of adult fish can be garnered from fishing statistics obtained after the Second World War, when the intense fishing in the North Sea was halted. During the war years the total fish biomass in the North Sea about doubled, but most fish species showed minor or no responses in catch per unit of fishing effort or in growth (Holden, 1978; Richards et al., 1978). If indeed food was limiting, growth should have decreased.

One of the major commercial species in the North Sea is the plaice, which can be an example of the lack of density-dependent control of growth in adult fishes (Bannister, 1978). Before World War II, fishermen harvested perhaps 23% of the plaice in the North Sea each year. When the fishing resumed at the end of the war, the stock had about doubled in number—yet the mean weight of postwar plaice had increased by 50%. Recruitment of plaice after the war was almost double that of prewar years. Different analyses of the pre- and postwar catch data yield different results, but the general conclusion is that the plaice did

TABLE 4-3. Density Dependence of Survival, Growth, and Recruitment in Young and Adult Fish Stocks.[a]

	Survival	Growth	Recruitment	Sources
YOUNG FISH				
Herring (Gulf of St. Lawrence)	—	—		Winters (1976)
Herring (North Sea)	—	—		Iles (1967, 1968)
Plaice (Waddensea)		Weakly−		Rauck and Zijlstra (1978)
Plaice (North Sea)	—	—		Cushing (1975)
				Hubold (1978)
				Harding et al. (1978)
				Lockwood (1979)
				Rijnsdorp and van Lceuwen (1992)
Sole (Waddensea)		Weakly−		Cushing (1975)
Sole (North Sea)		—		de Veen (1978)
Winter flounder (Mystic River Bay)	—			Pearcy (1962)
Halibut (N.E. Pacific)		—		Southward (1967)
Sardine (California Current)	—	—		McGregor (1957)
				Marr (1960)
				MacCall (1979)
Gadoids (North Sea)		—		Raitt (1968)
Salmon (North Pacific)[b]	—	—	0	Ellis (1977)
				Peterman (1980)
				Rogers (1980)
				Gulland (1970)
Southern bluefish tuna (N.W. Australia)		—		Jenkins et al. (1991)
Haddock (North Sea)[c]		—		Frank et al. (1992)
(Grand Branks)		—		
ADULT FISH				
Cod (North Sea)	0	0	Weakly+	Daan (1978)
Cod (Arcto-Norwegian)		+	0	Garrod and Clayden (1972)
Plaice (North Sea)	0	Weakly−	0	Bannister (1978)
				Cushing (1975)
				Rijnsdorp and van Leeuwen (1992)
Sole (North Sea)		0		de Veen (1978)
Herring (Gulf of St. Lawrence)		—	Very weakly−	Winters (1976)
Herring (North Sea)	0	0	0	Iles (1967, 1968)
				Saville (1978)
				Cushing (1981)
Sprat (Gullmarsfjörd)		0		Lindquist (1978)
Mackerel (North Sea)		Weakly−	0	Hamre (1978)
Mackerel (Gulf of St. Lawrence)			0[d]	Winters (1976)

TABLE 4-3. (*Continued*).

	Survival	Growth	Recruitment	Sources
Pacific halibut		0	0	Southward (1967) Cushing (1981)
Sardine (California)	0	0		Marr (1960)
Haddock			0	Cushing (1981)
Whiting (North Sea)		Weakly−		Jones and Hislop (1978)

[a] 0, no significant relationship to abundance; + or − indicates positive or negative relationship to abundance.
[b] Several species of salmon involved. These data include individuals up to 5 years of age so are not only young fish; salmon spend lengthy periods in freshwater before entering the ocean, and density-dependent growth of salmon is less in the marine environment than in freshwater (Cushing, 1981). West Coast salmon mature when they are ready to return to freshwater and reproduce just once.
[c] Two-year old fish show the relationship most clearly; the higher growth rates occurred in the stronger cohorts, in which years the food supply may have been considerably larger than average.
[d] Significant peak in recruitment during year with intermediate abundance of stock.

not show a very marked response to the increase in abundance, and neither did other species such as haddock (Gulland, 1970; Jones and Hislop, 1978). The implication is that at the time these species existed at an abundance considerably below the carrying capacity of the North Sea, and hence competition for food must not have been important for stocks of the adult fish.

There is an instance where more marked effects of density on growth of adult fish have been measured. Growth of herring in the Gulf of St. Lawrence (Table 4-3) is lower at higher densities. When the stocks of pelagic fish decreased by about 83% in the 1970's, the size of herring increased by about 53%. This area of the northwest Atlantic, along with the Grand Banks and Georges Bank, has supported a sizable fishery for a long time, known to Breton, Irish, and Scandinavian fisherman for centuries. These places of high fish density, therefore, appear not to be representative of the sea in general; the high abundance of fish was remarkable enough to justify long, hazardous voyages in small vessels.

The herring in the St. Lawrence demonstrate that adult marine fish can be numerous enough to tax their food resources and hence produce some degree of density-dependent growth. The effects of density are less marked in the vast majority of other reported cases of stocks of adult fish. Perhaps more and better data will show otherwise, but for now we have to conclude that although marine fish *can* tax resources, such instances are infrequent. Because of this, growth of adult marine fish seems not to be density dependent.

4.4.3 Effect of Abundance on Recruitment

The recruitment* of young fish into an adult population is not clearly related to the density of the adult fish (Table 4-3). The scatter of the data is too large to detect trends (Fig. 4-10) even though there are numerous published instances of models fearlessly fitted to such data. Koslow (1992) calls for a reexamination of the application of such models.

The lack of density dependence is not due to lack of variation in the recruitment from year to year. This variation is quite large; in Norwegian herring, for example, recruitment may vary over two orders of magnitude over just a few years (Cushing, 1981). This may be an extreme example, but variability in recruitment is the rule although the causes of the variability are largely unknown.

The evidence on mortality, growth, and recruitment obtained from commercial fisheries leads to the conclusion that the densities of larval young in pelagic fish may be high enough for competition for food to be important. Adult fish exist at densities low enough that the carrying capacities of their environment are seldom taxed. Abundance of these fishes—and that of the reef fishes discussed earlier—is thus not completely controlled by factors that depend on density of the population. Rather, the control, if it does exist, is exerted by density-dependent mortality and growth during early life. Recruitment is not dependent on adult density, so there is little evidence that adult abundances are controlled by density-dependent mechanisms. In spite of large adult fecundities, mortality suffered by the early stages may be so intense that very few recruits survive and are available to enter the breeding population. If the mortality of early stages depends on many factors—as seems likely—the sum of these factors may impose what may in the aggregate appear to be random variation on mortality rates. Thus, for many fishes the number of recruits would vary randomly, and may not be higher when adult numbers are low or lower when adults are abundant.

The above considerations are a caution that even though in many species the density is probably determined by density-dependent factors, density-independent processes may be very significant in population dynamics. There is still, however, as discussed in Chapters 8 and 9, no agreement on the extent to which competition establishes the abundance of species in nature (Schoener, 1983; Bailey and Houde, 1984).

---

*For demographic purposes recruitment refers to the maturing of individuals into the adult age classes. In fishery publications, recruitment is defined as the appearance of a cohort into the catch due to their having grown large enough to be caught given the mesh size of the fishing gear.

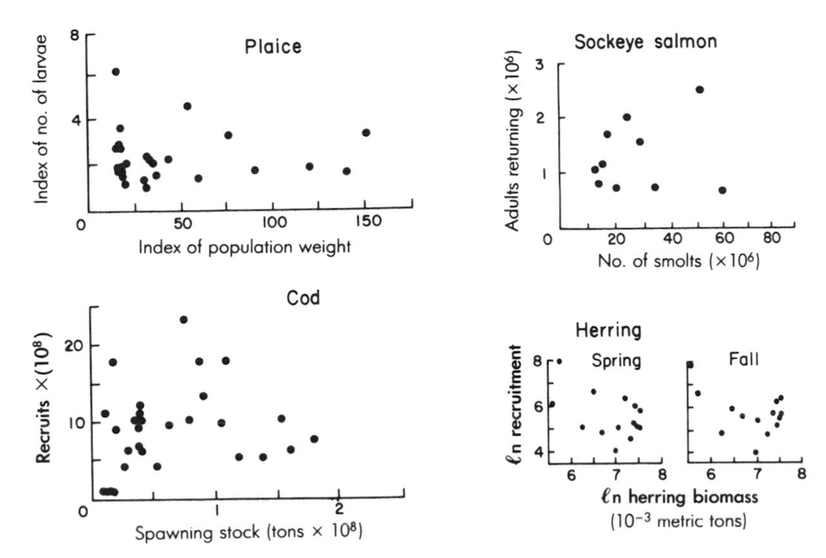

FIGURE 4-10. Recruitment in fish populations in relation to density. Top left: Recruitment of plaice in to the North Sea fishery in relation to stock density. The values are dimensionless indices obtained from catch statistics [adapted from Cushing (1975)]. Bottom left: Recruitment of cod in Arcto-Norwegian waters (1940–1969) in relation to the abundance of spawners [adapted from Garrod and Clayden (1972)]. Top right: Recruitment of sockeye salmon in Skeena estuary, British Columbia [adapted from Ellis (1977)]. Smolts refer to young fish leaving rivers for the sea. Bottom right: recruitment of herring in southern Gulf of St. Lawrence, Canada. Adapted from Winters (1976).

4.5 Resource Partitioning

There is a long history of examples, primarily in birds, in which species that live together are described as partitioning resources. Such sympatric species divide up limited resources by behavioral, morphological, or ecological mechanisms. Lack's (1947) study of Darwin's finches in the Galapagos Islands is just one prominent example. Many of these examples are ambiguous, due to faulty data and interpretation, but the remarkable degree of partitioning of resources that exists in all environments suggests (although it cannot demonstrate, since the appropriate controls are lacking) that competition has been a pervasive factor during the evolution of present species.

Different aspects of habitat or food resources may be apportioned differently in some way by the species present in one place. The ubiquity of this partitioning gave rise to what has been said to be one of the few "laws" of ecology, the "competitive exclusion" or "Gause's" principle: complete competitors cannot coexist. This idea has a venerable history (cf. Krebs, 1978) but a contentious development, with opinions ranging from "one of the chief foundations of modern ecology" to "a trite maxim." The point to remember is that if enough effort is spent, it is possible to find differences in habitat selection or food partitioning between any two species. On the other hand, it is impossible to complete a search for ways in which potential competitors could differ, since there are innumerable dimensions in which differences could be found. The circularity of the competitive exclusion principle makes it impossible to disprove, and hence less than useful as empirical science.

There are lots of examples, though, of cases where partitioning of resources is circumstantially suggested. Partitioning of resources can occur either at one time and site, or may be accomplished at different times of day or year, and competing species may diverge morphologically enough to exploit different portions of resources.

Slightly different parts of a habitat can be used either by different age classes within a species or by several species. An example of the former is the distribution of the limpet *Patella compressa* on the surface of the kelp *Ecklonia* (Fig. 4-11). *Ecklonia* is almost the only place where this limpet is found. Juvenile *P. compressa* aggregate on folds on the bases of the fronds; older individuals move onto the broad flat "hand," while the oldest individuals establish themselves in shallow cavities in the stipe. The base of the shell of large individuals even grows concavely to adapt to the cylindrical shape of the stipe. The separation among the age groups may partition space and food, presumably minimizing intraspecific competition.

Interspecific partitioning of a habitat by several species occurs in all environments. The assemblage of fish species found in New England salt marshes (Fig. 4-12, left histograms) partition the use of habitats

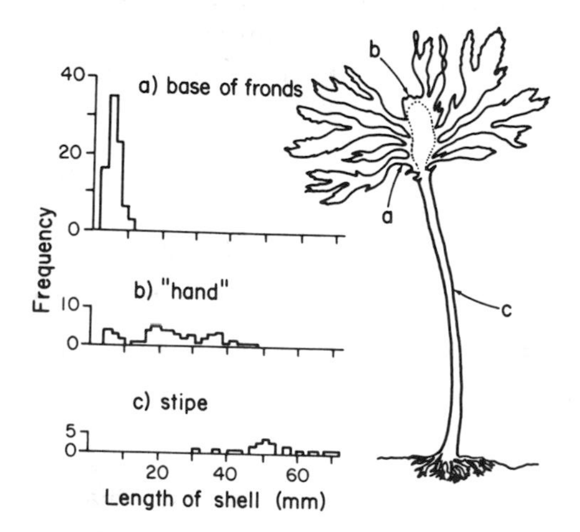

FIGURE 4-11. Location of size classes of the limpet *Patella compressa* on the kelp *Ecklonia*. The histograms show the frequency of individuals of various sizes. This kelp may reach 10 m in height. Adapted from Branch (1975a).

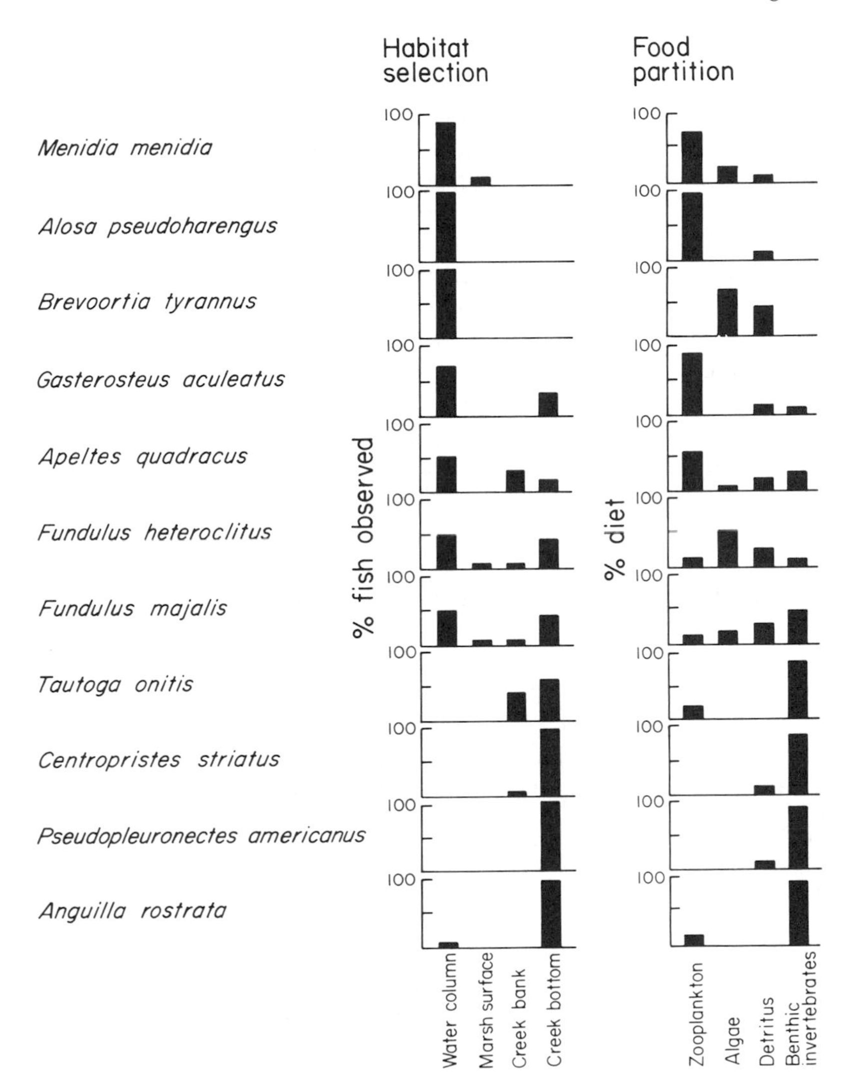

FIGURE 4-12. Partitioning of habitat and food by species of salt marsh fish. These results were averaged over an entire year and through many tidal cycles. The common names of the fishes, in order, are silversides, alewife, menhaden, three-spined stickleback, four-spined stickleback, common mummichog or killifish, striped killifish, tautog, sea bass, winter flounder, and American eel. Adapted from Werme (1981).

within the marsh. Silversides, alewives, and menhaden are found primarily in the water column. The two sticklebacks are less stereotyped in use of habitat, and are found in the water column as well as on the bottoms of creeks. The two killifish are the least specialized in habitat choice. The last four species are found mainly on creek bottoms, with tautog, sea bass, and winter flounder using sandy bottom while eels were mainly found on muddy bottoms.

The partitioning of food among size classes of one species and other aspects of food selection are more fully discussed in Chapter 6. Here we can emphasize the extent of food partitioning by looking at food habits of salt marsh fish (Fig. 4-12, right histogram). Of the three species found in the water column, the menhaden eat mainly algae and silversides feed on larger zooplankton than the alewives (Werme, 1981). The three-spined sticklebacks eat much more plankton than the four-spined sticklebacks. The killifish are the most generalized feeders of all the fish species, but even so show differences in food selection. The sheepsheads feed principally on algae and detritus, the striped killifish on benthic invertebrates found on creek bottoms, and the common killifish on algae and invertebrates found on the surface of the marsh, obtainable only during the time when the marsh in flooded. The remaining species of fish feed on benthos, and partition their food by using different combinations of size classes of prey.

Species often partition resources by using prey populations at different times. The three-spined stickleback is found primarily in the spring (April–June) while the four-spined is in the marsh for longer periods (April–October). The common killifish (and the related sheepshead minnow, *Cyprinodon variegatus*, not included in Fig. 4-12) feed primarily during high tide, while the striped killifish have no tidal feeding rhythm.

Partitioning of food and habitat may take place by changes in morphological characters in competing species (Brown and Wilson, 1956), although some of the original evidence for such character displacement has been interpreted in other ways (Grant, 1972). One example of character displacement has been claimed for two species of shallow-water marine isopods whose body sizes diverge when the two species are sympatric (Table 4-4). Size in isopods is important in various ways, one of which is that individuals of a given size occupy protective crevices of a given size. Perhaps displacement of size avoids competition among the two species for crevices.

Fenchel (1975) found another possible example of character displacement while surveying the size of two snails, *Hydrobia ulvae* and *H. ventrosa*, in estuarine sites off the Danish peninsula of Jutland. Shell lengths of the two species are similar in sites occupied by only one species, while in the 16 sites of sympatry there is a consistent divergence in shell length (Fig. 4-13, top left). Changes in size may have many ecological implications, one of which is that snails of different size feed on different size particles (Fig. 4-13, left bottom). This results in differences in the distribution of particle sizes ingested by sympatric snails compared to allopatric snails. The *Hydrobia* data

TABLE 4-4. Average (± Standard Deviation) Body Length of the Isopods *Sphaeroma rugicanda* and *S. hookeri* in Four Areas of Shallow Water on the Danish Coast.[a]

Site	Species	In body length (mm)	
		Sympatric	Allopatric
Sønderjylland	*S.r.*	1.64 ± 0.17	1.61 ± 0.17
	S.h.	1.38 ± 0.18	1.49 ± 0.16
Slien	*S.r.*	1.67 ± 0.17	1.61 ± 0.17
	S.h.	1.44 ± 0.20	1.49 ± 0.16
Storstrømmen	*S.r.*	1.68 ± 0.16	1.53 ± 0.21
	S.h.	1.43 ± 0.17	1.50 ± 0.16
Mariager	*S.r.*	1.88 ± 0.18	1.73 ± 0.20
	S.h.	1.32 ± 0.15	

[a] Adapted from Frier (1979).

suggest that character displacement is a real phenomenon and may result in reduced competition (Kofoed and Fenchel, 1976). Pacala and Roughgarden (1982) showed that in lizards the level of competition was inversely related to the degree of resource partitioning, as we would expect from the *Hydrobia* data. Schoener (1983) reviewed available results and concluded that overlap is inversely related to competition.

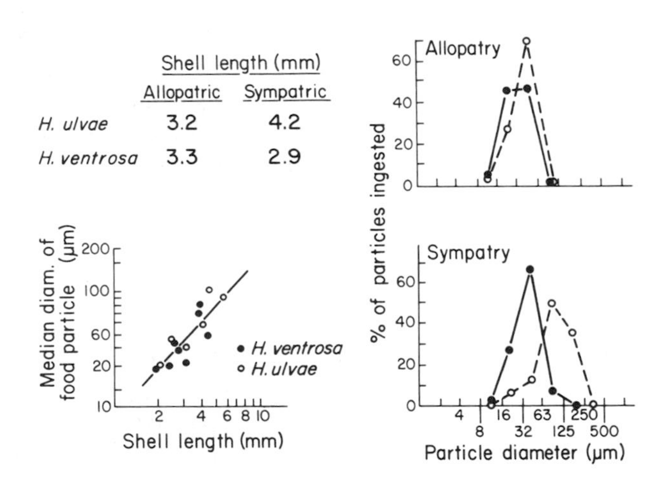

FIGURE 4-13. Character displacement and its consequences for feeding in *Hydrobia*. Top left: length of shells in two species of *Hydrobia* where they occur together and separately. Bottom left: relation of length of shell and median size of particles eaten by *Hydrobia*. Right: distribution of particles sizes ingested by the two species of *Hydrobia* where they occur together and separately. Adapted from Fenchel (1975).

Nonetheless, it is hard to prove that character displacement results from competition. Connell (1980) has pointed out that the localities studied by Fenchel varied in salinity, water turbulence, and particle size of sediment, so that it is not possible to prove that competition alone led to character displacement. Furthermore, in any one site the species may occur together 1 month and alone the next, so there may not be enough time for displacement to occur in species that disperse so actively. Character displacement remains a suggestive but unproven consequence of competition.

Static documentation of food and habitat partitioning and character displacement only constitutes circumstantial evidence of competition; it is not possible to attribute cause and effect. Competition could cause partitioning, but also a low degree of partitioning or displacement could reduce competition. In contrast, where experimental manipulation such as in Connell's work has been done, we were in a position to say that competition was responsible for the phenomena observed. Of course, it is difficult to carry out experiments on processes that take place over relatively long periods of evolutionary time, such as character displacement, but other aspects of competitive interactions are subject to an experimental approach.

4.6 Niche Breadth and Species Packing

Resource partitioning due to interspecific competition implies that very specialized species should be found in habitats containing many species. Much space has been devoted to this topic in the ecological literature. If we accept that the maxim "jack-of-all-trades, master of none" is applicable to how organisms use resources, we can argue that more specialized species will make more effective use of whatever resources are available. Hence more species should be able to be packed into an environment containing specialists. This tendency to increased specialization can go on until constrained by two factors. First, the variability in the resource being partitioned sets a limit. If a specialist becomes dependent on a specific resource, but the resource is not available in sufficient abundance or during a certain time, that specialist species is more liable to disappear. Second, intraspecific competition will favor some degree of nonspecialization so that not all individuals or age classes within a species exploit the same resource. Recall, for example, habitat partitioning due to intraspecific competition in limpets (Fig. 4-11).

The above arguments are completely hypothetical, with precious little evidence to support them. There are several reasons for the lack of evidence. First, it is extremely difficult to define a niche. Hutchinson (1957) first developed the concept of the niche as a multidimensional hyperspace, comprising all the variables important to a species. While interesting, this idea is not testable. The best that can be done is to isolate a few potentially important variables and see how species align themselves along these few dimensions of the niche. The range of values of the chosen variables that is tolerated by the species is then called the niche width and is measured in various simple ways (Colwell and Futuyma, 1971; Pianka, 1973) or by using multivariate statistics (Green, 1971). While the variables chosen may appear very important to the ecologist, they may be less critical to the organisms under study. Second, there is always the problem of scale of measurement; we may measure the temperature range in terms of a few degrees while the organisms may respond simply to hot or cold, while being far more sensitive to minute changes in another variable. Third, some dimensions of niche breadth are more relevant to some taxonomic groups than to others, and comparisons are difficult even within similar ecosystems. For example, Kohn (1971) finds remarkably high degrees of specialization in food habits of *Conus* snails in Pacific coral reefs. On the other hand, Sale (1977) concludes that the diet of coral reef fishes is rather generalized. The diets of species of *Conus* tends not to overlap, while those of fishes do. As Anderson et al. (1981) point out, however, this does not mean that the fishes are complete generalists, since reef fishes have quite specialized requirements in terms of habitat, for example. Further, Roughgarden (1974) shows that some reef fishes do show partitioning with regard to prey size (cf. Chapter 6). Such considerations have made it difficult to generalize about how niche breadth should vary in different kinds of environments.

The study of niches and how species are "packed" into an environment requires a more dynamic understanding of what a species does than is provided by an account of items in the diet. The morphology and behavior of an assemblage of sympatric predators, for example, determine the kinds and sizes of their prey, and how closely apportioned the prey are (Schoener, 1974 and others). To see what is involved in the acquisition of prey, it is necessary to learn a considerable amount about the feeding process and feeding selectivity involved, the subject of the next two chapters.

Chapter 5
Feeding and Responses to Food Abundance

5.1 Introduction

5.1.1 Modes of Feeding in Marine Consumers

Marine heterotrophs have evolved a very large variety of feeding mechanisms to obtain their rations, and any given species usually makes use of several feeding mechanisms (Pandian, 1975). Heterotrophs may use organic matter dissolved in water or on larger particles. Dissolved food may be taken up through the body or cell surface, as occurs in microorganisms, many invertebrate parasites, and pogonophorans. Other consumers take up fluids through their mouth, as in the case of some nematodes, trematodes, leeches, parasitic copepods, and young mammals. Dissolved food is also obtained by some heterotrophs from their endosymbionts, as in the case of zooxanthellae in corals and other cnidarians, and sulfur bacteria in some pogonophorans and bivalves.

5.1.1.1 Uptake of Dissolved Organic Matter

Heterotrophic uptake* of dissolved organic matter from water is primarily carried out by bacteria, fungi, and other obligate heterotrophs. Facultative heterotrophic uptake may also be carried out by algae and some invertebrates.

There are many measurements of heterotrophic uptake of dissolved organic carbon (DOC) by algae, usually done by experimental addition of high concentrations of organic substrates. If uptake of DOC by bacteria can take place at low (nano- or micromolar) concentrations of

---

*The kinetics of uptake of dissolved organic compounds can often be described by Michaelis–Menten kinetics (Parsons and Strickland, 1962; Wright and Hobbie, 1965). This implies, as in the case of nutrient uptake reviewed in Chapter 2, that there is a maximum uptake rate at some concentration of substrate, and that perhaps the number of transport sites in each cell sets this rate.

DOC, one would suppose that in nature bacteria do nearly all the heterotrophic uptake, since they would easily maintain the DOC concentrations below the level where algal uptake can take place. This supposition has not been conclusively verified. The best evidence to support the idea that bacterial heterotrophy is much more important than algal heterotrophy comes from studies in which uptake of organic substrates by populations of living microorganisms is fractionated using filters. About 90% of the heterotrophic assimilation of tritium-labeled organic compounds is due to organisms smaller than $5\,\mu m$ (Azam and Hodson, 1977), most of them smaller than $1\,\mu m$ and largely bacteria.

In sediments, where the concentration of specific organic substrates tends to be in the micromolar to millimolar range rather than the nanomolar to micromolar range, there is some evidence that some benthic diatoms can take up certain low-molecular-weight organic compounds in competition with bacteria (Saks and Kahn, 1979). The presence of diatoms with no photosynthetic pigments in sediments certainly supports this idea (J. Hobbie, personal communication). Conceivably, in very rich interstitial water and in detritus, heterotrophic uptake by algae could be significant.

There is a growing literature on the uptake of organic substances by invertebrates—reports by Costopulos et al. (1979), Stewart and Dean (1980), and Slichter (1980) are but a few of the good examples. Uptake takes place against very considerable gradients, and seems to be carried out by saturable processes describable by Michaelis–Menten kinetics. Measurements of uptake, however, have usually been carried out in the presence of bacteria, which has confounded the measurements, since bacteria can take up organic compounds very rapidly (Siebers, 1979). The measurements have also been done at extremely high substrate concentrations relative to ambient values, and so have not shown convincingly that invertebrate heterotrophy was important in nature. Recent developments of analytical techniques have made it feasible to demonstrate that uptake does take place very quickly at ambient concentrations as low as 38 nmoles amino acid liter^{-1} in the blue mussel *Mytilus edulis* and in other invertebrates (Manahan et al., 1982). Concentrations of DOC are higher in interstitial water of sediments (5–16 μM of amino acids, for example). Meiofauna (Meyer-Reil and Fauvel, 1980) and polychaetes may take up amino acids at these concentrations (Jørgensen and Kristensen, 1980, 1980a). The actual extent to which invertebrates can use heterotrophic uptake to complete their ration is not known. It seems likely that this is a small source of dietary requirements.

5.1.1.2 Symbiosis

There are many instances of heterotrophs making use of DOC released by closely associated autotrophs (Trench, 1979). Such cases of symbiosis range from very loose to very close associations. An example of a loose symbiotic arrangement is that of a species of *Flavobacterium* found associated with cells of diatoms (Jolley and Jones, 1977). Presumably these symbionts exchange organic substances which only one of the partners manufactures. In anaerobic protozoans, fermentation may yield end products of use to the bacteria found within them; the protozoans in turn may depend on the bacteria for cellulases (Fenchel et al., 1977). Symbiotic associations are common in corals and other cnidarians that contain dinoflagellate phytosynthetic inclusions—zooxanthellae—that fix and are thought to release DOC used by the heterotrophic coral host. Some host species, such as corals and some bivalves with reduced digestive systems (Cavanaugh et al., 1981) may be able to survive without symbionts, but there are obligate symbiotic relationships where the host has no independent source of energy. Examples of this are gutless pogonophorans (Southward and Southward, 1980; Cavanaugh et al., 1981) and some oligochaetes (Giere, 1981).

Some deposit or suspension feeders that live in environments with abundant sulfur resources have facultative symbionts that use the sulfur to drive the reactions of carbon fixation. Examples of the latter have been found in the spreading centers in the deep ocean, where geothermal sulfur is released, and in coastal environments where anoxic muds containing sulfide are common (Cavanaugh et al., 1981).

The best-studied cases of symbiotic exchange of DOC involve corals and other cnidarians. Lab experiments demonstrate that zooxanthellae provide significant amounts of carbohydrates for the metabolism of sea anemones. When particulate food and light are available to anemones with symbionts, the respiratory quotients (CO_2 evolved/O_2 consumed) of the anemones are between 0.9 and 1, indicating carbohydrate metabolism (Fig. 5-1, left). Anemones without symbionts that have no particulate food available switch to metabolism of fat in a short time, evidencing that their carbohydrate reserves are depleted (Fig. 5-1, right).

The inclusion of organisms in the tissues of other organisms has an extremely long evolutionary history, and Margulis (1981) believes that

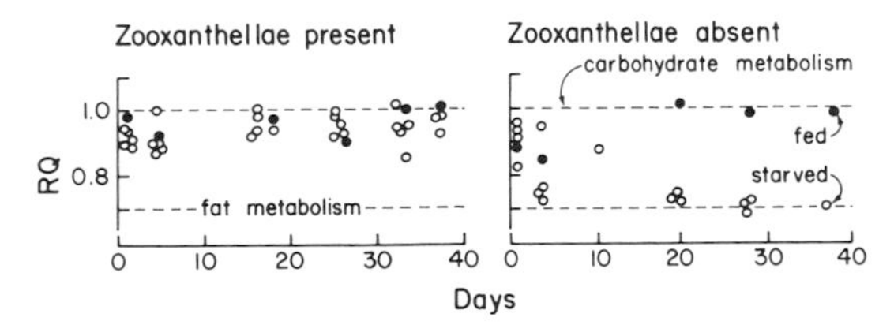

FIGURE 5-1. Respiratory quotient (RQ = CO_2 evolved/O_2 consumed) in the anemone *Anthopleura elegantissima* where zooxanthellae were present or absent. Anemones were either fed shrimp or starved. Adapted from Fitt and Pardy (1981).

many of the radical steps in the early evolution of organisms are based on symbiotic associations. Many of the organelles of cells, for example, are the result of incorporation of previously free-living species into a host species and subsequent modification of the symbionts. The evolution of symbiotic relationships is widespread and the ensuing adaptations are extremely varied. The acquisition of such symbionts has made possible behavioral and morphological adaptations that demonstrate that for such species symbiotic procurement of energy is as or more important than predation.

The importance of maintaining the symbiotic relationship can be seen in the evolution of behavior and morphology designed to facilitate symbiosis. Certain free-swimming medusae with zooxanthellae have daily vertical migrations that take them to strata of water where there is enough ammonium to satisfy the needs of the zooxanthellae (Muscatine and Marian, 1982). Certain sessile sea anemones in coral reefs may have their zooxanthellae and nematocysts located in different organs. The anemones expand the organs bearing zooxanthellae during the day (Fig. 5-2, left diagrams), when light is available; the organs bearing nematocysts are expanded at night (Fig. 5-2, right diagrams) when the zooplankton prey are most active.

In such anemones and in corals, the zooxanthellae provide carbon compounds to the host, while the nematocysts, through the prey they provide, furnish the nitrogen needed by the symbiotic algae (Trench, 1974). Seawater around coral reefs tends to be very nutrient poor, and the assurance of a ready supply of nitrogen is an important asset for producers. It is therefore not surprising that much of the primary production in reefs is carried out within corals.

Corals are only partially heterotrophic. Between 70–161% of the energy needed to support respiration may be furnished by symbiotic carbon fixation in corals (Muscatine, 1990). Among Caribbean reef-building corals, species more adapted to live autotrophically seem to have the smallest polyp sizes (Porter, 1976; Wellington, 1982a,b), since large polyps seem best suited for capture of zooplankton.

Evidence of the dependence of corals on photosynthetic symbionts is that the ratios of stable carbon isotopes of the tissues of the host coral match those of their zooxanthellae, instead of the isotope ratio of the zooplankton (Land et al., 1975). The coral-zooxanthellae unit therefore functions mainly as a primary producer, with the internal exchange of DOC being an essential feature for the success of the arrangement. Some corals, however, are capable of satisfying energy requirements by feeding heterotrophically (Lewis, 1992). The ratio of autotrophy to heterotrophy depends on coral polyp morphology and availability of suitable prey.

Symbiotic relationships of a looser sort may also occur between corals and fish. Schools of grunt feed in seagrass beds during the

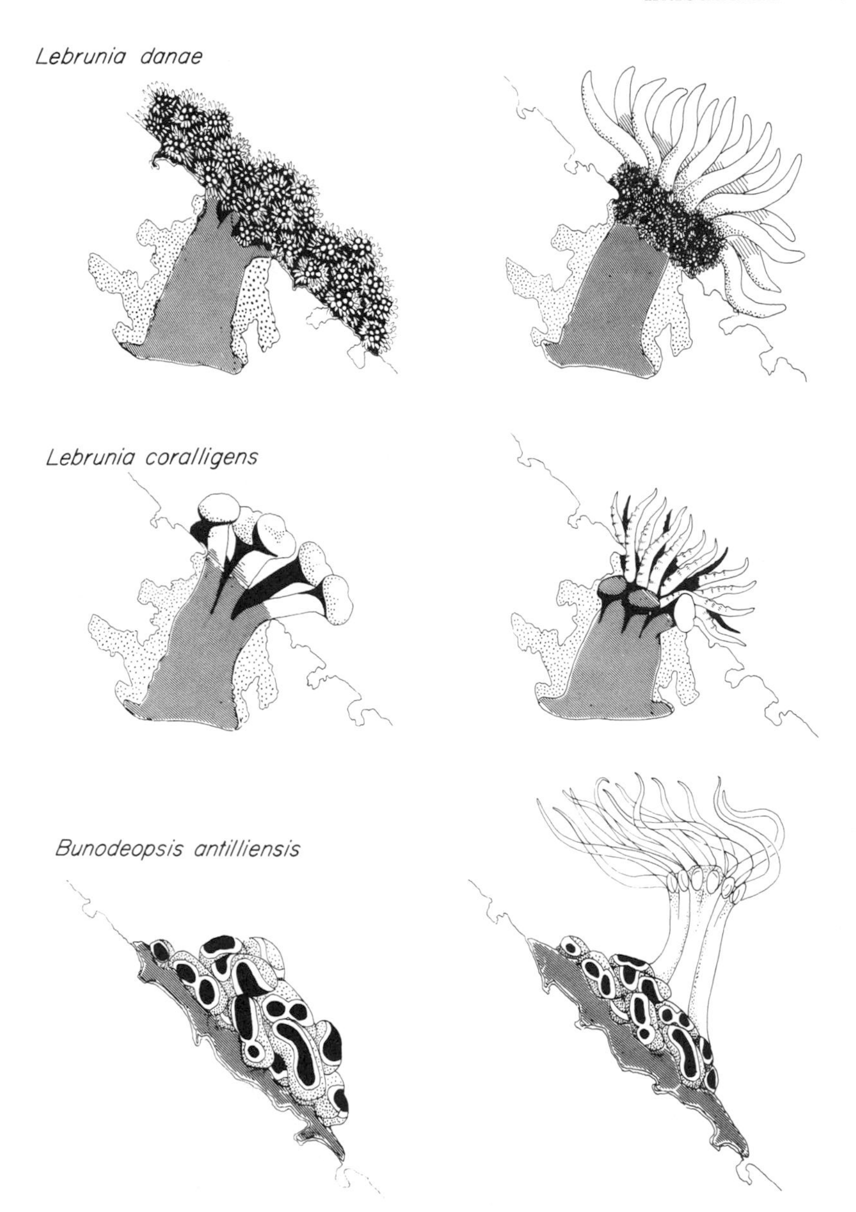

FIGURE 5-2. Appearance of sea anemones of a coral reef in Jamaica during the day (left diagram) and night (right diagram). Adapted from Sebens and De Riemer (1977).

night and spend the day near coral heads, perhaps a refuge from predators. The dissolved and particulate nutrients excreted by the fish are large enough to increase growth rates of the corals (Meyer et al., 1983).

5.1.1.3 Consumption of Food Particles

The majority of consumer species eat particulate foods. Food particles may be consumed by engulfing prey with pseudopods, as in, for example, radiolarians and foraminiferans. Suspension feeders collect food particles in various ways. Ciliates, sponges, bivalves, many annelids, and brachiopods create feeding currents by the beating of cilia or appendages to bring food to the animal. Tunicates and some gastropods produce mucus sheets that are used as sieves or sticky surfaces to capture food. Copepods and barnacles use setae- and setule-bearing appendages to capture particles much smaller than the consumer. Some fish strain particles by passing water through gill rakers, and whales use baleen for the same purpose. Deposit feeders such as many sea cucumbers, bivalves, polychaetes, and shipworms collect and ingest sediment that contains food particles, while raptorial feeders (gastropods, cephalopods, crustaceans, cnidarians, turbellarians, polychaetes, fish, birds, turtles, snakes, and some mammals) capture food particles that are relatively large compared to the size of the consumers. Feeding studies have been done mainly on species that feed on particles, so the remainder of this chapter focuses on particulate feeding.

5.1.2 The Study of Feeding Rates by Consumers

The mechanics of finding, capturing, and ingesting particles of food by consumers are similar whether they are carnivores, herbivores, or detritivores. The time and effort allotted to each of the components of feeding—search, pursuit, capture, and handling of food items—differs in a snail feeding on benthic microalgal mats compared to tuna pursuing mackerel. In both cases the food procurement process is made up of the same *components*, even though, for example, the time spent pursuing food differs. One approach to the study of feeding by consumers is to find out how each component is affected by the abundance of consumers and food. Once the effects of consumer and food abundance are understood, a synthesis of the process can be provided by models that incorporate these variables. The models, ideally at least, can then be used to assess the role of feeding by consumers in nature.

Models of the feeding process based on an analysis of components of feeding and their dependence on density of food and consumers have been constructed by Holling (1965, 1966) and others (Beddington, 1975; Beddington et al., 1976; Griffiths and Holling, 1969; Hassell et al., 1976; among many others). Since the work on these models is stated in terms of predator and prey, we will use these terms in this and the following chapter, but with the understanding that they may apply to herbivores, carnivores, or detritivores, and their food particles.

The rate at which consumers acquire food depends largely on their abundance and on the abundance of food and consumers. The following sections of this chapter examine the three major ways in which predators respond to abundance of their prey. These responses include (1) the functional response (Solomon, 1949), in which predators increase prey consumption as prey abundance increases; (2) the numerical response (Solomon, 1949), in which the number of predators increases as prey abundance increases; and (3) the developmental response (Murdoch, 1971), in which predators grow to a larger size faster when more prey are available.

5.2 Functional Response to Prey Density

5.2.1 Types of Functional Response

As prey numbers increase, predators eat more prey. This so-called functional response to prey density may take three different shapes (Fig. 5-3). In the type I response, the consumer eats an increasing number of prey items as prey density increases, in linear proportion to prey abundance, until a satiation point is reached. The satiation point occurs because the predator cannot handle prey any faster, so that the ingestion rate then remains approximately constant even if prey density increases further. A combination of two such linear relationships has been used to describe the functional response of both herbivorous and carnivorous copepods (Frost, 1977).

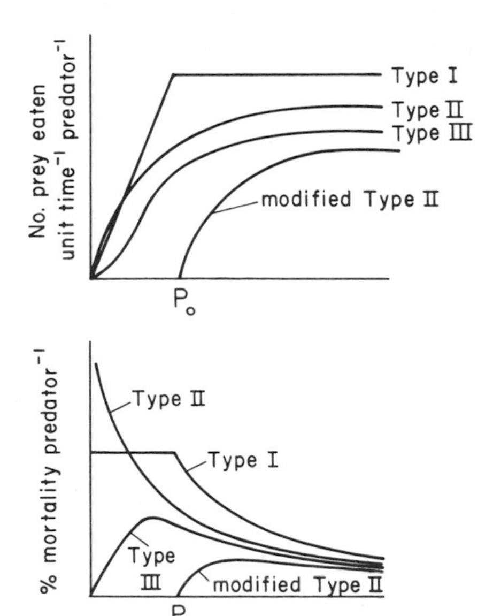

FIGURE 5-3. Shape of hypothetical functional response curves (top) and the percent mortality expected per predator by each of the types of functional responses (bottom). The height of the asymptotes in the top figure is not an important consideration. Adapted from Holling (1959) and Peterman and Gatto (1978).

Predators exhibiting a type II response increase consumption of prey at a decelerating rate as prey density increases. This results in a curve that rises to an asymptote, reflecting increased costs or constraints associated with the consumption of more prey per unit time. At increased prey densities either the mechanism of prey capture may become increasingly inefficient or the motivation for increasing feeding rate may be reduced.

The type II response has been described by many authors (Ivlev, 1961; Parsons and LeBrasseur, 1970; McAllister, 1970; Frost, 1972; Corner et al., 1972; Robertson and Frost, 1977; among others). Holling derived it in the equation

$$I = aP/(1 + abP), \tag{5-1}$$

where I is the ingestion rate at prey density P, a is the rate at which prey are encountered, and b is the time required to capture and eat a prey item. Ivlev (1961) adapted a curve derived earlier by Gause:

$$I = I_m(1 - e^{-dP}), \tag{5-2}$$

where I_m is the maximal rate of ingestion and d is a constant specifying the rate of change in I with respect to P. Cushing (1978) and Crowley (1973), arguing by analogy to Michaelis–Menten uptake (cf. Section 1.4.2.1), predicted ingestion rate using

$$I = I_mP/(K_d + P), \tag{5-3}$$

where K_d is the prey density at which ingestion equals $I_m/2$. These three expressions of a type II functional response are conceptually and arithmetically related (Parker, 1975) and the major point is that the acquisition of food items by the type II response can be represented by a variety of equations that describe a process that becomes increasingly saturated at higher densities of prey.

Parsons et al. (1967, 1969) claimed that in zooplankton no feeding took place below some low concentration of particles. They therefore modified Ivlev's equation to allow for such a threshold (Fig. 5-3):

$$I = I_m[1 - e^{d(P_0 - P)}], \tag{5-4}$$

where P is the prey density and P_0 is the threshold prey density below which no feeding takes place. Others reached similar conclusions (McAllister, 1970; Adams and Steele, 1966), and Steele (1974a) used equations incorporating this feeding threshold as a cornerstone of a model of the dynamics of grazing by copepods on phytoplankton in the North Sea (cf. Section 8.31). The attraction of the threshold idea in the type II response is that if grazers cease exploiting prey below a threshold density, this provides a sanctuary in which a few phytoplankton can survive to regenerate the population when grazers are no longer present near the depleted patch of phytoplankton. Inclusion of the hypothetical feeding threshold in the model thus furnishes a mechanism for recovery of a depleted patch of phytoplankton. The modeled populations of phytoplankton are thus not exterminated by grazers.

It is not clear whether this threshold is a general phenomenon in the sea. *Calanus* can remove an average of 92 cells day^{-1} even at the relatively low cell density of 270 cells liter^{-1} (Corner et al., 1972). In these experiments the copepods still fed at 50 cells liter^{-1} but counts were too variable to allow conclusions. Frost (1975) found that the rate of filtration by *Calanus* was depressed at about 200 cells ml^{-1} (type I response, Fig. 5-3) and some filtration still took place at about 100 cells ml^{-1}. In fact, when copepods were fed on larger cells,* filtration went on even at lower densities, and ingestion did not decline, let alone stop, until cell densities dipped below 4–11 cells ml^{-1}. These cell densities, while low, are in the range of concentrations of oceanic phytoplankton (Fig. 1-2). Thus, zooplankton, while showing lowered ingestion rates at low food densities, can still obtain food even at very low densities of phytoplankton.

Type III responses to increasing prey density initially show an accelerated rate of ingestion (Fig. 5-3). It is not clear how this acceleration is brought about; one thought is that the accelerating attack rate is perhaps related to the buildup of a search image (Tinbergen, 1960). As encounters with a given kind of prey are more frequent, the predator perhaps "learns" that that particular type of prey is adequate and concentrates feeding on that kind of prey. Several other mechanisms that could accelerate the density-dependent portion of the type III functional response have been proposed (Hassell and May, 1974; Crawley, 1975; Hildrew and Townsend, 1977; Hassell et al., 1977; Cook and Cockrell, 1978) but are not well established. Explanations of how these hypothetical alternative mechanisms might work are lengthy and can be found in the references cited above.

After the accelerating portion of the type III functional response there is an inflection point beyond which the rate of change of prey eaten per predator slows down, much as in the type II functional response.

Behavioral adjustments in feeding rate by predators can bring about all three types of functional responses. Since behavioral responses can be immediate, the functional response, takes place with no significant lag after a change in prey abundance.

---

*Note that cell size and cell density are not independent: this is the first of many examples where we will see that although we discuss components separately as an expository device, the components may be inextricably related.

Of the three types of functional response, type III would seem to have the greatest potential to regulate prey numbers, since it generates a negative feedback on prey density (Fig. 5-3, bottom): the percentage of the prey eaten by the predator *increases* as density of prey increases, at least for the lower range of prey densities.

Work on functional responses of predators has emphasized theory, sometimes verified in laboratory situations, and dealt mainly with terrestrial species. How generally applicable are functional response curves to marine organisms? In Fig. 5-4 we have the results of selected experiments with a variety of marine invertebrate species where the number or amount of food eaten was recorded in situations where density of food varied. In all cases, there is some sort of functional response: all the invertebrate species consumed more prey when prey density increased. There is, however, considerable scatter in the results, so that for most cases it is not possible to decide if curves of type I or type II responses best fit the data. In fact, where (Fig. 5-4, lower right) the fit of three models (two straight lines, Ivlev's and the Michaelis–Menten equations) has been explicitly tested, no one model fitted better than any of the others. Given the variability in even the best available results, it is difficult to decide which model fits best, let alone infer the mechanisms of feeding by the fit to a specific curve.

Examples of type III functional response curves from the field are harder to find in the literature, but predation by shore birds on bivalves may follow this pattern (Fig. 5-5, top). Feeding by certain fish and starfish in the laboratory also may show type III functional responses (Fig. 5-5). In the experiments with starfish, after exposure to the single prey species for 10–12 weeks, the entire predator population had "learned" that these were adequate prey and spent less time in investigative behavior and fed more readily on the prey offered. The number of prey eaten at any one density increased, and the curve changed to a type II response whose shape was largely due to the relative efficiency with which prey were processed. In cases where there is one or just a few prey species, the learning phase may thus be a temporary phenomenon. The term "learning" is used here in a very broad sense, including not only the strictly behavioral notion of changed behavior due to experience but also physiological adaptation such as induction of enzymes and changes in assimilative efficiency.

Feeding by some ctenophores on copepods (Reeve et al., 1978) also shows sigmoid (type III) functional response curves. Because of the mechanisms involved in ctenophore feeding (passive capture of particles in mucus-covered tentacles), it is difficult to imagine a way that learning can produce the sigmoid curve. This once again points out that using theoretically derived curves to infer mechanisms is an uncertain business. Feeding mechanisms are better studied directly, such as by *in situ* observations using SCUBA (Hamner et al., 1975), filming

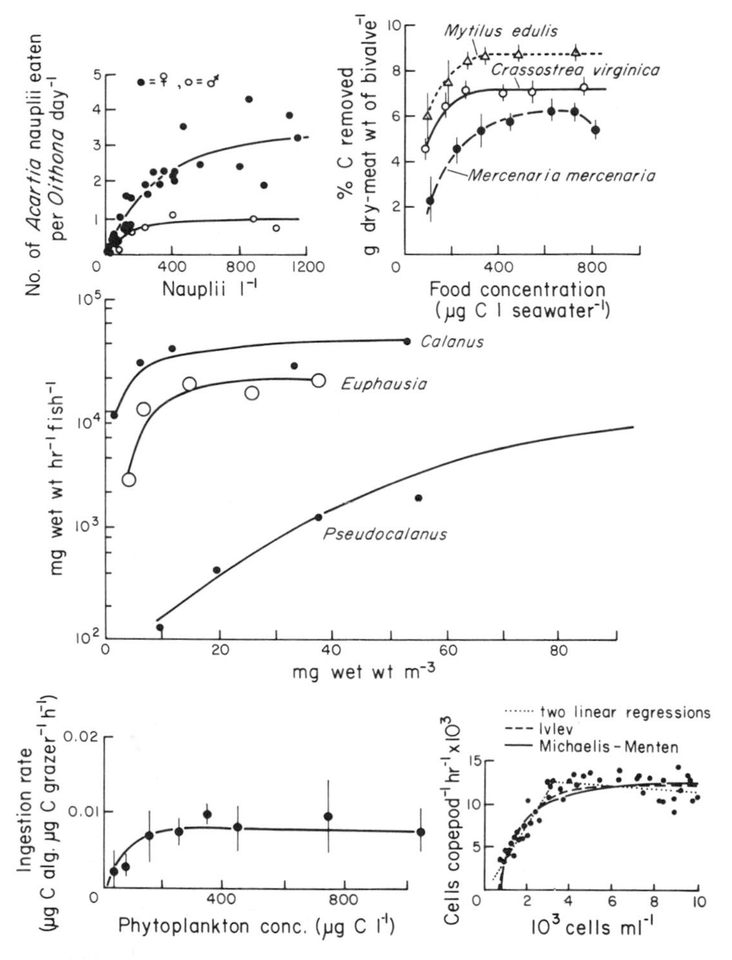

FIGURE 5-4. Functional responses of various marine invertebrates to increases in density of food. Top left: Predation by *Oithona nana* on nauplii of *Acartia* (Lampitt, 1978). Top right: Amount of particulate carbon removed from suspension by three species of bivalves (Tenore and Dunstan, 1973). Middle: Ingestion of three prey zooplankton species of varying size by juvenile pink salmon (Parsons and LeBrasseur, 1970). Bottom left: Ingestion rate of phytoplankton carbon by copepods (McAllister, 1970). Bottom right: Ingestion of the alga *Thalassiosira* by the copepod *Calanus*, adapted from Frost (1972) and Mullin et al. (1975), with three alternative descriptions of the functional response.

methods (Alcaraz et al., 1980), and radioactive tracer experiments (Roman and Rublee, 1980).

There are several aspects of feeding responses to increased food abundance that have not as yet been considered but may be important.

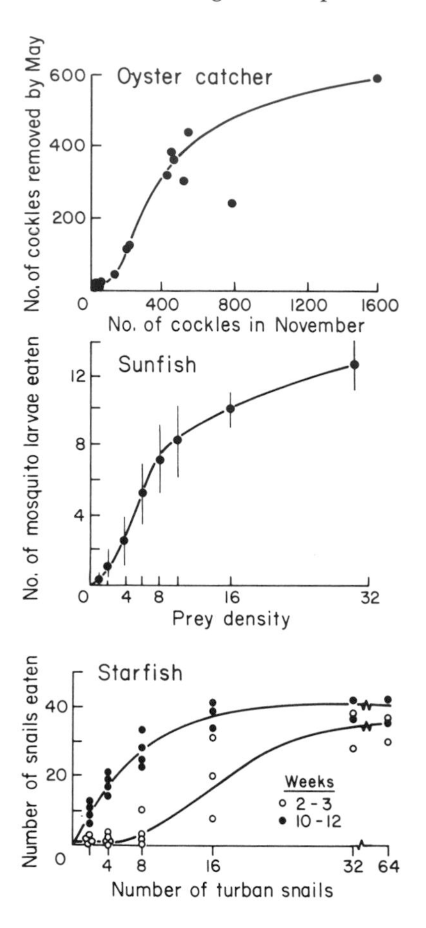

FIGURE 5-5. Type III functional responses. Top: Oystercatchers (*Haematopus ostralegus*) feeding on cockles (*Cerastoderma edule*) in Llanridian Sands, Burry Inlet, South Wales. The densities of cockles were measured in November and again in May, from 1958 to 1970. Missing cockles are inferred to have been eaten by the overwintering population of oyster catchers. Curve is fitted by eye, ignoring the one outlying point. Adapted from Horwood and Goss-Custard (1977). Middle: Feeding by bluegill sunfish (*Lepomis macrochirus*) on mosquito larvae in the laboratory. Data of R.C. Reed in Murdoch and Oaten (1975). Bottom: Feeding by starfish on turban snails (*Tegula*) in the laboratory. The data shown are those from weeks 2–3 and 10–12 in a 3-month-long experiment. Data of D.E. Landenberger in Murdoch and Oaten (1975).

First, none of the responses discussed above predicts that ingestion may be lower at very high prey densities, even though some of the data suggest lowered ingestion rates at the highest prey densities (Fig. 5-4). It is likely that in many cases there are additional mechanisms that lead to lower feeding efficiency or clogging at very high densities of food particles (Harbison and Gilmer, 1976). Such mechanisms have not yet been included in models of the functional response.

Second, there are instances where there is no asymptote in the feeding rate of copepods as the density of food increases (Mayzaud and Poulet, 1978) so none of the three responses described above may apply. Feeding by adult *Mnemiopsis* (a ctenophore) is carried out by extension of mucuscovered lobes, and does not become saturated at even the highest densities found in the natural environment (Reeve et al., 1978). This contrasts with the evidence described above and makes it clear that short-term experiments may not provide the entire picture of the functional response of predators to prey density.

Third, since the density of prey varies through a season, species of zooplankton may show different functional responses at different times of year. For example, the rate at which *Calanus pacificus* filters cells out of the water varies over the season (Runge, 1980), with a fivefold increase between early summer and the colder months. Most of the experiments discussed above involve collecting predators and placing them in a container with a range of food densities for short periods of time. Such experiments generally show that the functional response is saturated at higher densities. Given long enough periods of time, however, zooplankton may adjust their functional responses to the ambient prey density and composition. Several species of copepods in Canadian coastal waters acclimate their feeding rates to seasonally varying food densities, and feeding increases as ambient concentration of food particles in the water column increases (Mayzaud and Poulet, 1978). Both behavioral and physiological acclimation took place, since not only were more prey taken at higher prey densities but also there was higher activity of various digestive enzymes in the zooplankton. Short-term experiments carried out at different times during the year yielded type II saturated feeding curves, but the density at which the asymptote was reached varied with ambient food density. Although such acclimation may not occur in all predators (For and Murdoch, 1978), it is important to consider such changes before interpreting field data or developing theories of grazing in the sea.

5.2.2 Components of the Functional Response

5.2.2.1 Components Related to Prey Density

The components of the functional response that are related to prey density were listed by Holling (1965) as the rate of successful search for prey, the time that the predator spends searching, the time required for the predator to handle prey, the degree of hunger in the predator, and the manner in which the prey manages to inhibit the predator. We will examine each of these in turn.

5.2.2.1.1 Rate of Successful Search. The rate of successful search is determined by three subcomponents, including the relative mobility of predator and prey, the size of the perceptual field of the predator relative to the density and size of prey, and the proportion of attacks that result in a successful capture of a prey.

The relative mobility of predator and prey defines in part what prey are available to the predators. It also may reflect whether the predator uses stalk or chase tactics. Predatory fish attack prey using bursts of speed (about 10 lengths s^{-1}) that may be over three times the cruising speeds. If the predator is much larger (say 15 times) than the prey, bursts of speed are not needed: herring of 25–30 cm do not need to accelerate when feeding on *Calanus* (0.2–0.3 cm in length). The herring may need to find over 1,000 *Calanus* day^{-1} to provide daily food requirements, so a very large proportion of each day must be spent seeking copepods. On the other hand, a cod (1 m in length) may only need to feed on three whiting (about 20 cm in length) per day, but to capture the whiting the cod needs high chase speeds (Cushing, 1978). Thus, there are tradeoffs that species have made in how they have adapted to find food. Predators that use easily overtaken prey, necessarily small-sized, must spend long periods feeding to satisfy daily requirements. Chase predators may use large prey, spend only a fraction of their time in actual attack, but must have physiological and behavioral adaptations for acceleration and pursuit. Many species may alternate hunting strategies depending on available food density.

Herring larvae, for example, change swimming mode at different densities of prey (Fig. 5-6). The functional response to prey density (Fig. 5-6, top) is brought about by relatively active swimming at low to intermediate prey densities (Fig. 5-6, middle). At higher prey densities asymptotic attack rates are maintained, even with modest swimming activity. The increased attack rates are achieved not just by overall increases in foraging rates, but by different foraging behavior. The larvae feed mainly by fast bursts toward prey, and accordingly show more frequent bursts of high speed when prey are at relatively low density (Fig. 5-6, bottom). When no prey are present, burst frequency is low (the larvae carry out a few attacks on detrital particles). At highest densities, the time spent in bursts is very short, but prey are apparently so dense that prey capture is near the maximum calculated on the basis of gut capacity and rate of digestion (0.9 attacks min^{-1}) (Fig. 5-6, top).

Predators may use even more complex behaviors to modify the rate of successful search. Humpback whales round up zooplankton prey either by herding (Howell, 1930) or by the use of "bubble curtains" (Jurasz and Jurasz, 1979). Bubbles are released by a whale while swimming in a tight horizontal circle. The curtain of bubbles then rises around the clusters of prey. The zooplankton move away from the bubble walls, and as a result concentrate toward the center of the circle. The whale then harvests this temporarily increased density of zooplankton by swimming up the cylinder of seawater contained within the bubbles. Such complex behavioral adaptations may be the exception rather than the rule for most predators.

Another example of a predator that reduces search time is provided by siphonophores that bear tentacles that resemble copepods, mimicking even the presence of long antennae (Purcell, 1980). The "copepods" are even contracted at variable intervals that resemble the darting swimming of copepods. Examination of the gut contents of these

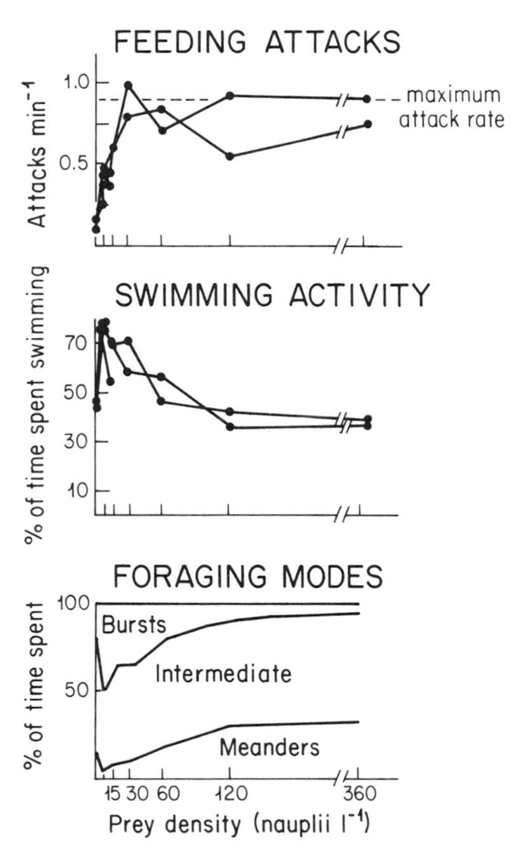

FIGURE 5-6. Feeding activity of larvae of herring (*Clupea harengus*) feeding on different densities of copepod nauplii. Top: Rate of feeding attacks. Maximum attack rate estimated based on gut capacity and rate of digestion. Middle: % of time spent swimming. A minimum of about 40% of the time spent swimming is needed for the larvae to counter sinking. For top and bottom graphs, data from 3 trials are shown as the black line joining points. Bottom: Partition of swimming activity into 3 modes: bursts of speed associated with feeding, intermediate speeds, and slow speeds in which larvae meander, perhaps in a digestive pause, maintaining vertical position. Adapted from Munk and Kiørboe (1985).

siphonophores shows crab larvae, large copepods, and euphausiids, all of which eat copepods. Other siphonophores have specialized organs resembling fish larvae.

The size of the perceptual field of the predator relative to the density and size of the prey (Kerr, 1971a) also influences whether a predator can find a prey. Visual predators are more likely to detect and eat prey nearby than further away (Fig. 5-7), and for any given visual acuity,

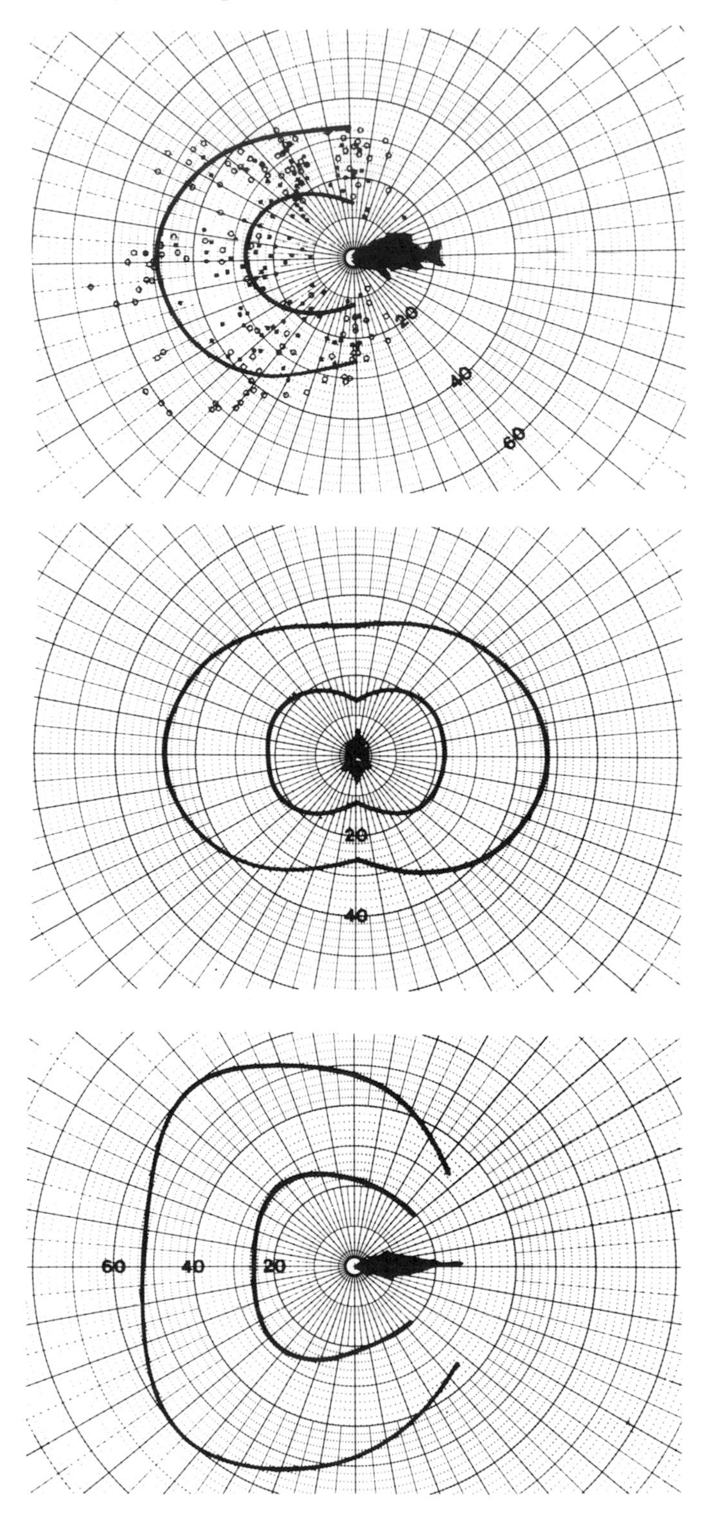
20
40
60
20
40
60
40
20

larger prey will be detected at further distances (O'Brien et al., 1976). This may result in the consumption of proportionately more large than small prey (cf. Section 7.4.3.3). Prey that are sparse will require considerable searching, and if they are small, may not compensate for energy expended. The shape and size of perceptual fields (Fig. 5-7) depend on the anatomy of the predator, and may be quite variable. We have discussed only visual means of prey detection, but mechanoreceptor, chemosensory, or other means may be used to detect prey (cf. Section 6.3.2).

The proportion of attacks that result in a successful capture of a prey varies widely for different prey and can be affected by many variables. For example, the hunting success of blackbellied plovers hunting in mud flats for benthic invertebrates is lower in the cooler parts of the year. Presumably, warm temperatures encourage prey organisms to be active nearer to the surface of the sediment, where they may be more susceptible to predation (Baker, 1974). Further, whether or not a capture attempt is successful determines what happens in subsequent intervals of time. Once larval anchovies find a patch of food organisms, they remain feeding within it (Hunter and Thomas, 1975). Very similar behavior is shown by other predators. When a prey is successfully captured by a foraging blackbellied plover, the bird mainly stays where it is and continues hunting. When a peck is unsuccessful, however, the plover on the average takes one to several steps and therefore moves to a new area (Baker, 1974). Thus capture success leads to apportionment of time to further feeding, while failure leads to nonfeeding behavior, designed to move the predator in search of patches with higher densities of prey. In situations such as mud flats where feeding may be restricted to certain hours of the days during favorable stages of the tide, it is essential for the predator to use the available time carefully. The result of the pattern of feeding is that the predators aggregate on patches where prey is more abundant.

5.2.2.1.2 Search and Handling Time. The time spent searching and handling prey can be partitioned into subcomponents: (1) time spent pursuing and subduing a prey, (2) time spent eating the prey, and (3) time spent in a digestive pause during which the predator is not

←

FIGURE 5-7. Prey detection field for a bluegill sunfish at about 5,900 lux. The bold lines indicate the zone within which 95% (closest to the fish) and 5% of the prey present were detected. Top: Lateral view, with black points showing sites where prey were detected and open points indicating where prey were placed but not located by the fish. Middle: Front view, points omitted. Bottom: Top view, points omitted. The numbers show centimeters away from the fish. In all cases the prey were 2-mm-long zooplankton (*Daphnia magna*). Adapted from Luecke and O'Brien (1981).

hungry enough to renew hunting activities. Differences in the nature of the predator and prey may require different apportionment of time spent in each of these three subcomponents. Different predators may use very different tactics in finding prey.

The time that predators spend handling each kind of available prey may vary substantially. The overall capture method used by a predator broadly sets limits on time spent handling prey. For predatory snails that bore through thick shells of bivalve prey by rasping and chemical secretions, handling time of a prey specimen may be on the order of days. Starfish feeding on mussels may take hours to process one prey. At the other end of the spectrum, handling time may be less important for intermittent suspension feeders such as copepods or fish larvae (Munk and Kiørboe, 1985). Suspension feeders may have very short handling times, on the order of seconds or less (T. Cowles, personal communication).

Handling time may also vary with the type of prey, even though a predator may be using the same overall method of capture. For black-bellied plovers hunting in mud flats, some clams and polychaetes take more time to capture than smaller prey (Baker, 1974). Handling time increases for larger predators feeding on prey within a similar range of prey sizes (Zwarts and Wanink, 1993). For shorebirds foraging on coastal flats, longer feeding intervals are detrimental, because they may not be able to satiate hunger before tides cover the feeding flats and because longer feeding increases exposure to birds of prey. Larger shorebirds would probably seek areas with appropriately larger prey.

5.2.2.1.3 Hunger Level. The level of hunger of the predator is determined by the rate at which food is digested and assimilated, and by the capacity of the gut of the predator. Availability of food is usually not continuous, so feeding, and hence the relative fullness of the gut, may vary over time. Fluctuations in feeding behavior may involve changes over months (Jones and Geen, 1977) or they may have much shorter time scales, days or hours (Elliot and Persson, 1978; Munk and Kiørboe, 1985). Fish that feed early in the morning do not feed again until the meal is digested, so that the fullness of stomachs and the consumption of food have a bimodal pattern over the course of a day. In fish, mechanical stimuli provided by the relative gut fullness may be the determinant of hunger level. In many arthropods there may be no feeding until the peritrophic membrane lining the gut is regenerated (Reeve et al., 1975). This membrane is involved in processing food in the midgut and is excreted as the lining around fecal pellets. Whatever the specific mechanism involved, hunger level varies over time. The quality of food available, gut capacity, and digestive efficiency of the predator then determine how quickly the next meal is sought.

The size of the prey relative to the stomach volume of the predator also is important in determining frequency of attack. Larger prey take longer to digest (Jones and Geen, 1977), so that hunger levels rise more slowly when the prey are large and therefore the next attack is delayed. Gelatinous zooplankton fishing with mucus nets capture very small particles, so they need to feed almost continuously. On the other hand, many other species show very intermittent feeding. Of course, expenditures of energy are required for the acquisition of prey, so that effort involved in capturing either very small or very large prey have to be metabolically balanced relative to the ration obtained. This topic is discussed more fully in Chapter 7, and Lehman (1976) furnishes a model that incorporates feeding, ingestion, and digestion in filter feeders. There are other factors that affect rates, such as temperature. In cod, for instance, ingested food may disappear from the gut in 25 hr at 10°C, while at 2°C food lasts about 60 hr (Tyler, 1970). Thus ambient conditions also affect feeding rates.

5.2.2.1.4 Inhibition of Predation by Prey. Prey can reduce their susceptibility to being eaten by behavioral or morphological adaptations. Such inhibition of predation by the prey is very well established in marine molluscs (see citations in Harrold, 1982), where there are many stereotyped responses to starfish predation. One example is provided by two kelp forest snails (*Tegula pulligo* and *Caliostoma ligatum*) that are eaten by the sea star *Pisaster giganteus*. Both snails twist their shells violently on contact with *Pisaster*, crawl away as fast as possible, and release their hold on the substrate and tumble away if contact persists. *Calliostoma* also covers its shell with the mucous secretions of its foot, and apparently this hinders efforts of *Pisaster* in grasping the shell. In addition, *Calliostoma* may aggressively bite *Pisaster* when the star has made contact with the snail. *Tegula* neither uses mucus nor biting for protection, and as a result, is eaten by *Pisaster* three times more frequently than *Calliostoma* even though it is encountered by *Pisaster* much less often. The difference is due to the effectiveness of the defensive behavior, not in preferences by *Pisaster*, as shown by a field experiment with narcotized snails (Harrold, 1982). Although many marine organisms have no way to actively defend themselves from predators, not all are just passive prey.

Another mechanism for inhibition of predators is through morphological adaptations,* a phenomenon known as mimicry. If a prey species is morphologically recognizable, and unpalatable, predators may learn to avoid it. This is related to density in that a certain frequency of encounter between predator and prey species is needed before learning by predator takes place. Simulation studies by Holling

*There is also the intriguing but largely unstudied possibility of chemotactic mimicry in the case of prey of predators that use chemotactic means of detecting prey.

(1965) showed that Batesian mimicry—where a palatable prey species resembles an unpalatable species—could provide considerable protection to the palatable prey even though predators may feed on some individuals of the palatable prey species, or even if the palatable prey species was more numerous than the unpalatable model. In the case of Mullerian mimicry—where relatively unpalatable species resemble each other—the resemblance developed between the species helps both prey species because the predatory burden is shared. The model further predicts that this second kind of mimicry also helps the predator because it allows a more effective avoidance of distasteful species, since just one image has to be learned as unpalatable. Holling (1965) also identified a third kind of mimicry where a group of edible species could share the burden of predation by evolving to resemble each other, or at least not evolving obvious differences in appearance.

The proportions of any one species consumed by predators would be smaller than it would be in the absence of the mimicry (Van Someren and Jackson, 1959). This protection is provided particularly at high densities of prey species. High densities could be the result of high absolute numbers of the mimic species or of local high densities resulting from schooling behavior (Brock and Riffenburgh, 1960). Below these high densities, predation favors high diversification of prey genotypes (and phenotypes) in Holling's model, a feature actually demonstrated for a predaceous echinoderm and the butterfly clam by Moment

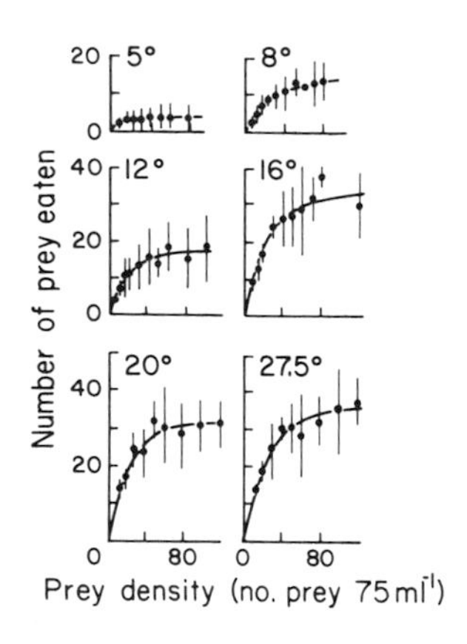

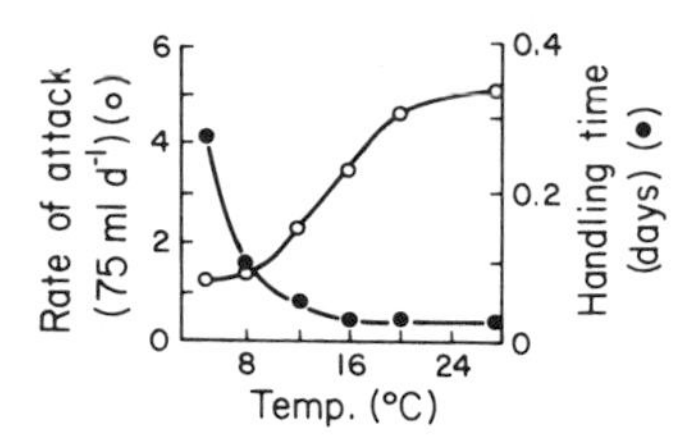

FIGURE 5-8. Top: The effect of temperature on the functional response (mean ± standard error of number of prey eaten) of a dragonfly naiad (*Ishnura elegans*) feeding on a cladoceran (*Daphnia magna*). Bottom: The effect of temperature on the time to handle prey and on the rate of attack by *Ishnura* on *Daphnia*. Adapted from Thompson (1978).

(1962). Presumably the presence of many different phenotypes would reduce the effectiveness of the learning component of the functional response to prey density.

Throughout this discussion of predation, we have focused on biological mechanisms. There are also nonbiotic variables that are important in determining the overall rates of attack. For example, a larger number of prey may be eaten (Fig. 5-8, top) at higher temperatures, because warmer temperatures allow faster movement of cold-blooded invertebrates and therefore the time they spend handling their prey (Fig. 5-8, bottom) is much shorter. This in turn makes it possible for a faster rate of attack and a more marked functional response to the density of the prey. Thus, temperature is important at least in affecting handling time. Similar effects can be expected on many of the other components and subcomponents of predation.

5.2.2.2 Components Related to Predator Density

There are additional components of predation that are strongly influenced by the density of the predator. These include the impact of social facilitation and avoidance learning by prey, intensity of exploitation, and interference among predators.

5.2.2.2.1 Social Facilitation and Avoidance Learning by Prey. As the number of predators increases, the competition for a limited number of prey may increase, and the probability of any one predator finding an unattacked prey decreases. In some cases, there may be an increase in the number of prey eaten as density increases if social facilitation of attack exists. Examples may be a group of predatory fish shoaling a school of smaller prey fish, or groups of swordfish coursing through fish schools and feeding on the frightened or distracted prey. Escape from one predator may make the prey temporarily more susceptible to another predator (Rand, 1954; Charnov et al., 1976). In such instances, increased contact with conspecifics leads a predator into eating more, searching more effectively or capturing prey more readily. On the other hand, as predator numbers increase, some species of prey may, through associative learning, avoid a specific predator encountered (and survived) before. The greater the density of the predator, the greater the probability that each prey will have acquired a successful way to avoid attack (Holling, 1965).

5.2.2.2.2 Intensity of Exploitation. In general, the intensity of exploitation of prey by a predator is complicated by the fact that predators and prey are usually patchily distributed over space (cf. Chapter 11). Thus, it is difficult to define the relevant densities of prey and predators, and therefore the distribution of predator attacks. Although Griffiths and Holling (1959), Hassell and Varley (1969), Free et al. (1977) treat such

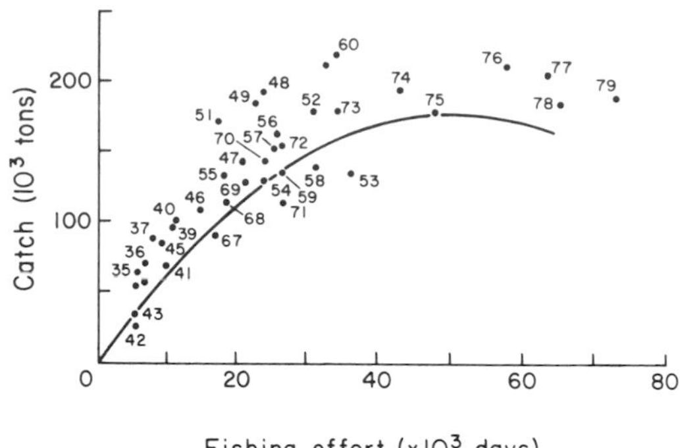

FIGURE 5-9. Relation between catch of yellowfin tuna (*Thunnus albacores*) in the tropical Pacific between 1967 and 1979 and fishing effort in days. The line represents the theoretical equilibrium yield curve for the species. Adapted from Inter-American Tropical Tuna Commission (1980). Values from 1935–1960 from Schaefer (1957).

complications in models and certain laboratory situations, it is not clear how these features apply in the field, nor are there field studies where such difficulties are dealt with.

The extensive work on the effect of fishing effort on fish stocks does not usually consider spatial distributions but is the only body of data available on the effect of various levels of exploitation on prey. Much theoretical fisheries work has centered on the idea that as the intensity of exploitation or fishing effort increases, the predators (fishermen) obtain an increased harvest until a peak yield is reached. This peak may be sustainable for some time if fishing effort is maintained, but any increase in predation pressure beyond this maximum yield decreases the amount of food harvestable by the predator. This sort of response of yield to exploitation pressure can be observed in laboratory studies (Silliman, 1968) and in commercial catches in the sea (Fig. 5-9). The mechanism behind such an optimal yield curve* may vary from

*The maximum yield curve has been a key concept for management strategies and models of commercial fisheries (Cushing, 1975; Ricker, 1975). There are many reservations as to its usefulness in actual management of fish stocks (Larkin, 1977), since (1) the concept is not helpful in dealing with the not infrequent catastrophic declines in stocks or in allocating fishing to different geopolitical areas so that no one local stock is depleted, thus maintaining genetic variability of the species; (2) maximum yield ideas have not been developed to include multispecies stocks, so do not include interactions among various species in one area; and (3) maximum yield may not be practically sustainable and may be economically undesirable, since perhaps only a lower supply of certain species may be marketable.

prey to prey, but it is thought to involve compensatory adjustments by the exploited populations through changes in growth, survival or fecundity. At some point, the removal rate is greater than the rate at which the prey can compensate demographically, and both the yield to predators and the stock of prey are reduced.

5.2.2.2.3 Interference Among Predators. As density of a predator species, predators may mutually interfere with each others' activity: as the number of nudibranch predators increases, the number of polyp prey eaten per nudibranch is lowered, regardless of the density of the prey (Fig. 5-10). The nudibranchs are typically solitary and at high densities the increased encounters with conspecifics somehow disrupt normal feeding behavior. Predator interference can also be interspecific. Terns and bluefish often feed simultaneously on schools of small fish along the NE coast of the United States. The frequency of successful dives by terns is reduced by about a third by the presence of bluefish (Safina, 1990). The bluefish appear to make the prey less visible or accessible to terns.

There is a second kind of interference due to the functional response: patches of dense prey are exploited more intensely than low density patches (Free et al., 1977). The result of this is that at high densities of predators it is harder to find patches of prey that are not already exploited (Rogers and Hassell, 1974; Beddington, 1975). This indirect effect of predator density reduces the number of prey available over time to the average predator. The relative importance of the two types of interference is not well known.

There is a further complication in that schooling may protect individuals of a predator species from their own predators (Brock and Riffenburgh, 1960). Schooling by predators, however, may reduce foraging efficiency if area of discovery of prey among members of the school overlaps as a result of the schooling. This would reduce the

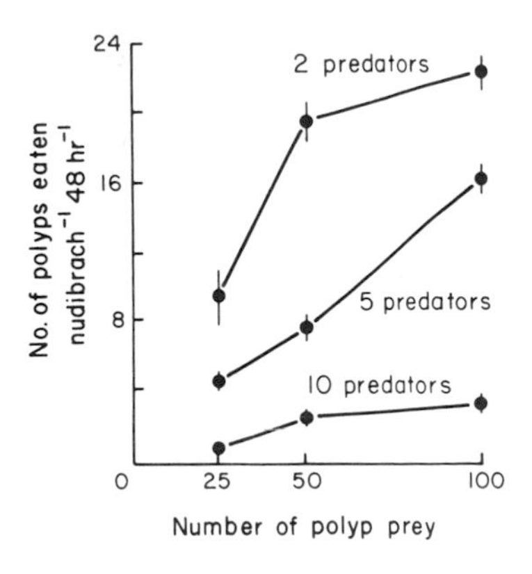

FIGURE 5-10. Predation rate (mean ± standard error) by the nudibranch *Coryphella rufibranchalis* on polyps of the hydroid *Tubularia larynx*. The experiments were run with sets of 2, 5, and 10 nudibranchs present to compare the effect of increasing density of the predator. Adapted from MacLeod and Valiela (1975).

prey available per predator, and it may be advantageous for schooling predators to disperse to an extent while feeding (Eggers, 1976). School of hungry sticklebacks (*Gasterosteus aculeatus*) (Keenleyside, 1955) and jack mackerel (*Trachyrus symmetricus*) (Hunter, 1966) are less compact than after feeding or when well fed*. The cost of schooling may, theoretically at least, be less when prey of schooling fish occur in very dense patches, since searches would then be rewarded by groups of many prey. Perhaps if the prey are very patchily distributed the trade-offs of schooling by predators may become favorable (Eggers, 1976).

The information available is insufficient to evaluate how widespread or important mutual interference by predators is in nature. One serious problem is that the operational densities of predators in the actual environment may be much different than in the very simplified laboratory settings where interference has been demonstrated. This is a basic problem needing further research.

5.3 Numerical Responses by the Predator to Density of Prey

5.3.1 Means by Which Predators Respond Numerically to Prey Density

When density of food particles increases, predators can increase their numbers per unit area by aggregation, increasing fecundity, and lowering mortality.

5.3.1.1 Aggregation

The first way in which abundance of predators may increase where prey are more numerous is by means of aggregation behavior, where the predators simply seek and move to areas of high density of prey. Aggregations of predators are usually found in sites with high densities of prey. As discussed earlier in relation to feeding by larval sardines and blackbellied plovers, predators search for areas with abundant prey. In mudflats, predatory wading birds are generally most abundant where benthic organisms are denser (Fig. 5-11). Overwintering sanderlings (*Calidris alba*), a common shore bird, defend feeding territories in beaches, and the area of the territories is inversely proportional to

---

*This response of predators to scarce prey is quite general and applies to a variety of quite different predators. The Athapaskan Indians in the Pacific Northwest of North America lived inland and lacked the abundant food supply of the coastal tribes. When hunting the Athapaskans dispersed into small groups to maximize chances of encountering game.

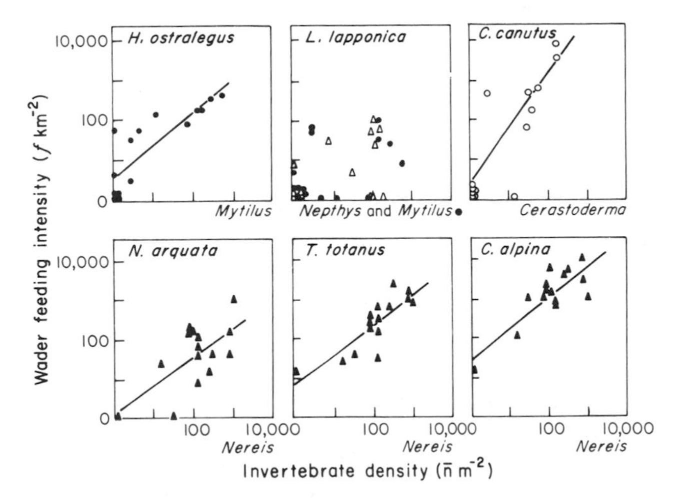

FIGURE 5-11. Relationship of hours of feeding by shorebirds km^{-2} (f km^{-2}) and density (average number prey m^{-2}) of invertebrate prey on intertidal flats. The names of the shorebirds are shown inside the panels, and are, in order, oystercatcher, curlew, bar-tailed godwit, redshank, knot, and dunlin. The invertebrate prey are named below the panels. A regression line is included where the relationship was significant. Adapted from Bryant (1979).

density of the food organisms. In some cases the reduced territories may simply be the result of sufficient food being available even in the smaller territory. In the specific case of sanderlings, the relationship of territory size and density is not due directly to the effect of the food density itself but occurs because it is so costly to defend a larger territory from the increased number of intruder sanderlings that are attracted to places of high prey density (Myers et al., 1979). The density of foraging penguins and seals (Hunt et al., 1992) and shorebirds (Goss-Custard et al., 1993; Zwarts and Wanink, 1993) are correlated to abundance of their principal foods.

Additional nonmarine examples of aggregation of predators in areas of high prey density are reviewed by Hassell et al. (1976). All such aggregations are the result of interaction of the feeding behavior of predators with densities of prey and yet again make evident the artificiality of our attempts to discuss the effects of various components of predation separately.

The behavior that results in aggregation of predators requires only a small time lag before the population of predators adjusts its local abundance to the abundance of prey. As long as predators are mobile and available, aggregative numerical responses are well suited to exploit, and may limit the number of prey organisms in some sites; of course, for the aggregative response to be effective there have to be enough predators at large to aggregate. In many instances there are no such dispersed populations, and there the aggregative response is less important than the fecundity response discussed below.

5.3.1.2 Increased Fecundity at High Prey Density

The second mechanism that may bring about a numerical response to prey density is the production of more young when food particles are more abundant. When the copepod *Calanus* is fed on cultures of *Skeletonema* of increasing cell density, the result is a nearly linear increase in fecundity (Fig. 5-12, top). Beddington et al. (1976) review other aquatic and terrestrial examples and provide analysis and a model of the coupling between fecundity and prey supply.

This fecundity response is of interest because more predators are produced even if the initial population of predators dispersed at large was low in density—as for example, at the start of the growing season in temperate water columns. Note (Fig. 5-12, bottom) that there is a time lag involved in this response: in the copepod, it takes several days for the full response of fecundity to increased availability of algae. This lag is important in that it may restrict the ability of grazers to limit populations of phytoplankton. Some predator species may have slower fecundity responses, perhaps requiring months, while others require

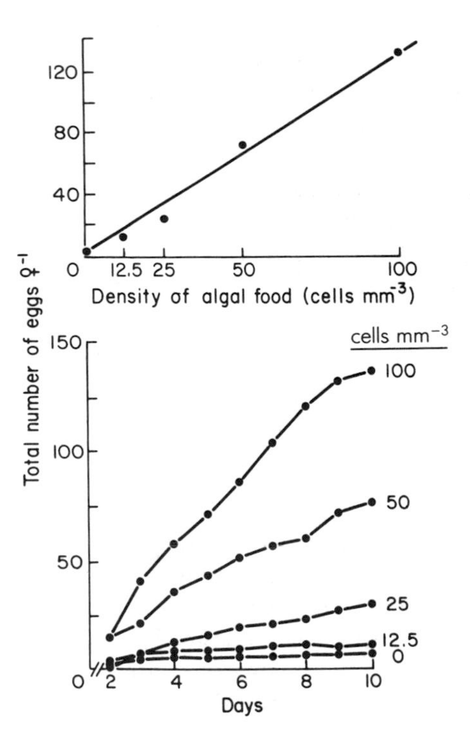

FIGURE 5-12. Numerical response of *Calanus* fed in the laboratory on *Skeletonema*. Top: The total production of eggs as influenced by the density of *Skeletonema* after 10 days. Bottom: The development over time of the numerical response to density of *Skeletonema*. Adapted from Marshall and Orr (1964).

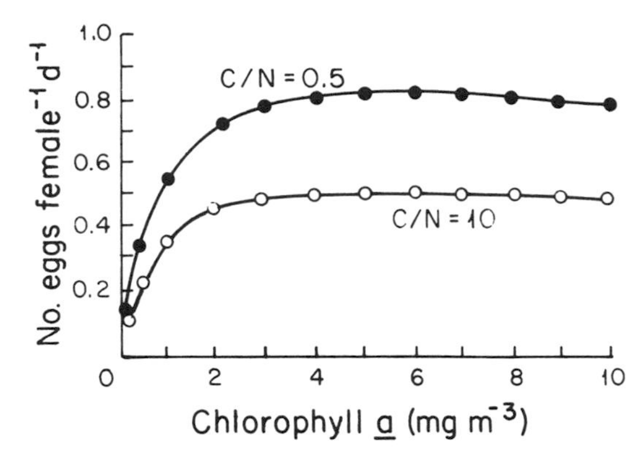

FIGURE 5-13. Fecundity response of the copepod *Acartia tonsa* to different amounts of algal food (measured as chlorophyll *a* > 5 μm), with different carbon to nitrogen ratios. Data of J.W. Ambler, adapted from Smith (1991).

less time. In the laboratory *Paracalanus*, a small copepod, can increase production of eggs after only 12 hr of feeding on high concentration of algae.

The fecundity response can be strongly affected by quantity as well as quality of food (Fig. 4-8). Egg production by *Acartia tonsa*, a coastal copepod, increases as more food is available, and is always higher when food particles contain more nitrogen relative to carbon, regardless of the quantity of food (Fig. 5-13). Selection for higher food quality can therefore affect number of offspring produced. The plasticity of the fecundity response evidences once again that the repertoire of predators can be extremely varied, and that there are many slightly different tactics that predators employ to obtain and make use of available food.

5.3.1.3 Increased Survival at High Prey Densities

Well-fed predators show better survivorship (Beddington et al., 1976; Cushing, 1975), and since population size is the balance between birth and death rates (emigration and immigration are part of the aggregative response), any decrease in death rate may lead to a relative increase in abundance. The implications of this specific mechanism have not been explored.

5.3.2 Numerical Responses in the Sea

Most of the data discussed in the previous paragraphs are laboratory results, but numerical responses to changes in food abundance have actually been documented in the sea. The phytoplankton abundance in an area of the North Sea varied over time and was measured during a

TABLE 5-1. Food Availability and Egg Production of *Calanus helgolandicus* in a Large Patch in the North Sea.[a]

Cruises	Algal stock (mm^3 liter^{-1})	Daily ration/body wt. × 100	Egg production (Eggs female^{-1} day^{-1})
5–7	0.5–0.6	370	18.5
7–10	0.01–0.2	26	2.0
10–13	0.003–0.004	0.8	—

[a] The data were obtained in a series of 13 cruises over several months. Algal stock was assessed by volume of cells in a standard sample. Adapted from Cushing (1964). Paffenhöffer (1971) believes that the high food consumption rates (third column) are overestimates; corrected values show about 195% of body weight of *Calanus* consumed per day with dense phytoplankton, but the point still remains that food is consumed at remarkably high rates.

series of cruises (Table 5-1). The copepod *Calanus helgolandicus* fed on the phytoplankton, and the daily ration available to the copepod varied from almost 400% to less than 1% of the copepod's body weight. Production of considerable numbers of eggs by *Calanus* only occurred while food supply was very high. Consumption of large quantities of phytoplankton may be needed to enable the copepod to achieve a reasonable portion of its potential fecundity. Perhaps some trace element or essential vitamin is required for maintenance, growth, and egg production, and the amounts needed are only obtainable by eating many algae.

5.3.3 The Numerical Response and Control of Prey

We have referred earlier to the fact that the type II functional response, the most common of the three types, did not seem to offer possibilities of controlling prey populations, because rate of attack did not accelerate as density of prey increased (Fig. 5-3, bottom). A combination of the type III functional response and the numerical response (specifically the aggregation of predators plus the increase in fecundity) do provide at least a theoretical mechanism for such a negative feedback (Hassell et al., 1976; Beddington et al., 1976), although the arguments supporting this statement are too detailed to elaborate here. Presumably, a numerical response plus either type I or type III functional responses may also enhance the chances of control of prey population. The role of the numerical response in nature has not been studied enough, and there is a very large gap in studies about the numerical response in general and in its integration with other aspects of predation.

In the field, the numerical response must be very important, for instance, in the case of overwintering populations of zooplankton in temperate or cold seas. These populations are usually in low abundance and must undergo a numerical response during spring blooms of phytoplankton. Whether or not the zooplankton control phytoplankton depends on whether the zooplankton densities achieved by aggregation or reproduction are high enough. Where zooplankton population density is low on a large spatial scale, aggregation probably is less important than the fecundity response. The lag inherent in the fecundity aspect of the numerical response is critical, as will be discussed in Chapters 8 and 15.

We have emphasized the control of prey by predators so far. We should recall, however, that prey may also influence the numerical response of predators. These two types of controls are discussed in more detail in Chapter 8.

5.4 Developmental Response to Prey Density

When food density is high, individual predators may, because of their functional response, consume more prey than at low prey densities. When animals can eat more, they become larger (Fig. 5-14). In turn, the resulting larger predators may eat more prey than smaller predators (Murdoch, 1971; Beddington et al., 1976). This phenomenon is the developmental response.

The growth rate of consumers is related to density of food [Fig. 5-15; cf. Chapter 7 and Brockson et al. (1970); Fenchel (1982)]. Since at least some of the food that the predator eats needs to be expended for maintenance of metabolism, there is a threshold of prey density below which the predator will not grow, and in fact growth can be negative in such circumstances (Fig. 5-15, top). Increased food supply may also

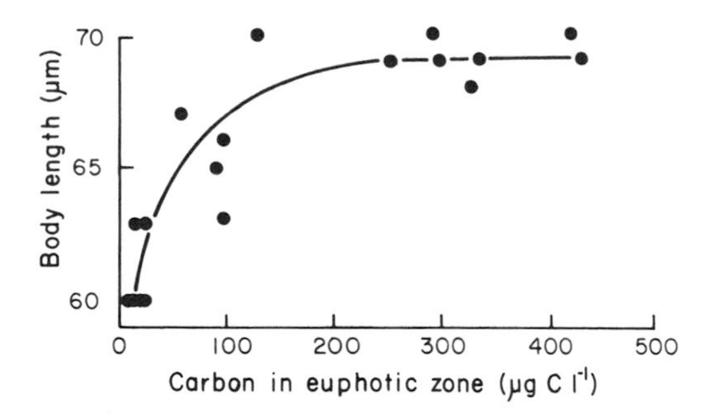

FIGURE 5-14. Prosome length of *Calanus pacificus* in Puget Sound versus an estimate of food supply, the phytoplankton carbon concentration in the euphotic zone during 1965. The carbon was estimated using a time-varying carbon to chlorophyll ratio. Adapted from Frost (1974).

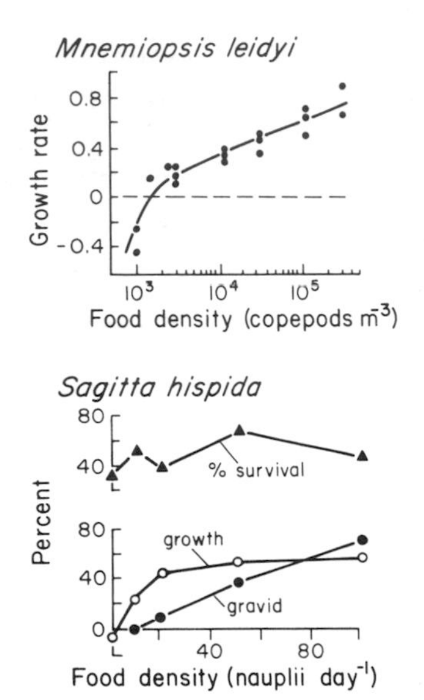

FIGURE 5-15. Laboratory studies of developmental responses to increased density of food particles in two marine invertebrates. Top: Growth rate of the ctenophore *M. leidyi* under increasing abundance of copepods. The growth rate is expressed as a growth coefficient k obtained by solving the equation $W_T = W_0e^{kt}$. A growth coefficient value of 0.8 implies that the organism is more than doubling (actually 2.2 times) in size each day. Adapted from Reeve et al. (1978). Bottom: Growth rate (expressed as increase in length) for the chaetognath *S. hispida,* including for comparison the percentage survival and percentage of the individuals that were gravid, two components of the numerical response to prey density. Adapted from Reeve (1970).

accelerate the completion of developmental and shorten generation time stages, thus hastening the numerical response to prey density. This is yet another example of the interconnections among components of predation, one that brings us to the relationship of food supply and growth, a topic that is considered further in Chapter 7.

The relative importance of the developmental versus the other responses to prey density has not been assessed for populations in the field. Laboratory studies suggest that the increase in survival component of the numerical response due to larger size may not be as marked as that of the fecundity component, as least for chaetognaths (Fig. 5-15, bottom).

The presence of a developmental response provides another potential way by which predation can exert density-dependent control of prey populations. In computer simulation studies, Murdoch (1971) concluded that the type II functional response by a predatory marine snail could not by itself control prey numbers. When the developmental response was added, however, there was a range of prey densities (toward the lower end of the range studied) where the percentage of prey attacked increased as density of prey increased. Beyond this density, prey numbers escaped control by the predator.

As prey are more readily available, the predators achieve larger sizes, and, as we will see in Section 6.3.1, larger predators are proportionately more effective at feeding. Larger size also often leads to greater reproductive output (Fig. 4-5), so there are very close couplings among the functional, numerical, and developmental responses to prey density. The way in which these three major components articulate with each other varies with each instance of predator and prey species. An enormous variety of combinations of density, availability, and quality of prey exist. Each species of predator can make use of the many components and subcomponents of predation in somewhat differing combinations to exploit available prey in the best manner.

The foregoing analysis of feeding on particles painstakingly went over many components and subcomponents of the process. Many further details could have been added. Holling's analysis of components ultimately intended to combine mathematical representations of the components in a model that synthesized how predation took place. The analysis has identified many mechanisms, but unfortunately, the details needed to specify the model exceed current knowledge about most predator-prey relations. Efforts at synthesis of predation have therefore focused on partial aspects of the predation process. Some of these models are discussed in Chapters 8, 9, and 15. In spite of the difficulties of synthesis, component analysis has made evident the mechanics involved in the consumption of particles in a systematic way.

Chapter 6
Food Selection by Consumers

6.1 Introduction

In the last chapter we focused on the importance of food quantity. It turns out, however, that not only abundance but quality of food items must be considered. Moreover, even though there may be sufficient abundance of food items of adequate quality, they may not be easily vulnerable to predators. In this chapter we take up the quality and availability of food particles. First, we need to consider the cues used by consumers in finding and selecting food of appropriate quality and the role of properties of the food and consumer, particularly size and chemical makeup, in food choice. We then consider the relative vulnerability of food items, mainly in reference to the morphology of prey and the physical structure of the habitat. Last, we consider consequences of food selection, especially in regard to changes of prey populations.

6.2 Behavioral Mechanisms Involved in Finding and Choosing Food

The behavioral mechanisms by which consumers find food vary widely. Marine microbes are attracted to a wide variety of stimuli, but in most cases it is not clear whether such tropisms result in finding food. Larger vertebrate and invertebrate consumers have better-known mechanisms with which to seek food. Many marine consumers are visual, so that size, color, and shape of food items are important.

Chemosensory organs are commonly found in many marine organisms. Cod, for example, can detect chemical compounds with organs on their pelvic fins and the barbel (Brawn, 1969), although they often use both chemoreception and sight for locating invertebrate prey. Shrimp can also use chemical cues to track and find prey (Hamner and Hamner, 1977). Experiments in which copepods were fed microcapsules containing phytoplankton homogenates show that chemosensory cues can be used by these organisms since the ingestion rates on such particles are considerably higher than ingestion of untreated particles. The specific compounds involved are not well known, but there is considerable specificity in feeding cues; for example, in *Acartia clausi* L-leucine stimulated feeding while L-methionine or glycolic acid prompt no response (Poulet and Marsot, 1980). Chemoreceptors in lobsters (*Homarus americanus*) are sensitive to 35 individual compounds tested, many of which—particularly the nitrogenous compounds—are in high concentrations in prey of lobsters (Derby and Atema, 1982). The lobsters were able to detect remarkably low concentrations of several compounds, in the range of 10^{-6} to 10^{-14} M.

In copepods, chemosensory detection of algae may be the primary mechanism used in feeding by herbivorous species (Friedman and Strickler, 1975), while mechanoreception may be used principally by raptorial feeders. Copepods may use mechanoreceptors located on their first antenna (Strickler and Skal, 1973) to sense hydrodynamic disturbances caused by moving prey (Strickler, 1975). Removal of the first antenna does not affect the rate of feeding on phytoplankton but sharply reduces the rate of capture of moving prey (Landry, 1980). Chaetognaths also respond to mechanical vibrations with pursuit and capture behavior (Feigenbaum and Reeve, 1977). Certain sharks can detect electromagnetic fields generated by life activities of their flatfish prey and can use this sensory ability to capture prey buried in sediments (Kalmijn, 1978).

6.3 Factors Affecting Food Selection by Consumers

The selection of food items by consumers is a complex phenomenon whose details are only beginning to be understood. The size and chemical composition of the item, however, are probably the two major cues that tell a consumer that it has found adequate food. Other properties of prey, such as specific morphological features and foraging methods, also are important in prey selection.

6.3.1 The Influence of the Size of Prey and Consumer on Selection of Food Particles

6.3.1.1 Sizes of Prey Particles Used by Consumers

Some predators attack prey larger than the predator itself, as in the case of the predaceous ctenophore *Beroe nata* feeding on the larger ctenophore *Bolinopsis* (Hamner et al., 1975).* The other extreme is the

*A curious and unusual feeding pattern is for a predator to attack prey of similar size to itself but to eat only a portion of the prey. Examples of this are typhloscolecid polychaetes feeding on heads of chetognaths (Feigenbaum, 1979) and young flatfish feeding on the siphons of bivalves (Trevallion et al., 1970). In both cases, the prey survive and regenerate the lost parts.

feeding by consumers on particles measuring less than 1 to a few μm, usually free-living bacteria. Particles of this size may constitute 25 to 50% of the daily ration of the larvacean *Oikopleura dioica* (King et al., 1980); bivalves and some crustaceans can remove particles of bacterial size from suspension (Peterson et al., 1978). Pelagic tunicates, tens of centimeters in length, secrete mucus webs that may function as passive filters and can retain particles as small as 0.7 μm in diameter (Harbison and Gilmer, 1976). The size of preferred prey is usually roughly 1/10 to 1/100 relative to the size of predators (Fenchel, 1988).

Most consumers fall between the extremes illustrated above, but the size of the food particle relative to the size of the predator is usually an important criterion. Although predators use many kinds of cues to choose prey, much prey selection is accomplished using the size of the prey as the discriminating variable.

6.3.1.2 Size Selectivity in Consumers and Some Consequences

Work on freshwater fish (Ivlev, 1961; Brooks and Dodson, 1965) first highlighted the fundamental importance of size of prey for predators. Ivlev made use of an index of electivity* to assess the degree of selection:

$$E = (r_i - p_i)/(r_i + p_i), \tag{6-1}$$

where r_i is the proportion of item i in the diet and p_i is the proportion of the same item available in the environment. Positive values of E show selection, while negative values show relative rejection by the predators. Using E, Ivlev could demonstrate that fish show clear preferences in the size of prey they elect to feed on (Fig. 6-1). Prey selection patterns such as those found by Ivlev imply that predators exploit certain size classes of prey population more intensely than others and that at least one way by which predators can partition a food supply is to have different preferred size classes. A further implication is that the sizes below and above the size classes preferred by predators are refuges from predation pressure, and that a prey may thus outgrow its vulnerability to predation.

Although the phenomenon of size selection was first studied in fish, it is widespread among various taxa and quite often selection for larger items is evident. Predatory birds such as the redshank, a common Eurasian shorebird, preferentially eat prey at the larger end of the size distributions available on mudflats. This can be seen by comparing the histogram of sizes of amphipods eaten by the birds relative to those found in the mud (Fig. 6-2, top), and by the regression of average size of amphipods eaten and available in the mud (Fig. 6-2, bottom).

---

*Jacobs (1974) and Cock (1978) assess Ivlev's electivity index and suggest modifications.

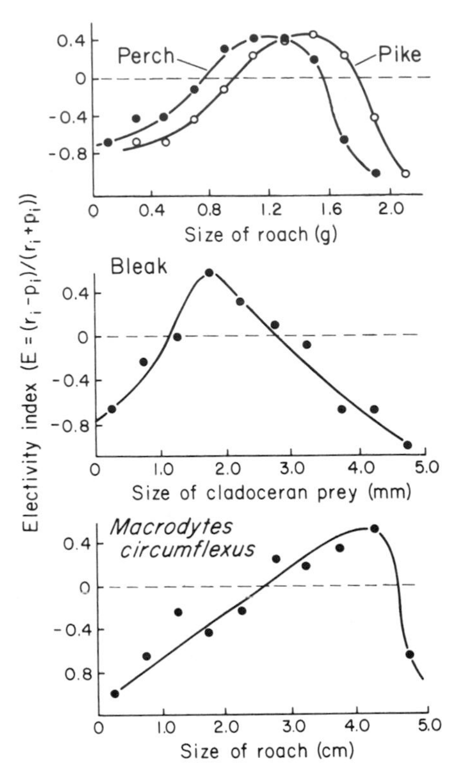

FIGURE 6-1. Selectivity of prey by predaceous fish in the laboratory. Top: Perch (*Perca fluviatilis*) and pike (*Esox lucius*) feeding on roach (*Rutilus rutilus*). Middle: Bleak (*Alburnus alburnus*) feeding on several species of cladocera of varying size. Bottom: Larvae of *Macrodytes circumflexus* feeding on small roach. Adapted from Ivlev (1961).

The preference for larger particles may also result in faster feeding rates when food of larger size is found. Filter-feeding fish such as menhaden filter more actively (or, in other terms, show increased functional responses) when large prey are available (Fig. 6-2). Unlike

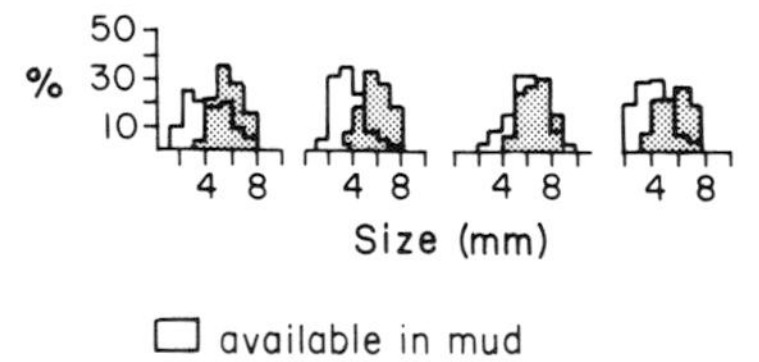

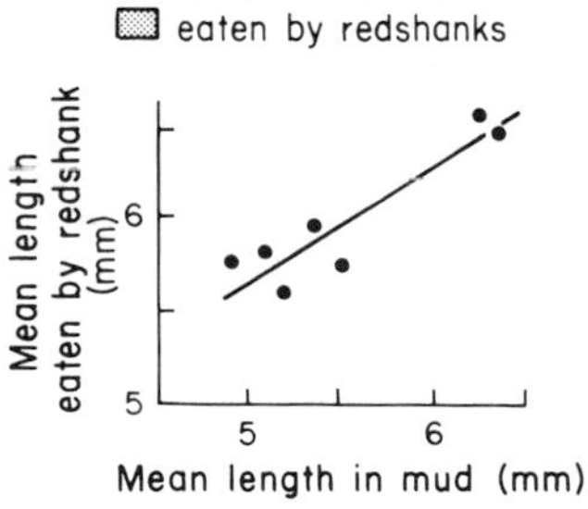

FIGURE 6-2. The sizes of the amphipod *Corophium volutator* taken by redshank (*Tringa totanus*) in comparison to the sizes available in intertidal flats, Ythan estuary, Scotland. The top histogram shows size classes for four sites, while the lower graph relates mean lengths of amphipods in redshank diets and in the mud. Adapted from Goss-Custard (1969).

FIGURE 6-3. Filtration rates of menhaden (*Brevoortia tyrannus*) on plankton particles of increasing size. Note that the top graph has a horizontal scale that comprises only a very small portion of the left part of the bottom graph. In the regression equation fitted to the data F = filtration rate and r = correlation coefficient. Adapted from Durbin and Durbin (1975).

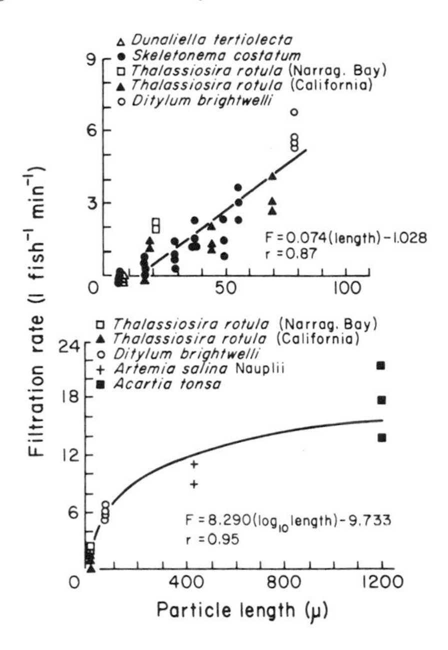

Ivlev's fish (Fig. 6-1), menhaden do not show a lowered feeding at the largest sizes offered (Fig. 6-3, bottom). This is due to the relative size of gape of the mouth to size of prey given in the experiment; if increasingly larger prey were offered to menhaden there would be some point where feeding would decrease. It is therefore important in studies such as these to offer particles that span size spectra usually found by the consumer in the field.

As consumers grow and increase in size, they feed on relatively larger food items. Larvae of the plaice (a flatfish of great commercial importance) feed on various species of small zooplankton, but in certain years their diet in the North Sea consists almost entirely of appendicularians. As the larvae of the flatfish grow (Fig. 6-4), the modal size of prey in the guts of the larval fish increases in size: the diet of the larval fish changes and eventually matches the available food supply.

Thus, increased size of predator is followed by increases in the preferred size of prey. Starfish (*Pisaster giganteus* and *P. ochraceus*) are common predators of the rocky shore in California. The starfish increase in radius as they age and feed on larger specimens of mussels (*Mytilus californianus* and *M. edulis*) (Fig. 6-5, left). The predators do not discriminate so much between species of mussels, but rather among the different sizes of the mussels. Fishes typically also feed on larger prey as they grow (Fig. 6-5, right). Again size of prey rather than species of prey is the more important variable. As herring larvae grow, they prefer larger prey, so there is a fairly narrow range of sizes of prey

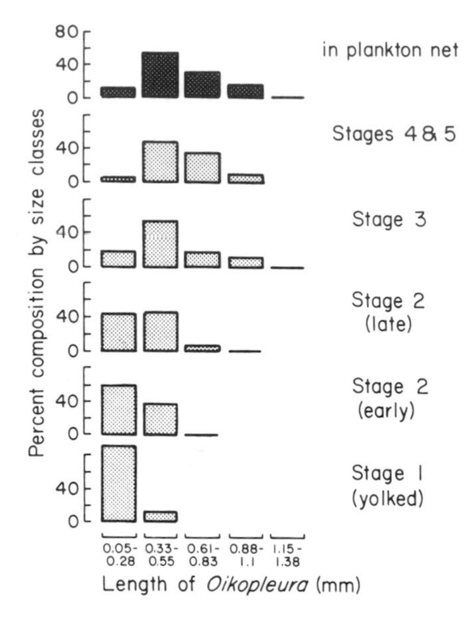

FIGURE 6-4. Size composition of prey (the appendicularian *Oikopleura dioica*) eaten by larvae of plaice (*Pleuronectes platessa*) in different stages of development (lower five histograms) and captured in plankton nets (top histogram). Adapted from Shelbourn (1962).

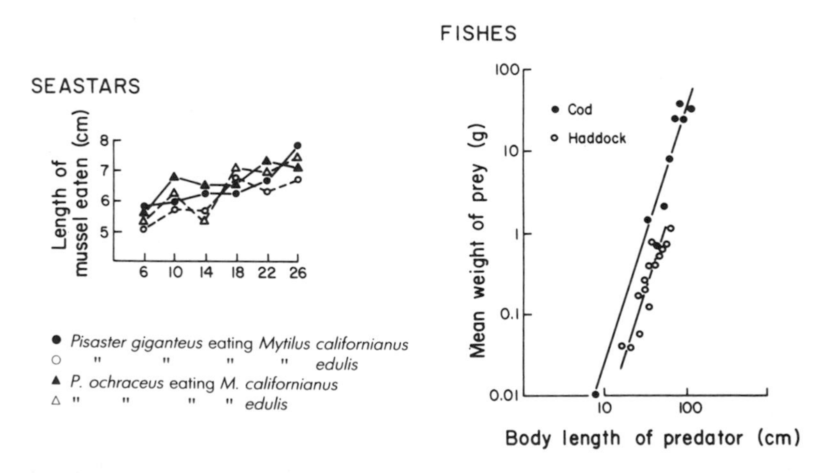

FIGURE 6-5. Relation of predator and prey sizes in invertebrates and vertebrates. Left: Size of mussels eaten in relation to the size of the starfish predator. Mussels were offered in the laboratory in a size range of 3–10 cm in shell length. The starfish only fed on mussels of 5–8 cm. Adapted from Landenberger (1968). Right: Length of two fish species in relation to mean weight of prey found in their stomachs. Adapted from Jones (1978).

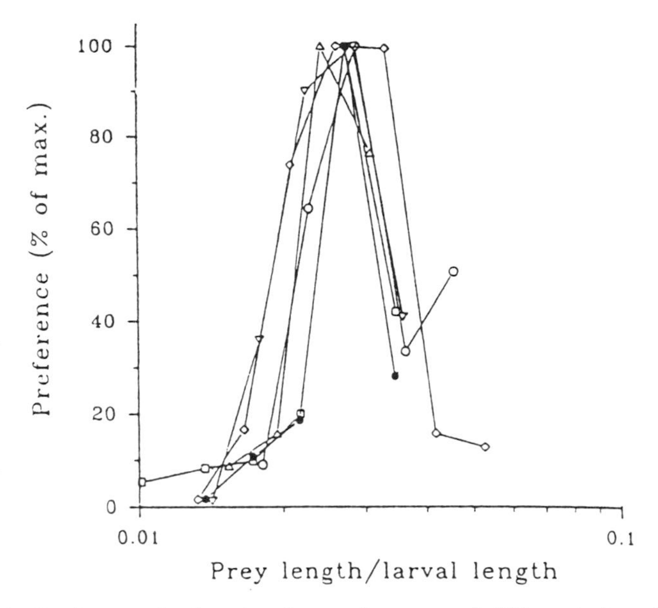

FIGURE 6-6. Preference by herring larvae for prey of different sizes (expressed as % of maximum preference) for different ratios of prey to predator length. Data obtained using groups of herring larvae of different length (open circles: 37 mm, inverted triangles: 30 mm, squares: 26 mm, triangles: 23 mm, diamonds: 22 mm, and black circles: 18 mm. From Munk (1992).

eaten relative to larval size. The ratio of preferred prey size to predator size is therefore reasonably rigid (Fig. 6-6). Many species of bottom-feeding (demersal) fish have pelagic larvae; as the young fish grow, they require larger prey; at some point they are unable to find large enough prey in the plankton and become demersal.

Larger predators do not merely eat larger prey, they also become more effective predators by adjusting feeding rates, and by being able to process food items more effectively. Medusae and combjellies increase rate of water clearance when prey are larger, and so do copepods (Stoecker and Capuzzo, 1990; Purcell, 1992). The filtration rate of some pelagic tunicates, for example, increases exponentially with size of the tunicate (Fig. 6-7, left). A larger tunicate, therefore, is capable of capturing more particles per unit time. In fact, the retention efficiency of the gelatinous sieves used by the tunicates depends on the size of the tunicate, with larger specimens being able to retain larger particle sizes more effectively (Fig. 6-7, right). Small herring larvae are most successful when feeding on the smallest class of copepods offered (Fig. 6-8, top); larger herring larvae are poor at feeding on small prey copepods, and get better when feeding on larger copepods. Note that Fig. 6-8 (top) deals only with the ascending leg of a predator/prey size

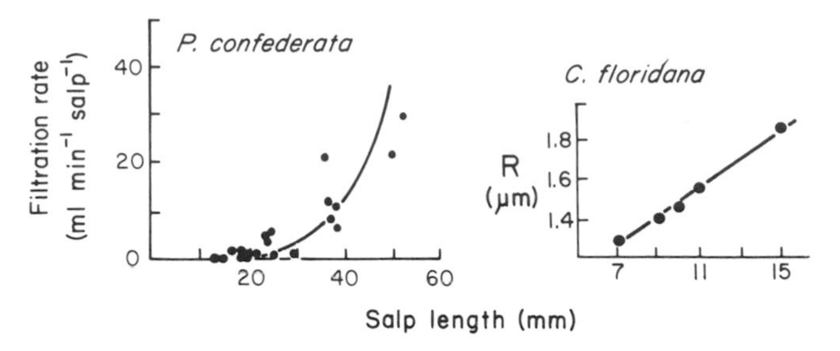

FIGURE 6-7. Left: Filtration rate as a function of body length for the salp *Pegea confederata* [adapted from Harbison and Gilmer (1976)]. Right: Retention index (R = particle size retained with 50% efficiency) as a function of body length of the salp *Cyclosalpa floridana* [adapted from Harbison and McAlister (1980)].

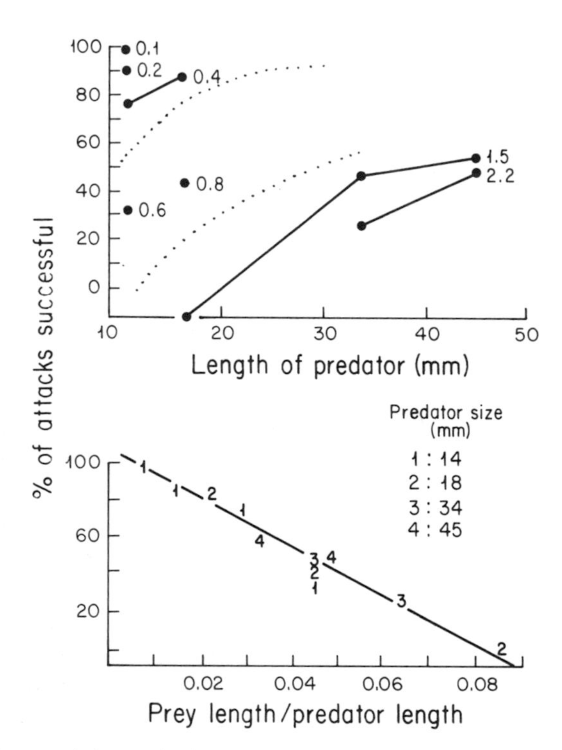

FIGURE 6-8. Successful attacks by herring larvae on copepod prey in relation to size of predator and prey. Top: Effect of predator length for different sizes of prey. Copepod length (mm) shown as number besides points. Dotted lines are extrapolated contours for the 0.5 and 1% success rates. Bottom: Effect of ratio of prey to predator size. Numbers (1–4) are size classes of herring larvae, defined in mm in the inset. Adapted from Munk (1992).

TABLE 6-1. Some Properties of Feeding by Three Species of Copepods on *Thalassiosira fluviatilis* in the Laboratory.[a]

	Mean dry body weight (μg)	Maximum ingestion rate (cells $cop^{-1} hr^{-1}$)	Cell conc. at which 90% of max. ingestion rate is achieved (cells ml^{-1})	Max. ration as fraction of body carbon
Pseudocalanus sp.	10	1,356	1,515	1.13
Calanus pacificus	170	12,652	4,489	0.62
Aetideus divergens	47	1,201	6,598	0.21

[a] From Robertson and Frost (1977).

relationship. For relatively larger prey, there must be a descending leg. In the case we are dealing with, the interrelation of prey and predator sizes can be shown independently of actual sizes by plotting attack success versus the ratio of prey to predator size (Fig. 6-8, bottom). Feeding attacks by herring larvae are most successful when prey are smaller relative to size of the predator, regardless of size of the larvae.

The performance of consumers of different size can be compared in Table 6-1. *Pseudocalanus* sp. and *Calanus pacificus* are both primarily herbivorous suspension feeders, but *C. pacificus* is about 17 times heavier. *Pseudocalanus* require only about 1,500 cells ml^{-1} to reach 90% of its maximum feeding rate, while *Calanus* needs to feed on phytoplankton about three times as dense to do so. *Calanus*, however, can filter at a maximum rate which is over nine times that of *Pseudocalanus*. The end result is that *Calanus* can obtain about twice as much carbon per unit time as *Pseudocalanus*. Data on *Aetideus divergens* are included (Table 6-1) show that even though larger than *Pseudocalanus*, a species with feeding appendages ill-adapted to handle small particles will perform poorly.

6.3.2 Selection of Food on the Basis of Chemical Cues

The palatability of food items depends on chemical composition. There are two major aspects to this notion. The first is the chemical defenses used by a potential food or prey to deter feeding by a consumer, and the second is the presence of cues that convey the nutritional adequacy of the food.

6.3.2.1 Chemical Deterrents

The use of chemical signals to deter consumers is widespread among organisms. The presence of chemical deterrents to feeding by herbivores is especially well established in terrestrial plants (Rosenthal and Janzen,

1979, for example), and it is becoming evident that marine producers also have such defenses (Hay and Fenical, 1988).

Dinoflagellates have well known toxins, but it is hard to know what the target for these compounds is, since many invertebrate grazers are not affected. Vertebrates can be severely disabled by the toxin, so perhaps dinoflagellates originally evolved such compounds to keep from being consumed by grazing fish (White, 1981). Some brown and red algae contain deterrents that prevent grazing by herbivorous snails (Geiselman and McConnell, 1981). Certain phenolic compounds in salt marsh grasses make tissues less palatable to geese (Buchsbaum et al., 1982). Extracts of eelgrass leaves containing phenolic acids inhibit both grazing by amphipods and microbial decay (Harrison, 1982). Cyanobacteria produce neuro- and hepatoxins (Demott and Moxter, 1991).

The deterrent substances are collectively known as secondary plant compounds to suggest that for the most part they are not involved directly in the primary metabolic pathways of producers. In marine producers these secondary deterrent compounds may include (a) phenolics: cinnamic acids in vascular plants, tannin-like compounds in brown algae, and simple, often halogenated compounds in red algae; (b) nitrogen compounds: carotenoids in diatoms, pyrrole alkaloids in red algae, alkaloids in monocots; (c) terpenoids: halogenated terpenoids in red algae, saponins in monocots; and (d) acetogenins typical of red tide dinoflagellates and toxic lipids in some green algae (Glombitza, 1977; Hoppe et al., 1979; Swain, 1979; Geiselman and McConnell, 1981; Fenical, 1982; Valiela and Buchsbaum, 1989; Hay and Fenical, 1988).

Organisms with sessile growth habits, regardless of whether they are a macroalga, vascular plant, or animal, often acquire chemical defenses against consumers. In the sea, where many taxa have a sessile, plant-like growth habit, there are many examples of chemical defenses against browsers. Marine sponges (Green, 1977) and ascidians (Stoecker, 1980) are just two examples of taxa that have such defenses, and chemical deterrents are widely distributed in many taxonomic groups (Jackson and Buss, 1975). In the Great Barrier Reef of Australia, 75% of common coral reef invertebrates (including four phyla and 42 species) are toxic to fish (Bakus, 1981). The importance of chemical deterrents is demonstrated by the presence of some structural adaptation against predation by fish in most of the species of invertebrates not protected by chemicals. In addition, of the invertebrates that usually are hidden in some protective microenvironment, and therefore are less exposed to being eaten, only 25% are toxic to fish.

The production of chemical deterrents by marine primary producers not only affects herbivores but also detritus feeders. Many of the substances that deter herbivores remain after the plant dies and their presence in detritus inhibits feeding by detritus feeders. For example, the consumption of salt marsh plant detritus by amphipods and snails is inhibited by high concentrations of ferulic acid, a phenolic compound that reduces palatability of marsh grasses (Valiela et al., 1979). Thus, a device used by plants to reduce herbivory has consequences for decay and the regeneration of organically bound nutrients, since the inhibition of consumption by detritivores slows decay of detritus.

6.3.2.2 Cues of Nutritional Quality

The nutritional quality of food is of obvious importance (cf. Chapter 7), and evolution must have provided for a correlation between the palatability of a particular food item and its nutritive value to a consumer. The need to maintain a diet of sufficient quality and containing specific dietary requirements while feeding on materials of uneven or low-quality forces grazers to display remarkable feeding selectivity. Selectivity may result from a preference hierarchy among food types based mainly on chemical cues. The cues that determine the hierarchy must include both cues of quality as well as deterrent cues. In some instances the nutritional quality of a food may override the effect of deterrent substances. Feeding on detritus by salt marsh snails (*Melampus bidentatus*) is inhibited by enhanced concentrations of ferulic acid in the food. When the nitrogen content of the detritus is experimentally increased, however, the inhibition of feeding by ferulic acid disappears (Valiela and Rietsma, 1984). The cues that determine palatability are hierarchically arranged; in the salt marsh snail, nitrogen content can overcome chemical deterrents; other variables such as pH and salinity have places lower in the hierarchy. In other species of consumers the hierarchy of cues may differ, or other cues may be more important.

The importance of food quality is made evident by the many unusual adaptations in feeding behavior that lead to improvements in the quality of food eaten. For instance, the green turtle (*Chelonia mydas*) is a herbivore that feeds on turtle grass (*Thallassia testudinum*) in tropical shallow coastal waters. The tissues of this plant are low-quality food, especially mature leaves that are rather high in fiber and low in protein. Turtles partially improve their diet by maintaining grazing plots where they repeatedly crop young leaves richer in nitrogen (Bjorndal, 1980).

For organisms that ingest sediment particles the abundance of the microbial flora growing on the particles may furnish a nutritional cue. In the archiannelids *Protodrilus symbioticus* and *P. rubriopharyngens* the more bacteria per gram of sediment the more attractive the sediment (Gray, 1966, 1967).

6.3.3 Other Properties of Food Items

Food choice may be affected by toughness or calcification of tissues. The herbivorous snail *Dolabella auricularia* prefer to feed on softer species of algae when fronds are offered intact, but is less choosy when fronds are ground up before feeding experiments (Pennings and Paul, 1992). In certain coral reef algae, calcified tissues add protection from herbivorous fish (Paul and Hay, 1986). Calcification deters feeding of *D. auricularia* at $CaCO_3$ concentrations of 28 and 50% of biomass, but not at 16%.

The combined effects of deterrents, nutritive quality, and toughness can be seen in compilations of data of the consumption of different producers by grazers (Fig. 6-9). Grazers, not surprisingly, are more effective consumers of high-quality, low-deterrent, tender producers. Phytoplankton consumption varies enormously, and there is no evident characteristic grazing intensity. Macroalgae are exposed to rather

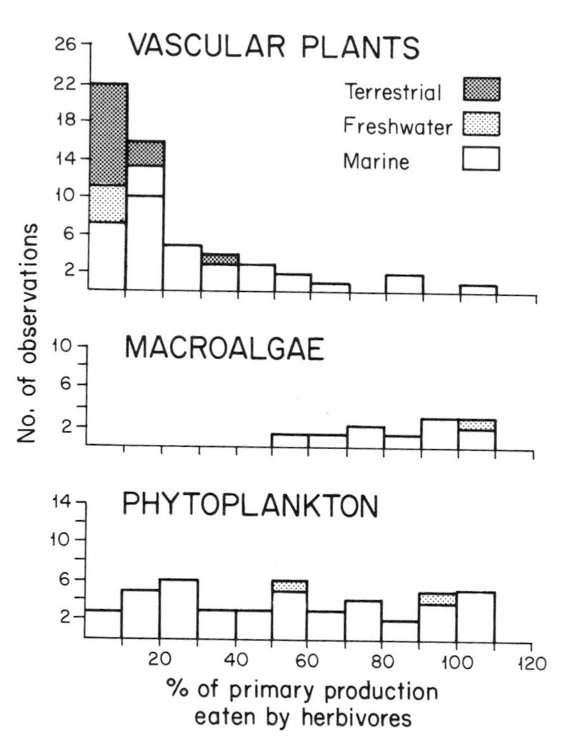

FIGURE 6-9. Frequency distribution of % of net primary production that is consumed by herbivores in vascular plants (top), macroalgae (middle), and phytoplankton (bottom). Data from terrestrial, freshwater, and marine environments are also shown. All values >100% are shown in one category. From compilations in Valiela (1984), Cyr and Pace (1993), and Cebrián and Duarte (1994).

intense consumption, even though they are often chemically defended (Hay and Fenical, 1988). There seem to no clear differences in exposure to grazing between fresh and marine producers. The chemically defended, tough vascular plants are seldom heavily attacked by herbivores. There seem to be no differences in grazing pressure among terrestrial and aquatic plants.

Other less obvious properties of particles may also be important, at least for benthic animals feeding on soft deposits (Self and Jumars, 1978). Many worms collect particles by means of long, sticky tentacles. The transport of these particles down the tentacles into the mouth by ciliary movement is less effective for heavier particles, so that ingestion is therefore "selective" for particles of lower specific gravity. Other mechanisms of particle transport in polychaetes also result in sorting of particles within the gut, since particles of low specific gravity have shorter residence within the gut. Other characteristics, such as surface texture and shape, are also likely to be important in ingestion by consumers, and are beginning to receive some study.

6.3.4 Morphology and Foraging Tactics of Consumers

Copepods, fish, and other consumers may have several ways to find and eat prey. By slight variations in the combinations of components of predation (or consumption in general), there are possibilities for widely varying ways in which consumers can partition food resources. The anatomy of a consumer is of course adapted to the kind of food it eats: the morphology of the mouth of nematodes, for example, varies, and details of musculature and teeth betray the species that grasp prey from those that feed by bulk ingestion of tiny microbial cells (Tietjen and Lee, 1977).

The species of consumers in any one community usually differ morphologically and display different foraging tactics. One example of just such an assemblage of predators is the group of terns that use Christmas Island in the Central Pacific as a nesting ground (Table 6-2) and forage around the island and offshore for food to feed their young. All the species of terns feed on prey captured very near the surface of the sea. Other predators that forage in the same area, for example, the shearwater *Puffinus nativitatis*, differ from the terns by feeding deeper below the surface. Except in the case of the blue-gray noddy, prey for the terns become available principally when shoaled by tuna or other predatory fish. There are, incidentally, two other species of tern nesting on Christmas Island. *Thalasseus bergii* is ecologically isolated from the others since it feeds primarily on beaches, and *S. lunata*, a small relative of *S. furcata*, that is not well studied.

The size of the terns varies (Table 6-2), with the sooty and the brown noddy being similar and larger than the others. The blue-gray noddy is

TABLE 6-2. Foraging Tacties of Species of Terns Nesting on Christmas Island, Pacific Ocean.[a]

Species	Wgt (g)	Diet	Modal fish length (cm)	Modal squid mantle length (cm)	Approximate fishing range (km)	Web area in feet (cm)
Sooty tern (*Sterna fuscata*)	173	Fish, squid	2–4	4–6	>100	4.8
Brown noddy (*Anous stolidus*)	173	Fish, squid	2–4	4–6	80	13.8
White tern (*Gygis alba*)	101	Fish	2–4	2–4	—	3.4
Black noddy (*Anous tenuirostris*)	91	Fish	0–2	2–4	8	10.9
Blue-gray noddy (*Procelsterna cerulea*)	45.4	Fish, invertebrates, some squid	0–2	0–2	8	10.1

[a] Adapted from Ashmole (1968).

the smallest while the remaining two are intermediate in weight. The taxonomic composition of the diet of the terns is similar, except that the blue-gray noddy eats a larger proportion of tiny crustaceans and marine insects (*Halobates*).*

The two species intermediate in size—the white tern and the black noddy—feed on fish and squid, with the noddy using less squid. The white tern feeds mainly at dawn, so it has access to a different group of species of fish than the noddy, since nocturnal and diurnal fish species are not the same. The noddy returns to its nest site at night, feeds on maller fish than the white tern, and uses dipping tactics more frequently.

The blue-gray noddy is the smallest of the species and can use only the smallest fish and squid, since its small bill cannot hold large fish. It fishes near the island, with lots of surface dipping and pursues whatever small arthropods it can get.

The pair of large species is about the same weight and feed on similar kinds and sizes of foods. Although they can be found fishing together, the sooty has a far larger foraging range than the brown noddy. Its search time therefore must be considerably longer. The sooty, furthermore, largely relies on shallow dives from the air to capture prey, while the brown noddy also uses the better developed

* A survey of knowledge on marine insects is given in Cheng (1976).

webbing on its feet to engage in dipping from a swimming position on the surface of the sea. The difference in attack probably provides different prey, since the speed, detection distance, and acceleration of prey and predator differ in diving and dipping.

Although these tern species resemble each other in morphology, feed largely on similar taxa, and use only a few foraging methods, there are very substantial differences in the details of how they obtain food. Thus the differences in species of prey in the diets of these predator species arise primarily because of slight differences in anatomy and foraging patterns. We can generalize that species of predators that coexist sympatrically have evolved enough differences in foraging tactics (and/or prey selectivity) so that the food particles used by the various species of predators are on the average different. The very diverse approaches to consumption of particles, made possible by the variety of components of the process of predation, thus allows for the partitioning needed by groups of similar species.

The example of the Christmas Island terns should have reminded us of the competitive partitioning of resources in salt marsh fish (Chapter 4). The similarity should indicate that competition and predation often are intimately linked, and both potentially are important factors affecting what consumers are present are what they eat. The relative role of competition and predation in thus structuring communities are evaluated in Chapters 8 and 9.

6.4 Examples of Feeding Mechanisms at Work: Suspension Feeding

Feeding by plankton on suspended particles has received a great deal of attention. An examination of some of the details of suspension feeding highlights many of the specific mechanisms we have discussed above and how they interact.

The feeding of copepods had long been thought to involve the use of appendages to direct feeding currents through sieves. The sieves were thought to be the array of setae on certain mouthparts, particularly the second maxillae (Fig. 6-10). The setae have setules set at specific distances apart, and water was thought to be forced past the setal sieves. Particles above a certain size would be retained in the sieves and could then be ingested. This filtration process has been called passive filter-feeding.

Recent work using slow-motion microcinematographic records of feeding by individual copepods shows that we need to revise our view of copepod feeding. Purely passive mechanical particle capture seldom occurs; copepods use their mouthparts to chemotactially select and

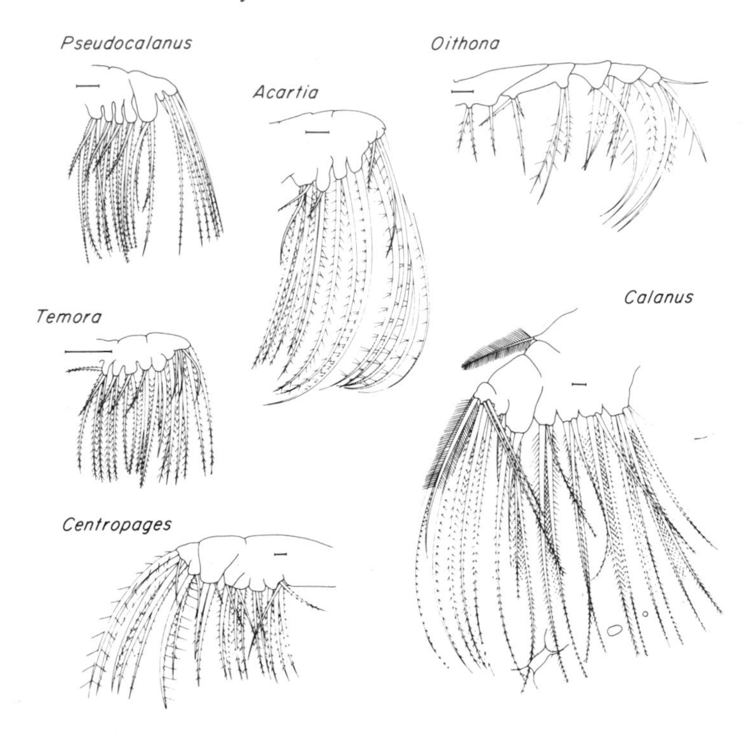

FIGURE 6-10. Feeding appendages (second maxillae) of some common copepods, showing setae and setules. Silhouettes of phytoplankton cells of differing size are superimposed on the second maxillae of *Calanus*, including, in order of increasing size, a microflagellate, a coccolithophorid and a diatom. The horizontal bars show 20 um. Redrawn by Masahiro Dojiri from Marshall and Orr (1955, 1966) and Steele and Frost (1977).

capture algal cells out of the feeding currents (Paffenhöfer et al., 1982). Kohl and Strickler (1981) point out that the viscous nature of water at the small size scales of feeding appendages would make sieving of water very difficult, and they conclude that in most cases particles such as algal cells are detected and actively captured individually by feeding appendages. The postulated action of feeding appendages is supported by the abundant chemo- and mechanosensory organs found in mouthparts (Friedman and Strickler, 1975).

It seems, however, unlikely that copepods can actively capture particles smaller than 5 μm individually, since they would be difficult to detect. Yet, they do consume such particles, so perhaps some bulk processing of water resembling passive filtering may be used by copepods when only small food particles are available (Paffenhöfer et al., 1982). In any case, it is clear that the term filter-feeders is inadequate: we therefore will refer to suspension feeding, a broader term that includes the possibility of other feeding behavior besides passive filtering.

Many suspension feeders also can pursue and capture relatively large individual prey by grasping. Any of a number of appendages are used, but certain species have mouthparts that are stronger and may be better suited for raptorial behavior (cf. *Oithona* in Fig. 6-10). Most zooplankton species appear to readily switch from one mode of feeding to another, depending on the abundance of small and larger prey. *Calanus pacificus*, for example, prefers to feed on copepod nauplii, but if phytoplankton are abundant enough it will feed preponderantly on algal cells (Landry, 1981). Euphausiids and very likely many other plankton consumers can also switch from feeding on phytoplankton to raptorial capture where necessary (Parsons and Seki, 1970).

Switching from one prey type to another is common in some species of copepods, who often feed on whatever particle size is available in the water. If the size distribution of particles changes over time (for example, compare the June to the December data, Fig. 6-11, top row), the particle distribution ingested by the copepods present (Fig. 6-11,

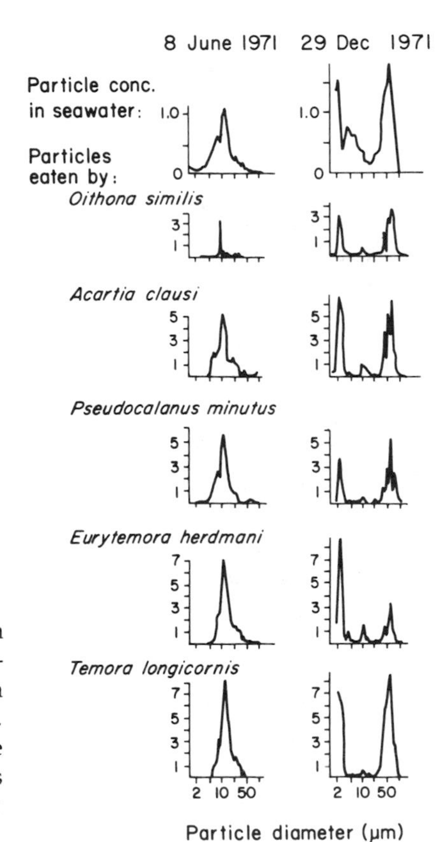

FIGURE 6-11. Particle size spectrum in sea water (ppm by volume) and consumed by several copepod species in Bedford Basin, Nova Scotia, Canada, on two dates. Particle consumption (the vertical axis of the bottom five rows of graphs) is in mg hr^{-1} $copepod^{-1}$ $\times$ 10^{-4}. Adapted from Poulet (1978).

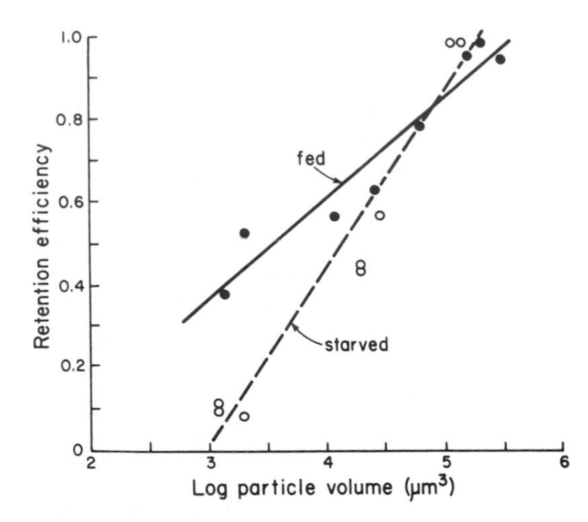

FIGURE 6-12. Retention efficiency of female *Calanus pacificus* feeding on the diatom *Thalassiosira.* Experiments run with previously fed and starved females to measure the effect of increased hunger level. Adapted from Runge (1980).

bottom five rows) changes accordingly (Richman et al., 1977; Poulet, 1978; Cowles, 1979).

In many other studies of suspension feeders there is, in contrast, evidence of a clear preference for larger particle sizes (Richman et al., 1977; Donaghay and Small, 1979). Some copepods, for example, retain and eat larger algal cells more efficiently than smaller cells (Fig. 6-12). When hungry, such copepods show more marked discrimination than fed copepods, and also favor large particles (Fig. 6-12). The latter is a counterintuitive result, since it might be supposed that a hungry individual would accept food in any size.

We thus have that size selective behavior is very variable from one suspension feeder to another. Some copepod species are more flexible with regard to the particle sizes on which they will feed than are other species (Fig. 6-13). *Centropages typicus*, for example, shifts its feeding in response to the abundance of suspended particles. On the other hand, *Undinula vulgaris* ingests particles toward the small end of the spectrum of available sizes in spite of the abundance of smaller and larger particles. Perhaps certain environments provide more constant supplies of particles than others and the feeding behavior of copepods may be adapted to such patterns of abundance. It is not yet clear whether the abundance and distribution of food particles in certain environments could favor a flexible or a stereotyped strategy in selection of food sizes. This topic could sustain further interesting research.

We have seen repeatedly that the specific feeding tactics of different predators can differ markedly and that feeding behavior can change depending on circumstances. Such changes take place by adjustments in the many components of predation and there are innumerable changes potentially available to even very similar predators. The resulting plasticity of hunting behavior of different species studied is no doubt responsible for the divergent results often obtained by different researchers.

There is a further ambiguity in regard to size selectivity which is caused by the way in which feeding selectivity is measured. Most work on particle selectivity in zooplankton has been done using electronic particle counters that measure the distributions of volumes rather than linear dimensions. Sieves such as hypothesized for some suspension feeders, however, remove particles in proportion to linear dimensions rather than particle volumes. Copepod sieves and electronic counters may thus not sort particles in the same fashion (Malone et al., 1979). Harbison and McAlister (1980) passed suspensions of particles through man-made sieves and obtained results similar to earlier work interpreted by others as selective feeding involving behavior. The shape of the phytoplankton used as food, whether spherical or elongate cells, or occurrence of cells furnished with spines or elaborate architecture, seriously affects the retainment by passive sieves, perhaps independently of the volume of the cells. All this further emphasizes the need to make direct observations of feeding behavior (Alcaraz et al., 1980; Paffenhöfer et al., 1982).

The different results of studies of size selectivity in suspension feeders and the ensuing controversies in the literature thus are fueled by the use of species with different hunting tactics and by the lack of a method free of artifacts. We have come full circle with regard to technology: it was to a large extent because of the advent of the Coulter counter in the 1960s that so much attention could be focused on grazing on small particles by zooplankton. Recent direct observations have shown that, at least for some copepods, there is a complex

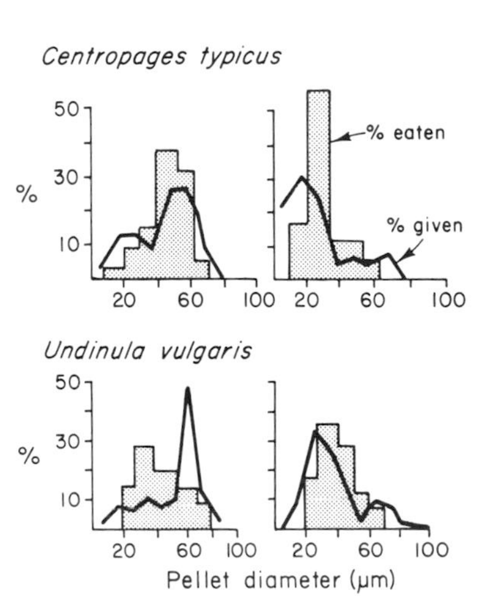

FIGURE 6-13. Shipboard experiments where the frequency of particles (plastic pellets) added to seawater were skewed to the large and small end of the size spectrum of suspended particles. The experiments with *C. typicus* were carried out with specimens from a station in the coastal shelf (200 m deep) off Cape Cod in February. The *U. vulgaris* data were obtained in the Sargasso Sea, at a 3,000-m deep station during September. Adapted from Adams (1978).

feeding behavior whose study will modify the passive-sieve interpretation. It remains to be shown how wide-spread this is and whether or not it is more quantitatively important than filtering through sieves.

A better example of passive filter feeding is provided by the gill rakers of certain fish. These organs serve as retaining nets for particles carried into the mouth of the fish by swallowed water. The size of the pores in the sieves involved are considerably larger than those of copepods and viscosity of water is less of a problem. Sardines, anchovies, herring, and other fish can thus filter feed by swimming with an open mouth, but they also pursue and capture individual prey of a larger size. When offered plentiful food of a large size, anchovies—and many other marine animals, including copepods (Runge, 1980)—go into a brief feeding frenzy and consume prey at higher rates. The rate of food consumption by anchovies decreases as the maximum gut capacity is approached. Filter feeding by sardines on smaller particles is not an effective way to obtain a ration, so that even after 1 hr of feeding only two thirds or so of the maximum capacity of the gut is filled. Leong and O'Connell (1969) conclude that anchovies could not obtain sufficient rations by filter-feeding alone. These fish must supplement their diet with raptorial feeding on large particles. Perhaps the feeding frenzy is an adaptation to exploit as much as possible of a patch of large particles as quickly as possible.

The mechanisms for detection and capture of prey do not always detect and provide consumable prey. Some prey will be too big, some too small, some will be distasteful or have too many protective spines, and these will be rejected. There is, therefore, considerable selectivity even in suspension feeders.

6.5 Optimization in Food Selection by Consumers

Throughout its life, a predator has to decide repeatedly where to find prey and what item of food to eat. In fact, predators have a very well-developed ability to discriminate among prey items and adjust their feeding accordingly. Goss-Custard (1970) allowed captive redshanks (a common European shorebird) to feed on a series of prey placed out sequentially in a row of depressions bored on a wooden plank. The prey were cut sections of mealworm larvae, and each trial provided the redshanks with "small" (one segment) and "large" (eight segments) prey in a ratio of 1 large: 3 small prey (Fig. 6-14). The procedure was repeated, and as a bird began to feed , it took 30–90% of small prey during the early trials. Within five to six trials, however, having encountered and evidently preferring large prey, the redshank searched for and ate large prey, missing only 1.9% of the large prey. Note that during this stage, virtually all small prey were neglected. So

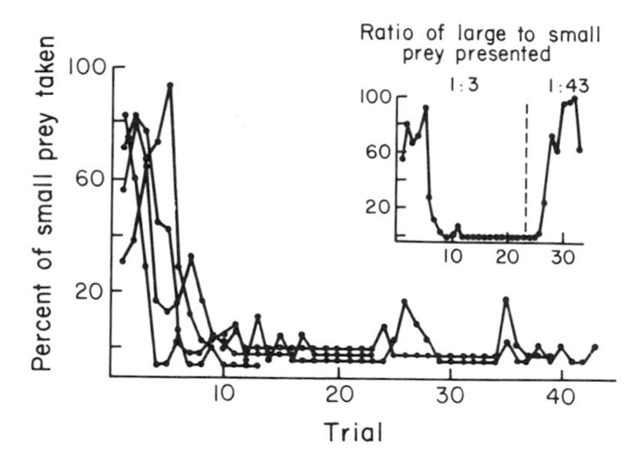

FIGURE 6-14. Proportion of small prey taken by redshanks (*Tringa totanus*) in a series of trials where they were allowed to feed sequentially on large and small prey offered in a ratio of 1:3. The inset shows the switching due to changing the ratio of large to small prey from 1:3 to 1:43. Adapted from Goss-Custard (1970).

far, all we have is a demonstration of size-selective predation, with rejection of less desirable prey. If the ratio of large to small prey in the trials was 1:43 rather than the 1:3 (inset, Fig. 6-14), the bird perceived the change in ratio and in just a few feeding trials adjusted feeding so that the less desirable but much more abundant small prey were taken. This may be an extreme example of a change in feeding to suit available food, but makes the point that predators are quite capable in this respect. It is thus not farfetched to think that a predator may choose prey items so as to maximize the net energy intake per unit foraging time (Emlen, 1968; Shoener, 1971; Pulliam, 1974; Charnov, 1976; Comins and Hassell, 1979; and others). Experimental evidence for such optimal foraging behavior is available for a variety of predators (Charnov, 1976; Emlen and Emlen, 1975; Kislalioglu and Gibson, 1976; Werner, 1977; Werner and Hall, 1976).

Predators have to spend time searching, subduing, handling, and eating prey. As yet no one has put together a complete accounting of the cost in time of each of these components of predation relative to the benefits in terms of energy or biomass ingested, but in certain predators and prey one or two components are the most important and a cost/benefit analysis is easier.

Carcinus maenas is the common green or shore crab of the coasts of both sides of the Atlantic, and feeds on a variety of prey, including sessile mussels (Elner and Hughes, 1978). Mussels are detected by chemoreceptors in the antennae and mechanoreceptors in the walking legs. The crabs examine an individual mussel for 1–2 s and then either reject it or proceed to try to crush the shell, turning it around until a weak spot is found, and if they are successful they then feed on the flesh. These activities may be time-consuming, on the order of 10^2–10^3 s for breaking and 10^2–10^3 s for eating mussels.

Both of these component activities require more time as the size of the prey increases (Fig. 6-15), and in addition, the size of the predator affects their duration (Fig. 6-15, right). A rough estimate of value of the prey versus the cost of pursuing it can be obtained by a ratio of the grams of biomass eaten relative to time invested in handling (Fig. 6-14, top). Because of the effect of size of predator on handling time, larger prey are of more value to larger predators than to smaller predators. Clearly, there are many other factors that potentially may affect cost/benefit, including time to eat, different assimilation efficiency of crabs of different size, and nutritive quality of mussels of different size. The use of time to break as the cost, however, gives a good approximation to the actual consumption of mussels (Fig. 6-16, bottom). Green crabs do seem to feed on a size distribution that optimizes the weight eaten relative to the time it takes to handle the prey. Lawton and Hughes (1985) show similar results for other crabs.

If crabs do not often encounter mussels of the optimal size, they may shift to a different, much more time-consuming method of attack on larger mussels, involving grinding away the edges of the valves until a chela can be inserted into the shell. Prey of larger than optimal size are taken if they are in high abundance relative to "optimal" mussels. The crabs at all times attack prey of all sizes; the size of the crab, the size of

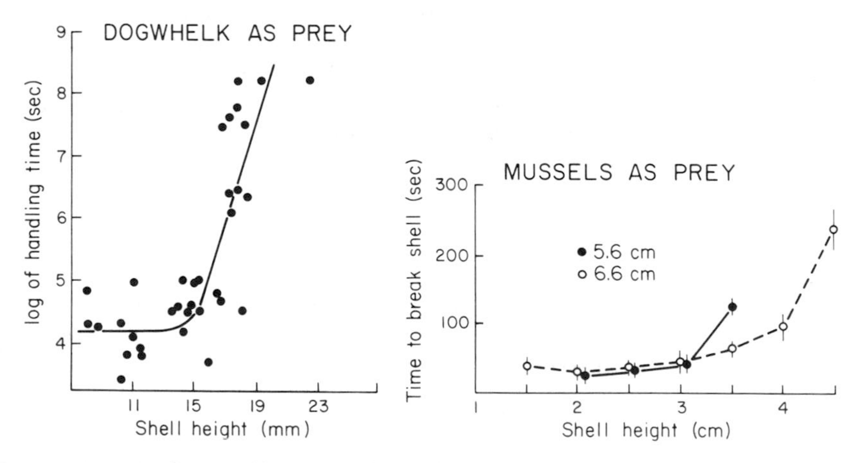

FIGURE 6-15. Left: Handling time for crabs (*Carcinus maenas*.6–7 cm in carapace width) feeding on dogwhelks (*Nucella lapillus*) of varying size. From Hughes and Elner (1979). Right: Shell breaking time for *C. maenas* of two sizes (5.6 and 6.6 cm carapace width) feeding on the ribbed mussels, Unpublished data of J. Wagner.

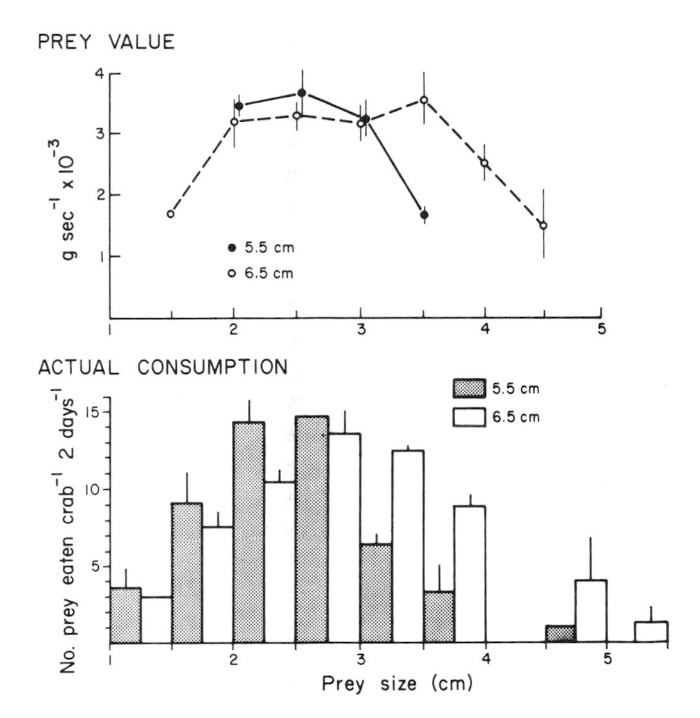

FIGURE 6-16. Top: Prey value, expressed as grams of mussels eaten by green crabs of two sizes (5.5 and 6.5 carapace width) per second. Bottom: Actual consumption when crabs were offered a choice of prey sizes. Unpublished data of J. Wagner from material collected in Great Sippewissett Marsh, Massachusetts.

prey, and their density determine the modal frequency with which a prey of given size is eaten.*

If prey are very rare, even very large mussels may be opened by crabs given enough time and hunger. There may be some size classes of prey that cannot be successfully attacked by crabs, no matter how long the crab persists. Starved crabs may persist in attacking invulnerable prey for up to 15 min, but for less hungry crabs with no other food, the handling time is usually cut short and the crab goes on to seek more suitable prey. For rare prey with invulnerable size classes, the optimal tactic may in fact be to cursorily examine the prey, eat the easily handled prey, and not persist too long on the more difficult items.

* Similar relationships have been proposed in models of zooplankton grazers (Lam and Frost, 1976; Lehman, 1976). Taghon et al. (1978) and Doyle (1979) extend these models to benthic animals feeding on particles of sediment, where the food is the microflora growing on the sediment particles.

In addition to the choice of specific prey size, foragers also have to choose where to feed. It would be most profitable to find patches where density of food is higher than the average density of food items in the entire habitat. Once feeding has reduced the density within a patch to near the average for the habitat, the foragers should move on. Mathematical expressions of this aspect of optimal foraging are available (Hughes, 1980; Krebs et al., 1981) and can be coupled to optimization of food obtained and time spent.

Optimal foraging theory is general enough to be applicable to many instances of food gathering by consumers. One difficulty with the theory is what common currency to use in optimizing food acquisition and expenditures during foraging. Most of the work on this area has been done on the basis of energy content, but it should be clear from our examination of properties of food items (Section 6.3) that other criteria could also be the subject of optimization. Redshanks may, for their own reasons, actually prefer energy-poorer amphipods to energy-richer worms (Goss-Custard, 1981); other consumers may optimize energy and nitrogen content while at the same time avoiding certain chemicals (Glander, 1981). Time is also a constraint of importance to many predators. Time of exposure to predation might have to be balanced against time spent searching for more favorable prey. It is therefore difficult to define just what should be optimized in an optimal foraging model. A second difficulty is that multiple factors may be optimized simultaneously by foragers. Food choice in the field, then, could differ substantially from that in the laboratory, where only one variable may have been considered in the experimental study of optimization. Ydenbeng and Houston (1986) address trade-offs among competing needs. It may be very difficult if not impossible to test multiple optimization theory in the field because of the complex observations needed (Zach and Smith, 1981). Optimal foraging theory has been criticized, for the above and other reasons. Pierce and Ollason (1987) conclude that optimality is not testable and assert that optimal foraging *theory* is "a complete waste of time." We can, however, say that *facts* such as the ratio of amount eaten to time spent vary, and we can describe how such ratios vary in relation to other variables, such as prey size (Fig. 6-16). Such empirical information is useful, and conveys some idea of how some predators may behave.

6.6 Vulnerability and Accessibility of Food Items

While some prey may be unavailable because of deterrents, low quality, or inappropriate size, there may be perfectly adequate prey that still may not be easily available to predators. We have seen earlier that one way a prey may escape predation is merely to hide untill it has outgrown its predator. This is a common occurrence and is well demonstrated in a whelk, *Nucella lapillus*, that escapes predation by green crabs after it reaches a shell height of about 15 mm (Hughes and Elner, 1979). In the field the susceptible size classes hide under stones or in crevices and are thus less available to green crabs. There is thus an evolutionary race between the searching ability of predator and the hiding ability of prey.

A somewhat different kind of spatial refuge from predation protects *Balanus glandula*, a common barnacle found primarily high in the intertidal in the coast of Washington. Whelks (*Thais* (= *Nucella*) spp.) eat all the *Balanus glandula* that settle toward the low end of the intertidal range. *B. glandula* does not grow to a size where they are safe from predators, but high in the shore the period between low tides (during which *Thais* is active) is too short to allow the whelks to complete drilling into the barnacles, and so above a certain height in the intertidal range the barnacles are in a spatial refuge (Connell, 1970).

Another kind of sanctuary from predation is that afforded by the complexity of the arena in which predation has to take place, as demonstrated in a classic paper on mites by Huffaker (1958): increased complexity of the habitat decreases the relative vulnerability of prey. This idea has broad application, and can be demonstrated to occur in salt marshes on the east coast of North America, where there are two major vegetated intertidal habitats, low marsh (dominated by cordgrass, *Spartina alterniflora*) and high marsh (dominated by marsh hay, *S. patens*). Many types of benthic organisms are found in the sediments of these two habitats and are eaten by killifish during high tide. Laboratory experiments where killifish were allowed to feed on amphipods show that the functional response to amphipod density differs in the two habitats (Fig. 6-17, top). In fact, only in the low marsh was it possible for fish of large size to evidence their superiority as predators (Fig. 6-17, bottom). The explanation for this has to do with the physical character of the two habitats. Low marsh plants are more sparsely distributed than high marsh plants. On average, the distance between nearest neighbor stems is 1 mm in high marsh, while stems in low marsh are spaced almost five timesas far. The dense stems of high marsh make it difficult for the fish to enter the habitat, and at the same time provide hiding places for prey. The prey are therefore less vulnerable to attack by fish in high marsh. Since it is more difficult for larger fish than for smaller fish to find their way into a complex, dense habitat, this explains the lack of effect of predator size on the number of prey eaten (Fig. 6-18, bottom).

The above experiments are laboratory results but there is evidence of the importance of habitat complexity in the field. We can compare the distribution of prey sizes and abundance in salt marsh plots where fish had access to prey with results from areas where they were excluded

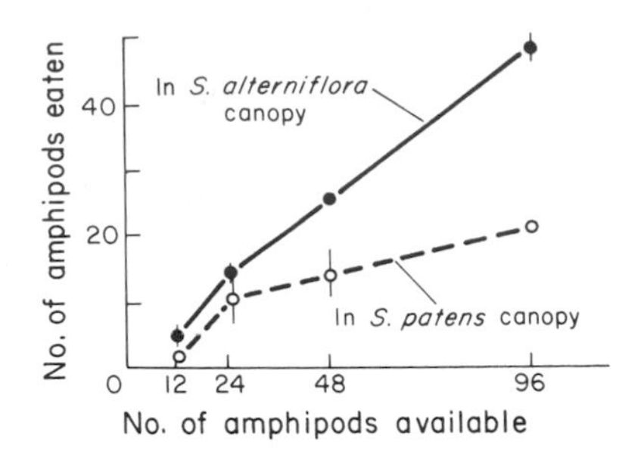

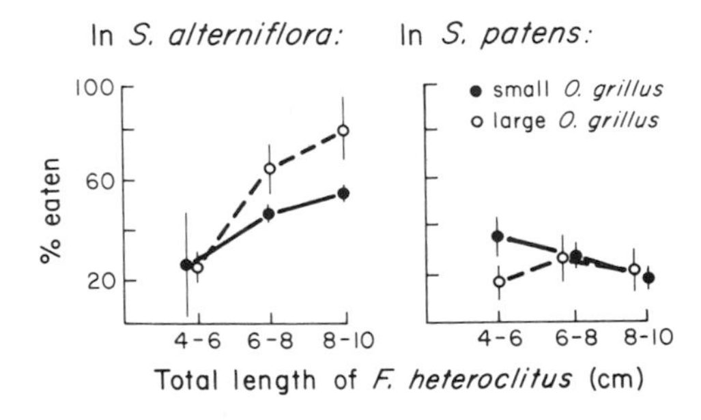

FIGURE 6-17. Functional response of *Fundulus heteroclitus* to densities of amphipod prey in two habitats with different physical structure. Bottom: Percent of amphipods prey eaten by *F. heteroclitus* of differing size in each of the two habitats. Adapted from Vince et al. (1976).

by means of fences (Fig. 6-18). The number (top right within each graph) of available prey in samples from low marsh was much lower than those from high marsh. Where fish were present, fewer prey were found than where fish were excluded. From our discussion of prey size, we know that predation is usually size specific, and the snail can escape predation by growing to an invulnerable size (Vince et al., 1976). In low marsh the smaller size classes of snails are removed where predators are present. The size distribution of snails in high marsh where fish were excluded resembles that of low marsh with no fish. The more remarkable result, however, is that the size distribution of prey in high marsh also has a large proportion of small individuals. This corroborates the laboratory results, and shows that predation by fish is so reduced in the more complex habitat that it cannot affect abundance or size distributions of prey.

Very similar conclusions emerge out of the data of field distributions of the amphipod (Fig. 6-18), the difference being that unlike the snail, the entire size distribution of amphipods is subject to predation and, in fact, fish prefer larger amphipods (Vince et al., 1976). The results of these experiments demonstrate that the physical architecture or structure of the habitat in which predation takes place has critical consequences for both predator and prey.

It is not clear that such structure is important to predation in planktonic communities. We tend to see the water column as relatively homogenous but there is an increasing amount of evidence that there are important heterogeneities in temperature, salinity, density, nutrient content, and also the motion of water. These heterogeneities pervade all spatial scales (cf. Chapter 11) and may actually provide a complex structure to the water column, with still to be explained consequences for predator-prey interactions. Ongoing research is Chesapeake Bay indeed suggests that it is just such physical features that lead to food item aggregation and to growth of estramine fish (W. Boynton, pers. comm.).

Accessibility of prey of suitable size may determine the numerical response of predators. The density of foraging shorebirds, for instance, is closely correlated to abundance of bivalves that not only fall into suitable size classes but that are also accessible (Fig. 6-19, top). Potential prey change their vertical position over the year in response to temperature changes. Access to these prey is determined by the depth at which bivalves are buried, relative to length of bill of the wader that consumes the bivalves (Zwarts and Wanink, 1993). Benthic organisms

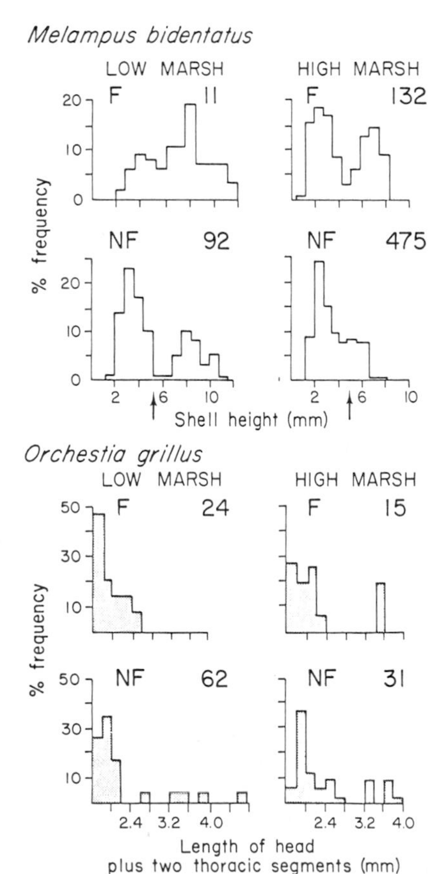

FIGURE 6-18. Abundance (numbers on top right of each histogram, in individuals per 2 cores of 9 cm radius) and size distribution of the prey species (histogram) in two salt marsh habitats. Area labeled "F" are exposed to natural predation by *F. heteroclitus* during high tide, while these fish were excluded from the "NF" areas by means of wire mesh fences. The arrows show the threshold above which *M. bidentatus* is not eaten by *F. heteroclitus*. Adapted from Vince et al. (1976).

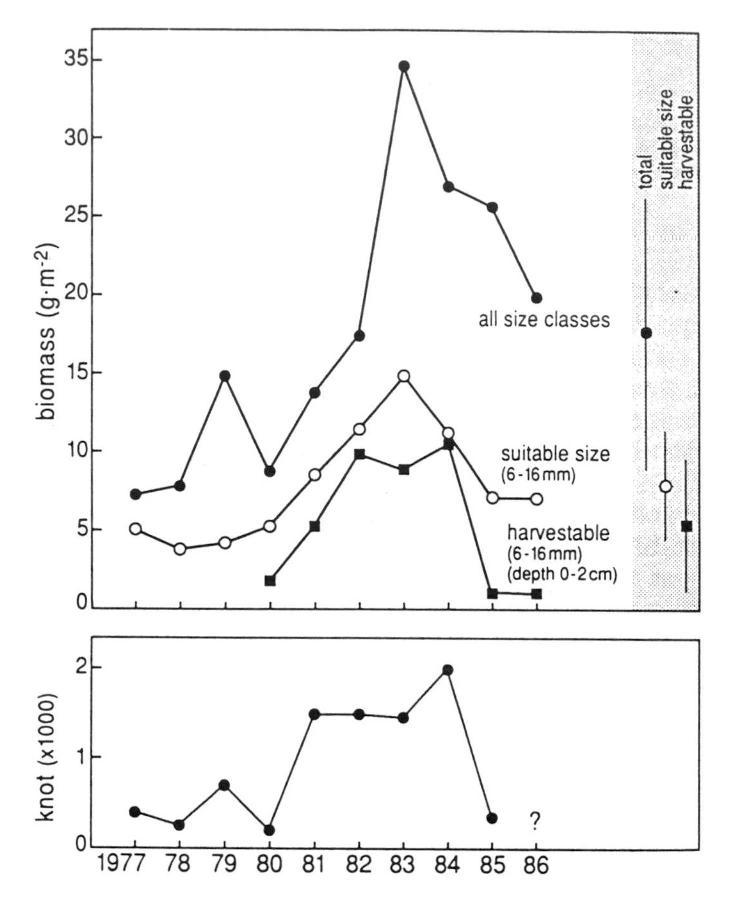

FIGURE 6-19. Top: Biomass of the bivalve *Macoma balthica* during August for the years 1977–1986 in three different categories. The categories include total biomass, biomass of individuals whose size makes them suitable as wader food, and harvestable biomass of individuals of appropriate size and depth in the sediment. Averages plus or minus standard deviations are given on the right. Bottom: Peak numbers of knot (*Calidris canutus*) during the span of years in the same areas. From Zwarts and Wanink (1993).

move deeper in winter and become inaccessible (Fig. 6-20). The lowered accessibility of prey during winter could in part explain why shorebirds have to migrate. Since prey access varies year-to-year (Figs. 6-19; 6-20), the degree of aggregation of predators in sites with many accessible prey also changes (Fig. 6-19, bottom). The data of Figs. 6-19 and 6-20 are only for two species of predator. Other predators with different bill lengths could exploit different benthic prey, or make use of the same prey at different times of year.

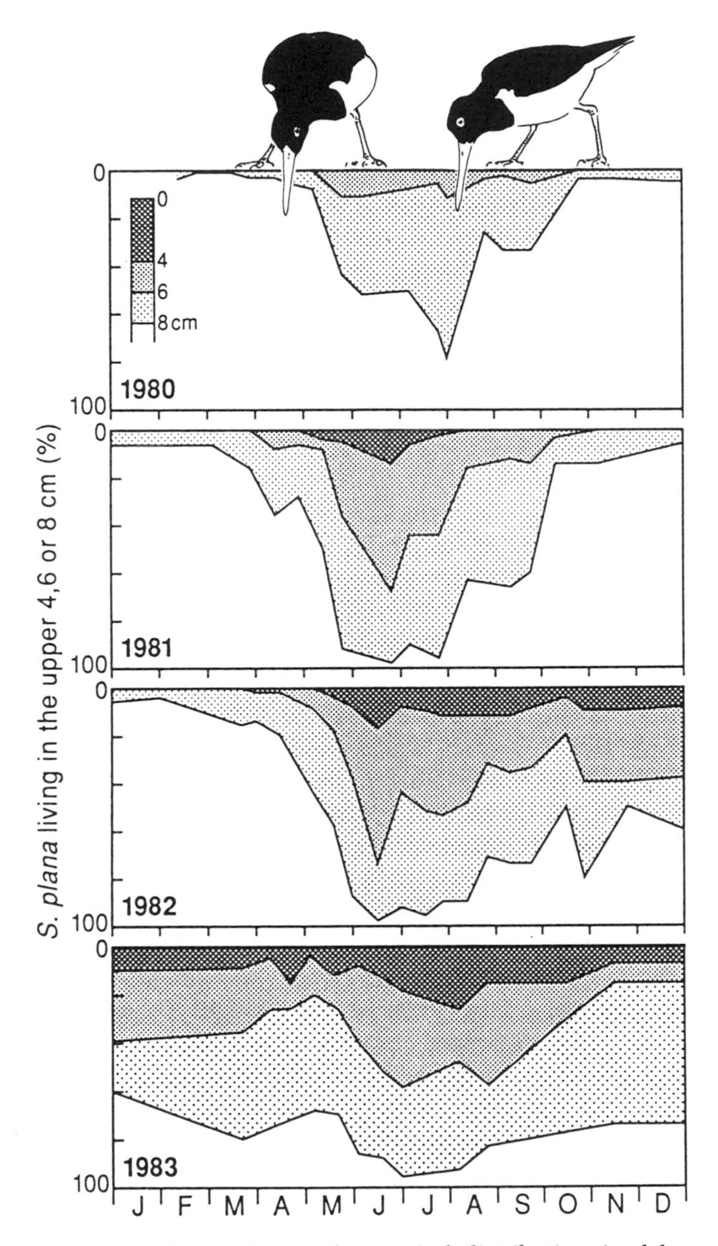

FIGURE 6-20. Seasonal variation in the vertical distribution (and hence accessibility to the oystercatcher) of the bivalve *Scrobicularia plana* over 4 years. The proportion of the population living in the upper 2, 4, 6, and 8 cm of sediment is shown. The feeding oystercatchers indicate the approximate depth to which these predators can find prey. From Zwarts and Wanink (1993).

6.7 The Importance of Alternate Prey

6.7.1 Prey Switching by Predators

In discussing optimal foraging we concentrated on the size of individuals within a species of prey. Prey choice, however, also involved choice among different species. If predators can feed on several specios, they might consume the most abundant prey first, perhaps by the buildup of a search image, as mentioned earlier. Then when numbers of this species are lowered, they might concentrate on whatever other prey species is most numerous at the time. This feeding behavior, similar to the patch exploitation pattern hypothesized in optimal foraging, has been called "switching" (Murdoch, 1969). Switching has received attention because such behavior on the part of the predator could exert maximum predation pressure on the most abundant species, and thus prevent any one species from monopolizing an environment, an important feature that may explain how multispecies communities are maintained (Elton, 1927).

Predators with strong preference among possible prey, such as carnivorous snails, seldom switch regardless of relative abundance of the prey species (Murdoch, 1969). In fact, certain sea slugs starve rather than feed on seaweed species not on their preferred list (Trowbridge, 1991).

Switching has, however, been found in consumers that have less marked preferences among the alternate prey available. Switching occurs in copepods (Landry, 1981), for instance, presumably because although suspension feeders are size selective they may not have strong preferences among prey of different species. Pelagic food webs, in fact, have been described as "unstructured" (Isaacs, 1973) precisely because any predator is thought, within constraints imposed by its size, to be able to eat most other species in the assemblage of species present.

In species where preference among alternate prey species is not too marked, the relative density of the alternate prey species may be important. The snail *Lepsiella vinosa* preys on two species of barnacles that settle on pneumatophores of Australian mangroves. Switching by the snail takes place provided that the density of the one slightly preferred prey species is not too high (Bayliss, 1982).

It is not clear exactly how switching takes place. It may not just be a reflection of relative abundance of alternate prey types, and learning and rejection may not be involved in every case. Possible mechanisms for switching might involve changes in foraging methods, predation on high density patches of different prey species, and the differential benefit of alternate prey to the predator.

Changes in foraging method. Switching may take place due to changes from one foraging method to another (Akre and Johnson, 1979).

Damselfly naiads are a very common predator in freshwater ponds and lakes and themselves are subject to visual vertebrate predators. To avoid detection by fish and amphibians, naiads use the less risky ambush mode of hunting as long as it results in sufficiently frequent encounters with prey, in this case two cladoceran species. There seems to be no clear preference by the predator for one or another type of cladoceran. One of the prey species is more easily captured using ambush methods, but if it becomes too rare the hunger level of the naiad reaches a point where the naiads start hunting by searching. Since the second species of prey is more easily encountered while searching, this type of prey then becomes more abundant in the diet. In this instance, no learning or rejection of less preferred prey is involved. Rather the hunger level relative to prey density is the driving force that changes feeding tactics.

Predation of high-density patches of different species. Switching may take place in predators that concentrate foraging in patches of high density of prey (cf. Section 5.3). If different patches where prey are found at high density are dominated by different types of prey, as the predator moves from one patch to another the result is effectively switching, with no rejection or learning behavior (Murdoch and Oaten, 1975). One important consequence is that this sort of predation leads to a type III functional response, as does the shift in hunting behavior seen in the naiads. This sort of switching could be important in thus allowing a density-dependent acceleration of prey consumption.

Differential benefit of alternate prey to the predator. Sea otters select prey species that provide them the most energy per time spent foraging and handling (Ostfeld, 1982). As a given prey becomes rarer, search time increases, and thus the otters would probably switch to a more abundant prey where the ratio of energy obtained to time spent may be more favorable. The criterion used to switch would be a function of profitability rather than merely depend on abundance of alternate prey.

The diversity of ways in which feeding interactions can be carried out demonstrate the manifold ways in which animals can carry out similar activities, both by flexible behavior within one species and by different adaptations by several species. This is a major lesson to be learned from this section, and is one of the reasons why we examined the plethora of examples set out in this chapter. A second major point is that the flexibility in behavior seen in the above examples, plus a weak initial preference when alternate prey are about equally abundant are preconditions to switching when prey densities may become unequal. As such, one may imagine that if prey preference is very marked there are potentially few situations where the mechanism of switching may be important in affecting stability of prey numbers. The degree of prey preference—and the related idea of profitability—are key elements in how feeding on alternate prey may be important in governing abundance of prey populations.

6.7.2 The Impact of Preference in Multiple-Prey Situations

In the field, the usual condition is for a foraging organism to be faced with choices among a variety of very different food types. A situation where a predator had a choice among various prey was examined by Landenberger (1968), in experiments in which two species of carnivorous starfish fed in an arena where seven species of prey molluscs were available. Their prey included three species of mussel, two species of whelk, a turban snail, and a chiton. Mussels were the preferred food of the starfish. The data are reported as to the percentage of mussels in the first, second, third, etc., groups of 30 prey eaten (Fig. 6-21, left panels). The first point to be made [one made earlier by Holling (1959)] is that the presence of alternate prey reduces the predation pressure on all prey, even the most preferred: mussels made up rather less than 100% of the first group of prey. Thus, prey gain some protection from predators in situations where alternate prey of similar size are present. An example of this for fish larvae is documented by Gotceitas and Brown (1993).

The second point is that as the predators gained experience with the second, third, etc., groups of prey eaten, the proportion of preferred prey (mussels) in the diet increased, so that after five weeks, starfish

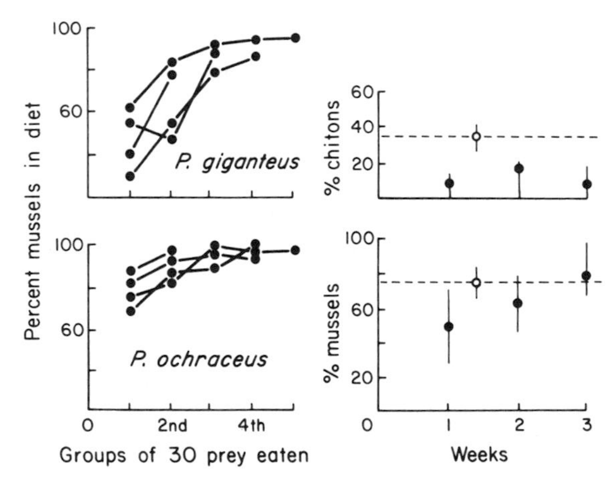

FIGURE 6-21. Effects of alternate prey on predation by two species of starfish (*Pisaster giganteus* and *P. ochraceus*). Left panels: Percentage of the preferred prey (mussels) in the diet, expressed as the 1st, 2nd, etc., groups of 30 prey eaten. Right panels: Percentage of alternate prey (mean ± range) eaten by starfish trained for 3 months to eat turban snails. For comparison, the dashed line and open circle show preference of untrained control starfish. Adapted from Landenberger (1968).

consumed mussels almost exclusively. For these predators the effects of alternate prey are not permanent.

A third point (not obvious in Fig. 6-21) is that the predators were less discriminating when more kinds of prey were available. Only broad taxonomic categories (e.g., pelecypods, gastropods) were recognized by starfish in situations with many alternate prey species, while when an experiment allowed the starfish a choice of only two species of mussels, starfish were capable of distinguishing species, and showed preference for one of the two species offered.

Since past experience with given species of prey seemed to be an important feature, Landenberger then tested the impact of alternate prey on starfish fed for three months on turban snails (Fig. 6-21, right panels). When the alternate prey were chitons (a prey not liked by starfish), only 10% or so of the prey of the snail-conditioned starfish were chitons, the rest being turban snails. When mussels were the alternate prey, there was an initial consumption of 50% turban snails, still to an extent reflecting their long conditioning on turban snails. As time passed, the starfish switched to feeding on mussels: the predators overcame previous history and used the most preferred food items.

In Landenberger's study, as in Ivlev's (1961) work on feeding by fish, the rate of feeding increased with time. The predators were also most selective when feeding at high rates.* With experience, the starfish learned how to be more effective predators on the array of prey available.

6.8 Interaction of Mechanisms of Predation

In the field, the various mechanisms of predation that we have been discussing interact in complex ways. Not all mechanisms are important everywhere, nor are they and their interactions constant over time. One example of such interactions is the well-studied and widespread phenomenon of vertical migration in microplankton, which currently is largely explained as a complex combination of prey size preferences, aggregative responses, differential predatory modes and cues, use

---

*We have already seen this counterintuitive effect of hunger level in Section 6.4 when discussing copepods. It is not clear exactly how generalizable is the relation of greater selectivity to feeding rate. In blackbellied plovers search times are shorter when the birds feed on a more varied diet (Baker, 1974). Shorter search times would allow faster feeding rates, so that the less selective the predator, the faster the feeding rate. The relation of feeding rate, alternate prey, and selectivity needs further study.

of refuges, and effects of alternative prey items, all altering the demographic characteristics of multiple predators and prey (Ohman, 1990).

In Dabob Bay, an arm of Puget Sound, there are impressive seasonal changes in the abundance of a small phytoplankton-feeding copepod, *Pseudocalanus neumani*, whose abundance seems to be followed by seasonal changes in abundance of its main nonvisual predators, two larger species, a copepod, *Eucheta elongata*, and a chaetognath, *Sagitta elegans* (Fig. 6-22, top). In addition, there are fish that visually hunt for prey in the upper illuminated photic zone. The fish prefer the two larger prey, but also feed on *P. neumani*. Predation pressure on *P. neumani* is therefore intense.

P. neumani feeds on phytoplankton, and without predator pressure, this copepod would aggregatively respond to food abundance by occupying the upper 20 m of the water column, where food is most abundant from April to November (Fig. 6-22, middle).

Ohman (1990) made use of stations where fish abundance differed to study how differences in predation pressure by fish affected vertical migration by plankton. Where fish were relatively more abundant (Fig. 6-22, bottom), there were few *E. elongata* and *S. elegans*, and *P. neumani* that were present migrated down during the day, and thus avoided risk of visual predation during the time of year when fish are active. Where there were fewer fish, there were relatively more *E. elongata* and *S. elegans*, and these predator species themselves avoided fish by migrating down-ward during the day (Fig. 6-22 middle, histogram insets). Since these two species do not use visual cues to find prey, they can feed in the upper layers during night. Where *E. elongata* and *S. elegans* are abundant, *P. neumani* reverse their daily migration: at night they are found deeper in the water column, where *E. elongata* and *S. elegans* are less abundant. Reverse migration reduces risk of being eaten by the non-visual predators, which, at least in this station, appears to be a greater priority than risk of consumption by the few fish that are present in the upper layer. The nonvisual predators do not depend exclusively on *P. neumani*, and find alternate prey in the upper layers of Dabob Bay.

The changes in vertical migration are unlikely to be learned behavior. It is more likely that differences in genotypic frequencies in migration behavior can be selected by predation pressure. *P. neumani* has 8–9 generations per year, so different gene frequencies could replace each other fairly quickly. Life table calculations show that if predator avoidance diminishes mortality by as little as 12%, migrants will replace nonmigrants. Demography of this species can therefore respond to the pressures brought by complex array of predation-related mechanisms.

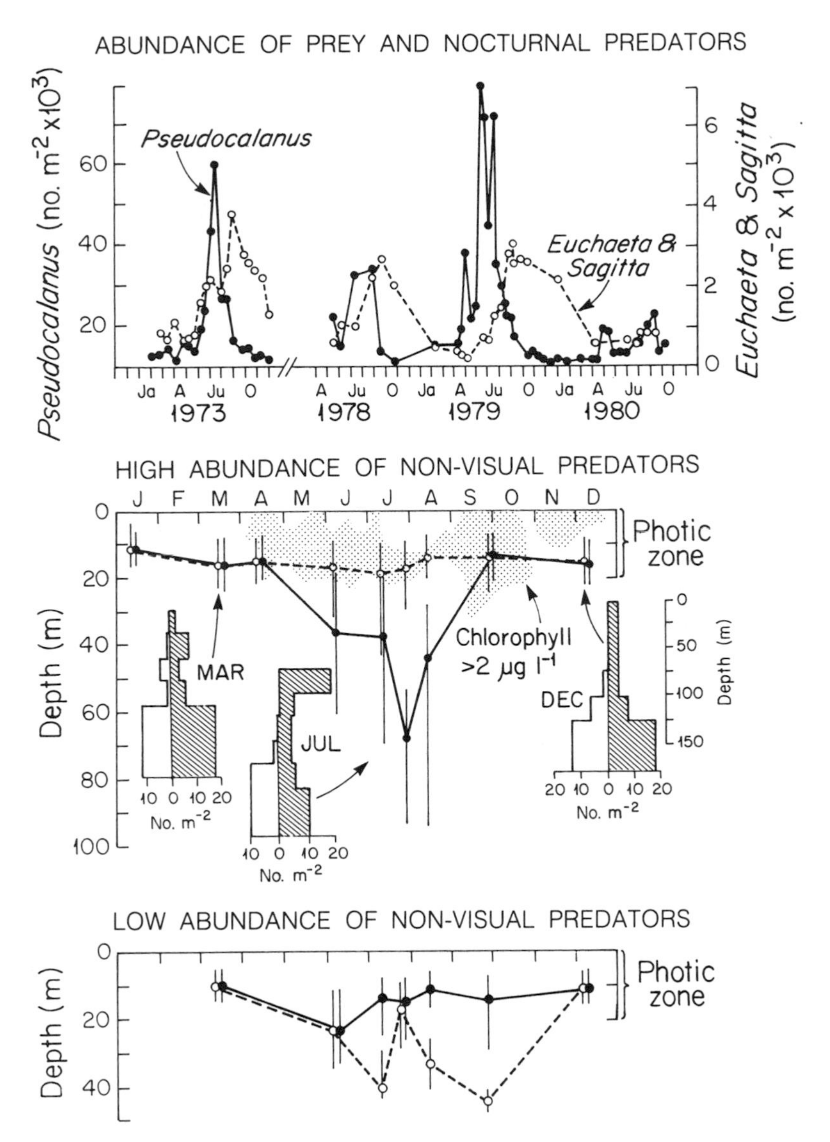

FIGURE 6-22. Top: Seasonal abundance of the copepod *P. neumani*, and two of its main nonvisual nocturnal predators, the copepod *E. elongata*, and the chaetognath *S. elegans*, in a station in Dabob Bay, Washington, where the predators of *P. neumani* were abundant. Middle and bottom: Seasonal change in median depth where female *P. neumani* were found, during the day (dashed lines) and night (black lines), in sites where nocturnal predators were abundant (middle), or rare (bottom). Histograms show the abundance and vertical location of female *E. elongata* at 25 m intervals, during day (white), and night (black). Stippled areas show the position of abundant phytoplankton (>2 μg chlorophyll l^{-1}). Adapted from Ohman (1990).

6.9 Predation and Stability of Prey Populations

In our discussion of Chapters 5 and 6 we have so far ignored whatever coevolution has taken place between predator and prey. We continue to do so, but evolutionary processes are clearly involved. Recall the discussion of chemical feeding deterrents and mimicry. We have already noted, for example, that some polychaetes may secrete chemical deterrents active against some predators. There may be morphological adaptations. Zooplankton may grow spines or processes that prevent ingestion. The shells of many marine snails show structures and sculpture that seem to reflect the presence of many molluscine predators through the evolutionary history of snails (Vermeij and Covich, 1978). The architecture of snail shells is much less ornate and more conservative in freshwater environments where typically there are fewer predators specializing in snails than in marine habitats.

Predators show adaptive responses of their own. In nature we find that, by and large, there is an interplay between prey and predator whereby both continue to be present, at least in ecological if not evolutionary time scales. Predators could potentially eat all prey, and prey could potentially fill the environment with their progeny, yet on average neither of these situations takes place.

What potential ecological mechanisms does the process of predation provide by which control of prey numbers can take place? In terms of preventing prey from reaching excessive numbers, the functional response does not seem like the most important contributor, except where a type III curve is found; we have seen that this may not be a very common or permanent feature. We cannot be sure that the rarity of type III functional response is not a laboratory artifact, but switching, patchiness of prey, and presence of alternate prey favor this response. The much more common type II response is by itself less likely to provide the means by which predators control prey. The functional response does not act alone, however. Predators also respond to changes in prey density or to prey patchiness by aggregation. This numerical, almost lagless response, combined with the increased recruitment of young predators, could make predators effective enough to control prey numbers, since both increase markedly as prey density increases. The developmental response, although involving a lag in time, as we have seen, could add to the effectiveness of predation in the control of prey.

On the other hand, complete extirpation of a prey species is prevented by the existence of refuges, of habitat complexity, and by the growth of prey to invulnerable size classes. These devices allow survival of certain proportions of the prey population, making extinction less likely. Another way to escape control by consumers is for prey to grow so fast as to overwhelm the ability of predators to respond to prey density. This aspect is discussed in Chapters 8 and 15 in regard to phytoplankton and their grazers.

In the complex phenomenon of "predation"—including animals feeding on plants, animals, microbes, and detritus—there are a variety of mechanisms that could lead to the control of prey numbers. A very rich array of predation tactics exists and similarly, a variety of prey defenses, and each case may have its own distinctive features. The generalizations that we can make are thus not about the specific ways in which consumers obtain food particles. Rather, the generalities concern the components of the predation process. Time is the most limited resource. Each predator must allot time to detect, pursue, capture, and handle prey. The apportionment of time among these various phases of predation, the deployment of the components of predation to carry out the processes, and the properties of the prey determine what a predator does and how important it may be in structuring the abundance of other species in the community in which it lives.

Chapter 7
Processing of Consumed Energy

7.1 Flow of Energy Through Consumers

In the previous two chapters we have discussed the various means by which consumers obtain sufficient rations. In this chapter we focus on the various fates of ingested material and how this apportionment affects consumers. In this section we will look at the overall scheme of flow of ingested material; in the following section we examine the processes that affect the various fates of matter ingested by consumers and some consequences of this partitioning. In many instances the rates of each process vary so widely that it is necessary for comparative purposes to use ratios between two of the processes, which we refer to as efficiencies. After completing the examination of partitioning of food in specific populations, we consider how the partitions are put together in energy budgets for populations and ecosystems.

It is customary to consider the partitioning of specific elements, usually carbon, nitrogen, or phosphorus (Butler et al., 1970; Corner and Davies, 1971; Kuenzler, 1961; Jordan and Valiela, 1982), or the energy* contained in ingested food. Our emphasis will be the fate of consumed energy in a consumer (Fig. 7-1). A certain portion of the energy of ingested material is not assimilable and is released back into the environment as unused feces (*F*); the remainder of the ingested food is assimilated (*A*). Some of the products of metabolic reactions are energy-rich nitrogenous compounds such as urea and uric acid; these compounds are voided (*U*) and are usually measured together with fecal losses (*F*) in a measurement of egestion (*E*). Some of the energy in ingested food is lost in the deamination of amino acids not incorporated into consumer protein, and there are further losses of energy in the

*There has been much research on energy in consumers and many measurements of energy content in terms of calories of biological materials (Paine, 1971; Cummins and Wuychek, 1971). In general, the ash-free caloric content of consumers lies between lower values typical of carbohydrates (3.7 $kcal\,g^{-1}$ for glucose, 4.2 $kcal\,g^{-1}$ for cellulose) and higher values of lipids and fatty acids (9.4 $kcal\,g^{-1}$). Most consumers have a far narrower range of caloric content, averaging about 5–6 $kcal\,g^{-1}$ (Holme and McIntyre, 1971).

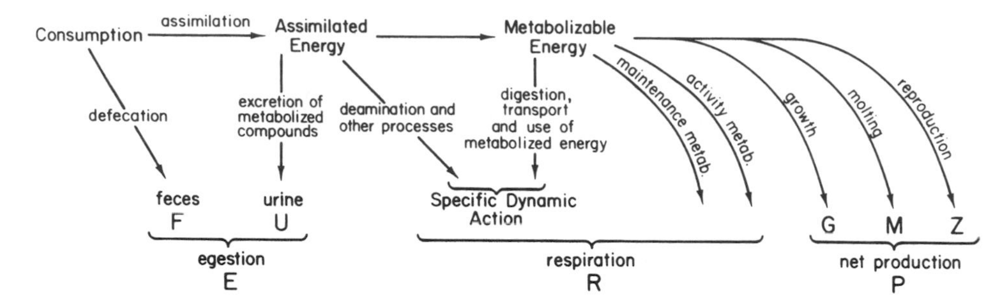

FIGURE 7-1. Scheme of the various fates of consumed energy and the terms used in discussing processing of energy. Many marine species release mucus in large quantities. This is not shown in this figure but could be a significant loss of metabolizable energy that should be included in net production for such species. Modified from Warren and Davis (1967).

digestion, transport, and metabolism of food. The sum of these losses of assimilated and metabolizable energy is referred to as the specific dynamic action and is evidenced as an increase in oxygen consumption soon after feeding occurs. The major losses of ingested energy are due to expenditures required to maintain metabolic processes and to sustain levels of activity sufficient for life activities. The sum of specific dynamic action plus metabolic and activity losses is respiration (R). Respiration is the oxidation of organic matter and results in the release of energy and CO_2 [cf. Eq. (1-2)]. Respiration is thus the release of energy fixed by producers.

The energy that is assimilated but not respired is devoted to production (P). Production consists of the energy invested in reproductive products (Z), and in growth (G). If the species has the kind of development that involves shedding of outer layers, there is a further loss due to molting or release of mucus (M) that is also included in production. The sum of these various terms is the energy consumed (C),

$$C = P + R + E. \tag{7-1}$$

C for field populations has often been estimated by some version of Eq. (7-1), usually using the assimilation efficiency (A),

$$A = [(C - E)/C]100, \tag{7-2}$$

which can be measured in the laboratory with known rations (cf. Section 7.2.1).* Then, by substitution into (7-1),

---

*There is some ambiguity in Eq. (7-2) in that $E = F + U$ (cf. Fig. 7-1), and most of F is material that has not been assimilated. It is generally not practical to measure this nonassimilated fraction separately.

$$A \cdot C = C - E = P + R, \qquad C = (P + R)/A. \tag{7-3}$$

The scheme of Figure 7-1 was derived for consumers that feed on particles and are fairly large. With some modification it could be applicable to smaller consumers such as aerobic heterotrophic microorganisms. The modifications would include release of vacuolar waste products instead of feces, secretion of mucopolysaccharide sheaths instead of molting, and reproduction by fission instead of release of reproductive products.

7.2 Assimilation

7.2.1 Measurement

The simplest measurement of assimilation would be to obtain the ash-free weight of food and feces, but it is often difficult to weigh feces, and reingestion of feces and losses of soluble materials will affect the measurement. Animals may also assimilate the ash portion of food (Cosper and Reeve, 1975); prawns, for example, can assimilate over 30% of the ash in their food (Forster and Gabbott, 1971). Feeding of radioactively labeled food, followed by measurement of labeled CO_2, feces, and soluble compounds, is a more acceptable approach. This procedure requires adequate controls to assess bacterial uptake of label and a long enough interval of time during which the consumer is exposed to the label so as to allow tracer content in the body of the organism to reach equilibrium.

There may be a substantial amount of organic matter lost during feeding by consumers. This means that ingestion is not equal to the removal of food organisms. The loss during such "messy" feeding by zooplankton can be up to 30% of the ingested food (Williams, 1981). ^{14}C studies of feeding by a marine isopod on *Calanus* spp. showed that perhaps 25% of the biomass of the prey was lost during feeding. Such losses, if not considered, may lead to overestimates of assimilation and assimilation efficiency. Inert radioactive tracers such as ^{51}Cr (Calow and Fletcher, 1976) can also be useful in measuring assimilation.

7.2.2 Factors Affecting Assimilation

It is usual to refer to the ratio of food assimilated to food ingested as the assimilation efficiency. Measurements of assimilation efficiencies for marine species can vary significantly even within one species. For example, the assimilation efficiency of *Calanus helgolandicus* may be anywhere from 12.4 to 61%; that of *Centropages typicus* 16 to 79%; *Temora stylifera* 0 to 91% (Gaudy, 1974). Such variation in assimilation efficiency may be due to a variety of factors including food quality, amount of food, and age of the consumer.

7.2.2.1 Food Quality

It is obvious that the higher the quality of a food, the higher the assimilation efficiency. Food quality has many dimensions; a relatively simple property is the percentage of inorganic ash: the higher the percentage ash in food, the lower the assimilation efficiency (Fig. 7-2).

Differences in the food habits of various species lead to significant differences in assimilation efficiency. Estimates of assimilation efficiency obtained in various ways for a very wide variety of marine, freshwater, and terrestial animals are compiled in Fig. 7-3. The classification of species into detritivore, herbivore, and carnivore is based on the actual food types eaten by the consumers (dead plant material plus microbial flora, living producers, and living animals, respectively) during the measurement.

Individual measurements of assimilation efficiency are subject to many experimental errors and are very variable, but considering the data set as a whole, there are clear trends. Carnivores characteristically show high assimilation efficiencies; their food is of high quality and readily digestible, since it lacks cellulose and other undigestible compounds. Herbivores span a wide range of assimilation efficiencies, but peak in intermediate values of assimilative ability. As discussed in Sections 6.3.2 and 8.2, producers may be inadequate food; for example, many plants often contain secondary chemical compounds that bind proteins, lowering the quality of food used by grazers, and this low food quality results in low assimilation efficiencies. Single-celled algae, yeasts, and bacteria tend to be assimilated with higher efficiency than macroalgae and higher plants, although not as efficiently as animal food. Several short-lived, opportunistic macroalgae, referred to as ephemerals in Section 8.3.1.1 are also consumed with high efficiency. This agrees with what we know about the lack or presence of structural or chemical defenses in these kinds of producers. The result is that single-celled or ephemeral species are often the preferred food of herb-

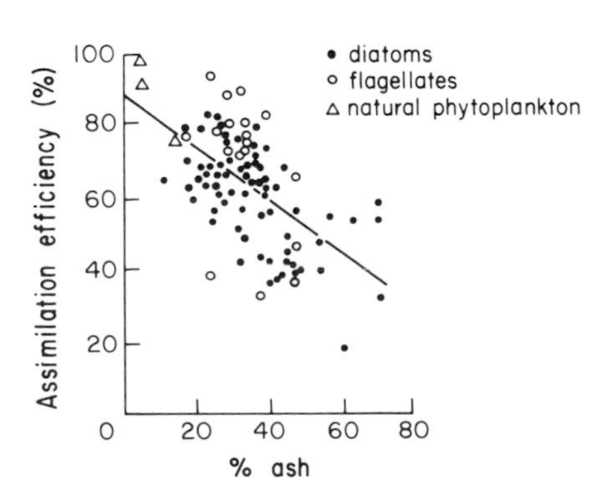

FIGURE 7-2. Assimilation efficiency in *Calanus hyperboreus* feeding on phytoplankton of different percent ash content. Adapted from Conover (1966).

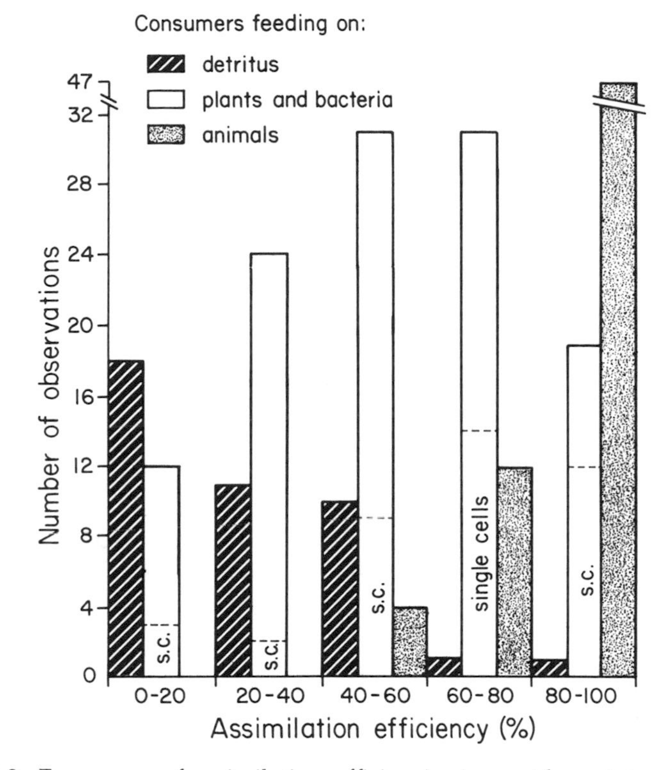

FIGURE 7-3. Frequency of assimilation efficiencies in a wide variety of animals that feed on detritus, plants, and animals. Herbivores that feed on single-celled algae, bacteria, or yeasts are grouped in the lower portion of the open bars. One species may have been included more than once if experiments with different food types were done. Values from many sources, including the compilations in Welch (1968), Conover (1978), Forster and Gabbott (1971), Dickinson and Pugh (1974, vol. 2), Krebs (1976), Winter (1978), Ricklefs (1979), Kiørboe et al. (1981), Horn (1989); Werner and Hallibaugh (1993), and many individual papers.

ivores. Other measurements of high assimilation efficiency in Fig. 7-3 were obtained with terrestrial herbivores that feed on seeds or sap; seeds have typically high protein contents and are high quality food, while feeding on sap avoids many of the secondary defensive compounds, since these are to a large extent associated with structural plant tissues. Herbivores that feed by ingesting tissues of vascular plants tend to have lower assimilation efficiencies, since these plants tend to have well-developed structural tissues and chemical defenses.

In most ecosystems grazing does not consume all the plant matter (Fig. 6-9), so that senescent algal and plant tissues largely become detritus. Even in oceanic water columns, where grazing can consume all that is produced by the phytoplankton, much of the organic matter is returned to the environment by the grazers in the form of fecal matter (cf. Chapter 13). Detrital matter is therefore plentiful and the majority of species in most environments consume detritus. Yet, detritus is poor food; detritivores are subject to both nitrogen and carbon deficiencies in their diet. Animals require that carbon be in roughly a 17/1 ratio to nitrogen (Russell-Hunter, 1970). Detritus is often lower in nitrogen; C/N values of 20–60/l are not unusual in detritus from vascular plants in salt marshes and 15–30 in eelgrass (Godshalk and Wetzel, 1978). The C/N ratio of particulate matter in the oceanic water column in the Atlantic is in the order of 7–21 (Gordon, 1977; Knauer et al., 1979). Particles at 0–250 m in depth are richer in N (C/N = 7.5) while at depths of 3,000–4,000 m the nitrogen is relatively lower (C/N = 15.1) (Gordon, 1977). In coastal areas off California. C/N varies between 8.8 and 15, depending on depth and intensity of production near the surface, while in the open Pacific the C/N values lie between 13 and 29, the high values associated with deeper water. The factors that govern the ratio of C/N in the water column are not well known, but differences in sampling and analytic methods may be responsible for part of the observed variability.

Sediments have C/N of 10 or more (Degens, 1970). Even though bulk C/N ratios may not exceed the desirable ratio of 17/1, much of the nitrogen in particulate organic matter is unavailable to suspension or deposit feeders due to binding of nitrogen compounds with clays, lignins, and other phenolic compounds, including nitrogen-containing humic acids (Odum et al., 1979; Rice, 1982). This immobilization of nitrogen means that even where C/N of organic matter are low, benthic deposit feeders or other detritivores generally do not get enough nitrogen (Tenore et al., 1979).

Fecal pellets are abundant in soft sediments, and they have C/N = 9–15 (Honjo and Roman, 1978), a more favorable ratio than the 17/1 presumably required by consumers. Fecal pellets are nevertheless a poor food. *Calanus helgolandicus* fed on fecal pellets develop more slowly from the third copepodid to adult stages (20–22 days) than specimens fed on phytoplankton (8–15 days). Survival is also lower in pellet-fed (57–71%) than in phytoplankton-fed copepods (66–99%) (Paffenhöfer and Knowles, 1979). Growth is reduced when detritus is included in diets, and copepods on detrital diets have higher respiration rates as more detritus is included in the ration (cf. Fig. 7-9).

The caloric content of detritus—mostly derived from carbon compounds—is a rough index of the potential energy available to detritivores. The polychaete *Capitella*, for example, incorporated more detritus of *Spartina* when it had a higher caloric content, while consumption and assimilation were not increased by increased nitrogen content (Tenore et al., 1982). Caloric content may thus be more impor-

TABLE 7-1. Assimilation Efficiencies in an Amphipod, *Hyalella azteca* (Hargrave, 1970), a Polychaete, *Nereis succinea* (Cammen et al., 1978), a Holothurian, *Parastichopus parvimensis* (Yingst, 1976) and oysters, *Crassostrea virginica* (Crosby et al., 1990), Feeding on Various Foods.

Species of consumer	Food	Assimilation efficiency (%)
Hyalella azteca	Bacteria	60–83
	Diatoms	75
	Blue-green bacteria	5–15
	Green algae	45–55
	Epiphytes	73
	Leaves of higher plants	8.5
	Organic sediment	7–15
Nereis succinea	Microbes	54–64
	Spartina detritus	10.5
Parastichopus parvimensis	Microorganisms	40
	Organic matter	22
Crassostrea virginica	Bacteria	52
	Detritus	10

tant than nitrogen content in some cases (Tenore, 1983). Detritivores therefore not only face the problem of obtaining sufficient nitrogen, but also have a problem obtaining enough carbon and energy. Even though detritivores may have the carbohydrases needed to digest detritus (cf. Section 10.1.3.4.3), little of the detritus may be assimilable (Table 7-1). Cammen et al. (1978) calculated that the polychaete *Nereis succinea* can only obtain 26–45% of its carbon requirement by feeding on detritus. Prinslow et al. (1974) found no growth for the killifish *Fundulus heteroclitus* on a diet of detritus alone, even though detritus was often found in their gut.

As we would expect, then, detritivores show lower assimilation efficiencies, lower for the most part than herbivores (Fig. 7-3). Detritivores can efficiently assimilate food of high quality, for example, microflora (Table 7-1), even though their assimilation of detritus may be low. Detritivores are therefore not physiologically restricted to low assimilation efficiency; it is the quality of the food that determines the low assimilation. The low quality of detritus as food for consumers is probably due to two features of detritus. First, if the detritus is derived from a vascular plant, it bears some of the antiherbivore compounds produced by the living plants. Although the soluble fraction of secondary compounds may be leached out of detritus, a large enough portion remains bound to cell walls in detritus to affect palatability, digestibility, and assimilability of detritus by detritivores (Harrison, 1982). Second, there is a marked loss of soluble nitrogen and other compounds from detritus soon after senescence of the parent plant (cf. Chapter 13);

this loss decreases the quality of detritus as food for consumers.

In view of the low quality of detritus as food, it seems likely that detritus feeders do not merely consume bulk detritus. They may be facultative predators, enhancing their diet with an occasional high-quality prey item ingested along with the detrital particles, or they may also selectively feed on nitrogen-richer fractions of detritus (Bowen, 1980; Odum et al., 1979).

The results compiled in Fig. 7-3 were calculated on the basis of organic matter, energy, or carbon, but the general results also apply to other specific nutrients. For example, the predaceous pteropod *Clione limacina* assimilates 99% of the nitrogen it consumes (Conover and Lalli, 1974), while the mussel *Modiolus demissus*, feeding on a mixed diet of phytoplankton and detrital particles, assimilates only 50% of the nitrogen consumed (Jordan and Valiela, 1982). In general then, ingestion of high-quality food leads to high assimilation efficiencies of most components of the food eaten.

7.2.2.2 Amount of Food

Assimilation efficiency is reduced as the amount of food ingested increases (Fig. 7-4). Excess feeding may waste time, energy, and food, and most consumers probably reduce feeding rates rather than simply ingest in excess and defecate what they do not use. For example, the rate of filtration by bivalves and zooplankton increases sharply as food abundance increases, up to some relatively high level of filtration determined by the maximum feeding effort exerted by the animal (Lam and Frost, 1976; Winter, 1978). If abundance of food particles increases further, the filtration rate remains at the constant maximum rate and provides a linearly increasing amount of food that can be ingested. At some higher food density, ingestion rate reaches the maximum number of particles that can be ingested and can pass through the digestive tract; if food density increased beyond this threshold, the rate of filtration is lowered, and the amount ingested remains near the maximum rate of ingestion.

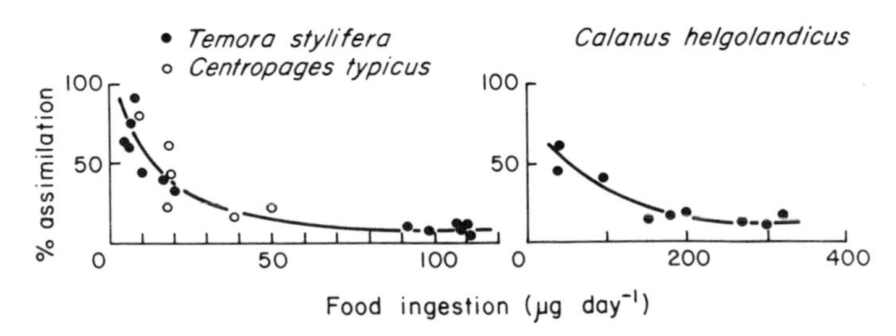

FIGURE 7-4. Assimilation of ingested food by the three copepod species in relation to the amount of food ingested. Adapted from Gaudy (1974).

Assimilation efficiency depends on the abundance of food particles up to the point where the maximum ingestion rate is reached; at higher densities of food, assimilation rate may be independent of food density, as occurs in copepods (Conover, 1978) and bivalves (Winter, 1978; Navarro and Winter, 1982). In general, there is probably some upper limit to the rates at which food can be assimilated, and at higher ingestion rates assimilation decreases and rate of defecation may increase. The hydroid *Clava multicornis*, for example, defecates 26% of food eaten when food is scarce and 39% when food is abundant (Paffenhöfer, 1968).

7.2.2.3 Age of the Consumer

Younger (or smaller) individuals generally show higher assimilation efficiencies than older specimens (Table 7-2). The exception to the

TABLE 7-2. Age-Specific Assimilation Efficiencies of Various Consumers.

		Assimilation efficiency (%)
Fiddler crab (*Uca pugnax*)[a]	1 yr (juveniles)	44.3
	2 yr (juveniles)	40.2
	3 yr (peak reproduction)	42.5
	3 yr (past peak reproduction)	34.5
Amphipod (*Orchestia bottae*)[b]	9 mg	50.6
	80 mg	30.4
Scallop (*Patinopecten yessoenis*)[c]	1 yr (prereproductive)	79
	2 yr (reproductive)	66
	3 yr (reproductive)	68
Crab (*Menippe mercenaria*)[d]	Zoea stages 1	42
	2	40
	3	49
	4	50
	5	71
	Megalops larvae	85
Blue whale (*Balaenoptera musculus*)[e]	Suckling	86
	Pubertal	79–83
	Adult	80
Fin whale (*Balaenoptera physalis*)[e]	Suckling	93
	Pubertal	79–83
	Adult	80
Sea urchins (*Strongylocentrotus intermedius*)[f]	1 yr	70
	2 yr	69
	4 yr	68

[a] Krebs (1976).
[b] Sushchenya (1968).
[c] Fuji and Hashizume (1974).
[d] Mootz and Epifanio (1974).
[e] Gaskin (1978), data of Lockyer.
[f] Fuji (1967).

trend (crab in Table 7-2) may be due to a switch from an algal to a carnivorous diet; choice of food thus confounds the age comparison. The need for high-quality food (and relatively greater quantities, cf. Sections 3.5, 7.4) may be an important factor in life histories and habitat use of marine animals. Many coastal fish, for example, use food-rich salt marshes and estuaries as nurseries (Werme, 1981), since this provides the juveniles with a better and more ample supply of food than is available in deeper waters.

7.3 Respiration

7.3.1 Measurement

Respiration of organisms is usually measured by the rate of oxygen consumed in either the water or the air that surrounds the organism. For species of very small size, a Warburg or Gilson respirometer, or oxygen microelectrodes can be used; in larger specimens the amount of oxygen consumed over a given time interval may be determined by an oxygen electrode or a Winkler titration (Edmondson and Winberg, 1971).

Respiration, food consumption, and energy use are roughly interconvertible using so-called oxycalorific coefficients. Very approximately, one liter O_2 consumed equals approximately 5 kcal of energy. With a respiratory quotient ($RQ = +\Delta CO_2/-\Delta O_2$) of about 1, 1 mg dry weight of food is equivalent to about 5.5 kcal of energy (Parsons et al., 1977). Elliot and Davison (1975) review oxycalorific coefficients of different substrates. Rates of respiration can be converted to carbon by using the relations

$$\text{mg } O_2 \text{ consumed per unit time} \times (12/34) \times RQ = \text{mg C used per unit time,} \quad (7\text{-}4)$$

$$\text{ml } O_2 \text{ consumed per unit time} \times (12/22.4) \times RQ = \text{mg C used per unit time.} \quad (7\text{-}5)$$

The values of *RQ*—determined mainly in mammals—vary from above 0.7 in organisms using fat for energy to 0.8 and 1 for consumers metabolizing proteins and carbohydrates, respectively.

7.3.2 Factors Affecting Respiration Rates

Respiration is an extremely variable process for two reasons: (a) it is markedly affected by many external factors and (b) its four components (Fig. 7-1) may all vary simultaneously and in different ways. Here we are concerned mainly with external factors such as temperature, size of the organism, level of behavioral activity, and biochemical composition of the source of energy.

7.3.2.1 Temperature

Transient increases in temperature prompt increases—often exponential—in respiratory rates (Fig. 7-5). Such increases are a very common effect of temperature on many biochemical and biological processes (Somero and Hochachka, 1976). The actual respiration rate at a given temperature varies with the species involved (Fig. 7-5, top left), but there are many examples of the general increase in respiration with temperature (Teal and Carey, 1967; Smith and Teal, 1973a,b, for example). There are temperatures beyond the range of acclimation of a species (Fig. 7-5, top left and top right), where temperatures are detrimental.

The response of a process, in this case O_2 use, to temperatures (t) can be represented by

$$Q_{10} = [r_1/r_2]^{[10/(t_1-t_2)]}, \tag{7-6}$$

where r_1 and r_2 are O_2 consumption rates at temperatures t_1 and t_2. The Q_{10} is an indication of the dependence of the rate in question on

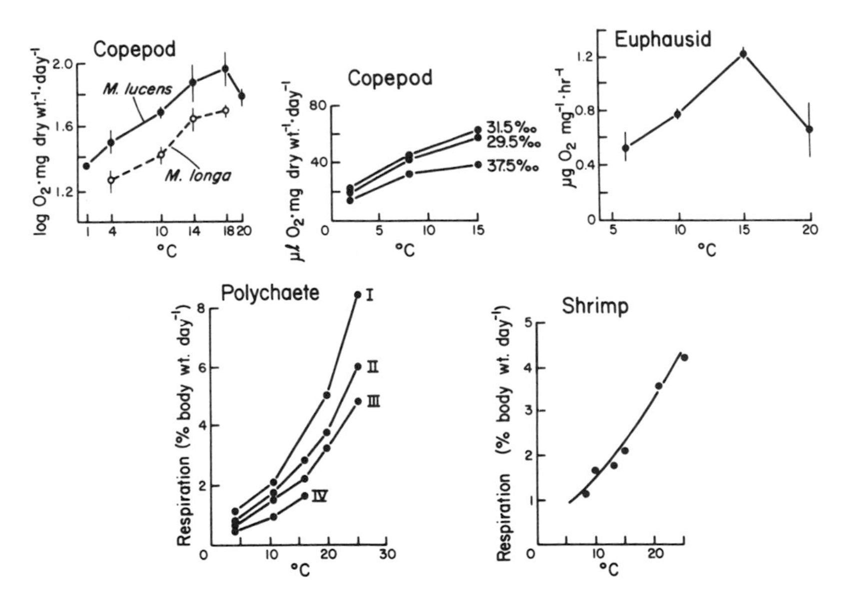

FIGURE 7-5. Respiration rate per unit biomass relative to temperature. Copepods: *Metridia lucens* and *M. longa*, from Haq (1967), and *Calanus finmarchicus* at three salinities, from Anraku (1964). Euphausiid: *Euphausia pacifica*, from Paranjape (1967). Polychaete: *Nereis diversicolor*, four size groups, I being the smallest, from Ivleva (1970). Shrimp: *Leander adspersus*, from Ivleva (1970).

temperature; it can be interpreted as the increase in reaction rate for a 10° change in temperature. Most biological rates have a Q_{10} between 2 and 3. In addition to the Q_{10} relationship, there are several other empirical formulas that nicely fit actual data (McLaren, 1963), among which is Belehradek's (1935) equation

$$\text{metabolic rate} = a(t + c)^b, \quad (7\text{-}7)$$

where t is the temperature, and the constants, a, b, c, are obtained empirically. A scale correction (c) shows the temperature at which the rate is 0, b determines the curvature of the line (in other words, the degree of dependence at different temperatures) and a is determined by the units in which the rate is measured.

The marked dependence of respiration on temperature stimulated comparisons of metabolic rates in species from cold and warm environments. Early measurements of metabolic rates of polar and tropical fish (Scholander et al., 1953; Wohlschlag, 1964) led to the conclusion that species of cold latitudes had a 5- to 10-fold larger respiratory rate. This elevated metabolic expenditure was taken to be the cost of maintenance at cold temperatures, and since maintenance costs were thus high, growth rates had to be low. This fitted well with the low rates of growth often recorded in polar species. It turns out that polar fish are rather sensitive, and the procedures involved in the measurements resulted in overestimation of standard respiratory rates* (Holeton, 1974). Subsequent careful experimental work has shown that standard metabolic rates of fish (Holeton, 1974) and invertebrates (Everson, 1977) of polar regions are very low. Other aspects of energy use—growth, activity, reproduction—are also low, so that the total food consumed is relatively low compared to that of similar species in warmer waters (Clarke, 1980). Freezing is avoided in Antarctic fishes by the presence of "antifreeze" metabolites (DeVries, 1988) and by other adaptations to low temperatures (Somero, 1991).

Other factors besides temperature affect respiration. Since osmoregulation may be an important metabolic expenditure, salinity of the medium may also affect respiration (Fig. 7-5, top middle). Pressure may increase respiration rates if an animal is deeper than its normal depth range. Invertebrates within their normal depth range have respiration rates unaffected by pressure, even if they live at great depths (Smith and Teal, 1973b).

7.3.2.2 Body Size

Total respiration rates increase as size of the organism increases (Fig. 7-6, top three graphs), but if respiration is expressed on a weight-

---

* See Section 7.3.2.3 for definition.

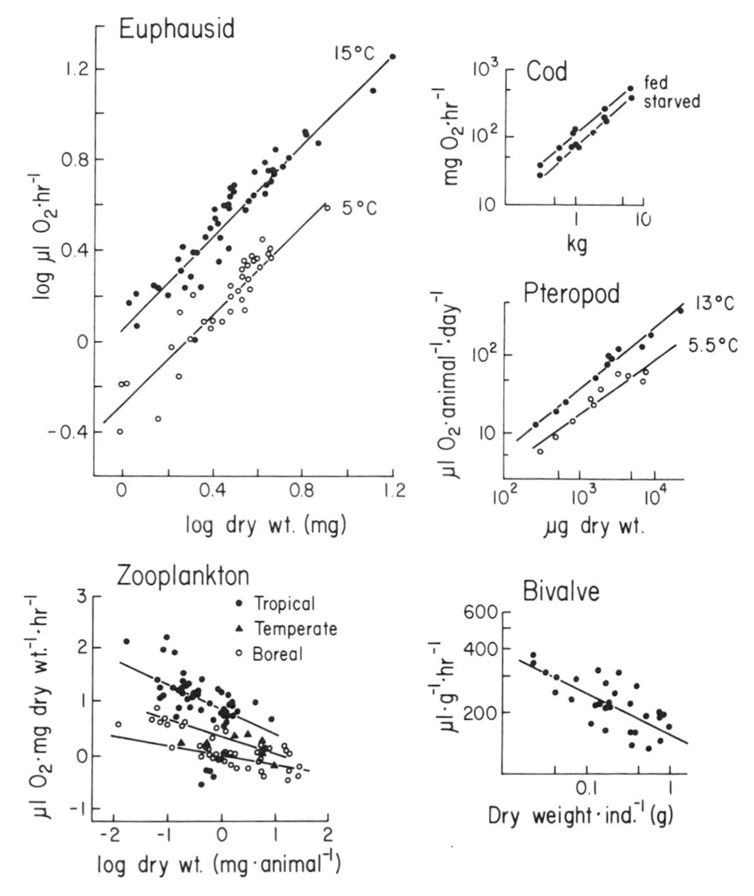

FIGURE 7-6. Respiration rates relative to size of organisms. The units in which respiration is expressed in the literature are perversely varied, as can be seen in these examples. The top three graphs show respiration rate on a per-animal basis; the bottom two graphs show respiration rate on a per unit weight basis. The euphausiid, *Euphausia pacifica*, at two temperatures, from Paranjape (1967). The pteropod *Clione limacina*, at two temperatures, from Conover and Lalli (1974). Cod (*Gadus morhua*), in recently fed and in starved individuals, from Saunders (1963). Seventy-five species of zooplankton from the Pacific, sorted out by geographic area, from Ikeda (1970). Bivalve data (*Scrobicularia plana*) from Hughes (1970).

specific basis (Fig. 7-6, bottom two graphs), smaller organisms have higher weight-specific respiration rates. Although there are exceptions to this rule, it is true for most species. The effect of size on respiration rates applies interspecifically (Fig. 7-6, bottom left) and intraspecifically (Fig. 7-6, bottom right). Smaller organisms may also be more subject to the effect of temperature, as evidenced by the divergence of the regression lines for the three geographic regions of Figure 7-6, bottom left.

The size dependence of respiration (R) for most organisms (Banse, 1976) can be described by

$$R = aW^{0.75}, \tag{7-8}$$

where a is a constant that may have different values for different taxa, and W is the size or weight. The exponent may vary somewhat for different taxonomic groups, but generally the values fall between 0.7 to 0.8. In fish, for example, this relation of metabolism to weight is (Winberg, 1971)

$$\text{rate of } O_2 \text{ consumption} = 0.3W^{0.8}$$

in ml O_2 hr^{-1} at 20°C. Moloney and Field (1989) review values for a and the exponent for various taxa.

In animals where growth takes place by stages that differ morphologically and physiologically the usual respiration-size relation may not apply. For example, in lobsters (*Homarus americanus*), late larval stages undergo very rapid changes, and to accommodate this metamorphosis increased energy expenditures are needed (Capuzzo and Lancaster, 1979). The respiration per unit weight in these larvae peaks during the later, larger stages when changes are most demanding. It is not clear whether this is unique to lobster larvae or more general.

7.3.2.3 Level of Activity

In warm-blooded species there is a "basal metabolic rate" set by the homeothermic mechanisms. For cold-blooded species there is no such thing, since body temperature depends largely on external heating, and it is usual to refer to the metabolic rate of unfed animals as the "standard metabolic rate." Short-term increases in behavioral activity (such as are produced by changes in prey abundance [Section 5.22]) markedly increase respiration above either basal metabolism or standard rates (Fig. 7-7). In the Pacific sardine (*Sardinops caerulea*) bursts of swimming activity may increase oxygen consumption by two or three times the rate observed when the fish is cruising slowly. Activity generally exceeds standard metabolic levels; in young fish an activity level of at least 1.5 times the standard metabolic level is required to obtain a maintenance ration (cf. Fig. 7-14, left). In king penguins the metabolic activity needed to forage for food to feed chicks plus maintain the parent bird is about two to three times the basal metabolic rate (Kooyman et al., 1982). In cod the average swimming speed depends on prey availability (expressed as food intake in Fig. 7-8). At low prey abundance and consumption, cues that stimulate hunting are few, so swimming is slow. Speed increases as encounters with prey increase.

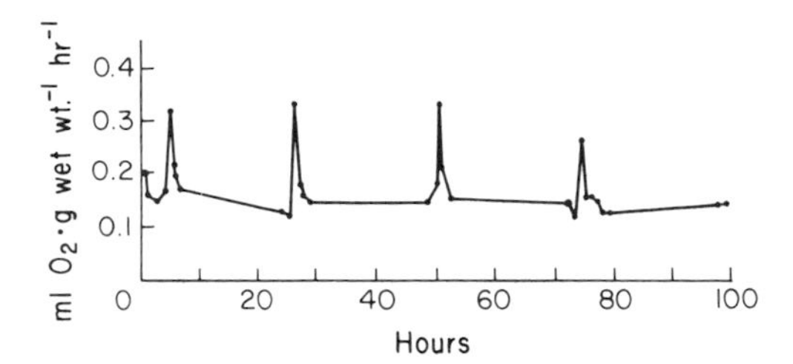

FIGURE 7-7. Rate of oxygen consumption by a single Pacific sardine (*Sardinops caerulea*) at normal cruising speed (baseline) during and after feeding (sharp peaks) on brine shrimp introduced into the tank. Prey introduced into the tank were quickly consumed, so that feeding lasted only a few minutes. Adapted from Lasker (1970).

At high prey densities, satiated cod swim slowly. The energetic costs of swimming parallel changes in swimming speed, and appear to be significant compared to standard metabolic expenditures.

Respiration rates also increase after feeding, but rates do not immediately return to basal metabolism after feeding is ended; there is lag

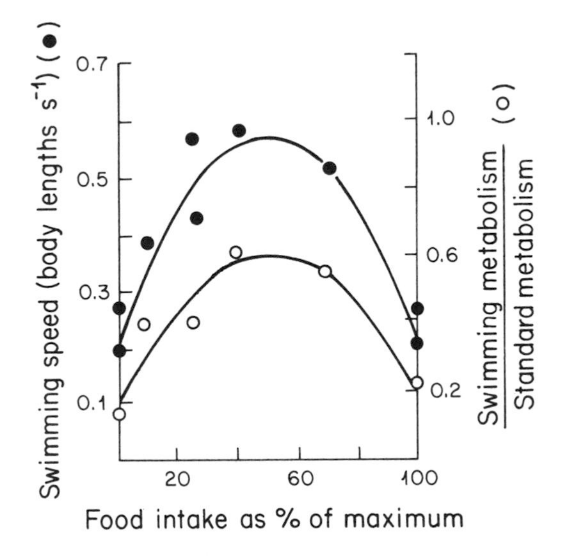

FIGURE 7-8. Average swimming speed and relative metabolic cost for cod (*Gadus morhua*) in experiments done in large tanks. Black circles: Swimming speeds averaged over periods during which cod were supplied with a range of different prey densities. Open circles: Ratio of rate of metabolic experditures during swimming to rate during standard metabolism (80 mg O_2 kg^{-1}h^{-1} for conditions during the experiment) versus food intake normalized to maximum intake. Adapted from Björnsson (1993).

of several hours, during which metabolic rates remain elevated, a phenomenon referred to as the specific dynamic action of food. Similar differences due to recent feeding can be found in many animals (Fig. 7-6, top right), and in fact, in certain fish that feed very intermittently the effects of a single feeding may last for 2 days (Conover, 1978). Calculations of the energy requirements of a species therefore requires some estimate of the level of activity and specific dynamic action over the time of interest.

Processes related to activity, such as the functional response to food abundance, take place at temperatures within the range of acclimation of the species. At low temperatures, for example, the activity of cold-blooded organisms or microbes is curtailed. Copepods in early spring may thus be unable to respond to algal densities, and do not consume many algae until the water warms up later in the season.

7.3.2.4 Source of Energy

The chemical makeup of the substrates used by animals to obtain energy can to some extent also affect the rates of respiration, since the respiratory quotients for different organic compounds vary. The oxidation of carbohydrates supplies the animal about 4 kcal g^{-1}, protein about 4–4.5 kcal g^{-1}, and fat about 9.5 kcal g^{-1}.

It is common to consider the rate at which O_2 is used relative to excretion of nitrogen (as ammonium) as a standardized way to compare metabolic rates of zooplankton (Conover, 1978). Use of protein should yield a O:N of about 8, while oxidation of average marine organic matter should produce a ratio of 17. Values of O:N for various marine invertebrates compiled by Conover (1978) range from 1.3 to 15.6. The low values in this range are probably due to the use of inadequate methods.

There are seasonal patterns of O:N ratios that convey a consistent picture: when zooplankton or bivalves (Bayne, 1973) are rich in lipids due to plentiful food during the spring bloom, respiration of protein is low and therefore O:N is highest. In winter, food is less available and lipid reserves are depleted, and the ratio decreases. Thus, within one species and one season, the relative respiration rate and O:N ratio can be higher or lower depending on the substrate within the organism used to supply energy. The ratio may also depend on the stage of development in invertebrates, if lipids become less important as a primary source of stored energy as the animal grows (Capuzzo and Lancaster, 1979).

Since internal lipid resources can be exhausted, the quality of the diet can effect respiration rate within relatively short periods of time. For example we have seen (Section 7.2.2.1) detritus is a low-quality

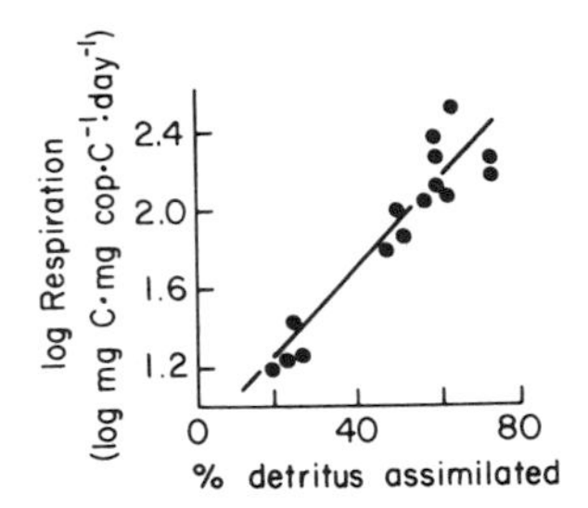

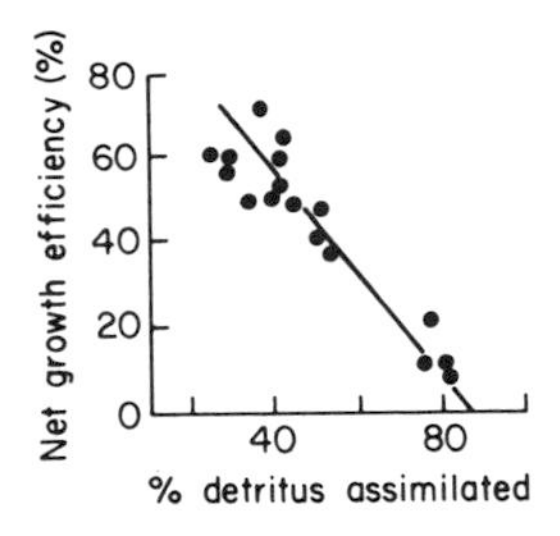

FIGURE 7-9. Effect of increased assimilation of detritus on the respiration and net growth efficiency of copepods. Adapted from Chervin (1978).

food, and the more detritus in the diet of the copepod, presumably the lower the lipid and protein content of the ration and the more energy expended to collect a sufficient ration (Fig. 7-9, left). It is not surprising then that respiration increases as more detritus is included in the diet.

7.4 Growth

7.4.1 Measurement

The growth of individual microbial consumers in the field is not measurable currently with any accuracy (Brock, 1971), but that of larger consumers has been measured extensively. The relative ease of measurement of individual growth of animals is very taxon specific. Increases in dry weight over intervals of time are the most direct measurement, but linear measurements can often be more easily done—for example, the width of carapace in decapods. Growth layers in mollusc shells, polychaete mandibles, whale teeth and earplugs, fish scales, and otoliths have all been used to determine the age and growth of individuals (Rhoads and Lutz, 1980). Such criteria for size and age are related to weight by an equation such as $w = ql^b$, usually in the linearized form

$$\log w = \log q + b \log l, \qquad (7\text{-}10)$$

where w = weight, l = length (or some other size criterion) and q and b are constants determined empirically; growth rate can be calculated from the slope of the line.

Many quantitative descriptions of growth have been based on equations of von Bertalanffy

$$L_t = L_\infty[l - e^{-k(t-t_0)}] \qquad (7\text{-}11)$$

where L_t = length at time t, L_∞ is an asymptotic or maximum length typical of the species, k is a constant that indicates the rate at which L_∞ is approached, and t_0 is the time at which growth starts. With slight modification, weights can be used in Eq. (7-11) instead of lengths, for instance, in the Gompertz equation

$$w_t = w_0 e^{G(1-e^{-gt})}, \tag{7-12}$$

where w_t and w_0 are the final and initial weights within the period of interest, G is the instantaneous rate of growth at t_0 and when $w = w_0$, and g is the instantaneous rate of decrease of the instantaneous rate of growth. The Gompertz curve allows for sigmoid growth while the von Bertalanffy model is a decelerated monotonic curve. At the inflexion point of the Gompertz curve g equals the instantaneous rate of growth. The differences among the various growth models such as Eqs. (7-11) and (7-12) are slight when compared to the variability of field data. These expressions assume constant weight changes over time, an assumption that may be unrealistic; nevertheless, the fit of such models to actual growth data is quite good (Ricker, 1975). Some examples that show how well theoretical growth curves such as Eq. (7-11) can fit actual data are included in Fig. 7-17 (top left).

7.4.2 Growth Efficiency

In comparisons within a species, growth rate is closely related to the net growth efficiency (K_2)* (Calow, 1977; Kiørboe et al., 1981), the proportion of assimilated food that is incorporated into growth,

$$K_2 = [G/(C - R)] \times 100. \tag{7-13}$$

Net growth efficiencies for many species of cold-blooded consumers range very broadly, but most consumers apportion 20–40% of assimilated energy to growth (Fig. 7-10).

Warm-blooded species are not included in Fig. 7-10. Their expenditures for maintenance are larger than for cold-blooded species, so that the proportion of assimilated food that is devoted to growth is much less than in fish and invertebrates. Marine birds and mammals therefore put relatively less assimilated energy into growth than other marine taxa. Cold-blooded species are thus better candidates for mariculture, since the interest there is to grow as much animal protein as possible with as little food as possible.

---

*Net growth efficiency is often labeled "K_2" in the literature; "K_1" is the gross growth efficiency, equal to (G/C) 100, and sometimes called "conversion efficiency."

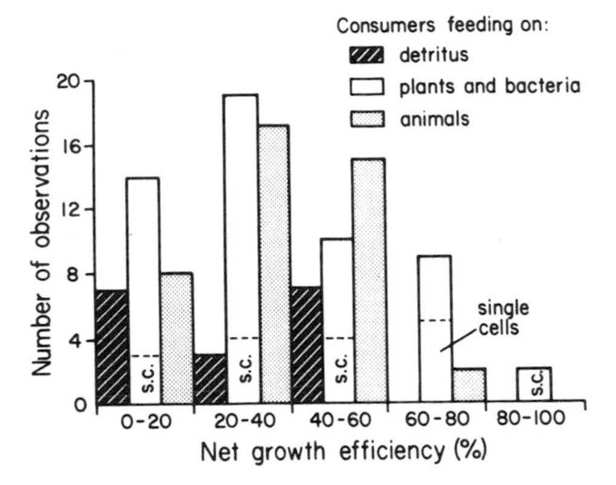

FIGURE 7-10. Net growth efficiencies for many species of consumers, including insects and other invertebrates, and vertebrates both in the field and in the laboratory. Most entries are for marine organisms, but some freshwater and terrestrial species are included. Single-celled foods are indicated in the lower part of bars in the plant category. Values from many individual papers and compilations by Paffenhöffer (1968), Trevallion (1971), Krebs (1976), Hughes (1970), Jones (1978), Shafir and Field (1980), Bougis (1976).

Bacteria as consumers are also not included in Fig. 7-10. Bacterial net growth efficiency does vary within and among substrate types (Table 7-3). Bacteria synthesize biomass more effectively in situations where they metabolize estuarine, rather than oceanic dissolved organic matter, and phytoplankton-derived particulate matter, rather than macroalgal and plant particulate matter.

TABLE 7-3. Ranges in Net Growth Efficiencies for Bacteria Consuming Different Organic Substrates.

Organic substrates	Net growth efficiency
Dissolved organic matter	
Amino acids	34–95
Sugars	10–40
Mussel exudates	7
Ambient DOM, estuaries	21–47
" ", oceanic	2–8
Particulate matter	
Phytoplankton	9–60
Macroalgae	9–15
Plants	2–10
Feces	10–20

[a] From Ducklow and Carlson (1992).

7.4.3 Factors Affecting Growth

7.4.3.1 Temperature

Growth, like most other physiological processes in animals, is markedly influenced by temperature (Fig. 7-11). Growth rates are low at the lowest temperatures, peak at more intermediate temperatures, and decrease again if temperatures are so high as to require large energy expenditures for maintenance (Fig. 7-11), salmon and clam examples). Differences in growth rate result in individuals of differing size.

Lower temperatures—within limits—may allow a reallocation of energy from respiration to growth. The colder water temperatures usually found at depth may be responsible for increases in size in taxa found deeper in the oceanic water column (Tseytlin, 1976, for example).

Higher temperatures abbreviate the duration of life history stages in cold-blooded organisms. In general, the warmer the water, the faster the development of eggs and juveniles, and the shorter the generation times (Fig. 7-12). The effects of temperature are quantitatively significant: for instance, an increase of 6°C can reduce the generation time of a chaetog-nath by nearly 2.5 times (Fig. 7-12, top right).

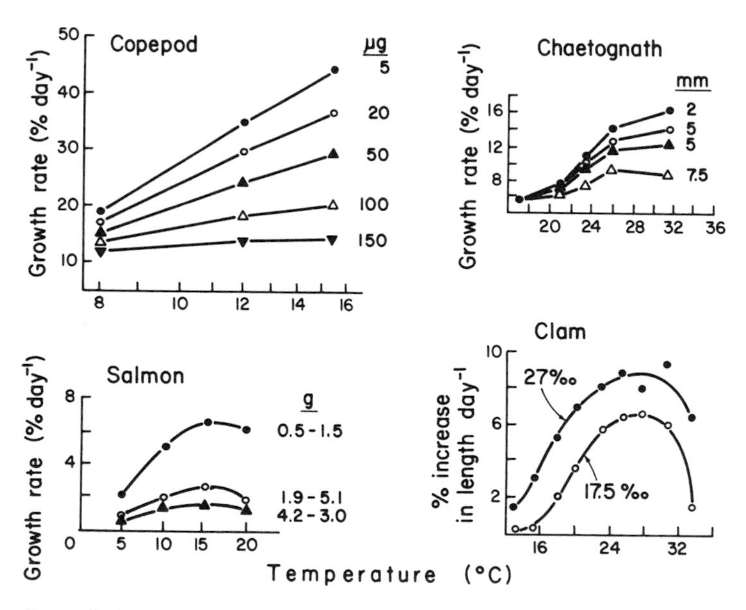

FIGURE 7-11. Influence of temperature on growth rate. Data for copepod (*Calanus pacificus*) from Vidal (1980), chaetognath (*Sagitta hispida*) from Reeve and Walter (1972), sockeye salmon (*Oncorhynchus nerka*) from Shelbourn et al. (1973), and quahog clam (*Mercenaria mercenaria*) at two salinities from Davis and Calabrese (1964). Data for salmon include values for fish of different weight. Data from chaetognath includes values for individuals of various lengths.

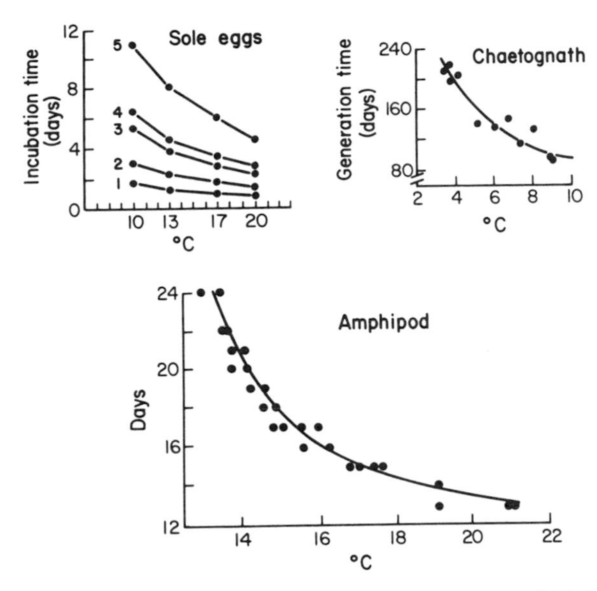

FIGURE 7-12. Effect of temperature on rate of development of life history stages. Incubation time (time elapsed before hatching) in five different stages in the development of eggs of sole (*Solea solea*), from Fonds (1979). Generation time of chaetognath (*Sagitta elegans*) from Sameoto (1971). Intervals between successive molts in adult female amphipods (*Gammarus zaddachi*) from Kinne (1970).

The result of all these temperature effects is that developmental and numerical responses by consumers such as zooplankton depend on temperature at times of temperature changes such as during spring. There are of course limits; if temperatures exceed the range to which a species is adapted, other species better suited to warm temperatures take over from the more cold-adapted species.

7.4.3.2 Abundance of Appropriate Food

Nutritive quality of food affects gross growth efficiency. In bacteria, for example, gross growth efficiency is at a near-threshold value of 40–50% when the C/N of the substrate is >6/1, and increases to 94% if the C/N of substrate is 1.5/1 (Goldman et al., 1987). Food type seems to have less of an effect on net growth efficiency (K_2) than it does on assimilation efficiency. There are no obvious differences in K_2 values for species feeding on animals or plants (Fig. 7-9). If food provides enough of a ration, the biochemical transformations of assimilated compounds are the same regardless of the origin of the compounds. Growth rates do however, depend on food quality. Relative increases in nitrogen content, for example, raise growth rates of deposit feeders (Fig. 7-13). In fact, growth is not sustained below a certain nitrogen

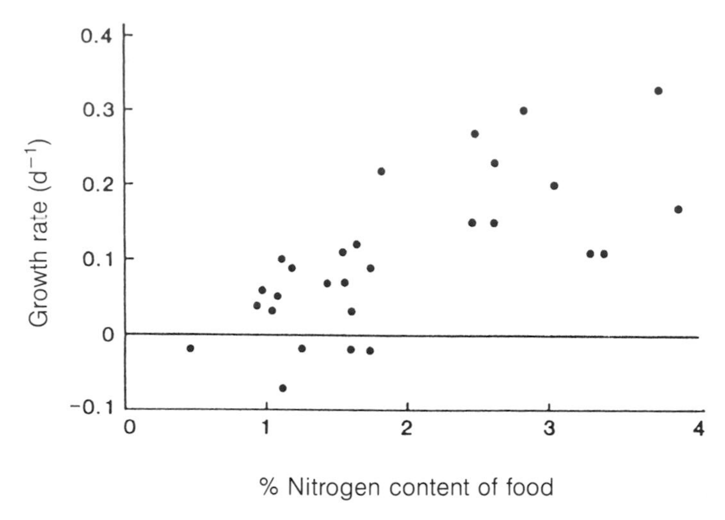

FIGURE 7-13. Growth rate of the deposit feeding polychaete *Capitella capitata* grown on foods containing different % nitrogen. Adapted from Smith (1991), data of K. Tenore.

content of food. Human detrital diets may not provide sufficient nutrition, as suggested by the somewhat lower K_2 values. When the growth efficiency of copepods was measured in experiments where the diet contained increasing amounts of detritus, K_2 was reduced (Fig. 7-8, right). Other studies with zooplankton confirm the reduction of growth under detrital diets as well as slower development and reduced survival (Paffenhöfer and Knowles, 1979). The nutritional inadequacy of detrital diets may be caused by insufficient nitrogen content, particularly the low supply of essential amino acids, such as histidine and phenylalanine (Marsh et al., 1989).

We have already discussed some aspects of the effect of ration size on growth under the guise of the developmental response to prey density (Section 5.4). Increases in rate of food consumption result in increased growth rates of consumers (Fig. 7-14), with a deceleration at higher rations. This is a measure of the cost of maintenance of larger body sizes.

Growth efficiency is negatively related to ration in laboratory experiments. Within any one size class of young salmon, the growth efficiency is lower at higher rations (Fig. 7-15). Fish that have more food items available tend to be more active (Kerr, 1971).* Since the energetic cost

---

*We have already seen this pattern in Section 5.2.2, when discussing search modes created by different prey densities. An extreme manifestation of this phenomenon for hungry predators is the "feeding frenzy" that takes over many predators when suddently exposed to a very dense aggregation of prey.

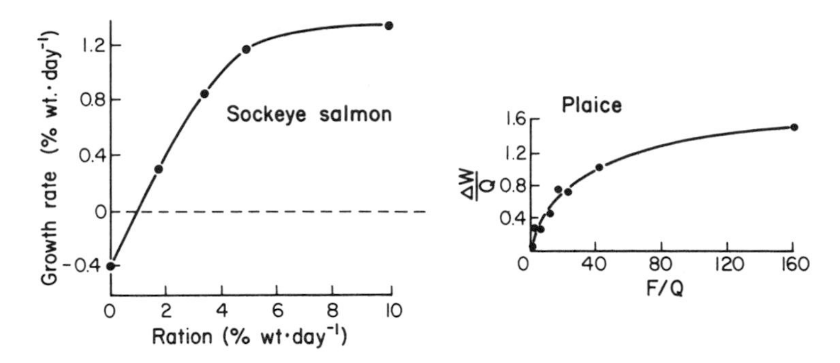

FIGURE 7-14. Growth rate in relation to food consumed. Left: Sockeye salmon (*Oncorhynchus nerka*) in culture. Adapted from Brett et al. (1969). Right: Young plaice (*Pleuronectes platessa*) feeding on siphons of the bivalve *Tellina tenuis*. Growth (ΔW) and food consumption are both normalized to Q, the standard metabolic rate. Adapted from Edwards et al. (1970).

of swimming in fish varies approximately as the square of the velocity (Fry 1957), the respiratory requirements increase more than proportionately to level of activity. It is not clear, however, whether this phenomenon of lower growth efficiency at higher prey densities occurs in the field.

As consumers increase in size, the ratio of growth to consumption is lower and the effect of ration on growth efficiency is considerably less (i.e., the slopes of the so-called "K_2-lines" flatten out at lower K_2 values, Fig. 7-15). This is probably due to the greater total metabolic costs of large individuals.

It should be evident that curves such as those on Figure 7-14, based on single food types in the laboratory, do not convey the complete picture of how growth is regulated. Size (Section 6.31) and spatial

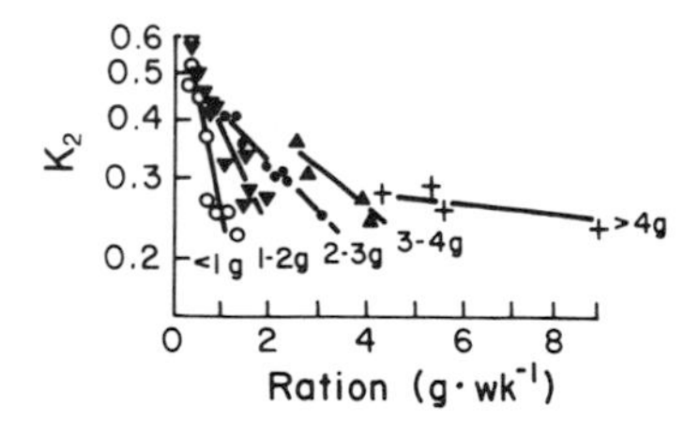

FIGURE 7-15. Net growth efficiency (K_2) of chum salmon (*Oncorhynchus keta*) in the laboratory at different weekly food rations. The straight lines refer to young salmon in 1-g size classes. These "K_2-lines" have been obtained only for the right side of the curves; there must be extensions to the left that have a positive slope, since K_2 is 0 when ration available reaches a certain minimum threshold. Adapted from Kerr (1971).

distribution (Section 11.4) as well as abundance of food particles play some role in setting growth rates.

No one has documented all the variables affecting growth efficiency, but computer simulation studies show that based on what is known of fish feeding, growth—expressed as K_2—is best when relatively large prey are available, even if they are rare (curve for 100 g prey, Fig. 7-16, left). Small prey can support growth of the predator if the smaller prey are relatively abundant (compare the curves for prey of 0.1 g, Fig. 7-16, left). A diet of small and rare prey demands increased metabolic expenditures because of the high level of activity needed to gather enough prey. In such circumstances energy is therefore devoted to maintaining activity rather than to growth, and growth of the predator is therefore curtailed.

7.4.3.3 Size of Consumer

Growth of individuals decelerates as individuals age (Fig. 7-17). In both vertebrates and invertebrates there are physiological mechanisms that reduce growth per unit weight as time passes. Other physiological controls make for differences among sexes (Fig. 7-17, bottom). The effects of size have been thought to apply not just to stages within the life span of a species but interspecifically as well, since in general, small species grow faster than larger species (Fenchel, 1974). Growth efficiency, however, does not vary with size in comparisons among species (Banse, 1979), although it is not clear why this is so.

In many crustaceans, as animals reach adulthood, the yearly growth may be considerable but is converted to reproductive products and cast molts (Krebs 1978), so the net annual growth for the individual is close

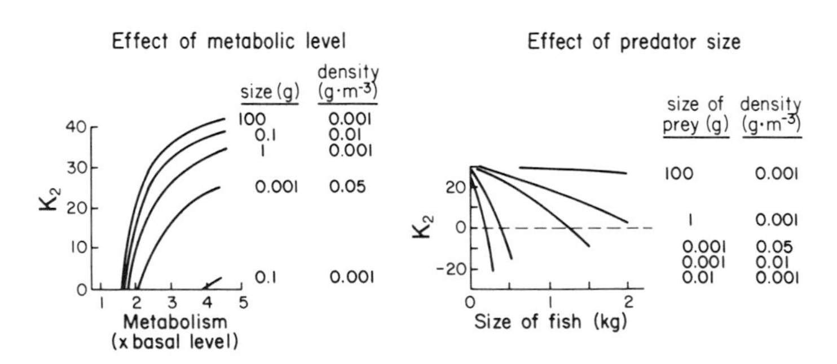

FIGURE 7-16. Results of computer simulation studies on the effects of metabolic level and size of brook trout (*Salvelinus fontinalis*) on net growth efficiency (K_2). Left: K_2 as function of metabolic level for a 1-kg trout for different sizes and abundance of prey. Right: Changes in K_2 as a predator grows, for different combinations of size and abundance of prey. Metabolic level is assumed to be 2.5 × basal metabolic rate. Adapted from Kerr (1971a).

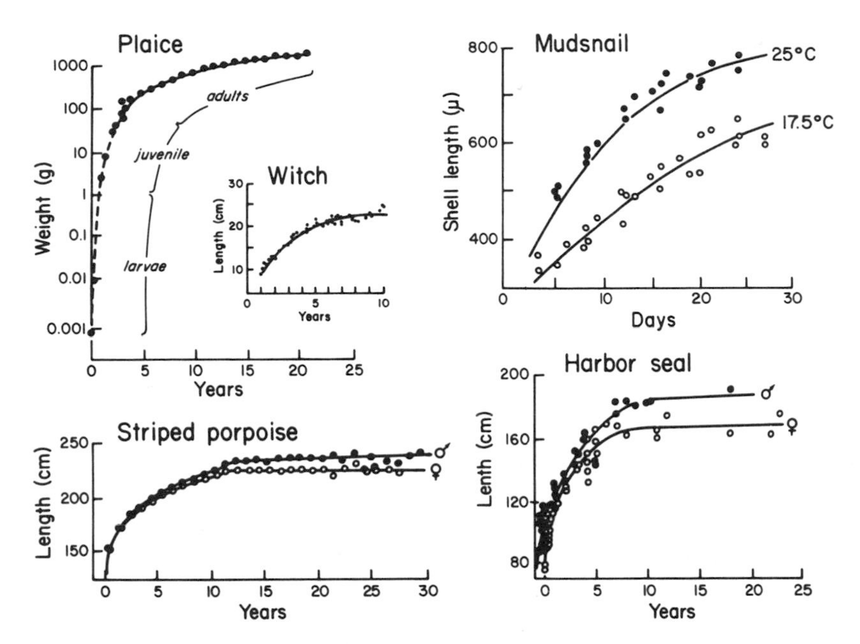

FIGURE 7-17. Growth of consumers over time. For plaice (*Pleuronectes platessa*) the curve fit by von Bertalanffy equation is shown as a complete line; smaller specimens do not fit equation. Adapted from Cushing (1975) after Beverton and Holt. Witch (*Glyptocephalus cynoglossus*) data for female fish only, fitted by a von Bertalanffy equation by A.B. Bowers. Mud snail (*Nassarius obsoletus*) from Scheltema (1967). Striped porpoise (*Stenella coeruleoalba*) from Miyazaki (1977). Harbor seal (*Phoca kurilensis*) from Naito and Nishiwaki (1972).

to nil. In many animals the reduction in growth rate with age within a species may come about by the increased food energy required to supply maintenance expenditures by larger individuals (Section 7.322).

7.4.3.4 Interactions Among Variables

In Chapter 2 we discussed the interacting variables that determine rates of primary production. For consumers there are also important interactions among the variables discussed in this chapter.

Temperature and size. Smaller specimens of a species generally grow proportionately faster than larger individuals as temperature increases. This can be demonstrated in serpulid polychaetes (Dixon, 1976), and can be seen in the growth rates of mud snails of various sizes exposed to different temperatures (Fig. 7-17, top right).

Temperature and ration size. At low temperatures growth rates are low (Fig. 7-18). As temperatures increase so do growth rates, but the increase is greatest where food is abundant; loss of weight may occur

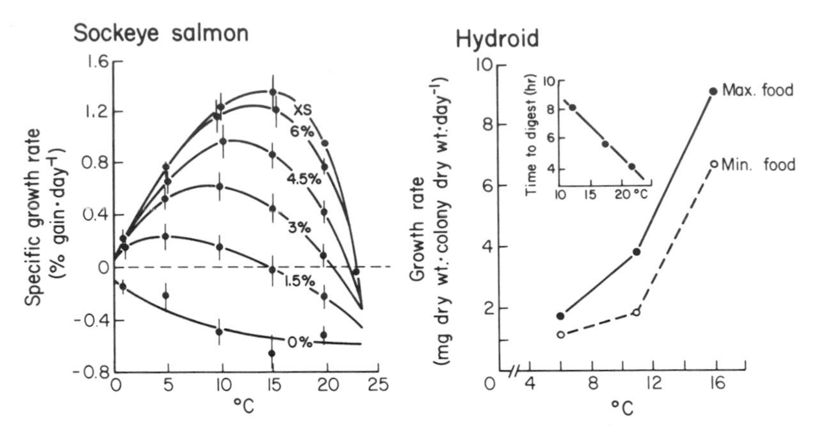

FIGURE 7-18. Combined effects of temperature and food supply on growth rate. Left: Young sockeye (*Oncorhynchus nerka*) grown in the laboratory at rations of 0–6% of body weight per day and at excess (xs) ration. Means ± standard deviation. Adapted from Brett et al. (1969). Right: Growth rate of the colonial hydroid *Clava multicornis* in the laboratory at various temperatures and a maximum level of consumption and at the lowest level that allowed growth. Adapted from Paffenhöfer (1968) and Kinne and Paffenhöffer (1965).

where food is scarce (Fig. 7-18, left). The increase in growth at higher temperatures reflects the general increase in metabolic activity as, for example, the increased rate at which digestion takes place in warmer temperatures (Fig. 7-18, right, inset). The combined effect of ration size and temperature producs a peak growth rate which shifts to the right in peak growth at each ration. This shift is an expression of the balance between the increased growth due to higher physiological activity of higher temperatures and the greater amounts of food needed at higher temperatures. At the highest temperatures, the resulting increased respiration reduces growth.

Consumer size in relation to prey size and density. Kerr (1971a) followed changes in growth efficiency in growing fish in his simulation study (Fig. 7-16, right). For the smallest fish, growth is possible (K_2 is positive) even on diets of small and rare prey. As growth takes place, the fish move toward the right, and the "K_2 lines" slope downward. There is a point where a fish needs to seek either larger or more abundant prey to maintain a positive K_2. The largest fish can only sustain growth on large prey. This is the metabolic basis behind the pervasive tendency for size selection in the feeding behavior by predators (Chapter 6).

Temperature, ration size, and size of consumer. Growth rates of *Calanus pacificus* increase with temperature and decrease as body size increases (Fig. 7-11, left). Note, however, that the slopes of the lines decrease with larger copepods, so that the growth response of large copepods to temperature is much reduced. This may be the result of the amount of food required to sustain maximum growth increases with temperature and size of copepods. Food densities capable of sustaining maximum growth for small copepods are therefore insufficient to support maximum growth of large copepods. It may be, therefore, that at low food concentrations the growth rate of small copepods is not determined by food but by temperature, while in large copepods food supply determines growth rate (Vidal, 1980). In fact, larger copepods may attain their largest size at low temperatures if enough food is available. This very different response of large and small copepods to temperature gradients may in part explain why there have been reports of reduced body sizes as temperature increased (McLaren, 1963; Sameoto, 1971), while other studies show no such relation (cf. review in Vidal, 1980).

Oligotrophic tropical waters have mainly small species of zooplankton while rich temperate waters hold communities of larger zooplankton. It may be that small species predominate in depauperate environments, where they can grow despite low food supplies, but only do so at high temperatures. Larger species may be excluded from oligotrophic warmer water and may do best where food supply is high, even if temperatures are low (Vidal, 1980).

7.5 Production

Net production rate of a population is the number of individuals present times the net growth in these individuals. In microorganisms that divide by fission, production is just the increase in number of cells, since growth of individual cells is difficult to measure. More complicated modes of growth and reproduction add further components to production, as pointed out in Section 7.1. In crustaceans, for example, net production is the sum of growth, reproduction, and molting. The measurement of production must therefore include the relevant components, and the factors that affect production are those that apply to its component parts.

7.5.1 Measurement of Microbial Production

The field of microbial ecology has burgeoned over the past decade, stimulated by the availability of new methods to measure abundance and activity of microbes, new knowledge about heretofore poorly or unknown groups of organisms (cf. Section 1.1), and new ideas as to the role of microbes in aquatic ecosystems (cf. Chapter 9). The methodological advances are reviewed in great detail by Karl (1986) and in the handbook edited by Kemp et al. (1993).

Hobbie et al. (1977) provided the first method that showed that bacteria were considerably more abundant than previously thought. The epi-illuminated counts of appropriately stained and fluorescent cells furnished data that demonstrated far larger natural abundances of field populations than estimates obtainable by previous methods that depended on culture media. Other stains have also been used (Porter and Feig, 1980).

Several approaches that have been used to assess rate of bacterial production (Karl, 1986; Kemp et al., 1993). The principal approach has been to add a radioactively labeled substrate, and measure uptake by the microbes. Since all species of bacteria do not use a single substrate, it was not possible to find a convenient analog to the ^{14}C-bicarbonate incorporation method used for algae. Instead, specific substrates have been used, such as ^{3}H-thymidine, a precursor of DNA in bacteria, and ^{3}H-leucine, which is converted to protein.

The thymidine and leucine incorporation methods have produced most of the newly available estimates of bacterial production. There methods are based on the knowledge that proteins make up a fairly constant 60% of bacterial biomass, so that rates of protein synthesis can be converted to biomass production, but there is still controversy about these methods. For the thymidine method, there are still questions as to how much of the thymidine may be synthesized internally rather than originate from outside the cells, how many compounds other than DNA may take up label, and there is concern about the large effects of errors inherent in the calibration factors required to convert uptake of thymidine to carbon production. Potential problems with the leucine method are that leucine may be manufactured intracellularly from other componds, and that there may be protein synthesis and degradation independent of growth. In addition, the leucine method shares the difficulty of finding appropriate conversion factors.

Thymidine and leucine uptake have been applied mainly to aerobic situations, but seem applicable also to anaerobic conditions (McDonough et al., 1986). The needed conversion factors, however, are likely to differ from those used in aerobic conditions.

The thymidine and leucine methods have also been used to assess bacterial production in sediments (Findlay, 1993). The principal problems in sediments are the need to disturb the sediments to introduce the label and suitable extraction of DNA or protein from the sediments.

A completely different approach is to calculate population growth rates from censuses of dividing bacerial cells (Hagstrom et al., 1979). Bacteria divide by fission, and it is relatively easy to distinguish cells that are dividing. The proportion of cells that are dividing, relative to the total population, provides a measure of growth of the population. This method has produced estimates similar to those obtained using thymidine and leucine uptake methods (McDonough et al., 1986; Pedrós-Alió et al., 1993).

7.5.2 Methods for Animal Production

Reviews of methods to measure biomass and production of invertebrates and fish can be found in (Greze, 1978; Southwood, 1978; Holme and McIntyre 1971), which provide the material below.

7.5.2.1 Cohort Methods

Where distinct cohorts are evident, in any given time interval $t_0 - t$, production (P_{t-t_0}) is the increase in biomass due to growth ($B_4 - B_{t_0}$) plus the biomass lost due to predation and other mortality (B_m)

$$P_{t-t_0} = B_t - B_{t_0} + B_m. \tag{7-14}$$

This general approach was first used by P. Boysen-Jensen and involves measurement of biomass at different times. B_m is calculated as the sum of the products of the number of individuals at t_0 by the weight of the individuals at t. This provides the biomass of animals not present at the time of measurement.

K.R. Allen and V.N. Greze independently improved the Boysen-Jensen method, and calculated production of a cohort as

$$P_{t-t_0} = (n_t - n_{t_0})[(w_t - w_0)/2], \tag{7-15}$$

where n and w are the numbers and average weights at the beginning (t_0) and end (t) of the interval of measurement.

The time over which production is measured needs to be short relative to changes in the life history of the species, since both growth and mortality can be exponential and the use of linear interpolations such as ($n_t - n_0$) or ($w_t - w_0$) would not adequately assess rates over relatively long periods of time. Parsons et al. (1977) recommend that measurement intervals should be about 10% of the generation time of the species.

7.5.2.2 Cumulative Growth Methods

Where there are no clear cohorts and reproduction is continuous or repeated, production can be calculated based on the growth curve of the species. The basic idea is to obtain the product of number of individuals and average growth increment over a certain interval, and then sum over the entire life history (up to the kth age or size classes):

$$P = \sum_{i=1}^{k} n_i g_i, \tag{7-16}$$

where g_i = the mean increase in weight during the interval for each age class. Values of g_i should be obtained by direct measurement. Values of g_i can be calculated from studies of net growth efficiency (K_2) and respiration (R) as

$$g_i = R[K_2/(C - K_2)], \tag{7-17}$$

where $R = aw^b$, but this is less desirable. This latter procedure based on efficiencies should be only used as a check on production rates determined by other methods, since R and K_2 are rarely constant over all the age groups. The calculation of growth is more realistic if the effects of temperature are included by using curves such as those in Figure 7-11 or 7-18. The cumulative growth method is more frequently applicable than cohort methods (Kirmmerer, 1987).

7.5.2.3 Life Table Methods

Elster (1954), Edmondson (1960), Hall (1964), and others have used life table methods to calculate the growth of populations of freshwater zooplankton based on number of eggs and their development time relative to ambient temperature. The calculation of number of eggs produced by the population (cf. review by Paloheimo, 1974) and their rate of development allow the estimation of N_t, the potential increase in numbers of the population within a period of time $(t - t_0)$. The difference between the potential (N_t) and actual ($N_a = N_t - N_0$) increase in the population is the mortality (N_m), which can be calculated as

$$N_m = N_t - N_a = [n_{eggs} \times 1/\text{development time}] - N_a. \tag{7-18}$$

Multiplication of N_t by mean weight per individual provides estimates of total production. Heinle (1969) applied this approach to measure production in copepods, including naupliar and copepodid stages.

7.5.2.4 Energy Budget Approach

Another way to measure production is to measure consumption, respiration, egestion, and excretion, and from this calculate production using Eq. (7-1). This procedure is cumbersome, time-consuming, and subject to large errors. The inherent variability is large because the error of the estimate of production is the cumulative error associated with measurement of each of the components. There are additional ambiguities, such as whether to enter reproductive products separately or as part of growth. In most cases the growth term will have included biomass of reproductive products as the eggs and sperm developed within the adults. As with all the other methods, many assumptions need to be made as to conversion coefficients and the environmental conditions, for example, temperature, during the time interval of interest. Identification of the diets of consumers in the field are seldom well known, and many assumptions usually have to be made prior to the calculation of consumption.

A rough way of estimating production is to measure O_2 consumption or CO_2 production—in other words, to measure respiration—and then calculate production using an empirically derived relationship to production, such as that developed by Engelmann (1961)

$$\log R = 0.62 + 0.86 \log P, \tag{7-19}$$

or by McNeill and Lawton (1970)

$$\log P = 0.8262 \log R + 0.0948. \tag{7-20}$$

where R and P are respiration and production.

This was originally done for terrestrial species but fits marine animals tolerably well (Hughes, 1970). The application of log-log relationships such as Eqs. (7-19) and (7-20) needs to be interpreted with care, since such straight lines drawn with logarithmic scales may hide large absolute deviations at the upper end of the scales. The usefulness of respiration as a predictor of production is in part due to respiration consuming a very significant proportion (70–80%, cf. Section 7.6) of the consumed food. A relation between production and respiration thus seems reasonable and may be useful as a measure of total consumer activity. This approach has been applied to faunas in sediments of very deep oceans, where no other method to measure production can be applied (Smith and Teal, 1973a,b).

7.5.2.5 Comparisons of Methods to Measure Secondary Production

Each of the various methods discussed above is subject to large errors, and it is necessary to make many assumptions when using all the methods. In fact, consumer production measurements are some of the most variable and error-prone of the variables in ecology. Other approaches to measure consumer production have been suggested, for example, use of ^{14}C-labeled food (Chmyr, 1967), but these also require many assumptions and ancillary measurements. Methods of measure primary production apply to the whole assemblage of species engaged in photosynthesis, in contrast to methods to estimate secondary production, which generally measure production of a specific population.* This requires that ancillary data on numbers, size specific growth rates, etc. be gathered. This adds to the difficulties, since the reliability of the production measurement, in addition to the problems inherent in each procedure, also depends to a large extent on the quality of the additional data sets needed. In spite of all these caveats, Greze (1978) after reviewing a few studies where production was measured on the same population by two or three different methods, concluded that the

---

*One exception to this is that measurement of O_2 consumption or CO_2 production can provide assessment of respiration by the assemblage of consumer populations present. This approach has been used to measure "community respiration" in marine benthic communities, as will be seen in Chapter 13.

results were roughly comparable. Moran et al. (1989) reached similar results. They show that appropriate sampling during periods of high production is crucial. When choosing a method it is necessary to consider carefully all the methods suitable for the population to be studied, and then to carefully design a research program based on the idiosyncrasies of the specific case.

7.5.3 Biomass and Production of Consumers

7.5.3.1 Planktonic Consumers

Epifluorescence microscopy methods (Hobbie et al., 1977) plus the use of electron microscopy have provided estimates of bacterial number and biomass for seawater and sediments (Meyer-Reil, 1982). The range of abundance is relatively narrow (10^4 to 10^6 cells ml^{-1} in seawater, equivalent to 1–200 μg C $liter^{-1}$). Most cells (80–90%) are probably free bacterioplankton, while the rest are attached to particles. Most of these bacteria are taken to be heterotrophic. Other methods to estimate bacterial biomass involve measuring the abundance of compounds specific to bacterial cells, such as certain lipopolysaccharides (Watson et al., 1977) or muramic acid (Moriarty, 1977),

Ducklow and Carlson (1992) reviewed many measurements of bacterial production in marine waters and concluded that the different methods yielded similar values, as also found by Cole et al. (1988). Their survey also showed that although individual estimates ranged over seven orders of magnitude, from 0.0004 to over 2,000 $mg\,C\,m^{-3}\,day^{-1}$, average values for different habitats spanned a narrower range, 2–119 $mg\,C\,m^{-3}\,day^{-1}$.

Bacterial production is on average, only 10–30% as large as autotrophic production (cf. Fig. 7-19). In a few instances, bacterial production is 2–25 times as large as phytoplankton; this suggests that in those sites there is an additional external carbon source. In marine environments dominated by macroalgae, bacterial production may amount to 6–33% of macroalgal production (Newell et al., 1981; Stuart et al., 1982).

Production by the larger planktonic consumers varies tremendously (Fig. 7-19). Some rough estimates of the biomass and production of microzooplankton are provided in Table 7-4. Note that production is higher in the nutrient-richer coastal waters than in more oligotrophic deeper waters. To make comparisons it is convenient to take the ratio of production to the average biomass; this is often not a reliable measure, since, for example, there are large seasonal variations in biomass. Nonetheless, comparisons of P/B are instructive, since this ratio can be thought of as the turnover rate of these populations (Fig. 7-19).

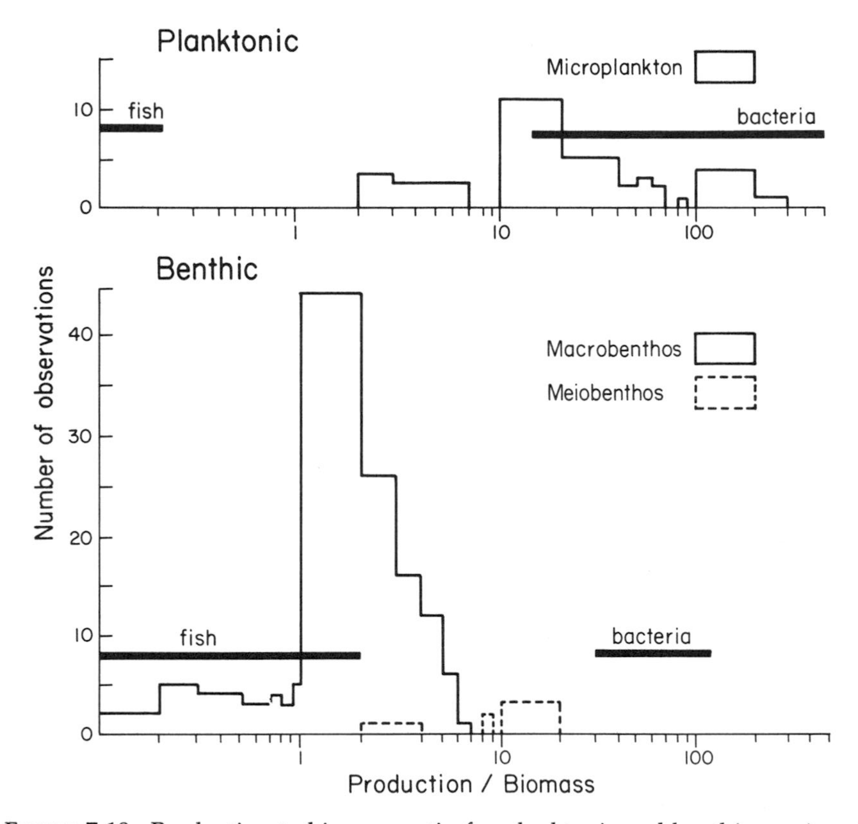

FIGURE 7-19. Production to biomass ratio for planktonic and benthic species. A few freshwater species are included; their *P/B* did not differ from those of related marine species. The ranges of the few available estimates for bacteria and fish are included as bars to show how they compare with invertebrates. Data compiled from Allan et al. (1976), Ankar (1977), Banse and Mosher (1980), Berkes (1977), Burke and Mann (1974), Cederwell (1977), Conover (1974), DucKlow and Carlson (1992), Fleeger and Palmer (1982), Feller (1977), Gerlach (1978), Hibbert (1977), Josefson (1982), Peer (1970), Siegismund (1982), Mullin (1969), Lasker et al. (1970), Warwick and Price (1975, 1979), Sanders (1956), Mills (1980), and Warwick (1980).

The *P/B* of the few available—and controversial—measurements for bacteria indicate a turnover of 100–400 times a year (Fig. 7-19, top). This is faster than most turnover rates for zooplankton, including copepods, cladocerans, and euphausiids. Zooplankton have a wide range of *P/B*, but the modal turnover rate is about 10–20 times a year. Unfortunately there are no measurements for the planktonic ciliates and flagellates; their *P/B* should lie closer to the bacterial values than to the larger zooplankton. Measurements for pelagic fish are few, in spite

TABLE 7-4. Biomass and Production of Microzooplankton and Fish from Various Marine Regions.[a]

	Depth[b] (m)	Biomass (g dry wt. m^{-3})	Production (g C m^{-2} yr^{-1})
Zooplankton			
Inshore waters	1–30	122	15.3
Continental shelf	30	25	6.4
Shelf break	200	108	5.5
Open Sea	200	20	5.7
Benthos		(g C m^{-2})	5.3–17
Estuaries	0–17	5.3–17	0.7–12
Coastal seas	18–80	1.7–4.8	2.6
Continental shelf	0–180	23	2.4
Continental slope	180–730	18	—
Deep Sea	More than 3,000	0.02	
Pelagic Fish			
Continental shelf	0 – 180	2.6	0.3
Continental slope	100–730	10.6	1.3
Demersal Fish			
Continental shelf	0–180	8.6	0.3
Continental shelf	180–730	4	0.2

[a] Adapted from Mann (1982), Mills (1980), and Whittle (1977). Note the difference in units in biomass of zooplankton and other entries. Fish production calculated roughly from catch statistics. Demersal fish live and feed near the sea floor.
[b] Depth to which sampling took place.

of the extensive fishery work. Production by pelagic fish species is relatively low (Table 7-4) and *P*/*B* values are small (Fig. 7-19, top).

7.5.3.2 Benthic Consumers

Bacteria are more abundant in sediments (10^{10}–10^{11} cells ml^{-1}) than in the water column although the biomass of sediment bacteria turns over more slowly (Fig. 7-19 bottom) than that of planktonic bacteria. Most sediment bacteria may be relatively inactive. This needs further study. Other microbes such as fungi and yeasts also need study.

Consumers larger than bacteria (3–4 to 45 μm) are called microbenthos; these are mainly ciliates and some flagellates. Organisms between 45 and 500 μm in diameter (nematodes, foraminifera, harpacticoid copepods, some ostracods, and polychaetes) are called meiofauna, while the macrofauna are larger than 500 μm—or 300 μm if we wish to include more of the immature stages. Macrofauna are very diverse, with amphipods, various annelids, and bivalves being typical representatives. These size categories are to an extent arbitrary, but Schwinghamer (1981) shows that the biomass of marine benthic organisms peaks at size ranges of 0.5–2 μm (bacteria), 8–32 μm (meiofauna), and 32–250 μm (macrofauna). Thus, the classification does have some real meaning.

As in the plankton, benthic biomass and production are very variable (Table 7-4). Biomass and production are higher in shallower waters and decrease at great depth. The few available *P* to *B* ratios of meiofauna are considerably larger than those for the larger macrofauna (Fig. 7-19 bottom); the meiofauna turn over about 10 times per year, while the modal macrobenthic turnover rate is one to two times a year. When data become available, the microbenthos and bacteria *P*/*B* will probably lie to the right of both macrobenthos and meiobenthos. Vernberg and Coull (1974), for example, estimate that the turnover rates of ciliates in shallow marine sediments is between one and four orders of magnitude greater than that of the meiofauna and macrofauna. Some *P*/*B* ratios of fish associated with the benthos are included in Figure 7-19, bottom; these species turnover rather slowly, perhaps once a year, and production is smaller (Table 7-3) than that of smaller-bodied consumers. The larger *P*/*B* ratios for fish included in Figure 7-19 are from rich estuaries; fishes of deeper water have very low *P* to *B* ratios.

The faster turnover—the result of growth and mortality—of smaller organisms (Fig. 7-21, left) means that although the biomass of small-sized species may be smaller than that of larger species, the higher specific production—what we referred to above as the *P*/*B*—of smaller species often makes them proportionately more important producers than larger species (Fenchel, 1969; Gerlach, 1978; Vernberg and Coull, 1974).

7.5.4 Factors Affecting Production

Anything that affects growth, reproduction, metabolism, or molting affects production. Temperature, size of individuals, and food quality and quantity are therefore important since they affect growth, as we have seen earlier in this chapter. For example, weight-specific excretion increases with temperature (Mayzaud and Dallot, 1973; Ikeda, 1974), so that it is not surprising to find also that production depends on temperature (Fig. 7-20).

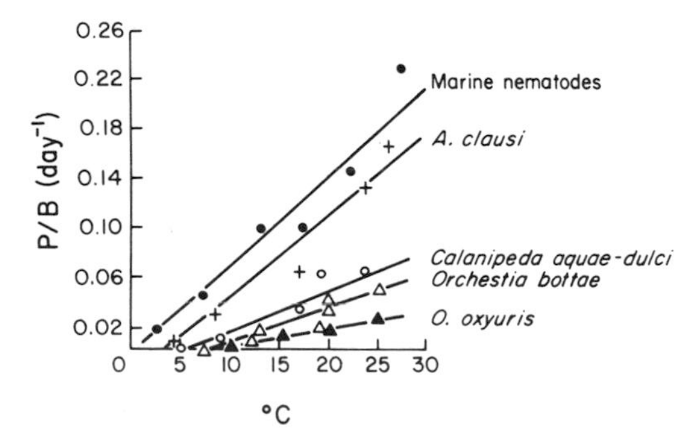

FIGURE 7-20. Relation of specific production (production/biomass) to temperature in various marine species, including nematodes, two copepods, and amphipods. The ratio *P*/*B* is used here to normalize production in the case of species with different biomass. Adapted from Zaika and Makarova (1979).

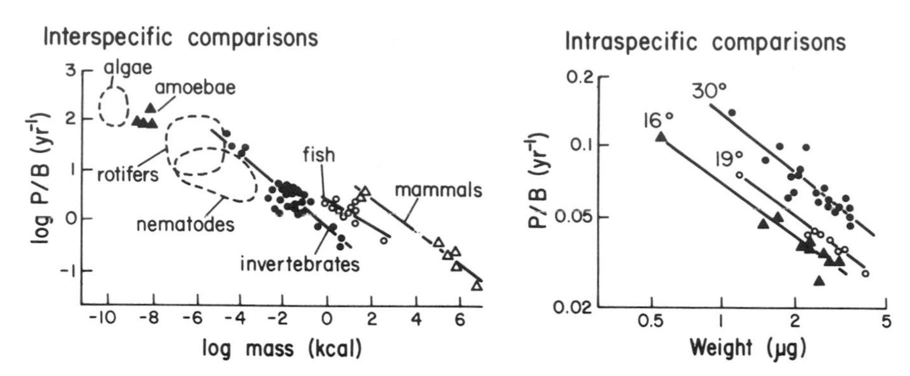

FIGURE 7-21. Specific production per year in relation to body size (expressed as energy content or weight—1 kcal is roughly 1 g wet weight). Left: Comparisons for various taxonomic groups: Fast-growing single-celled algae, soil amoebae, rotifers, nematodes, other invertebrates, fishes, and mammals. Dashed lines surround inferred ranges. Adapted from Banse and Mosher (1980). Right: Intraspecific comparison for the copepod *Acartia clausi*, Azov Sea, at three temperatures. This copepod strain is adapted to rather warm temperatures. After Kinne (1970), modified from V.E. Zaika and L.M. Malowitzkaja.

Respiration rates are size dependent, and therefore production, as mentioned above, is inversely related to size of the consumer [Fig. 7-21, left, and Warwick (1980)]. The effect of size on production holds both for comparisons of species of differing adult size and for comparison of smaller and larger or younger and older specimens within one species (Fig. 7-21, right). There is a general pattern of higher specific production with smaller size;* there are also slightly different relations for taxonomic groups, whose slopes are somewhat offset from each other. Sheldon et al. (1972) review production rates of many species, using time to double numbers as an index of production. They find that microbes and algae of 1–100 μm in diameter double in less than 10 to about 100 hr. Invertebrates (primarily zooplankton 100–10,000 μm in average diameter) double numbers in less than 100 to 1,000 hr. Fish, 10^4 to 10^6 μm, take the longest time to double their number, about 10^3 to 10^4 hr.

Both the quantity and quality of foods affect production. In terms of quality, we have seen earlier that nitrogen content of food is important to herbivores; in experiments with a detritus-feeding polychaete the production of biomass increased as the percentage of nitrogen in detritus increased (Fig. 7-22). In terms of quantity of food, in the same

---

*The production-size relationship is good enough to have encouraged Sheldon and Kerr (1976), based on the dimensions and food availability in Loch Ness, plus more speculative guesses as to size, to predict that there may be 10–20 monsters present in the Loch, assuming they are top predators.

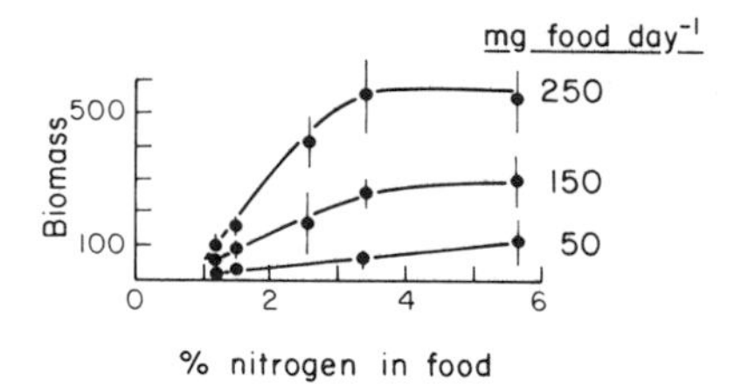

FIGURE 7-22. Biomass of the polychaete worm *Capitella capitata* (mg ash-free dry weight/0.1 m^2 tray) when fed on detritus from a series of foods of different percent nitrogen (*Spartina alterniflora, Zostera marina,* mixed cereal, *Fucus* sp., and *Gracilaria* sp., respectively, in the graph). The diets were offered at three dosages, shown in the right of the graph. Adapted from Tenore (1977).

experiments, production expressed as biomass accumulation was clearly higher where the amount of detritus was larger (Fig. 7-22). Notice, however, that when enough nitrogen was available, there was a threshold beyond which production was no longer stimulated. Some other essential constituent of detritus, probably essential amino acids, or trace elements (Marsh et al., 1989), limited production by the polychaetes (Tenore et al., 1982).

7.5.5 Production Efficiency

The production efficiency of a population ([*P*/*A*]100) is the proportion of assimilated energy devoted to production. Humphreys (1979) surveyed many studies and concluded that birds and mammals had production efficiencies of 1 to 3%. In fish and invertebrates the production efficiencies were 10–25%. This difference is probably due to the higher metabolic costs of warm-bloodedness mentioned above, and is evident in a graph of production versus respiration (Fig. 7-23). For any one

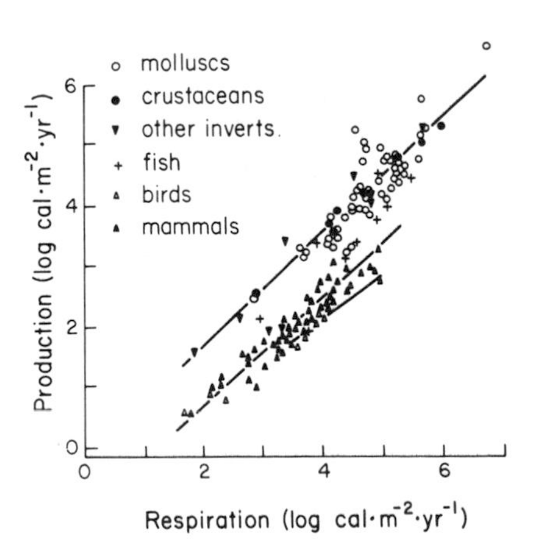

FIGURE 7-23. Relation between production and respiration in some major taxonomic groups. Species from marine, freshwater, and terrestrial environments are included. The regression lines for non-insect invertebrates, mammals and birds are included. Adapted from Humphreys (1979).

level of respiration, warm-blooded birds and mammals are less production. The magnitude of production or respiration did not affect production efficiency in any of the taxa, since the slopes of all regressions were equal to one. There was also no effect on size of the organism and only meager evidence that food quality affected production efficiency.

7.6 Energy Budgets for Populations

There are many studies where the various fates of matter consumed by a population have been identified and the fluxes measured. Specific values for each component vary enormously with local conditions, and comparisons among species are sometimes difficult because not all components are separated in each study. Rather than list individual measurements, it is instructive to look at the frequency distributions of measurements of the percentage of assimilated energy invested in production, respiration, reproduction, egestion, and molting.

Producers use light energy, consumers use energy stored in organic matter; respiration produces energy from stored products. The common currency in all these transformations is energy, so that it has become customary to measure them in energy units.*

Different species allot their assimilated energy in widely different ways, but there are some recognizable patterns (Fig. 7-24). Most consumers convert about 0–30% of their assimilated energy into production and about 40–80% into respiration. (Note the contrast to the 10% of total production lost via respiration in producers (Fig. 1-15); this difference shows the energetic cost of being a consumer versus a producer.) Investment in reproduction in consumers is low, generally considerably lower than 10% of assimilation. In organisms with ecdysis, molting also generally consumes less than 10% of assimilation. Egestion is hard to interpret since both feces and urine are included and fecal material is not assimilated, while urine is the waste product of assimilated food; in any case, 10–20% of assimilation is the usual extent of this modified "excreted" loss. For respiration, the standard maintenance levels are usually comparable to the losses due to specific dynamic action, while active metabolism may account for another third of total respiration. These proportions, however, are extremely variable.

---

*The units generally used are calories. The content of 1 g of phytoplankton carbon is about 15.8 kcal, 1 g of dry weight phytoplankton is about 5.3 kcal. Consumption of 1 ml of O_2 by an animal provides about 3.4–5 cal; taking the latter value, and an *R.Q.* of 1, 1 mg of dry food is about 5.5 cal. Energy content is also expressed in joules (J), and 1 cal = 4.19 J.

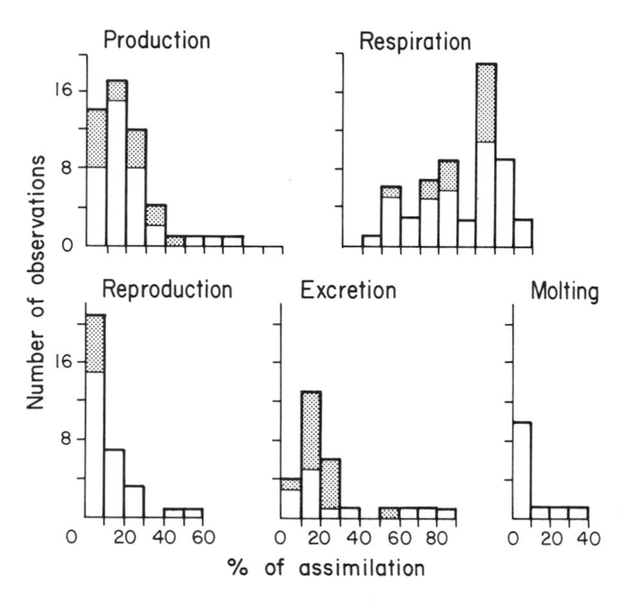

FIGURE 7-24. Partition of assimilated food into the various parts of the energy budget of populations of marine invertebrates (open part of bars) and fish (stippled part of bars). Many of the measurements of respiration do not include values for active expenditures. The excretion values for fish include both feces and urine. From data compiled by Conover (1978), Sushchenya (1970), Parsons et al. (1977), Baird and Milne (1981), and Matin and Pincell (1992).

Of course, the processing of energy by populations changes over time, both through seasonal patterns and through aging. Consumption by most species has a strong seasonal signature (Fig. 7-25), increasing in the warmer months, as we expect from our prior discussions of respiration, assimilation, growth, and from seasonal changes in food abundance. Seasonal fluctuations of growth and standing crop may not necessaily be correlated over time (Fig. 7-25), since mortality and migration also play a role in determing standing crop.

The other effect of time is through the aging of individuals comprising the population. As individuals grow through the stanzas of their life history their metabolic abilities change, as discussed above. The result is that the ratio of assimilation/consumption decreases as animals mature, and relatively more of the assimilation goes to respiration at the expense of production (Fig. 7-26). In addition to respiration, excretion also increases at the expense of production. Larger animals also defecate larger proportions of consumption. For example, a 35-mg mussel (*Mytilus californiensis*) egests an amount of feces equivalent to 15.2% of assimilated energy, while a 10-mg mussel only releases 4.5% (Elvin and Gossor, 1979). The net result of all this is

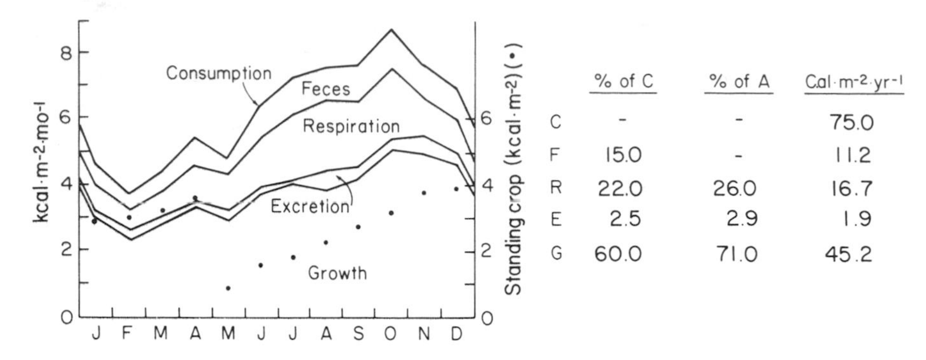

	% of C	% of A	Cal·m-2·yr-1
C	-	-	75.0
F	15.0	-	11.2
R	22.0	26.0	16.7
E	2.5	2.9	1.9
G	60.0	71.0	45.2

FIGURE 7-25. Partitioning of energy consumed through the year by the polychaete *Nereis virens*. Reproduction is included in growth. To the right are the percentage of total consumption (C) or assimilation (*A*) that is devoted to *F*, *R*, *E*, or *G*. The values for respiration seem low, probably because the calculations were done assuming a diet of high quality prey rather than detritus; further, metabolic expenditures for activity may be low if this polychaete were sedentary in behavior. This polychaete is also capable of anaerobic metabolism and of uptake of dissolved organic compounds, so both losses and consumption may thus be different than calculated. The average standing crop through the year is 28 kcal m^{-1}. From data of Kay and Brafield (1973).

that specific production is lower for larger individuals, and if older cohorts predominate in a population the level of production by the population may change.

Production of biomass (*P*) in a given trophic link relative to food consumed by that link (C) cannot be very high, since there is so much

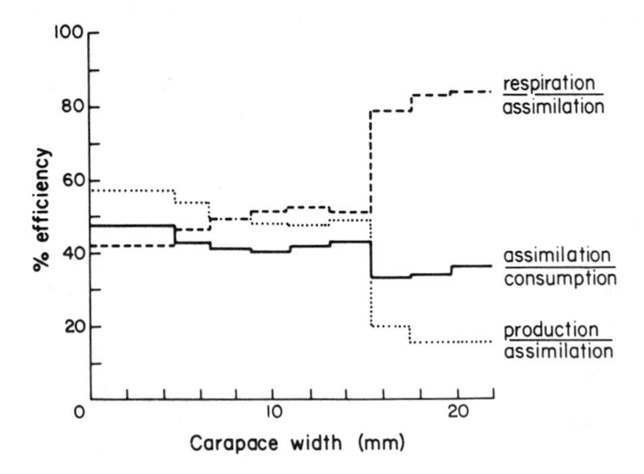

FIGURE 7-26. Changes in metabolic efficiencies during the life history of the fiddler crab *Uca pugnax* in Great Sippewissett Marsh, Massachusetts. Data from Krebs (1976).

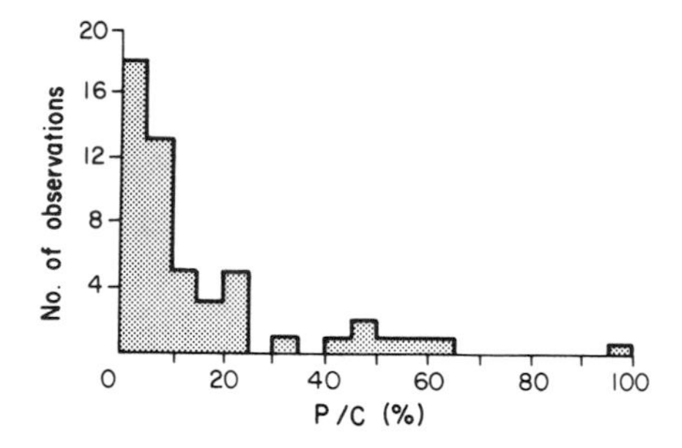

FIGURE 7-27. Ecological efficiencies (production by consumer/consumption by consumer) for invertebrates. Data were obtained from Conover (1978), Shafir and Field (1980), Tenore et al. (1973), Krebs (1976), Reiswig (1974), Teal (1962), Petipa et al. (1970), and Baird and Milne (1981). Data include a wide variety of taxa, trophic habitat, and size.

dissipation of energy by metabolic demands (Fig. 7-1). The ratio of P to C is called "ecological efficiency" and is important because it yields some notion of what proportion of consumed energy is available to the next link in the food web, whether it is another predator or a fleet of fishing vessels. This is a fundamental property of food chains, and is of interest to fisheries and to understand how marine ecosystems work. Steele (1974a), following Ryther (1969) concluded that ecological efficiencies in the water column should be about 10–20%. Actual measurements of ecological efficiencies in many types of animals range broadly, but most measurements are lower (Fig. 7-27). There is a scatter of high values, some too high to be reasonable, in view of what we know of respiration and egestion. In general, the ecological efficiency of a trophic link is most frequently less than 10%, some values reach 25%, and there is a scattering of not totally convincing larger values.

Part III
Structure and Dynamics of Marine Communities

The terms "structure" and "dynamics" are overused in many fields, so it is appropriate to be specific as to their use here. In our context, structure of communities is conferred by the way organisms in a community are linked trophically in food webs, by the species composition of the assemblage of organisms present, by the spatial arrangements of the organisms, and by the appearance or replacement of species over time. By "dynamics" we mean how the structural elements (food webs, species diversity, patchiness, colonization and succession) change over time and are controlled. Chapters 8, 9, 10, 11, and 12 cover these topics at the level of marine communities.

Chapter 8
Trophic Structure 1: Controls in Benthic Food Webs

8.1 Defining Food Webs

Food webs are descriptions of what consumes that. Data for construction of food webs come from inferences from anatomical structures, observations of consumption, gut analysis, and more recently, from stable isotope signatures.

The architecture of organs involved in capture and eating may hint what the species eats. In nematodes, for example, the construction of the buccal apparatus has been taken to indicate food habits (Platt and Warwick, 1980). Unfortunately, this approach is useful only for a minority of marine consumers.

Observations on what individuals of a population eat (or defecate) has been used to construct food webs. In rocky shores, for example, seastars and predaceous snails can be seen eating specific prey items (Paine, 1980). This works well for relatively large species that eat one prey at a time in a dilatory fashion.

Analysis of gut contents can furnish a record of recent meals. This method has been applied mostly to study feeding habits of fish (Simenstad and Caillet, 1986). It has drawbacks, since duration of residence of specific items in guts depends on their digestibility, many fragments are unrecognizable, and the method thus favors identification of foods that are relatively large, and have hard parts. Nonetheless, much has been learned from data gathered using this approach (cf. Fig. 4.12, for example).

Stable (nonradioactive) isotope signatures have in recent years been proposed as a better, albeit more technically demanding method to identify trophic links (Peterson and Fry, 1987). This approach makes use of the principle that "you are what you eat"; the results record longer-term nutrition, in contrast to gut content analysis, which provides a snapshot of feeding habits. The stable isotope method takes advantage of differences in the relative abundance of different isotopes of an element. Organic materials bear a stable isotope "signature," owing to differences in fate of the lighter and heavier isotopes during reactions that created the organic material. The fractionation recurs because biological transformations most often leave behind the heavier isotope, and the resulting differences create signatures that are incorporated into consumers that assimilate the organic matter (Peterson and Fry, 1987; Fry, 1988).

Stable isotope signatures for carbon, nitrogen, and sulfur can provide information about sources of materials for different links in food webs, and identify origins of allochthonous organic matter. Carbon isotope ratios have been used to establish the sources of organic matter that support food webs (Fig. 8-1). Carbon isotope ratios in plankton, land plants, and salt marsh grass differ clearly (stippled areas in Fig. 8-1), owing to different stable isotope signatures in the carbon sources

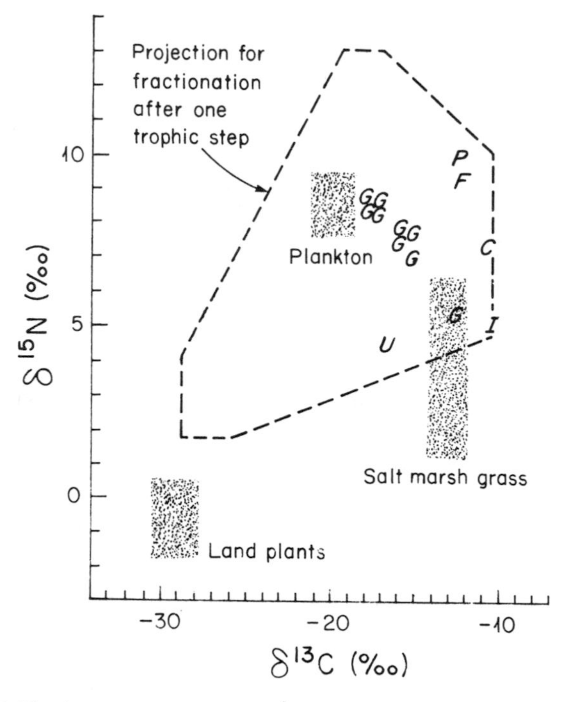

FIGURE 8-1. Stable isotope values for nitrogen and carbon in producers and consumers of Great Sippewissett Marsh, Massachussetts. The stippled areas show values for three major potential sources (plankton, salt marsh grass, land plants) of organic matter for consumers. The dashed polygon shows where possible values for consumers could fall given the isotopic signature of the producers, a fractionation of 3.5‰ for N and 1.5‰ for C resulting from one trophic step. The consumers are ribbed mussel, *Geukensia demissa* (G), blue mussel, *Mytilus edulis* (M), fiddler crab, *Uca pugnax* (U), mud snail, *Illyanassa obsoleta* (I), sheepshead minnow, *Cyprinodon variegatus* (C), killifish *Fundulus heteroclitus* (F), and grass shrimp, *Palaemonetes pugio* (P). Adapted from Peterson et al. (1985) and D'Avanzo et al. (1991).

(carbon dioxide in air, and carbonate in water) for land plants and plankton, and to reuse of CO_2 in C_4 plants, which leads to heavier ratios in salt marsh grasses.

Each time that stable isotopes are metabolized by organisms there is some degree of fractionation, since isotopes of different atomic weight do not participate equally in reactions. This isotopic fractionation can provide information on a species' position in a trophic web. It is well established that there is a fractionation of about 3–4‰ for nitrogen isotopes, and about 1.5‰ carbon isotopes, at every trophic step (Fig. 8-1). Knowing the potential sources of organic matter in a given site and the expected fractionation for a trophic step, we can delineate the projected area (enclosed by the dotted line in Fig. 8-1) where consumers of the available producers may fall. Indeed, the several species of consumers do fall in the projected area, but their specific position yields additional information of which producers were consumed. Mussels must have a diet consisting of a mixture of marsh grass particles and plankton, with mussels in deeper water apparently consuming relatively more plankton than marsh grass (Peterson et al., 1985). The proportion of organic matter from land sources assimilated by mussels seems constant, regardless of where they are, and land sources were less important. Land plants are more important to animals that live in environments such as streams (Findlay et al., 1991). The fiddler crabs use salt marsh grasses and land plants, the mud snail, and sheepshead minnow assimilate marsh grasses. Plankton seems somewhat more important for the shrimp and killifish.

Stable isotope studies have shown that organic matter can be transported into distant environments (Fry and Sherr, 1984; Suchanek et al., 1985). Thresher et al. (1992), for example, used stable carbon data to show that fish larvae living on the continental shelf off Tasmania contained more carbon from allochthonous seagrass detritus than from phytoplankton.

The stable isotope approach is not infallible; signatures may vary not only because of differences in sources and processes, but also because of short-term (Simenstad et al., 1993) and seasonal (Cifuentes et al., 1988) changes in isotopic composition in organisms and the environment. Taxa that are simultaneously predators and prey [for example, squid in Antarctic food webs (Rau et al., 1992), can show stable isotope signatures that have problematic interpretation.

We can depict a food web by applying the above approaches, but the picture of food webs obtained, however, is never unambiguous (Paine, 1988; Goldwasser and Roughgarden, 1993). Many species select prey on the basis of size rather than type, are omnivorous, or change food type as individuals grow (Section 7.4.3.4), and thus confuse or alter the picture of the web. There are many organisms that feed on detrital matter of originates from various sources. Food webs are further confused by not infrequent trophic "loops," in which a species feeds on what should have been its consumers, a problem that Isaacs (1973) has referred to as "unstructured" food webs. Furthermore, components of food webs are often treated at different level of categorization ("flagellates" for example, are never identified to species, while large species such as whales are almost always so identified). Not all taxa are equally observable or charismatic; less interesting, cryptic, or intermittently present species are less likely to appear in food web diagrams. The concept of food webs (and the related "tophic levels") is riven with too many unanswered aspects, ambiguities, and forces researchers to make too many suppositions, even with the most detailed descriptive work (Goldwasser and Roughgarden, 1993) and latest techniques (Rau et al., 1992).

Over the last decade there has been a flurry of publications on "food web analysis," where properties of the geometry of food webs (number of trophic links, relative specificity of trophic pathways, number of trophic steps) are studied. The premises and findings of food web analysis seem compromised by the uncertainty and incompleteness of trophic link data in virtually every food web that has been described, plus the tautological nature of the approach (Paine, 1988; Peters, 1988, 1991). Food webs exist, but study of webs per se seems surprisingly intractable.

Recent evidence shows that communities and food webs are not tightly integrated entities. The truly spectacular number of invasions by exotic species in many marine environments (Carlton and Geller, 1993, and cf. Chapter 16) shows how remarkably "open" marine communities are. Buzas and Culver (1994) examined the geological record of marine foraminiferal assemblages over 55 million years, and found that the "community" of species that reinvaded areas was highly variable. The species that happened to find themselves together again after a local extiction was a variable subset of a larger regional pool of species. These findings do not suggest that communities are strongly linked, interdependent assemblages of species. Below we therefore focus on mechanisms that control *population* abundances and distributions in specific settings rather than on abstract properties of food webs or community structure.

8.2 Controls of Community Structure

Even if we had an unambiguous depiction of a food web, we would still be a long way from understanding how these assemblages of organisms function, what controls flow of matter and energy from one link to the next, and what determines abundance of different populations. Description of the abundance and kinds of organisms, and the food web linking the species, is not tantamount to understanding the mechanisms that control function and structure of a community.

Terrestrial ecology has had the benefit of a seminal paper (Hairston et al., 1960) that inquired into the mechanisms that determine the abundance of populations of species in different trophic levels. Hairston et al. (1960) made a series of observations and drew conclusions as to how abundances of producers, herbivores and carnivores are determined. Even though some of their points are now outdated, their arguments are still of interest:

1. Since the accumulation of organic matter in environments is negligible in comparison to the amount produced by photosynthesis, decomposers must be limited by their food supply. If they were not, organic matter would accumulate.
2. Depletion of green plants or their destruction by natural catastrophes is exceptional. Producers are therefore not limited by herbivores or catastrophes. They must be limited by their own exhaustion of nutrients.
3. Where carnivores have been excluded by extrinsic factors, herbivores do deplete vegetation. Herbivores are therefore limited by predators rather than by their depletion of their plant foods.
4. If carnivores limit herbivore populations they are limiting their own resources, and therefore carnivores must be food limited.

These enormously broad generalizations attracted their full share of attention and criticism (Murdoch, 1966; Ehrlich and Birch, 1967; Slobodkin et al., 1967; Wiegert and Owen, 1971). They also served to stimulate a large amount of research. Much of the heated argument that ensued dealt with nuances of interpretation and semantics, but several aspects of Hairston et al. (1960) do need updating.

The classification of consumers into decomposers, herbivores, and carnivores is easy to make on paper but difficult when dealing with real animals. The typology stems from the idea of trophic levels, an idea that has heuristic value but is not easily applicable to natural assemblages. This is because most detritus feeders are facultative predators, most predators can feed occasionally on dead organic matter, and many herbivores can use animal prey at certain times.

Biotic and abiotic mortality factors seldom act separately, and resources and predation may also work in tandem to limit population growth. Hairston et al. (1960) were aware of these difficulties but isolated the effects as a polemical device to focus the arguments.

The fact that on land the "world is green," as put by L. Slobodkin, suggests that terrestrial vegetation is not largely consumed by herbivores (cf. Fig. 6.12), does not mean that herbivores have an unlimited food supply. Recent work, in part stimulated by Hairston et al. (1960), has shown that plants produce a multitude of compounds that can act as antiherbivore agents (cf. Section 6.3.2.1). Further, there is evidence that plant tissues, although plentiful, can be nutritionally inadequate (McNeill, 1973; Caswell et al., 1973; Mattson, 1980) for herbivores. Thus either because of poor quality or because of the presence of chemical defenses, green plants may not be adequate or usable food for terrestrial herbivores. This new information requires modification of the statement by Hairston et al. (1960): terrestrial herbivores can be limited by the quality of their food resources.

This modification of the original formulation helps solve an awkward difficulty in Hairston et al. (1960) having to do with the fact that they had to conclude that limitation due to restricted food supply—in other words, competition for food—would not be important to herbivores, since plant food was not limiting. This seems unreasonable in view of the many examples of sympatric species of terrestrial herbivores that clearly partition food resources (Bell, 1970, for example). The revised version allows for competition among all groups of consumers, which corresponds to what seems to take place in nature.

There have also been criticisms of some of the data on which Hairston et al. (1960) based their statement that herbivores can be controlled by carnivores (Caughley, 1966), but other examples still support that conclusion.

The ideas of Hairston et al. (1960) also stimulated research in freshwater communities, whose results led to discussion of "bottom-up" controls (by supplies of resources to populations), and "top-down" controls (by the action of consumers on their food populations) (Carpenter et al., 1985; McQueen et al., 1986). Perhaps the speculations can be summarized by saying that resource supply may set overall potential level of abundance of a population, which is modified by the action of consumers.

There might also be effects that "cascade" down food webs: a consumer might reduce abundance of its food items, thus releasing pressure on the next lower trophic link. Such ideas have led to the suggestion that stocking piscivorous fish may prevent zooplankton-feeding fish from depleting herbivorous zooplankton, and so would maintain water clarity in lakes. The hoped-for end result is that zooplankton would maintain low concentrations of phytoplankton and so keep the water clear (Andersson et al., 1978; Elliot et al., 1983; Carpencer et al., 1985), although it may be that influence of nutrient loading overcomes the expected top-down effect (Lehman, 1988).

Cascade effects may not necessarily involve consumption. We discussed earlier how increases in dissolved nitrogen in water led to increased growth of epiphytes on the surfaces of eelgrass. If epiphyte grazers are present, however, they may reduce epiphyte biomass sufficiently to free eelgrass from light limitation imposed by epiphytes

(Howard and Short, 1986; Williams and Ruckelshaus, 1993). Moreover, larger predatory fish and birds may feed on the herbivores, and provide trophic cascade effects that further complicate non-trophic controls (Power, 1992). In fact, there are opportunities for so many interactions of possible control processes that unraveling the top-down or bottom-up alternatives may become unfeasible (Williams and Ruckelshaus, 1993).

8.3 Control Mechanisms in Benthic Communities

Populations can be affected by action of consumers, and by supply of resources, the top-down and bottom-up notions described above. Resources such as nutrients affect marine producers (Chapter 2), and producers affect consumers (Chapters 4, 5, 6, and 7). Consumers also alter abundance of their nutritional resources (Chapters 4, 5, 6, and 7).

Nonbiological features may also be important, in at least two ways. First, certain variables may modify or overwhelm other controls of populations. For example, desiccation and wave action may be crucial in determining what species are present at different elevations in the rocky intertidal; models that do not include appropriate effects of temperature on demography of zooplankton do not fit field data (Davis, 1987). Second, vagaries of weather and currents can powerfully alter conditions for the assemblage of species in a community (Connell, 1985). For example, recruitment of intertidal barnacles to California rocky shores depends on (a) southerly winds that stop the upwelling that moves water away from shore (Farrell et al., 1991), and (b) hydrographical vagaries that determine abundance of larvae-eating fish in the kelp forest through which barnacle larvae must pass on their way to settling on shore (Gaines and Roughgarden, 1985, 1987). The gauntlet barnacles traverse is dependent on physical controls. Such non-biological factors may be more or less important from one area to another. Menge (1991) showed that differences in recruitment caused by vagaries of weather accounted for less than 11% of the variability in species composition in New England rocky shores, but 39–87% in Panama. Comparisons among many sites, however, did not show a clear relation of recruitment success on community composition (Menge and Farrell, 1989).

Even though physical factors and meteorological features may therefore have important effects on structure of communities, we focus our attention on resource and consumer interactions. To evaluate the relative role of such top-down and bottom-up controls we will examine case studies of controls of structure in marine benthic communities.

8.3.1 Case Studies of Top-down Controls

8.3.1.1 Grazers in the Rocky Intertidal

The vegetation of the upper half of the rocky intertidal in New England is a mosaic of various species. In tide pools the extremes vary from almost pure stands of the opportunistic green alga *Enteromorpha intestinalis* or of the perennial red alga *Chondrus crispus* (Irish moss), to situations where many different types of algae coexist.

The most obvious herbivore in the upper intertidal zone is the snail *Littorina littorea*, and its density varies from pool to pool. *L. littorea* has clear-cut preferences among the algal foods available in the New England shore: *Enteromorpha* is a highly desirable food, while *Chondrus* is not (Fig. 8-2, bottom). *Enteromorpha* is abundant where there are few snails and *Chondrus* is the dominant species in pools with lots of snails. The relative ease with which manipulative experiments can be carried out in the rocky intertidal has made this environment one of the best understood in regard to competitive and predatory interactions among the constituent populations. To see if the differences in seaweed species composition were caused by the different herbivore abundances, Lubchenco (1978) carried out experimental alterations of snail density (Fig. 8-2). All snails were removed from a pool where *Chondrus* was the

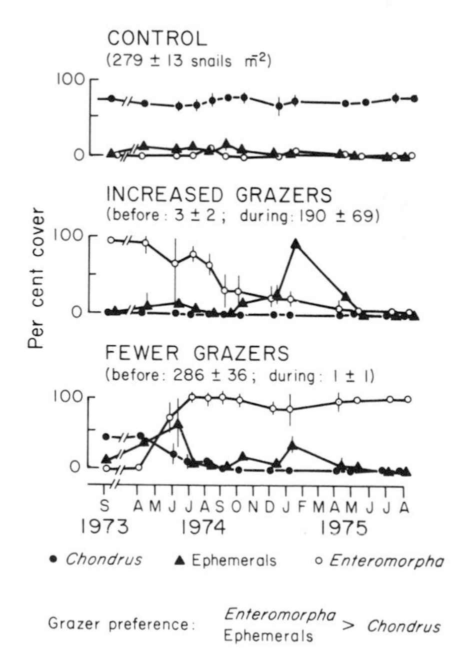

FIGURE 8-2. Experimental manipulation of a grazer snail (*Littorina littorea*) in tide pools in the higher reaches of the New England rocky intertidal zone. Adapted from Lubchenco (1978).

dominant producer, while snails were added to a pool dominated by *Enteromorpha*. In an untreated control pool, the abundance of *Chondrus* remained high through the 1½ years of the study (Fig. 8-2, top).

In the pool where snails were abundant, their grazing gradually reduced the abundance of *Enteromorpha*. Some "ephemerals" (fast-growing, short-lived algae) became abundant in winter, the time when snails were less active.

In the pool with no snails, *Enteromorpha* and several ephemeral species settled quickly, but after a short period of time *Enteromorpha* became the most abundant species in the pool. *Enteromorpha* actually settled on *Chondrus*, grew and shaded the fronds of the Irish moss, and after one summer of the experimental manipulation only the holdfasts of *Chondrus* remained.

These experiments show that grazers control the composition of the producer community in tidal pools. The first-order effect of grazers is the virtual elimination of preferred species such as *Enteromorpha* and other ephemerals. The second-order effect of grazers is that by removing *Enteromorpha*, the best competitor for space, other species may become the dominant components of vegetative cover.

The effect of herbivores is therefore considerably more complicated than merely a removal of a more or less important proportion of the plant biomass.* Selective feeding by the herbivores leads to differential grazing pressure on the various species of producers present in the environment, a feature of fundamental importance in determining what species make up the community. The amount of biomass of the surviving species, the ones not preferred by herbivores or competitively dominant, is then determined by other factors, principally light and nutrients.

8.3.1.2 Role of Predators in the Rocky Intertidal

Predation (Paine, 1966; Connell, 1972) and physical disturbances (such as waves and floating logs) (Dayton, 1971) are important as causes of mortality and maintain competing species of prey at relatively low densities. In such situations, competitive interactions are less intense and competitive exclusion is usually avoided. Since predation and other disturbances periodically provide open space (the resource most limiting in the rocky intertidal environment), recruitment of new individuals can take place.

---

*Herbivores may have many additional effects on vegetation, including influencing plant morphology. For example, some algae of the upper intertidal zone in rocky shores exist as upright morphs during the part of the year where grazing pressure is low, while crustose or boring morphs are dominant when grazers are more active (Lubchenco and Cubit, 1980).

A good example of these interactions is provided by experimental studies by Lubchenco and Menge (1978) in the rocky intertidal zone of New England, where the predatory intertidal whelk, *Thais lapillus,** and the sea star, *Asterias forbesi,* feed on the barnacle *Balanus balanoides* and the blue mussel *Mytilus edulis,* as well as affect their competitor for space, the macroalgae *Chondrus crispus.* The experiments included clearing areas on the rock surface and following the course of events in the areas (Fig. 8-3, top four graphs). In cleared areas with no further manipulations (called "control" in Fig. 8-3) *Mytilus* settled but did not survive predation for very long, and *Chondrus* became the dominant species in the site. Where stainless-steel cages excluding *Thais* and *Asterias* were attached to cleared rock surfaces, *Mytilus* survived and was able to out-compete *Balanus* in a short time. In other cleared areas with cages, the *Mytilus* were removed from the cages; this allowed *Balanus* to settle and eventually to grow into the most abundant species. In cages where *Mytilus* were removed and *Balanus* were reduced in density, *Chondrus,* after some initial coexistence with *Balanus,* became dominant. Thus, at least as colonizers of bare substrates, the competitive hierarchy is (1) *Mytilus,* (2) *Balanus,* (3) *Chondrus.*

There are places where predators are naturally absent, such as in sites exposed to severe wave action, and in such habitats *Mytilus* is the most abundant species. In protected sites, predators are not swept away or damaged by waves, and their presence prevents their preferred prey (*Mytilus* and *Balanus*) from monopolizing space. *Chondrus* can then colonize and establish itself.

Another set of experiments was started in already-established stands of *Chondrus crispus* (Fig. 8-3, bottom two graphs). The exclusion of predators allowed recruitment of *Mytilus* into the stand, and growth of the mussels eventually resulted in exclusion of *Chondrus* from the experimental site. Then, regardless of whether *Chondrus* was established or not, mussels, the dominant competitors, replaced the alga, but only if predators were absent. Notice that the mechanisms involved, preferences by the consumer and competitive dominance, are very much the same as discussed for herbivores.

The outcome of predatory and competitive interactions can be markedly modified by physical-chemical features. The clearing experiments of Figure 8-3 were done at 0.3 m above mean low water. Similar experiments done higher in the intertidal showed that *Chondrus* does not cover all the space, probably because it cannot survive long exposures to drying.

---

* The genus of these snails has recently been changed to *Nucella*. We retain the earlier usage for convenience.

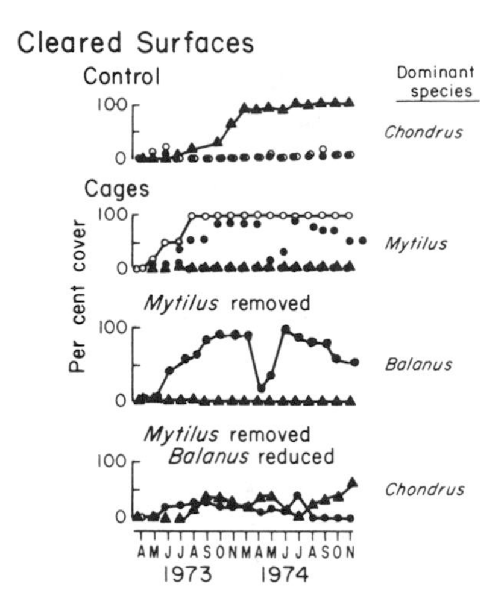

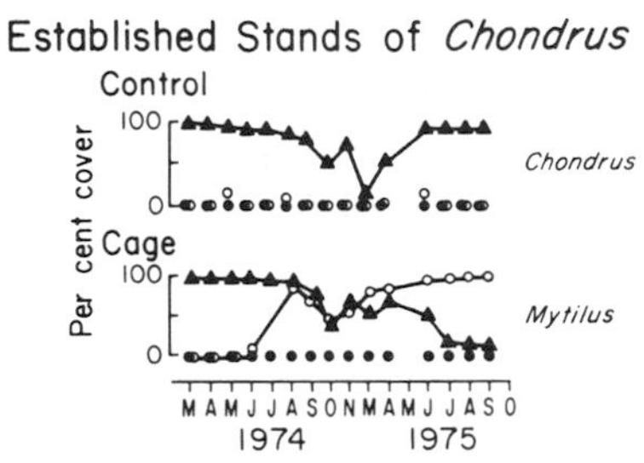

FIGURE 8-3. Experiments on the role of predation and competition in the lower rocky intertidal in New England. The abundance of each of the three major species, *Mytilus edulis, Balanus balanoides,* and *Chondrus crispus,* is expressed as percentage cover. One set of experiments (top four graphs) was carried out starting with experimentally cleared surfaces, while the second set (bottom two graphs) was done on already well-established stands of *C. crispus.* Cage controls (not shown) consisted of cages with no sides and show results similar to controls. Adapted from Lubchenco and Menge (1978).

The results of these experiments suggest how the rocky intertidal community of New England is structured. In habitats protected from severe wave action, predators can severely limit the abundance of the dominant competitor (mussels). Feeding by predators in protected areas produces open space that is colonized by algae. Competition among the various species of algae ensues, and *Chondrus* eventually excludes other algal species. In exposed habitats the intense wave action may remove predators, since they are not permanently anchored in the substrate. In such habitats the competitively superior mussels monopolize available space.

The results of work on intertidal rocky shores by Connell, Paine, and Lubchenco and Menge provide an appealing scheme of how community structure is determined, one that has been applied to a wide variety of other environments. The scheme may not, however, be entirely generalizable to other marine environments, or even perhaps to all hard substrates. The outcome of competitive relationships among assemblages of invertebrates on the underside of foliose corals seems determined by competition rather than by predation, and competitive exclusion is prevented by reversals in competitive superiority from one site to another (Jackson, 1977a). Predation also does not seem to determine the structure of fouling communities that settle on hard surfaces suspended in the water column (Sutherland and Karlson, 1977). In spite of such exceptions, the rocky intertidal experiments have produced the basic model from which hypotheses have been derived to be tested in other environments.

8.3.1.3 Grazers in Kelp Forests

The subtidal kelp forests off California consist of the canopy-forming giant kelp (*Macrocystis pyrifera*), and two understory kelp species (*Laminaria dentigera* and *Pterygophora californica*). A luxuriant growth of foliose red algae covers the rock surfaces below the kelp canopy. Along the seaward margin of the forest is a band of the brown alga *Cystoseira osmundacea* and the canopy-forming bull kelp *Nereocystis leutkeana*.

Sea urchins are widespread grazers of marine macroalgae (Lawrence 1975), including kelp. *Strongylocentrotus franciscanus* prefers *Macrocystis* over all other species (Leighton, 1966), but other species of kelp are eaten when *Macrocystis* is not available. If no kelp at all are available (years 1974–1975, Fig. 8-4) urchins feed on detritus, microscopic plants, newly settled juvenile plants, or whatever else may become available.

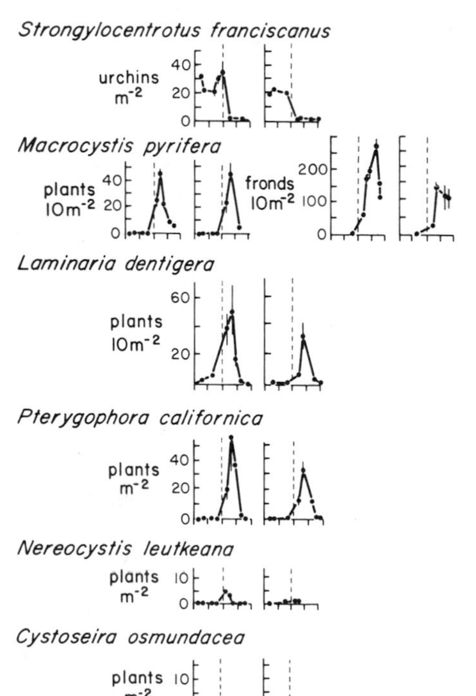

FIGURE 8-4. Changes in vegetation of kelp forest after collapse of sea urchin population off California. The vertical dashed lines show the time of occurrence of the sea urchin mortality. Adapted from Pearse and Hines (1979).

A natural experiment that showed the importance of grazer control of producers took place in 1976 on the seaward edge of a giant kelp forest when an unknown disease decimated the population of the sea urchin *S. franciscanus*, with remarkable consequences for the vegetation (Fig. 8-4). Soon after the near-disappearance of urchins off California, the density of *Macrocystis* increased markedly (Fig. 8-4, second row), and by 1977 only about 1% of the light at the surface reached the bottom. This light intensity is about the lower limit at which *Macrocystis* can achieve positive net photosynthesis (Neushul, 1971), so that intraspecific competition through self-shading was probably responsible for the decline in numbers of *Macrocystis* during 1976–1977. The number of fronds on the surviving kelp increased, so that the total biomass of *Macrocystis* remained significantly higher after the removal of sea urchins. Shading by the large established kelp prevented further recruitment of young kelp.

The two understory kelp species (*Laminaria* and *Pterygophora*) increased rapidly after the urchins were gone (Fig. 8-4, third and fourth rows). In subsequent years, both these species decreased in abundance, with few live individuals left by 1977. The decrease in *Laminaria* and *Pterygophora* was most probably due to shading by the *Macrocystis* canopy; to test this hypothesis, a plot of $200\,m^2$ was cleared of large *Macrocystis* and the vegetation was examined 3 months later. The increased light due to the clearing allowed increases in density of *Laminaria* and *Pterygophora* as well as of juvenile *Macrocystis*. The biomass of foliose red algae also increased. As the kelp forest recovers following the disturbance, the biomass per unit area of *Macrocystis* will presumably be lowered to the average of about 70 fronds $\cdot\,10\,m^{-2}$ of undisturbed kelp forests. This lower density may allow enough increased light penetration so that the understory kelp species can reestablish themselves.

Sea urchins are not usually found grazing on the upright kelp, since wave action dislodges them from the fronds. Instead the urchins feed on kelp fronds kept down near the sea floor by the combined weight of many clinging urchins. The ability of urchins to affect kelp forests is related to urchin abundance. In kelp forests off Nova Scotia, sea urchins* have to exceed a density of $2\,kg\,m^{-2}$ before these grazers can consume enough kelp to reduce the area covered by kelp forests (Breen and Mann, 1976). If the urchins are abundant enough, they can graze out a kelp bed (Bernstein et al., 1981). Then it is rather difficult for the kelp community to reestablish itself, since the urchins remain, feeding on detritus, algae, and benthic fauna. The remaining urchins slowly

---

*In Nova Scotia, the main kelp is *Laminaria longicrurus* and the sea urchin is *Strongylocentrotus droebachiensis*.

lose weight and newly recruited urchins may not obtain maximum size; dense populations of stunted urchins may occupy the area of a former kelp bed for many years.

The specific species and details of the grazer-producer interaction may vary from site to site. Off the coast of California, *Nereocystis* and *Cystoseira* were not influenced by the absence of urchins or competition among algae (Fig. 8-4, bottom two rows), although elsewhere they may be subject to both (Paine and Vadas, 1969; Foreman, 1977). Nonetheless, although the specific *result* of grazing in rocky shore and kelp forests may differ, the role of grazing and competitive *processes* evident in the kelp case history are very similar to that of the rocky intertidal shore. Grazing pressure may be intense enough that if applied to competitively dominant species of producers it may result in competitive release of other producer species. This conclusion may be general enough to apply to other marine environments.

In the kelp forest case history, we have a species, *Macrocystis,* that for some reason has not evolved effective phytochemical defenses and is therefore susceptible to overgrazing by herbivores, much like *Enteromorpha* on the rocky shore. It turns out that *Macrocystis* is a species of fundamental importance to the entire kelp forest ecosystem, so that the impact of grazers is critical in determining what the community looks like. Just what is so particular about the phytochemistry of *Macrocystis* or the ability of urchins to ignore what feeding deterrents may be present in this plant is not known. In actuality, what might be a deterrent to a given herbivore may not inhibit another herbivore. For example, one amphipod species significantly consumed a brown alga that presumably contains substantial chemical defenses, while ignoring the perhaps chemically underfened filamentous green alga, while another species of amphipod did the reverse (Duffy, 1990).

In both our macroalgal case histories, the producer–herbivore interaction is primarily influenced by the relative palatability of the array of producer species, which is in turn determined by the chemistry of the tissues. The impact of herbivores is greatest on the most palatable species. Once palatable species are removed, the success of the less palatable producers is determined by competitive interactions (for light in the case of the kelp forests). The interplay of these features, plus restrictions due to wave action, and other factors of local importance, determines the make-up of the producer community.

8.3.1.4 The Impact of Predators in Kelp Forests

We saw earlier (Section 8.3.1.3) that grazing by sea urchins can destroy kelp forests if urchin densities are high enough. Removal of these herbivores by experimental manipulation (Paine and Vadas, 1969), by oil spills (Nelson-Smith, 1968), or by disease results in rapid recolonization by marine vegetation. The presence or absence of predators also has important consequences for urchin populations and for kelp.

The sea otter (*Enhydra lutris*) once occupied a range from northern Japan through the Aleutian Islands and south to Baja California. At present, the sea otter occurs mainly in certain islands off Alaska, as a remnant population in Central California, and in small populations that have been reintroduced to the coasts of Oregon, Washington, and British Columbia. The near extinction of the sea otter was due to the hunting pressure for the highly prized pelts and occurred after the coming of Europeans to the northwest of North America.

At Amchitka Island (Aleutian Archipelago, Alaska), sea otters are present at high densities (20–30 individuals km^{-2}) (Estes and Palmisano, 1974; Estes et al., 1978). On average, an otter weighs 23 kg and consumes 20–30% of its body weight in food daily. Otters feed mainly on benthic invertebrates and some fish, and as a result of their high consumption rates, have a major role in structuring the near shore community. The population in Amchitka has evidently recovered from the near extirpation of the early 20th century, but recolonization has not yet taken place on Shemya Island, 400 km to the West. At Shemya, where sea otters are absent, there is no subtidal algal cover, and the density of sea urchins (*Strongylocentrotus polyacanthus*) is much higher than that in Amchitka (Fig. 8-5, top and bottom left). Other herbivorous invertebrates are also scarcer and smaller (Simenstad et al., 1978). As usual with many of the predators we have examined, otters prefer to eat the largest urchins (over 32 mm diameter). Only small urchins remain where otters are present (Fig. 8-5, bottom right).*

In Amchitka algae almost completely cover the substratum (Fig. 8-5, top left), while in Shemya space is occupied by urchins, chitons, mussels, and barnacles. Climate, sea state, tidal ranges, and substrates were similar in both sites so that the marked differences in the two communities are likely due to the presence or absence of sea otters.

Confirmation of the importance of otters is also evidenced at other sites where otters have become reestablished (Table 8-1). Otters were reestablished in Surge Bay, Alaska following transplantation into adjacent areas (Duggins, 1980). After some time, the otter population became very numerous and presumably reduced urchin populations (cf. Deer Harbor data, Table 8-1). This led to large enough increases in

---

*The fact that large urchins are present in high densities in Shemya seems unusual. Other studies referred to earlier show that stunted populations were more often the result of crowded, overgrazed conditions where the key predator was not present. Perhaps in Shemya there is enough horizontal transport of food from the rich rocky intertidal zone into the area where urchins are found (Simenstad et al., 1978) or cannibalism of young urchins so that stunting is prevented.

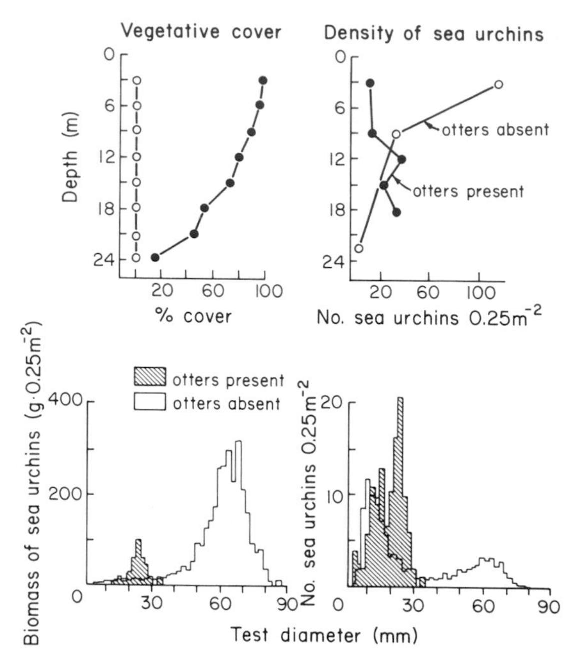

FIGURE 8-5. Interactions among sea otters, sea urchins, and vegetative cover in kelp beds off the Alaskan coast. Sea otters are present in Amchitka Island and absent in Shemya Island. Symbols on top left are the same as top right. Size of sea urchins is shown as the diameter of the test not including spines. Adapted from Estes and Palmisano (1974).

fast growing annual kelp species that local fishermen reported the area so congested with the annual kelp *Nereocystis* that some boats were unable to make their way through the kelp beds. In Torch Bay, where no otters were present, *Nereocystis* and several other annual kelp became the dominant primary producers one year after Duggins (1980) experimentally removed urchins from an area devoid of kelp. The annuals were subsequently outcompeted and replaced in the experimental quadrats by the perennial *Laminaria groenlandica*. These experimental results agree nicely with the data of Table 8-1 and demonstrate that the presence of otters fundamentally alters the nearshore community.

The impact of the otters on the kelp forest community has several other probable consequences. Estes and Palmisano (1974) and Simenstad et al. (1978) believe that the low abundances or lack of certain fish (rock greenling, *Hexagrammos lagocephalus*), harbor seals (*Phoca vitulina*), and bald eagles (*Haliaetus leucocephalus*) at Shemya are related to the significant lack of algal cover, although the specific linkages to sea otters are not clearly known. Islands with sea otters have near-shore fishes that depend on and use sublittoral macroalgae for cover and spawning.

TABLE 8-1. Mean Number m^{-2} (± Standard Deviation) of Key Species in 1-m^2 Quadrats in Southeastern Alaskan Kelp Forests.[a]

	No otters (Torch Bay)	Otters present less than 2 yrs (Deer Harbor)	Otters present more than 10 yrs (Surge Bay)
Urchins			
S. franciscanus	6 ± 7	0.03 ± 0.03	0
S. purpuratus	4 ± 13	0.08 ± 0.06	0
Kelp			
Annual species[b]	3 ± 7	10 ± 5	2 ± 5
Laminaria groenlandica	0.8 ± 5	0.3 ± 0.6	46 ± 26
Percentage samples with no kelp	69	0	0

[a] From Duggins (1980).
[b] Annual kelp include *Nereocystis leutkeana, Alaria fistulosa, Cymathere triplicata,* and *Costaria costata*. These algae are opportunistic species that temporarily use space free from urchin grazing.

These fish feed on epibenthic organisms such as mysids and amphipods that consume detrital macroalgal particles (Simenstad et al., 1978). There are fewer near-shore fish at islands lacking otters, and the species that are present are mostly species that feed on the open water. These manifold consequences of the activity of sea otters demonstrate that they are what Paine (1969) calls keystone predators, that is, species whose activities structure the surrounding community. Otters also feed on the older, less productive age classes of fish such as the rock greenling (Simenstad, unpublished data). Thus, otters not only structure the near-shore communnity but also may increase the secondary production by the effect of their selective feeding.

Removal of the keystone predator in kelp forest communities may have occurred more than once, as shown by animal remains in kitchen middens of aboriginal Aleuts and pre-Aleuts at Amchitka (Simenstad et al., 1978). During time intervals when sea otters, harbor seals and fish remains were abundant in the middens (about 580–80 B.C. and a period around 1,080 A.D.), sea urchins and limpets were not, and vice versa. The Aleuts may therefore have supplanted the otter as the keystone species. The Aleuts had the technological know-how with which to hunt or even exterminate sea otters* but for whatever reason Aleut hunting of otters varied in intensity over long periods of time.

*During the enslavement of Aleut hunters by Russian fur traders, the Aleut nearly eliminated the sea otter population (Kenyon, 1969). Thus, when they had to, these hunters could bring about a marked decrease in abundance of sea otters.

When hunting pressure increased, the otters became scarce and the kelp forest community reverted to its sea urchin-dominated state.

Elsewhere in the world where there are kelp forests, other predators such as crabs, lobsters, or fish may perform the keystone role. Kelp forests off Nova Scotia have declined in area, with reduction of up to 70% of the habitat in St. Margaret's Bay (Breen and Mann, 1976; Wharton and Mann, 1981). At the same time, the lobster catch per unit fishing effort declined by 50% between 1959 and 1973, probably due to over-fishing of lobsters. Subsequently, sea urchins have increased and have destroyed kelp beds at a high rate. The picture of lobsters as the keystone predator controlling sea urchin numbers is not entirely clear because there are substantial numbers of crabs and fish that prey on urchins and the fish have declined in number (Wharton and Mann, 1981; Bernstein et al., 1981). The crabs in turn are highly preferred by lobsters over urchins (Evans and Mann, 1977). There may also be important effects of diseases or parasites (as we have seen in Section 8.3.1.3) in changing urchin abundance.

8.3.1.5 Grazers of Attached Microalgae

Single-celled algae that grow on surfaces are found in the sediments of the shallow coastal zone, as epiphytes on floating surfaces and on surfaces of rocky seashores. The abundance and distribution of attached microalgae are very markedly affected by grazers.

An example of the impact of grazers can be found in the intertidal zone of Southern Oregon, where dark brown carpets of attached diatoms grow on the sandstone surfaces during the colder months. When the weather warms up, the diatoms are removed by grazing herbivorous snails (*Littorina scutulata*). Castenholtz (1961) installed wire-mesh covers over artificial tide pools and varied the density of the littorines within these cages. As the density of grazers increased, the carpet of diatoms was reduced in percentage cover, until it disappeared above a threshold density of grazers. A density of about 3 snails dm^{-2} kept an area free of diatoms, while algal patches developed below this density of grazers. In the summer natural densities of grazers were equal to or several times the threshold values so that during the warm season grazers eliminated algal patches.

The major effect of grazers on attached microalgae may be due to several factors. The elimination of attached microalgal producers by grazers suggest a lack of grazing deterrents in the attached algae. In addition, while phytoplankton cells are widely dispersed, attached microalgae offer many cells in close proximity to grazers. The effort (search time, energy expenditures) involved in feeding is therefore much smaller in the case of attached microalgae, and this may be enough to allow consumption rates by grazers to exceed growth rates by populations of attached microalgae, and hence limit abundance of these producers.

8.3.1.6 Effects of Consumers in Soft Sediment Benthic Communities

Predators, particularly fish, are major influences on the competition and abundance of organisms of soft sediments. There is correlational evidence for this: the abundance of benthic invertebrates in the Baltic Sea, for example, increased after overfishing depleted flatfish during the 1920s (Persson, 1981). Better evidence can be obtained by caging experiments where an area of sea floor is protected from predation. Early experiments showed that such caging led to 60-fold increases in the densities of prey on the sea floor (Blegvad, 1925). Experiments using cages or exclosures are now commonplace, and many of them have been done on soft sediments.*

In soft organic sediments of intertidal creeks of New England salt marshes the major macrofaunal taxa include amphipods, an oligochaete (*Paranaius litoralis*), hydrobiid snails, and anemone (*Nematostella vectensis*) and a diverse group of annelids, including *Streblospio benedicti*, *Capitella* spp., *Manayunkia* spp., and *Hobsonia florida*. During warm months of the year, fiddler (*Uca pugnax*) and green (*Carcinus maenas*) crabs and fish (principally the killifish *Fundulus heteroclitus* and *F. majalis*) forage through the surface sediments and detritus, ingesting a variety of foods, including the fauna. Experiments in which cages were used to exclude predators show that in the absence of predators, the biomass of benthic macroinvertebrates remains high through the growing season or increases a bit. Where predators have access—in natural sediments and in control cages—biomass of macroinvertebrates decreased markedly after an initial increase during spring. Predation by fish and crabs thus seems responsible for removal of a very large proportion of prey biomass (Werme, 1981; Wiltse et al., 1984; Sardá et al., 1995). Recruitment and growth of macroinvertebrates take place in early summer but the predators quickly remove the added biomass, so that by August or so the standing crop has been reduced by at least

*During the 1970s it became fashionable to question the validity of caging experiments in soft sediments, since caging is claimed to lead to artifacts, including changes in the sediments and fauna within the cages (McCall, 1977; Virnstein, 1977; Eckman, 1979). There is no question that there are cage effects, but many of these changes are no doubt the result of changes in the fauna brought about by the lack of predators since organisms can markedly change the density and physiochemical properties of sediments (Aller, 1980a). The caging approach is too powerful a tool to dismiss too readily, especially when not many alternatives are available to directly study the effect of consumers. Cages should be used with careful attention to minimizing artifacts, and cage effects should be assessed using partial cages.

one order of magnitude. It is clear that the predators are indeed key members of this community and that their activity structures the benthic community.

Similar caging studies have also been carried out in a variety of lagoons and estuaries with unvegetated soft sediments. In all these studies, protection from predation increased the density of the macroinvertebrate prey two- to three-fold (Peterson, 1979). There is reduced growth of individuals, so competition for food exists, but species seldom disappear (Peterson, 1992). This suggests that, as in the other marine environments we have examined, predation can control composition and density of the fauna of soft sediments.* We should note that none of these experiments distinguish between the effect of ingestion of prey by the predator from mortality due to the disturbance created by the activity of the "predators." Disturbance of sediments by large animals can disrupt the life activities of benthic animals and may, if severe enough, kill or smother smaller species.

Caging studies on soft sediments, in contrast to the studies on rocky substrates, show little evidence of competitive exclusion once predators are removed (Peterson, 1979). In fact, more species of invertebrates were often found within the cages. The conclusion from experiments done in rocky intertidal environments that predation, by preventing competitive exclusion, fostered the presence of multiple species, is therefore not completely extrapolable to soft sediments.

The effectiveness of predators in determining abundance of prey species varies from one habitat to another. A major feature that affects the success of predators is the physical complexity of the habitat where predation takes place (cf. Section 6.6). This complexity is provided by topography, rock fragments, cobblestones, reefs, or vegetation. Experiments using cages to exclude predators have also been carried out in estuarine or lagoon sediments on which vegetation was present. The results of these experiments (Peterson, 1979) are that the exclusion of predators has very slight effects on the macroinvertebrate community of such habitats, since the vegetation already excludes large predators or permits prey to hide. The fauna in vegetated habitats is usually more numerous compared to similar but unvegetated sediments and far more species are present. In these relatively predator-free habitats there is also no tendency for competitive exclusion.

The experimental work on soft sediments thus fails to show competitive exclusion, but the mechanisms that prevent exclusion are not clear (Peterson, 1979). Perhaps one explanation is that in soft sediments

---

*The supposition is that small predators about the same size as the macroinvertebrate prey are not effective as predators, as discussed in Chapter 5. This assumption needs some more careful examination, especially in soft sediments and in plankton communities.

food rather than space is the limiting resource (Peterson, 1992). Since food is renewable, there is some reasonable chance that less able competitors could gain access to food sometime, somewhere.

Soft sediments also do not provide the hard substrate where competitors can be crushed, pried off, or overgrown (cf. Chapter 4). Thus, the more obvious mechanisms of interference competition, leading to direct elimination of one species by another, may be lacking in soft sediments. Animals that live in soft sediments may compete mainly by exploitative competition. Another explanation may be that in soft sediments there are frequent small scale local disturbances that may alter competitive advantages. Such disturbances may thus prevent competitive exclusion and maintain multi-species assemblages. There are many possible mechanisms that may disturb patches of soft sediments, including physical and biological mechanisms (Thistle, 1981). Van Blaricom (1982), for example, showed that bat rays (*Myliobatis californica*) repeatedly disturb the sea floor by digging pits in the sand. This markedly affected the fauna of the disturbed pits and resulted in a sequence of various species colonizing the disturbed patches. On average, about 23% of the sea floor was in some phase of recovery following disturbances by bat rays; during periods when rays were very active this figure reached 100% discuss below.

Another reason for the lack of exclusion may lie in the nature of "predation" in soft sediments. In many instances, as in the salt marsh example of Figure 8.8, the "predators" are really omnivores. Salt marsh fish and fiddler crabs, for example, ingest detritus, algae, and meiofauna as well as macrofauna. Exclusion of the fish and crabs results in more abundant algae (Fig. 8.8, top). If these algae are suitable alternate food for the prey macrofauna within the cage, exclusion is less likely to occur.

Yet another possible mechanism that prevents saturation, and hence exclusion, may be a lack of sufficient larvae available to settle and saturate available space. Whatever the factors involved, the coexistence of many competing species is a major unresolved issue in the understanding of communities in soft bottoms.

The lack of competitive exclusion seen in short-term experiments does not mean that competition has not occurred on different spatial or time scales. There is suggestive, indirect evidence of character displacement, resource, and habitat separation among benthic organisms (Rees, 1975; Fenchel and Kofoed, 1976; Levinton, 1979; Frier, 1979, 1979a).

Although there is room for considerably more study, it seems from the evidence on soft sediment, foliose coral, and fouling communities that the entire picture of the role of predation and competition may not be directly transferable from the rocky shore to other environments.

8.3.1.7 Summary of Impacts of Consumers in Benthic Food Webs

The examples discussed above highlight several important aspects of the herbivore-producer interaction. First, feeding preference, mediated primarily by chemical composition in the case of grazers and size in the case of predators, is a key element. Deterrents in the producer, quality of producer biomass as food, and sizes of prey and predator often determine how much and which of the producer species will remain in the environment.

Second, the differential effect of herbivores on different species of producers alters competitive relationships. The activity of consumers determines which array of food species are present, depending on whether the preferred food is one species or another, as evident in the cases of the nonpreferred *Chondrus crispus*, and in the preferred *Macrocystis pyrifera*.

Third, the effect of consumers often depends on quantitative factors. To control producers, there must be enough herbivores, and their feeding rate must be faster than the growth of producers; otherwise, the producers will escape control by herbivores and be limited by some other factors. In the case of kelp forests, a minimum density of urchins was needed to cause kelp forests to recede; in the plankton, the number and feeding rates of zooplankton may be insufficient to cope with reproductive rates of phytoplankton, at least during blooms. Where algal cells are close together, as in benthic microalgae, feeding is expedited, and control by herbivores may take place. Every consumer can potentially exhaust its food supply; whether this happens depends on the rates of consumption and renewal of the food supply.

The above three points update the ideas of Hairston et al. (1960) on the control of producers by grazers, but still are far from a complete statement of interaction between herbivores and their food. There is still much to be learned in this field, both as to actual consequences of feeding in different environments and in regard to food-consumers coevolution. For example, Owen and Wiegert (1976, 1981), based on some observations on terrestrial cases, hypothesize that there is a complex mutualism between grazers and plants advantageous to both; this has been disputed (Silvertown, 1982; Herrera, 1982), but there are many other examples of coevolution between herbivores and their food plants (Rosenthal and Janzen, 1979). One such example is the apparent induction of higher concentrations of secondary compounds in leaves of oak trees that were previously defoliated by larvae of gypsy moths (Schultz and Baldwin, 1982). In fact, Mattson and Addy (1975) argue that the long coevolutionary history of consumers and plants has resulted in feeding rates inversely proportional to the productivity of forest vegetation. This relationship may allow herbivores to actually regulate production in forests, in spite of the low consumption typical of forest herbivores. Much of the research in this field is being done in terrestrial situations, but many aspects are extendable to marine environments. In both terrestrial and marine situations much remains to be learned about such interactions.

8.3.2 Case Studies of Bottom-up Controls

Experiments in which nutrients were added to salt marsh parcels show that increased nutrients lead to more salt marsh grass biomass (Fig. 2-27), and to shifts in species composition of salt marsh plants (Fig. 15-11). In addition, the changes brought about by increased resource supply also alter the composition of salt marsh consumer assemblage (Fig. 8-6). The mechanisms underlying the responses of the food web to such bottom-up controls involve the chemistry of producers.

As seen in the rocky shore and kelp forest case histories, the preference by the grazer for one species of producer over another affects the composition of producer communities and raises the important question as to how certain plants deter feeding by herbivores.

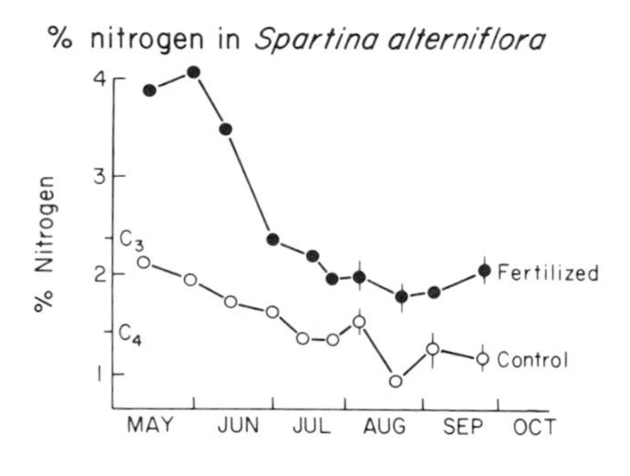

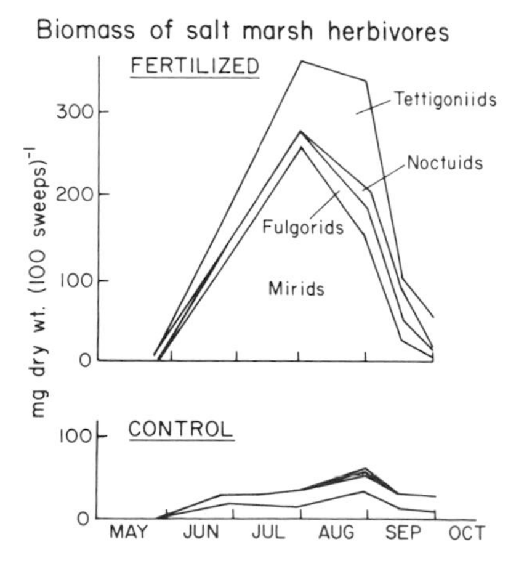

FIGURE 8-6. Effect of increased nitrogen in plants on the herbivore fauna in a Massachusetts salt marsh. Top: Percentage nitrogen through the year in control and experimentally nitrogen-enriched plots of cordgrass (*Spartina alterniflora*). C_3 and C_4 refer to average concentration of nitrogen in C_3 and C_4 plants at harvest, comparable to September values for cordgrass. Middle and bottom: Biomass of herbivores in fertilized and control plots. Unpublished data, I. Valiela and S. Vince.

Deterrents are involved in establishing feeding preferences. Terrestrial plants protect themselves by means of spines, hairs, and chemical defenses (Section 6.3.2). Marine macroalgae have not been studied as thoroughly, but as discussed in Section 6.3.2, many species do have a variety of chemical defenses against herbivores.

The quality of certain marine vascular plants as food for herbivores is also important. Enrichment studies in salt marshes and mangrove swamps (Section 6.3.2) show that there is close coupling between nutritive quality of plants and palatability to herbivores. The biomass of herbivores and the effects of herbivores on the plants are also influenced by nutritive quality. Increased nitrogen content of salt marsh plants* (Fig. 8-6, top) leads to increased biomass of herbivores (Fig. 8-6, middle and bottom).

Herbivores do not respond merely to the nitrogen content of the food plant. In these enrichment experiments, the fertilization also decreased the concentration of certain secondary plant substances (phenolic acids), making the plants more palatable to herbivores. The relatively low consumption of salt marsh plants by grazers is the result of the low nutritive quality of these plant tissues as well as by the relatively high concentrations of secondary plant substances. In salt marshes and in mangroves, which also provide relatively poor food, herbivores do not usually deplete vegetation; they are limited not by the quantity, but probably by the quality of their food plants.

The same holds for many marine grazers that feed on algae growing in kelp forests and intertidal pools. These grazers can eliminate the edible species of macroalgae. The remaining vegetation is composed of a few species—such as Irish moss—that are not consumed by grazers, perhaps because of chemical deterrents. Herbivores in such environments can and often do completely consume the highly palatable or edible food supply; the remainder is undesirable food, so that these herbivores face a limiting supply of food. In both the kelp forests and intertidal rocky shore case studies we lack information as to the particular bottom-up factors that determine the amounts of biomass of the plants that successfully escape grazers. These details will most likely involve nutrients or light.

---

* *S. alterniflora* is a species that shows C_4 metabolism. The name C_4 derives from the four-carbon compound that is the first product of carbon fixation rather than the three-carbon compound typical of the more usual Calvin cycle metabolism. Such C_4 species have a number of remarkable biochemical, physiological, and ecological properties (Black, 1971, 1973). One property of C_4 plants relevant here is that they are relatively free of herbivores (Caswell et al., 1973), and one reason for this is their relatively low nitrogen content (Fig. 8-6, top). The experimental fertilization increases the percentage nitrogen of *S. alterniflora* to that of the C_3 plants, and the ensuing response of the herbivores shows one reason why C_3 plants are more attractive to herbivores.

Researchers who work on soft sediments and rocky shores have placed different emphasis in their studies. There are many studies that show evidence of both bottom-up and top-down controls on structure of communities on soft marine sediments. A list of studies addressing top-down controls (principally predation and competition) in rocky shores would be even longer, but studies of nutrient control on hard-bottom communities are remarkably few. Perhaps it is harder to manipulate nutrient supply than predator abundance in rocky shores, but bottom-up controls do exist, as revealed by a few studies that examined effects of nutrient additions by bird guano in tidal pools (Wootton, 1991; Bosman et al., 1986), and episodic nutrient supply by physical mixing to kelp forests (Zimmerman and Kremer, 1984). It is also conceivable that nutrient supply could have powerful indirect effects on species living on rocky shores. For example, we saw above that nutrients stimulated epiphytic growth on seagrasses, and thus indirectly limited the seagrasses. Nutrients could similarly stimulate epiphytic growth on key species of rocky shores. Enhanced epiphytic growth could lower rates of growth and reproduction in mussels (Dittman and Robles, 1991), mortality in barnacles (Farrell, 1991), as well as increase frequency of dislodgement of mussels (Witman, 1987) and large algae (D'Antonio, 1985). It is therefore plausible that increased nutrient supply could restructure the composition of species assemblages on hard substrates. Further work on the role of increased nutrient supply on hard bottom communities would be welcome, since eutrophication of coastal waters is increasing, and it would be useful to be able to predict consequences of increased nutrients.

8.3.3 Studies of Both Kinds of Controls

We have so far discussed top-down and bottom-up controls of community structure as though they act separately, but in reality, the assemblage and abundance of species in any environment is the net effect exerted by both types of controls (Hall et al., 1970; Young and Young, 1978; Foreman, 1985; Duggins et al., 1989; Williams and Ruckelshaus, 1993; Geertz-Hansen et al., 1993). It has been claimed that top-down and bottom-up mechanisms are about equally important in freshwater systems (Hunter and Price, 1992), and it may well be that the same holds for marine communities. The balance and interaction between the two is likely to vary over the seasons, and probably varies from one marine habitat to another. The task remaining is to understand how the balance of top-down and bottom-up is struck in different environments.

One example of a study in which simultaneous manipulation of nutrient supply and predator pressure were carried out was done in salt marsh creek sediments. Experimental addition of nutrients increases abundance of benthic microalgae for most of the year (Fig. 8-8, top graph). The effects of enrichment are superimposed on seasonal patterns influenced by light and temperature (discussed in Chapters 2 and 15). The bottom-up effect of nutrient supply propagates up the food web, increasing abundance of meiobenthos (Fig. 8-8, middle graph). The meiofauna increase in abundance in response to the microalgal crop created by the larger nutrient supply (Fig. 8-8, middle graph, black lines). When predator access was reduced by installation of cages, the abundance of meiobenthos increased, more markedly so where nutrients had been added (Fig. 8-8, middle graph, dashed lines). The "predators" also consumed some chlorophyll, since caged marsh sediments maintained higher chlorophyll standing crops than uncaged sediments (Fig. 8-8, top graph). The macrofauna showed a seasonal peak in abundance in late spring, and decreased markedly as a result of intensified predation by fish (Fig. 8-8, bottom left). Growth of the predators was closely tied to prey abundance: for fish of any size, growth was higher in early summer, and decreased as supply of benthic macrofauna decreased (Fig. 8-7). Fish leave the marsh after late summer as food supply dwindled (Werme, 1981). Abundance of macrofauna then increased again (Fig. 8-8, bottom left). The importance of top-down control during late summer is made evident by cage exclusion experiments in which abundance of macrofauna remained high even

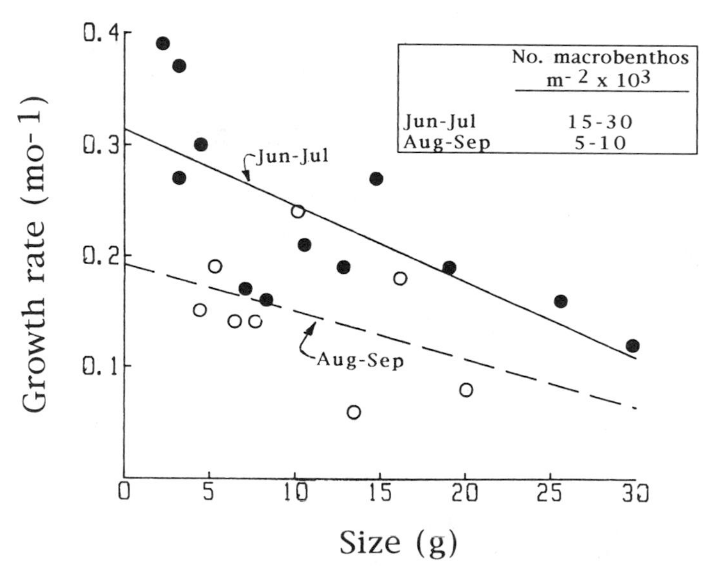

FIGURE 8-7. Growth rates of *Fundulus heteroclitus*, the principal predatory fish in Great Sippewissett salt marsh, for periods of the year when availability of prey of suitable size differed (inset box). Data shown for fish of different length. Data from Valiela et al. (1977), and Wiltse et al. (1984).

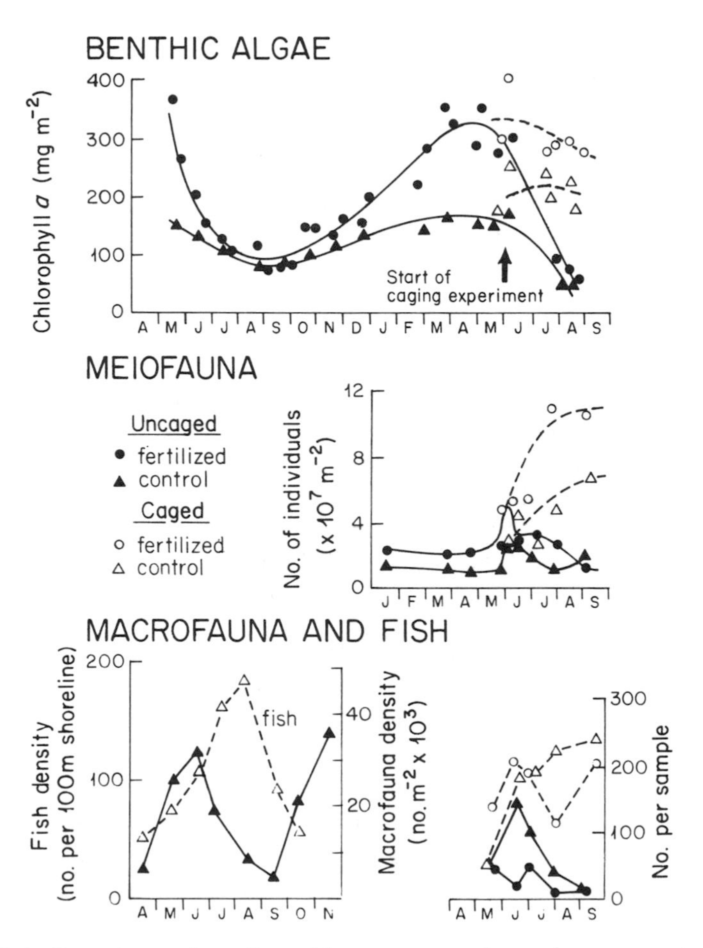

FIGURE 8-8. Experimental studies of bottom-up and top-down controls in soft sediments of Great Sippewissett salt marsh. Top: Seasonal changes in benthic chlorophyll in nutrient-enriched and control sediments and in sediments inside and outside predator exclosures (from Foreman, 1989). Middle: Abundance of meiofauna in the same experiments (from Foreman, 1989). Bottom left: Abundance of fish and macrofauna in control salt marsh sediments (from Werme, 1981; Wiltse et al., 1984). Bottom right: Abundance of macrobenthic fauna in the experimental treatments described for top two graphs (from Wiltse et al., 1984).

during late summer, the time when macrofauna in unprotected sediments decreased (Fig. 8-8, bottom right).

The salt marsh experiments show that abundance of various groups of benthic animals is determined by a tight intertwinning of bottom-up and top-down controls. The relative importance of the controls changes over a season, and depends on changing key features (nutrient supply, abundance of fish and prey organisms, light, temperature). Abundance of prey and predators, and their growth rates, are tightly coupled, even though continually changing in response to the external variables (light, temperature), and to the natural history of the organisms.

The simultaneous manipulation of bottom-up and top-down controls in salt marsh sediments demonstrate that these overall controls do affect communities in the field. We have to add, however, that not all populations showed the same responses; among the meiofauna, for example, the nematodes and copepods did respond to nutrient supply and predator exclusion, while the abundance of foraminifera increased in cages that were not enriched, and abundance of ostracods was diminished by nutrient enrichment (Foreman, 1985). Aggregate variables such as "total ind. m^{-2}", and "mg Chl. m^{-2}" might at best represent trends followed by dominant species, at worst a meaningless aggregate of disparate responses by different species of more equitable abundance. This is a general issue; in planktonic studies, for instance, Davis (1987) showed that changes in abundance of coastal zooplankton were best simulated when species-specific effects of driving variables were included in the model. Key responses were species-specific (Davis, 1984; Davis and Alatalo, 1992); responses to temperature differed among species, and, for example, *Pseudocalanus* spp. and *Paracalanus parvus* were not food-limited, while *Centropages typicus* was. Consideration of populations, particularly prominent species, are therefore desirable, whenever possible, for field and model studies, and probably more informative than aggregate variables.

Chapter 9
Trophic Structure 2: Components and Controls in Water Column Food Webs

9.1 Food Webs in Marine Water Columns

No topic within marine ecology and biological oceanography has changed more in the last decade than our notions about components and structure of planktonic food webs. Knowledge about marine water column food webs has been considerably enlarged, and made much more complex, by recent findings about the existence and role of smaller organisms, release and reuse of dissolved organic matter, and reassessment of the function of certain larger organisms. The changes in our understanding have been significant enough (Pomeroy, 1974) to amount to a new "paradigm", a signal advance in understanding (Kuhn, 1970).

The classic planktonic food web included diatoms, dinoflagellates, and other microalgae, which provided food for microzooplankton (mainly copepods), which in turn were eaten by larger consumers. Several lines of evidence have emerged since the early 1980s, findings that showed that earlier notions of planktonic food webs needed revision (Pomeroy and Wiebe, 1988). First, the advent of new methods, as already discussed, permitted researchers to identify new groups of important microorganisms, and showed that microbes were far more abundant than previously thought; perhaps half of the biomass in the plankton was contributed by microbes (Section 1.1.1.3). Second, small organisms were rather more active than was realized: production by small chlorophytes and coccoid bluegreens often exceeds that by diatoms and dinoflagellates. Third, bacterial activity was also high: bacterial respiration may consume much of primary production. High rates of bacterial production required large amounts of available organic matter in the water, which posed a problem, since sufficient sources of dissolved organic matter were unknown. The answer to this dilemma was provided by studies that showed release of organic materials by phytoplankton (by exudation, breakage during grazing by consumers, and more recently by viral lysis), bacteria (by release of exoenzymes, incomplete assimilation of the attacked organic matter, and viral lysis), and animals (via molts, feces, and mucus release). Fourth, bacterial numbers in seawater were higher than estimated by older plate culture methods, but their abundance fell in a relatively narrow range, which suggested that some strong controls were restraining bacterial numbers.

The array of new facts prompted Azam et al. (1983) to propose a new view of the planktonic food web that included a "microbial loop", in which organic matter was cycled through microbes before entering the classic food web, in addition to the linear arrangement of the classic food web. The new view continues to expand; every passing month brings a new trophic relationship among microbes to light, so that the neat microbial loop initially proposed is becoming far more involved (Sherr and Sherr, 1988) and is best described as the microbial food web.

Since so much of primary production is carried out by small organisms, it is of some importance to ascertain whether any of the carbon that courses through the microbial web makes its way up to larger consumers. Larger animals can consume smaller organisms such as bacteria (Crosby et al., 1990; Langdon and Newell, 1990), but there has been disagreement about the quantitative fate of microbial carbon (Pomeroy and Wiebe, 1988).

Our concepts of the role of larger animals in water column food webs, species large enough to swim in inertial conditions, have also been markedly transformed. Studies on larger consumers found in water columns demonstrate that many gelatinous taxa were more abundant and active than was thought.

Below in this chapter we update notions of the workings of water column food webs. Since size (Section 6.3.1) is of great importance to consumers, we divide the subject into three size domains. First, we discuss microbial food webs, webs whose components range from pico- to nanoplankton sizes. Second, we review knowledge of the classic food web, where the organisms range from micro- to mesoplankton. Third, we speculate about controls of nekton food webs, whose organisms are large enough to live in an inertial rather than in a viscous world. Throughout this treatment, we identify the components, trophic links, and assess what is known about control mechanisms.

9.2 Microbial Food Webs

The principal groups that fix carbon through photosynthesis in the plankton are prochlorophytes, coccoid cyanobacteria, flagellates, dinoflagellates, and diatoms. These groups span a size range between the pico-and mesoplankton. As already mentioned, the contributions to total biomass (Section 1.1.1.3) and production (Section 1.4.1) by smaller taxa are as or more important than those of larger producers. During most of the year smaller size groups dominate production. Larger microplankton such as diatoms become prominent during seasonal blooms of short duration (cf. Section 15.3.1).

Phytoplankton are consumed, usually in a size-specific fashion, by a myriad kinds of organisms. The major consumers in water column communities, and their role in control of their food organisms, are discussed below.

9.2.1 Viruses

Viruses are likely to play important roles in marine food webs. Viruses are exceedingly small, 0.02–0.2 μm in diameter. Recent studies show 10^6 to 10^7 virus particles per ml in ocean surface waters, and up to 10^8 in richer coastal waters (Suttle et al., 1990; Fuhrman and Suttle, 1993). Virus infections are largely host-specific, and are therefore remarkably diverse. Viral populations infect a wide variety of host cells, including bacteria, cyanobacteria, and eukaryotic phytoplankton, but most are bacteriophages (Cochlan et al., 1993).

Depending on the turnover rate of virally infected cells, viruses could potentially reduce phytoplankton and bacterial populations. Experimental additions of viruses obtained from seawater reduced densities of cultures of diatomas, cyanobacteria, and flagellates by about one order of magnitude (Suttle et al., 1990). Further, the viral additions lowered primary production in the cultures by 78%. About 2–16% of marine bacteria may contain mature phage particles (Proctor and Fuhrman, 1990; Heldal and Bratbak, 1991). Such infection rates could lyse 2–24% of the bacterial population per hour. Suttle and Chen (1992) estimated that 4–13% of bacteria are infected daily by viruses; if bacteria double once a day, viruses are responsible for 8–26% of bacterial mortality. Similar calculations suggest that viruses may be capable of causing a mortality of 5–10% of cyanobacterial populations daily (Fuhrman and Suttle, 1993).

Virally caused mortality of phytoplankton may be the mechanism behind heretofore unexplained lysis of cells (Cottrell and Suttle, 1991) and losses of inorganic carbon in excess of primary production (Proctor and Fuhrman, 1990), and could cause significant seasonal changes in abundance of phytoplankton. Viral-induced mortality could account for 25–100% of the mortality suffered by a bloom of the coccolithophore *Emiliania huxleyi*, which led Bratvak et al. (1993) to conclude that the viruses may have caused the end of the bloom.

Virus-caused algal mortality increased in experiments where phosphate was made available (Bratvak et al., 1993). Low supply of phosphate, but not nitrate, lowered the impact of the viral infection. The controlling effect of phosphate is probably related to the lower N/P (which largely reflects the ratio of protein to nucleic acid) of viruses relative to algal cells. Viruses are of course mainly nucleic acids, and lack protein.

The role of viral lysis in the sea remains to be evaluated, particularly since some initial results have been hard to reproduce, but viruses could be involved in many important processes, including terminating phytoplankton blooms, release of phytoplankton and bacterial DOC as a result of cell lysis, fostering formation of aggregation of DOC into amorphous particles, and maintainance of diversity of hosts (Fuhrman and Suttle, 1993). Such potentially key points are discussed further below, and need further study.

9.2.2 *Bacteria*

Bacteria are important in the metabolism of aquatic ecosystems. Carbon demands by bacteria, for example, are equivalent to 40–60% of carbon fixation by producers (Cole et al., 1988).

Bacteria obtain organic matter to support their metabolism from at least three sources. Uptake of dissolved organic compounds released by other organisms, principally those in the classic plankton food web (cf. Sections 13.2.3, 13.2.4.2.2), increases in nutrient-richer systems, but supports less than half the carbon required to support bacterial growth (Baines and Pace, 1991). Bacteria also release exoenzymes to lyse particulate organic matter; this is probably as important quantitatively as uptake of already-dissolved organic matter, but of course requires organic particles. A few bacteria attack living larger organisms (flagellates, diatoms, and dinoflagellates) to obtain organic matter (Imai et al., 1993); these predatory gliding bacteria are not common and require surfaces.

Bacteria require nutrients as well as dissolved organic matter, but it is difficult to sort out the effects of nutrients and organic matter. The latter may be released or produced by phytoplankton, but phytoplankton depend on nutrient supply, so supplies of nutrients and organic matter, and abundance of bacteria and phytoplankton, are linked (Banse, 1991; Ducklow and Carlson, 1992). There are experimental, within-system, and cross-system conparisons that show the relative role of nutrients and organic substrates, and the coupling of bacteria and phytoplankton.

Nitrogen enrichments done in MERL tanks increased bacterial abundance and productivity (Hobbie and Cole, 1984). Other enrichment experiments showed in early spring, when coastal waters are rich in nutrients, bacterial productivity increased when an organic substrate was provided (Fig. 9-1, top). During summer, when nutrients have usually been depleted by phytoplankton, bacterial production increased when additional nutrients were added (Fig. 9-1, bottom).

Karner et al. (1992) took advantage of different nutrient supply to waters along a transect in the Adriatic to show that the more chlorophyll in the water, the higher the rates of bacterial production. Water from eutrophic waters in a Belize reef contained more bacteria than oligotrophic waters, but, curiously, bacteria from nutrient-poor waters grew better in response to microplanktonically generated organic supplements (Peduzzi and Herndl, 1992).

Compilations of data from many aquatic systems show that bacterial production increases as phytoplankton production increases (Fig. 9-2*). Points from marine systems tend to lie below the points from freshwater systems, where presumably there are larger quantities of organic matter in the water owing to nearby terrestrial sources. Bacteria are more productive relative to phytoplankton (up to 5:1 in terms of carbon) in nutrient-poor waters, and become relatively less productive (less than 1:1) in richer waters where there are larger populations of phytoplankton (Fig. 9-2). The experimental enrichment studies and the cross-system compilations therefore suggest that bacterial activity in marine waters is strongly affected by availability of nutrients and

---

*Comparative studies such as Fig. 9-2 provide useful views across a wide range of systems, but interpretation of cross-system data should be conditioned by certain caveats. First, cross-system comparisons unavoidably confound effects of variables of interest (bacteria and phytoplankton abundance, in the case of Fig. 9-2) with effects of other variables inherent in comparisons made across different places or times. In the case of Fig. 9-2, the large scatter typical of such comparisons stems from the multiplicity of variables that affect phytoplankton and bacteria across the different sites. Temperature, for example, affects bacterial abundance and activity (White et al., 1991); in cold waters, for example, bacteria are less effective at using organic substrates for growth (Wiebe et al., 1992). The second caveat is that conclusions from cross-site comparisons apply to the entire data set, and may not apply to any one place in particular. For example, while for the aggregate data from all marine sites in Fig. 9-2 there is a positive relation between phytoplankton and bacteria, we should not assume that the same holds for a given parcel of water. Third, correlations do not necessarily identify causes; it is difficult to be sure that phytoplankton is the governing agent in the relationship being studied. Manipulative experiments that include or exclude factors to be tested are better suited for study of how specific communities function. Manipulations do have drawbacks: they may be difficult to carry out, usually can only be applied to small spatial scales, and sometimes create unrealistic conditions. Nonetheless, if well-conceived, manipulations have the advantages of establishing causality and testing specific questions unconfounded by nuisance variables. The ideal situation is to establish mechanisms and causes by experimentation, and to extend the range of application by cross-system comparisons.

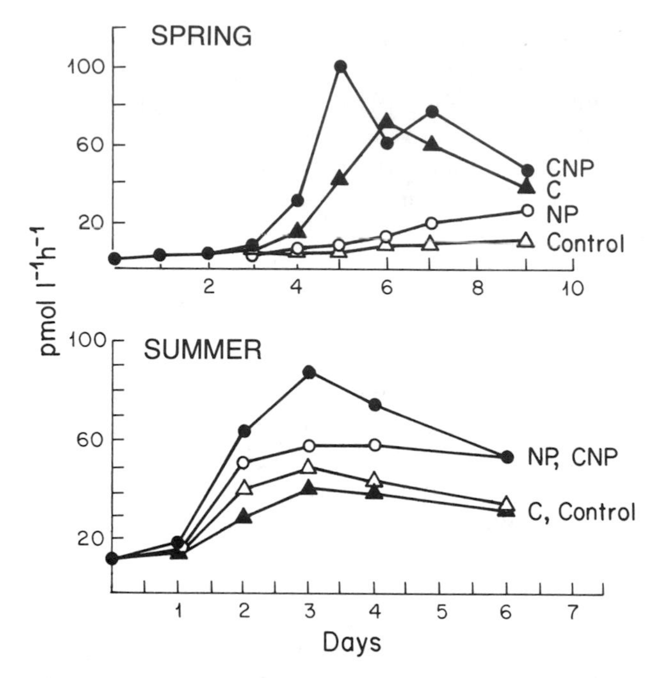

FIGURE 9-1. Thymidine incorporation rate in Baltic Sea bacteriplankton, in batch enrichment experiments done in early spring (top), and summer (bottom). Enrichments consisted of addition of sucrose (C), NH_4Cl (N), or KH_2PO_4 (P). control batches received no additions. Adapted from Kuparinen and Kuosa (1993).

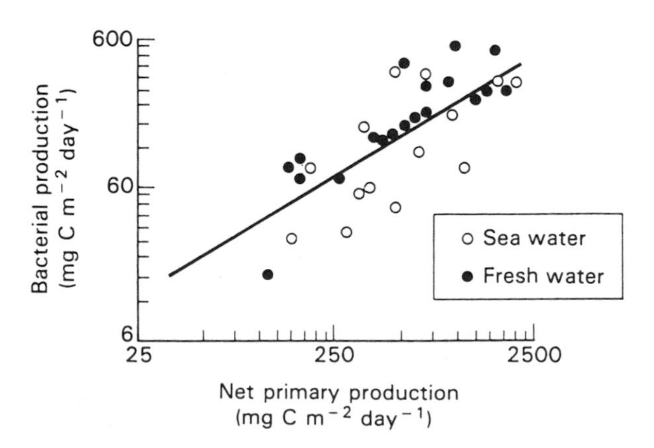

FIGURE 9-2. Relationship of bacterial and net primary production in many different bodies of freshwater and marine environments. Adapted from Cole et al. (1988).

organic matter, and that phytoplankton are a major source of the organic matter.

9.2.3 Heterotrophic Nanoflagellates

Heterotrophic nanoflagellates* feed on viruses, bacteria, and presumably also on other pico- and some nanoplankton. In one study, viruses provided up to 9% of the carbon, 14% of the nitrogen, and 28% of the phosphorus consumed by nanoflagellates in cultures where there were 10^6 bacteria ml^{-1} and 10^7 to 10^8 viruses ml^{-1} (Gonzalez and Suttle, 1993). Bacteria thus furnish most of the food for nanoflagellates, but abundance of heterotrophic nanoflagellates is weakly correlated to that of bacteria (Sanders et al., 1992; Gasol and Vaqué, 1993). Densities of bacteria in different waters range from 10^5 to 10^8 cells ml^{-1}, and those of nanoflagellates from 10 to 10^5. In specific systems there are usually on the order of 1,000 bacteria per 1 nanoflagellate (Fenchel, 1986; Sanders et al. 1992).

Nanoflagellates are major consumers of bacteria. For example, if the doubling time of bacteria is 9 h, and the rate of nanoflagellate grazing is $2.6\,d^{-1}$, for a bacterial population of 4.6×10^5 cells ml^{-1}, the nanoflagellates will remove 63% of the bacteria daily (Bautista et al., 1994). Other such calculations range up to 250%, perhaps high enough to match bacterial division rates and prevent increases in bacterial population numbers (Andersen and Fenchel, 1985; McManus and Fuhrman, 1986).

Evidence on top-down control of bacterial abundance by nanoflagellates in the field is inconclusive (Ducklow and Carlson, 1992; Gasol and Vaqué, 1993). Cross-system comparisons have concluded that there is (Sanders et al., 1992) and there is not (Gasol and Vaqué, 1993) a relationship between bacterial and nanoflagellate abundance in aquatic environments. Similar comparative studies also show that rates of consumption of bacteria by nanoflagellates are correlated to rate of production by the bacteria (Fig. 9-3), and that on aggregate, since the points fall around the 1:1 line, there seems to be a rough balance between production by and consumption of bacteria by nanoflagellates. Note, however, that any one system may be "out of balance" by up to one order of magnitude (Fig. 9-3).

A few manipulative studies have been done to ascertain the importance of bottom-up and top-down control of bacteria. Kuparinen and Bjørnsen (1992) tested whether bottom-up or top-down mechanisms controlled Southern Ocean bacterial abundance by adding nutrients and excluding different-sized consumers in 100 l containers.

---

*There are many nanoflagellates that are both heterotrophs and autotrophs (Estep et al., 1986). These species contain chlorophyll and can also engulf particulate prey.

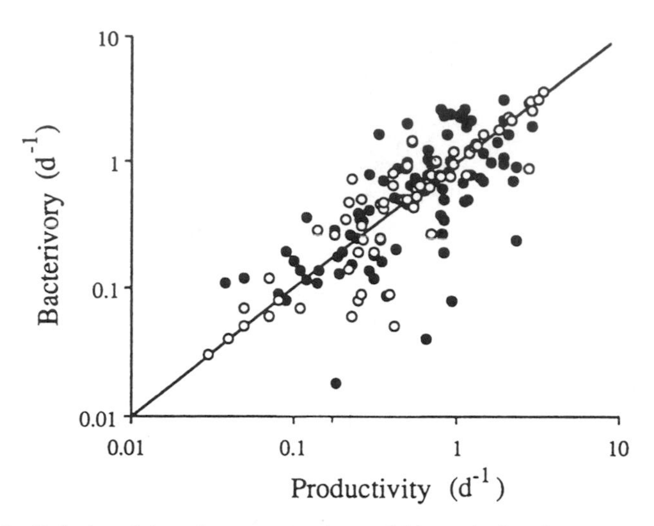

FIGURE 9-3. Relationship of consumption of bacteria by heterotrophic nanoflagellates and production by bacteria. The line shows the 1:1 relationship. From Sanders et al. (1992).

Bacterial abundance was less affected by increased nutrients (and presumably organic substrates) than by consumption by heterotrophic nanoflagellates. In treatments where larger dinoflagellates were present, they consumed nanoflagellates and allowed more bacterial growth. Nanoflagellates therefore have the potential to control bacteria, but it may be that in natural waters larger predators hold down the abundance of nanoflagellates, creating a cascading that releases bacteria from control by nanoflagellates.

Experiments in fresh water show that bacterial abundance depends more on resources than on nanoflagellate activity (Kirchman, 1990; Pace and Funke, 1991; Weisse, 1991). There is a also a trophic cascade effect, in that the nanoflagellates are kept in check by larger predators, and do not control bacterial abundance.

There are few estimates of the grazing impact of nanoplankton in the field. Fenchel (1982a) measured the abundance of six species of flagellates in waters of Limfjord off Denmark. The flagellates ranged in size from about 0.5 to 10 μm. Using estimates of the volume of water cleaned per organism, Fenchel calculated that, on average, 20% of the water of Limfjord was filtered by the microflagellates per day. The flagellates thus could consume 20% of the bacteria daily. In the Baltic, bacterial generation times may vary between 10 to 100 hr (Hagström et al., 1979). Fenchel (1982a) calculates that if the flagellates consumed 20% of the bacteria daily, and had a gross growth efficiency of 35%, the bacteria could support a growth rate of the flagellate population of about one doubling per day. This is a high rate and, if true, implies that grazing has to be an important factor affecting bacterial abundance.

Comparative and experimental results show that in the absence of larger predators, nanoflagellates can effectively control bacterial abundance, but that where larger predators are present, as occurs in most field situations, control of bacteria by nanoflagellates is less likely. Cross-system comparisons suggest that nanoflagellates may be more able to control bacteria numbers in nutrient-rich waters, but may be less effective in low-nutrient environments, where bacterial numbers may be controlled by resources (Sanders et al., 1992). Similar comparisons led Gasol and Vaqué (1993) to conclude, in contrast, that abundances of bacteria and nanoflagellates were more closely linked in oligotrophic waters. The interplay of top-down and bottom up controls on bacteria needs more study.

9.2.4 Nano- and Microplanktonic Protozoans

Various microplanktonic protozoan groups, mainly ciliates and heterotrophic dinoflagellates, consume bacteria, nanoplankton, and microphytoplankton (Lessard and Swift, 1985; Verity, 1991; Caron et al., 1991). Some phytoplankton may engage in bacterivory to acquire nutrients during periods when dissolved nutrients are in low supply in the water (Nygaard and Tobiesen, 1993). Larger protists such as ciliates do not consume bacteria too effectively. McManus and Fuhrman (1986), for instance, report that only 0.3% of the bacterial crop was removed daily by ciliates.

Top-down control by protozoans seems likely in many cases, but sorting out the ability of protozoans to control the abundance of their prey is made difficult because not only do protozoans engulf smaller cells, but they also release nutrients, which can stimulate growth of their pico- and nanoplankton prey (Ferrier and Rassoulzadegan, 1991).

9.3 The Classic Microplankton Food Web

9.3.1 Size Classes Within Microzooplankton

Consumers of the classic planktonic food web and their particulate prey occur in the water column in a very wide variety of sizes. To deal practically with this very large size range, a series of categories has been proposed, as discussed in Section 1.1.1.2: picoplankton, smaller than 2 μm; nanoplankton, chiefly small plankton between 2 and 20 μm; microplankton, consisting of large phytoplankton and small zooplankton about 20–200 μm; macroplankton, made up of large zooplankton ranging in size between 200 and 2,000 μm; and megaplankton, the remaining plankton larger than 2,000 μm. There are consumer species in every one of these categories, and each consumer species eats smaller species. Research has focused on grazing by macroplankton since measurements are easier, but grazing by nanoplankton and microplankton is important (Pomeroy, 1974; Williams, 1981). The larger size categories (2 cm to 20 m) often consist of active swimmers (nekton) in addition to the more passively drifting plankton. Such distinctions are not always clear, and different classifications have been forwarded (Sieburth et al., 1978).

A large proportion of phytoplankton standing crops seems available and palatable to herbivores, even though there is some evidence from freshwater studies that certain phytoplankton such as blue-green bacteria are unpalatable (Arnold, 1971; DeMott and Moxler, 1991) or undigestible (Porter, 1973, 1976) for zooplankton. Certain dinoflagellates such as *Amphidinium carterae* produce choline-like substances that may be grazing deterrents (Wangersky and Guillard, 1960) and are rejected by their tintinnid grazers (Stoecker et al., 1981). There are many examples of preference by grazers for one phytoplankton species over another, but there is little information on the cues used by grazers to establish preferences. Most algal cells seem to be relatively palatable to some herbivore.

The standing crop of algae is always much lower than that of terrestrial plants, although production rates may be comparable (Chapter 1). In the English Channel and in the South Atlantic, standing crops may be on the order of 2–3% of the annual production. In sharp contrast to the green terrestrial world, the ocean is blue, largely devoid of green cells. In part, the low standing crop of phytoplankton typical of seawater is due to low nutrient supply, and to consumption of producers by grazers.*

Most copepods, one of the dominant components of the classic food web, are omnivorous rather than herbivorous or carnivorous, as assumed in the classic notion. Copepods readily consume autotrophic or heterotrophic nanoflagellates (Caron, 1991), and protozoans (Stoecker and Capuzzo, 1990). Contrary to earlier notions, therefore, copepods feed not only on phytoplankton, but on a large variety of foods, selected largely on the basis of size rather than taxonomic categories.

Some microzooplankton can consume cells smaller than 2 μm. Bivalve larvae can ingest *Synechococcus* cells $1 \times 0.5\,\mu$m in size (Gallager, 1988)

---

*In terrestrial plants extensive support and vascular systems are needed to keep photosynthetic tissues exposed to adequate light and transport nutrients; in the water column water movement supplies these functions. The extensive, nongrowing tissues of terrestrial plants slow turnover rates (Table 2-3). This contrast is also important in chemical defense, since many of the deterrents are usually chemically associated with structural tissues.

and are capable of assimilating bacterial carbon (Douillet, 1993). Most microzooplankton, however, do not feed effectively on particles in the small nanoplankton range (2–5 μm, mainly heterotrophic flagellates), although certain rotifers, cladocerans, and even larger animals such as pteropods, salps, doliolids, and larvaceans can do so (Stoecker and Capuzzo, 1990). Most suspension-feeding copepods, rotifers, and cladocerans, gelatinous mucus-net feeders, and raptorial fish larvae feed more effectively on nanoplankton within the 5–20 μm range, and best of all on microplankton-sized prey.

9.3.2 *Trophic Transfers Between Microbial and Classic Food Webs*

If we have a microbial food web in which there are several trophic steps, there must be a significant loss of carbon at each transfer. The efficiency with which bacteria, for example, transform carbon in substrates to cells has rough bounds at about 10–50% (Table 9-1). Earlier estimates were higher, but are likely to have been overestimated (King and Berman, 1984). Protozoan yields average 32%, with large variation (Caron and Goldman, 1990). If we assume an unrealistically high 50% transfer at any trophic step, 12% of the carbon passes 3 steps, and only 6% survives four steps (Cole et al., 1988; Pomeroy and Wiebe, 1988).

Actual measurements in planktonic systems show insignificant carbon transfer from bacteria to crustaceans larger than 10 μm (Ducklow et al., 1986), except in a few environments dominated by consumers adept at capture of bacteria, such as cladocerans in lakes (Wylie and Currie, 1991). The conclusion is that there is likely to be little transfer of carbon from microbial to classic planktonic food webs.

In the benthos there is a possibility that microbial transfer is more important. Free-living bacteria at most supply only 13 and 16% of the C and N assimilated by clams, but 3/4ths of bacteria are associated with

TABLE 9-1. Estimates of Yield Efficiency of Bacteria Growing on Different Organic Substrates.[a]

	% conversion of substrate carbon to biomass carbon	
Source of Organic Carbon	Dissolved OC	Particulate OC
Plants	4–53	2.5–37
Macroalgae		6–43
Phytoplankton	10–36	10–58
Seston	31	8.8–20

[a] Summary of a compilation from various sources by Pomeroy and Wiebe (1988) and Alber and Valiela (1994). Seston includes all particles in water.

particles (Werner and Hollibaugh, 1993). Use of both free and attached bacteria could double the importance of bacteria as clam food. Moreover, perhaps half the bacterial production could also be made available as nanoflagellates and ciliates, both of which can serve as food for clams (Werner and Hollibaugh, 1993). These are speculations, but they do suggest that it is too soon to disregard the role of the microbial food web as a source of food for larger organisms, at least in the benthos.

9.3.3 Controls of the Relationship of Phytoplankton and Microzooplankton

The relationship between classic phyto- and zooplankton has been simulated in many models (Riley, 1947, 1963; Steele, 1974a,b; Wroblewski and O'Brien, 1976; Landry, 1977; Davis, 1982; among others). Newer models introduce size-based dynamics (Moloney et al., 1991), reflecting new knowledge about food webs. Although progress has been made, there is still uncertainty about the formulation of the links between consumer and food. To better define empirical relationships of food to consumers, correlational and experimental approaches have been applied.

Bottom-up controls of the classic food web are manifest by a variety of results. Maps of global distribution of zooplankton show peaks closely tied to peak phytoplankton production (Bardach and Santerre, 1980). Surveys of herbivore biomass, and grazing rates, show both increase as primary production increases (Cyr and Pace, 1993; Cebrián and Duarte, 1994). Standing crops of zooplankton are positively correlated to chlorophyll concentrations (Beers and Stewart, 1971; Smith et al., 1981). A causal connection may not underlie such correlations, but experimental manipulations do show that increases in phytoplankton (as a result of nutrient enrichments) produce more zooplankton (Fulton, 1984; Nixon et al., 1986).

Top-down controls seem more difficult to find in the classic food web. Tenfold increases of chlorophyll concentration followed removal of zooplankton from oceanic seawater in simple shipboard experiments (Thomas, 1979). Zooplankton may graze phytoplankton effectively in nutrient-poor waters, or at times of year during which low nutrients reduce phytoplankton division rate. Some zooplankton can graze most of the phytoplankton production (cf. Fig. 6-9); most of these examples come from warmer oligotrophic ocean waters, where primary production rates may be lower than in coastal areas, and grazers are active most of the year (cf. Section 15.2.1).

9.4 Speculations as to Control of Prey Populations by Larger Predators in the Marine Water Column

9.4.1 The Kinds and Sizes of Predators in Pelagic Food Webs

Planktonic predators often feed on each other, and usually have very variable diets. There are also marked changes in plankton communities even over relatively short periods of time. These features make it difficult to define planktonic food webs. Just as there are herbivorous species in most size groups of zooplankton (Section 9.3), there are a plethora of predaceous species of every size. In the microzooplankton, for example, there are ciliated protozoa, including tintinnids and oligotrichs, that consume naked ciliates, as well as graze on dinoflagellates. Macro- and megaplankton, and nekton, including copepods, euphausiids, bony fishes, sharks, mammals, birds, squid, and some of the larger gelatinous plankton are usually nearer the top of pelagic marine food webs.

The small plankton consume even smaller particles and are themselves the prey for still larger predators. Oligotrophic waters may have more size classes of zooplankton than eutrophic waters. Predation is size-dependent (Chapters 5 and 6), so the more size classes of plankton that there are, the more kinds of predators that can be present. It is therefore possible to have a larger number of links in oligotrophic oceanic than in coastal or more eutrophic food webs.

Pelagic food webs tend to be "unstructured" (Isaacs, 1973) because there are many predaceous species that are only slightly larger or smaller than other potential prey species. Most zooplankton have the capacity to feed on most everything within a certain size range. Since few species of marine zooplankton are therefore free of some predation, it may be difficult for any one species to build up its own abundance in response to the abundance of its prey. Thus, the complex feeding relationships among consumers themselves may prevent planktonic predators from completely determining the number and kinds of prey, because not many predators are themselves free of predation pressure.

In contrast, freshwater food webs do not often consist of as diverse a fauna as marine planktonic food webs, and tend to be more unidirectional (more "structured"). A freshwater predator may be able to respond more clearly to changes in prey density and, rather than be limited by its predators, it is more likely limited by its own food supply.

There are many major unanswered questions about the role of the various groups of planktonic predators. In particular as already mentioned there is a notable lack of information on the abundance and impact of nano- and microzoo-plankton, squid, and gelatinous zooplankton, and on their importance as consumers. The problem is mainly the lack of adequate methods to estimate densities of, for example, gelatinous zooplankton. Observations during *in situ* diving by R. Harbison and L. Madin and associates (Harbison et al., 1978) indicate that gelatinous zooplankton may be as important consumers as copepods. In some areas, jellyfish are major predators of zooplankton. Matsakis and Conover (1991) suggest that medusae exert considerable control over zooplankton in Bedford Basin, Nova Scotia; medusae consumed 10–59% of the zooplankton daily, a rate that may exceed growth rates of zooplankton. Gelatinous species consume 2–78% of the zooplankton (Purcell, 1992), and 20–40% of fish eggs and larvae (Cowan and Houde, 1993) in Chesapeake Bay. Gelatinous nekton are therefore abundant enough to consume significant amounts of prey, but it is not evident yet whether they exert control of prey populations. Predation among gelatinous species could diminish the controlling role of these species (Matsakis and Conover, 1991; Cowan and Houde, 1993), by cascading effects. Other predators still need study. Squid, for example, are often abundant, and are uniformly predators, but their ecological function is poorly known.

Throughout the discussion on controls we have focused on trophic relationships, but there are other mechanisms whereby some organisms restrict the abundance of others. Release of exocrine substances is one example of nontrophic controls. In Narragansett Bay, Rhode Island, the red tide flagellate *Olisthodiscus luteus* is abundant following blooms of *S. costatam*; when the flagellate density decreases, *S. costatum* increases again (Pratt, 1966). Water previously inhabited by high densities of *O. luteus* inhibits growth of the diatom, evidence that substances released by *O. luteus* may prevent its replacement by the diatom. Similarly, high densities of *O. luteus* reduce growth and increase mortality of two tintinnid grazers (Verity and Stoecker, 1982). Other examples reviewed in Johnston (1963a) also suggest that replacement of one species by another may be mediated by antimetabolites. Red tide dinoflagellates provide another instance of release of toxic substances, although it is not clear just what is the target organism for the very powerful dinoflagellate toxins; the toxins do not always inhibit invertebrate grazers such as shellfish, although they may deter vertebrate grazers (White, 1981). These compounds may be anticompetitive (allelopathic) devices. Allelopathic relations are characteristic of situations where resources such as water (on land) or surfaces (in marine hard-bottom, coral reefs etc.) are in short supply.

There are other more exotic mechanism that are used by organisms to prevent activity of competitors or predators. Certain crabs avoid predation by being well camouflaged when in its chemically protected host macroalga (Hay et al., 1990a). Tube-building amphipods construct tubes using fragments of a fairly rare brown alga that is chemically protected by diterpenes that deter consumers (Hay et al., 1990); the tubes containing the alga also protect the amphipods. Sea slugs and snails acquire the defensive chemical armory synthesized by seaweeds, sponges, bryozoans, ascidians, cnidarians, and other mollusks on which seaslugs feed, and transfer the defensive substances to body parts exposed to predators (Todd, 1981; Hay, 1990a; Avila, 1995). Other seaslugs eat coelenterates and ingest nematocysts. The seaslugs somehow manage to keep the nematocysts untriggered while they are relocated to the seaslug mantle. The relocated nematocysts then serve as defense against predators of seaslugs.

In spite of the long history of research in the sea and lob of specific examples, such as those cited above, there is very little known about the role of most invertebrate planktonic consumers in the sea.

There is just as little known of the role of larger predators, such as fish and mammals, as controls of marine food webs, even though bony fishes are the most studied of all marine predators, primarily due to the impetus of commercial fisheries. Marine birds and mammals are also important and obvious predators that have attracted attention and some data on these groups of consumers are available. In the sections that follow we examine what is known of the role of fish, mammals, and birds as potential predators in pelagic food webs. We treat the fish separately from the mammals and birds because, as will be seen, most fish use a reproductive stategy very different than mammals and birds. This fundamental difference (typically large numbers of tiny young in fish versus few, large young in mammals and birds) leads to very different results in their role as predators.

9.4.2 The Role of Fish in the Sea

Cross-system comparisons suggest that fishery yield increases as the average primary production rate of aquatic systems increases (Fig. 9-4). Yield is a proxy for fish abundance; it may or may not reflect actual abundance in the sea but is the kind of data most available. In the case of marine fish we are dealing with populations exploited by the fishing fleet, so cascade effects might prevent fish from more closely controlling their prey, rather than reflecting food abundance. Nonetheless, the comparisons show considerable bottom-up influence: the more nutrients are available, and the greater producer activity, the greater fish harvest. The comparisons also show that estuaries are among the most productive aquatic systems, and that for some reason, on average, freshwaters yield lower fish crops than marine waters.

Although on aggregate the comparative data show evident bottom-up effects, we know from Chapter 8 that fish are effective predators in specific shallow water marine environments, and exert considerable

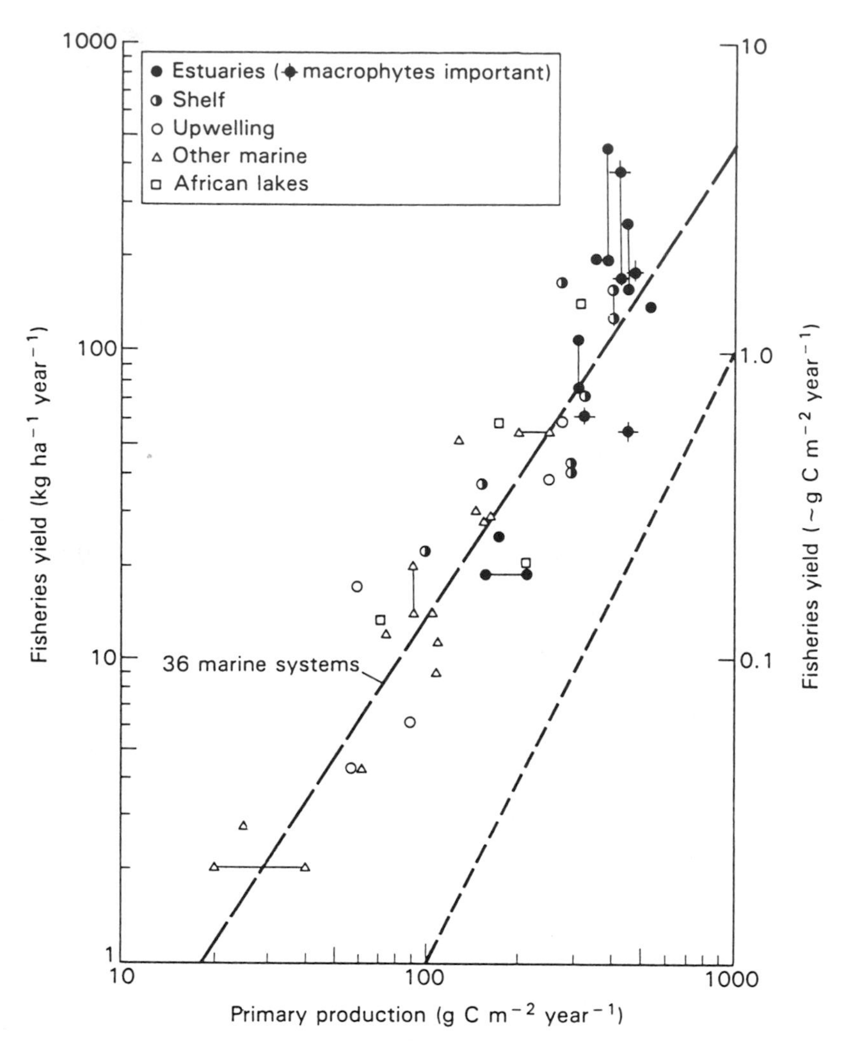

FIGURE 9-4. Annual fisheries yield per unit area as a function of annual primary production for many different aquatic ecosystems. The dashed line shows best-fit line for 15 freshwater bodies other than the large African lakes. Adapted from Nixon (1988).

control of prey abundance in the specific environments in which they live. Below we examine top-down effects of feeding by fish, and discuss some implications of the food/consumer relation at the population level, especially in the case for pelagic marine fish.

9.4.2.1 Some Relevant Evidence from Freshwater Environments

There are several examples of the impressive influence of predation on communities in freshwater environments. Smith (1968) records the remarkable changes that have taken place in the fish fauna of Lake Michigan as a result of changes in intensity and selectivity of predation, including introduction of sea lamprey and changes in fishing effort and gear. Brooks and Dodson (1965) document the consequences of introduction of alewives (*Alosa pseudoharengus*) in Connecticut, U.S. lakes: the larger zooplankton species were eliminated after the appearance of the alewife. Zaret and Paine (1973) describe the results of the invasion of *Cichla ocellaris*, a predatory fish, in Gatún Lake, Panama, where its feeding changed the food webs of the lakes, with a marked reduction of prey organisms and the appearance or increase of nonprey species. In the case of *Cichla*, the depletion of prey also meant that densities of other predators that fed on the same prey as *Cichla* were also much reduced. Since, as we know, predation is generally size specific, the prey requirements of the new fish species led to a very different community of herbivorous zooplankton and presumably to a very different distribution of sizes and species of phytoplankton.

Similar conclusions emerge from experimental manipulation of fish densities in replicated freshwater ponds by Hall et al. (1970), who also manipulated densities of invertebrate predators. Invertebrate predation did not seem to have nearly as important an effect as vertebrate predators. The fish were larger in size than the invertebrate predators, more mobile, and evidenced better-developed functional and developmental responses. Nutrient concentrations were also experimentally manipulated but did not seem to have as much of an effect on the composition of the zooplankton as did fish but did lead to higher growth of the fish predators, as we saw in the case of salt marshes (Section 8.3.3).

The theory that emerges from these studies (Hall et al., 1976) is that as fish density increases, size-selective predation leads to decreases in the size of the dominant zooplankton. The resulting community of abundant, but small zooplankton species may, however, be just as productive as the community dominated by the larger species. In the absence of fish, the smaller species cannot outcompete the larger species because the larger species have higher grazing and food processing efficiencies (cf. Chapter 5).

Freshwater communities thus seem to be thoroughly dominated by the top predators. The supply of nutrients and phytoplankton may have less of an effect on the composition and production of freshwater zooplankton than do the fish. Most freshwater phytoplankton are available to grazers, although some species may not be readily palatable or assimilable. Many freshwater benthic macrophytes, for example, do have what may be chemical defenses, such as alkaloids in water lilies. The many different compounds found in a variety of species are summarized in Hutchinson (1975).

Grazers are much more vulnerable to fish predation in the water column than in habitats with benthic vegetation. Since the protection provided by the structure of macrophytes is lacking, the impact of predation on planktonic species is not damped as it may be in vegetated benthic habitats.

We have no data with which to evaluate the impact of predation of pelagic marine fish comparable to those available for freshwater. The kind of manipulative experiments that allowed the role of fish in lakes to be studied are very hard to do in the open sea. Cases of invasions or extinctions of well-defined parcels of ocean are also not easily studied. Thus a scheme about how predation by fish may effect pelagic marine communities can only be speculatively pieced togather using various other lines of evidence.

9.4.2.3 Fish as Carnivores in Marine Pelagic Communities

Fish can be effective consumers, but it is difficult to say whether they control the number and size of prey in the open sea. Fishery biologists traditionally believe that stocks of adult fish rarely overexploit their food resources (Beverton, 1962; Gulland, 1971; Cushing, 1975). For example, the average-year class of haddock (*Melanogrammus aeglefinus*) is thought to consume only a fraction of its potential food supply, and only when the occasional very strong year class appears is the potential food resource used more extensively. The 1962 and 1967 age classes of haddock in the North Sea were about 20 times as large as those of other years. Such large increases in abundance must to some extent have taxed food supply, as indicated by the small decrease in the average size of fish entering the fishery (Jones and Hislop, 1978). The 1958 autumn age class of herring in the Gulf of St. Lawrence was 44 times as large as the smallest age class recorded (1969) and the 1959 spring age class was over 60 times as large as the 1969 age class. These contrasting abundances provide some idea of the tremendously different recruitment success from year to year, reflecting how variable water column conditions and plankton communities can be over time. There is little evidence of how the peak abundances due to strong recruitment relate to the carrying capacity of marine water columns. It is clear, though, that primary production does not change by 60X from year to year.

To really assess how close populations of marine fish are to the carrying capacity of their food supply, we need an idea of the food resources and demands by fish. There are very few instances where both abundance of food and requirements for growth of fish have been measured. Burd (1965) compiled data on the size of adult (3-, 4-, and 5-year-old) herring in the southern North Sea and the average density of *Calanus*, a favored food item for herring. Herring mature at about 3 years of age. The size of adult herring is *not* related to food availability at the time of collection of the sample, but, if size of the 3-year-old herring is plotted versus the average density of *Calanus* during the 3 years of life of the fish (Fig. 9-5), a response in growth during early life to prey abundance is evident. These data are certainly not conclusive, since many other factors could have affected fish length, and because older fish grow more slowly than younger fish. The example does corroborate the evidence seen earlier (Table 4-2) that growth in young fish is more closely coupled to abundance of food than is growth of adult fish, which was also concluded by Heath (1992).

The importance of a growth response to prey density such as seen in Fig. 9-5 is that larger young fish have better survivorship, mature faster, and become more fecund adults (Sections 4.4, 7.4). Thus, the developmental response discussed in Section 5.4 is a major potential mechanism whereby population numbers may be adjusted to food supply. This growth response has been consistently observed in freshwater, where fish predators effectively control prey populations (Le Cren, 1965). In marine environments the abundance of adult fish seldom seems to affect the development response, implying a looser coupling of numbers of pelagic marine fish to their food resources.

Pelagic bony fish have generally adopted a reproductive strategy very similar to the complete metamorphosis of terrestrial insects: their larvae are very numerous and use very different resources than the adults. Other large marine predators, of course, use other strategies. Mammals, sharks, skates, and birds produce few, large offspring that by size and feeding habit enter the ecosystem at a similar position to the parent. Gelatinous zooplankton use both the few-large and many-small strategies, depending on the species.

The reproductive strategy used by pelagic bony fishes has a major drawback: producing lots of larvae necessarily means that they will be small, on the order of 1 mm or so. The problem with being small in the plankton is that the steps in the food chain are primarily based on size

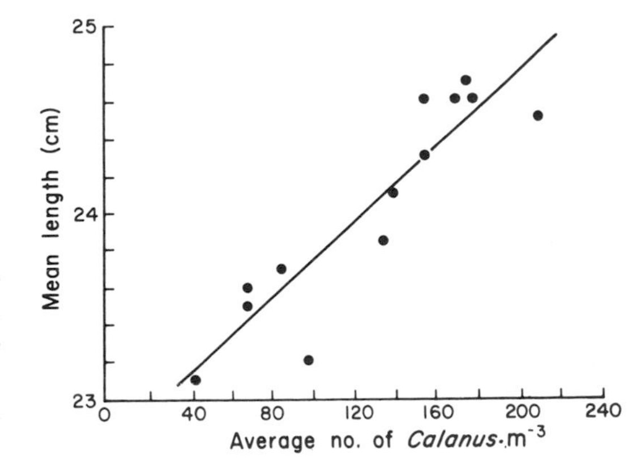

FIGURE 9-5. Length of 3-year-old herring in the southern North Sea in relation to the abundance of *Calanus*, a favored food item. Data from Burd (1965).

of predator and prey [cf. Section 6.3.1 and Ursin (1973)], and a growing fish larva essentially has to grow its way up the food web, and is exposed to a gauntlet of predators at each stage (Bailey and Houde, 1989).

A further problem is that the larvae are true plankters and drift with the water. During the period of larval drift the small larvae are very susceptible to haphazard meteorological and hydrographic events which may take the larvae to unsuitable areas (Heath, 1992). These multiple detrimental factors not only lead to a marked mortality of young fish but also impose what appears to be a random variability to rates of larval survival. The variability is large and haphazard enough to overwhelm the influence of density of parent and cohort stocks at some point along the survival curve of the cohort.

Fish larvae live in a habitat populated by numerous other larvae and zooplankton of similar size, all feeding on similar food items of appropriate sizes. The influence of cohort density is significant in early life, as evidenced by the density-dependent mortality and growth of young fish (Section 4.4). Density-dependent competition for food is therefore a dominating feature of larval life (Section 4.4). The larvae that are most successful at feeding grow faster, stay healthier, are more active, and more sucessfully escape predators.

Densities of larvae fall very markedly as the larvae age and competition may become less intense. The density dependence lessens rapidly with age in most fish species due to the random nature of the predation and hydrographical events that affect mortality and dispersal. This uncoupling of survival of young individuals from adult density makes it difficult to find density-dependent recruitment to adult populations in most pelagic species (cf. Section 4.4.4). The lack of such a relationship suggests that adult fish stocks rarely overexploit their food resources. There is also no clear relationship between growth rates and densities of adult fish (Section 4.4).

Another telling fact that suggests that adult fish populations are generally low relative to resources is that many fish stocks are maintained mainly through sporadic but exceptionally strong year classes (Cushing, 1975). As pointed out earlier, in general, perhaps only 10 out of every million fish larvae survive to adult size. There are so few larvae that make it through the tight bottleneck of larval and juvenile mortality that adult stocks are seldom abundant relative to resources. On occasion the severe mortalities that affect young fish may be slightly and fortuitously relaxed, by chance, there is a good match of a ready cohort, availability of food resources, and lack of potential predators (Cushing, 1975). Even a small increase in survival rate may result in a very strong cohort. Unusually strong currents took the 1987 year class haddock to waters where they survived well. This year class became one order of magnitude more numerous than other year classes along the New England shelf (Polachek et al., 1992).

Strong-year classes can be achieved even with minimal egg production; for example, the large size class of herring of 1958–1959 in the Gulf of the St. Lawrence (Fig. 4.10, left) arose out of a very small egg production.

Recruitment of strong-year classes is, by all appearances, a common but haphazard event that depends on meteorological and hydrologic factors. Often strong-year classes of particular species may occur simultaneously over much of a large area such as the North Atlantic, suggesting that a large-scale climatic event is responsible.

Stunting of individual fish within a strong cohort is rare and, in fact, strong cohorts usually are characterized by faster growth of young individuals (Gulland, 1971). This is unusual since mortality and growth of larvae and young fish are generally density-, and by extension, food-dependent. It is as though larval mortality due to predators or other factors had been reduced at the same critical time that larval resources increase. The common occurrence of the strong-cohort recruitment pattern and high growth rate of the cohort throughout the subsequent years suggests that there may be a surplus of food not used by the extant, average-year fish stock.

The strong-cohort reproductive pattern may be a result of what might be called an *r*-selected, exploitative, or "many-small" reproductive strategy (cf. Section 3.4) and of life in water columns subject to high variability. Conditions in the water column are extremely heterogeneous both in space and in time. Fish release as many larvae as possible into this hazardous and variable environment, and every so often circumstances allow greater survival of a cohort. Since the extant population is usually below carrying capacity, the strong cohort can proceed to grow well.

9.4.2.4 Role of Fish in Aquatic Environments

The observations discussed in the previous paragraphs suggest that substantial increases in abundance of pelagic fish do not lead to substantial changes in mortality, growth, recruitment, or fecundity of adult stages. Some indication of incipient resource limitation such as reduced growth are evident only in environments where very high densities are found. The body of evidence as a whole suggests that these predators may be limited by their food resources as young fish but that adults are not limited by and therefore do not limit their food resources. This contrasts with the marked effectiveness of freshwater and coastal fish in controlling prey populations.

Oceanic food webs often have five steps or more, while in freshwater three steps may be more typical. This is an as yet unexplained but significant qualitative difference between the two environments. We may speculate that freshwater fish larvae are therefore exposed to fewer types of predators, and are perhaps less subject to predatory mortality. In freshwater there may therefore be a better chance for the effect of parent density to be reflected on recruitment, and therefore a greater likelihood of a density-dependent response by the total fish population to changing abundance of the prey. In part this may be why freshwater fish are more effective at controlling the size of prey populations.

The contrasts in performance between pelagic and freshwater fish might more likely be due to the large differences in density of food and fish between these two habitats. It is not easy to make comparisons of the representative densities of marine and freshwater fish, and, in fact, there are few methods available to do so directly. Horn (1972) estimated in a very rough way the number of species and average number of individuals per species in fresh and seawater. The final point of his calculations was that each individual marine fish had about 10 to 10,000 times more volume of water to deal with than each freshwater fish. Clearly this calculation needs to be taken lightly due to the assumptions needed to make the calculations, but it does suggest that marine waters are dilute not only in the nutrients, algae, and grazers, but also in regard to fish.

We then have a situation in the sea where both fish and their food are in low abundance compared to freshwater environments. Yet pelagic fish, at least adult stocks, seem, in general, not to be food limited. Why don't marine fish tax their food supply? It may be that the crucial bottleneck is the larval stage. Mortality of larval fish in the sea may be so severe due to catastrophes, competition for food, and predation that, as said above, there is seldom a chance for the numbers of adults to accumulate.

Why do marine fish not make more eggs? Actually marine fish do produce more eggs than freshwater fish. The modal marine bony fish produces about 10^6 eggs per spawning (with a range of 100 to 3×10^8), while freshwater bony fish produce a mode of about 10^4 eggs per spawning (with a range of 10 to 10^7) (Altman and Dittmer, 1972, pp. 149–150). Since marine and freshwater fish are not too different in size (marine fish *on average* are probably only a bit larger than freshwater species in spite of swordfish, tuna, marlin, and other giants), the production of more eggs means that the eggs are smaller. Perhaps there is a lower limit to the size of a viable egg; smaller eggs may just have too many trophic steps to grow through, or perhaps there are anatomical and physiological constraints.

Alternatively, why do pelagic marine fish not evolve better larval survival ability? Some have: sharks, for example. Bony fish may have taken irreversible evolutionary steps in embryology so that larger larvae are not feasible. In addition, perhaps planktonic conditions are so unpredictable that the *r* strategy of opportunism has proven favorable in the long term.

Freshwater fish may be freer to respond to food abundance because larval mortalities in freshwater fish are lower, perhaps because there are fewer trophic steps and less variability. If this is true it might explain why there is a greater density dependence of growth and of age of first reproduction in freshwater than in marine fish (Cushing, 1981).

This section was headed as "speculations," and most of the above discussion was just that. Much critical research remains to be done on this subject. We need to add that fishes of certain marine communities (coral reefs, salt marshes, estuaries) may achieve very high densities, and may be less exposed to variable hydrography. Many fishes of coastal environments tend to take better care of their eggs, either by actual brooding or care in selection of a potential site for oviposition. Not all species engage in these activities, but in species that do so, larval mortality is less random or intense. The marine fish of some coastal environments perhaps are thus more likely to limit prey abundance (cf. Section 8.3). In the open ocean, the evidence available suggests that populations of adult fish are not likely to control abundance of their prey.

9.4.3 Competition and Predation by Mammals and Birds in Pelagic Communities

Mammals and birds, in contrast to most fish, use a *K*, saturation selected, or few-large reproductive strategy (cf. Section 3.5), and this may have fundamentally different consequences for the role of those animals in the exploitation of their food resources. Unfortunately, there is little concrete information on the way in which pelagic communities may be structured by mammal and bird activity. The only data available by which we can infer some of the effects of mammals and birds as predators on pelagic communities are provided by the consequences of the intensification of whale harvesting in the southern oceans since the late 1920s (Fig. 9-6) due to the development of the exploding harpoon, faster whaling vessels, and factory ships.

The whalers behaved as a typical predator, first harvesting the largest whales accessible to their gear, focusing their attention in the early years on the blue whale. When stocks of blues began to dwindle, the whalers turned ("switched"), again in classical predator fashion, to two other large whales, the fin and the humpback. In the 1960s, as the larger whales became very scarce, the whale catch fell steeply and the remaining whaling fleets turned to the considerably smaller sei and

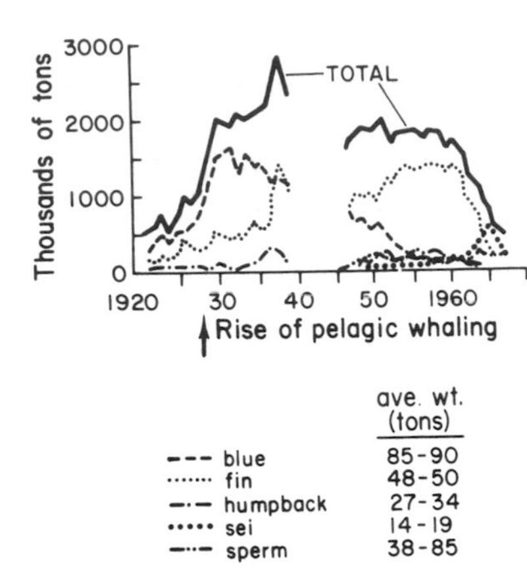

FIGURE 9-6. Tonnage of Antarctic whale catches over time. The whale species include the blue (*Balaenoptera musculus*), fin (*B. physalus*), sei (*B. borealia*), minke (*B. acuturostrata*), humpback (*Megaptera novaeangliae*), and sperm whale (*Physater catadon*). The latter feeds on squid and fish, a diet rather unlike the copepod, krill, and small fish eaten by the ballen whales represented by the other two genera. Adapted from MacIntosh (1970); weights from Laws (1977a,b) and Lockyer (1976).

even to minke whales. The stock of plankton-feeding whales was sharply reduced after only 30 years of industrialized whaling.

By the late 1960s stocks of baleen whales had been lowered in the Southern Ocean to about a third of their former numbers. Stocks that initially were around 46×10^6 tons of whale biomass more recently were 8×10^6 tons (Laws, 1977a,b). Blue and humpback whales were the most affected, having been reduced to 3 and 5% of the estimated initial stock (Laws, 1977a,b). The sperm whale was thought to be half as abundant as formerly.

Such large changes in abundance of major consumers altered food use; especially consumption of krill in the Antarctic pelagic food web (Table 9-2). The total tonnage of whales is now perhaps one fifth of

TABLE 9-2. Rough Estimates of Consumption of Antarctic Krill by Major Groups of Predators in the Southern Ocean in 1900 and 1984. Numbers are Based on Published Data, Guesses Based on Information from Various Sources, and Back-Projected Population Estimates.[a]

	Annual consumption ($\text{tons} \times 10^6$)	
Consumer type	1900	1984
Whales	190	40
Seals	50	130
Birds	50	130
Fish	100	70
Squid	80	100
Total	470	470

[a] From Laws (1985).

what it was. Before the pelagic fishery reduced whale stocks, the annual migration of whales in and out of the Southern Ocean to more northern waters drained over six times the amount of biomass away from the Antarctic compared to present losses.

If production and biomass of krill, fish, and squid are about the same over the years, the reduction in whales leaves considerable amounts of food "unconsumed" (Table 9-2). In terms of krill alone, the tonnage left "unconsumed" by the decrease in whales reaches 150×10^6 tons. We can perceive the enormity of this number by comparison with the total catch of marine fisheries, which is about 75×10^6 tons y^{-1}. There is, unfortunately, little direct information about whether the densities of krill, squid, and copepods have actually increased in response to increased food. We need, therefore, to look for indirect evidence of such a response by predators still present in Antarctic waters.

Available data on current whale stocks suggest that availability of food per whale has increased, and this increased food availability has affected the surviving stocks as shown by changes in variables known to be density and food dependent (cf. Chapter 4). The reduction of whale densities has, in fact, resulted in increased growth rates of the remaining individuals (Laws, 1977a,b). The percentage of the blue, fin, and sei females that are pregnant has also increased since the 1900s; there is evidence of earlier maturity in fin, minke, and sei whales. The halt in whaling during World War II allowed increased survivorship and lowered food availability for a short time, reducing the reproductive performance of fin and blue whales. The hiatus in whaling during the war may also have resulted in a brief period of reduced growth of average individuals, since smaller fin whales were taken while whales were more abundant (1945–1946) than during a period of lower whale standing stocks (1962–1963) (Fig. 9-7). In sperm whales, average length at maturity has increased for males, and females mature earlier (Kasuya, 1991).

The increase in growth and earlier maturation when more food is available per capita* have also been recorded on other environments for harp, fur, and elephant seals (Sergeant, 1973), so it seems that in general, marine mammals are food limited. In contrast to the fish populations discussed earlier, whale populations seem very closely coupled to their food resources.

---

* Such improvements in the "standard of living" after density has been lowered seem to be a rather general phenomenon. In the Baltic Sea, growth of flatfish increased after those populations were thinned out markedly in the 1920s (Persson, 1981). The phenomenon is not limited to marine organisms: food supply, wages, and quality of life in general improved for European people as a whole during 1350–1550, a period following severe outbreaks of the Black Death (Braudel, 1981, p. 193).

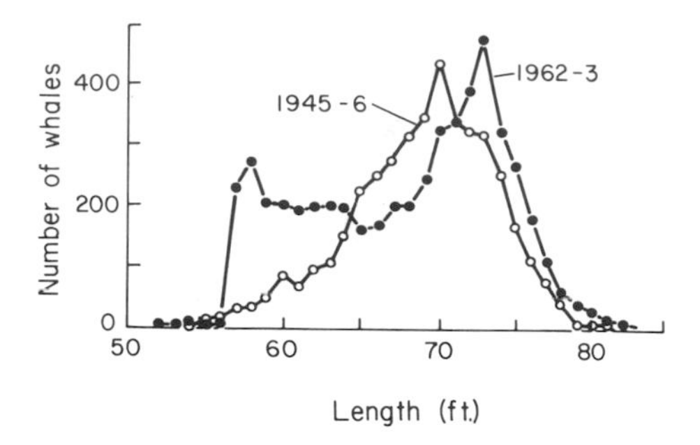

FIGURE 9-7. Length composition of female fin whales caught in the Antarctic in the 1945–1946 and 1962–1963 seasons. The small mode in the 1962–1963 data is an artifact of the well-known "elastic" tape measure used by whalers on smaller whales after the introduction of the 57-ft lower size limit. Adapted from Gulland (1971).

All the above changes in the whale populations suggest that whales were limited by food resources prior to industrial whaling. Where there were fewer whales, more food was available, and this led to faster growth, earlier maturity, and successful pregnancies. These increased pregnancy rates turned into more recruitment of young since survival rates of young are so much greater than those of most fish (Chapter 3). Such reproductive changes in the sei whale population are especially convincing because the response took place before the whalers began to exploit sei populations extensively (Gambell, 1968). The general conclusion from all these observations is that whales remaining after exploitation responded as if there were an increase in food abundance.

Other vertebrate predators in the southern oceans also seem to have responded to the apparent increase in availability of food, and probably consume the krill "left" by the whales (Table 9-3). Data on the several penguin and six seal species present in Antarctica are very difficult to get because penguins and seals are sparsely distributed, often are difficult to see and count among the ice floes, and spend much time diving. Surveys of nesting colonies show that four of the five major penguin species in Antarctica have increased in abundance. The fifth (the emperor) is a species that feeds inshore and probably was seldom in competition with whales. There is information on two primarily krill-feeding seal species (Table 9-3), and for both there is some evidence of increased numbers, especially for the fur seal. The seals presently in the southern oceans have increased to the point where their estimated consumption of krill, fish, and squid exceed that of whales (Table 9-3).

Although the above observations are not very convincing individually, in the aggregate they do suggest that other open-ocean predators currently in the southern oceans have responded to the

TABLE 9-3. Recent Changes in Penguins and Seal Populations in Antarctica.[a]

	Principal foods	Changes in popuation
Penguins		
Emperor (*Aptenodytes forsteri*)	Fish	No significant increase
King (*A. patagonica*)	Mainly squid	Marked increase (5% y^{-1})
Adelie (*Pygoscelis adeliae*)	60% krill, 40% fish and other	Local increases (2.3% y^{-1}) in whaling areas
Chinstrap (*P. antarctica*)	Krill	Marked increase, extended range
Gentoo (*P. papua*)	Benthic fish, some krill	Some increases
Macaroni (*Endyptes chrysolophus*)	75–98% krill, 2–25% fish	(Increases of 9% y^{-1})
Seals		
Crabeater (*Lobodon arcinophagus*)	94% krill, 3% fish, 2% squid	Earlier maturity, increase in numbers (7.5% y^{-1})
Fur (*Arctocephalus gazella*)	34% krill, 33% fish, 33% squid	Population explosion (14–17% y^{-1}) especially in overlap with range of baleen whales; appearance of new colonies

[a] Data from Conroy (1975), Stonehouse (1975), Laws (1977a,b), Payne (1977), Øritsland (1977), Croxall and Prince (1979), Hinga (1979), Laws (1985), and Cooper et al. (1990).

greater food availability due to removal of whales by increasing their own abundances. The species involved (mammals, birds) do not use the "many-small" reproductive strategy of fish, and perhaps thus avoid the loose coupling to resources typical of populations of adult bony fish. The Antarctic surveys imply that prior to industrialized whaling the abundance of whales had increased to the level of being limited by and limiting their food supply. After the advent of industrialized whaling, other predators (seals, penguins) used the no-longer-consumed food resources, and their populations, in turn, have increased.

Scarcity of whales, public opinion, and the exigencies of economics, especially fuel costs, have caused the whaling effort to dwindle, and soon there may be little if any large-scale hunting of whales. Will they return to their earlier abundance? Perhaps some, such as blue whales, have too few individuals left to do so,* but other species certainly can

*Actually blue whales are not rare in certain parts of the ocean (W. Watkins, personal communication), and since they have notable ability to move for very long distances, there is a good chance of recolonization of locally depleted areas if whaling pressure stops.

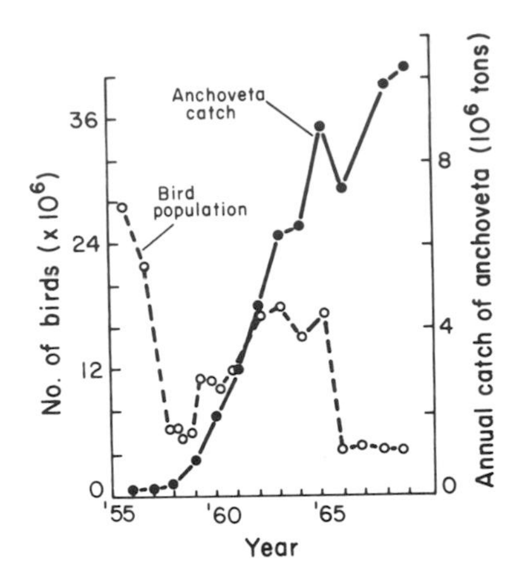

FIGURE 9-8. Commercial catch of anchoveta (*Engraulis ringens*) off Peru and census of guano birds during the same span of time. Adapted from Schaefer (1970).

recuperate earlier abundances. Human ability to affect natural environments, however, never ends: Russian efforts to harvest krill in the Antarctic are already under way (Omori, 1978). With so many protein-short human populations, it seems inevitable that krill will be exploited.

Whether whales can recover their earlier abundance in the face of a new, efficient competitor for krill, we can only speculate at present, but it is certain that a new balance will take place. There is a very pertinent comparison available in the interactions among birds, men, and anchovies in the Peruvian upwelling region (Fig. 9-8). Sporadic natural episodes of El Niño* usually lead to massive reductions in anchoveta. The lack of fish causes populations of the fish-eating guano birds to crash, as occurred in 1957 and 1965. While the birds recovered readily from the earlier crash, competition from men produced a different result after 1965.

By the 1960s the human harvest of anchoveta had become quite large, about 7.5×10^6 metric tons, compared to $0.2–0.3 \times 10^6$ tons commercially harvested during the late 1950s. The birds in the late 1950s ate about 2.5×10^6 metric tons of fish. In the late 1960s the sustainable yield of anchoveta was calculated to be about 10×10^6 metric tons; the birds took 0.7×10^6 metric tons. The fishermen had by then developed the capacity to take at least the remaining 9.3×10^6 metric tons. The result is that guano birds did not therefore have the surfeit of fish on which to base a return to earlier abundances. This

* El Niño (The Child) episodes start toward the end of the calendar year (hence the connection to the Nativity), when warmer water intrudes into coastal areas of Peru, preventing upwelling of cold, rich water. The result is markedly lowered primary production and massively reduced abundance of anchoveta. The hydrodynamics of El Niño are connected to events going on at surprisingly distant areas of the tropical Pacific (Wyrtki, 1975). The El Niño phenomenon is further described in Section 16.2.2.2.

may be the situation that may await the whales if the krill fishery becomes well established.

The Peruvian upwelling is a second example of a pelagic ecosystem where the species occupying the upper trophic positions seem to be closely coupled to food availability. As in the case of the whales in the southern oceans, the predators that control prey abundance in the Peruvian upwelling are mammals or birds. These predators, as mentioned already, use a *K*-selected reproductive strategy which may be best suited to control populations of prey or to increase numbers to closely match available resources.

All species of marine mammals have single offspring, even though they belong to four different orders whose nearest terrestrial relatives have variable and larger clutch sizes (Estes, 1979). In marine species the strongly developed parental care reduces mortality of the young, especially from predation. This reproductive strategy may have set up a situation where the numbers of marine mammals would be limited by resources, rather than by predators. If marine mammals are resource-limited, they in turn are likely to have a strong influence on the structure of the communities of organisms on which they feed.

Whales, seals, penguins, and most marine birds thus seem to exist near the carrying capacity of their environment, and show *K* or saturation selected traits such as iteroparity, few offspring, parental care, delayed age of first reproduction, and long lives. Such life history characteristics make a population easily susceptible to over-exploitation since there is little slack in the demography of such species to absorb predatory losses [cf. Chapter 4 and Estes (1979)]. Whale stocks were lowered by whaling, as were the exploited stocks of elasmobranchs, another group of *K* strategists, described in Section 3.5. Exploitation of sea otters led to extinction over most of their range (Estes et al., 1982); exploitation of the Great Auk, a marine bird ecologically similar to penguins, led to its complete extinction.

We should be careful to note that virtually all the evidence we have examined comes from exploited populations. Fishing has reduced populations of fish and other top predators over much of the sea. Fishing mortality, for example, can be 58% of all mortality in cod (Daan, 1975). The many-tiered food webs of the marine water column have at present relatively reduced populations of many top predators. For example, during and after the late fifties the tuna stocks were drastically reduced by fishing in the North Sea (Tiews, 1978). The annual consumption of fish by bluefin tuna in certain areas of the North Sea during the maximum *recorded* tuna abundance (1952) has been roughly estimated at 210,000–256,000 metric tons of mackerel and herring (Tiews, 1978). If bluefin tuna consume 75% herring, this means that about 150,000–180,000 tons of herring were eaten by tuna during their annual stay in the North Sea. In 1952 the stock of herring in the area in question was about 140,000 tons (Burd, 1978), so that consumption of herring by tuna could have removed the entire stock of herring. The reduction of tuna abundance, due to fishing, may have therefore been one reason for the very marked increase in herring (Daan, 1978) and other pelagic fish in the North Sea during the 1950s.

Such comparisons are very rough but do bring up an important notion. What we see today in the food web of the North Sea—and elsewhere in the sea [for example, in the gulf of Maine (Whitman and Sebens, 1992), or in the Caribbeans coral reefs (Hay, 1984)]—is severely affected by the ultimate top predator, the fishing fleet. Virtually all large top predators in most marine environments have been much reduced by fishing and man may now have replaced these as the keystone predator of marine food webs. We are in the same position that a sentient otter may be in interpreting what might be going on in the coastal communities of Amchitka Island. It may be hard to separate our effects from those of other control mechanisms.

The fishing activity of man has been clearly instrumental in affecting the community structure of every marine environment. The increased stocks after the respite of World War II, the lack of herring in recent years in the North Sea, the near disappearance of tuna and whales, are among the many examples of the real impact of fishermen as the top predator. It is not yet clear how this affects our conclusion as to the role of pelagic fish in regard to their food supply.

The studies reviewed in this section provide some tantalizing glimpses at some processes that organize open water marine communities. This entire field needs innovative approaches, especially including experimentation and study at the multispecies and ecosystem levels. In coming decades, exploitation of resources in the sea will increase, and the evaluation of the consequences and the management of marine resources will require considerably improved knowledge as to the dynamics and organization of marine communities. The speculations in this section merely probe the surface of the matter.

Chapter 10
Taxonomic Structure: Species Diversity

10.1 Introduction

Terrestrial ecologists long ago remarked on the richness of the floras and faunas of tropical environments relative to colder climates. The diversification within many specific taxonomic groups is clearly greater in the tropics than in temperate latitudes, both in terrestrial and in marine environments (Fig. 10-1). Low numbers of species are also typical of severe and disturbed habitats. Such observations have spawned an abundant and contentious body of publications that have dealt with three major problems: first, how to quantify the clearly observable differences in diversity, second, how are such differences in taxonomic richness of communities generated and maintained, and third, what do such differences mean ecologically.

10.2 Measurement of Diversity

10.2.1 Diversity Within a Habitat

The assessment of number of species in a habitat is not a straightforward matter, for it is hard to know when to stop sampling additional individuals. Rare species will be missed if large numbers of individuals are not included. To solve this uncertainty, the number of species (S) can be graphed against the total area sampled (A) (Fig. 10-2, top) or the number of individuals included in the sample (Fig. 10-2, bottom). Such a curve is described by

$$S = CA^z, \tag{10-1}$$

where C and z are constants that can be estimated if a series of values of S and A are available and a regression using logarithmic transformations of the numbers of species and individuals is used:

$$\log S = \log C + z \log A. \tag{10-2}$$

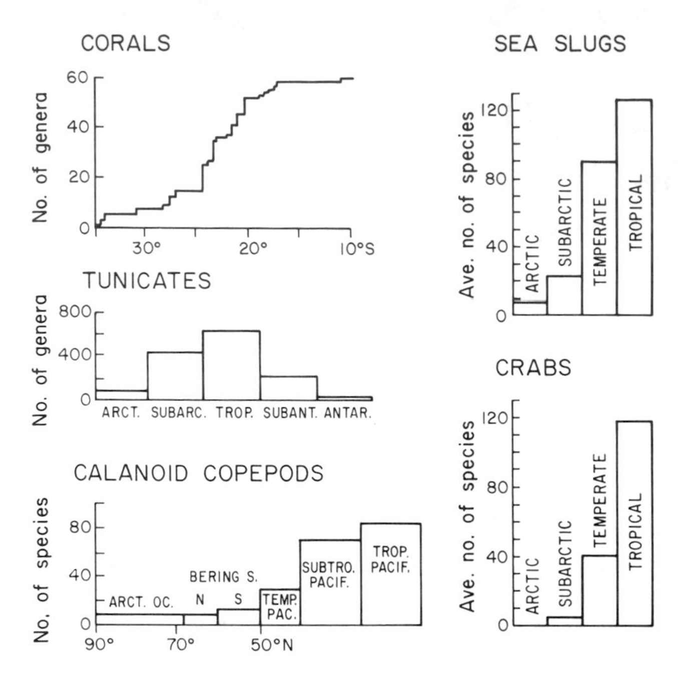

FIGURE 10-1. Diversification in selected marine taxa along geographical gradients. Corals from the Great Barrier Reef, Australia; the copepods are from the Pacific and its northward extensions, while the remaining data are from all oceans. Adapted from Thorson (1957) and Fischer (1970).

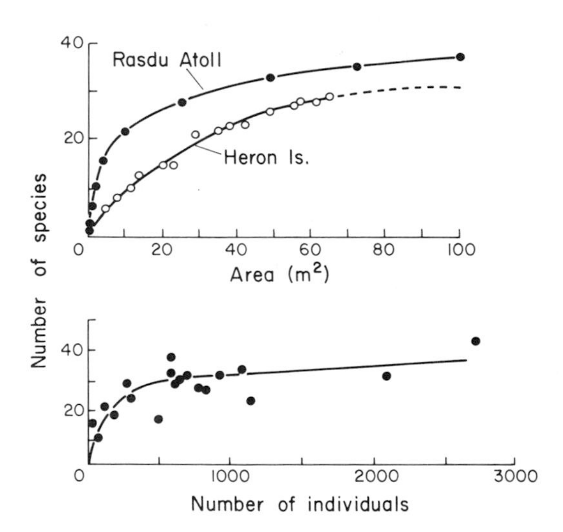

FIGURE 10-2. Top: Species-area curve for corals in coral reefs on Rasdu Atoll, Maldives, and on Heron Island, Great Barrier Reef. Adapted from Scheer (1978). Bottom: Relation of number of species and number of individuals in a sample, based on twenty samples of benthic invertebrates collected from Buzzards Bay, Massachusetts, during a 2-year period. Adapted from Hessler and Sanders (1967).

The number of species in any area can then be obtained. Such species–area curves generally depict deceleratingly increasing numbers of species as area or number of specimens increase. The asymptote eventually reached gives some idea of the total number of species in the environment.

Such an estimation of the total number of species is still not a completely satisfactory way to measure diversity, since the distribution of individuals among species also needs to be considered. For example, a sample containing two species, each represented by 50 individuals seems intuitively more diverse than one consisting of 99 individuals of one species and 1 of another species. Diversity, then, has two components, species richness and equitability of abundance of the species. To deal with these two components a number of approaches using either probability or information theory have been made. We deal with these only briefly; anyone interested in measuring diversity should consult the reviews by Poole (1974), Peet (1974), and Pielou (1975), the major sources of the following discussion.

Simpson's index. Simpson (1949) suggested the use of an index of diversity that was based on the probability that two individuals drawn from a population of N individuals belong to the same species. This index, modified more recently (Pielou 1969), is

$$D = 1 - \sum_{n_i} n_i(n_i - 1)/N(N - 1), \qquad (10\text{-}3)$$

where n_i is the number of individuals of the ith species and N is the total number of individuals. D can also be used as the probability of interspecific encounters in random meetings of the individuals of the population.

Logarithmic series. Fisher et al. (1943) proposed that the frequency of species with a given number of individuals was described by a logarithmic series, and derived an equation for S, the number of species in a sample:

$$S = \alpha \log(1 + [N/\alpha]), \qquad (10\text{-}4)$$

where α is a fitted constant that can be used as an index of diversity.

Log-normal distributions. If a variable is affected by many independent factors, the distribution of the variable will be normal. Since the occurrence of individuals is affected by such an array of factors, in a community there ought to be mainly intermediately abundant species, with few abundant or rare species as in normal distributions. This abundance pattern can be fitted by a log-normal distribution where data are plotted as the number of species in steps (called octaves) of one to two individuals, two to four individuals, etc. (Fig. 10-3). The steps are equivalent to taking logarithms of the abundances to the base

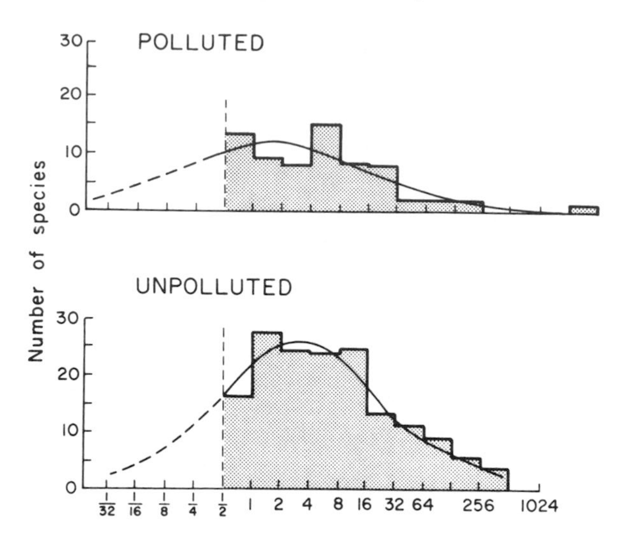

FIGURE 10-3. Frequency distribution of species of diatoms settling on glass slides in a polluted and an unpolluted stream. The histogram shows the number of species in each of the octaves (doubling intervals) of abundance. The line shows the idealized log-normal distribution as if it were a continuous distribution. Since frequencies of less than one individual cannot occur, the dotted part of the distribution cannot be seen left of the dashed vertical "veil line" located left of the first octave. Adapted from Patrick et al. (1954).

2. Such log-normal distributions fit many populations (Preston, 1948, 1962). The number of species (s) in the Rth octave is

$$s = s_0 e^{-(aR)^2}, \qquad (10\text{-}5)$$

where a is a value estimated from the data and s_0 is the number of species in the modal octave. The total number of species (S) in the environment from which the range was taken can be obtained from

$$S = s_0\sqrt{2\pi\sigma^2} = 2.5\sigma s_0, \qquad (10\text{-}6)$$

where sigma is the standard deviation of the log-normal distribution expressed in octaves. The accuracy of this calculation of S is not as good as desirable. Further, S is an index of species richness and does not reflect equitability. To include both richness and equitability, Edden (1971) suggests the use of the ratio s_0/σ. Because of Eq. (10-6), the ratio equals $S/\sqrt{2\pi\sigma^2}$. The ratio, then, has both the elements of species richness (S) and equitability, since the variance reflects the frequencies with which individuals are distributed. Values of Edden's index ratio calculated for phytoplankton correlate fairly well with those calculated using the Shannon–Weaver index (see below).

The use of the log-normal requires that the mode be clearly identified, so that the distribution should be moved to the right as far as possible

(Fig. 10-3). For the mode to move one octave to the right, a doubling of the number of individuals is needed, so the use of the log-normal requires quite large censuses.

Information theory indexes. The use of diversity indexes derived from information theory were first suggested by Margalef (1957). If we are examining a collection of specimens from an environment, and we find that all the individuals examined so far belong to species A, we feel reasonably sure that the next individual will also be an A; we therefore gain little information by looking at that specimen. If, in contrast, the sequence so far has been A, N, B, D, X, Z, we cannot easily predict what species will be represented by the next individual, and its identification does produce new information about the collection. An increase in the number of species and a more equitable distribution therefore increase uncertainty, and the average uncertainty can be calculated in various ways.

The information theory index most often used is the Shannon–Weaver expression*

$$H' = -\sum_{i=1}^{s} P_i \log p_i, \tag{10-7}$$

where S is the number of species and $p_i = n_i/N$, for the ith species. Actually Eq. (10-7) is a biased estimator, and at least one correction term $[(S - 1)/2N]$ should be subtracted from the right-hand side of Eq. (10-7) (Hutcheson, 1970). The variance of H' can be obtained from the series

$$\mathrm{Var}(H') = \frac{\sum_{i=1}^{s} p_i \ln^2 p_i - \left(\sum_{i=1}^{s} p_i \ln p_i\right)^2}{N} + \frac{S - 1}{2N^2} + \ldots \tag{10-8}$$

For large samples the first term of Eq. (10-8) is usually sufficient (Poole, 1974). The calculation of Eq. (10-8) makes it possible to subject the values of H' to statistical tests (Hutcheson, 1970).

In some instances involving sessile, colonial, or vegetatively growing organisms (bryozoans, tunicates, salt marsh grasses are examples) it is not possible to define individuals. For such cases Pielou (1966) suggests that the area of study should be divided into many quadrats, and the diversity of a randomly chosen quadrat calculated using Brillouin's index

$$H = 1/N \log \frac{N!}{N_1! N_2! N_3! \ldots N_s!}, \tag{10-9}$$

---

* The choice of the base of logarithm is arbitrary; if base 2 is used, the units will be in bits; if base e is used, the units are bels.

where N is the total number of individuals, and $N_1, N_2 \ldots N_s$ are the abundances of each species.

Brillouin's index applies where all individuals in a collection or a community can be identified and counted. Since H measures the entire community it has no standard error, and values of H calculated for different communities that are not the same are therefore significantly different.

The area covered by a species or perhaps the weight of a plant within a quadrat are taken as N_i. The procedure is then repeated but using values of N_i from the first quadrat pooled with those of a second randomly chosen quadrat. Then a third quadrat is added to the first two, and H estimated for the pooled data, and so on. The cumulative values of H are graphed versus the number of pooled quadrats, and there will be a point where the calculated pooled diversity reaches an asymptote. An estimate of the pooled diversity (H'_p) of the whole area is

$$H'_p = \frac{1}{z - t + 1} \sum_{k=t}^{z} \frac{M_k H_k - M_{k-1} H_{k-1}}{M_k - M_{k-1}}, \tag{10-10}$$

where z is the total number of quadrats sampled, t is the number of quadrats at which the pooled diversity reaches an asymptote, M_k is the summed area or weight of all species in k quadrats, and M_{k-1} is the summed area or weight of all species in $k - 1$ quadrats.

Many other indices of diversity exist; none, including the ones listed above, have a well-documented underlying biological mechanism; all the justifications given above are analogies or logical speculation. All have some drawbacks (Hurlbert, 1971; Taylor et al., 1976). Further, they each respond somewhat differently to aspects of diversity, for example, to changes in the importance of the rarest or most abundant species (Peet, 1974; May, 1975; Taylor et al., 1976). The result is that there may be a great deal of discordance in the trends shown by the different indices even when applied to the same data (Hurlbert, 1971). It should always be kept in mind that the diversity we more or less quantify using indices is our construct. There is no evidence that diversity has any meaning to the organisms involved.

Many other problems or difficulties need to be considered, not the least that indices used should be independent of sample size. The avalanche of new indices and papers on diversity has currently subsided, and the attitude prevails that perhaps the number of species or a simple index of species richness (R), such as Margalef's (1951)

$$R = (S - 1)/\log N \tag{10-11}$$

will suffice for most purposes.

One of the reasons diversity and diversity indices became so popular in the 1960s and early 1970s was the growing need to provide some

TABLE 10-1. Effects of Experimental Eutrophication on the Diversity of Higher Plants and Benthic Diatoms of Salt Marsh Plots.[a]

	Higher plants	Diatoms		
	No. of species	No. of species	H′ (bits ind^{-1})	Percentage *Navicula salinarum*
Control plots	9–11	129–196	3.8–4.1	5–9
Fertilized plots	5	92–117	3.1–3.4	20–23

[a] Fertilized plots received additions of nitrogen or mixed fertilizers throughout the growing season. H' is the diversity index of Shannon–Weaver. The right-most column is the percentage of the total number of cells made up of *Navicula salinarum*. Adapted from Van Raalte et al. (1976a) and unpublished data.

assay of the effect of pollution on natural communities. It was obvious that the number of species decreased in polluted circumstances (Fig. 10-3 and Table 10-1) and that the more severe the level of pollution the greater the reduction (Fig. 10-4). An index such as H' contains information on the abundance and species composition of an array of species (Fig. 10-4). It also depicts the dominance by an opportunistic, tolerant species (the polychaete *Capitella capitata* in Fig. 10-4, bottom) in more

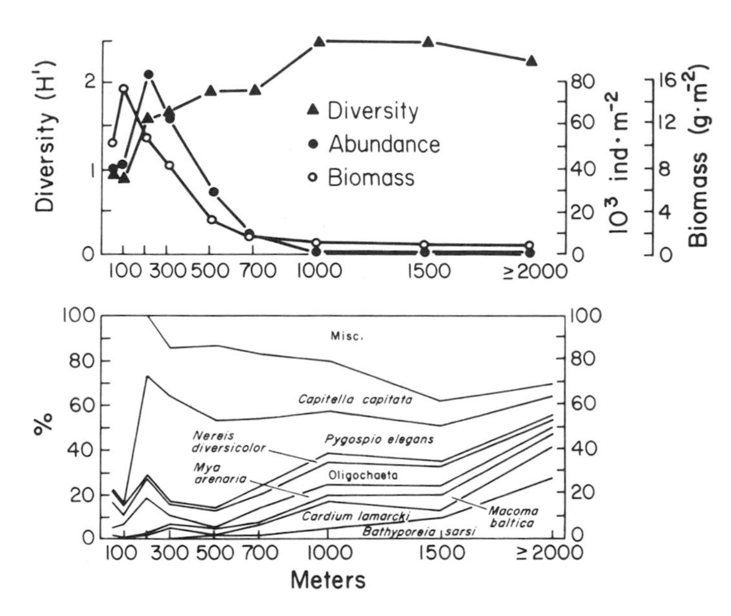

FIGURE 10-4. Benthic fauna at a depth of 3 m in the Baltic Sea. Top: Abundance, biomass, and diversity in samples collected at various distances away from an outfall releasing untreated sewage. Bottom: Percentage composition of benthic species along the same transect. Adapted from Anger (1975).

polluted waters, and the more diversified assemblage away from the source of pollution. Applied scientists thus seized on the apparently comprehensive, easy-to-calculate indices as a way to subsume a lot of complex data.

As another example, a comparison of the number of species and variance of the distribution of Fig. 10-3 shows that there were fewer species and that these fewer species were relatively more abundant in a polluted that in an unpolluted environment. Pollution simplifies the assemblage of diatoms and prompts the numerical dominance of a handful of species. The Shannon–Weaver index, because of ease of calculation, has been a favorite (Fig. 10-4). In addition to an ambiguous ecological interpretation, this use of diversity indices delivers less than it promises. Assemblages of species are variable, and so are the indices calculated from samples of such assemblages. This variability, plus the statistical idiosynchracies of the index used, may prevent the indices from being sensitive enough to detect incipient pollution, a role that would be of great benefit. By the time a diversity index reflects changes in the assemblage of species, there are other clear and more dramatic symptoms of pollution. Judicious use of the indices can, however, serve as a shorthand way to show change in species composition, and the results should be used keeping in mind their limitations.

10.2.2 Diversity Between Habitats

If larger and larger samples of organisms are taken there will be increases in the diversity of the total census that are due to the inclusion of different habitats with different faunas (MacArthur, 1965). This can be seen in Fig. 10-5, where the number of species from environments that contain a variable number of habitats is shown. This between-habitat component of diversity has also been termed beta-diversity, in contrast to alpha- or within-habitat diversity (Whittaker, 1972). The distinction of within- and between-habitat diversity is to an extent ambiguous since it depends to a large degree on the size of the organism and the size of habitat patches.

A variety of indices can be used to measure the compositional similarity of assemblages of species from different sites (between-habitat diversity), including some from information theory (Field, 1969; Williams et al., 1969). A commonly used index of overlap among samples containing two arrays of species is that of Morisita (1959)

$$C_D = \frac{2\sum_{i=1}^{s} x_i y_i}{(D_x + D_y)XY} \tag{10-12}$$

where species i is represented x_i times in the total X from one sample, while y_i times in the total Y from another sample. D_x and D_y are

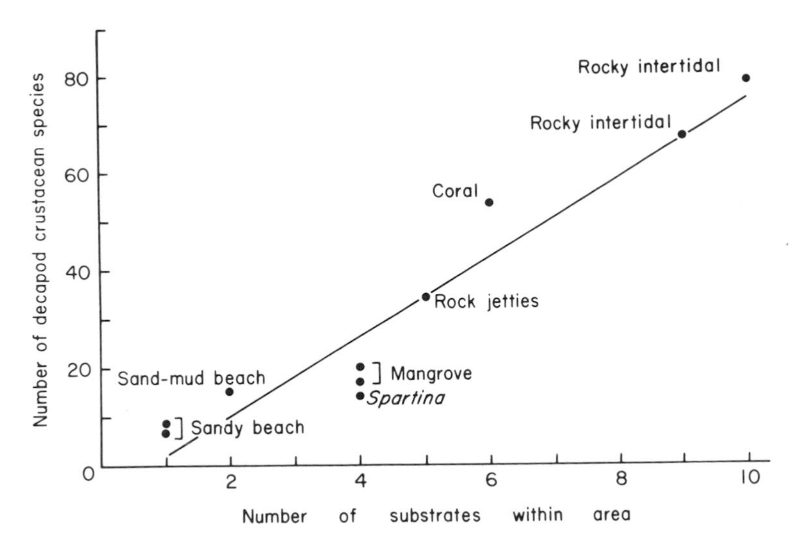

FIGURE 10-5. Effect of between-habitat diversity on decapod species present. The number of different substrates in each of the habitats was more or less arbitrarily designated from experience. The species counts are probably underestimates in all cases but serve for comparative purposes. From Abele (1982).

Simpson's index [Eq. (10-3)] for the two samples. When the samples are completely different $C_D = 0$. The value of C_D increases to about one when species are in the same proportions. Some pitfalls of similarity indices are discussed by Bloom (1981). Simplified versions of Morisita's index and others adapted from information theory are provided by Horn (1966). Plant ecologists have developed many other simple empirical indices of similarity that can be used for such comparisons (Goldsmith and Harrison, 1976). Much more complex analyses are available for the identification of relations among assemblages of organisms from many different sites (Whittaker, 1973).

10.3 Factors Affecting Diversity

The tropical–temperate gradient in species diversity mentioned above has been taken as an example of the observation that diversity varies from one environment to another and has thus attracted an enormous amount of attention. There are many hypotheses put forward to try to explain the gradient and differences in diversity (Rhode, 1992). In very general terms, the various hypotheses can be summarized as follows (Sanders, 1968):

The time hypothesis. Communities diversify with time, and older communities are therefore more diverse.

The spatial heterogeneity hypothesis. The more heterogeneous and complex an environment the more diverse the flora and fauna.

The competition hypothesis. In benign environments selection is controlled by biological interactions rather than physical variables. Interspecific competition is therefore important, species have evolved narrower niches and hence there is higher diversity because more species can partition the available resources.

The environmental stability hypothesis. The more stable the environmental variables, such as temperature, salinity, oxygen, etc., the more species are present, since it is less likely that species will become extinct due to vagaries of environmental variables.

The productivity hypothesis. All other things being equal, the greater the productivity, the greater the diversity, since more resources allow more species.

The predation hypothesis. Predators crop prey populations and thus lower competition that may lead to exclusion of many species.

Detailed descriptions of each of these are available in Pianka (1966), Ricklefs (1979), Pielou (1975), and Huston (1979). Most of the papers that proposed these hypotheses dwell on the evolutionary origin of species diversity and often disagree on factors that might affect species diversity. These hypothesis are either circular reasoning or are not supported by evidence (Rhode, 1992). This is a field where logical argument exceeds factual content and, as such, is far from settled in spite of great effort.

We do not have space to scrutinize the evidence for each hypothesis, but the various hypotheses listed above are a mix of evolutionary considerations and factors that affect diversity on a local scale. We will discuss these two aspects separately to provide some idea of why most of the hypotheses falter.

10.3.1 Evolutionary Considerations

Sanders (1968) combined features of the several hypotheses listed above into an integrated statement that he called the time-stability hypothesis. This hypothesis incorporates evolutionary considerations, and was based on the most thorough study of marine benthic organisms available. Sanders (1968) concluded that variable, stressful shallow habitats had lower diversities than the more constant, benign abyssal environments. To account for these differences the time-stability hypothesis holds that diversity in communities increases during evolutionary time. Competitive pressures result in competitive displacements; the assemblage of species becomes relatively more specialized over time. The increase in diversity, however, is restricted by the level of physiological stress and environmental variability that the assemblage of species has to endure. Thus, diversity is the net balance of a process in which the more specialized an assemblage of species, the more subject it is to vagaries of the environment.

The ambiguities and difficulties of schemes to explain diversity are represented by those of the time-stability hypothesis (Peters, 1976). Unfortunately it is impractical to follow the diversity of communities over evolutionary time, marking it difficult to test directly whether diversity does indeed increase over time, the first part of the time-stability hypothesis. The second element of the hypothesis, specialization, is very difficult to define in most situations, since species may have narrow preferences for one dimension of their needs, while showing broad tolerances for another. Branch (1976) put together a subjective index, based on the occurrence of a species in one or more habitats, and on morphological traits, and applied it to an assemblage of limpet species found on rocky shores of South Africa. In this one example there is a clear relation of specialization of the assemblage and the number of species (Fig. 10-6). This agrees with the prediction of the time-stability hypothesis, but it does not necessarily mean that the relationship arose as per the hypothesis.

Because of the difficulty of measurement over time and difficulty in the definition of specialization, the tests of Sanders' hypothesis have to focus on stress, the remaining element. Stress, however, is usually defined by its effect so there is danger of circular reasoning, and it is not easy to state a falsifiable hypothesis about the effect of stress. Thus, the time-stability scheme does not have the clear predictive quality that defines scientific hypotheses. Perhaps tests can be carried out if stress can be thought to be related to the extent to which variables fluctuate in different environments. Comparison of communities of similar age

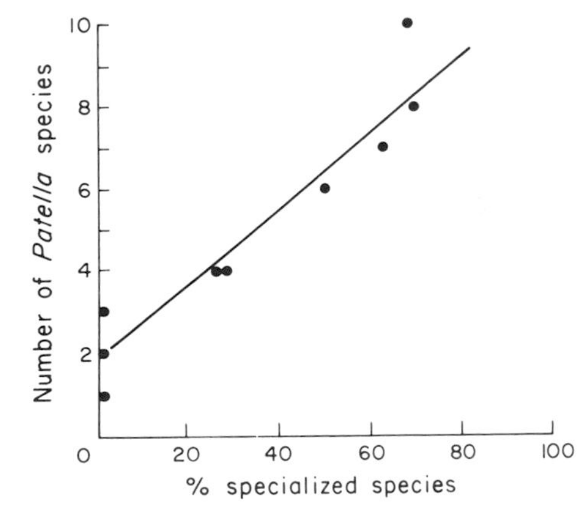

FIGURE 10-6. Relation of number of *Patella* species in a South African shore and the percentage of specialized species. Adapted from Branch (1976).

but differing amplitude of variation could then test of the effect of "stress" on diversity. Coral reefs of approximately the same age (Abele, 1976) contain more species of decapod crustaceans (55 vs 37) in reefs with more variable conditions (temperature ranges of 17–29°C vs 28–29°C, salinity ranges of 22–36% vs 31–32%). This, although a very inadequate test, does not seem to support the time-stability hypothesis in regard to higher diversity in constant, benign environments.

The high diversity of benthic fauna found by Sanders (1969) in the deep sea floor and the very uniform conditions at such depths inspired the time-stability hypothesis. The high diversity no doubt was facilitated by the long evolution in relatively constant conditions. Part of the explanation, however, may be attributable to other factors. The increased diversity of the deep-sea benthos compared to shallower areas may to a large extent depend on the inclusion of a series of heterogeneous habitats in the very lengthy tows used to sample the larger areas in the deep sea. The data may thus include substantial between-habitat diversity. There may also be sampling artifacts due to the use of different collection devices (Osman and Whitlach, 1978; Abele and Walters, 1979). Regression analysis of the species–area curve for some of the data collected by Sanders (1968) and colleagues showed that perhaps 80% of the variation in the species numbers down to about 2,000 m could be accounted for by the effect of area alone. More extensive analysis of benthic data including data down to about 5,000 m shows that the diversity of several taxonomic groups decreases below about 3,000 m (Rex, 1981) (Fig. 10-7). Since the largest area of sea floor is at or below this depth, the effect of area alone cannot explain the gradients in diversity. Nevertheless, the devices used to sample the deeper parts of the sea floor, such as epibenthic sleds and anchor dredges, are towed over the sea floor for several kilometers, while shallow stations were sampled in Sanders' data by coring devices only a few centimeters in diameter. The sampling is thus at different scales

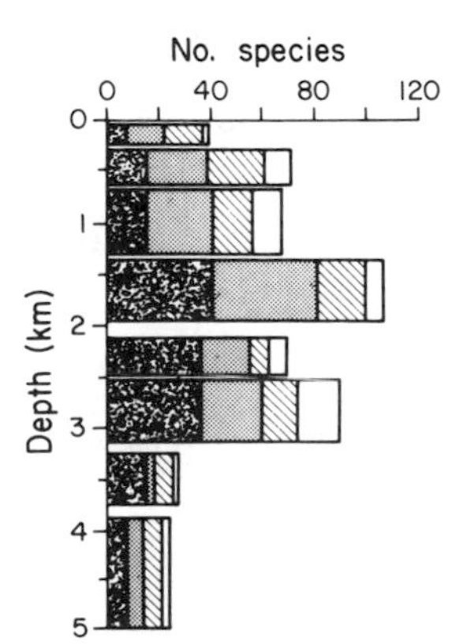

FIGURE 10-7. Species richness in the sea floor off New England, at different depths. The taxa are, from left to right, echinoderms (irregular stippling), fish (dotted), decapods (diagonal lines), and other groups (blank). The collections were done by otter trawls during several cruises, and all the data are pooled. Adapted from Haedrich et al. (1980).

and must affect the number of species obtained. Newer approaches to deal with such spatial problems are discussed by Jumars (1976, 1978).

The difficulties in interpretation and methods associated with the time-stability hypothesis are representative of other evolutionary arguments to explain taxonomic diversity. On the whole, these arguments seem unlikely to advance empirical knowledge. We discuss them here because they have been given much attention in the past, and knowledge about these issues is part of the development of our field.

10.3.2 Factors Affecting Diversity on a Local Scale

Questions about trends in diversity of biotic assemblages over evolutionary time and on regional geographical scales have not yet been resolved, in spite of sustained efforts. On the other hand, work on the local, short-term control of diversity has been more successful. The local diversity of a community can be affected over relatively short periods of time by at least four kinds of factors: (a) the concentrations of deleterious substances or physiologically severe conditions in the environment, (b) the abundance of key resources, (c) the abundance of key consumers or disturbances, and (d) specific features of the local environment.

10.3.2.1 Deleterious Substances or Conditions

Few species survive in environments that contain large quantities of deleterious substances. These substances include natural and man-made compounds, such as sulfide, ammonium, or allelochemicals, and petroleum and chlorinated hydrocarbons. The effects of intense levels of pollution on assemblages of organisms are clear cut [for example, in the case of oil in marshes, Sanders et al. (1980), and Figs. 10-3, 10-4]. Despite the widespread nature of marine pollution little is known about the effects of low-level, chronic contamination by petroleum and polychlorinated biphenyls now found dispersed everywhere in the sea. Since few marine environments remain uncontaminated, deleterious substances may play a role in determining the diversity of many marine systems. We should note that such substances affect species populations, rather than diversity per se.

10.3.2.2 Abundance of Key Resources

A low supply of an essential resource may be insufficient to maintain many species, and thus diversity may be low. Lowered diversity of benthic polychaetes and bivalves is related to reduced oxygen concentrations in upwelling areas off southwest Africa, for example (Fig.

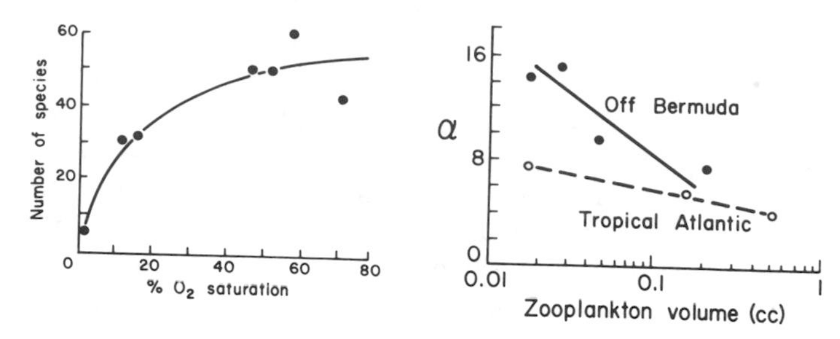

FIGURE 10-8. Left: Number of species of polychaetes and bivalves in a transect of benthic stations off Walvis Bay, Namibia, and percentage saturation of oxygen in the water overlying the sediments. Adapted from Sanders (1969). Right: Diversity (alpha, the index derived from the logarithmic series) of zooplankton in two areas, versus abundance of zooplankton. The volume of zooplankton between the two data sets cannot be compared because different nets and sampling methods were used. Adapted from Sutcliffe (1960).

10-8, left). This should bring to mind the adversity-selected species discussed in Section 3.5.

Low as well as high abundance of resources may lower diversity. The very low supply of food for benthic invertebrates at depths below 3,000 m may be responsible for the lower diversity at such depths (Rex, 1981). It may be that there has to be a certain supply of food resources before much partitioning among sympatric species is feasible. On the other hand, very large amounts of resources may also lower diversity as, for example, in diatoms and higher plants in experimentally enriched salt marsh plots (Table 10-1). In this case, in the presence of excess resources, one opportunistic (*r* or exploitatively selected) species proliferates. This also happens in phytoplankton communities of upwellings, where a few species become dominant, and diversity is lowered (Margalef, 1978). The response of diversity of resource abundance seems to be, therefore, a humped curve, with peak diversity at intermediate resource abundance.

There have been speculations as to the relation of diversity and productivity. This relation should also follow a humped pattern. Unfortunately, the examples available usually only span the lower or upper slopes of the curve, and have led to diverging conclusions in regard to the productivity hypothesis (Sanders, 1968; Huston, 1979). For example, the diversity of zooplankton near Bermuda and in the tropical Atlantic decreases the more zooplankton there are (Fig. 10-8, right). Since standing crops of zooplankton are correlated to productivity (cf. Section 9.3.1), higher production is correlated to lower diversity. Presumably there is a left-most portion of this relation, missing in Figure 10-8, where diversity increases as production increases.

10.3.2.3 Abundance of Consumers and Disturbances

We have already discussed (Section 8.3.1) experiments by Lubchenco (1978) where she manipulated the density of grazing snails on tidal pools and on the rock surface of a rocky shore. These experimental changes in density of the herbivores resulted in alteration of the diversity of the macroalgae. In the tidal pools with few snails (Fig. 10-9, top) *Enteromorpha* outcompeted other algal species, and diversity is therefore low. At intermediate densities of snails, the abundance of *Enteromorpha* and the few other preferred food species is reduced; this prevents competitive exclusion by *Enteromorpha* so that many algal species, both ephemeral and perennial, can coexist. At high densities of snails all edible algal species are consumed, leaving low diversity stands consisting mainly of the inedible *Chondrus*. Similar results, with a dome-shaped response of species richness to increased herbivore pressure, are found in cases of sea urchins feeding on macroalgae (Paine and Vadas, 1969), the sea urchin *Diadema antillarum* feeding on algal mats (Carpenter, 1981), and in damsel fish feeding on coral reef algae (Hixon and Brostoff, 1983).

The tide pool results (Fig. 10-9, top) reflect the combined impact of competition and selective grazing. This is not the case on the exposed rock surface. Disturbances such as ice damage in winter can be frequent enough in the rock surface (Fig. 10-9, bottom) that the competitive dominance of the perennials is repeatedly thwarted. If snails are absent, diversity is high, since at least 14 species of ephemeral macroalgae grow on the disturbed patches and on the perennials. As snail density increases, the ephemeral species decrease, since they are preferred

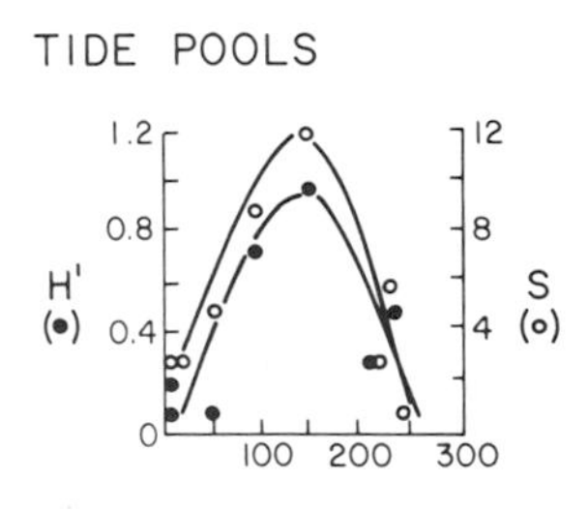

FIGURE 10-9. Effect of density of the grazing snail *Littorina littorea* on the diversity of algae in two habitats of the rocky shore of New England. H' is the Shannon-Weaver diversity index and S is the number of species. Adapted from Lubchenco (1978).

food of *Littorina*. Since only the few species of perennial macroalgae remain, the canopy is one of low diversity (Fig. 10-9, bottom).

Note in Fig. 10-9 that the information theory index H' provides a similar depiction as species richness. In this, and other cases, the use of the more complex index does not furnish any new information.

A more concrete example of the role of physical disturbances on species richness is provided by a study on the species growing on boulders on Californian beaches. Boulders of different size are turned over in a frequency proportional to their weight. An examination of the diversity of boulders of various sizes (Table 10-2) thus provides a gradient of disturbance frequency or intensity, since the overturning of boulders damages the attached organisms. Small boulders are populated by few species, mainly opportunistic early colonizer species, such as the alga *Ulva* and the barnacle *Chthamalus*. The largest, least disturbed boulders tend to be covered by a late colonizer, the red alga *Gigartina canaliculata*. The surface of boulders disturbed at intermediate frequency are covered by a combination of early and late colonizers and also show bare areas. The differences in flora are not due to the size of the boulders, since small rocks experimentally attached to the sea floor supported an array of late colonizing red algal species, and eventually *G. canaliculata* became dominant. In boulders where disturbance was frequent, only the few quick-growing opportunists are present; where disturbances seldom occur, the competitive dominants exclude other species. Maximum diversity is found at intermediate disturbance frequencies.

Thus, it is clear that not just grazers, but any agent—such as predators or disturbances—that prevents competitive dominant species from monopolizing resources may affect diversity. In rocky shores

TABLE 10-2. Species Richness (Average Number of Species ± Standard Error) of Macroalgae on Boulders Whose Size Resulted in Different Frequencies of Disturbance.[a]

	Incidence of disturbance		
	Frequent	Intermediate	Seldom
Nov. 1975	1.7 ± 0.18	3.3 ± 0.28	2.5 ± 0.25
May 1976	1.9 ± 0.19	4.3 ± 0.34	3.5 ± 0.26
Oct. 1976	1.9 ± 0.14	3.4 ± 0.4	2.3 ± 0.18
May 1977	1.4 ± 0.16	3.6 ± 0.2	3.2 ± 0.21

[a] The size classes are grouped into three frequency of disturbance classes: frequent, that rock size that required less that 49 Newtons (N) to move horizontally; intermediate, 50–294 N; seldom, more than 294 N. From Sousa (1979).

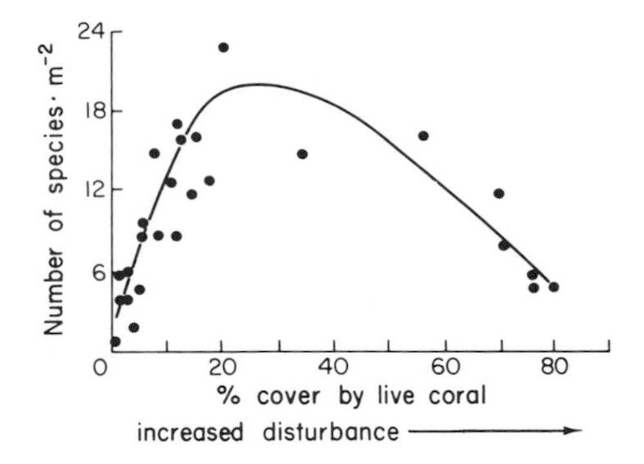

FIGURE 10-10. Species richness in coral reef of Heron Island, Queensland, versus the percentage of the area covered by live corals. The values are numbers of species of corals in samples of $1\,m^2$ in surface area. The area covered by live corals is an index of the degree to which predators and storms have been active, since starfish and other predators feed largely by scraping the live tissues off the coral skeletons, and storms break and damage corals. Adapted from data of Joseph Connell.

predaceous sea stars are often such "keystone" species and their presence may increase diversity (Paine, 1966), and wave action can also play a key role.

In coral reefs predators and storms may remove live tissue from coral skeletons, and the percentage of live coral cover can be taken as an index of disturbance. Species richness has a dome-shaped response to increases in such disturbances (Fig. 10-10). Much as in the case of grazers, some disturbance or predation prevents competitive dominance and leads to increased diversity.* Beyond a certain predation pressure or disturbance frequency only the few less desirable prey or tolerant species remain, and species richness is low.

10.3.2.4 Specific Features of the Local Environment

Much of what constrains within-habitat diversity has to do with the physical nature of the habitat (MacArthur, 1965). In soft sediments the activity of certain animals can provide just such a within-habitat structure (Fig. 10-11). Sea cucumbers (*Molpadia oolitica*) live head down in the mud of Cape Cod Bay, and ingest sediment particles. Feces are expelled out the anus onto the water-sediment surface and form a

---

*This applies to the sum of predators in the community but it may not be true for all species of predators. For example, the crown-of-thorns starfish (*Acanthaster planci*) feeds preferentially on rarer species of corals, and its feeding thus fosters the growth of common, fast-growing species. *Acanthaster* could thus lower species richness (Glynn, 1974, 1976).

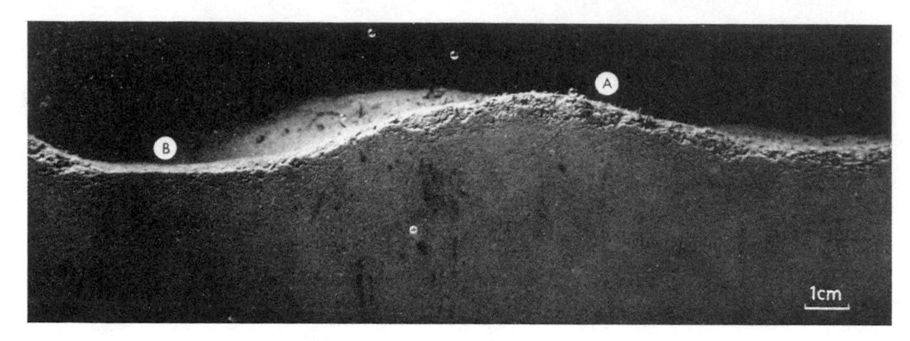

FIGURE 10-11. Photograph of a profile of surface sediments in Cape Cod Bay, Massachusetts. Defecation by the sea cucumber *Molpadia olitica* produces the relatively stable fecal cones (A) populated by a variety of species of suspension feeders. The cones provide habitat heterogeneity and increase and diversity compared to that of the more unconsolidated mud (B). From Rhoads and Young (1971).

cone-shaped mound whose surface is considerably more consolidated than the flocculent, unconsolidated sediment surrounding the fecal mound. The mounds are colonized by tube-building polychaetes (*Euchone incolor, Ninoe nigripes,* and *Spio limicola*). These tubes in turn make it possible for the caprellid amphipod *Aeginina longicornis* and the bivalve *Thyasira gouldi* to be present. Thus the microtopography provided by the feeding of *Molpadia* increases species richness.

Other features of the habitat may also affect diversity. The particular geometry of coral heads may attract differing groups of species. Even more exotic mechanisms may be important, such as the increased diversity of coral reef fish associated with the vicinity of territories of the cleaner wrasse (Table 10-3). The wrasse removes parasites off other fish, and this is apparently an important enough service to attract many individuals of many other fish species. In fact, where cleaner wrasses were absent, 26 of the 49 fish species found in the reef were not found and an additional 12 species were less abundant.

TABLE 10-3. Effect of Presence and Absence of the Cleaning Wrasse (*Labroides dimidiatus*) on Fish Diversity on a Red Sea Coral Reef.[a]

	Number of species per habitat	Number of individuals per transect	Number of species absent	Number of species with reduced abundance
Labroides absent	0–6	1–10	26	12
Labroides present	7–31	18–310	3	8

[a] From Slobodkin and Fishelson (1974).

10.4 Integration of Factors Affecting Diversity and Some Consequences

Given that there are enough resources available, the diversity of communities seems, from the examples above, to be a balance between a tendency toward competitive exclusion and the prevention of this tendency by disturbances, either due to consumers, natural accidents, or pollution. Increased resources favor species with rapid growth rates and these may become dominant enough to reduce diversity; the dominance will increase up to the level allowed by the disturbances or consumers. The way in which a consumer affects the diversity of its food items may depend on its selectivity. These trends can be altered where toxic materials eliminate certain species or where particular features of the habitat play a significant role. These simple statements cannot yet account for the notable geographical gradients in diversity with which we started this chapter.

Even if concrete facts as to how diversity is determined within a habitat are available and reasonable ways to measure diversity are used, there still remains the matter as to its biological significance. No one has found an answer as to what it means for an organism to live in a high- or low-diversity environment. Clearly the ratio of intra- to interspecific encounters in each of these two kinds of environments may vary (Lloyd, 1967) but there is no evidence as to the significance of this. Perhaps diversity matters more to ecologists than to organisms.

Another area of ambiguity is the much-discussed relation of diversity and stability of the assemblage of species. Ecological wisdom followed Charles Elton and others who held that a community with many species would be more stable than one with few species. The more diverse system would be less likely to disappear due to some haphazard event since it would be more likely that some of the species present would tolerate the disturbance. A more diverse system would also be better able to adjust to changes in resources, since each component species is bound to possess slightly different characteristics. The community as a whole could thus deal with a wider variety of changes.

In the early 1970s, a spate of theoretical publications showed, using simulation models, that in fact stability was reduced in diverse systems (Gardner and Ashby, 1970; Smith, 1972; Hubbell, 1973; May, 1973; among others), only to be followed by others that showed that by changing the models diversity could increase stability (Jeffries, 1974; De Angelis, 1975; Gilpin, 1975; and others). In part, the problem is in the definition of stability, a very vague concept that has been variously defined and interpreted (Orians, 1975; Harrison, 1979). A second problem is the difficulty in formulating testable hypotheses due to the ambiguity of the concept and the often tautological nature of discussions of stability—"this community is stable because it is not perturbed by this change." The result is that there has been a great deal of modeling and reasoning and little experimental verification, a situation appropriately described by McNaughton (1977), who thinks that "the marked instability of attitudes regarding diversity-stability relationships in ecosystems arises from a low diversity of empirical tests of the hypothesis."

A few attempts at experimental evaluation of parts of the diversity-stability idea have taken place, and have come up with disagreeing conclusions (Abele and Walters, 1979; Smedes and Hurd, 1981) or failed to show meaningful relationships (Hairston et al., 1968). Studies on a few terrestrial environments show that diversity and stability increase together (McNaughton, 1977), as claimed by the earliest speculations. By and large there has been, despite much interest, little progress in recent years, and certainly no study documenting causal relationships, largely due to the difficulty in formulating testable hypotheses.

Explanation of species richness awaits new approaches. It may be that the explanations of differences in species richness are not ecological but rather are of genetic origin. Perhaps a first new step is understanding of the specific genetic and mating system mechanisms that lead to speciation in marine organisms (Palumbi, 1992). In sea urchins, abalone, and other groups, the proteins that mediate binding of sperm and egg show relatively rapid generation of considerable variation in sequences of amino acids. Such changes at the molecular level could originate reproductive isolation. Moreover, gene complexes that control selfing and breeding compatibility have hundreds of alleles, and recombination during inbreeding may produce new combinations that may reproductively isolate small local populations. Once we have identified the specific genetic or breeding system mechanisms that promote appearance of new species, we could ask if there are environmental factors that increase or decrease the effectiveness or frequency of speciation in one environment or another, and compare groups of organisms that show different degrees of species diversity.

Chapter 11
Spatial Structure: Patchiness

11.1 Scales of Patchiness

In general, organisms are not distributed uniformly over space; consider for example, the distribution of primary production over the world's oceans (Fig. 2-30). Moreover, such heterogeneity occurs at all spatial scales. We can use the size of a patch of higher abundance or the distances among such patches as a way to assess the scale of heterogeneity. There are significantly higher or lower rates of production in distances of 10^5 to 10^8 m within any linear transect in Fig. 2-30. Spatial variability is also pervasive at smaller distances: the variation in concentration of dissolved CO_2 in surface waters of the Gulf of Maine is evidence of spatially heterogeneous biological activity on a scale of 10^4–10^5 m, while the variability in concentration of chlorophyll in St. Margaret's Bay, Nova Scotia, varies on a scale of about 10^4 m (Figs. 11-1a,b). Measurements of numbers of zooplankton m^{-2} off the California coast show patches of tens of meters (Fig. 11-2c); a careful examination of spatial variability of phytoplankton production shows significant patches in a scale of 10^{-1} m, most clearly in samples from the Gulf Stream (Fig. 11-1d). Recent technical advances (for example, Davis et al., 1992) have provided data over spatial scales of microns to hundreds of meters, confirming the presence of aggregated distributions of organisms at scales of 20 cm to 200 m.

Such variability over space, especially on the smaller scales, has made field measurements notoriously variable and has long been considered as one of the chief nuisances of environmental research. Field measurements of densities of many kinds of organisms, for example, usually have associated variations of 10–30% of the mean. Much of this variability is due to unrecognized patchiness and it is only recently we have become more aware that such patchiness is the signature of some very powerful biological and physical processes. Although we are at a very primitive stage in our understanding of spatial distribution, future studies of spatial distributions may provide insight as to how the structure of marine ecosystems is determined, as well as provide a way to understand the variability of ecological measurements.

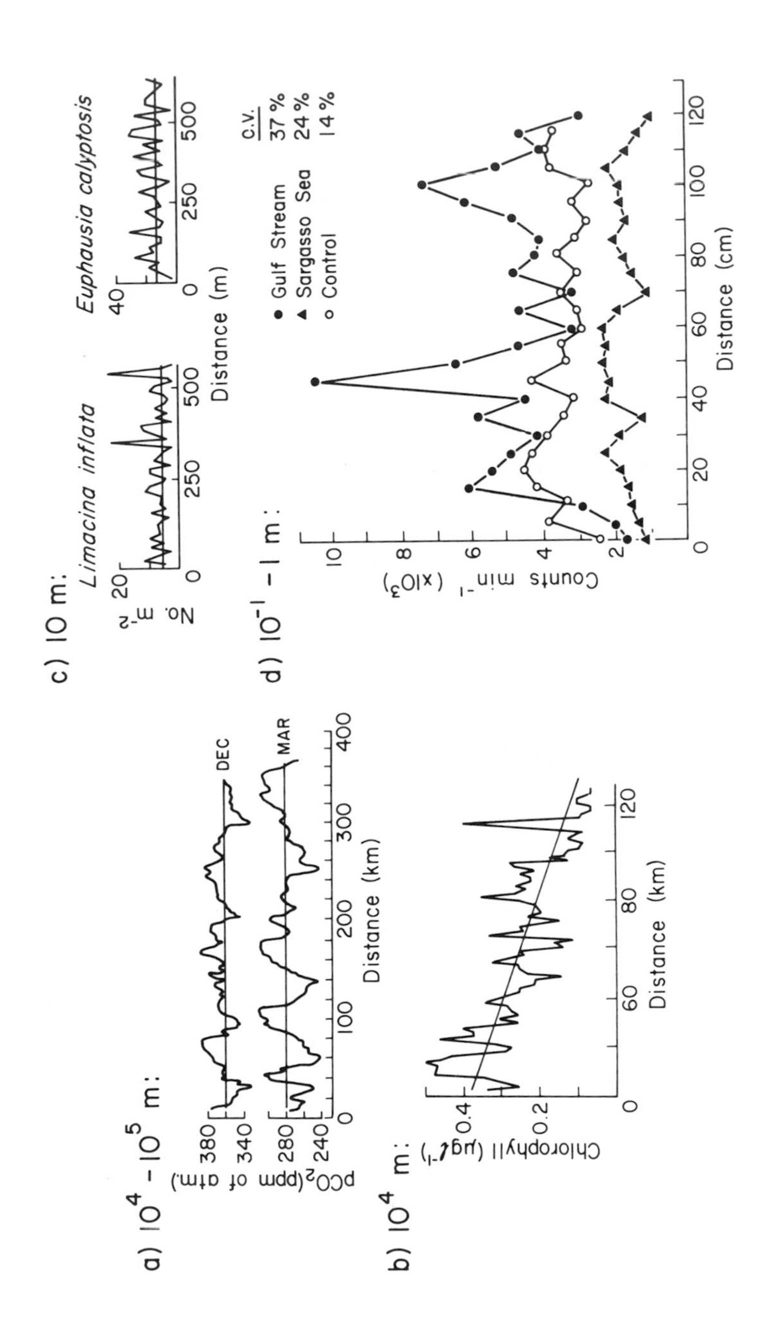
a) 10^4 - 10^5 m:
pCO2(ppm of atm.)
380
340
280
240
DEC
MAR
0
100
200
300
400
Distance (km)
b) 10^4 m:
Chlorophyll (μg l^-1)
0.4
0.2
0
60
80
120
Distance (km)
c) 10 m:
Limacina inflata
Euphausia calyptosis
No. m^-2
20
40
0
250
500
Distance (m)
d) 10^-1 - 1 m:
Counts min^-1 (x10^3)
10
8
6
4
2
0
20
40
60
80
100
120
Distance (cm)
Gulf Stream
Sargasso Sea
Control
C.V.
37 %
24 %
14 %

FIGURE 11-1. Patchiness on approximate scales of 10^5 to 10^{-1} m for abundance and activity of plankton. (a) Concentrations of CO_2 in surface waters for transects in the Gulf of Maine. Adapted from Teal and Kanwisher (1966). (b) Chlorophyll concentrations at 10 m along a transect in St. Margaret's Bay, Nova Scotia. The straight line is the regression of chlorophyll on distance. Adapted from Platt et al. (1970). Reprinted by permission of the Canadian Journal of Fisheries and Aquatic Sciences. (c) Abundance of a pteropod mollusc and an euphausiid in near-surface tows off California. Medians of entire data shown by the horizontal line. Adapted from Wiebe (1970). (d) Radioactive counts in a ^{14}C uptake experiment with phytoplankton in the Atlantic. The control consisted of shelf water that was thoroughly stirred to homogenize the distribution of cells. The height along the y-axis depends on level of productivity and is not relevant here; the important thing to notice is the variability over space. A measure of this variability is given by the coefficient of variation (C.V. = coefficient of variation = variance/mean). Data from Cassie (1962a).

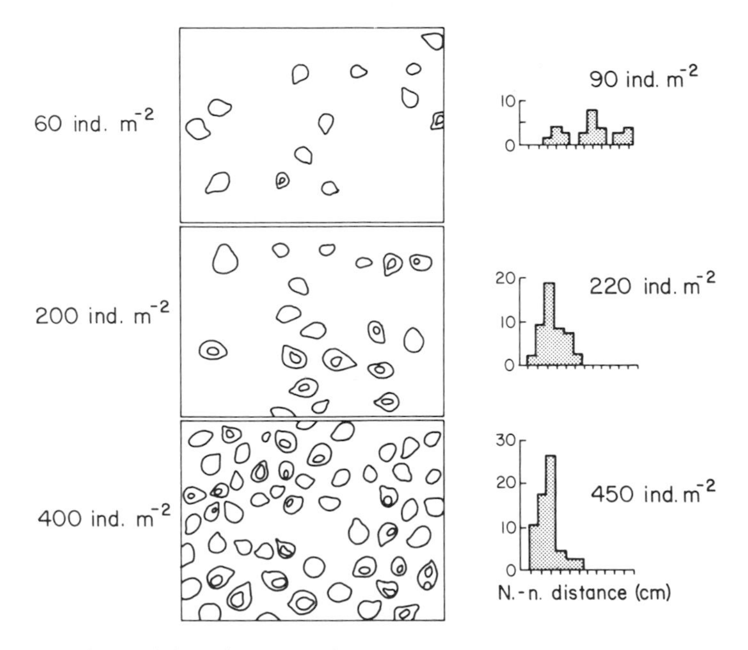

FIGURE 11-2. Spatial distributions of limpet (*Patella cochlear*) on a South African rocky shore. Left panels contain plan views of the distribution of limpets, which are shown as the tear-shaped objects. Small limpets settle on large limpets since territories on the rock surface are defended by adults, so that young limpets are prevented from settling on bare rock. Histograms on right show the frequency of distribution of nearest-neighbor distances (excluding those on the backs of others) for three densities of limpets approximating the densities of the diagram in the panels on the left. Adapted from Branch (1975a).

Spatial structure is evidenced by a wide variety of phenomena. We have already dealt with examples of one-dimensional spatial distributions (vertical zonation) within the intertidal zone (Section 8.2). In Section 2.1.1.2 we examined the mechanisms that provide a marked vertical structure to the distribution of phytoplankton. In this chapter we concentrate on the scale of spatial variations, on how spatial heterogeneity is described, and what some of the sources of spatial heterogeneity may be. Lastly, we will consider some consequences of spatial variability.

11.2 Description of Spatial Distributions

There are many ways to describe distributions over space. Patchiness can be described in terms of the distances that separated units of the populations being studied over a continuous space, or by assessing the occurrence of units of the population within discrete spatial units such as quadrats, tidal pools, or samples. Methods for both of these situations are included in Cassie (1962) and Poole (1974). Spatial heterogeneity has also been described by spectral analysis (Platt and Denman, 1975) and applied to seasonably long data series.

11.2.1 Dispersion Patterns Over Continuous Space

Spatial distributions are most easily studied in two-dimensional surfaces such as rock surfaces. The distribution of intertidal surface-dwelling limpets provide examples of various spatial patterns (Fig. 11-2, left panels). Measurements of nearest neighbor distances* are often compiled into frequency distributions (Fig. 11-2, right panels). The mean ($\bar{x}$) and variance (s^2) of these distributions are not the same in these three examples (Fig. 11-2, right panels). The ratio $s^2/\bar{x}$ can be used as an indicator of the kind of spatial distribution displayed by the population. In randomly dispersed populations, $s^2/\bar{x} = 1$. When spacing is fairly uniform, as in Figure 11-2, bottom right, $s^2/x < 1$. When spacing is very variable, and clusters of individuals are found, $s^2/x > 1$. This latter condition is generally the most common of all distributions, and is variously referred to as aggregated or contiguously distributed.

---

* Distributions of nearest neighbor distances are usually calculated by compiling distances on a plane from only one of the four possible 90° quadrants; this makes it possible to include distances between individuals situated in different aggregations. Otherwise, nearest neighbors would always be within aggregation and the distribution of nearest neighbor distances would not include any reflection of the distribution of patches,an important aspect of spatial structure.

Clark and Evans (1954) used nearest neighbor distances to calculate an index, R, that also can distinguish among the three kinds of spatial distribution,

$$R = \frac{\sum_{i=1}^{n} r_i}{N} 2\sqrt{p}. \qquad (11\text{-}1)$$

In this expression, r_i is the distance between an organism and its nearest neighbor, where there are n such measurements. N is the number of organisms, and p is the density of the population (individuals/unit area). Values of this index between 0 and 1 indicate that the population is aggregated, and the more aggregated, the closer to 0. Values of the index between 1 and 2.15 correspond to uniform distributions, while a value of 1 indicates a randomly spaced population. Values of R calculated for a limpet population (Fig. 11-3) show that at low density individuals are aggregated, while at high densities, as we surmise from Figure 11-2, limpets are closely and uniformly spaced. Such a pattern may be a feature of species whose major limiting factor is space.

Many other populations share the tendency to aggregate when their density is low and be either uniform or randomly spaced at high densities (Hairston, 1959). It is possible, however, for rare and abundant populations to show any of the spatial distributions. For example, judging by s^2 to $\bar{x}$ ratios, most species of crustaceans, molluscs, and echinoderms in the deep sea off California have mostly random spatial distributions, with some occasionally aggregated species (Jumars, 1976). The most dense populations tended to show the most aggregation. In the example of limpets (Fig. 11-2) it just so happened that increases in density were correlated to changes from aggregated to uniform distributions; the actual response is apparently species specific.

The size of the sample or quadrant needs to be considered carefully since the choice of sample size can artificially create one or another

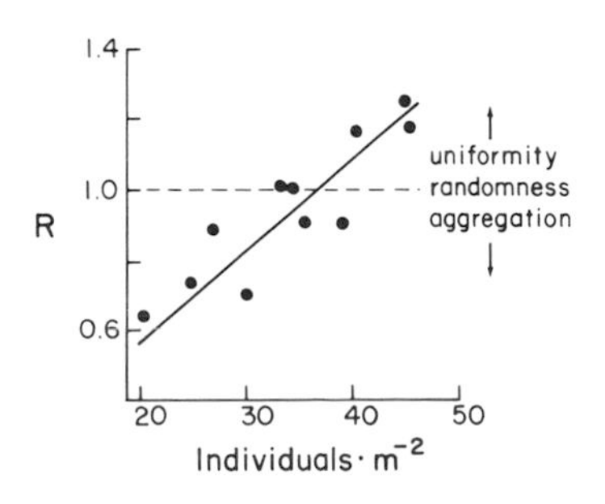

FIGURE 11-3. Relation of aggregation (R is the Clark-Evans index) to density of *Patella longicosta*. The horizontal dashed line shows the value of R that indicates a random distribution. Adapted from Branch (1975a).

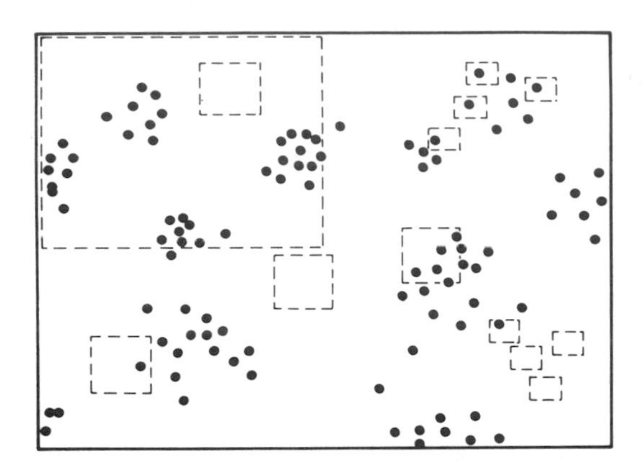

FIGURE 11-4. Hypothetical distribution of a population over space. The dashed squares indicate the changing effect of the use of different sample sizes.

distribution. Consider a population distributed as in Figure 11-4, and sampled using quadrats of three sizes. An array of the largest quadrats provide a more or less uniform number of individuals, and misses the smaller-scale patchy nature of the distribution. A collection of individuals obtained by use of the intermediate-size quadrats shows the aggregated distribution quite well, while the smaller quadrats either provide a very uniform distribution or show that the species is very rare, depending on where the quadrats are placed.

11.2.2 Dispersion Patterns in Discrete Units

The frequency of occurrence of individuals in discrete units or in samples from an environment can be described in a fashion similar to that used for nearest-neighbor distances. Figure 11-5 shows a generalized scheme of the frequency distribution associated with uniform, random, and aggregated populations. There are well-established statistical descriptions of these distributions (Cassie, 1962; Poole, 1974). The simplest assumption that can be made is that any point in space is equally likely to be occupied by an organism. The expected frequency distribution of samples with 0, 1, 2, 3, etc., individuals is given by successive terms of the binomial series $(q + p)^k$. In this expression k is the maximum number of individuals in a sample, p is the probability of occurrence of an organism in a sample, and $q = 1 - p$. For the binomial distribution, the mean $\mu = kp$, and the variance $= \sigma^2 = kpq = \mu - [\mu^2/k]$. If k is very large, $\sigma^2 \simeq \mu$. Such a frequency distribution, where each organism is equally likely to appear in a sample, and the number in the sample is very small compared to the total population, is called a Poisson distribution, a special case of the binomial, and both are random distributions. Based on the Poisson relationship, $\sigma^2 = \mu$.

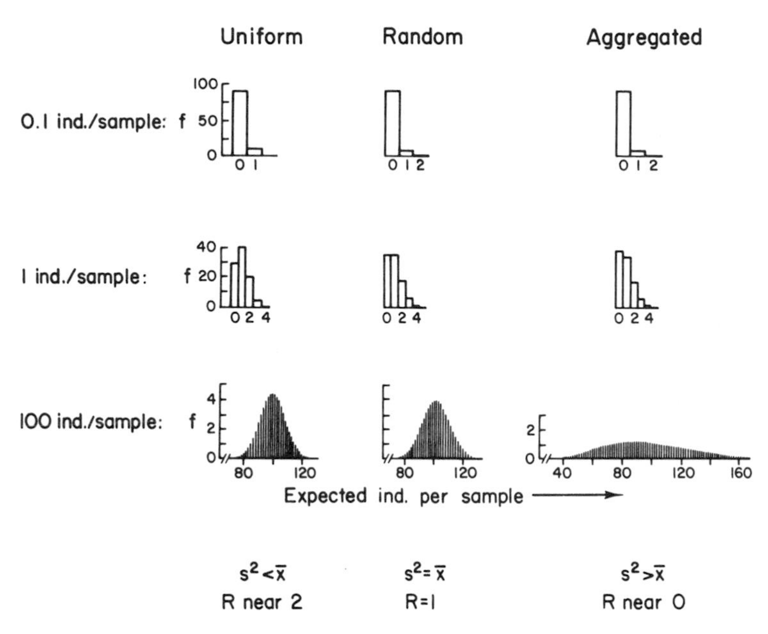

FIGURE 11-5. Generalized frequency distributions for uniform, random, and aggregated distribution of occurrence in discrete units of sampling. Each distribution is shown as it would look with mean number of 0.1, 1, and 100 individuals per sample. The relative size of variance and mean, and the Clark-Evans index of aggregation (R), two criteria that can be used to identify each of the three distributions, are shown at the foot of the figure. Adapted from Cassie (1962).

Two other kinds of distributions—uniform and aggregated—can be distinguished by comparison of values of R or by using $s^2/\bar{x}$ (Fig. 11-5, bottom). An example of the latter procedure is possible with data on the occurrence of a planktonic mysid in samples taken by net tows (Fig. 11-6). In this particular case, increased abundance is correlated to increased aggregation, while less dense populations tend to be randomly distributed. This pattern seems to be typical for plankton (Barnes and Marshall, 1951; Clutter, 1969). In plankton, it seems that uniform distributions are rare: there are few points where the variance is smaller than the mean (Fig. 11-6).

The abundance of a population has a significant impact on spatial arrangements and on our ability to detect various spatial patterns. Only when density is relatively high is it possible to see the differences between uniform and aggregated distribution. At low densities (Fig. 11-5, top row), all three distributions are highly skewed and it is very difficult to distinguish one distribution from another. Of course, rarity in the samples may be due to the smallness of the sampling unit, and

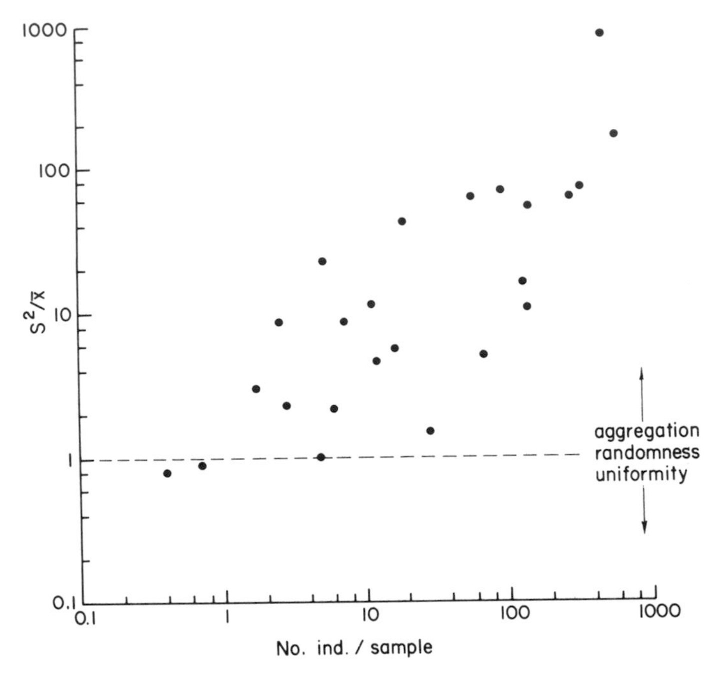

FIGURE 11-6. Index of aggregation ($s^2/\bar{x}$) for a variety of zooplankton samples containing the mysid *Metamysidopsis elongata*. Data from Clutter (1969).

an increase in the number or size of samples may allow the identification of the spatial arrangement.

It is clear from the foregoing that the ratio of variance to mean partly depends on the number of individuals per sample rather than being due solely to the spatial distribution. The aggregation of Figure 11-6 could be a function of density, for example. It may be common for plankton data to move diagonally across Fig. 11-5, from "uniform" distributions at low densities to "aggregated" distributions at high densities. If it is of interest to avoid the effect of density, coefficients of aggregation that are not affected by abundance may be used (Cole, 1946; Pielou, 1977).

11.3 Sources of Patchiness

11.3.1 Social and Demographic Mechanisms

11.3.1.1 Associations of Individuals in Mobile Groups

Many marine organisms form schools, clusters, or swarms (Fig. 11-7) including fish, euphausids, mysids, and copepods, among other taxa. These patches are due to social and reproductive behavior. The social

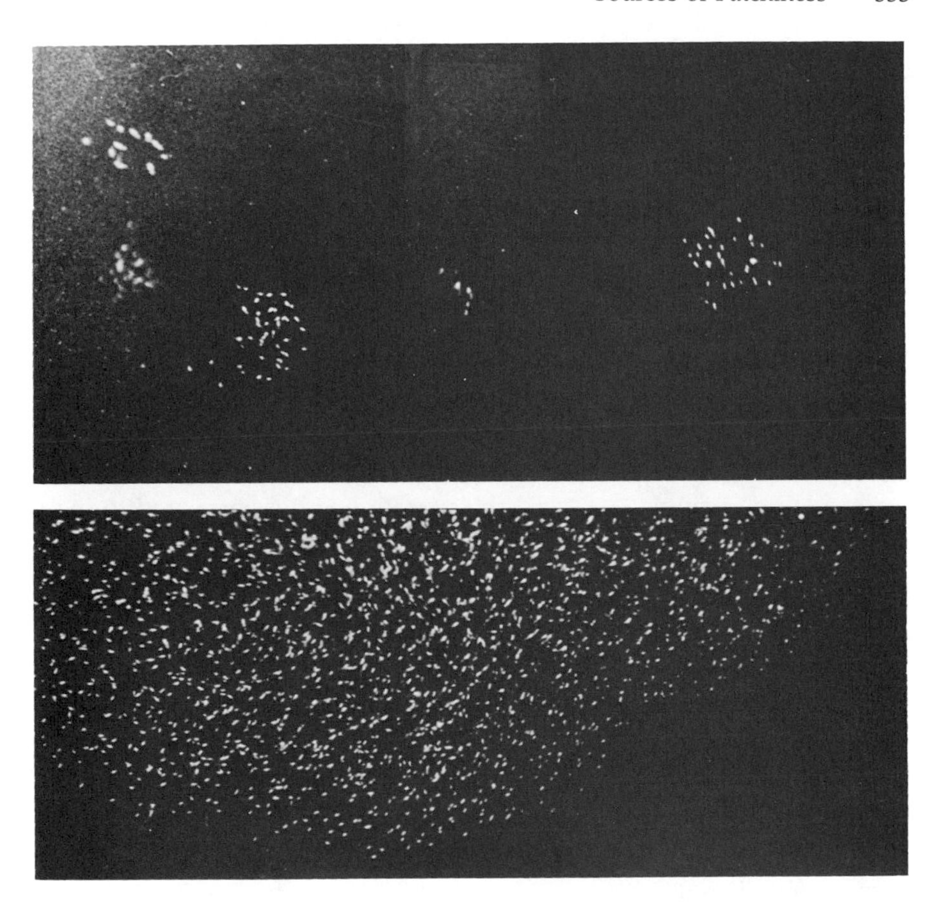

FIGURE 11-7. Schools or swarms of the mysid *Acanthomysis* sp. photographed at a depth of about 6 m in the Gulf of California. Length of each specimen is about 8 mm. From Clutter (1969).

nature of such aggregations is evident by the cohesion of movement of individuals within the clusters and by the uniform distances among members of a swarm or school.

Patches of the krill *Euphausia superba* move ". . . almost as if there was some leader in command . . ." (Hardy and Gunther, 1935). The shape and size of clusters are very fluid but there is a strong tendency for social plankton to reconstitute clusters even after disturbances (Kamamura, 1974; Clutter, 1969). There is clearly a strong impetus to remain in the swarm, as shown by motion pictures of swarms of pelagic mysids (Cluter, 1969). Individuals that lag behind their swarm make rapid leaping movements to regain their place in the school, and even overcompensate for their temporary isolation by swimming well into the swarm on their return. Smaller groups of mysids coalesce with larger groups if they happen to be less than about 25 cm away from the larger warm.

Underwater observations and motion pictures of mysid swarms show that, in spite of great variability in the size of the swarm (13,000–32,000 individuals), the average nearest neighbor distance is quite constant (1–4 cm) and is maintained at less than 4 mysid lengths (Clutter, 1969).* Similar small distances are maintained by schooling fish and euphausiids. *Euphausia pacifica* swarms with an average distance of 1–2 cm among individuals (Komaki, 1967), while the density of *E. superba* can be 1 individual cm^{-3} (Marr, 1962). The distribution of intraswarm distances is one of the few instances of uniform spatial spacing in planktonic populations. This spatial pattern is of course destroyed when samples are taken with nets. Certain high-density zooplankton patches are mating aggregations. These characteristically are made up of single taxa, and their location is not correlated to high density patches of other species (Davis et al., 1992).

Some aggregations are due to reproductive phenomena, not necessarily involving social behavior. For example, eggs and larvae may be patchy due to the occurrence of reproduction in one area. Such aggregations may last for considerable time. Copepod and mysid (Clutter, 1969) clusters are made up primarily of individuals at specific stages in the life history or at similar stages of sexual maturity. In *E. superba* high numbers of cast molts are often collected in specific net hauls (Marr, 1962), suggesting that molting of individuals within a cluster is sychronized. Such observations indicate that clusters are often made up of individuals whose growth and reproduction are to some extent linked.

The sensory mechanisms used to maintain motile swarms or schools are not clear. In mysids, visual means are probably important, since eyes are well developed and swarming is lessened at night (Clutter, 1969). Since copulation takes place at night, however, other senses, including mechano- and chemoreception must also operate. Social mechanisms operate over small spatial scales, perhaps a centimeter to a meter for larger zooplankton; the reproductive patterns, however, can synchronize populations and create patches over larger spatial scales, up to diameters of perhaps 10^3 km (Haury et al., 1978).

11.3.1.2 Spacing of Individuals in Sessile Clusters

Territoriality is another social mechanism that affects the distribution of individuals over space; in this case, behavioral mechanisms set the

---

*We saw earlier that mysid distributions tended toward aggregation at higher densities (Fig. 11-6). The process of sampling collects swarms or parts of swarms so that the individuals appear aggregated. This is matter of size of the sample unit, as discussed in reference to Figure 11-4.

distances among individuals within a cluster or aggregation. We have already discussed territorial spacing in shorebirds (Chapter 5) and in limpets (Chapter 4 and Fig. 11-2). Territorial defense by individuals is a proximate mechanism that provides for some ultimate need, such as sufficient supply of food or suitable breeding area. The garibaldi, a pomacentrid fish common off the California coast, offers a good example of a population whose spatial distribution is dominated by territoriality. The population clusters on areas of the sea bottom that are appropriate as a habitat (Clarke, 1970). Territorial behavior then more or less uniformly spaces the individuals within the dense cluster (Fig. 11-8). Outside the cluster the density of the garibaldi population is very reduced, so that on a larger scale the fish are patchily distributed.

The spacing is primarily defined by aggressive behavior of nesting males, who seldom leave their territory. Male garibaldi have unique nest-building behavior in which a thick growth of red algae is somehow slowly fostered in a 15–40 cm diameter patch within their territory. This patch is the nest, made more visible during the nesting season when the males clear away all vegetation from a band 5–11 cm wide around the nest. These algal nests require 2 or more years to culture and are used for several years, quite unlike other fish whose nests are

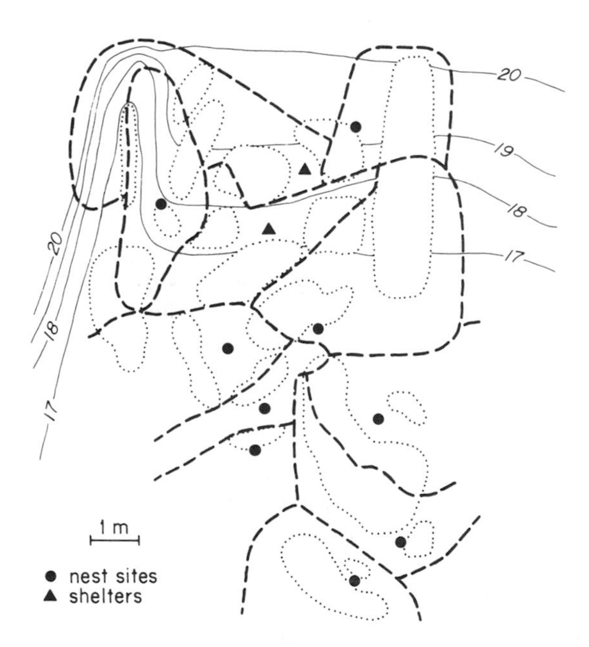

FIGURE 11-8. Territories of eleven garibaldi (*Ilypsypops rubicunda*) off La Jolla, California. Nine males had nests within their territory (dashed lines). Depth contours (thin lines) are in meters; rocks protruding from the sea bottom are indicated by the dotted lines. Adapted from (Clarke (1970).

abandoned after breeding. Thus the investment in a garibaldi nest is very high, and nests are defended throughout the year. Any nests that become vacant are taken over by other males at any time of the year. Because continuous defense of the nest is needed, each territory must also be large enough to provide food and shelter. In the case of garibaldi, both food abundance and reproductive needs define spatial distribution. Successful reproduction apparently requires some minimum space (van der Assem, 1967); in the garibaldi, the minimal average distance among nests is about 1.9 m.

Settlement of barnacle larvae on surfaces may or may not be aggregated. Some barnacles settle in patches preferentially near adult conspecifics (Bayliss, 1993); apparently, some advantage overcomes the potential for intraspecific competititon. Other barnacle species settle randomly (Satchell and Farrell, 1993).

11.3.2 Physical Mechanisms

The properties and motion of seawater are the most important mechanisms affecting spatial distribution of plankton in most spatial scales. Most of the spatial variability in physical properties that can be observed in the sea is created by processes at larger scales. The variations are simultaneously transmitted to smaller and smaller scales, and dispersed over larger scales. Regional-scale events such as upwellings take place; the energy contained in the water is then redistributed to larger and to smaller spatial scales. The patches of upwelled water both expand outward into larger areas, and are twisted and stretched into smaller and smaller features. Such redistribution of variation, in particular the "turbulent cascade" (Mann and Lazier, 1991) down to smaller scales, provides the physical basis for much of the biological patchiness we observe. Eventually, the cascade transfers turbulence down to the viscous-dominated spatial scales, where the energy is dissipated as heat.

If plankton were just a passive tracer of water movement, we would expect plankton abundance to be well correlated to passive tracers (such as salinity and temperature) of water masses. One way to examine whether the patchiness of phyto- and zooplankton is a result of passive movements with water masses is to examine variance spectra in instances where long series of measurements along a spatial transect are available (Fig. 11-9). Temperature, an indicator of water properties, and fluorescence, a measure of chlorophyll, can be collected while a ship is underway, with measurements taken at intervals over transects covering many km. The spatial variation of temperature, and of phytoplankton are therefore recorded. Spectral analysis (Platt and Denman, 1980) can be used to quantify the amount of variation (shown as they axis in Fig. 11-10) that can be assigned to different spatial distances (the

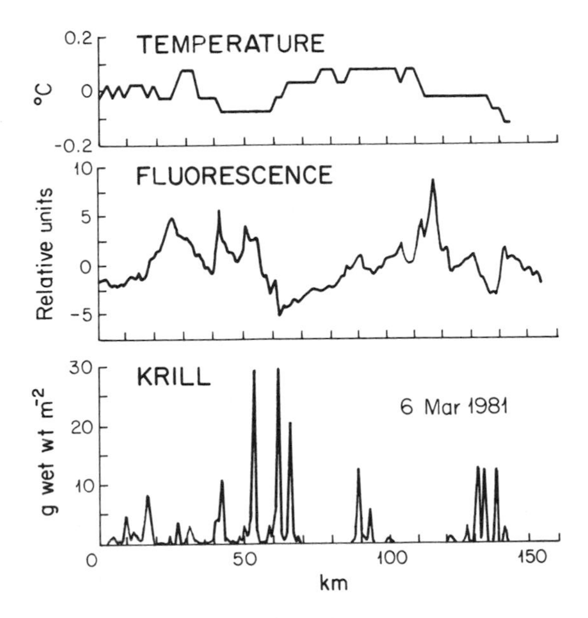

FIGURE 11-9. Horizontal transects of surface seawater temperature, fluorescence, and krill density in the Antarctic Ocean south of South Africa. Adapted from Weber et al. (1986).

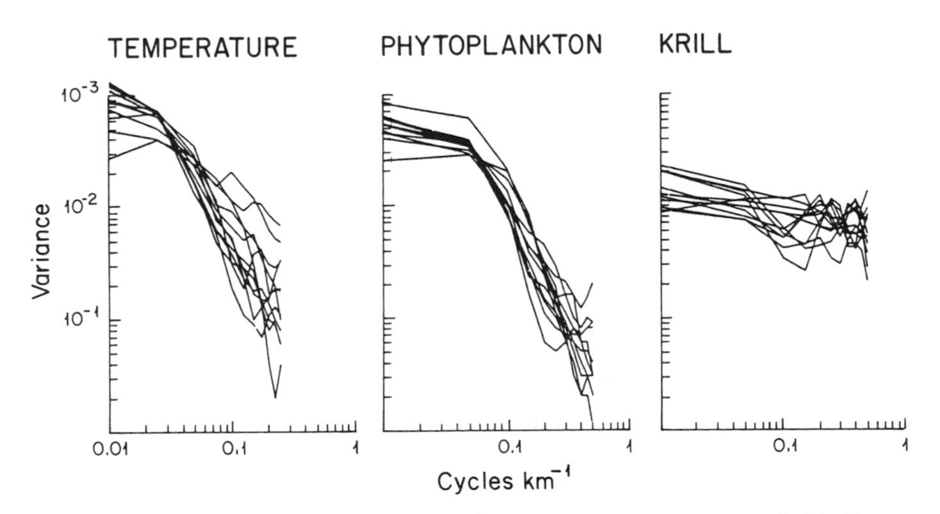

FIGURE 11-10. Variance spectra for surface seawater temperature (left), fluorescence (center), and krill biomass (right) in 12 transects (indicated by the different lines) from the Antarctic Ocean south of South Africa. The x axis depicts spatial scale, represented by the number of measured peaks ("cycles") in each transect normalized to length of the transect in km. Adapted from Weber et al. (1986).

x axis in Fig. 11-10). If spatial distribution of phytoplankton were only affected by changes in water mass dynamics, chlorophyll data would follow the curve for temperature. In fact, the spectral lines show that spatial distributions of temperature and chlorophyll, although variable, are similar. In addition, most of the variability in temperature and chlorophyll occurs toward the left side of the spectrum, that is, at the smallest spatial scales over which measurements were taken (note that the axes are logarithmic).

Another way to examine the relation of water physics and phytoplankton is to correlate measurements of the two variables taken over the same spatial scales. Small-scale variation in water salinity and temperature can be related to variability in chlorophyll concentrations (Duarte et al., 1992). Such correlations could arise in two ways. First, mixing and transport by water movements could merely redistribute existing chlorophyll. Second, hydrographical variation could create conditions that alter abundance of chlorophyll; for example, hydrographic variability can be linked to differences in properties that determine growth, such as nutrient concentrations. For both the above reasons, it is reasonable to think that variation in abundance of all plankton depends to some extent and at some scales on hydrography.

Zooplankton may aggregate in spatial scales of tens of meters due to local movements of water in near-shore waters. For example, density of zooplankton in the lee of a point in the Queensland coast reaches up to 40 times the density in adjoining water (Alldredge and Hamner, 1980). The actual mechanism by which this aggregation is accomplished is not known.

One of the better-known physical causes of patchiness of microplankton distributions is Langmuir circulation. Wind blowing over the surface of the sea causes the surface water to flow in Langmuir cells (Pollard, 1977). Langmuir cells extend up to 400 m in length, and can be several meters in diameter. These cells consist of water moving in horizontal helical paths in the direction of the wind. Wind of about 3 m sec^{-1} is required to form Langmuir cells, but higher wind speeds may disrupt them. Many parallel spirals are usually formed, and adjoining spirals roll in opposite directions, so that there are alternate parallel lines of convergence and divergence along a field of Langmuir cells. Such flow patterns can cause accumulations of floating particles on the sea surface in the areas of convergence between cells and also may affect the spatial distribution of plankton in the water within the spirals (Stavn, 1971; Evans and Taylor, 1980).

There are many other larger-scale physical mechanisms that can be important (Haury et al., 1978). Intrusion of water masses from the English Channel or from Dutch coastal waters determines whether either *Oikopleura dioica* or *Fritillaria borealis* are present in large patches of the southern North Sea (Wyatt, 1973). The impact of such movements of large parcels of water on the distribution and abundance of organisms is not unusual. Over half the variation in abundance of copepods in certain areas of the North Pacific may be due to variation in such advective transport (Wickett, 1967).

Another example of physical processes that operate at a very large scale are the consequences of large rings that spin off meanders of major western boundary currents such as the Gulf Stream (Wiebe, 1982), the Kuroshio (Tomosada, 1978), and the East Australia Current (Nilson and Cresswell, 1981). In the northwest Atlantic rings can contain a core of warm or cold water, depending on whether a portion of continental slope or Sargasso Sea water has been pinched off during the formation of the meander and ring. About five to eight cold core rings are formed per year, and each may reach a diameter of 250 km and depths of over 1,000 m (Lai and Richardson, 1977). Rings last for 2, perhaps 3 years. There may be 10–15 rings in the northwest Sargasso at any one time, covering a substantial part of the area (Richardson, 1976). Warm core rings are fewer and smaller. The formation of rings thus inserts large patches of nutrient-rich, cold waters of the continental slope into the Sargasso Sea and depauperate, warm water into the slope water. In addition, the organisms within the rings can be very different from those in the surrounding waters (Wiebe et al., 1976a), even in the case of mobile species such as fish. In warm-core rings in the Tasman Sea, for example, warm-water species dominate the fish fauna, while the cold water surrounding the rings holds a different array of cold water fish species (Brandt, 1981). Warm and cold core rings are therefore a major feature affecting the spatial distribution of organisms. Their study may clarify many questions of zoogeographical distribution as well as contribute to understanding heat transfer and productivity of the North Atlantic.

The very largest scales of patchiness such as shown in Fig. 2-34 are also primarily due to hydrography, as discussed in Chapter 2. Nutrients in large measure determine regional primary production and are made available by the transport and mixing of water masses. It is therefore clear that the physics of water is behind much of the spatial structure of planktonic distributions.

11.3.3 Physiological and Behavioral Mechanisms Related to Consumption of Resources

Nutrient uptake, grazing, and predation are very likely to be involved in determining spatial distributions, but we are a long way from understanding their effect in the sea. We have already mentioned current ideas on the potential importance of small nutrient patches for phytoplankton in oligotrophic water in Chapter 2.

Many of the regional-scale heterogeneities in biological properties are related to nutrient enrichment, which in turn depends on physical forcing (Chapter 2), such as upwellings, fronts, estuarine circulation, and other processes (Mackas et al., 1985; Mann and Lazier, 1991). Such links make it difficult to separate the influence of physical from resource-related mechanisms influencing patchy distributions.

Feeding activities may furnish numerical responses (Section 5.3) that aggregate individuals. Foraging in fish larvae (and probably other predators) is episodic; larvae swim for relatively long distances, seeking patches with higher prey densities. Once in such a patch, larvae swim shorter distances, and turn more often (Hunter and Thomas, 1974). These responses to prey density work to have larvae remain in high-density patches. Many benthic species are often found aggregated into patches, which may be a numerical response to food abundance or a response to predators. Mortality rates of queen conch (*Strombus gigas*) found outside aggregations are higher than those of conch in high density patches (Stoner and Rav, 1993).

There has been considerable theoretical work in relation to patchiness in plankton. Kierstead and Slobodkin (1953) derived a relationship between the size of a patch of phytoplankton and the rate of growth of the algae required to counter the losses of cells by diffusion out of the patch. Their model shows that patches smaller than 10 km could not be maintained against turbulent diffusion. Modification of the model by Steele (1974) brought the critical patch size down to a few kilometers, but we know that spatial heterogeneity smaller than this scale exists (Fig. 11-1). The addition of an Ivlev equation (Section 5.2.1) to include grazing in the Kierstead–Slobodkin model results in the repeated appearance of patches of all sizes (Wroblewski et al., 1975). Simulation studies using models thus suggest that the fate of phytoplankton patches depends on growth rates of the phytoplankton, grazing rates by zooplankton, spatial distribution of grazers, limiting nutrients, and turbulent diffusion.

The relative importance of different mechanisms as explanations of distributions of pico-, microplankton, and nekton depends on the scale at which we make observations. Take, for example, the distribution of bacteria in an upwelling or estuarine zone. If we wish to examine the mechanisms that create patterns of regional distribution of bacteria in these waters, we will readily conclude that regional physical process are involved, because physical tracers will correlate well to bacterial abundance. If, instead we focus at the scale of microns, we find that physical processes are less important, since at this scale viscosity dominates, so only biological mechanisms are available to create heterogeneous distributions.

In the case of larger plankton and nekton, specific inertial physical mechanisms (Owen, 1989) could potentially account for the widespread aggregation (Mann and Lazier, 1991), but there are many biological sources of patchiness as well. Note, for example that the variance spectrum of krill biomass (recorded by acoustic methods) is quite different from the temperature and chlorophyll spectra (Fig. 11-11, right); krill are far patchier (i.e., there are larger proportions of the variance) at larger scales than would we predicted on the basis of water mass properties alone. The distribution of krill cannot be explained as passive responses to movement of small or intermediatescale water masses (Mackas et al., 1990; Trathan et al., 1993). Active behavior provides a better account of distribution of such larger organisms.

11.3.4 Statistical Processes

Statistical properties of biological processes can generate nonrandom distributions. Suppose the settling of larval limpets on each quadrat has a random (Poisson) distribution. If survival of the settled limpets within the quadrat is also random, the resulting spatial pattern of the surviving limpets will be aggregated and will have a distribution called Neyman's type A (Poole, 1974), even though both processes that contributed to it were random. If survival is instead distributed logarithmically, another common aggregated distribution, the negative binomial results. There are also other statistical ways to generate negative binomial distributions (Cassie, 1962). Since the distribution may then be of statistical rather than have a direct ecological origin, presence of a specific distribution should be regarded at best as an empirical description rather than as evidence of the action of specific processes.

11.3.5 Artifacts of Measurement

We measure spatial patterns through procedures or methods that act as a filter that often imposes its own constraints on the data. This should be clear from our discussion of the effect of quadrat and sample size on the resulting frequency distributions. This is true of all the different areas of marine ecology that we have examined, in fact of all of science. The problem is especially obvious in dealing with the statistical descriptions of spatial arrangement as, for example, in studies carried out by Haury et al. (1979) for zooplankton and Duarte and Vaqué (1992) on bacterial distributions.

In Cape Cod Bay, Massachusetts, tidally generated internal waves (waves moving within the ocean rather than at the surface) with a period of 6–8 min and a wavelength of about 300 m propagate over a shallow bank (Fig. 11-11). Such internal waves are important physical mechanisms that may produce patchiness at intermediate (10–100 m)

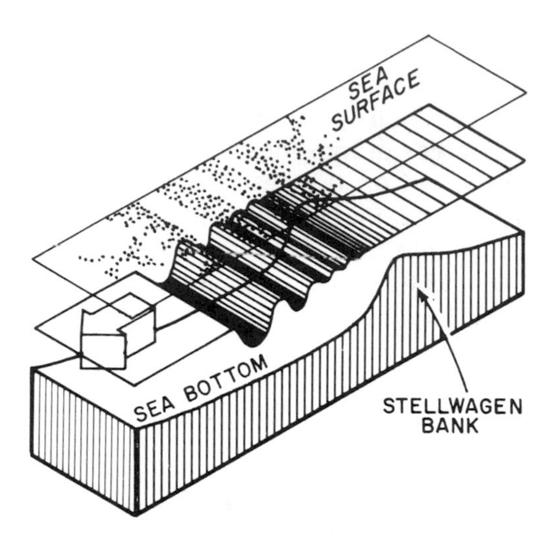

FIGURE 11-11. Schematic diagram of internal waves generated in the lee of Stellwagen Bank, Cape Cod Bay. While the waves can be quite steep at depth, there are only minor surface expressions. Adapted from Haury et al. (1979).

spatial scales (Haury et al., 1978). Data collected by net tows* at a depth of about 17.5 m reveal some effects of these waves (Fig. 11-12). The pattern of the internal waves is depicted by the excursions recorded in temperature and salinity. Deeper water was colder and saltier (Fig. 11-12, left). At sample number 20, for example, there was a wave peak that brought deeper water up. On the right side of Figure 11-12 are biological data collected at the same time. The deeper water contained more chlorophyll, fewer appendicularians and amphipods, and more copepods. The results of these tows are an excellent example of the ambiguous nature of patchiness and the potentially artifactual effects of the sampling procedure.

Consider the data on the right-hand side of Figure 11-12. These transects could be taken to be excellent examples of patchiness; they are quite comparable to transects such as shown in Figure 11-1. Should we consider the high densities true patches or are they just artifacts due to the collection of a transect at a uniform depth of 17.5 m? Further, are the measured patches meaningful to organisms, that is, do organisms (other than the ones in Fig. 11-12) cross the very clear boundaries between colder and warmer water? If the species present always remain in their own water mass, the patches do not have much

---

*These samples were collected with a modified Longhurst–Hardy plankton recorder (LHPR). This device enables the collection of a series of samples as it is towed through the water. Ribbons of mesh are rolled across the cod end of the net so that the plankton from different sections of the tow are caught on successive portions of the mesh. New versions of the LHPR reduce difficulties of net avoidance by fast-swimming copepods, extrusion of specimens through the mesh and the hang up of specimens along the net (Haury et al., 1976). A number of other instruments were used to measure temperature, salinity, and chlorophyll (Haury et al., 1979).

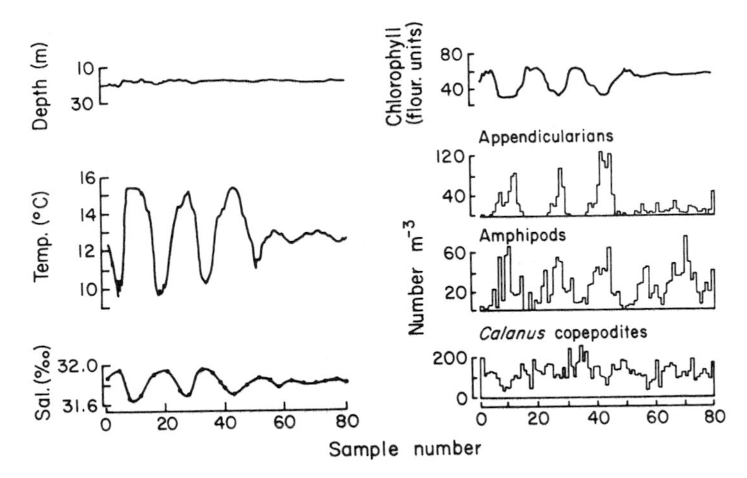

FIGURE 11-12. Data from a horizontal tow taken at about 17 m through a wave packet such as shown in Figure 11-11. Samples were taken every 14.6 m, and the total length of the tow was 1,155 m. Adapted from Haury et al. (1979).

biological significance. On the other hand, if a whale, for instance, were feeding while swimming horizontally at 17 m, prey would be patchy in abundance.

It has been noted that abundances of bacteria in seawater are fairly uniform (Azam et al., 1983; Fenchel, 1984), and from this observation, it was inferred that bactivores impose top-down controls on bacteria. It may be that the uniformity is in part a feature of volumes sampled. Bacterial abundances depend on the scale of measurement: abundances were higher, and reasonably uniform, in samples with volumes >5 ml (Fig. 11-13, right). The patchiness of bacterial cells occurs at small-

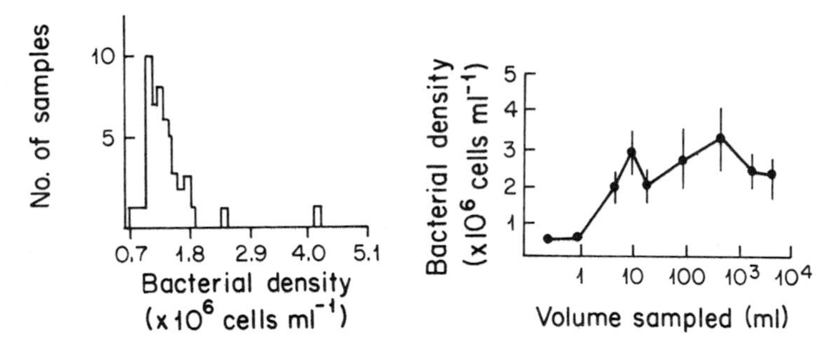

FIGURE 11-13. Distributions of bacteria in subsurface water from Alfacs Bay, Ebro Delta. Right: Average (± std. dev.) bacterial density in a series of measurements (4 obs. per mean) in different volumes of sampled water. Left: Frequency distribution of bacterial density in 50 samples of 1 ml from Alfacs Bay. Adapted from Duarte and Vaqué (1992).

enough spatial scales that the 4 replicates of volume <5 ml simply missed the high-density patches. This was shown by a larger sampling of 50 samples of 1 ml (Fig. 11-13, left), where there were enough samples to reveal the infrequent but high-density patches. Larger (100–1,000 ml) samples, such as routinely used, provide uniform abundances merely because they would most likely include some high-density patches. Such samples, however, may not provide information about abundances at spatial scales most relevant to bacteria or their consumers.

The nature of "patches," in the two examples discussed above and many other instances, depends on the point of view and scale of measurement. This is a pervasive problem that needs careful attention. The physics of the water masses and the *in situ* behavior of the organisms are important, but scale of measurement may mask other effects.

11.4 Ecological Consequences of Patchiness

The pervasive patchy distribution of both resources and consumers has a variety of ecological consequences. Some aspects particularly linked to material in earlier chapters merit some note.

Feeding. We have already discussed the aggregative response of predators to patches of high prey abundance. This may occur at every scale; on small scales, distance among prey items is less than that among isolated individuals, so feeding in patches is more efficient for the consumer. For example, carp feeding on the benthos of the Volga River only achieve their full rations when they feed on highly aggregated food items (Ivlev, 1961). Laboratory populations of *Calanus pacificus* reach sizes comparable to those in the field only when densities of phytoplankton higher than natural concentrations are available (Anderson et al., 1972). As predators move about their environment, they often remain to feed in patches where prey items occur at high densities, and so accumulate in places where food is more abundant. Sperm whales, for example, gather in this fashion when feeding on patches of food thousands of miles long (Cushing, 1981). Notice that aggregations of feeding predators are not the same as predators hunting while in schools. Eggers (1976) has shown using simulation studies that hunting in schools may reduce the prey consumption per predator, except when prey densities are very high or the prey are patchily distributed. Since the latter situation is very common, we find that many predators—tuna, swordfish, porpoises—do feed while in schools.

Reduction of risk of being eaten. The behavior that leads to aggregation into schools and the maintenance of uniform distances within a school seem designed to avoid risk of predation (Brock and Riffenburgh, 1960; Vine, 1971). The chances of one individual being eaten are reduced by such spatial distributions, even though the school may provide predators with a patch of high density.

Demographic effects. Copepods in an oceanic water column feed on patchily distributed food particles, so they must necessarily encounter areas where food items are scarce.

Feeding on patchily distributed food is the norm for copepods, and if patches are frequent enough, even "food-sensitive" species such as *Centropages typicus* can grow, reproduce, and subsist on patchy diets (Davis and Alatalo, 1992). Seeking patchily distributed food may demand energy expenditures that compete with energy devoted to reproduction: rates of egg production in *Acartia tonsa* were about two and a half times smaller when the same food density was distributed patchily than when it was homogeneously distributed (Saiz et al., 1993).

Turbulence changes prey patchiness, and alters frequency of encounters with prey (Davis et al., 1991; Marrasé et al., 1990). Modest increases in turbulence increase feeding success in cod larvae and feeding rate in anemones (Fig. 11-14), growth of copepods (Davis et al., 1991) and growth and survival of anchovy larvae (Wroblewski, 1984; Wroblewski et al., 1989). Very high turbulence, as may occur during

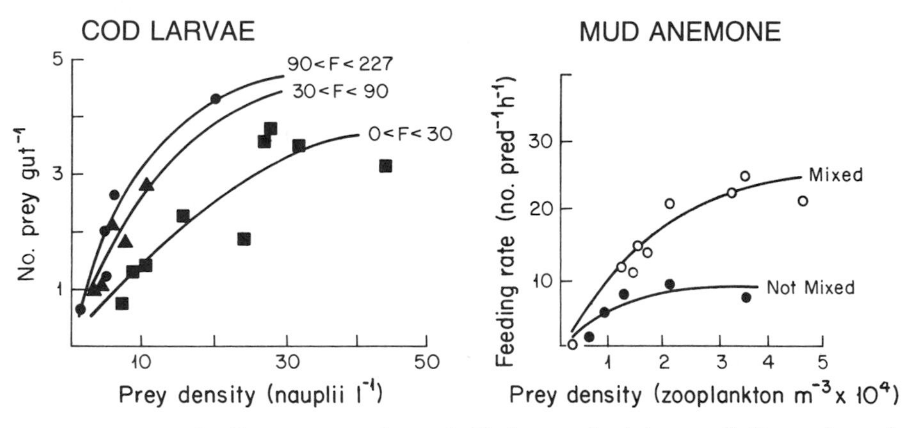

FIGURE 11-14. Feeding success in cod (*Gadus morhua*) larvae (left), and mud anemones (*Ceriantheopsis americanus*) (right) in different turbulence regimes. Left: The different curves result from different ranges of the turbulence parameter $F = W_8^3/N_{5-25}$, where N_{5-25} is the average Brunt-Väisälä frequency in the water column between 5 to 25 m in depth, W_8^3 is the average cube of wind speed during 8 h previous to sampling [high values of F reflect higher turbulence; cf. Mann and Lazier (1991) for further explanation]. Adapted from Sundby and Fossum (1990). Right: Curves refer to different treatments in MERL mesocosms, in which the mesocosms were or were not mixed vertically by mechanical means. Adapted from Sullivan et al. (1991).

storms, can be detrimental: growth of menhaden larvae is lower after such disturbances (Maillet and Checkley, 1991). High turbulence may disperse patches of prey and make encounters less frequent. Models suggest that even small differences in growth significantly affect reproductive performance, and that effects through growth are as important as mortality in determining recruitment of copepod populations.

Since wind speed has such a primary influence on turbulence, much of the demography of plankton is subject to vagaries of weather. Such chance meteorological events, coupled to the enormous larval mortality, could easily uncouple recruitment to fish populations from the reproductive output of adults, as discussed in Chapter 9.

Habitat use and selection. Since different patches will differ in what they offer as resources, individual animals will show patch or habitat selectivity. We have already seen this when discussing the aggregation response to prey density and in the discussion of searching time. Both an active choice of habitat and a simple response to resource abundance can bring about such "habitat selection."

Social organization. In species where resources are allocated by interference competition, the most favorable of patches in a habitat are occupied first. This leads to a uniform distribution of individuals within the favorable areas. If there are more individuals than can fit in the choice habitats, less favorable areas are used. In the garibaldi, the nesting males take up the best habitats; other fish, including females, have less desirable areas and there is a far less intense defense and attachment to these poorer territories. In many other cases, the areas of habitat needed for nesting and obtaining food differ widely. In terns, for example, as we saw in Chapter 5, foraging areas are necessarily far larger than nesting areas; the limited amount of suitable nesting habitat required by the population forces a colonial type of social organization. Depending on the spatial distributions of the patches suitable for a species, other forms of social behavior may occur. In whales tuna, and in other animals, long-distance migration of groups to winter feeding grounds is common. These organisms form pods, herds, or flocks that migrate from one area to another.

Dispersal. Dispersal patterns are frequently bimodal (Wiens 1976), with many individuals moving short distances within a favorable patch, and some moving long distances from one patch to another. Species that specialize in rapid colonization of newly available habitats, often called "fugitive" species (Chapter 12), show propensities for long-distance movements. Both the distributions of patches of suitable habitat and the frequency of change in a habitat affect the need or advantages of long-distance dispersal.

The above list of ecological consequences does not exhaust the possibilities; rather it serves to show the fundamental implications of spatial heterogeneity for almost every aspect of ecological organization. Awareness of the importance of patchiness goes back into the history of ecology, but concerted work on this topic is relatively recent. Many critical questions remain to be answered, such as: what is the effect of patchiness on consumption and growth? How does patchiness affect the ability of a consumer to limit its food resource? Does patchiness prevent a consumer from exterminating prey? How do biological mechanisms interact with the powerful effects of physical mechanisms?

It is to an extent paradoxical that in this section of the book where we are supposedly dealing with community and ecosystem levels, most of the examples have involved populations or individuals. Our knowledge of spatial distribution needs to be carried further, and we need to ask questions at the level of the community. For instance, are some environments likely to contain more species distributed in aggregated fashion? As communities age, are randomly distributed species replaced by aggregated or uniformly distributed populations? Does spatial arrangement reflect resource availability? These are among the many difficult questions raised by the existence of spatial heterogeneity. Bright young ecologists will no doubt find ways to answer them experimentally in the future.

There is no doubt that spatial distributions reflect powerful influences of important processes. It is troubling, though, that the definition of scale of interest can be so critical. This is yet another field in which it is necessary to make sure that we make use of operational terms and do not confuse researchers' assumptions with real mechanisms.

11.5 The Problem of Upscaling

We have so far established that there is patchiness at most spatial scales, and that spatial heterogeneity has multiple sources and powerful consequences. In marine ecology we often take measurements at one scale—say, primary production in 1 liter containers, and we then use such measurements to try to calculate primary production for much larger units, say the Subarctic Pacific. Are such extrapolations appropriate? There are too few evaluations of such upscaling. Here we examine three instances, involving recruitment rates to coral reefs, seasonal cycles of production in the open ocean, and nutrient export in estuaries.

Fowler et al. (1992) recorded the recruitment of *Chaetodon rainfordi*, a dominant fish, at different sites within one reef, several reefs within one region, and at several regions of the Great Barrier Reef. The measurements were repeated over three years to assess interannual variation at the three different spatial scales.

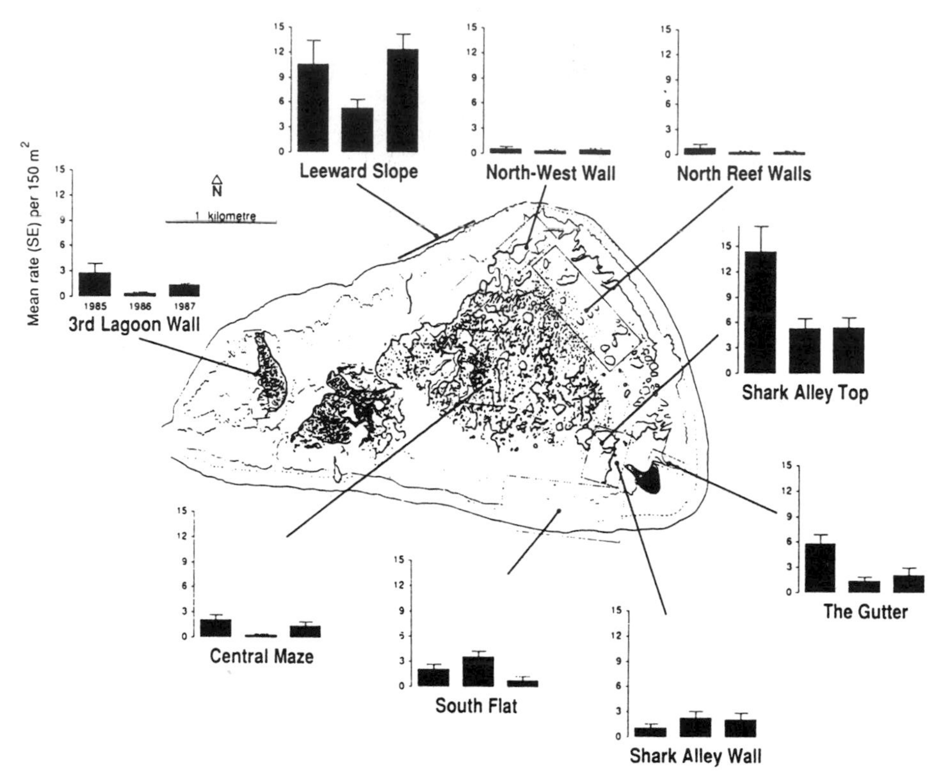

FIGURE 11-15. Mean (+ std. error) rates of recruitment of *Chaetodon rainfordi* for each of 3 years to 9 sites within One Tree Reef. From Fowler et al. (1992).

There was notable intersite variability within one reef during the three years (Fig. 11-15). At this scale of hundreds of m, there were no coherent interannual patterns; recruitment was apparently governed by local site characteristics. The mean recruitment rates for different reefs within one region (scales of 100s of km, and averaging over several sites in each reef) in contrast, showed only slight among-reef variability, and quite coherent variation over years (Fig. 11-16). Somehow, the local within-reef variation is averaged out when data from entire reefs are pooled, and the means show coherent interannual variation, probably imposed by regional hydrographic processes. Lastly, recruitment rates varied somewhat among reefs within a region (Fig. 11-17), and the interannual variation was coherent, but in addition, important North–South large scale geographical variation is evident.

The second comparison is for phytoplankton cycles in the North Atlantic. Colebrook (1979) compiled much data collected by plankton recorders towed by commercial vessels in the North Atlantic since 1948. These results, obtained from volumes of water at a reasonably

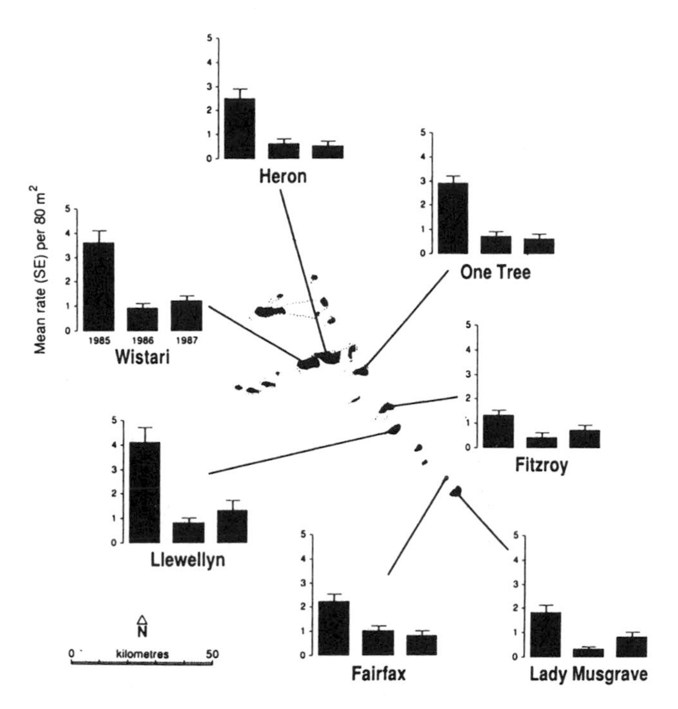

FIGURE 11-16. Mean (+ std. error) rates of recruitment of *Chaetodon rainfordi* for each of 3 years to the reef slopes of each of 7 reefs of the Capricorn/Bunker Group. From Fowler et al. (1992).

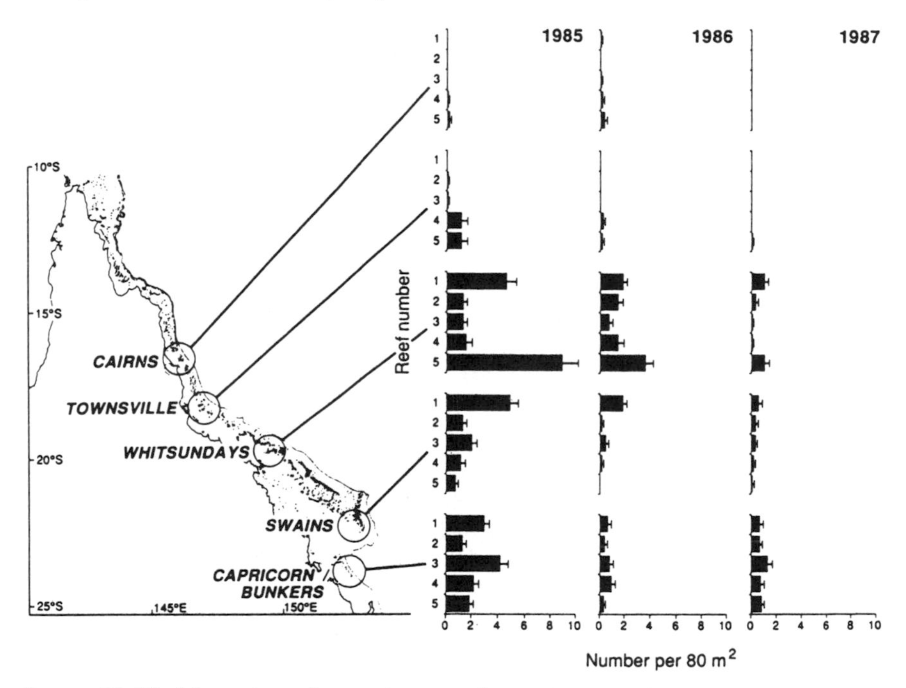

FIGURE 11-17. Mean (+ std. error) rates of recruitment of *Chaetodon rainfordi* for each of 3 years to the reef slopes of each of 5 reefs of 5 regions of the Great Barrier Reef. From Fowler et al. (1992).

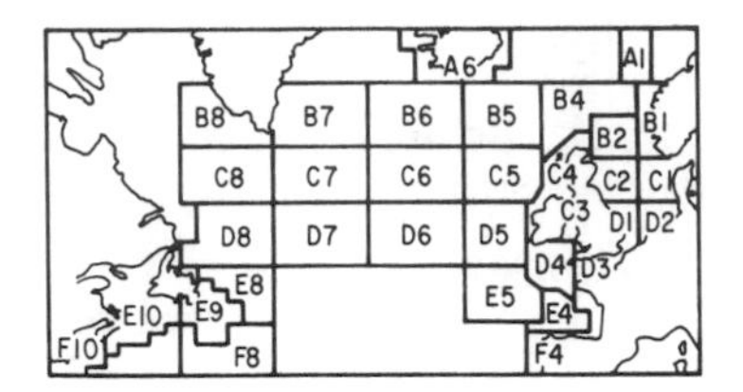

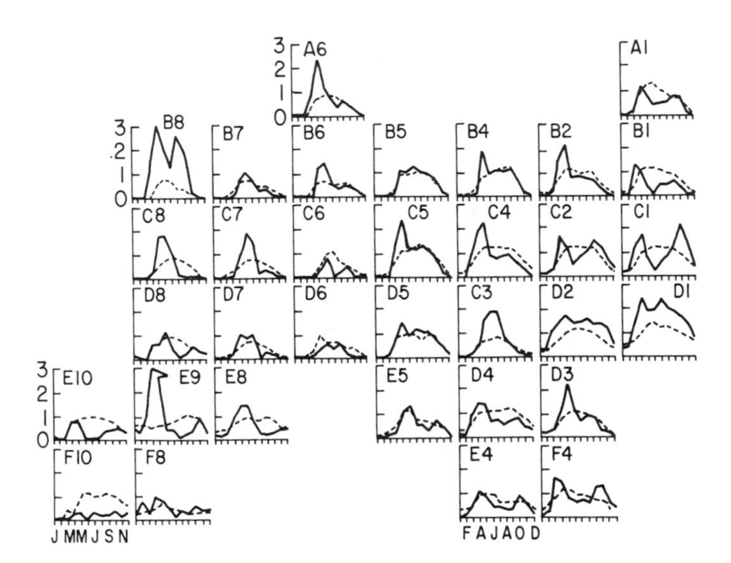

FIGURE 11-18. Seasonal variation (bottom) in abundance of phytoplankton (whole lines) and zooplankton (copepods) (dashed lines) in different areas of the North Atlantic (top). Phytoplankton are in arbitrary units of greenness and copepods in log means of numbers per sample. From Colebrook (1979).

small scale, are grouped according to geographic areas where they were collected in Fig. 11-18. In each zone there was a somewhat different seasonal cycle of abundance of phytoplankton, some areas showing cycles that were unimodal, others bimodal (cf. Section 15.2.1 for further discussion of seasonal cycles).

Banse and English (1994) used median values of pigments calculated from NASA Global Ocean Data Set, averaged per month for areas about 77,000 km^2, to describe seasonal plankton cycles. Figure 11-19 shows selected data from Banse and English (1994) that approximately correspond to three regions of the Colebrook (1979) plankton recorder results. Even though both data sets were averaged over many years, and over reasonably large areas, the seasonal cycles of phytoplankton obtained using data collected at regional scales do not evidently match the results from smaller water sample spatial scales. Differences in methods, and among years of the two studies could be responsible for

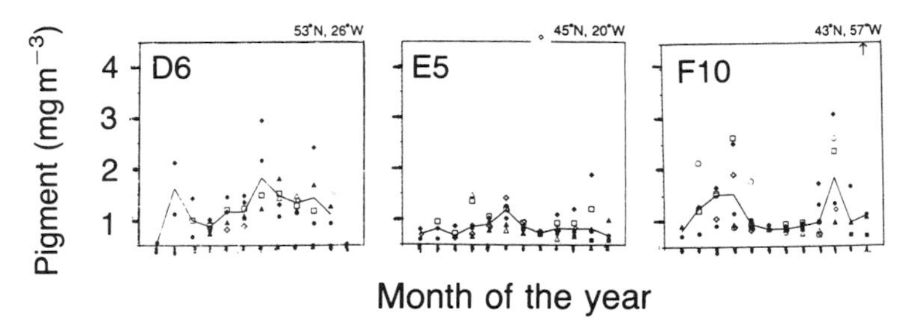

FIGURE 11-19. Seasonal variation in median pigment concentration (black lines) for regions centered on the coordinates shown on top right of each graph. These regions approximately correspond to areas designated as D6, E5, and F10 in Fig. 11-18. Data from NASA Coastal Zone Color Scanner, integrated for areas of about 77,000 km^2, medians calculated from records obtained during nine different years (different symbols) between 1978 and 1986. Tics on x-axis show months of the year from Jan–Dec. Adapted from Banse and English (1994).

the lack of match, but it may also be that, as in the case of the coral reef fish recruitment rates, local effects led to features in the smaller scale results that were smoothed out in the larger-scale data.

A third example of issues raised by the upscaling problem is one frequently faced in estuarine studies. We can measure processes such as nitrogen transport (from land through the estuary to the sea) in one system, say Waquoit Bay (Fig. 11-20, top right), and wonder whether the measurements are applicable to Chesapeake Bay (Fig. 11-20, left). Note that the spatial dimensions in these two systems differ by three orders of magnitude.

There are two ways we could think of comparing data from two such bays. First, we could think of coastal environments such as Waquoit Bay as representative of the myriad estuaries (Fig. 11-20, bottom right) of size similar to Waquoit Bay that make up the "skin" of Chesapeake Bay. In this case, we could multiply results from Waquoit Bay by the number of similar landscape units fringing Chesapeake Bay. In this concept, we conceive that the necklace of Waquoit Bay-like systems is interposed between land and Chesapeake Bay, and whatever losses or gains in nitrogen take place in the "skin" of Chesapeake Bay can be estimated from data obtained in the smaller system.

Second, we may ask the more usual upscaling question, can we simply directly apply results from one bay to the other? In this case, we need to find a way to make the comparisons in spite of the differences in scale. Clearly, we need to consider the differences in volumes, but a moment's thought shows that we should also think of the rate at which those volumes are being renewed within each estuary. The rate of

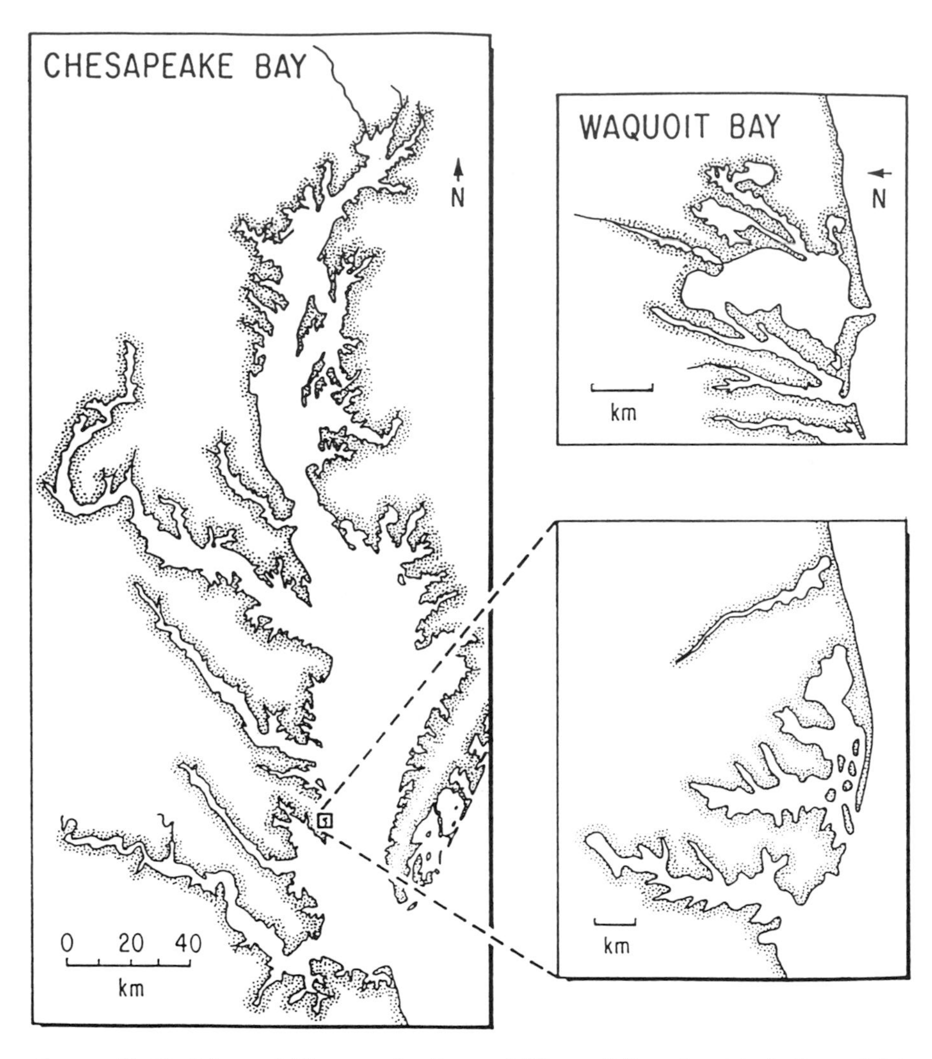

FIGURE 11-20. Maps of Chesapeake Bay and Waquoit Bay. Inset in bottom right shows an expanded detail of a smaller estuary along the coast of Chesapeake Bay that closely resembles Waquoit Bay in size and shape.

nitrogen export could be related, therefore, to the residence time of water within estuaries (Fig. 11-21). Estuaries approximately align themselves along a curve that suggests that while similar biogeochemical processes govern transformations of nitrogen in different estuaries, export depends on water residence time (which itself depends on volume, and freshwater and tidal exchanges). Although the idea is an untested notion so far, we include it here to show an example of scaling factor that may be needed to make comparisons across broad

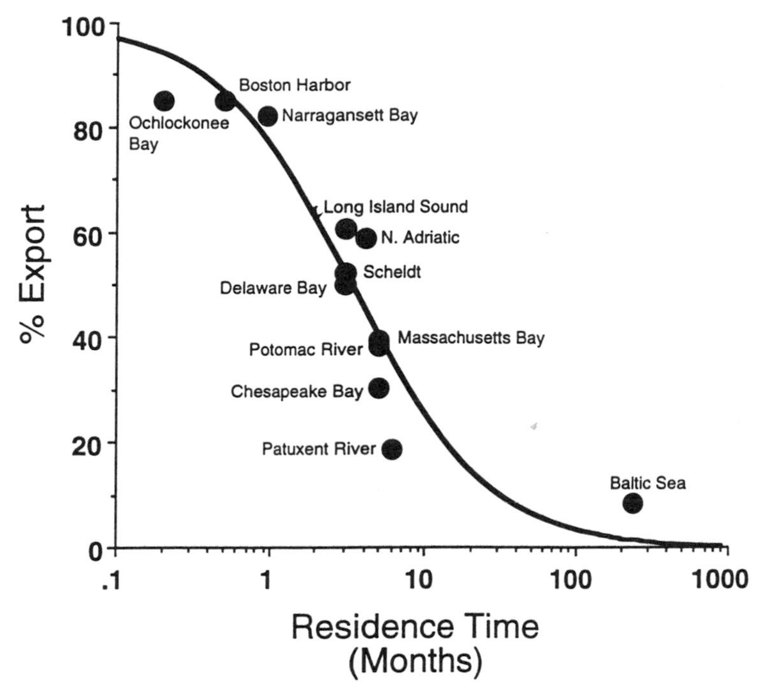

FIGURE 11-21. Relationship of % export of nitrogen and water residence time in various estuaries. The equation of the line shown is % export = 1/1 + kR), where k is a fitted coefficient with a value of 0.3 in this case, and R is residence time. The expression is used to describe yields of industrial digestors in chemical engineering. Waquoit Bay would presumably lie towards the top left in the graph. Courtesy of S. Nixon, D. Di Toro, and W. Boynton.

ranges on spatial scales. Residence time is a scaling factor that makes comparisons possible, as well as reveals the action of control processes.

These three examples demonstrate that extrapolation from local sites to layer geographical zones requires caution, and that the action of different controls may or may not be manifested at different spatial scales. Asking the null hypothesis about upscaling may deepen our understanding of ecologically important processes, but testing the hypothesis will demand detailed studies at different spatial scales for a variety of systems.

Chapter 12
Development of Structure in Marine Communities: Colonization and Succession

12.1 Introduction

Any newly available patch of habitat—a newly upwelled parcel of water, or a recently cleared surface—is subject to colonization by organisms. There is nearly always a supply of propagules from an array of species, ready to colonize the unoccupied environment. Colonizers continually threaten to invade any habitat. The composition of the assemblage of species present in any one environment is a composite of invasive opportunity and ability vs. propensity of the species already present to prevent or further invasion. Almost invariably, there is some replacement of species over the course of time, which has been called succession. Because the species that may have propagules available to colonize at any one time may vary, and because of meteorological or hydrological events, the resulting array of "settled" species, and the timing of replacements, all have an element of uncertainty.

The concept of succession was developed in terrestrial ecosystems (Odum, 1969), and although much of succession theory has been criticized as tautological (Peters, 1991), the idea is still widely applied. In marine systems, succession has been applied to two kinds of phenomena. One is the sequence of plankton species that develops in nutrient-rich water, be it newly upwelled parcels, or late winter water. The plankton blooms that result involve more or less recognizable sequences of taxa (Johnston, 1963a; Margalef, 1967; Smayda, 1980). Such sequences may take place over days or months, repeated each time nutrients are renewed in stratified water, or when light triggers a new seasonal production cycle (more on seasonal cycles is found in Chapter 15).

The second marine application of the concept of succession is to the process by which species appear and are replaced on new or disturbed surfaces. In this chapter we focus on this second aspect, because its mechanisms have been much better established.

We will first examine a few case histories that show the processes involved in colonization of a newly available parcel of habitat, and then consider mechanisms that maintain or alter the composition of the assemblage of species present on that parcel. This latter type of sequence of species is closer to what terrestrial ecologists call succession.*

12.2 Colonization Processes

12.2.1 Propagule Supply

Although reproductive activity by organisms may be closely timed by diverse cues, the vagaries of meteorological and hydrographical events, plus uncertainties regarding presence or activity of predators, lead to a more or less variable supply of propagules ready to colonize and survive on any new surface. The timing of propagule arrival imparts a specific history to the assemblage. This historical feature is a critical component of recruitment: postsettlement processes can only work on the species that in fact have arrived.

Many sessile species can be considered as arrays of isolated local populations connected genetically and demographically by the mixing of larvae during the pelagic phase of the life history. In species with such "open" life cycles the reproductive effort of a local population cannot easily limit or regulate local population size, since locally produced propagules largely end up elsewhere. Instead, it has been argued that externally-determined supply of larvae determine density and presence of species, and thus of the structure of communities on rocky surfaces (Roughgarden et al., 1988). Borrowing a term from recent economic controversies, "supply-side" interpretations suggest that increases in numbers of larvae result in increased local density.**

The link postulated by supply-side ecologists seems unlikely to be a strong one. Postsettlement mortality, for instance, could dilute the effects of larval supply. In a colonial ascidian, 50% of newly settled colonies disappeared after 1 d, and 90% after a week (Stoner, 1990). Under such intense mortality rates of newly settled recruit, the

---

* In terrestrial ecology primary succession is the sequence of species that colonize a new substrate; secondary succession takes place on previously inhabited surfaces that have been cleared of biota by some disturbance.

** If mortality of settled larvae is density-independent, recruitment-limited assemblages will retain the densities imposed by the vagaries of larval availability, and will show abundant cohorts following periods of high recruitment. We have already discused this idea as the "strong cohort" strategy characteristic of many marine fish; it is widespread in marine organisms (Hughes, 1990).

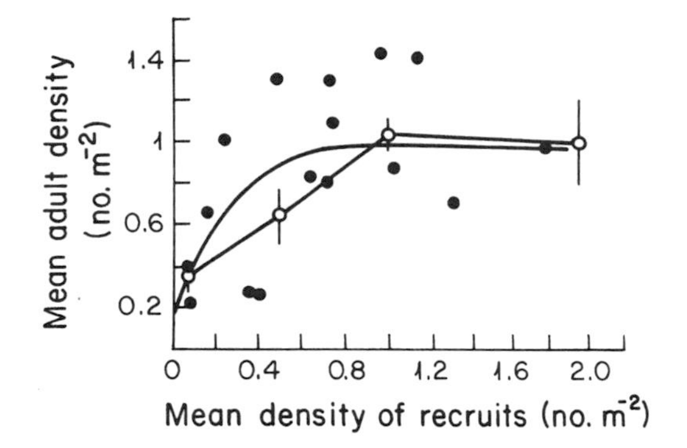

FIGURE 12-1. Relationship of density of recruits vs. resulting density of adults of coral reef fish, in experiments in which recruit density was manipulated (open circles) and in surveys done in different areas (black circles). The curved line was fit to the survey points. Adapted from Jones (1990).

quantitative effects of propagule supply on adult density, even at high recruitment rates, would be easily lost within natural variability. The link between recruitment and adult densities may be more significant in species that suffer less intense mortalities. In certain coral reef fish, for example, resulting adult densities increase to an asymptote after higher larval recruitment rates (Fig. 12-1), and mortality rates of settled juveniles are lower at low adult densities (Forrester, 1990). Postsettlement larval mortalities in such fish are indeed less intense (55–90% after 10 mo, Forrester, 1990) than was the case for the ascidians. Nonetheless, the large scatter in the relationship between adult density and larval recruitment (Fig. 12-1) suggests that other factors play a major role in determining densities of survivors. It appears that postsettlement events (Doherty and McWilliams, 1988), modified to a modest extent by propagule supply, may largely determine the abundance and assemblage of species present.

12.2.2 Mechanisms of Settlement

Settlement by new propagules involves active selection and is strongly influenced by water movement (Butman, 1987). Propagules colonize specific sites by means of active selection. Larvae of the polychaete *Capitella* in still water preferentially settle onto estuarine mud rather than on glass beads (Fig. 12-2). Water movement reduces the number of settled larvae; part of the effect of water flow is passive, judging from the parallel reduction of inanimate spheres found on the sediments. Regardless of whether effects are passive or active, rate of flow plays a role. In some species (Eckman et al., 1991; Pawlik and Butman, 1993) larval settlement peaks at intermediate flow velocities, because

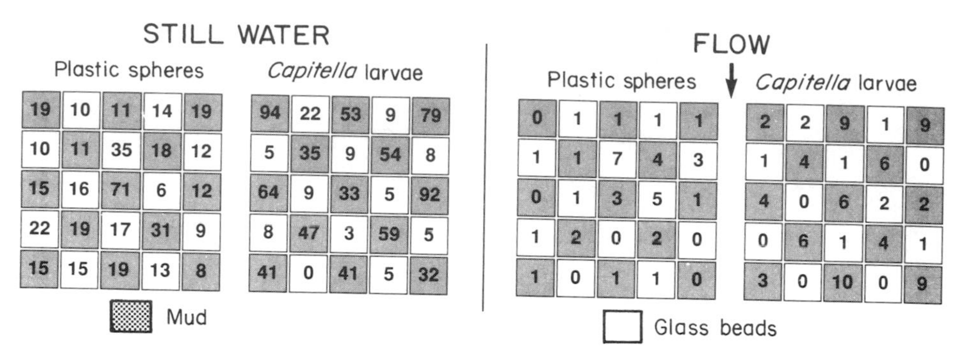

FIGURE 12-2. Number of larvae *Capitella* and plastic spheres found on assays of containers holding either estuarine mud from New Bedford Harbor or glass beads. Experiments done in laboratory with still water (left), or in flumes with flowing water (right). From, Grassle and Butman (1989).

high velocities erode larvae from surfaces, and because perhaps sites with low velocities are unsuitable for adults, and are avoided. Specific selective behavior is superimposed onto the hydrodynamic controls. In the *Capitella* experiments, even in flowing water there is selection for the estuarine mud. The selection is not merely a passive process, because live larvae behaved quite differently from the inanimate glass beads (Fig. 12-2), regardless of hydrographic regime.

To find a site for attachment, propagules have to descend through the water column and attach to a solid surface. Their distribution in water is likely to depend on physical and biological processes over a range of spatial scales characterized by different physical properties, as we already found when discussing solute uptake (Chapter 2, Fig. 2-12). Water velocity is zero at the seabed surface, because of the "no-slip" condition of water molecules, is minimal within the viscous layer, and increases logarithmically up to the "free-stream" velocity beyond the upper limit of the boundary layer (Fig. 2-12).

Water velocity greatly affects larval settlement. Currents above the bottom frequently flow at speeds considerably faster than swimming speeds of larvae. Larvae are generally small, with accordingly relatively slow swimming speeds (cf. Fig. 2-13). To actively explore sites on the seabed, larvae in the water column may have to sink or swim to some height above the seabed, below which flow speeds may be slow enough to allow the larvae to maneuver effectively.

Butman (1986) used data of tidal velocities in a site where tidal currents reached 5–10 $cm\ s^{-1}$ twice daily to calculate vertical profiles of flow velocities above the sediments (Fig. 12-3). *Mediomastus ambiseta* is a common, small polychaete in Buzzards Bay sediments; its larvae face the problem of finding appropriate places to settle in a hydrodynamic

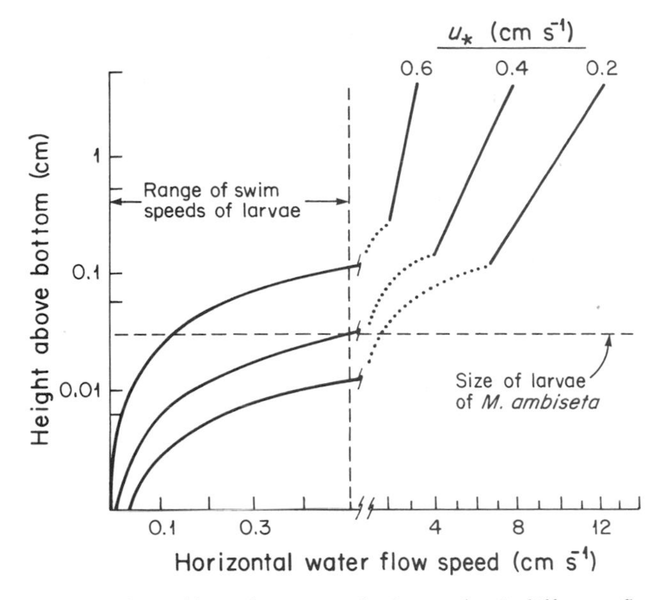

FIGURE 12-3. Vertical profiles of water velocity under 3 different flow regimes. U_* is the friction velocity, a variable that describes the viscous stress between adjacent layers moving at different velocities (Mann and Lazier, 1991). Note that this graph is a semilog transformation of Fig. 2-12, and that there is a scale break between the left and right sides of the x axis. The lines on the right side correspond to the logarithmic part of Fig. 2-12, the curves on the left side correspond to the viscous-dominated linear layer of Fig. 2-12. Adapted from Butman (1986).

environment depicted by curves of Fig. 12-3. The dashed lines show constraints within which active horizontal swimming to select a settlement spot is feasible. The horizontal dashed line shows the approximate size (300 μm) of larvae of *M. ambiseta* that are ready to settle. Below the dashed line, the larvae "hit bottom," so they are unable to swim horizontally. The vertical dashed line shows maximal speeds achievable by polychaete larvae. In currents whose horizontal components are to the right of the vertical line, the larvae would not be able to swim against the current. Only in the region to the top left of Fig. 12-3 could larvae actively swim exploring sites suitable for settling. From field data, Butman (1986) calculated that conditions that permitted such behavior would be limited to perhaps 40% of the tidal cycle, and probably much less considering other hydrographic processes. Larger larvae would find constraints on horizontal swimming even more stringent. It seems unlikely that searching within the boundary layer is the way larvae find settling sites.

The most common pattern of settlement may be that larvae fall through the water column in velocities proportional to their size. Swim velocities are in the same range as sinking speeds; polychaete larvae may have to swim actively just to maintain their vertical position in the water column. Once larvae fall to the seabed, they can decide whether or not to stay in the site where they are; if not, upward swimming or turbulent water motion could take them up again to a sufficient height where additional water motion would move them horizontally, and the procedure may be repeated until a suitable settlement site is found. A combination of passive sinking plus active choice during exploration and attachment seem the most likely way in which propagules deal with the physical processes in boundary layers to select settlement sites (Butman, 1986; Mullineaux and Butman, 1991).

12.3 Case Histories of Colonization and Succession

12.3.1 Colonization and Succession in the Rocky Intertidal Shore

There are many studies on rocky shores that describe the development of the community of sessile organisms attached to the rock surface. We will focus on a study designed specifically to describe succession and to evaluate just how species replacement takes place. This study describes the colonization and establishment of algae and animals on the top

TABLE 12-1. Species Composition of Early, Middle, and Late Successional Communities on Boulders in an Intertidal Boulder Field in California.[a]

Species	Successional stage		
	Early	Middle	Late
Ulua spp. (green alga)	73.1 ± 15.6	10.8 ± 8.4	1
Chthamalus fissus (barnacle)	18.1 ± 14.7	4.7 ± 7.3	1
Gigartina leptorhynchos	1	15.8 ± 10.4	3.5 ± 5.6
Gelidium coulteri	1	4.9 ± 5.7	2.4 ± 4.8
Gigartina canaliculata	1	47.1 ± 19.0	91.8 ± 6.2
Centroceras clavulatum	1	3.7 ± 6.2	1
Anthopleura elegantissima (sea anemone)		1	1
Corallina vancouveriensis		1	1
Gastroclonium coulteri		1	1
Laurencia pacifica		1	
Porphyra perforate		1	
Rhodoglossum affine		1	
Bare area	5.4 ± 4.8	9.7 ± 6.9	1.9 ± 2.8
Number of species	6	12	10
Diversity ($e^{H'}$)[b]	1.6 ± 0.5	4.1 ± 1.3	1.5 ± 0.3

[a] Species are red algae except where indicated. Values are mean ± std. dev. of percent cover by each species on 30 boulders. Adapted from Sousa (1979, 1980).
[b] Includes only species with >1% cover. H' is the Shannon–Weaver diversity index.

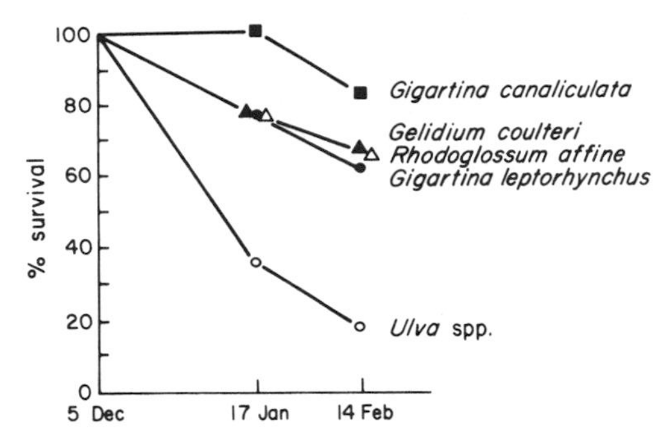

FIGURE 12-4. Survival for five species of macroalgae over a 2-month period during a period of harsh conditions (low tide in afternoon with the algae exposed to dessication). Thirty plants of each species were tagged on 5 December 1976. Adapted from Sousa (1979).

surface of intertidal cobblestones in certain parts of the southern coast of California (Table 12-1).

Species typical of early succession on these cobblestones are generally cosmopolitan, produce many flagellated motile spores for much of the year, settle densely, and grow quickly. *Ulva* spp. and *Enteromorpha,* both green algae, are representative of pioneer species, and occupy much of the space made available by disturbances (Table 12-1). Pioneers do not persist for at least two reasons. First, they suffer greater mortality from physical disturbances such as desiccation than do later species (Fig. 12-4). Second, pioneer species such as *Ulva* are preferred foods of grazers (Fig. 8-2 and Sousa, 1979a).* Over time disturbance and grazing reduce the amount of space occupied by the pioneer species.

In the absence of disturbances or grazing, *Ulva* spp. do retard colonization by other species, since they are good interference competitors for space. When *Ulva* spp. are removed, whether experimentally (Fig. 12-5, top) or naturally, the species of the later successional stages increase in abundance. These taxa of later stages are often much more seasonal in their release of young, produce fewer young, settle more sparsely, and grow relatively slowly. *Ulva* spp. can outgrow them, and only when *Ulva* spp. are removed by grazers, or by a disturbance, does succession go on.

Once other species are established, they in turn inhibit the settlement and growth of additional species. For example, removal of

*This seems typical of species of early succession. In terrestrial situations, palatability of plants of early succession is also greater than that of species typical of the later stages (Cates and Orians, 1975). Pioneer plants appear not to invest much effort in chemical or morphological defenses against herbivores.

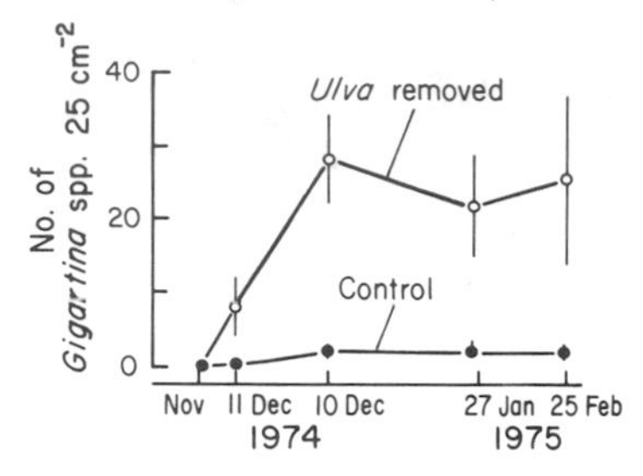

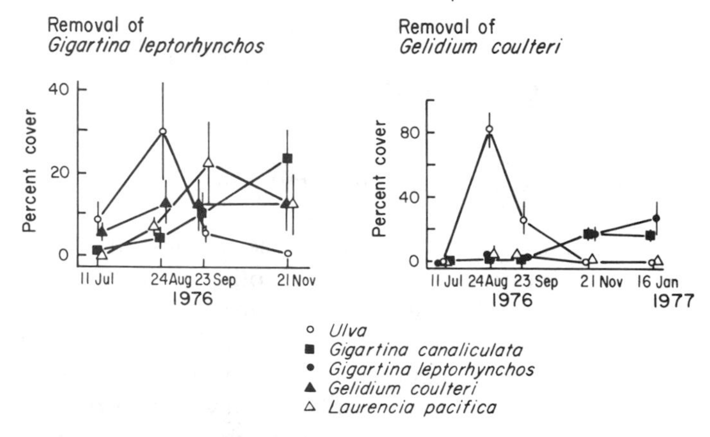

FIGURE 12-5. The effect of experimental removal of early and middle-succession species. Removal was repeated at each sampling interval. The controls for the experiments with *Gigartina leptorhynchos* and *Gelidium coulteri* showed no significant changes and are not shown here. Adapted from Sousa (1979).

Gigartina leptorhynchos and *Gelidium coulteri* (Fig. 12-5, bottom) leads initially to an increase of other species on the newly opened space. *Ulva* spp. may first become abundant, but as time goes on they are replaced by late-succession species. In all these examples, each species can inhibit the appearance of species characteristic of later stages in the succession.

Where late succession species were removed, *Ulva* spp. temporarily took over because they grew so much faster than any other plant on the cobblestones. The slow-growing late-succession species can tolerate the presence of *Ulva* spp., and eventually, as they grow, they replace the early colonists, perhaps through shading or by surviving grazing pressure.

In addition to desiccation and grazing, there is another important disturbance on the cobblestone community: during episodes of strong wave action, the cobblestones may be overturned. The frequency of disturbance (cf. Section 12-3) also affects whether early- or late-succession species are present on the boulders. The boulders that are most frequently turned over only hold a few opportunistic species. Disturbance therefore halts succession. If rocks are not disturbed, middle- and late-succession species colonize the surface, and in time, *Gigartina canaliculata*, the competitive dominant, monopolizes the surface of the boulder (Sousa, 1979a).

Disturbances are important in many habitats. Strong wave action, for example, can provide open spaces in intertidal rocky shores dominated by mussels (Paine and Levin, 1981). The disturbed patches devoid of mussels vary in size from the space occupied by one mussel to $38\,m^2$. The rate at which these patches appear is enough to turn over the mussel population in several years, so although at any one time the open patches occupy less than 5% of the rock surface, patches are quantitatively important over several years. Small patches disappear rapidly due to movement of mussels, but large patches (on the order of square meters in area) require settlement by new individuals. In such patches succession takes place. A major portion of space is occupied by a rapid growing red alga, *Porphyra pseudolanceolata*. Herbivore limpets prevent this species from completely covering all the space during the first year or so. A herbivore-resistant red alga, *Corallina vancouveriensis*, grows during the second year. Other species of various and unknown relationships continue the succession process. Thus the patches produced by disturbance in rocky shores are an essential feature of the community.

Many of the same features of colonization and species replacement seen in Sousa's study of cobbles can be seen in other rocky shores. One additional feature that merits notice is the occurrence of indirect effects on the course of succession. In the Oregon rocky intertidal barnacles indirectly delay succession by impairing in some fashion the ability of limpets to graze colonizing seaweeds (Farrell, 1991). On the other hand, limpets, when present, consume colonizing seaweeds, and so indirectly favor colonization by barnacles (Van Tamelen, 1987). Consumers can therefore hasten, as in the New England rocky intertidal (Lubchenco, 1983), or restrain species replacement. Indirect effects can be even more complicated, and unrelated to consumption. In many rocky intertidal sites of the NW coast of the United States, available space is in short supply, and mussels out-compete turf-like seaweeds, which in turn outcompete brown algae known as sea palms (Paine, 1988a). If mussels are not present, space is occupied by turf algae; the sea palms can only successfully colonize bare space after it is made available by removal of mussels from rock surfaces by storms or other disturbances.

12.3.2 Colonization of Hard Substrates by Fouling Communities

Ship bottoms, outfall pipes, offshore oil derricks, and any object that people place in the sea are soon colonized by organisms. This fouling of hard surfaces has had great economic interest and has been the subject of numerous studies. The species that make up these communities are those that would settle on any hard surface subtidally; fouling communities therefore cover a sizable proportion of the coastal sea bottom.

Microbial invasion is the initial process (Mitchell, 1978) that corresponds to terrestrial primary succession. Within 24 hr a new substrate may be invaded by short, rod-shaped bacteria; within 2–4 days a mixed population of bacteria and detritus particles is enmeshed in a polymeric matrix of insoluble, high-molecular-weight polysaccharides (Dempsey, 1981). By 4–7 days there may be a thick film of bacteria mixed with pennate diatoms and within 2 weeks protozoa are grazing on the micro-flora. The polysaccharides and glycoprotein secreted by microbes may be chemically involved in prompting settlement of larger

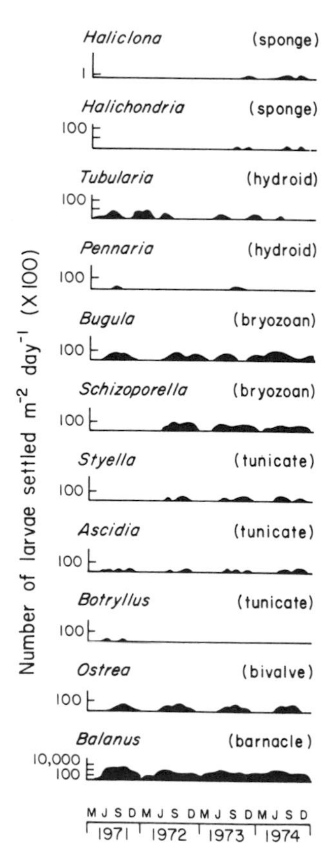

FIGURE 12-6. Larval recruitment over time for the 11 most abundant taxa in the fouling community in North Carolina. Adapted from Sutherland and Karlson (1977).

plants and animals (Kirchman et al., 1982), but there may also be just a sticky effect of the film, since where bacteria have formed a film the "settlement" of inanimate latex particles is also enhanced (Di Salvo and Daniels, 1978).

There are data that suggest that the bacteria that grow on the surfaces of different species of macroalgae differ in distribution and chemical signature (Johnson et al., 1991). The distinctive chemistry of the bacterial populations may be derived from the host frond, and the distinctive bacterial flora may be a cue used by the larvae of many invertebrate species to select a place to settle. Such a chain of mechanisms may couple the host, bacterial films, and invertebrate larvae in distinctive assemblages, but this has not been studied sufficiently.

The development of invertebrate communities on previously bare surfaces is rather unpredictable, principally because the availability of larvae, and hence larval recruitment, varies markedly within and among years (Fig. 12-6). The initial composition of the assemblage is therefore variable. In the coast of North Carolina new patches of bare surface are provided by feeding of the sea urchin (*Arbacia puntulata*). Larvae of the bryozoan *Schizoporella unicornis* colonize these bare patches, and once established, the adult bryozoans can to some extent prevent the settlement of other species. The larvae of the tunicate *Styela plicata*, however, are good competitors for space, and slowly settle on the bryozoan patches, replacing them after 2 years or so. The adults of the tunicate are also good interference competitors and

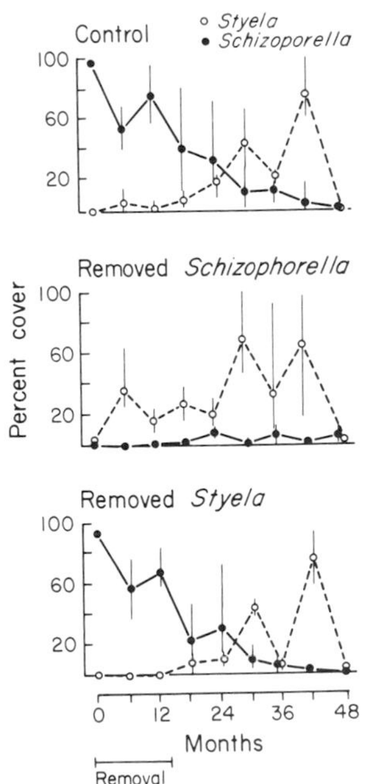

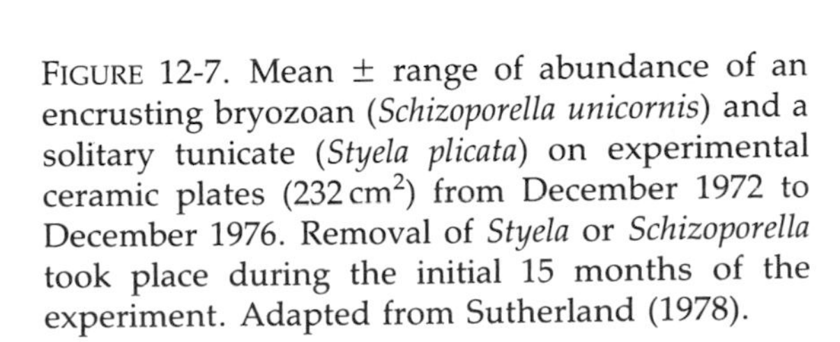
FIGURE 12-7. Mean ± range of abundance of an encrusting bryozoan (*Schizoporella unicornis*) and a solitary tunicate (*Styela plicata*) on experimental ceramic plates (232 cm^2) from December 1972 to December 1976. Removal of *Styela* or *Schizoporella* took place during the initial 15 months of the experiment. Adapted from Sutherland (1978).

prevent other species from colonizing (Sutherland, 1978), at least for the life span of the specimens present on a surface.

In experiments where adult bryozoans were removed from plates (Fig. 12-7, middle) the colonization of the plate surface by *Styela* was hastened; when the tunicate was removed, the results were very much as in the untouched control plates: tunicate larvae settled on and overgrew the bryozoans (Fig. 12-7, bottom). These and other examples from fouling communities demonstrate that inhibition of one species by another in a colonizing or successional sequence is a common phenomenon.

In another study of a fouling community on the coast of Delaware, mussels (*Mytilus edulis*) settled more abundantly on panels previously occupied by tunicates or hydroids than on bare panels (Dean and Hurd, 1980). Thus, in contrast to the many examples of inhibition reviewed above, in this case the presence of early-succession species facilitated settlement by a later species. Once established, *Mytilus edulis* became dominant and resisted invasion by any further competitors for space, as it does in the rocky intertidal shore.

In rich temperate coastal waters, where there are many larvae available to occupy surfaces, bare surfaces are nearly filled after a few months (Schoener and Schoener, 1981). Colonial species expand rapidly by asexual reproduction and cover most of the space. Solitary species settle also, but the space they occupy depends on the rate of growth of individuals. After about a year, however, the solitary species manage to cover most of the space.

The pattern of settlement in nutrient-poor tropical coastal waters is apparently different (Jackson, 1977). Colonization of bare surfaces is slow, perhaps due to low numbers of larvae available to settle, and slow growth due to low food supply. Even after 6 months half the surfaces may still be bare. Solitary species predominate early in the colonization and are replaced by colonial species through lateral overgrowth.

12.3.3 Species Replacement in Coral Reefs

The marvelous complexity of coral reefs (Fig. 1-12), with myriads of species vying for space has long attracted attention. These very diverse communities are continually changing, with multiple small-scale adjustments where one species with a small competitive advantage gains space or position over another while perhaps losing space to a third (Porter, 1974; Jackson and Buss, 1975; Connell, 1978).

The changes in allocation of space over time are largely under biological control. Macroalgae grow on bare surfaces such as dead corals. Corals and other invertebrates, once they are well established, can actively prevent fast-growing filamentous algae from growing and

TABLE 12-2. Responses by Benthic Seaweeds and Grazing Fish to Experimental Removal of the Herbivorous Sea Urchin, *Diadema antillarum* in St. Thomas, U.S. Virgin Islands.[a]

Responses	Treatment	
	Sea urchins removed	Controls
% cover by upright algae	98	1–5
% cover by crustose algae	4	68–82
% of transplanted plants of *Thalassia testudinum* eaten by herbivorous fish	22	6–7

[a] Data from Hay and Taylor (1985).

taking over surfaces. The filamentous algae, however, are a preferred food of sea urchins and herbivorous fish, and grazing controls growth of the algae and other early colonizers (Birkeland, 1977; Hay and Taylor 1985); sometimes territorial behavior of some fish species prevents herbivorous fish species from consuming algae within territories (Vine, 1974). Experimental removal of herbivorous sea urchins allows upright macroalgae to overgrow crustose macroalgae, and the increased abundance of palatable upright algae leads to increased grazing by fish (Table 12-2). The reef fish and urchins do not eat the young corals, and may in fact, enhance recruitment of corals by reducing the biomass of competitors for space (Birkeland, 1977; Sammarco, 1980). Once established, the corals (the late-succession species in this environment) dominate the community because they are good interference competitors as mature colonies or may even feed on newly settled larvae of other species. There is thus a rich pattern of interrelations that determines how far succession goes.

Any agent that removes or kills a patch of corals—storms, predators—can restart the cycle of replacements, and depending on what larvae are available at the time, a very different assemblage of species may colonize the patch. Thus succession is not simply a predetermined, linear sequence of replacements as portrayed in the classical idea of succession. The combination of species present in an area reflects the history of different patches within the area. After a patch is disturbed, the available larvae and environmental conditions in that patch may prompt the development of any of several assemblages that may retain their species composition over considerable time, as shown by Sutherland (1974) in fouling communities. A new disturbance either allows the previous species to return or causes further replacement with new species. The term "disturbance" is sufficiently vague to include any catastrophe, change in nutrient supply, appearance of new consumers or superior competitors, etc. This idea of a community as a patchwork of small parcels of slightly different history of disturbance is not just applicable to marine systems but is also emerging as the current view of the development of communities in terrestrial situations (Loucks, 1970; Denslow, 1980).

In diverse communities such as coral reefs the vast assemblage of species has a long geological history that allows survival of these species as long as the vagaries of the environment fall within their past evolutionary experience. Disturbances that occur in reefs are slight enough to allow the development of the finely adjusted biological detail characteristic of coral reefs. These communities are especially susceptible to many of the novel perturbations brought about by human activities—contamination by petroleum, heavy metals, agricultural chemicals, and disturbance of sediments by activities such as dredging—since these disturbances are mostly new to the species making up coral reefs. This contrasts with other marine environments such as salt marshes, where the climatic and geochemical settings naturally provide some exposure to heavy metals, eutrophic water, and severe changes in salinity and temperature. Salt marsh organisms, as a consequence, are somewhat more tolerant of many kinds of contamination and disturbance than coral reef organisms.

12.3.4 Colonization and Succession in Soft Sediments

12.3.4.1 Characteristics and Effects of Early and Late Colonists

Physical disturbance of shallow marine bottom sediments is a common event. Tidal surges, storms, pollution, and predation can defaunate patches of soft bottoms. Dumping of spoils from dredging operations is another source of disturbance. The latter is a pressing problem in coastal zone management and has led to interest in the study of consequences of disturbance for bottom faunas.

The appearance of bare sediment almost invariably prompts massive invasion by a few opportunist species, especially polychaetes (Grassle and Grassle, 1974; Rhoads et al., 1978; Sanders et al., 1980; Thistle, 1981). The recolonization of a benthic habitat after abatement of a pollution source illustrates the process (Fig. 12-8). After the initial increase in density and subsequent crash or replacement of the few early species, there is an increase in diversity, when many of the less opportunistic species become relatively more prominent. Subsequent to this peak diversity, there may be some competitive exclusion, with resulting lower diversities, as we have seen in other environments. Over subsequent periods diversity may change further depending on season, vagaries of weather, and availability of larvae (Sanders et al.,

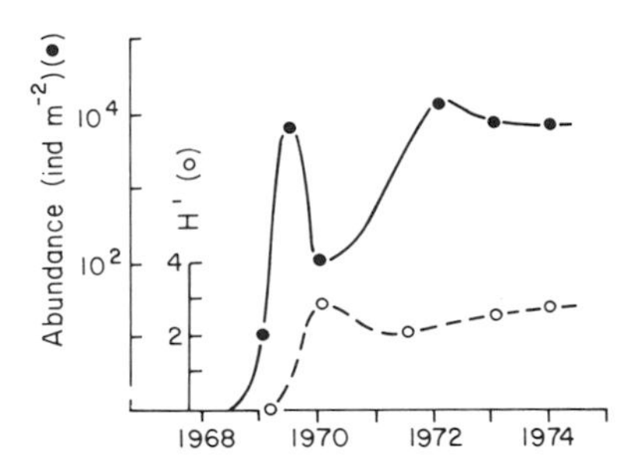

FIGURE 12-8. Abundance and diversity of benthic fauna after abatement of pulp mill pollution in Saltkallefjord, Sweden, allowed the recovery of natural communities. The peak in abundance is due to the early colonizers. Adapted from Pearson and Rosenberg (1978).

1980). Where density of larvae is lower, such as in the deep sea, colonization of bare sediments is slower and settlement more sparse than in coastal sediments, where densities of larvae are larger (Grassle and Morse-Porteus, 1987; Snelgrove et al., 1992). Colonization rates are strongly affected by the quality of organic matter available in sediments: the addition of only 1% weight of ground phytoplankton or *Sargassum* to sediments increased colonization by two orders of magnitude (Snelgrove et al., 1992).

Rhoads et al. (1978) studied the colonization of soft sediments by placing boxes of azoic sediment on the sea bottom and subsequently monitoring faunal changes. They also followed colonization of areas where dredge spoils were dumped. In both experiments bare muds were invaded by larvae.

The first colonizers were mostly small-sized (Fig. 12-9) species that had fast growth rates, reached very high densities, and achieved high production rates (Table 12-3). They also suffered high mortalities, so these populations turned over quickly. The high densities achieved by the opportunist species lead to depletion of resources and diminish the carrying capacity of their environment (Grassle and Grassle, 1974). The polychaete *Capitella capitata*,* is a typical opportunist. The abundance of this worm quickly increased in a box of mud placed on the bottom of salt marsh creeks and crashed after a month or so after reaching densities of over 400,000/m^2. While the polychaetes were crashing in some boxes, there was growth to about 250,000/m^2 in new boxes, so the decline in density must have been due to density-dependent depletion of food or to the accumulation of toxic substances. The former seems more likely, since at high densities there was a reduction of

**C. capitata* has subsequently been found to consist of a group of five sibling species (Grassle and Grassle, 1976).

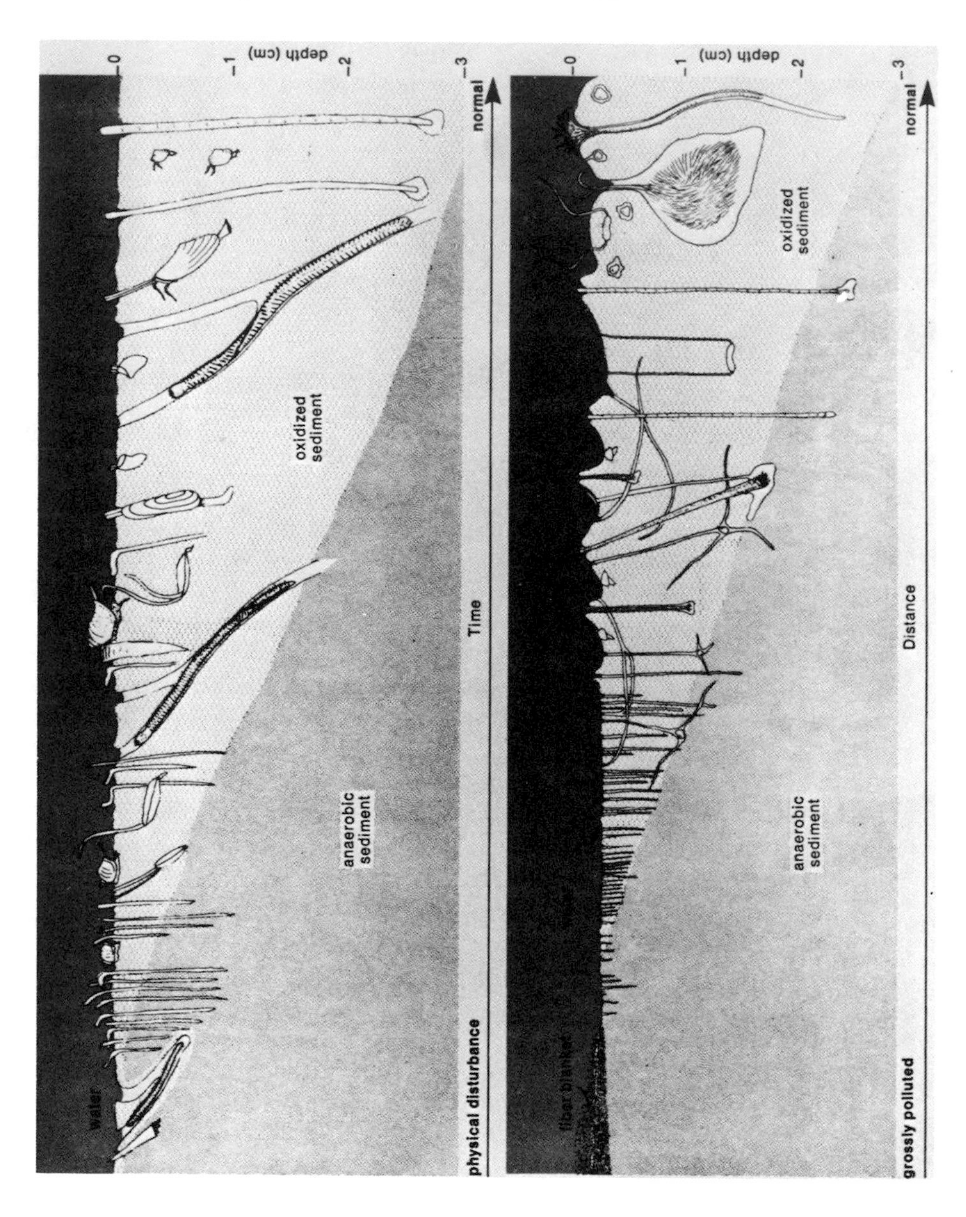

FIGURE 12-9. Idealized scheme of sequence of species invading bare sediment such as newly dumped dredge spoil (top). There is a close parallel to the gradient over space away from a grossly polluted site (chronic release of fiber from a cellulose factory) (bottom). Immediately after distubance or close to the source of pollution a few species of abundant, small and productive polychaetes

continued

the size of brooding females and of number of eggs per female (cf. Chapters 4 and 7).

The early colonizer species tend to feed on particles suspended above the bottom or on recently sedimented particles. They live very near the surface (Fig. 12-9) and may form dense aggregations of tubes. Since the depth of the tubes is slight, pumping of seawater only affects the geochemical properties of a very shallow layer of the sediment. Pioneer species are easily accessible to predatory fish and crustaceans since they live so near the surface, and this is in part why their mortality rates are high (Table 12-3).

The later colonizers are less abundant, larger, slower growing, and less productive (Table 12-3) than the earlier species. The late colonizers also suffer lower mortalities since they tend to live buried in sediment (Fig. 12-9) and are harder for predators to find. Late colonizers are typically deposit feeders whose burrowing and pumping of seawater through long, often U-shaped tubes (Fig. 12-9) change the chemical and physical condition of the sediments to a considerable depth (20 cm) (Aller, 1980). Pumping of seawater into tubes aerates the sediments and transports chemicals and particles. The movement, ingestion, and defecation of sediment by deposit feeders increase the fluffiness of sediments and up to a point foster growth of microbes on the particles.* The animals can then feed on the microbial biomass produced by this microbial "gardening" (Yingst and Rhoads, 1980; Hylleberg, 1975; Woodin, 1977).

The surfaces of soft sediment are often covered by fecal pellets, produced by both deposit and suspension feeders. New fecal pellets are less laden with microbes or microbial films than uneaten sediment particles and are readily colonized by microbes. Defecation by benthic animals may therefore maintain relatively high microbial activity.

Water moving over the sea floor can transport surface sediments. The roughness, cohesiveness, and specific gravity of the sediment

---

*The result of all these activities is often referred to as "bioturbation."

---

are found. These are followed, either over time or space, by suspension-feeding or surface deposit-feeding molluscs. The latter are replaced by large, slow-grazing species that live deeper in the mud, feed on buried deposits, and oxidize the sediment by their activities. The close equivalence of space and time is another example of the same principle we saw in the similarity of the early–late gradient of planktonic succession and coastal–oceanic gradients. These comparisons point out the commonality of the processes involved in organizing communities. From Rhoads et al. (1978) and Pearson and Rosenberg (1976).

TABLE 12-3. Life History Data for Representative Species Colonizing Experimental Sites Where Either Trays of Defaunated Mud were Provided or Where Dumping of Dredging Spoils Provided Opportunities for Colonization.[a]

	Days to peak abundance	Maximum abundance (m^{-2})	Size ($mg\,ind^{-1}$)	Generations yr^{-1}	Estimated production rate ($g\,m^{-2}\,day^{-1}$)	Mortality rate
Early colonizer						
Streblospio benedicti (polychaete)	10	420,000	0.15–50	3–4	0.57	High
Capitella capitata (polychaete)	19–50	80,000	0.15–50	5–8	0.27	High
Ampelisca abdita (amphipod)	29–50	10,000	0.5–1	2	0.06	High
Owenia fusiformis (polychaete)	—	—	0.5–1	—	—	High
Mulinia lateralis (clam)	—	—	2–10	1–2	—	High
Intermediate colonizers						
Nucula annulata (clam)	50	3,700	5–10	1–2?	0.12	Moderate
Tellina agilis (clam)	80	1,400	about 5	1?	0.04	Moderate
Pitar morrhuana (clam)	—	—	about 10	1	—	Moderate
Late colonizer						
Nepthys incisa (polychaete)	86	220	30–70	1	0.03	Low
Ensis directus (clam)	175	30	100–300	1	0.01	Low
Nassarius trivittatus (snail)	50–223	—	3–10	2	0.01	Low

[a] Experiments carried out in Long Island Sound. Adapted from Rhoads et al. (1978).

surface affect transport and are largely determined by the organisms of the benthos. There are numerous mechanisms by which animal activity can affect the integrity of the surface layer of sediment (Rhoads and Boyer, 1983; Rhoads et al., 1978a; Jumars et al., 1981; Taghon et al., 1978; Grant et al., 1982).

It is difficult to predict whether the presence of pioneer or late-colonizing species will stabilize or destabilize surface sediments. Rhoads and Boyer (1983) predict that although at low densities of pioneer species sediment surfaces may be destabilized, high population densities build up very quickly and increase the stability of surface sediments by stimulating microbial films and building tubes. On the other hand, late-succession species may tend to destabilize sediment because their activity aggregates sediments, and their feeding may deplete microbial populations. These activities make the surface sediments fluffier or more fluid and easier to disrupt. These speculations need further scrutiny, but if they are true, the late-colonizing species are very important in that their activity sets the stage for disturbance and transport of the unconsolidated surface of the sediment in which they live.

The animal–sediment interaction and its effect on sediment transport and stabilization have significant applied interest. Sediment transport is obviously important in navigation, maintenance of harbors, and erosion of shorelines. In addition, if pollutants are present, transport of particles along the sea floor is often accompanied by the movement of pollutant substances, since many are strongly adsorbed to particles. Particles bearing adsorbed pollutants would tend to be trapped in sediments populated by pioneer species, while further transport is more likely in sites at a later stage of succession.

The change in the oxidation state of sediment brought about by colonization and establishment of benthic fauna (Fig. 12-9) also changes nutrient dynamics (Aller, 1980). For nitrogen, the increased oxidation fosters a change from an ammonium-dominated to a nitrate-dominated phase.* Decay rates increase, and fungi may degrade some of the more refractory organic compounds. Nutrients in interstitial water are incorporated into microbial cells and are otherwise immobilized. In the case of phosphate, the more oxidized the sediment, the less soluble (and removable) the form of phosphorus (Section 14.1.3). For nitrogen, nitrate is not easily adsorbed, and is therefore easily transported, and

*This contrasts with some results in terrestrial environments, where it is believed that the succession moves from a nitrate-based to an ammonium-based nitrogen economy, since as succession proceeds, nitrogen may be supplied more and more by internal decay and recycling (mineralization of organic nitrogen) (Rice and Pancholy, 1972; Bormann and Likens, 1979). Such trends are not always found in terrestrial ecosystems (Vitousek et al., 1982).

the more aerobic the sediment the more that nitrogen is available for transport. Ammonium, however, can also diffuse or be pumped out of sediments (Section 14.2.3). It is therefore not clear which stage of succession leads to greater or lesser release of dissolved inorganic nitrogen or phosphorus. Quantitative measurements of nutrient regeneration in sediments inhabited by pioneer and late-succession assemblages will be usefull.

12.3.4.2 Mechanisms of Species Replacement in Soft Sediments

The rate of species replacement in soft sediments depends on interspecific mechanisms. The species present may retard succession, as when newly settled larvae are eaten by established species (Feller et al., 1979; Wilson, 1980). On the other hand, surface-dwelling suspension feeders typical of early succession may be buried when sediments are bioturbated by late-coming deposit feeders that burrow, ingest, and defecate sediment (Rhoads and Young, 1971; Brenchley, 1981).

There is also evidence that in soft marine sediments the presence of many species facilitates rather than inhibits the settlement of other species. Experimental manipulations in a mud flat in Puget Sound show that the presence of tube-building polychaetes increases the recruitment of other polychaetes, bivalves, and oligochaetes (Gallagher et al., 1983). The mechanism of facilitation may be related to the presence of tubes; artificial tubes cause aggregation of polychaetes and tanaids. Tubes may change the properties of flow over the sediment surface in some favorable way, although the specific benefits provided by the tubes remain to be identified.

The presence of some species is neither inhibited nor facilitated by other species. The abundance of the polychaete *Pygospio elegans*, for example, in a mudflat in Puget Sound is unrelated to the abundance of other species (Gallager et al., 1983). This species tolerates other species, and its rates of growth and reproduction are set by other factors.

12.3.5 Colonization and Sequence of Species in Salt Marshes

New England salt marshes are mosaics of short and tall forms of *Spartina alterniflora*, bare patches, *Salicornia europaea*, *Spartina patens*, and *Distichlis spicata* (Fig. 12-10). A series of long-term experiments have provided a sketch of how the balance among the vegetation types is struck (Valiela et al., 1985). In particular, the colonization of newly-bare patches of sediments in salt marshes by plants illustrates mechanisms of replacement and the nonlinear relationships of replacement within disturbed patches.

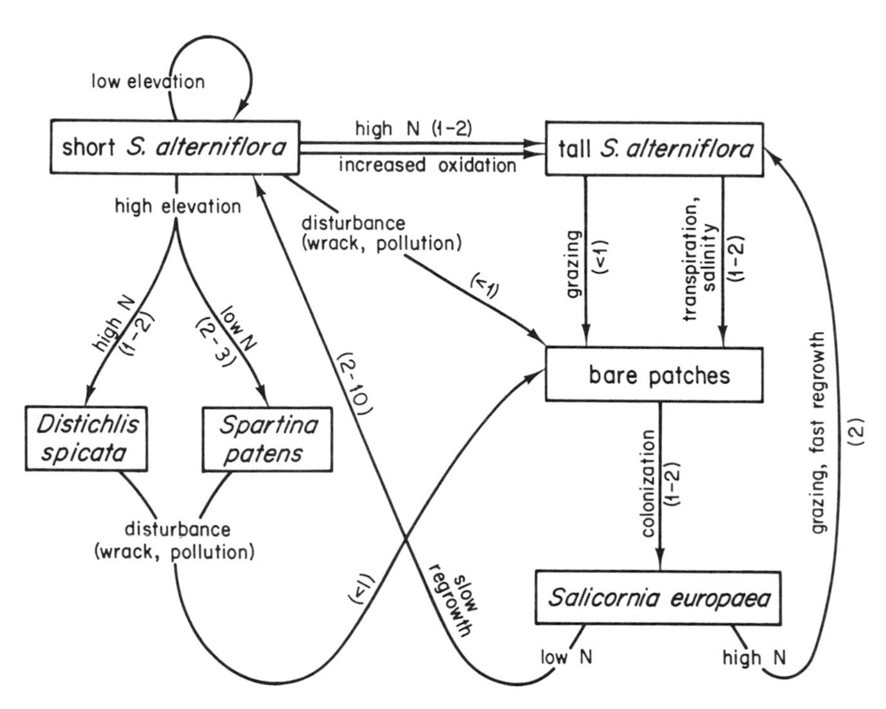

FIGURE 12-10. Elements (in boxes) and transformations (arrows) of the patches of salt marsh vegetation. The numbers in parentheses are duration of the transition in years. From Valiela et al. (1985).

Added supply of nitrogen (eutrophication), or increased oxidation of sediments due to increased drainage both foster conversion of patches of short to tall *S. alterniflora* within 1–2 years. The increased biomass of *S. alterniflora* could then increase salinity of the sediments, since more biomass leads to increased transpiration of fresh water. In relatively more eutrophic patches the tissues of *S. alterniflora* become enriched in nitrogen and are more readily attacked by grazers (cf. Section 6.322). Both increased salinity (Haines and Dunn, 1976) and grazer pressure (R. Buchsbaum and C. Cogswell, unpublished data) can thus lead to death of patches of *S. alterniflora* within one growing season or so.

The resulting bare patches in salt marsh sediments are colonized by an opportunistic plant, the glasswort *Salicornia europaea* within 1–2 years. This species is a poor competitor and does not do well except in bare patches. If it turns out that the sediment of the patch newly colonized by *S. europaea* is rich in nitrogen, the plants are eventually heavily attacked by herbivores and there is little regrowth of the glasswort. The high nutrients also foster the regrowth of tall *S. alterniflora*, either by new seedlings or by lateral vegetative growth. If the patch happens to have a lower nitrogen supply, the effect of grazers is less obvious, and slower regrowth of short *S. alterniflora* takes place.

If the patch in question is located at a low elevation within the intertidal range, the short form of *S. alterniflora* remains. If the patch is located at a higher elevation, above mean tide level, the short *S. alterniflora* is replaced by other species of plants. The replacement mechanism may be differential nutrient uptake. When nitrogen supply is high, *Distichlis spicata* takes over, while *Spartina patens* does so when nitrogen supply is lower.

Disturbances also create bare patches. Pollution events, such as an oil slick entering the marsh, or more commonly, ice-rafted sediments or layers of stranded sea wrack (Hartman et al., 1983) will kill patches of vegetation. Such disturbances might occur both in low and high tidal elevations and kill patches of vegetation in a short time. High sites tend to collect more wrack, so are more disturbed than low sites. These bare patches are then colonized by a variety of marsh plants. While the number of plant species (18) is greater in higher reaches of the marsh than in lower sites (8 species of plants), unlike the situation described for coral reefs and forests in Chapter 10, disturbance frequency seems unlikely to affect species richness (Valiela and Rietsma, 1994).

The succession of species on a particular patch of salt marsh is therefore not necessarily a predetermined, linear sequence [see den Hartog (1977) for a similar description for seagrass beds]. The sequence of events is affected by a variety of factors. "Regressive" steps, in which some change prompts reversion to other assemblages, are common. Such changes can be quite local, and add to the patchiness and between-habitat diversity within a salt marsh, and probably to all marine environments.

12.4 Interaction Among Communities at Different Stages of Succession

In nature no parcel exists by itself. There are linkages among adjoining environments. Margalef (1968) has pointed out that if the adjoining parcels are in a very different state of succession, the exchanges between the two will be asymmetric. Consider an early-succession assemblage composed of fast-growing organisms with few defenses against consumers. These pioneer organisms initially have a surfeit of resources, which they exploit inefficiently. They also have demographies suited to sustain high mortalities. In contrast, late-succession assemblages are, in theory, slow growing, have many defenses against consumers, are unable to sustain high mortalities, and are efficient in use of resources. In late succession, resources are in short supply, and whatever is made available is consumed by some species.

If two assemblages, one in an early, the other in a late stage of succession, exist next to each other, it does not seem far-fetched for the species of the late succession to venture into the adjoining early succession parcel, since there they would have the opportunity to exploit resources that are far more freely available and consumable. There are no data for such a phenomenon for adjoining parcels in early and late stages of succession within a given community, but we can consider adjoining pairs of communities in ecosystems, one resembling our description of early succession, and another more similar to late succession situations.

We have already discussed examples of such pairs of communities. These include the sedimentation of particles from the plankton which supports the benthos. There are other examples. Benthic fish of shallow waters often feed on zooplankton in the water column. The blacksmith (*Chromis punctipennis*), an abundant pomacentrid fish, lives in rocky reefs off California, feeds on zooplankton in the water column during the day, and shelters at night in rocky crevices on the bottom. The feeding pattern of just this one species results in the importation of $8\,g\,C\,m^{-2}\,year^{1}$ of feces (with a C/N ratio of 8) to the nocturnal shelter on the bottom (Bray et al., 1981). Such feeding behavior is also common in other species, and in fish and zooplankton of coral reefs (Porter and Porter, 1977), where, as we have seen, food availability is typically very low.

Another example is provided by salt marshes and adjoining coastal waters. Salt marshes are very productive systems with abundant dissolved nutrients, energy-rich reduced compounds, particulate matter, and very high densities of potential prey organisms. There are asymmetric flows of nutrients and particulates to less productive and less eutrophic coastal waters (cf. Section 14.2.4). The annual exports of dissolved nitrogen from marshes to coastal water, for example, can in some places amount to over 40% of the amount of dissolved nitrogen in the near-shore water column (Valiela, 1984). Further, fish of the deeper coastal waters exploit salt marshes and related tidal creeks in at least two ways: (a) coastal fishes use salt marshes as nurseries for their young (Werme, 1981), since the early life stages require relatively larger food rations and so are often found in salt marshes where food is more available; and (b) larger fishes enter tidal marsh creeks during high tides to feed on the dominant prey that live within the marsh.

There are thus asymmetric flows of resources from the loosely organized plankton to the more tightly organized benthos, and from eutrophic productive marshes to the less rich coastal water column.

12.5 Generalized Properties of Succession in Marine Environments

Succession is the composite result of many processes described in earlier chapters. There are several possible general types of interactions among species that determine whether or not replacement takes place (Connell and Slatyer, 1977; Van Tamelen, 1987), and at what rates it is accomplished if it does occur.

Depletion. The use of a resource by a species may prevent its continued presence; other species better suited to exploit resources in very short supply may then become dominant. This loose sort of interaction is prevalent in planktonic and benthic environments. Early colonizers may have lower assimilation efficiencies, or take up nutrients best at high concentrations. They therefore may be outcompeted by late-succession species. It would be of interest to compare assimilation and growth efficiency in early- and late-succession species and relate these physiological variables to demographic properties.

Tolerance. Some species that appear late in succession do so simply because their life history takes a long time to complete. Some late species may tolerate whatever the early colonizers do and, after inconspicuous development of the immature stages, may become dominant and outcompete pioneer species. We found examples of this in soft sediments, although it is probably unusual in nature. This is perhaps an odd entry here, since it essentially refers to a lack of interaction among species, at least during the early stages.

Facilitation. Some species of an early stage modify the environment so as to facilitate the appearance of species characteristic of late succession. Examples of this are bacterial films on hard substrates or certain activities of benthic animals that dramatically change the chemistry of sediments in mudflats so as to facilitate settlement of other species.

Inhibition. Earlier colonists may inhibit the appearance of later species. This delays succession, and may produce alternate assemblages that last for the life span of the species present, as we have seen in fouling communities. Many examples of this process are available in the environments surveyed above, where larvae of other or the same species are consumed, damaged, or expelled by the gauntlet of feeding appendages brandished by the animals already present on a substrate. The inhibitory interaction is frequently found in nature, especially in benthic situations, but it is only a temporary delay of succession. Open space for further colonization will still be made available by disturbances and by mortality of the early colonizers. The latter is especially applicable because early colonizers tend to be more palatable and have shorter lives, and so turn over faster than later species. This can lead to a slow change toward an assemblage of species characteristic of late succession.

Removal. Species of late succession may simply consume, overgrow, or disturb earlier species, and replace them in water, sediments, or surfaces. We have examples of this in coral reefs, rocky shores, and soft sediments.

Allelopathy. Early colonists may be replaced by a late-colonizer that can release ectocrine substances. Such substances could inhibit growth of a competing, earlier species. We have seen evidence of this in the plankton, in microbial communities, and in coral reefs.

Several of these mechanisms probably co-occur in most communities. It would be of interest to know if the preponderance of one or another of the mechanisms of succession depends on the type of community or affects the way the community is structured.

Early colonizers tend to be rapid growing, productive, *r* (or exploitatively) selected species.* Early colonizers tend to be more palatable as food for consumers—relatively free of chemical deterrents, spines, and other protections. It is no accident that the major crops of agriculture are grasses, typical early succession species of terrestrial ecosystems. In the sea also the more productive fisheries are also those characteristic of early succession stages. The classic examples are clupeid fisheries in upwelling areas, where the harvest is only one trophic step from the primary producers and few species of consumers share in the harvest.

As succession proceeds, the species—in most cases—tend to be larger, grow more slowly, be less productive, and have more complex morphology and special requirements, as we saw in the benthos. Such species use resources in short supply, and to point this out we have called them *K* or saturation selected species.

As time goes on, more species accrue, and diversity increases. Spatial heterogeneity may also increase during succession (as discussed in Chapter 11), since more patches accumulate through time due to disturbances.

Mosaics of heterogeneous patches containing a variety of species may replace the perhaps more homogeneous assemblages of early succession. In the later stages of succession, diversity may decrease if disturbances are infrequent enough so that competitive dominants can exclude some species. The diversity and patchiness of the environment reflect a balance between the advantages of specialization and the

---

*The terms opportunistic, fugitive, weedy, or pioneer are often applied to these species. The reproductive strategies involved were discussed in Chapter 4, where we also pointed out the *r-K* continuum was not a complete statement. Species adapted to live in difficult conditions should also be included.

degree of disturbance in the environment, at least in theory (cf. Section 10.3.1).

As succession proceeds, the complexity of biological interactions increases, featuring chemical inhibition, symbiotic, and behavioral mechanisms in addition to predation and competition. The activity of late-succession species often markedly changes their environment, with notable alteration of the chemistry and physics of the environment. In the water column many substances are either taken up or released by organisms, and as pointed out in Chapter 14, much of the nutrient supply in late succession may be regenerated by zooplankton. In sediments the bioturbation by benthic animals markedly changes the biogeochemistry and physics of the substrate. The marked effect of animals on soft sediments demonstrate how effectively late-succession assemblages may modify the properties of their environment (Aller, 1980). The domination of nutrient exchanges and sediment transport by animal activity are important but not well-enough known consequences of succession.

Any exploitation of a given community by consumers from late-succession patches or by man tends to halt or reverse the trends of succession (cf. Fig. 12-9, bottom). The more intense the degree of pollution or exploitation, the more succession is held back. At severe levels of contamination or exploitation, only a few of the more weedy species survive.

The complex and fragile biological interactions that govern communities of advanced succession make such assemblages of species more susceptible to many man-made alterations. For example, if the density of species of late-succession deposit feeders is reduced below the number needed to maintain the oxidized state of soft sediments, the whole benthic community may be altered. The fragility of late-succession assemblages derives from two features (1) late-succession species tend to markedly change their environment and (2) late-succession species may be more susceptible to novel disturbances. The management of environments that contain assemblages of late succession needs to consider these features.

Succession is seldom a predetermined, orderly replacement of one species by another. Rather, the assemblage of species may shift to any of a number of alternate states—as we have seen in the rocky shore, fouling, and salt marsh examples. The shifts from one to another state may be due to competition and predation among the species that happen to be present, and to external factors such as nutrient supply or disturbances. The shifts may be recursive, as happens when a plankton community is exposed once again to nutrient rich water: early-succession species reappear.

A habitat is a mosaic of patches, each of which may be at a different stage of succession—either because of different age of the patch or because the species that were available for recruitment differed. This provides a remarkable degree of heterogeneity in the spatial distribution and species competition of most habitats. The degree to which disturbances create new patches is responsible for the conservation of this heterogeneity. Where disturbances (predation by large mobile species, pollution, wave action) are very infrequent, top-down or bottom-up controls become more important.

There may be interactions among patches that are in different stages in succession. Generally, these interactions are asymmetrical, with species of the later stage exploiting resources in the earlier patch. This is just another way to describe where colonizers come from; the "colonization" can be short term, though, as in the coastal fish that only feed in a salt marsh during high tide and return to deeper waters at low tide.

Succession integrates most of the topics covered earlier in this book and reflects the processes that structure communities over time. Every exploitation of marine organisms or management of marine environments by people is necessarily done in the context of the stage of development of the community in question. Clearly, the mechanisms that govern communities are loose compared to, say, the tight organization due to endocrine regulation within individual organisms. Nevertheless, there are properties of species characteristic of different stages of community organization that can affect what we do to or what we get from any marine ecosystem.

Part IV
Functioning of Marine Ecosystems

Fecal pellet of a copepod, photographed at three different magnifications to show shape and contents. Photos courtesy of Susumo Honjo, Woods Hole Oceanographic Institution.

Chapters 4–7 dealt with transformations of organic matter and energy in individuals, populations, and in food webs, and expanded into physiological and behavioral aspects. It has been made clear that the chemical composition of organic matter is critical in many ways. Nutrients are involved in virtually all other aspects of marine ecology. To understand how marine ecosystems work, it is of vital importance to consider the role of essential nutrients. The specific mechanisms that link nutrients to primary production, consumers, and microbes have been documented in earlier chapters. In the next two chapters we delve into properties and transformation of just a few of the most important chemical elements involved in organic matter in the sea: carbon (Chapter 13), phosphorus, nitrogen, and sulfur (Chapter 14), and their linkages via stoichiometric relations of ecosystems.

To be entirely consistent with the orientation to processes in this book, the material ought to have been organized in terms of nutrient uptake, regeneration, sedimentation, and so on. Instead, each element is treated separately, simply because the information available for each process and for each element is not equivalent and also because each element has idiosyncracies of its own.

The separate treatment of elements should not obscure the intimate coupling among elemental cycles. This coupling is especially tight in the formation and decay of organic matter, since element contents and their transformations, as seen in earlier chapter, tend to follow certain ratios; these relationships can be referred to as ecosystem stoichiometry. Nitrogen, carbon, phosphorus, and sulfur are linked by various assimilative microbial processes. The point is that although for explanatory purposes the elemental cycles are here separated, this is by no means true in nature.

We then explore how seasonal (Chapter 15) and multiyear (Chapter 16) changes reveal much about processes controlling marine ecosystems. The long-term changes also highlight the interplay of anthropogenic and climatic factors, and allow a view as to how marine ecosystems are linked to global changes.

Chapter 13 The Carbon Cycle: Production and Transformations of Organic Matter

13.1 Inorganic Carbon

The sea contains a substantial amount of carbon compared to that present in the atmosphere* or in terrestrial organisms (Table 13-1); only rock deposits contain more carbon, but these resources are not involved in the carbon cycle at ecologically meaningful time scales.

Most of the carbon in the sea is present as dissolved inorganic carbon (Table 13-1), which originates from complex equilibrium reactions of dissolved carbon dioxide and water

$$H_2O + CO_2(aq) \rightleftharpoons H_2CO_3 \rightleftharpoons H^+ + HCO_3^- \rightleftharpoons 2H^+ + CO_3^{2-} \quad (13\text{-}1)$$

The properties and consequences of the CO_2-carbonate equilibrium depicted by Eq. (13-1) are of fundamental importance to the biology and geochemistry of the sea, but an adequate treatment of carbonate equilibrium is beyond the scope of this book. Reviews of the topic are available in Riley and Chester (1971) and Broecker (1974). More advanced treatments are provided by Stumm and Morgan (1981) and Skirrow (1975).

---

*The relation of atmospheric CO_2 to aqueous CO_2 in the sea has received a lot of attention recently due to the continued increase in CO_2 in the atmosphere attributable to the burning of fossil fuels (Keeling et al., 1976), and the consequent potential for disturbance of the world's climate. Recall that CO_2 is a major absorber of energy (Fig. 2-1), esepcially infrared radiation. Increases in CO_2 might increase the temperature of the earth's atmosphere by trapping heat in the so-called "greenhouse effect." The ocean holds about 50 to 60 times as much CO_2 as the atmosphere and is able to take up more CO_2. It is not certain just how much of the CO_2 added to the atmosphere will be taken up into the larger pool or what the effects of the increased temperature will be. In fact, some believe that the higher temperature will increase evaporation and will result in increased interception of energy such that temperatures will drop worldwide. The only conclusion agreed upon is that the effects of our altering the CO_2 content of the atmosphere will have major consequences, and that the ocean will play a large role.

TABLE 13-1. Approximate Size of Major Natural Carbon Reservoirs.[a]

Reservoir	$gC \times 10^{20}$
Atmosphere (about 1973)	0.000675
Ocean	
Inorganic carbon	0.38
Organic carbon	0.01
Detrital carbonates	0.0129
Terrestrial	
Organisms	0.0164
Organic C in sedimentary rocks	68.2
Carbonate rocks	183

[a] Compiled by Skirrow (1975) from several sources.

One major aspect of the CO_2-carbonate system is the buffering capacity it provides. Carbon dioxide or rather, carbonic acid, its aqueous form, is present mainly as the bicarbonate (HCO_3^-) in seawater, at concentrations of about 25 mg liter^{-1}. CO_2 is removed by photosynthesis and added by respiration. The amounts of CO_2 in seawater generally exceed photosynthetic demand, so the changes in pH that occur by Eq. (13-1) due to photosynthetic removal of CO_2 are relatively small. Aqueous solutions of weak acids and bases such as Eq. (13-1) resist pH changes, so that the pH of seawater is buffered between 7.5 to 8.5 by the CO_2-carbonate system, at least for time scales of tens to hundreds of years. The CO_2-carbonate system is not the sole buffering element in seawater; borate for instance, also contributes to buffering. Temperature and pressure also affect carbonate equilibria.

In anoxic water columns or in interstitial water within sediments, anaerobic metabolic reactions by microorganisms generate carbon dioxide in far larger concentrations than in aerobic seawater. These waters also have very concentrated dissolved salts. In anoxic situations, pH does not therefore depend solely on the CO_2-carbonate system.

Our too-brief mention of the CO_2-carbonate system here is just one way to show that organisms are intimately linked to the carbon cycle. Because of the chemical consequences of such biological control of the CO_2-carbonate equilibrium, affecting pH, buffering capacity, and the equilibrium reactions of many other elements, much of chemistry of the oceans is mediated biologically. There are other important biogeochemical processes, such as the deposition of carbonates on reefs and sea floor, that are also biologically influenced.

The uptake of inorganic carbon from seawater by organisms produces organic carbon, and this leads to complex transformations discussed below. In addition, there are transports of organic carbon by physical processes, transports that can have major consequences. The carbon cycles of aerobic and anaerobic environments are described separately below because many of the biological processes differ significantly in these two circumstances.

13.2 The Carbon Cycle in Aerobic Environments

13.2.1 Concentration of Organic Carbon in Seawater

Organic carbon in seawater occurs in dissolved forms (DOC) and in living or dead particulate forms (POC). DOC is taken to be the organic carbon that passes through certain filters (usually of a pore size of 0.2–0.45 μm) while POC is retained. Somewhat different methods of analysis are used for these two fractions. The separation into POC and DOC is not clearly justified, since there is a continuous size distribution of organic matter in seawater from 10^{-3}–10^{3} μm (Sharp, 1973). Some of the DOC is also colloidal rather than truly dissolved, but here we use the term DOC to comprise both.

The concentrations of dissolved organic matter in seawater are much larger than those of particulate organic carbon (Table 13-2); there is usually about a 100:10:2 ratio among DOC, dead POC, and living organic carbon (Parsons, 1963). The DOC is made up of many classes of compounds; metabolically some of the most valuable are low-molecular-weight carbohydrates, sugars, and amino and fatty acids. These compounds generally are present at concentrations of about 1–100 μg liter^{-1}.

In most aquatic environments that bulk of particulate organic matter is not free-swimming living organisms (Table 13-3). The so-called detrital matter could be morphous fragments or mucus from organisms,

TABLE 13-2. Range of Concentrations of Dissolved (DOC) and Particulate Organic Carbon (POC) and of Dissoved Organic Matter (DOM) in Marine and Lake waters.[a]

	DOC (mg C l^{-1})	POC[b] (mg C l^{-1})	DOM (mg l^{-1})
Open ocean, surface	0.4–2	0.1–0.5	0.5–3.4
Open ocean, deep water		0.01	
Inshore water			1–13
Lake	Trace–10	Trace–10	1–50

[a] POC includes living and detrital particulates. Adapted from Parsons et al. (1977), Fenchel and Blackburn (1979), and Wetzel (1975).
[b] The composition of POC is given in Table 13-3.

TABLE 13-3. Composition of the Particulate Matter Suspended in Seawater in Selected Areas of the Sea.[a]

	Particulate organic matter (mg C m^{-3} or μg C liter^{-1})	Percentage of total particulate organic matter			
		Phyto-plankton	Zoo-plankton	Bac-teria	Detri-tus[b]
Sea of Azov	750–1500	5–10	3–10	0.3–7	80–92
Arabian Sea	100–250	1–31	—	—	—
Black Sea	200–250	0.2–1	5–20	0.4	78–95
Tropical Atlantic					
15th Meridian	450–600	0.5–1.3	0.6	—	98–99
16th Parallel	100–250	0.6–1.3	0.7	—	98–99
Upwelling off					
S.W. Africa	70–900	30–43	4–14	—	9–14
Hudson River Estuary	660–2,250	2–72			40–93
New York Bight	200–840	12–51			38–90
Western Baltic Sea	492–505	23–27	33–35	—	41–43
Chesapeake Bay	11.5–84[c]	23	—	—	77
English Channel	950–2,500	15–17			
Aberdeen Bay,					
Scotland	200–3,400	8–10			
Wadden Sea,					
Netherlands	1,000–4,000	10–25			
Akkeshi Bay,		9.7	1.7		
Hokkaido			0.1[d]		
Baltic Sea	99–668	3–66[e]			51–94

[a] Adapted from Finenko and Zaika (1970), Hobson (1971), Chervin (1978), Lenz (1977), Van Valkenberg et al. (1978), Jorgensen (1962), Hogetsu et al. (1977), and Andersson and Rudehäll (1993).
[b] Detritus include attached fungi and bacteria.
[c] Particles ml^{-1} $\times$ 10^4. Totals and percents do not include zooplankton.
[d] Microplankton.
[e] Includes pico-, nano-, and microphytoplankton.

amorphous aggregates of previously dissolved organic matter, and complex of associated microrganisms (Alldredge and Silver, 1988). Phyto- and zooplankton contribute larger biomass to POC than bacteria (Table 13-3). Even in coastal water, where bacteria are relatively abundant at about 7 bacterial cells ml^{-1}, bacteria only contribute up to 4–25% of total plankton carbon (Ferguson and Rublee, 1976). Offshore bacterial densities may be less than 1 cell ml^{-1} (Ferguson and Palumbo, 1979), so the proportion of POC due to bacteria in open ocean water is very small.

13.2.2 External Sources of Organic Carbon

Carbon is introduced into the sea by various mechanisms (Fig. 13-1 and Table 13-4). In terms of the entire World Ocean, photosynthesis is the

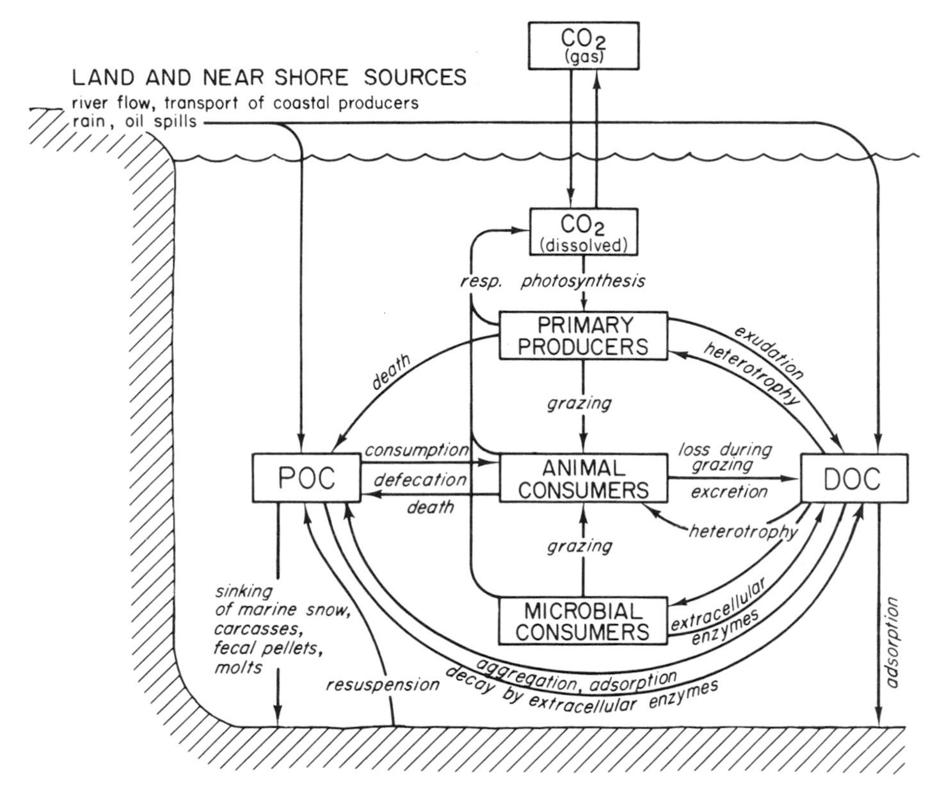

FIGURE 13-1. The transfer of carbon in aerobic marine environments. The boxes are pools while the arrows represent processes. The inorganic parts of the cycle have been simplified. "Marine snow" refers to organic aggregates and debris not shown in diagram. There is probably some release of DOC from sediments into overlying waters. Viral lysis of primary producer cells could be an additional mechanism that releases DOC.

TABLE 13-4. Estimates of the Amounts of Organic Carbon Transported to the Oceans by Various Mechanisms.[a]

	$g\,C\,yr^{-1} \times 10^{14}$
Primary production by phytoplankton	200–360
Rivers and streams	3–3.2
Groundwater flow	0.8
Plant volatiles and airborne particles	1.5–4
Petroleum hydrocarbons	0.046

[a] Adapted from Handa (1977), Duce and Duursma (1977), and Farrington (1980).

most important mechanism by which organic carbon is formed, by about two orders of magnitude (Duce and Duursma, 1977; Deuser, 1979). Fluvial and groundwater transport (Fig. 13-1) are important but smaller carbon sources (Table 13-3). The POC organic matter carried by rivers, often of vascular plant origin, is relatively decay resistant. Such terrestrially originated and fairly refractory material then finds its way to the sea floor and may be a quantitatively important source of sedimentary carbon in coastal areas. Studies on several chemical tracers—hydrocarbons (Gardner and Menzel, 1974; Farrington and Tripp, 1977; Lee et al., 1979), ratios of stable isotopes of carbon (Parker et al., 1972), and degradation products of lignin (Hedges and Mann, 1979)—show that there is considerable horizontal transport of terrestrial, estuarine, and coastal organic matter to deeper sediments. There is, however, evidence that some of the very resistant compounds—humic and fulvic acids—found in sea floor sediments are formed *in situ* (Nissenbaum and Kaplan, 1972).

Horizontal transport from land can be very large in certain locations. For example, in certain areas off the state of Washington the input of terrestrial matter may constitute 95% of the total transport of organic carbon to the sea floor (Hedges and Mann, 1979). Tidal flushing of estuarine rivers and creeks transports organic matter between coastal marshes and adjoining coastal waters, as discussed in Section 7.7.1. There has been some dispute as to the direction of the export (Nixon, 1980; Valiela, 1983), but recent evidence suggests that marshes export carbon. Hopkinson and Wetzel (1982) and Hopkinson (1985), for instance, found that a source of organic matter, most likely from adjacent salt marsh estuaries, was needed to balance carbon budgets of nearshore environments off Georgia. The heterotrophic activity of coastal organisms is considerably higher within 10 km from shore (Kinsey, 1981), and measured exports of organic matter and nutrients from estuarine salt marshes in Georgia seem large enough to support activity by the auto- and heterotrophs present (Turner et al., 1979). The presence of degradation products of lignins in sediments of the sea floor provides direct evidence that significant amounts of organic matter from land or coastal wetlands reaches shelf waters. Export of organic matter from mangrove swamps seems to support metabolism in Great Barrier Reef lagoons (Alongi, 1990). Export of organic matter from salt marsh vegetation in Georgia could contribute 36, 26, and 18% of the DOC in nearshore, inner, and outer shelf waters, based on calculations from data on lignin phenols (Moran et al., 1991). Moran and Hodson (1994) report that 11–75% of dissolved humic substances on the continental shelf off the SE United States originates from vascular plants, half from salt marshes, half brought by rivers.

The occurrence of detritus of seagrasses and other coastal plants is locally common in the deep sea (Wolff, 1976; Wiebe et al., 1976;

Staresinic et al., 1978 ; Thiel et al., 1988–1989). This source of carbon to the deep sea is very small, considering the area of ocean relative to that of coastal zones. The amounts of carbon needed by deep-water organisms are too large to be supplied by terrestrial and coastal export (Hinga et al., 1979).

Organic materials are also transported to the sea by wind and water transport of coastal and terrestrial carbon and manmade hydrocarbons (Table 13-4) (Duce and Duursma, 1977; Handa, 1977; Wakeham and Farrington, 1980). Lateral transport from one marine environment to another inject organic carbon into a water column such injection may increase oxygen consumption by heterotrophs in the affected layers of the water column, and lead to otherwise unexplainable zones of oxygen depletion (Craig, 1971; Skopintsev, 1972).

Within any one water mass it is the death of the producers that yields most of the nonliving POC (Fig. 13-1). Death and defecation by animals, and fragmentation of food during feeding, all add or modify organic particles, but of course these particles are derived from carbon fixed by plants. The aggregation and adsorption of dissolved organic carbon onto large complexes (cf. Section 13.2.4.3) are futher internal sources of POC (Fig. 13-1).

13.2.3 Internal Sources of Dissolved Organic Carbon

Dissolved organic carbon is excreted by consumers, released from broken cells during feeding by consumers or viral lysis, released by leaching of soluble organic carbon and by hydrolysis due to microbial extracellular enzymes, and exuded by primary producers (Fig. 13-1).

The largest single source of DOC in aerobic situations may be release (exudation) by producers (Fig. 13-1). Primary producers release DOC as a result of excretion of metabolic waste, secretion of compounds inhibitory to competitors or predators, release of carbohydrate end-products of photosynthesis carried out in excess of carbon needed for growth (Kelly, 1989), or as unavoidable diffusive losses through membranes in an aqueous medium (Bjørnsen, 1988).

Both single-celled and multicellular producers release some of the assimilated carbon, but there is some controversy as to how large this release is. Estimates of exudation by phytoplankton vary widely, from 7 to 62% of photosynthetically fixed carbon (Hellebust, 1967; Anderson and Zentschel, 1970; Choi, 1972; Smith et al., 1977; Lancelot, 1979; Larsson and Hagström, 1979). The *in situ* rates of exudation by natural phytoplankton assemblages may lie toward the low end of the range. Williams (1975) estimated that perhaps 10% of the net phytoplankton production is exuded in the open ocean, and Smith et al. (1977) calculated that about 9% of fixed carbon was excreted in the up-welling off northwest Africa.

The estimated rates of exudation by macroalgae are as variable as those of microphytes. They range from 1–4% (Majak et al., 1966; Brylinsky, 1977) up to 35–40% of the carbon fixed (Sieburth, 1969; Sieburth and Jensen, 1969; Hatcher et al., 1977). In coastal waters release of mucilage can amount from 14–25% and 50–60% of carbon fixed photosynthetically by macroalgae and corals, respectively (Linley et al., 1981; Crossland et al., 1980; Branch and Griffiths, 1988).

The extracellular products released as exudates by producers are mainly carbohydrates, but include various nitrogenous compounds, organic acids, lipids, phenolics, and many other kinds of organic compounds (Hellebust, 1974). This DOC is therefore potentially nutritionally valuable to heterotrophs.

The amount of DOC released may depend on growth conditions, light intensity, and species present. The growth condition of producers affect the release of DOC, although it is not yet possible to generalize. In cultures, cells in log growth phase release less organic carbon than those in a stationary growth phase. *Thalassiosira fluviatilis*, for example, releases only 5% of photoassimilated carbon when growing exponentially, while over 20% is released from cells in a stationary phase. Senescence or poor growth conditions, where cells are metabolically active but unable to divide, often lead to release of organic carbon. Thus, early in a bloom or in succession (cf. Chapter 15) in nutrient-rich water, the release of DOC per cell is lower than later when nutrients are depleted (Ittekkot et al., 1981). On the other hand, Lancelot (1979) found that phytoplankton producing at high rates release *more* DOC. This relation also applies when DOC release is expressed on a per chlorophyll unit basis, so that in high-production environments, more cells may be "leaky" than those where production is low. Overall, the % of primary production that is exuded is not a function of nutrient enrichment, at least in marine waters (Baines and Pace, 1991).

Light seems to have an impact only at high intensities. The release of DOC is proportional to the amount of photoassimilated carbon in cells over a wide range of light intensities (Fogg et al., 1965). At very high light intensities, where inhibition of photosynthesis occurs, there are unusually high releases of DOC.

The variability in rates of exudation from producers may be in part due to species-specific characteristics. For example, eelgrass exudes 1.5% of its net production (Penhale and Smith, 1977). In salt marshes, where cordgrass dominates, 9% of the net above-ground production is released into the water as dissolved organic matter (Turner, 1978).

Measurements of exudation such as reported above are rough, since the compounds released by exudation tend to be easily metabolized by microbes. Many measurements of exudation are hampered by the presence of bacteria and their rapid uptake of the released DOC.

Excretion by consumers (Fig. 13-1) can account for perhaps 3–10% of the carbon fixed photosynthetically in coastal waters off California

(Eppley et al., 1981). Excretion by zooplankton is perhaps of greater importance in regeneration of nutrients than of carbon (cf. Chapter 14).

Loss of DOC from cells damaged during grazing is probably small (Baines and Pace, 1991), amounting to 15–20% of the consumed carbon (Fig. 13-1). Lasker (1966) used ^{14}C-labeled carbon to show that about 16–21% of the carbon was released into the water during feeding by an euphausiid. Copping and Lorenzen (1980) used similar techniques to calculate that grazing by a copepod released 17–19% of the ingested labeled carbon. DOC release during viral lysis in potentially important but awaits confirmation of viral activity in seawater.

The release of DOC from dead particles is, of course, substantial. Up to 14–60% of the initial weight can be leached as DOC from recently dead organic matter in aerobic situations (Figs. 13-1 and 13-9). During senescence and the initial phases of decay macroalgae may release 30–85%, while vascular plants lose 10–30% of wet weight (Valiela et al., 1985; Buchsbaum et al., 1991).

13.2.4 Losses of Organic Carbon

13.2.4.1 Sedimentation

Algal cells are subject to sinking. Animals release fecal pellets, cast molts, shells, skeletal parts, and carcasses, all of which also fall down the water column (Fig. 13-1). This process of vertical movement of particles is crucial to one of the most debated problems in marine ecology—the source of organic matter for consumers living far below the photic zone.

Measurements of particulate carbon in seawater done in the 1960s showed that in many parts of the ocean concentrations decreased sharply below the surface layers (Fig. 13-2), and that the low concentrations continued unchanged to very great depths (Menzel and Ryther, 1970). Such vertical profiles suggested that labile carbon compounds were degraded by heterotrophs near the surface, and that only residual nondegradable carbon compounds remained at depth. In fact, Menzel

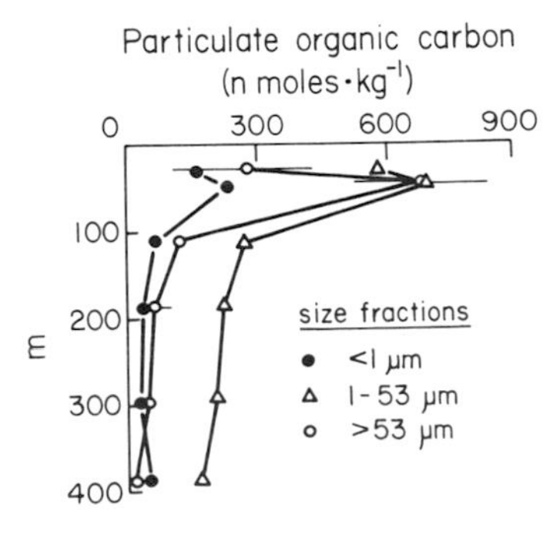

FIGURE 13-2. Organic carbon profile for three particle size fractions. Samples obtained with a large volume *in situ* filtration system in the tropical Atlantic. Adapted from Bishop et al. (1977).

and Goering (1966) incubated particulate matter from various depths and found that bacterial growth took place on POC from surface water but not on POC from 200 to 2,000 m. These results suggested that there was little if any vertical flux of organic carbon and that the carbon present at depth consisted mainly of biologically unavailable compounds. How, then, do the deep water and benthic faunas obtain food?

The water samples studied by Menzel and colleagues were obtained by collecting water with water samplers that closed at a desired depth and collected some tens of liters of water. This method of collection samples the small abundant particles that are more or less suspended in the water. It does not adequately sample the large, rare particles that sink through the water column, yet these large particles are largely responsible, as will be seen, for the carbon fluxes that supply the needs of the consumers found at great depths (Fournier, 1971, and many others).

Various approaches to measure vertical flux of large particles have been used. The various approaches have included comparison of production in the photic zone with consumption in the water column and benthos; others have calculated flux from vertical profiles of organic matter in the water column. Another approach has been to make direct measurements using sediment traps.* There is substantial variation in estimates of POC leaving surface waters using these various methods even in similar areas.

---

*Sediment traps are of remarkably variable design (Blomqvist and Hakanson, 1981). In general they are containers of various sizes and shapes open at the top and are moored at the desired depth to collect particles for certain periods of time. Decay of the collected organic matter is possible, especially at warm temperatures in shallow water, so poisons are often added to the traps to prevent decay. Animals that eat detritus may also be attracted to traps and may consume detritus; in poisoned traps, the attracted consumers may be killed, so that measurements of accumulation may be increased. The design and deployment procedures for sediment traps present several problems. The opening-to-height ratio changes the effectiveness of traps (Hargrave and Burns, 1979; Gardner, 1980, 1980a). The raising of traps to the ship may result in loss and alteration of the captured particles. Some sediment traps of different shapes and sizes may be especially susceptible to strong currents, since the containers themselves alter hydraulic flow lines such that under- and over-trapping can result, and the faster the current the larger the effects. Particles of different size react differently in any given velocity of current, so it is not a simple matter to apply a correction factor to the trap based on flow velocity. One way to reduce the effects of currents is to use drifting traps that move along with a water mass, collecting particles as they go (Staresinic et al., 1978). A preliminary report of recent intercalibrations (Spencer, 1981) show that different traps gave very different measures of flux for one site in the Panama Basin of the Pacific, ranging from a few to about 200 mg m^{-2} day^{-1} of material. The size and shape of traps seemed less important than other unidentified sources of variation.

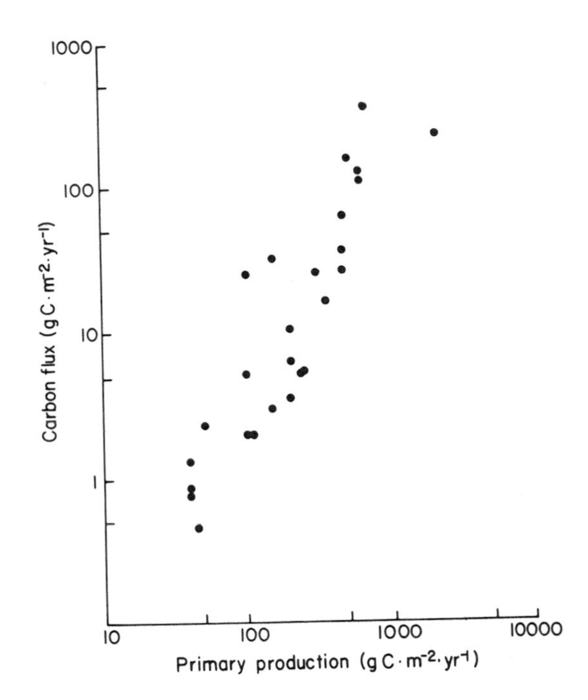

FIGURE 13-3. Rates of carbon flux from the surface layers to deeper waters in relation to the rate of annual production in the photic zone. Data from many authors complied by Hargrave (1984).

The range of values for export of carbon from oceanic photic zones to deeper water obtained using sediment traps is 1–100 mg C m^{-2} day^{-1}. The flux in nutrient-rich coastal waters is greater, ranging from 30 to 600 mg C m^{-2} day^{-1}. The intensity of production on the surface therefore affects the extent of carbon flux, as can be seen in Figure 13-3, where values of primary production are plotted against the resulting carbon fluxes from the surface layer.

Since intensity of production varies seasonally (cf. Chapter 15), so does the flux of carbon to the deep sea. Near Bermuda (Fig. 13-4) the flux varies from 20 to 60 mg of total particles m^{-2} day^{-1} (Deuser

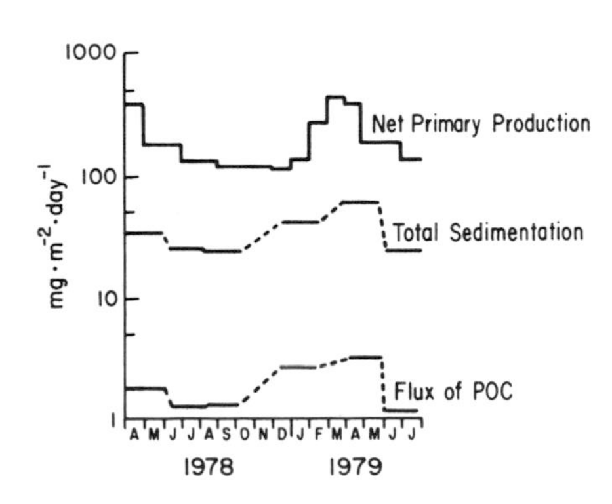

FIGURE 13-4. Seasonal pattern of net primary production [from Menzel and Ryther (1961) for the years 1957–1960], total sedimentation and POC flux into traps located at a depth of 3,200 m and moored at 1,000 m above the sea floor in the Sargasso Sea. Adapted from Deuser and Ross (1980).

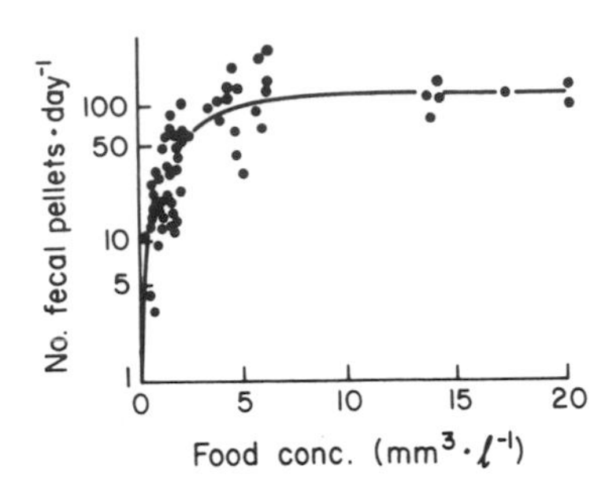

FIGURE 13-5. Egestion of fecal pellets by adult female *Eucalanus pileatus* feeding on various concentrations of phytoplankton. Adapted from Paffenhöfer and Knowles (1979).

et al., 1981). Such pulses provide one of the few variations that may be experienced by the fauna of the otherwise very constant deep sea environment. The more production, the more food available to zooplankton, and therefore the higher the rate of egestion of fecal pellets, up to a threshold beyond which feeding rate (Fig. 13-5) and release of fecal pellets do not increase further. Most of the time algal production rates are low and cell concentrations are such that increased production increases the release of fecal pellets.

Large particles such as fecal pellets may be only a small proportion of the suspended organic carbon in surface water [4% in the tropical Atlantic (Bishop et al., 1977)], but the *flux* of large particles may be large. The relatively greater specific gravity of large particles increases their fall velocity (Fig. 13-6); for example, a fecal pellet containing several coccolithophorid cells may sink 160 m day^{-1}, while a single coccolith sinks at 0.14 m day^{-1} (Honjo, 1976). The export of POC out of the photic zone due to the sinking of fecal pellets can be large; Honjo (1980) estimated that 92% of the coccolithophorid production in a station in the Equatorial Pacific was transported to the sea floor.

Larger particles (mm to cm in diameter) fall even faster through the water column.* Such large organic aggregates may contain thousands of fecal pellets and fall fast (43–95 m day^{-1}) through the water column. The fall of such particles may remove 3–5% of the POC and 4–27% of the particulate organic nitrogen daily from surface water (Shanks and Trent, 1980). Thus sinking of fecal matter and relatively large aggregates can lead to very significant losses of organic matter from the water column (Fig. 10-1).

There are other indirect lines of evidence that suggest that fast sinking by large particles is a major pathway by which organic matter

---

*Some of this so called "marine snow" does not sediment out of the water column and is perhaps the site of substantial primary production. In addition to algal cells such aggregates are also enriched microenvironments for bacterial and protozoan populations, which are five orders of magnitude more abundant in marine snow than in the surrounding water (Caron et al., 1982).

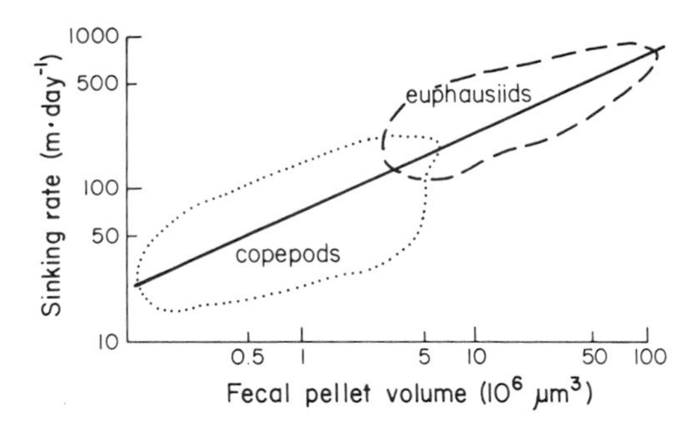

FIGURE 13-6. Relation of sinking rate to volume for fecal pellets of copepods and euphausiids in the Ligurian Sea. Adapted from Small et al. (1979).

reaches the sea floor. This evidence includes the synchrony of seasonal pulses of primary production at the surface of the sea and the flux of organic matter at great depth (Fig. 13-4). The very fast dispersal of recently manufactured industrial chemicals (polychlorinated biphenyls) to great depths (Harvey et al., 1973) also argues for transport of organic matter by adsorption onto fast sinking large particles.

13.2.4.2 Consumption

13.2.4.2.1 *Ingestion of Particulate Carbon.* Most of the organic matter within the photic zone and that leaving the photic zone is consumed (Fig. 13-1). Some calculations show that at least 87% of the organic carbon, 91% of the nitrogen, and 94% of the phosphorus is recycled above 400 m in the Equatorial Atlantic (Bishop et al., 1977). Degens and Mopper (1976) estimate that 93–97% of the flux of organic matter is recycled within the oceanic water column.

Grazers in different layers use particles that fall from the layer above. In deep water the "green" fecal pellets containing phytoplankton are far less abundant than "red" fecal pellets containing mainly mineral sediments (Honjo, 1980). This suggests that the zooplankton that produce pellets at different depths have very different foods available, and that only shallow-living zooplankton have access to phytoplankton. The peritrophic membranes surrounding fecal pellets slowly disintegrate as the pellets sink, probably due to microbial attack. The degradation is faster at warmer temperatures (Honjo and Roman, 1978; Iturriaga, 1979). The pellets are also consumed by the water column fauna.

Consumption of particles falling through the water column by microbes and fauna has not been measured directly but must be large, and is responsible for the high degree of recycling of organic matter and nutrients within the water column. These consumption processes produce an exponential decrease in the flux of carbon as a function of

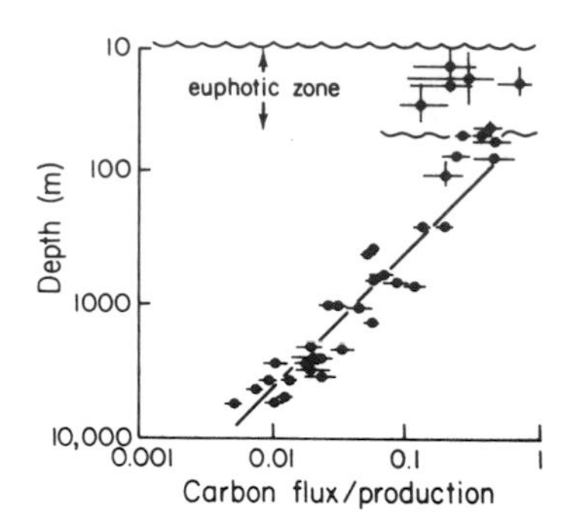

FIGURE 13-7. Carbon flux standardized to production in the photic zone in relation to depth, for a series of observations on a wide variety of areas. The equation that fits the points is $C_{flux(z)} = C_{prod}\ (0.0238\,z + 0.212)$, for depth z, and $z \geqslant 50$, $r^2 = 0.79$. Adapted from Suess (1980).

depth below the photic zone (Fig. 13-7). The decrease in POC down the water column is reflected in the markedly lower rates of oxygen consumption at 5,000 m (about 0.01–0.1 μM year^{-1}) compared to surface water (about 1–100 μM year^{-1}) (Deuser, 1979). In shallow water about 5–50% of the primary production reaches the deepest traps while only 1–3% is likely to reach the sea floor (Cole et al., 1987). Although delivery of particles may partially depend on differences in production in the photic zone, the diminution of delivery is mainly a function of depth (cf. Fig. 13-7, where flux is normalized to production). Depth can be equated to time exposed to consumption by animals and degradation by microbes, so the deeper the water column, the greater the disappearance of particulate organic carbon.

One result of differential sinking rates by particles of different size is that if zooplankton communities are dominated by smaller species their smaller pellets sink relatively more slowly and the sedimentation to the sea floor is smaller. For instance, 67% of primary production can be consumed within the water column over shallow shelf sites in the southeastern United States, but if small species of copepods are dominant, their small* pellets sink slowly, so that only 0.2% of the primary production reaches the sea floor (Hoffmann et al., 1981). This low rate of delivery of organic matter to the sea floor probably accounts for the sparse benthic fauna of the southeastern shelf of the United States (Hoffmann et al., 1981). This is an example of the structure of one community—the plankton—affecting the function of a very different but dependent community—the benthos.

There has been much discussion on the origin of the organic matter that supports benthic production. Some of the carbon fixed in the photic zone reaches the sea floor, regardless of depth, as anticipated

---

* Fecal pellets of small copepods such as *Paracalanus* and *Oithona* nauplii may be $2 \times 10^3\,\mu m^3$, while a large zooplankton such as the euphausiid *Meganyctiphanes norwegica* makes pellets of $10^5\,\mu m^3$ (Paffenhöffer and Knowles, 1979).

long ago (Agassiz, 1888). Most of the flux takes place via sinking of smaller particles. Estimates of the flux of carbon to the sea floor are variable, as can be expected by the nature of the methods used to obtain the measurements, but the estimated downward carbon fluxes may account for 59% of respiratory demand by benthic infauna (Smith, 1987). In addition, fall of larger particles, such as carcasses, may supply about 11% of respirating needs by the benthic (Smith, 1987). There is likely to be a modest additional lateral delivery of carbon from coastal environments (Thiel et al., 1988), but this source has evaded quantification. These are only very rough comparisons that do not always include the mobile epifauna, carbon losses to sediment, and other components of consumption besides respiration. Respiration, however, can be correlated to and is a large fraction of total metabolic demands (cf. Chapter 7). Respiration values are a lower estimate of carbon demand, and estimates of flux and consumption are about the same order of magnitude. Organic particle transport thus likely provides all the carbon needed to support benthic communities.

The variability of flux estimates is primarily due to the limitations of the ways in which we measure flux. That methodological variability masks what in fact may be a fairly tight coupling between production in the photic zone and activity and abundance of consumers in the benthos. We have some evidence of this in the results of the studies of the relation between zooplankton and benthos on the shelf off the southeastern United States, referred to earlier. More generally, the abundance of benthos can be related to the crop of phytoplankton during the spring bloom (Fig. 13-8, left). The respiration of the benthos, furthermore, is greatest at shallow depths (Fig. 13-8, right), because the flux of carbon to the sediment is higher at shallow depths (Fig. 13-7). Since respiration is higher (presumably because biomass of organisms is higher), more of the carbon flux is consumed, so even where the particulate carbon flux is high, there is little of the carbon that arrives at the sea floor that is not consumed by some heterotroph.* Cole et al. (1987) suggest that 85% of the organic matter reaching the deep sea floor is mineralized within a year.

One mechanism that assures very high degree of consumption even in shallow waters is the intense reworking of sediments done by benthic organisms. In Buzzards Bay, a shallow, relatively rich water body off Cape Cod, Massachusetts, Young (1971) found that the bivalve *Yoldia limatula* was capable of reworking the annual deposition of sediments 2.5 times per year. This is more impressive if we realize that

---

*In a few places sedimentation rate can so high that some organic matter is buried before being completely decomposed (Müller and Suess, 1979). The material that is buried, even in such cases, is probably mainly organic compounds high resistant to microbial attack.

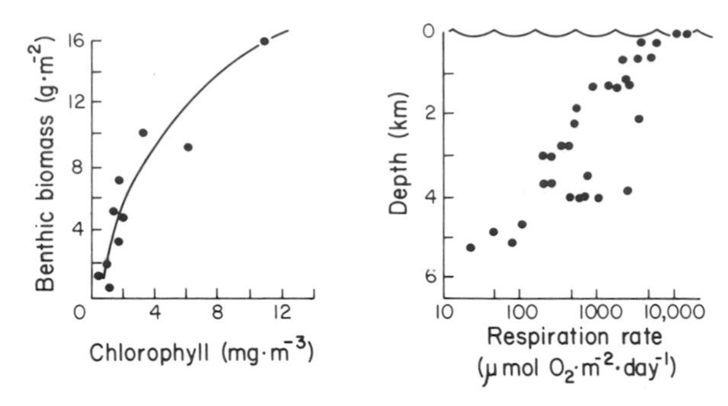

FIGURE 13-8. Left: Biomass of benthic macrofauna in relation to the chlorophyll concentration in the photic zone during March–May, in various areas. After Mann (1978), from data of B. T. Hargrave and D. C. Peel. Right: Respiration of infaunal bottom organisms at various depths. Compiled by Hinga et al. (1979) from various authors.

Yoldia make up only 0.25% of the animals in the sediments. The reworking maintains the surface layer in a flocculent state so that particles can be easily resuspended. Resuspension results in renewed colonization of particles by bacteria and the particles thus enriched by growth of microbes are reingested by animals.* This recycling of organic particles via resuspension amounts to 88–99.5% of the annual sedimentation.

It is evident, therefore, that there are powerful couplings by which processes taking place in the upper layers influence what takes place in the benthos below. In addition, in shallow waters, the benthos can influence what goes on in water columns above. These are a different kind of top-down and bottom-up controls—in this case actual vertical positioning rather than position in a food web. One example of such controls is provided by experiments that evaluated consequences of presence and absence of sediments in both nutrient-enriched and unenriched treatments in vertically well-mixed large containers (Sullivan et al., 1991). The presence of benthic consumers curtailed the response of the water column food web to nutrient enrichment. The results suggest that if water columns are well mixed, there is intense benthopelagic coupling, and consumers in shallow waters may dominate the flow of matter. Stratification of water columns may decouple the bottom-up processes, and only sedimentation may remain as a link between surface and bottom.

---

*This potential improvement of the quality of particulate organic matter due to microbial regrowth or colonization is important to many detritivores. For example, hydrobiid snails do not ingest fecal pellets until enough time has elapsed to allow microbial colonization (Levinton and Lopez, 1977).

13.2.4.2.2 Direct Uptake of Dissolved Carbon. Some of the dissolved organic matter exuded by producers or leached from detritus is readily usable by microbes (Fig. 13-1). Ogura (1975) found that about 10–30% of the DOC from phytoplankton is labile and is degraded rapidly, with a turnover rate of 0.045–0.25 per day. Brylinsky (1977) found that 20–30% of released carbon was assimilated by heterotrophic microbes within 2 hr. Others have found even faster assimilation of exudates (Larsson and Hagström, 1979; Lancelot, 1979). Twenty-nine percent of the DOC released from kelp is converted into bacterial biomass while only 11% of the particulate debris is used by bacteria (Stuart et al., 1981).

Some of the exuded DOC is fairly resistant to microbial attack. For example, the DOC released by brown algae such as *Fucus, Ascophyllum,* and *Laminaria* contains phenolics and is very similar to the refractory yellow humic substances (gelbstoff) found in fairly high concentrations in coastal waters (Craigie and McLachlan, 1964; Sieburth and Jensen, 1969; Kalle, 1966). Much of the DOC in seawater is made up of high-molecular-weight polymers including proteins, lignins, and humic acids (Williams, 1975), the bulk of which are not readily degraded by microbes.

The annual input of labile DOC released by plants and other sources is probably small relative to the total DOC pool (Larsson and Hagström, 1979; Wiebe and Smith, 1977). Since most of this total DOC may be refractory, turnover of this carbon may be slow, up to 3,400 years in some waters (Williams et al., 1979).

There is as yet no consensus on the importance of the producer: DOC: microbe pathway relative to other fates of fixed carbon. In coastal water Iturriaga and Hoppe (1977) calculated that microbial activity was high enough to metabolize the carbon exuded daily by phytoplankton. Turner (1978) measured carbon lost by salt marsh grass and found the amounts were high enough to sustain the respiration of plankton. In coastal waters off Stockholm, Larsson and Hagström (1979) estimated that bacteria assimilate 49 g $C m^{-2}$ of the total phytoplankton production of 110 g $C m^{-2}$, so that about 45% of primary production is channeled through the bacteria.* Andrews and Williams (1971) obtained similar results.

Smith et al. (1977) calculated that in the upwelling area off northwest Africa bacteria could remove carbon exuded by algae fast enough to maintain a low concentration of DOC. The production by hetero-

---

*Bacterial production was about 29 g $C m^2$ per year, so the growth efficiency was about 60%, a high value compared to most heterotrophs (cf. Fig. 7-10). For bacteria using organic matter derived from kelp, Robinson et al. (1982) calculated growth (or conversion) efficiencies of 28 to 91%, while Hoppe (1978) believes that the range is most often between 60 and 70%.

TABLE 13-5. Estimates of the Potential Contribution to Nitrogen Needs of Two Species of Bivalves by Feeding on Non-limited Supplies of Phytoplankton, Aggregates, Morphous Detritus, or by Uptake of Dissolved Organic Matter. Data Obtained by Feeding Experiments Using ^{15}N as a Tracer from the Different Food Sources.[a]

Source of organic matter	(Nitrogen incorporated/ N required) 100	
	Geukensia demissa	*Mytilus edulis*
Phytoplankton	83	49–96
Aggregates	11	7–14
Morphous particles	1	1–3
DOM	1	1

[a] From Alber and Valiela (1994, 1995).

trophic bacteria may be as much as two orders of magnitude lower than primary production, so that there is not much bacterial biomass available to other consumers in the food web of the sea.

Rhee (1972) believes that in nature bacteria are limited by the supply of organic carbon (recall the arguments of Chapter 9), while phytoplankton are nutrient-limited, as discussed in Chapter 2. Bacteria thus appear to have the potential to determine the chemical composition of seawater, at least in regard to DOC. Although the calculations of Larsson and Hagström (1979) and Smith et al. (1977) are only preliminary, they establish the potential importance of exudation and uptake of DOC as a significant pathway of carbon and energy flow. Future models of marine food webs will need to consider DOC dynamics.

Uptake of DOC by macrofauna is of limited quantitative significance. Such uptake, even when DOC supply is not limiting, can only supply a minor portion of nutritional needs of bivalves, for example (Table 13-5).

13.2.4.3 Aggregation of Dissolved Organic Carbon and Use of Aggregates

The action of surface charges and microbial activity prompts the formation of amorphous aggregates and flakes of marine snow from dissolved organic matter (Baylor and Sutcliffe, 1963; Riley, 1963b; Barber, 1966; Johnson and Cooke, 1980).

Some of the DOC released from benthic producers may be aggregated into amorphous organic particles. In the case of macrophytes, about 19% of the released DOC became aggregates (Alber and Valiela, 1994). C/N of these aggregates (4–12) are similar to those of bacteria

(4.5–6.7, Bratvak, 1985), suggesting that aggregates are made up of organic materials derived from producers but transformed by bacterial activity.

Aggregates are not only nutritively rich but also range in size from a few to 500 μm (Alber and Valiela, 1994). This size range makes aggregates available for feeding by a wide variety of consumers. Aggregates can be ingested and assimilated by suspension-feeding bivalves (Alber and Valiela, 1994), and other metazoa (D'Avanzo et al., 1991). Bivalves assimilate aggregates more effectively than morphous detrital particles (Table 13-5), owing to differences in nutritional value. Bivalves use aggregates with a probable conversion efficiency of 45%, so this trophic pathway may be potentially important in waters with a ready supply of DOM.

13.2.4.4 Decomposition

13.2.4.4.1 Phases of Decomposition. The decay of organic matter (POC and DOC) (Fig. 13-1) is a complex of several processes. The processes that affect the rates of loss of nonliving organic matter include leaching of soluble compounds, microbial degradation, and consumption by other heterotrophs (Valiela et al., 1985).

Strictly speaking, leaching is not a mechanism of decomposition but a way in which soluble material is transported. We have already referred to consumption by consumers. We include these processes in this section because in most cases they are very hard to separate from microbial degradation and thus have been studied together. Leaching, degradation, and consumption act at all times, but these processes may not be equally active during different phases of decay.

The phases of decay for organic matter derived from phytoplankton (Otsuki and Hanya, 1972), macroalgae (Tenore, 1977) and vascular plants (Wood et al., 1967; Godshalk and Wetzel, 1978) have been described. First, there is a short-lived phase during which leaching rapidly removes materials that are either soluble or autolyzed after death of cells (Fig. 13-9, top graph). The dissolved material lost from dead organic matter (detritus) is readily available to microbial heterotrophs for uptake and mineralization into CO_2 and inorganic salts (Otsuki and Hanya, 1972; Godshalk and Wetzel, 1978; Newell et al., 1981). The rates of loss during this early leaching phase are high and are most likely not mediated by microbes since losses occur at similar rates in sterile and nonsterile conditions (Cole, 1982). The leaching phase may last from minutes to a few weeks, depending on the detritus involved.

A second stage, the decomposition phase, with lower rates of loss, is longer lasting and occurs after the leaching phase. During this second phase (Fig. 13-9, first graph), microbial activity is the prime

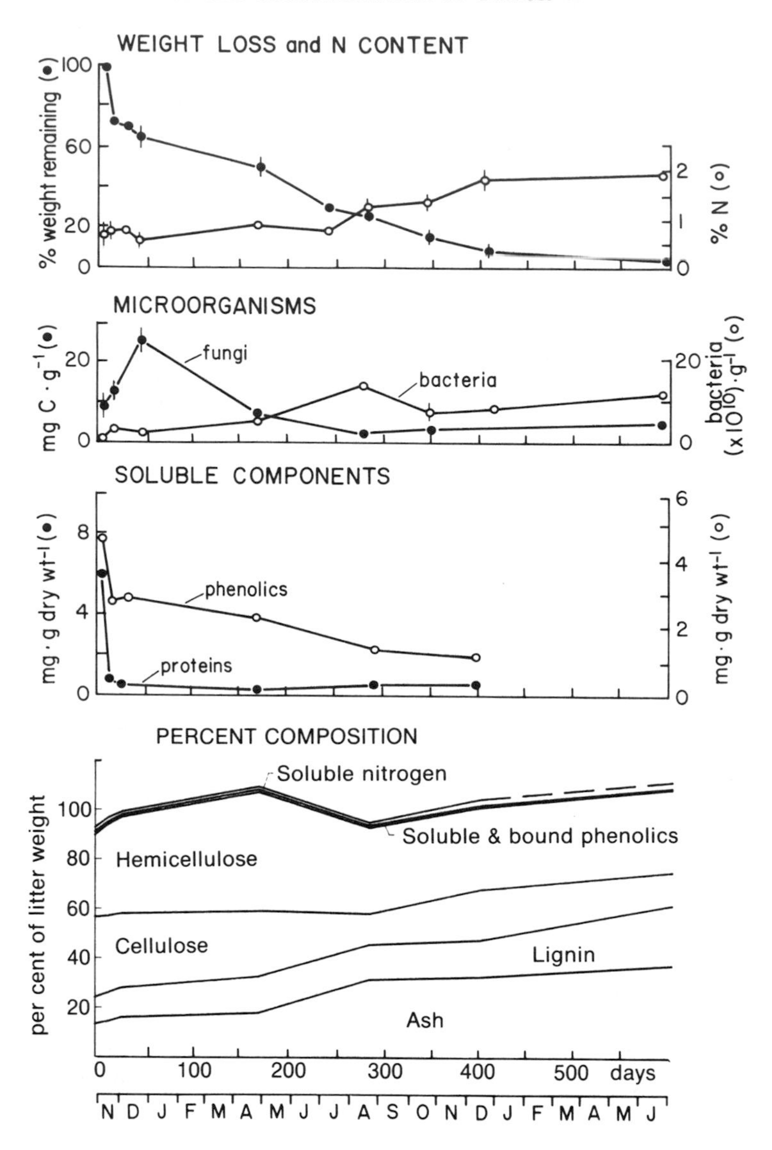

FIGURE 13-9. Decomposition history of biomass of *Spartina alterniflora* in a Massachusetts salt marsh. Data obtained by means of litter bags set out in November and collected at intervals. The two soluble components are small relative to other chemical components, and so the pattern of the third graph does not show in the bottom graph. Cellulose is a straight-chain polymer of glucose, and hemicellulose is a base-soluble cell wall polysaccharide closely associated with cellulose. Lignin is an amorphous phenolic heteropolymer. Data of I. Valiela, John Teal, John Hobbie, Tony Swain, John Wilson, and Robert Buchsbaum, and Valiela et al. (1985).

process that degrades organic matter (Figs. 13-1 and 13-9). Microbial degradation takes place mainly through hydrolysis by enzymes released from microbial cells, with subsequent uptake of the solubilized compounds by the microbes. Part of the partially broken-down organic matter is used by microbes for growth and part is respired as CO_2. Some of the detritus converted to DOC by the extracellular enzymes is lost by leaching away from the immediate vicinity of the microbes (Fig. 13-1).

Some compounds—sugars, some proteins—contained in detritus are susceptible to microbial attack, and are depleted (Stout et al., 1976; Velimirov et al., 1981; Cauwet, 1978) during the decomposer phase. Other compounds—cellulose, waxes, and especially certain phenolic compounds such as lignins—are less easily degraded, and thus during the decomposer phase the chemical composition of POC changes as a result of differential degradation (Fig. 13-9, bottom graph). The percent lignin, for instance, increases during decay. Such resistant compounds are present in vascular plants, but may be absent or less abundant in algae; hence, algal detritus disappears faster than vascular plant detritus (Tenore et al., 1982). The differences in rates among detritus of different taxonomic origin are not due to differences in the processes or phases of decay but rather in the chemical composition of the organic matter. The chemistry of decay of phytoplankton-derived detritus needs more detailed chemical study (Cauwet, 1978; Lee and Cronin, 1982).

Eventually, the remaining detritus primarily contains compounds that are degraded slowly or very slightly. This third or refractory phase may occur within several weeks in the case of phytoplankton detritus (Otsuki and Hanya, 1972) or may last months to years in detritus from vascular plants such as *Spartina* (Fig. 13-9, top graph) or *Zostera* (Godshalk and Wetzel, 1987a). The end result of the decay of both algae and vascular plants is organic matter with high contents of fulvic and humic acids. These are very refractory phenolic polymers and complexes (Nissenbaum and Kaplan, 1972), which eventually form what may be called marine humus or "gelbstoff." The differences in initial composition of algal and vascular plant detritus lead to subtle differences in the chemistry of the derived fulvic and humic acid (Gagosian and Stuermer, 1977; Stuermer and Payne, 1976). It might thus be possible to use these differences as indicators of the source of the gelbstoff. This last refractory phase may occur principally, but not exclusively, in sediments.

13.2.4.4.2 Chemical Changes During Decay and Internal Controls of Decay Rates. It is evident from the above discussion of the phases of decay that the relative abundance of certain types of organic compounds in organic matter can affect rates of decay. The more lignin initially

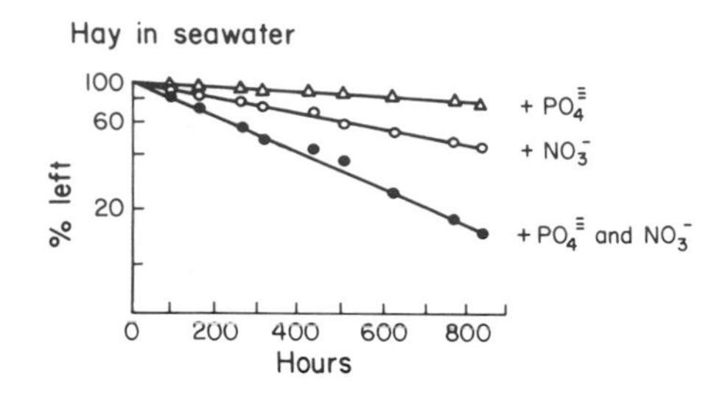

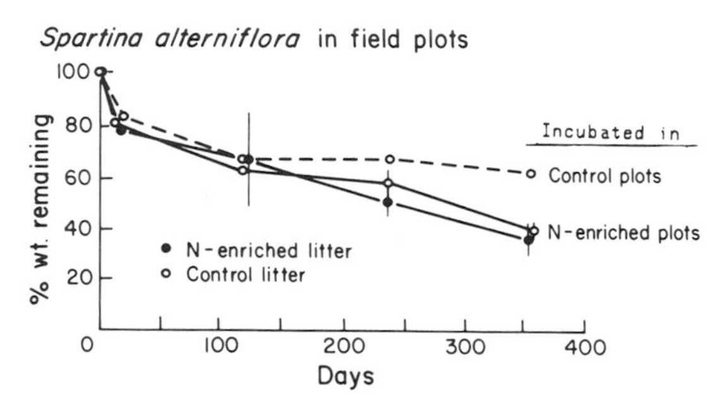

FIGURE 13-10. Effect of nutrient enrichment on loss of weight of dead organic matter. Top: ^{14}C-Labeled hay incubated in seawater in the lab with additions of nutrients (adapted from Fenchel 1973); bottom: *Spartina alterniflora* litter incubated in mesh bags in nitrogen-enriched and control salt marsh plots. Litter from plants grown in enriched plots has higher nutrient content (cf. Fig. 8-7); enriched litter incubated in control plots (not shown) decayed at the same rates as when incubated in enriched plots. Data of I. Valiela and John Teal.

present, for example, the slower the decay. The mineral elements in detritus also affect decay rates. During the course of decay of organic matter, bacteria and fungi take up mineral elements, principally nitrogen, but also phosphorus. The source of nutrients can be internal (contained in the tissues) or external (supplied by water). Increased internal nitrogen in detritus (Fig. 13-10, bottom) or nutrients supplied externally both may increase microbial activity and decay of salt marsh detritus (Fig. 13-10, top and bottom).

As we have seen above, particulate organic matter falling through the oceanic water column is to a large extent degraded by the time it reaches the deep sea floor, with significant changes in organic and mineral composition, including nitrogen content. For example, in some parts of the open ocean the POC in surface water may have C/N values of 14, while below 1,000 m the C/N may reach 29 (Knauer et al., 1979). Much of the organic matter in the deep sea may be refractory. Labile dissolved organic matter (such as provided by exudation) may constitute perhaps 15–25% of total organic matter. This "young" organic matter degrades rapidly, while the older fraction (perhaps 4,000 to 6,000 years old) are complex substances refractory to degradation (Degens and Mopper, 1976).

The rates of decay of the POC that does reach great depths may be slow for other reasons rather than just the chemical composition of detritus. This was brought into focus serendipitously by the accidental sinking of the research submarine *Alvin* in 1968. *Alvin* was recovered 10 months later with the crew's lunches largely intact (Jannasch et al., 1971). This evidence of obviously slow decay in the deep sea prompted a substantial effort (Jannasch and Wirsen, 1983; Morita, 1980) to identify the factors that inhibited decay rates in the deep sea. Enrichment studies (Jannasch and Wirsen, 1973) show no large increment in the rate of *in situ* decay of various organic substrates, even at high concentrations. Other factors have to be involved. The uniformly low temperatures (2–4°C) in the deep sea slow growth of even cold-adapted bacteria, and the high hydrostatic pressures further reduce bacterial growth. Temperature and pressure cannot be the complete answer, since there is considerable bacterial activity found in the guts of deep-sea cold-blooded animals, where nutrients are significantly concentrated but where temperature and pressure are the same as in the water. In addition, it is not clear why pressure would so hamper procaryote activity while apparently not really decreasing the activity of eucaryotic organisms. Perhaps free-living bacteria in the deep sea are adapted in a fundamental way to living at very low substrate concentrations. Such bacteria have been unfortunately named "oligotrophs" by microbiologists—needlessly confusing a term otherwise used by ecologists. Such bacteria, perhaps better called "oligocarbophilic," may be characterized by rather slow metabolic and growth rates (Poindexter, 1981) and may be very poorly adapted to using substrates in high concentrations, hence the negative results of enrichment experiments. Perhaps the free-living deep sea microflora is so dominated by oligocarbophilic species that the low rates of decay of organic matter associated with these microbes prevail in the deep sea. The finding of such slow decay of organic matter even when substrate may be added has applied importance, since organic wastes that find their way to the deep ocean are therefore likely to be degraded very slowly.

While the percentage carbon content in most kinds of dead organic matter generally decreases with age during decay as a result of respiratory losses of CO_2, in detritus of vascular plants the percentage nitrogen may increase. During the course of decay of vascular plant detritus bacteria and fungi immobilize nitrogen by incorporation into new cells, and this is in part responsible for the increased nitrogen content of detritus as it ages (Fig. 13-9, top). Microbial nitrogen is not enough to account for all the nitrogen in aged detritus (Cammen et al., 1978; Odum et al., 1979). Fungal biomass accounts for 12–22% of the nitrogen in salt marsh detritus, while bacteria amount to a tenth of the fungal contribution (Marinucci et al., 1983). Nitrogen may also be immobilized by the accumulation of microbial extracellular protein and nitrogen-containing exudates (Glenn, 1976; Hobbie and Lee, 1980). The third way by which nitrogen may be immobilized is through binding of proteinaceous substances by phenolic compounds in the detritus (Leatham et al., 1980; Tenore and Rice, 1980; Rice, 1982).

The actual amounts of immobilized nitrogen accounted for by these mechanisms has not been measured. As a result of immobilization of nitrogen, the carbon to nitrogen ratio of vascular plant detritus is progressively reduced. At C to N ratios of about 30:1 the rates of immobilization approximate the rates of mineralization; below 15:1 or 10:1, nitrogen in litter can be mineralized since such ratios provide N in excess of that needed by the microbes for cell building (Alexander, 1977).

In phytoplankton detritus the percent nitrogen does not increase during decay (Otsuki and Hanya, 1972a). In dead phytoplankton fungi are scarce, while in vascular plant detritus fungi make up one of the principal pools of nitrogen that increases over time. Further, the phenolic compounds that bind proteins are less abundant in phytoplankton than in vascular plants. The result is that immobilization of nitrogen may not be very important in phytoplankton detritus.

13.2.4.4.3 The Roles of Different Groups of Decomposers. Fungi and bacteria are the principal competitors for organic substrates. The fungi contain relatively less nitrogen relative to carbon than bacteria (Table 13-6) and do better on detritus that has a higher C to N ratio. Fungi are more efficient than bacteria at transforming carbohydrates into living tissue and may assimilate 30–40% of the carbon of organic matter, while bacteria only assimilate 5–10% (Alexander, 1977). Bacteria, in addition, may cause faster loss of carbon, since they require only 1–2 units of nitrogen for the decomposition of 100 units of carbon, while fungi require 3–4 units of nitrogen (Alexander, 1977). Fungi therefore tend to immobilize nitrogen more strongly than bacteria, so that mineralization may therefore be slower where fungi predominate. Organic matter such as derived from salt marsh grass, mangroves, or eelgrass has high C/N values (Table 13-6) and fungi are prominent. With detritus derived from phytoplankton, C/N values are lower (Table 13-6) and bacteria may be more important.

Further evidence of the role of fungi in immobilizing nitrogen is given by the low nitrogen content in anaerobic sediments such as found in salt marshes. Fungi are aerobic and are rare in marsh sediments. Organic matter in sediments has low nitrogen content, suggesting that mineralization of nitrogen has taken place.

Fungi are more important than bacteria in the degradation of cell wall material, especially in regard to complex molecules such as lignin,

TABLE 13-6. Approximate Carbon to Nitrogen Ratios in Some Terrestrial and Marine Producers.[a]

	C/N
Terrestrial	
Leaves	100
Wood	1,000
Marine vascular plants	
Zostera marina	17–70
Spartina alterniflora	24–45
Spartina patens	37–41
Marine macroalgae	
Browns (*Fucus, Laminaria*)	30 (16–68)
Greens	10–60
Reds	20
Microalgae and microbes	
Diatoms	6.5
Greens	6
Blue-greens	6.3
Peridineans	11
Bacteria	5.7
Fungi	10

[a] Data compiled in Fenchel and Jorgensen (1977), Alexander (1977), Fenchel and Blackburn (1979), and data of I. Valiela and J.M. Teal.

but the breakdown of such compounds generally requires aerobic conditions. Since fungi are rare or inactive under anaerobic conditions, lignins are not degraded in such situations.

Detritivores may directly assimilate some of the available organic matter, since cellulases and other needed enzymes have been detected in their guts (Elyakova, 1972; Hylleberg-Kristensen, 1972; Kofoed, 1975); most of these cases may involve symbiotic floras (Fong and Mann, 1980). Direct assimilation of ingested detritus by animals usually accounts for a small proportion of decomposition (Yingst, 1976; Lopez et al., 1977; Levinton, 1979) (see also Section 7.2.2.1) of organic matter that is "old." Newly added organic matter is consumed much more readily by detritivores (Kristensen et al., 1992). There are, however, important indirect effects of detritivore feeding, including the stimulation of microbial activity and changes in the microbial flora. Feeding by large consumers may stimulate microbial activity through the fragmentation (comminution) of detrital food particles. Comminution alters the surface to volume ratio of particles; where amphipods were allowed to feed on detritus there was a significant reduction of size of detrital particles (Fig. 13-11, left), with a 2–3-fold increase in surface area. The increased surface area led to more microbes and larger

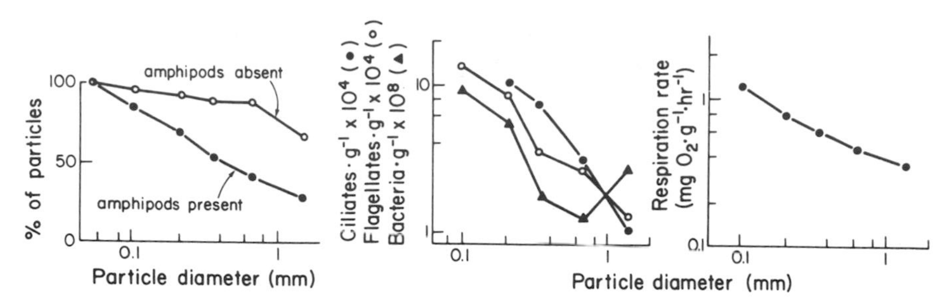

FIGURE 13-11. Left: Comminution of particulate organic carbon derived from the seagrass *Thalassia testudinum* by feeding of a detritivore, the amphipod *Parhyalella whelpleyi*. Middle and right: Effect of size of particulate organic carbon on abundance of organisms (middle) and on respiration by the biota (right). Adapted from Fenchel (1970).

standing crops of organisms that consume microbes, such as ciliates and flagellates (Fig. 13-11, middle). The larger surface area also enhanced activity of the biota, as measured by respiration rates (Fig. 13-11, right). The presence of amphipods increased the oxygen consumption of the entire system (detritus, amphipods, and microorganisms) two-fold in less than 4 days, compared to the respiration of detritus without the amphipods. The increase was about 110% of the respiration due to the amphipods, so the increase must have been due to microbial respiration.

The major consumers of microbes are small organisms—ciliates, microflagellates, nematodes, and foraminifera. The presence of these small organisms stimulates degradation of organic matter. Feeding by flagellates and ciliates can increase the percentage loss of weight of detritus of eelgrass by 46% over the loss that takes place when microbes alone are present, even though bacterial populations are reduced by one or two orders of magnitude (Harrison and Mann, 1975). A mixed protozoan fauna decreased the bacterial fauna of eelgrass detritus by 70%, while increasing the rate at which plant matter decayed (Fenchel and Harrison, 1976). In these cases the bacterial population is most likely turning over at a rapid rate, even though the standing population may be low.

There are important interactive effects of activity by large and small detritivores on rates of decay. Meiofauna may enhance use of detritus by large detritivores. Where ciliates are present, the polychaete *Nereis succinea* incorporates significantly more carbon from detritus of a red algae than from detritus with no ciliates (Briggs et al., 1979). When meiofauna are present 30–400% more ^{14}C from detritus of *Spartina* and *Zostera* is incorporated by polychaetes (Tenore et al., 1977). Such increases in uptake of detrital carbon are due both to enhanced microbial activity and to the consumption of small animals by the large detritivores.

Feeding by large detritivores may, in turn, greatly affect abundance and activity of small consumers of microbes. In field situations, the abundance of meiofauna can be mediated by the cropping pressure due to macrofaunal predators (Bell, 1980). The predation pressure by large species on small ones (another cascade effect) may be so large that the microbes increase in abundance (Morrison and White, 1980, Lopez et al., 1977) when macrofauna crop meiofauna. The increased abundance of microbes is probably due to the reduction of their principal predators, the micro- and meiofauna, to comminution, to increased water exchange in burrows, which renews supplies of O_2, CO_2, and nutrients (Kristensen and Blackburn, 1987), and to the mechanical stirring of detritus by the large animals. Animal activity makes nutrients released by mineralization more available to microbes. For example, activity of the large polychaete *Nereis* increases nitrification and denitrification in sediments (Kristensen and Blackburn, 1987). There are therefore important second- and third-order effects that link large detritivores, small microbe-feeding species, microbes, and rates of decay of organic matter.

There is also the possibility that feeding by detritivores affects the relative abundance of fungi and bacteria. Evidence from terrestrial soils shows that animal feeding on detritus reduces fungal biomass (Hanlon and Anderson, 1980; Hanlon, 1981). Apparently fungi may not grow fast enough to compensate for losses due to cropping by detritivores. Bacteria may grow faster, and the cropping thus results in a shift from a fungal to a bacterial microbial community. Morrison and White (1980) found such a shift in detritus cropped by coastal amphipods in the laboratory. So far such a shift has not been demonstrated in the field but detritus-feeder activity may be responsible for the shift from fungi to bacteria over time evident in Fig. 13-9 (second graph). The importance of this effect of detritivore feeding lies in the changes in chemistry and decay rates that may accompany a shift from fungi to bacteria. There may be less immobilization of nitrogen and, if lignin-containing detritus is involved, slower decay rates.

Throughout this section we have been discussing examples from coastal and marsh systems. Comparable data are not available for open-water environments, but we can hypothesize that the principles apply. What we have said need only be modified to account for the much more dispersed nature of detritus in the plankton, and for the differences in chemical composition between phytoplankton and coastal and marsh producers. The implications of these modifications need some concerted study.

13.2.4.5 Burial in Sediments

In general, in most oxic marine environments (Table 13-7), relatively little carbon accumulates, as suggested in Section 8.2. Rates of consumption and decomposition of organic matter by heterotrophs are such that, in general, only a small proportion of the organic matter produced by autotrophs accumulates in the sediments over most of the sea (Table 13-7). In the Buzzards Bay sediments mentioned above, permanent accretion to sediments amounts to about 0.5–2% of annual sedimentation. In MERL experimental ecosystems, about 4% of carbon inputs accumulated in sediments (Sampou and Oviatt, 1991). The organic matter that does accumulate in aerobic environments consists mostly of refractory compounds unsuitable for assimilation by consumers (Degens and Mopper, 1976).

The % of organic matter delivered to the sediment surface that becomes buried depends on rate of sedimentation. In situations of very high production and shallow depths, such as in the Baltic Sea and the Peruvian upwelling (Table 13-7) considerable proportions of annual primary production may accumulate in sediments. At fast accretion rates perhaps 10% of the organic matter is buried (Canfield, 1989).

Burial also increases with reduced conditions (cf. Fig. 13-14). Salt marsh sediments are anoxic, and peat forms out of partially decayed organic matter (Table 13-7).

The proposition of Hairston et al. (1960) (Chapter 8) that organic matter generally does not accumulate in terrestrial environments seems to be true for most marine systems. Since there is little excess organic matter available and what is available may be refractory, we conclude, as Hairston et al. (1960) did for terrestrial detritivores, that marine consumers that feed on dead organic particles are limited by the quantity and quality of their food supply.

TABLE 13-7. Accumulation of Organic Matter in Marine Sediments.[a]

	Percentage of primary production accreting in sediments	Sedimentation rate (mm yr^1)
Oceanic Sediments		
Abyssal plain, average of several sites	0.03–0.04	0.0001
Central Pacific	<0.01	0.002–0.006
Off N.W. Africa, Oregon, and Argentina	0.1–2	0.02–0.7
Peru upwelling, Baltic	11–18	1.4
Salt Marsh Sediments		
Cape Cod, USA	5.3	1
Long Island, USA	37	2–6.3

[a] Pelagic data compiled by Müller and Suess (1979) and Wishner (1980). Salt marsh data from various references in Valiela (1983).

13.3 The Carbon Cycle in Anoxic Environments

13.3.1 Occurrence of Anoxic Conditions

Reducing conditions occur in marine environments where the rate of consumption of oxygen exceeds the supply. The dissolved oxygen used in the oxidation of live and dead particulate matter is derived from photosynthesis and the atmosphere. The supply of oxygen from the water column thus depends on events within the surface layer of the ocean but there may be lateral movements of oxygen or oxygen-consuming organic matter. Such local phenomena ultimately may create a larger or smaller concentration of oxygen or carbon, but generally most of the oxidation that takes place in the sea goes on near the surface. The combination of supply and consumption rate at various depths results in characteristic oxygen profiles with a minimum concentration at some intermediate depth. Below the oxygen minimum, consumption rates are lower, and oxygen concentrations rise again up to perhaps 6 ml O_2 per liter (Deuser, 1975).

Where there is some impediment to water flow, such as a physical restriction or a stratification due to changes in density of different layers of water, water below the mixed layer may be devoid of oxygen. Such situations occur in the Black and Baltic Seas, the Cariaco Trench, and in some coastal lagoons and fjords.

There may also be low oxygen waters below photic zones with high rates of primary production. This occurs in certain large oceanic areas north and south of the Equator in the eastern Pacific (Deuser, 1975), regions where divergences create upwellings of nutrients, and as we will see in Chapter 16, in anthropogenically enriched waters.

The other major kind of situation where oxygen consumption exceeds supply is within sediments below productive water columns. In most coastal zones much organic matter reaches the sea floor, and since flux of interstitial water is slow, renewal of O_2 is very restricted. Deep-sea sediments receive much smaller amounts of organic matter, consumption of oxygen does not exceed supply rates, and interstitial water remain aerobic.

13.3.2 Sources of Organic Carbon in Anoxic Environments

The fixed carbon of anoxic environments is primarily of allochthonous origin. The major source is of course the rain of organic matter from above, but there are two other mechanisms that can produce small amounts of autochthonous organic carbon. As mentioned in Section 1.2, carbon can be fixed by anoxygenic photosynthesis carried out by bacteria (cf. Section 1.2). Another alternative source is the chemosynthetic reaction in which hydrogen gas is oxidized by bacteria that use

TABLE 13-8. Some Representative Reactions Illustrating Pathways of Microbial Metabolism and Their Energy Yields.[a]

		Energy yield (kcal)
Aerobic respiration	$C_6H_{12}O_6$ (glucose) + 6 O_2 = 6 CO_2 + 6 H_2O	686
Fermentation	$C_6H_{12}O_6$ = 2 $CH_3CHOCOOH$ (lactic acid)	58
	$C_6H_{12}O_6$ = 2 CH_2CH_2OH (ethanol) + 2 CO_2	57
Nitrate reduction and denitrification	$C_6H_{12}O_6 + 24/5\ NO_3^- + 24/5\ H^+ = 6\ CO_2 + 12/5\ N_2 + 42/5 H_2O$	649
Sulfate reduction	$CH_3CHOHCOO^-$ (lactate) + $1/2\ SO_4^= + 3/2\ H^+$ = CH_3COO^- (acetate) + $CO_2 + H_2O + 1/2\ HS^-$	8.9
	$CH_3COO^- + SO_4^= = 2\ CO_2 + 2\ H_2O + HS^-$	9.7
Methanogenesis	$H_2 + 1/4\ CO_2 = 1/4\ CH_4 + 1/2\ H_2O$	8.3
	$CH_3COO^- + 4\ H_2 = 2\ CH_4 + 2\ H_2O$	39
	$CH_3COO^- = CH_4 + CO_2$[b]	6.6
Methane oxidation	$CH_4 + SO_4^= + 2\ H^+ = CO_2 + 2\ H_2O + HS^-$	3.1
	$CH_4 + 2\ O_2 = CO_2 + 2\ H_2O$	193.5
Sulfide oxidation	$HS^- + 2\ O_2 = SO_4^= + H^+$	190.4
	$HS^- + 8/5\ NO_3^- + 3/5\ H^+ = SO_4^= + 4/5\ N_2 + 4/5\ H_2O$	177.9

[a] Energy yields vary depending on the conditions, so different measurements may be found in different references. The values reported here are representative. Adapted from Stumm and Morgan (1981), Fenchel and Blackburn (1979), and Martens and Berner (1979).
[b] This reaction is sometimes considered fermentation.

CO_2 as the electron acceptor. The result of this reaction is the release of methane (Table 13-8).

13.3.3 Losses of Organic Carbon in Anoxic Environments

13.3.3.1 Microbial Decomposition Processes

Carbon compounds can be degraded by dissimilative reactions in which the energy contained in reduced compounds is used by bacteria. In these reactions the reactants are not assimilated. Some chemosynthesis may be carried out based on inorganic reduced compounds (chemolithotrophy), including reactions such as the oxidation of H_2 with SO_4^{2-} or NO_3^-, or the oxidation of HS^- with NO_3^- (Table 13-8).

More relevant to our topic here are reactions in which reduced organic compounds are the source of energy (chemoheterotrophy). Examples of these reactions are fermentation, nitrate reduction, denitrification, and sulfate reduction (Table 13-8). In the latter three reactions some oxidant (NO_3^-, SO_4^{2-}) must be transported into a reduced environment from an oxidized environment. This usually means that the microbial reactions take place near the interface between oxidized and reduced environments.

Fermentation and anaerobic reduction reactions are the means used by bacteria to reoxidize the compounds—such as NADPH—that transfer reducing power in living cells when oxygen is not available. Both these types of reactions can be carried out with a variety of specific substrates (Fenchel and Blackburn, 1979; and Hamilton, 1979). Below we examine some general types of reactions involved in the anaerobic degradation of organic matter.

13.3.3.1.1 Fermentation. Fermentation evolved before the appearance of oxygen in the atmosphere. It has been retained in microorganisms living in anoxic environments and also occurs within many other higher organisms.

The specifics of fermentation in marine systems are poorly known. In this process, energy is transferred by the oxidation of part of an organic compound and the reduction of another part; there is no net oxidation of the substrate. The fermentative reactions are considerably less efficient at harnessing energy than aerobic respiration (Table 13-8).

The most common fermentation reactions involve sugars or other high-molecular-weight carbohydrates, with a variety of end products, including acetate, propionate, CO_2, and alcohols, among others (Table 13-8). Fermentation is important in anoxic environments, because it is the major mechanism by which organic matter synthesized by primary producers is broken down into organic compounds of lower molecular weight readily usable by other groups of microbes (Fig. 13-12).

13.3.3.1.2 Nitrate Reduction and Denitrification. When oxygen is depleted, bacteria use compounds other than oxygen as terminal electron acceptors. The energy yielded from the oxidation of organic matter by the reduction of nitrate (NO_3^-) or nitrite (NO_2^-) is comparable to the yield by aerobic respiration (Tables 13-8, 13-9) and NO_3^- and NO_2^- are thus readily reduced in anoxic environments. The further reduction through nitric (NO) and nitrous (N_2O) oxides to nitrogen gas (N_2) completes the process of denitrification (cf. Fig. 14-6). Denitrifiers generally obtain energy from organic compounds of low molecular weight (Fig. 13-12), but some also use H_2 and reduced sulfur.

The reactions that oxidize organic matter at the expense of nitrate can result in either complete conversion of the organic nitrogen to N_2 or to release of both N_2 and NH_3 (Table 13-9). In coastal sediments of Japan, for example, denitrification to N_2 accounted for 27–57% of the nitrate consumption, so both reactions may be important.

The relative magnitude of the oxidation of carbon compounds by nitrate and oxygen in marine environments is not well known (Fenchel

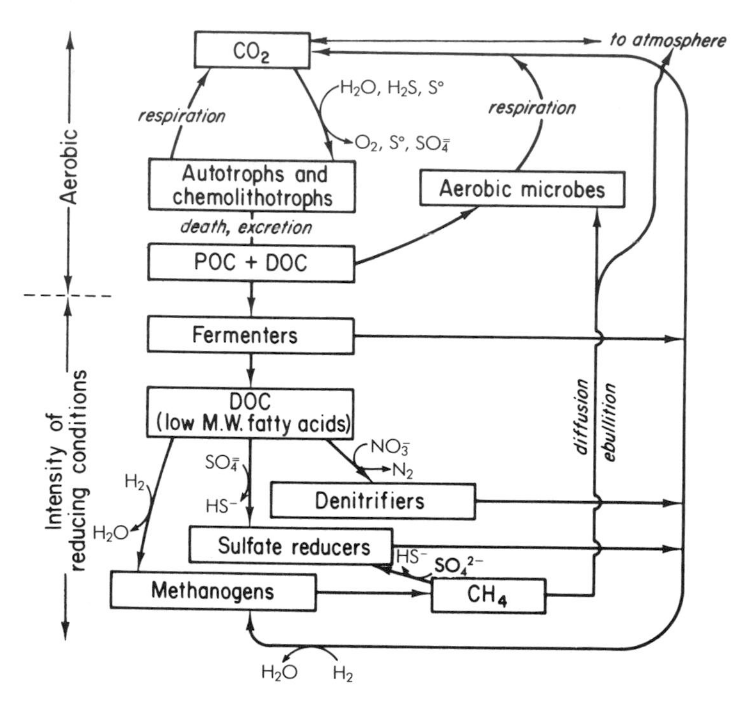

FIGURE 13-12. Carbon transformations in the transition from aerobic to anaerobic situations. The top part of the diagram is a simplified version of Fig. 13.1. The gradient from aerobic to anaerobic can be thought of as representing a sediment profile, with increased reduction and different microbial processes deeper in the sediment. Boxes represent pools or operators that carry out processes; arrows are processes that can be biochemical transformations or physical transport. Elements other than carbon are shown, where relevant, to indicate the couplings to other nutrient cycles. Some arrows indicate oxidizing and some reducing pathways.

and Blackburn, 1979). In part the difficulty lies in that it is hard to measure small changes in production of N_2 into water that already contains a very large pool of dissolved N_2. Further, it is not clear how much nitrogen is lost as N_2O, or the extent to which nitrite is denitrified. Depending on these factors, 2–5 carbon atoms could be oxidized for every 4 nitrogen atoms reduced. In a few anoxic marine environments where nitrate is available, nitrate reduction to nitrite can consume up to 50% of the photosynthetically fixed carbon (Richards and Broenkow, 1971). Such cases are the exception rather than the rule, and in most instances the oxidation of organic carbon by nitrate is small. In a coastal sediment off Denmark, Jørgensen (1980) calculated that 2% of the carbon was oxidized by denitrification. Denitrification in a New England salt marsh consumed less than 1% of readily available carbon (Table 13-10). We can conclude, then, that heterotrophic consumption of DOC via nitrate reduction and denitrification is usually unimportant as a mechanism of oxidation of organic matter (Table 13-10), even though the energy yields are quite favorable (Table 13-8). It simply turns out that in anaerobic environments there is little available nitrate, and so nitrate reduction and denitrification are therefore limited by the amount of oxidized nitrogen.

13.3.3.1.3 Manganese and Iron Reduction. Reactions involving the oxidation of organic matter by Mn^{4+} and Fe^{3+} (Table 13-9) doubtless go on, although it is not certain whether microorganisms directly reduce these metals. The reduction to Mn^{2+} and Fe^{2+} may be indirectly mediated by microbes, since the anaerobic conditions created by bacterial activity may reduce and solubilize Mn^{4+} and Fe^{3+} (Fenchel and Blackburn, 1979).

The reduction of manganese can co-occur with nitrate reduction, since the energy yield of these reactions is similar (Table 13-9), and both these processes may take place in sediments that are not completely anoxic (Froelich et al., 1979; Emerson et al., 1980, among others). After the labile nitrate and managanese are consumed, iron is reduced. The energy yield of reduction of ferric iron is lower than that of nitrate and Mn^{4+} (Table 13-9).

Dissimilatory Fe and Mn reduction have not been well quantified, but have been thought to be of minor importance in the loss of organic matter from marine sediments. The recent finding of bacteria that gain energy from these processes (Lovley and Phillips, 1988), measurements of rates (Aller, 1990), and evidence of fast reuse of Fe and Mn (Canfield et al., 1993) show that oxidation of organic matter by reduction of these metals may be more important than was thought, at least in some sites.

13.3.3.1.4 Sulfate Reduction. Bacteria capable of using SO_4^{2-} as terminal election acceptors were thought to be strictly anaerobic (Fig. 13-12). Recent work has shown that these bacteria have broad metabolic capacities, and can grow and reduce sulfate even in oxic environments (Jørgensen and Bak, 1991; Canfield and Des Marais, 1991). Nonetheless, most sulfate reduction takes place mainly in anoxic sediments, in anoxic microzones within aerobic sediments, or in anoxic waters. Even in aerated water columns there may be anoxic microzones within particles where diffusion of O_2 may be slow enough to produce anoxic conditions within the particles. Jørgensen (1977a) calculated, for example, that a detrital particle of 2 mm diameter was large enough to be anoxic at the center, even if it was surrounded by oxygenated water.

The energy yield of sulfate reduction is much less than that of nitrate reduction (Tables 13-8, 13-9), so that sulfate is consumed by microorganisms in anoxic sediments only after nitrate and perhaps after Mn^{4+} and Fe^{3+} are depleted. Since there is little nitrate in coastal

TABLE 13-9. Theoretical Stoichiometry of Major Reactions Involved in the Oxidation of Organic Matter (OM. defined as $C_{106}H_{236}O_{110}H_{16}P$) Through Reduction of Oxygen, Nitrate, Manganese, Iron, and Sulfate.[a]

Reactants		Products	Energy yield (kcal · mole OM^{-1})
$138\ O_2$ + 1 unit OM	$\longrightarrow$	$106\ CO_2 + H_3PO_4 + 122\ H_2O + 16\ HNO_3$	13,334
$94.4\ HNO_4^-$ + 1 unit OM	$\longrightarrow$	$106\ CO_2 + H_3PO_4 + 177.2\ H_2O + 55.2\ N_2$	12,665
$84.8\ HNO_4^-$ + 1 unit OM	$\longrightarrow$	$106\ CO_2 + H_3PO_4 + 148.4\ H_2O + 42.4\ N_2 + 16\ NH_3$	11,495
$236\ MnO$[b] + 1 unit OM + $472\ H^+$	$\longrightarrow$	$106\ CO_2 + H_3PO_4 + 366\ H_2O + 8\ N_2 + 236\ Mn^{2+}$	12,206–12,916[c]
$212\ Fe_2O$[b] + 1 unit OM + $848\ H^+$	$\longrightarrow$	$106\ CO_2 + H_3PO_4 + 520\ H_2 + 16\ NH_3 + 424\ Fe^{2+}$	5,559–5,894[c]
$53\ SO_4^{2-}$ + 1 unit OM	$\longrightarrow$	$106\ CO_2 + H_3PO_4 + 106\ H_2O + 16\ NH_3 + 53\ S^{2-}$	1,588

[a] Adapted from Froelich et al. (1979). The exact stoichiometry can be expressed in different ways (cf. Stumm and Morgan 1981; Emerson et al., 1980, for example), and the energy yield depends on the way the equation is written.
[b] The Mn^{4+} or Fe^{3+} compounds may differ; for example, 424 FeOOH may react with the same compounds, and 742 H_2O would be produced.
[c] The actual energy yield varies depending on the specific mineral phase of the reactants.

TABLE 13-10. Amounts of CO_2 Released by the Various Pathways of Mineralization of Organic Matter in Sediments of a New England Salt Marsh.[a]

Pathway	Rate of CO_2 release (mole C $m^{-2} yr^{-1}$)	% of total mineralization
Aerobic respiration[b]	30.1	44.9
Nitrate reduction[c]	0.4	0.6
Sulfate reduction[d]	36	53.7
Methanogenesis[e]	0.5	0.7
Total	67	

[a] Data from Howes et al. 1984, 1985.
[b] Even though the sediments are anoxic, the plants present have internal air ducts; there is also some movement of aerated water into and out of the sediment. These mechanisms support some aerobic respiration. Value obtaied by subtracting other terms from total.
[c] Based on Kaplan et al. (1979).
[d] The sulfate reduction is based on fermentation products, and the efficiency of use of these products by sulfate reduction is very hig. Fermentation does not therefore appear as a separate entry in this table.
[e] Based on $CH_3COO^- = CH_4 + CO_2$ (Table 13-8).

marine sediments, sulfate reduction becomes the dominant process by which organic carbon is oxidized. In some coastal sediments about half of the oxidation of carbon is due to sulfate reduction (Jørgensen, 1977). In salt marsh sediments about half the carbon dioxide that results from degradation of organic matter may be consumed by sulfate reduction and fermentation (Table 13-10). Since sulfate is plentiful in seawater but is rare in freshwater, sulfate reduction is far more prominent in the sea than in freshwater ecosystems.

Organic substrates of relatively low molecular weight usually serve as electron donors for sulfate reduction, and the sulfate reducers depend on fermenters to provide them. Lactate, acetate, hydrogen, and methane are oxidized, with H_2S, CO_2, and H_2O the end products (Table 13-8). The activities of sulfate reducers, fermenters, and methanogens are thus closely coupled. The H_2S and organic sulfides, primarily dimethyl sulfide, are the source of the "rotten-egg" odor of anoxic marine sediments.

13.3.3.1.5 Methanogenesis. Methanogens are strict anaerobes whose metabolic activity generates methane (Fig. 13-12). We have seen above that methanogenic bacteria can be a source of organic carbon (as methane, Fig. 13-12). To make matters confusing, some bacteria produce methane by degrading organic compounds, so some methanogens also act as decomposers of organic matter. The principal reactions involved in this latter role are the reduction of CO_2 (Deuser et al., 1973; Claypool and Kaplan, 1974) or the fermentation of acetate (Atkinson and Richards, 1967; Cappenberg, 1974, 1975) (Table 13-8). Methanogens use metabolic end products such as CO_2 or acetate released by the activity of other groups of bacteria (Table 13-8). This consumption of metabolic end products of other decomposers is important because accumulation of end products could otherwise inhibit activity of the other bacteria.

The energy yield of methanogenic reactions (Table 13-8) is low, and other potential electron acceptors (O_2, NO_3^- and $SO_4^=$) are used first (Oremland, 1975; Balderston and Paine, 1976). Methane production generally occurs in environments with no oxygen or nitrate (Fig. 13-12), such as anaerobic muds rich in organic matter, and in the guts of animals.

Methane concentration in reduced sediments are usually low where sulfate concentrations are high (Fig. 13-13, left). It has been suggested that this is because there is competition for substrates by sulfate reducers (Rudd and Taylor, 1980), or by inhibition of methanogens by sulfide produced by the sulfate reducers. Recent evidence suggests that the lack of methane is due to oxidation of methane by sulfate reducers (Table 13-8, Fig. 13-12). This oxidation may be responsible for the concave shape of the upper part of methane profiles (Fig. 13-13) (Martens and Berner, 1974). The methane that escapes the reduced part of the sediment may then be subject to aerobic oxidation. Methane is energy rich and provides a favorable energy yield for bacteria capable of its oxidation (Fig. 13-12, Tables 13-8, 13-9). Thus, very little of the methane produced within sediments escapes to overlying waters (Table 13-10).

In environments with high rates of production of organic matter and anoxic sediments, such as those of upwellings and marshes, methane from sediments can be found in overlying waters (Scranton and Farrington, 1977). The mixed layer of the open ocean, however, is so distant from the sea floor that transport of methane from sediments does not appear feasible, yet methane concentrations 48–67% in excess

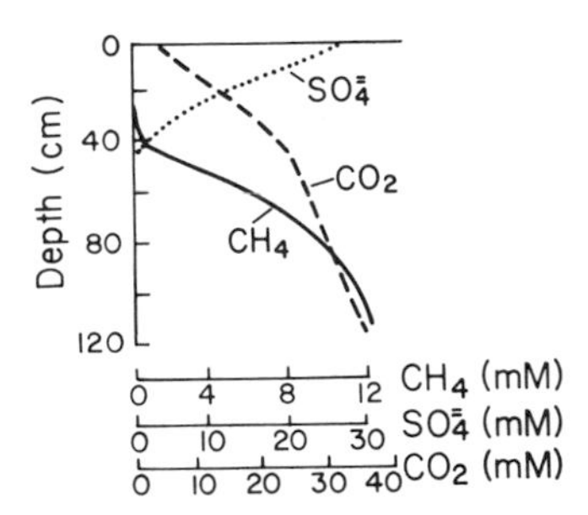

FIGURE 13-13. Schematic diagram of vertical profiles of sulfate, methane, and carbon dioxide in a reduced coastal marine sediment. Adapted from Rudd and Taylor (1980), data from Reeburgh (1976).

TABLE 13-11. Approximate Rates of Methane Production in Various Aquatic Environment.[a]

	Methane production ($\mu M\ m^{-2}\ hr^{-1}$)
Marine environments	
Seagrass beds	0.2–2
Coral reefs	0.02–0.5
Cariaco Trench	18
Santa Barbara Basin	12
Salt Marsh	3–380
Freshwater environments	
Paddy soils	570
Lakes	33–4,580

[a] Condensed from various authors in Rudd and Taylor (1980), and King and Wiebe (1978).

of equilibrium concentrations with the atmosphere have been measured (Scranton and Brewer, 1977). There is also a subsurface maximum where the methane concentration may exceed equilibrium concentration by more than twofold. This methane must be generated within the water column, either as a by-product of algal metabolism or in the guts of animals. Although the latter source at first sight may seem trivial, it is not: about 20% of the biologically produced methane that enters the atmosphere is produced in animal digestive tracts (Enhalt, 1976).

The rates at which methane is produced in freshwater systems are much larger than those in marine environments (Table 13-11). Freshwater systems have considerably less sulfate than marine systems (about 0.1 mM compared to more than 20 mM, respectively), so competition by sulfate reducers, inhibition of methanogens by sulfides, and anaerobic reduction of methane by sulfate reducers are less important in freshwater than in marine environments (Reeburgh and Heggie, 1977). The lack of sulfate thus allows methanogens to mineralize a larger proportion of the primary production. The large amounts of methane produced in freshwater muds are not anaerobically oxidized and may escape into the water, where aerobic oxidation takes place, or into the atmosphere.

Some very rich marine and brackish sediments resemble freshwater environments, with occasional depletion of sulfate near the surface of the sediment due to sulfate reduction (Martens, 1976). One of the reasons this can take place is that during the times of year where the sea floor is anoxic the benthic fauna are killed (Sansone and Martens, 1978). Tube building, feeding, and pumping of seawater by the benthic animals are prime mechanisms that transport seawater—and the dissolved sulfate and oxygen—into surface sediments. With no fauna there is little available sulfate, reduced activity by sulfate reducers, and the bubbling of methane out of anaerobic sediments then becomes a major source of methane to overlying water.

In seagrass beds, mangroves, and salt marshes the vegetation provides a large, intricate network of oxygen-containing roots (Oremland and Taylor, 1977; Teal and Kanwisher, 1966a). The oxidation associated with roots may provide for a significant amount of aerobic respiration of methane in these very productive environments; this may be part of the reason why marsh profiles of sulfate and methane differ from both marine and freshwater sediments in that while they are very productive and rich in sulfate, there is enough oxidation by sulfate reducers that little methane leaves the sediments. The patchwork of oxidized and reduced microzones could also probably allow the simultaneous occurrence of aerobic respiration and sulfide oxidation.

13.3.3.2 Vertical Distribution of Anaerobic Metabolic Processes

The reactions by which organic matter is decomposed by microbes have long been known to follow as a sequence of oxidant reductions (O_2, NO_3, Mn oxides, Fe oxides, SO_4, and CO_2) that extends over time or into redox gradients. The actual distribution of these reactions through depth of sediments depends on supply of oxidants and labile organic matter, diffusion of solutes, and sediment reworking by fauna or disturbances.

In coastal sediments, the oxic-suboxic reactions take place nearer the surface, while the lower layers support sulfate reduction and methanogenesis. In muddy shelf environments, for example, aerobic metabolism occurs within 0.1–1 cm, nitrate, Mn, and Fe metabolism occur between 1–10 cm, sulfate reduction mainly (but not exclusively, cf. Section 13.3.3.1.4) takes place within 10–100 cm, and methanogenesis at greater depths (Aller et al., 1986). In such sediments sulfate reduction is usually the dominant anaerobic metabolic pathway, although there are exceptions. For example in the Amazon shelf sediments there are deeper zones where Fe reduction is predominant (Aller et al., 1986). The actual mechanisms that set up the gradient are not all known, but are thought to be at least partly competitive. Iron reducers, for instance, can outcompete sulfate reducers for organic substrates, so that microbial iron reduction may restrict the vertical extent of sulfate reduction (Sørensen and Jørgensen, 1987; Hines et al., 1991).

In deep-sea sediments, where rates of supply of organic matter are lower, the vertical disposition of the various metabolic zones are displaced deeper into the sediment. Oxic zones may extent to more than 10 cm, nitrate and Mn reduction may take place deeper than 100 cm, and sulfate reduction may occur at great depth or not take place (Aller et al., 1986).

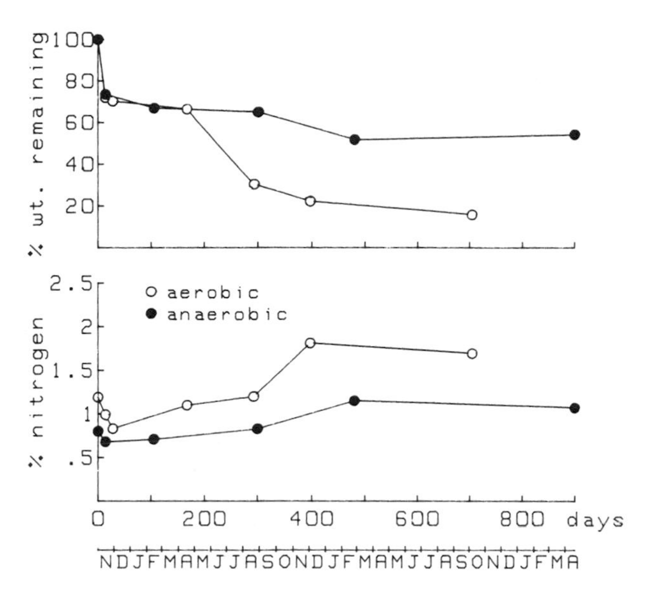

FIGURE 13-14. Course of decomposition of biomass and %N content of *Spartina alterniflora* exposed aerobically (litter bags on the marsh surface) and anaerobically (buried 5–10 cm below the marsh surface). Data of I. Valiela and John Teal.

13.3.3.3 Incorporation Into Anoxic Sediments

Only a small fraction of the organic matter produced by autotrophs accumulates in sediments (Table 13-7), due to consumption by animals and both aerobic and anaerobic decay. Where production is very high, such as in upwelling regions, salt marshes, and mangroves (Table 13-7), environments that characteristically have anaerobic sediments, organic matter or peat do accumulate. A certain proportion of organic material is resistant to decay under anaerobic conditions. Fungi are rare, as discussed earlier, so the abundant lignin fraction in detritus from vascular plants is not degraded and peat accumulates. This may explain the accumulation of organic matter in marshes, where decay of detritus under anaerobic conditions is slower than under aerobic conditions (Fig. 13-14).* This explanation may not apply where the organic matter originates as phytoplankton, since lignins and fungi are rarer in phytoplankton.

---

*This discussion applies to decay prompted by microbes during the decomposer phase; note in Figure 13-14 that the leaching phase is more or less the same under both conditions.

Based on analyses of chemical composition of water masses, Richards (1970) concluded that there was no evidence for slower decay of organic matter under reducing compared to oxidizing conditions. Actual experiments in which dead phytoplankton were incubated under anaerobic and aerobic conditions provide evidence that in fact anaerobic decay is somewhat slower than aerobic decay of phytoplanktin detritus (Otsuki and Hanya, 1972, 1972a). Perhaps the accumulations of toxic end products such as sulfides, ammonium, alcohols, or acids are responsible.

The rates of carbon oxidation carried out based on oxygen respiration are as fast or faster than those achieved based on sulfate reduction (Fig. 13-14 and cf. Fig. 14-18). The rates of oxidation supported by oxygen and sulfate metabolism become similar only when rates of sediment accumulation are high (Fig. 14-18). Under most field conditions, mineralization and release ("regeneration") of nutrients contained in organic matter (Table 13-9) take place more readily in anoxic environments than under aerobic decay. Fungi are rarely present to aid the immobilization of nitrogen. Further, nutrients in detritus are released in large quantities by sulfate reducers since they must oxidize more organic matter to obtain a given amount of energy (Table 13-8). This results in a higher carbon-to-nitrogen ratio of detritus under anaerobic conditions than under aerobic conditions. The percentage of nitrogen of anaerobic detritus is about half that of aerobic detritus (Fig. 13-14, bottom). Less nitrogen has been immobilized during decay, and more nitrogen is mineralized and available for transport out of the organic matter. This is probably one mechanism that produces the very high concentrations of inorganic nitrogen (principally NH_4^+) in waters of salt marshes, mangroves, and estuaries.

Chapter 14
Nutrient Cycles and Ecosystem Stoichiometry

Nutrients have been discussed throughout this book in terms of limiting factors for producers, consumers, and decomposers, as electron donors for microbial decomposers, and in a number of other roles. It is clear, then, that nutrients are inextricably linked to almost all other ecological processes. In this chapter we first look at the transformations, exchanges, and general dynamic of three nutrients: phosphorus, nitrogen, and sulfur. The dyanmics of these three elements illustrate how nutrients interact with sediments and water, and the great significance of oxidation state and the interactions with organisms. Then we examine the stoichiometry by which nutrient cycles are coupled to transformations of organic matter in ecosystems.

14.1 Phosphorus

14.1.1 Chemical Properties of Phosphorus

In aerobic environments inorganic phosphorus occurs virtually exclusively as (ortho-)phosphates* in which the P atom has an oxidation state of +5. The orthophosphate ions include phosphoric acid (H_3PO_4) and its dissociation products ($H_2PO_4^-$, HPO_4^{2-} PO_4^{3-}), and in ion pairs and complexes of these products with other constituents of seawater (Gulbrandsen and Roberson, 1973). HPO_4^{2-} is the major ion in seawater by about one order of magnitude over PO_4^{3-}, which in turn is one order of magnitude more abundant than $H_2PO_4^-$, while H_3PO_4 is negligible. Although polyphosphates are abundant within cells, only traces are found naturally in seawater (Hooper, 1973); high concentrations of polyphosphates are often used as an indication of pollution with waste

---

*Orthophosphate refers to any salt of H_3PO_4, phosphoric acid, which can also be described as $3H_2O \cdot P_2O_5$. Water molecules can be removed from orthophosphate; the various resulting condensed compounds are named polyphosphates (Van Wazer, 1958).

waters. Bacteria can reduce phosphate to phosphine (PH_3) and phosphide (P^{3-}) but such reduced compounds are of rare occurrence (Hutchinson, 1975); phosphate is by far the dominant form of phosphorus. The range of concentrations of phosphates in seawater is given in Fig. 2-19.

A number of organic compounds that contain phosphorus can be found dissolved in seawater. These are primarily phosphate esters that originate from living cells and can be hydrolyzed to phosphates once released into th environment.

There are at least two aspects of the chemistry of phosphate that are important in maintaining low concentrations of the dissolved forms and greatly complicate the dynamics of phosphorus: the facility of adsorption, and the propensity to form insoluble compounds with certain metals.

Phosphate adsorbs readily under aerobic conditions onto amorphous oxyhydroxides, calcium carbonate, and clay mineral particles (Krom and Berner, 1980; Stumm and Morgan, 1981). Adsorption onto clay particles takes place by bonding of phosphate to positively charged edges of clays and by substituton of phosphates of silicates in the clay structures (Stumm and Morgan, 1981). Because of the ease of adsorption under aerobic conditions, phosphate seldom travels far in sediments, except where transported by movement of particles. Phosphate ions also precipitate with cations such as Ca^{2+}, Al^{3+}, and Fe^{3+}. On a geological time scale phosphate in seawater is in equilibrium with phosphate minerals such as francolite ($Ca_{10}[PO_4CO_3]_6Fe_2$) or hydroxyapatite ($Ca_5OH[PO_4]_3$), but for ecological purposes these phosphate compounds can be considered almost insoluble.

The chemistry of phosphorus changes markedly in anaerobic environments. In anaerobic soils certain microorganisms may reduce phosphorus to oxidation states of +3 and +1 (HPO_4^{2-} and HPO_2^{2-}) (Silverman and Ehrlich, 1964), but these oxidation states have not reported in anaerobic marine sediments.

As far as phosphorus is concerned, a much more important aspect of anaerobic sediments is that bacteria and the H_2S present reduce ferric iron (Fe^{3+}) to ferrous iron (Fe^{2+}). Ferrous iron is much less effective at adsorbing phosphate than ferric iron (Krom and Berner, 1980), and the reduction of iron thus results in greater availability of dissolved phosphate in anaerobic environments. The reduction of ferric to ferrous iron causes the amorphous ferric oxyhydroxides to dissolve, making them unavailable to adsorb phosphates. This mechanism is the principal cause of the solubilization of phosphate, aided perhaps by inhibition of adsorption on clay surfaces due to the buildup of a coating of the abundant organic matter typical of anoxic sediments (Krom and Berner, 1980). The result of all this is that in reduced sediments the ratio of adsorbed phosphate (by weight) to the equilibrium concentration in the interstitial water is 1–5, while in oxic sediments the ratio is 25–5,000 (Krom and Berner, 1980).

Dissolved phosphate may leave anoxic sediments, but some of the phosphate may reprecipitate as $FePO_4$ at the oxic-anoxic interphase, and most is probably adsorbed onto amorphous ferric oxyhydroxides (Krom and Berner, 1980a). The phosphate adsorbed to ferric hydroxides is readily exchangeable and available as phosphate (Shukla et al., 1971; Golterman, 1973). In any case, the release of phosphate from anaerobic sediments is faster than the reoxidation and immobilization, and phosphate is regenerated from anoxic sediments into overlying water.

14.1.2 Transformations and Exchanges in the Phosphorus Cycle

Phosphorus in seawater is found in living organisms or as dissolved inorganic phosphorus (DIP), dissolved organic phosphorus (DOP), and particulate phosphorus (part. P).* In most aquatic environments the amount of part. P is much greater than that of DOP, which in turn is larger than DIP. In lakes the phosphorus is partitioned into about 60–70% part. P, 20–30% DOP, and 5–12 DIP. In seawater, where cell concentrations are lower than in lakes, the dissolved organic fraction (which comes from cell exudates) may be smaller than the inorganic fraction. In the English Channel most of the phosphorus present in winter is DIP, but in summer, when cells are most abundant and senescing, there is more DOP than DIP (Harvey, 1995). In eutrophic marine environments, such as the Baltic Sea, the amount of DIP may be 91% of the dissolved phosphorus (Sen Gupta, 1973).

The sources, losses, and pools of particulate, dissolved inorganic, and dissolved organic phosphorus are schematically represented in Fig. 14-1, which should be referred to for the remainder of this section.

Uptake by primary producers and bacteria is responsible for the low phosphate concentration typical of surface waters (Fig. 2-18; the kinetics of uptake are discussed in Section 2.2.1). Some coastal macrophytes in salt marshes, mangrove swamps, and eelgrass beds take up phosphorus from the sediments, as do some freshwater macrophytes (Carignan and Kalff, 1980). This pathway is not shows in Fig. 14-1 because in the sea the amounts of P involved are trivial relative to other exchanges.

---

*The standard reference for methods of analysis for nutrients in seawater is Strickland and Parsons (1968), now revised as Parsons et al. (1984), which includes procedures for the various forms of phosphorus as well as nitrogen and carbon. A critique of methods of measuring phosphorus is given by Chamberlain and Shapiro (1973), and for measurement under anaerobic conditions see Bray et al. (1973) and Loder et al. (1978).

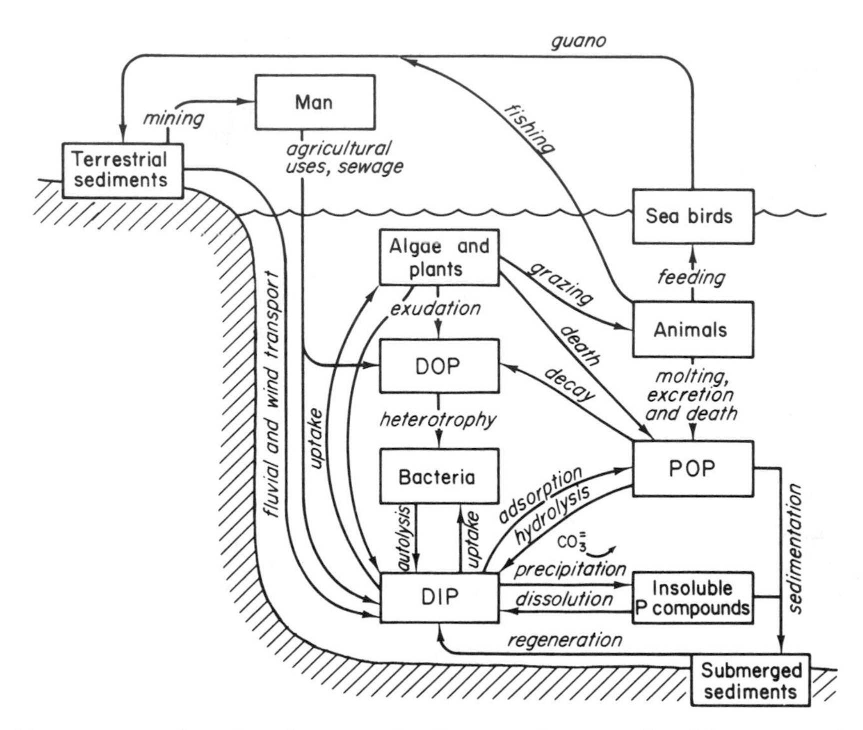

FIGURE 14-1. The phosphorus cycle. Boxes indicate pools of P, arrows show processes, including transport or transformations.

Some phosphate is excreted by bacteria, and DIP is also provided by microbial hydrolysis of the esters of DOP. This lysis is a very rapid proces that limits the persistence of DOP to a few hours (Van Wazer, 1973). Abiotic hydrolysis also occurs (Johannes, 1965), but is 10^3-10^4 times slower than where bacteria are active. The release of phosphate directly from macroalgae is very low (Kirkman et al., 1979; Carignan and Kalff, 1980) and the amounts produced by hydrolysis of DOP are very small. By far, phosphate is principally regenerated by the decay of particulate organic phosphorus (POP) and by animals. The zooplankton in the Central North Pacific Ocean, for example, may release 55–183% of the daily requirements of phosphorus by the phytoplankton (Perry and Eppley, 1981).

The processes that provide and remove DIP are rather fast; the residence time of dissolved phosphate in oligotrophic waters, for example, is just a few minutes (Hayes and Phillips, 1958; Pomeroy, 1960), mainly due to rapid uptake by algae and bacteria. The rapidity of uptake is a major reason why P in seawater is found principally incorporated in living particles. This can be demonstrated by considering the exchanges between the DIP, DOP, and particulate P as equilibrium

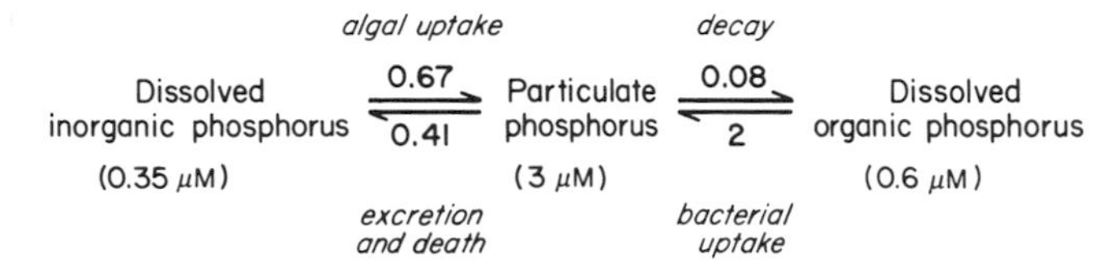

FIGURE 14-2. Relationship between DIP, Particulate P, and DOP in coastal waters off Nova Scotia. The concentrations of the three species of P are shown in parentheses. The arrows indicate the exchange rates in μM P day^{-1}. Adapted from Watt and Hayes (1963).

reactions (Fig. 14-2). The rates of transformation are such that the concentration of particulate P exceeds those of DIP and DOP.

Death, shedding, or molting of organisms plus adsorption of phosphate onto particles produce particulate organic phosphorus (POP) (Fig. 14-1). Some of the POP is released as DIP as particles decay in the water column, but some settle onto sediments. In the sediments further degradation of settled organic P to DIP can take place and some of this DIP is precipitated or adsorbed.

There is an extremely slow part of the phosphorus cycle in which submerged sediments are uplifted over geological time to provide phosphate rocks on land. The weathering of this rock is the geological process that supplies P back to the oceans by river run off and wind transport. Phosphate rocks are also mined, and much of the phosphate used as fertilizer finds its way down estuaries to the sea, transported on particles. After 1 year following use as fertilizer, 12% of the total P mined in the United States finds its way to aquatic environments, and 40% within 4 years (Griffith, 1973). Guano birds who feed on fish and defecate on land during the nesting season provide the only natural mechanism whereby some small amount of the P in the sea returns to land on a short-term basis (Fig. 14-1). Because of the chemistry of phosphorus, it is therefore clear that, at least for human use, phosphorus must be treated as a nonrenewable resource in limited supply.

14.1.3 Phosphorus Dynamics in the Marine Water Column and Sediments

In surface waters phosphate concentration is usually very low (usually less than 1 μM P, cf. Fig. 2-18), primarily due to uptake of phosphate by algae and bacteria. Algae are most likely the primary agent of phosphate removal (Fuhs et al., 1972; Cole, 1982) since their K_s concentrations are low (0.6–1.7 μM) and comparable to ambient concentrations (cf. Fig. 2-18). Values of bacterial K_s range from 6.7 to 11.3 μM, very high values compared to seawater concentrations. Th ability to store nutrients also makes algae a more important reservoir than bacteria. The importance of the uptake be algae is reflected by the sharp reduction of phosphate in surface layers of temperate coastal waters during the spring bloom. Algal growth produces an increase in particulate phosphorus and, as the algae senesce, release of dissolved organic phosphorus (Harvey, 1955; Strickland and Austin, 1960).

In the open ocean phosphate concentrations increase below near-surface waters to a peak near the permanent thermocline (Fig. 2-18). Peak concentrations are usually associated with the layer of minimum O_2 and maximum CO_2, where microbial and zooplankton activity regenerates both dissolved phosphorus and carbon. We have already seen that carbon fixed in the oceanic photic zone is regenerated within the water column (Chapter 13) and so is phosphorus. Measurements of the rate of release of phosphate (Pomeroy et al., 1963; Johannes, 1965) suggest that zooplankton may regenerate substantial amounts of the nutrients fixed by algae in oceanic wates. Butler et al. (1970) calculated that in *Calanus* about 17% of the P ingested was retained for growth, 23% egested as fecal pellets, and 60% excreted as dissolved phosphorus. The excreted P is accessible to organisms in the water column. Since most of the organic matter, including fecal pellets, that leaves the photic zone is degraded within the water column (Chapter 13), recycling of the contained P is very active. In coastal waters of California, for example, about 68–87% of the P used in surface production is regenerated above 75 m; above 700 m about 91–95% of the P is regenerated (Knauer et al., 1979). In oceanic waters of the Pacific, 87% of the P was regenerated above 75 m and 99% at 1,000 m (Knauer et al., 1979).

The rate at which zooplankton regenerate phosphorus is not constant. When algae are abundant, a larger amount of P is converted to eggs (cf. Chapter 3) and is stored as phospholipids. When little food is available, the stored lipids are used as an energy source and phosphorus is excreted (Martin, 1968), so that relatively more phosphorus is regenerated into the water.

If the water column is deep enough there is a more or less uniform concentration of phosphorus below the thermocline, probably maintained by vertical mixing, and at some depth nearer the bottom there is usually an increase in phosphate. Such gradients suggest that deeper waters are a common source of phosphate. In the Gulf of Maine, for example (Fig. 14-3, left), the concentrations are higher near the bottom, suggesting that vertical diffusion from the sediments supplies the water column. There are examples of horizontal advection of nutrient-rich water, but by and large the pattern of Fig. 14-3 is general.

Phosphate concentrations in sediments are quite variable. In anaerobic sediments there is a peak at some relatively shallow depth, presumably where reduced conditions combine to make phosphate salts

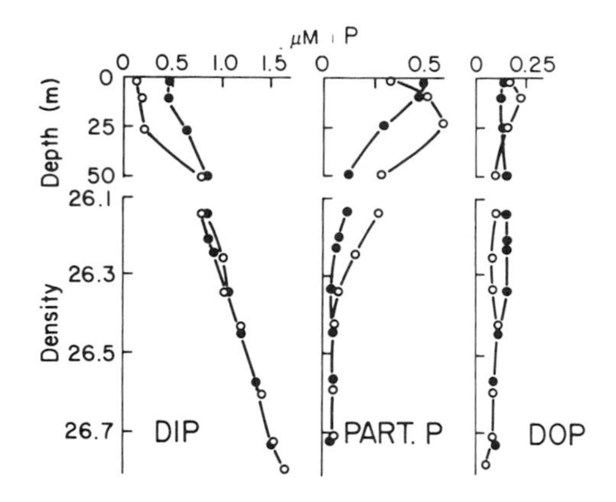

FIGURE 14-3. Vertical profiles of dissolved inorganic, particulate, and dissolved organic phosphorus in the water column of the Gulf of Maine. Below 50 m depth is expressed as density of the colder salties deeper water. Adapted from Ketchum and Corwin (1965).

most soluble (Fig. 14-4). Above the maximum there is a pronounced decrease toward the surface, suggesting that there is a significant diffusion of phosphate out of reduced sediments to the overlying water. This can be the source of DIP that diffuses upwards in Fig. 14-3, as shown by Klump and Martens (1981) for organic-rich coastal sediments. The gradients in the top few centimeters of the sediments of Fig. 14-4, if they are due to diffusion, could provide about 5% of the total P content of the water column above per week, a significant flux of phosphate.

More evidence that regeneration of benthic phosphate affects water column concentrations is given by the relation of the seasonal pattern of phosphate in water and the flux of phosphorus from sediments (Fig. 14-5). Regeneration of phosphate is temperature dependent, increases in the spring, and is followed by increases in phosphate in the water, at least in Narragansett Bay. The rates of release of phosphate by a wide variety of marine sediments range from −15 to 50 μmoles $m^{-2}u2-^{1}$ (Nixon, 1981; Fisher et al., 1982). Regeneration from sediments in Narragansett Bay can provide enough phosphate to support 50% of the primary production in the water column (Nixon et al., 1980). An additional 20% or so may be provided by DOP also released by the sediments. In various other coastal sediments the rates of regeneration of phosphate provides an average of 28% of phytoplankton

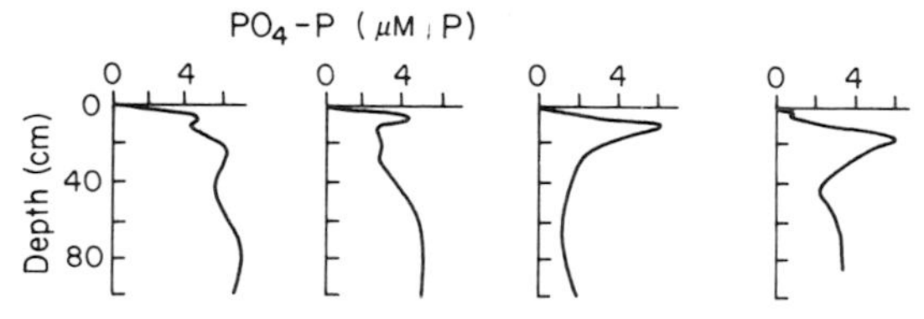

FIGURE 14-4. Vertical profiles of phosphate in the interstitial water of sediments at different stations within Chesapeake Bay. Adapted from Bray et al. (1973)

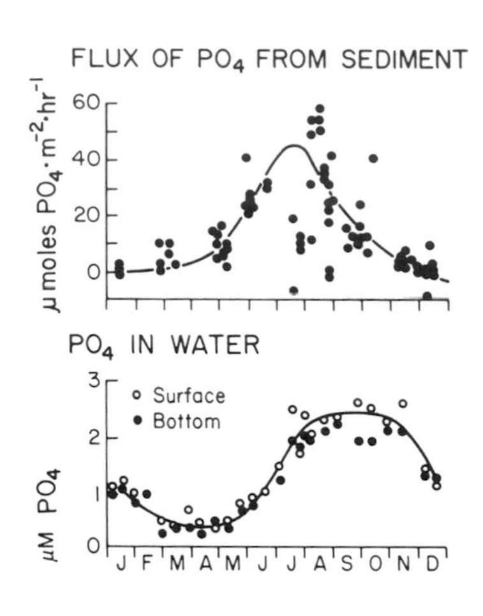

FIGURE 14-5. Flux of phosphate from sediments over the year and phosphate in the water of Narragansett Bay, 1972–1973. Lines drawn by eye. Adapted from Nixon et al. (1980).

requirements (Fisher et al., 1982). Regeneration of PO_4 by sediments is thus important in coastal water.

The regeneration process is so active that only a very small proportion of the P in coastal ecosystems is lost permanently to the sediments by burial. These losses to sediments may be balanced by inputs from terrestrial sources, including wind-borne dust particles and fluvially carried particles (Fig. 14-1).

Our discussion of phosphate in water and sediments demonstrates that its concentration is to a very large extent determined by biological activity. As a result, for example, oxygen uptake by organisms in water is well correlated to phosphorus (and nitrogen) concentrations (Stumm and Morgan, 1981), since nutrients are released during aerobic respiration of organic matter [Eq. (1-2)]. All the dissimilative reactions (Table 13-8) result in regeneration of phosphorus.

Biological activity by microbes and animals also controls the distribution of phosphorus and other elements in organic matter by altering the redox potential of sediments, and hence the chemical properties of phosphorus.

The stoichiometric equations of Tables 13-8 and 13-9 emphasize the point that cycles of individual nutrient are never isolated; the biological transformations link phosphorus, nitrogen, sulfur, and all elements present in organic matter. The reduction reactions involved in oxidation of organic matter are also important in that they not only alter redox conditions but also change alkalinity* (Emerson et al., 1980), and this

* Alkalinity is a measure of the bases that are titratable with strong acid. It can be thought of as the acid-neutralizing capacity. It is measured in milliequivalents, and in seawater is due primarily to the presence of bicarbonate (HCO_3^-). Various forms of borate and any other base in seawater add to the alkalinity. Edmond (1970) and Morel and Morgan (1972) discuss advanced methods to determine alkalinity in seawater.

buffers the pH of the medium. Sulfate reduction (Table 13-8) increases alkalinity, as does reduction of nitrate, Mn^{4+}, and Fe^{3+}. Since alkalinity has important consequences for chemical reactions, it is clear that the chemical status of water and sediments is thoroughly affected by biological activity. The CO_2-carbonate system is of course a major influence on water chemistry, but it is also affected by biological activity.

14.2 Nitrogen

14.2.1 Transformations of Nitrogen

Unlike the phosphorus cycle, the nitrogen cycle is dominated by a gaseous phase, and microbial transformations involving changes in oxidation state play important roles.

Nitrate (NO_3^-) is the most oxidized form of nitrogen; it is taken up in aerobic environments by algae, bacteria, and plants, and once within the cells, reduced by assimilation processes involving several enzymes [including nitrate reductase (Packard et al., 1971)] to the amine form, which is used in metabolic processes. The uptake kinetics can be described by Michaelis–Menten kinetics (Section 2.2.1).

Nitrate is also used as a terminal electron acceptor and is reduced by dissimilatory processes in a series of steps. The first step is the reduction of nitrate to nitrite (NO_2^-), then NO and N_2O, and finally to N_2 gas (Fig. 14-6). This pathway is called denitrification and requires a supply of organic compounds (i.e., reduced carbon) which is concomitantly oxidized (Table 13-8). Denitrification also requires anaerobic conditions (Reddy and Patrick, 1976), is most active at pH of 5.8–9.2 (Delwiche and Bryan, 1976) and is vey temperature dependent (Focht and Verstaets, 1977; Kaplan et al. 1977). Denitrifiers may reduce nitrate to ammonium via hydroxylamine intermediates (Fig. 14-6) but this is probably not a large source of ammonium. This source of NH_4^+ is probably not as important as the degradation of organic nitrogen.

Ammonium can be removed from water by uptake by plants, algae, and bacteria (cf. Section 2.2). Ammonium may also be oxidized by nitrifying bacteria (Fig. 14-6). Nitrification is a bacterially mediated process in which the first step is the oxidation of ammonium to nitrite by bacteria, usually *Nitrosomonas*. Nitrification continues with the further oxidation of nitrite to nitrate by *Nitrobacter* and other genera. Both reactions require oxygen and are energy yielding. These two genera of bacteria are autotrophic, in that they use the energy from reduced nitrogen compounds to fix carbon. Nitrification rates are often limited by supply of oxygen. Increase in aeration of sediments can increase nitrification, but nitrification can occur even at low levels of O_2 (Billen, 1975), due to the high affinity of nitrifiers for O_2. The importance of this is that some nitrate is produced even in sediments or

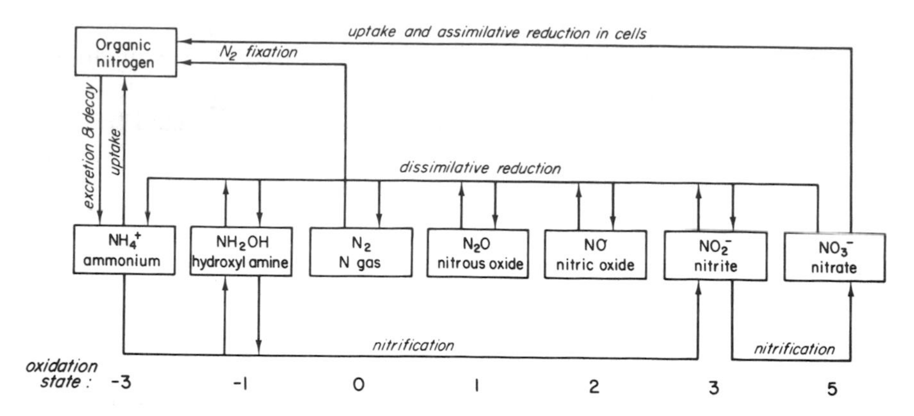

FIGURE 14-6. Transformations of nitrogen. The boxes show the various nitrogen species and arrows indicate the processes of transformation. Adapted from Fenchel and Blackburn (1979).

water low in oxygen. The production of nitrate than allows denitrification to take place.

The cycle of nitrogen also includes a pathway by which nitrogen in gaseous form can be fixed into organic compounds by certain bacteria, including blue-greens (Fig. 14-6). The microbes that fix nitrogen can be free-living (Carpenter, 1973; Marsho et al., 1975; Gotto and Taylor, 1976; Potts and Whitton, 1977), or symbiotic with primary producers (Head and Carpenter, 1975; Zuberer and Silver, 1978; Teal et al., 1979; Capone et al., 1977) or with animals [shipworms, Carpenter and Culliney (1975), and sea urchins, Fong and Mann (1980)].

Nitrogen fixation is much reduced if free ammonium is available (Goering et al., 1966; Van Raalte et al., 1974; Carpenter et al., 1978). Uptake of ammonium should be energetically favored over fixation since the assimilative reduction needed within cells when N_2 is fixed is an energy-demanding process. Anaerobic conditions are required by fixers because the enzymes involved in N_2 fixation are sensitive to oxygen. Most blue-greens that fix nitrogen have structures called heterocysts in which anaerobic conditions are maintained and in which fixation may take place even in aerobic water. Some cyanobacteria like the pelagic *Oscillatoria* lack such structures; they may create anoxic zones within bundles of associated cells (Carpenter and Price, 1976). Relatively large amounts of iron are required for growth of cyanobacteria with N_2 rather than NH_4^+ or NO_3^- as the nitrogen source. Iron is in low abundance in seawater, so it might restrict rate of nitrogen fixation, and might be a reason why fixation is less prevalent in the sea than might be expected in a nitrogen-poor environment (Rueter, 1982).

Nitrogen fixation requires energy, which in shallow sediments may be provided by the abundant organic matter. In the water column the DOC may be very low, so other energy sources are used. Fogg and Than-Tun (1960) postulated that perhaps light energy could drive fixation; this agrees with results that show that fixation in the water column by the blue-green *Oscillatoria* stops in the dark (Dugdale et al., 1961).

Large amounts of organic nitrogen are released after death of animals (Fig. 14-6). Most of these organic nitrogen compounds are probably decomposed by heterotrophs, ultimately to ammonium.

Excretion by animals also releases dissolved nitrogen. Zooplankton excrete free amino acids and ammonia, with urea making up some of the remainder (Corner and Newell, 1967). Fish excrete relatively more urea and other organic compounds. We have already discussed the potential importance of animal regeneration of nutrients in regard to primary production (Chapter 2). This is a topic needing further study. Even less is known about the release and transformations of organic than inorganic nitrogen compounds in seawater.

14.2.2 Distribution of Nitrogen Species

14.2.2.1 Water Column

Nitrate is generally the primary form of inorganic nitrogen in seawater (Table 14-1). It is usually most abundant in winter (Fig. 14-7) when it is not taken up by producers. Nitrate may disappear from the photic zone during algal blooms. Since active uptake by phytoplankton is restricted to the photic zone, nitrate concentrations are greater at depth (Fig. 2-18). Where there is a breakdown of stratification of the water

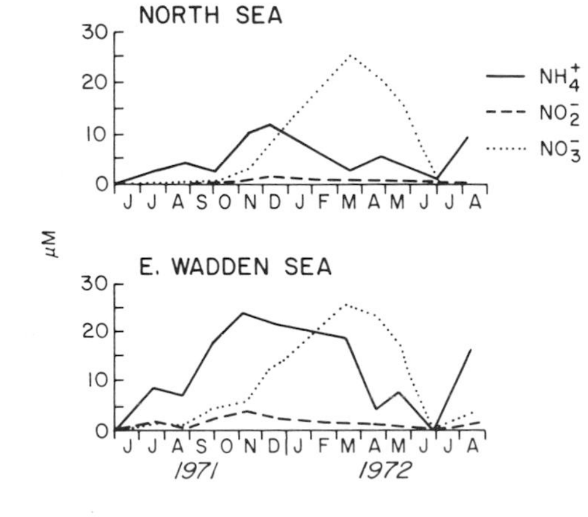

FIGURE 14-7. Seasonal changes in the concentrations of ammonium, nitrite, and nitrate in the North Sea and eastern Wadden Sea off Holland. Adapted from Helder (1974).

TABLE 14-1. Approximate Concentrations (μM) of the Major Forms of Nitrogen in Different Marine Waters. Oceanic Values are Approximate Means, Coastal and Estuarine Values are Approximate Ranges.[a]

N species	Oceanic waters: Surface (0–100 m)	Oceanic waters: Deep (>100 m)	Coastal waters	Estuarine waters
Nitrate	0.2	35	0–30	0–350
Nitrite	0.1	<0.1	0–2	0–30
Ammonium	<0.5	<0.1	0–25	0–600
Dissolved organic N	5	3	3–10	5–150
Particulate organic N	0.4	<0.1	0.1–2	1–100
N_2 gas	800	1150	700–1,100	700–1,100

[a] From Antia et al. (1991), data compiled by J. Sharp.

column, vertical advection of deeper water replenishes nitrate (and other nutrients) in surface waters (Table 14-1) (cf. Section 15.3).

The amount of nitrite present in the water column is low (Fig. 14-7). Ammonium may be abundant in very productive shallow environments (Fig. 14-7) and since it is generated in large measure from decay of organic nitrogen, it peaks when warmer temperatures favor microbial degradation of organic matter. Concentrations of ammonium range from traces in oceanic water to much higher in coastal areas (Table 14-1).

The amount of dissolved organic nitrogen (DON) (Table 14-1) is usually much larger than the amount of inorganic nitrogen. Amino acids and urea comprise only a small component of the DON. Amino acids range from 0.025 to 1.4 μM while urea concentrations reach 8.9 μM N but are usually much less (McCarthy, 1980). The remainder, or rather most of the DON, is not nutritionally suitable for oceanic phytoplankton (Thomas et al., 1971) or for estuarine microbes (Asiz and Nedwell, 1979). Mass balance and other calculations show only slight degradation of DON (Valiela and Teal, 1979; Nixon, 1981). Thus much of the DON may be resistant to decomposers. Most of the DON in seawater is still chemically uncharacterized and its chemical and biological properties are becoming better known (Antia et al., 1991).

14.2.2.2 Sediments

The concentrations of nitrogen compounds in interstitial water of sediments are much higher than in water (Fig. 14-8). This is due to degradation of the large concentrations of organic matter, low rate of percolation, and to the presence of active exchange surfaces in sediments.

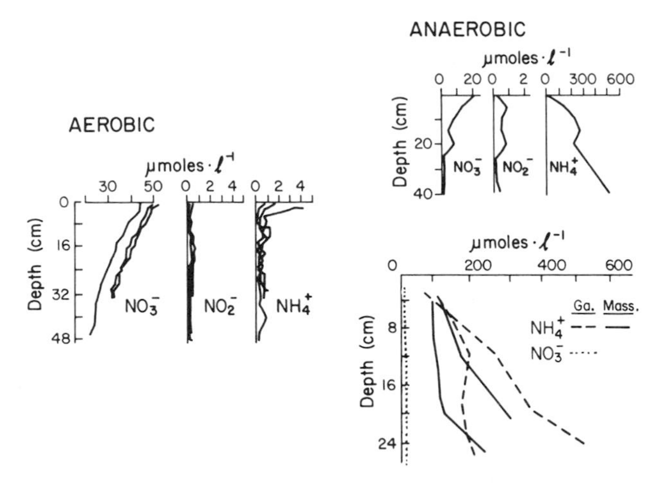

FIGURE 14-8. Vertical profiles of dissolved inorganic nitrogen in sediments. Left: Aerobic sediments in the East Pacific Rise [adapted from Emerson et al. (1980)]. Right top: Anaerobic sediments of Puget Sound [adapted from Grundmanis and Murray (1977)]. Right bottom: Anaerobic salt marsh sediment in two sites each in Massachusetts and Georgia. Concentrations of nitrite were very low. Adapted from Giblin (1982) and Haines et al. (1977).

The redox state of a sediment determines the relative abundance of inorganic nitrogen compounds [NH_4^+, NO_2^-, and NO_3^-], and their vertical distribution in sediment. In aerobic sediments (Fig. 14-8, left) organic matter accumulates near the surface due to the rain of particles from the water column; this leads to active decay of organic matter near the surface, so that most ammonium is released near the sediment surface. Benthic fauna are more abundant near the surface and their excretion also contributes to high ammonium near the sediment surface. Nitrification converts the ammonium into nitrate, and tube building and feeding activity of the benthos carry nitrate-bearing water down into the sediment column; both these processes create the nitrate gradient.

In anaerobic sediments (Fig. 14-8, right) ammonium is far more abundant than in aerobic situations due to the stoichiometry of anaerobic decay (cf. Table 13-8). The vertical profiles of ammonium in intertidal water show that ammonium diffuses upward to the overlying water. Oxidized inorganic nitrogen is seldom abundant (Fig. 14-8, bottom right) in anaerobic profiles, since denitrification, and less importantly, ammonification, remove nitrate. In some anaerobic sediments the activity of animals may transport nitrite and nitrate-bearing waters into the sediments (Fig. 14-8, top right). Models of vertical distribution of nitrogen species constructed on the basis of diffusion and the stoichiometry of reduction reactions (Table 14-1) may not completely describe the actual distribution. Stirring of sediment by animals (Fig. 14-8, top right) or ebullition of bubbles of gases up animal tubes may enhance vertical fluxes (Emerson et al., 1980; Klump and Martens, 1981). Dissolved and exchangeable ammonium are released and flux out of sediment to overlying waters at rates between 13 to 710 μmole m^{-2} hr^{-1} (Nixon, 1981). Thus, not only do biological reactions provide nutrients in proportions determined by the biochemical stoichiometry, but also biological activity by animals affects the distribution and transport of the products of these reactions.

Of the ammonium produced by decay, some remains dissolved in interstitial water, and one to two times as much can be adsorbed to sediments (Rosenfeld, 1979). Ammonium is adsorbed by two means: exchangeable ammonium is attached to surface of clays and organic matter by ion exchange; fixed ammonium is adsorbed within layers of the clay structure where it is not readily replaced by other cations.

14.2.3 Cycle of Nitrogen in the Sea

Nitrogen is transformed and transported in a complex pattern in marine environments (Fig. 14-9). We will not consider all components of the

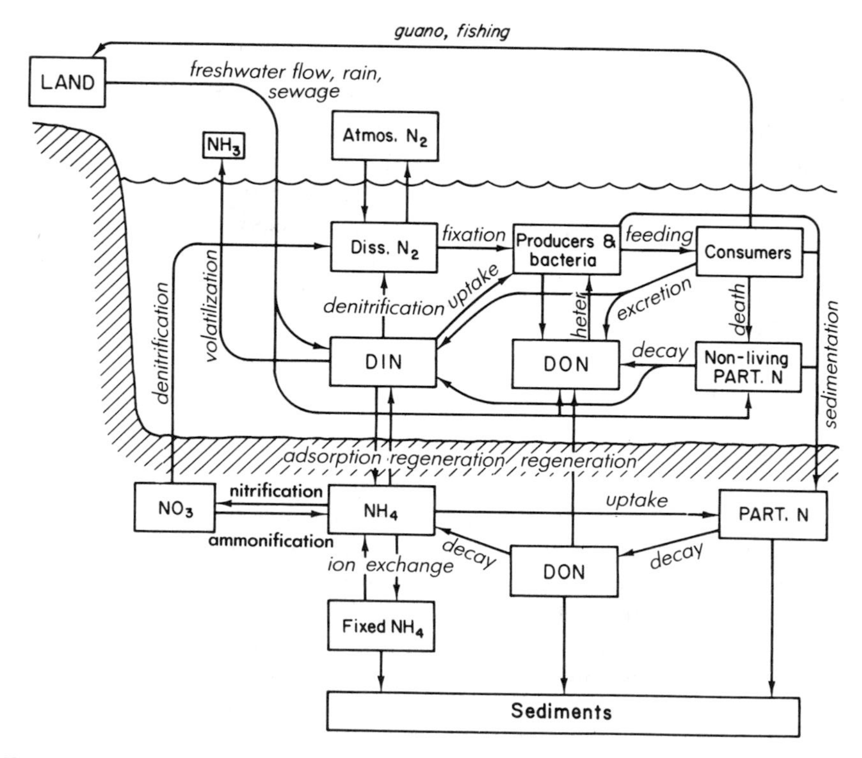

FIGURE 14-9. Simplified scheme of major transformations and transport of nitrogen in marine environments.

cycle but rather focus on the major sources of new nitrogen, mechanisms of uptake and regeneration of nitrogen, and on budgets of inputs and outputs. Referral to Fig. 14-9 will help through this dicussion.

14.2.3.1 Sources of New Nitrogen

14.2.3.1.1 Nitrogen Fixation. Gaseous nitrogen dissolved in seawater is generally in equilibrium with the atmosphere, except in special circumstances where water exchange is restricted (fjords, Black Sea deep water, interstitial water in sediments). The "fixation," meaning the incorporation of N from N_2 gas into organic nitrogen, is a source of new nitrogen for marine ecosystems.

Rates of nitrogen fixation are measured most easily by the acetylene reduction method. In this procedure the reduction of acetylene to ethylene is measured by gas chromatography, and the rates are transformed to N_2 fixation using a ratio of 3:1. This ratio assumes that the enzymes used by the microorganisms to attack the triple bonds in acetylene (HCCH) and N_2 are the same. The acetylene reduction method has drawbacks, and a better, albeit more cumbersome, method is apply ^{15}N tracer techniques (Painter, 1970; Hauck and Bremner, 1976).

The cyanobacteria are the principal organisms that fix nitrogen in aquatic environments. Bluegreens make up up to 50% of the planktonic biomass in freshwaters (Smith 1983), and this high abundance leads to relatively high rates of nitrogen fixation (Table 14-2). Such fixation of N_2 provides sufficient nitrogen for producers in freshwaters, whose growth and production, in consequences, becomes limited by the phosphorus supply. In contrast, abundance of nitrogen fixing bluegreens, and hence, rates of nitrogen fixation in marine waters are generally lower than in freshwaters (Table 12-2). The scarcity of bluegreens in seawater is curious, since the low supply of dissolved inorganic nitrogen in seawater would seem to make nitrogen fixation advantageous (Paerl, 1990). The abundant small marine cyanobacteria (Johnson and Sieburth, 1979; Waterbury et al., 1980) do not fix nitrogen.

The explanation for different rates of nitrogen fixation in fresh- and marine waters remains elusive. Many hypotheses have been forwarded. One intriguing alternative is that nitrogen fixation by marine bluegreens is limited by the availability of molybdenum, an element that is part of the enzymes involved in nitrogen fixation. Howarth et al. (1985) hypothesized that the abundant sulfate in seawater interferes with the uptake of molybdenum by fixers because of its steric similarity to molybdate, the form of Mo that is taken up by bluegreens. Abundant sulfate in water therefore could reduce the activity of nitrogen fixers. Some enrichment-based tests of the inhibition idea fail to demonstrate a clear effect of Mo at ambient seawater concentrations (Paulsen et al.,

TABLE 14-2. Rates (Mean and Range) of Nitrogen Fixation Inputs Via N Fixation As a % of Total N Loads, and N Fixation As % of the N Needed By Producers, in Various Aquatic environments.[a]

	Nitrogen fixation rate ($g\,m^{-2}\,y^{-1}$)	N fixation as % of total N loading	N fixation as % of N needed by producers
Water Columns:			
Marine systems			
Oceans	0.04 (0.02–0.09)	<0.04	0.4 (0.07–0.5)
Estuaries	0.7 (0.01–1.8)	12 (4.7–17)	2.2 (1.3–3)
Freshwater lakes			
Oligotrophic	0.3 (0–2)	16 (0.02–81)	3 (0.03–8.9)
Meso- to eutrophic	1.7 (0.2–9.2)	28 (5.5–82)	3 (0.5–7)
Sediments:			
Marine systems			
Oceans	0–0.0004		
Estuaries and coastal waters	0.01–2		
Seagrass beds	0.8–50		
Salt marshes	0.24–51		
Cyanobacterial mats	1.3–76		
Mangrove swamps	0.03–3		
Freshwater systems			
Lakes	0.001–0.3		
Freshwater marshes	0.01–6		
Peat bogs	0.05–2		
Cypress swamps	0.4–3		

[a] Adapted from Howarth et al. (1988).

1991). Measurement of uptake kinetics suggest that sulfate may inhibit fixation rates by 1–5% in freshwaters, 15–20% in seawaters, and more than 70% in highly saline lakes (Cole et al., 1993).

Another possibility is that marine systems are generally more turbulent than freshwater lakes and ponds, and the turbulence disrupts anoxic microzones near cells, reduces likelyhood of anoxia in general, and mixes bluegreen populations below critical depths (Paerl, 1990). All such effects are likely to reduce nitrogen fixation rates. This hypothesis is attractive, but experimentally manipulations of turbulence failed to reduce rates of nitrogen fixation in freshwater cyanobacteria (Howarth et al., 1993).

Other possible controls of nitrogen fixation rates might include insufficient supply of iron, as already mentioned in Section 14.2.1, or of appropriate organic sources of energy (Paulsen et al., 1991). In any case, we lack a compelling explanation for the different abundance of nitrogen fixing cyanobacteria in different kinds of aquatic environments. There is a tendency, however, for more eutrophic waters to support somewhat higher rates of nitrogen fixation (compare rates in estuaries vs. oceans, and eutrophic vs. oligotrophic lakes in Table 14-2). We know (Section 14.2.1) that high concentrations of ammonium inhibit nitrogen fixation, so we can speculate that perhaps the organic matter associated with enrichment might be responsible for raising fixation rates.

Nitrogen fixation tends to be modest in relation to nitrogen inputs from other sources in marine settings (Table 14-2, second column of numbers). Befitting the higher rates, fixation in freshwaters can furnish perhaps up to quarter of nitrogen inputs. In terms of supplying the nitrogen needs of producers, the differences between water types are smaller, and are small in any case (Table 14-2, third column of numbers).

Nitrogen fixation in and on sediments vary enormously (Table 14-2, lower part of first column of numbers). Sediments supporting vegetation usually show larger fixation rates than bare sediments. Controls of sediment fixation rates are poorly known. Sulfate inhibition is unlikely, since in most coastal sediments sulfate is reduced to sulfide, which is not sterically similar to sulfate. Interstitial concentrations of NH_4 are usually high enough to inhibit fixation.

14.2.3.1.2 Vertical Mixing. The second source of new nitrogen for the photic zone is vertical advection or mixing of nitrate-rich deeper waters. We have already mentioned how upwelling of nutrient-rich water or the breakdown of thermo- or haloclines can provide a source of new nutrients to the photic zone. This is by far the major source of nutrients for most pelagic systems (McCarthy and Carpenter, 1984), and is discussed further as a major influence on seasonal patterns of production in Chapter 15. The hydrographic provision of nutrients has already been discussed in Chapter 2 as the feature that generally determines the overall level of primary production.

14.2.3.1.3 Horizontal Transport of Nitrogen. In estuarine, coastal, and oceanic areas, there is considerable horizontal transport of nutrients, both in solution (nitrate and ammonium) and adsorbed to particles (ammonium). Shelf waters off Georgia, for example, are enriched by intrusion of nutrient-rich waters from the Gulf Stream (Hanson et al., 1981). Another example is provided in Section 15.3.1.2, were the importance of river outflow for phytoplankton production in coastal waters is discussed, but there are many other examples. The surface waters within the plume discharging from the Columbia River in the Pacific northwest of the United States contains up to $1.2\,\mu M\ NH_4^+$ even many miles off shore. Beyond the plume, ambient oceanic surface waters average about $0.1\,\mu M$ (Jawed, 1973). Much of the nitrogen transported by rivers originates as fertilizer nitrogen. In lakes and rivers of the midwestern cornbelt of the United States 55–60% of the

TABLE 14-3. Budget of Inputs and Regeneration of Nitrogen and Phosphorus for Narragansett Bay, Rhode Island.[a]

	Annual inputs (10^6 moles y^{-1})	
	Nitrogen	Phosphorus
Inputs		
Fixation	0.2	—
Rain	2.8	0.19
Runoff	16.2	0.8
Rivers	235	17.3
Sewage	278	21.7
Total	532	39.9
Regeneration		
Menhaden	0.8	0.1
Ctenophores	8.1	0.8
Net zooplankton	98.5	—
Benthos	264	41.1
Total	371	42

[a] Adapted from Nixon (1981).

nitrate in surface waters is due to agricultural use (Kohl et al., 1971); a substantial portion of this can flow to the sea.

While river transport is an obvious source of nutrients, there is increasing evidence of considerable intertidal and subtidal flux of groundwater carrying nutrients from land to coastal laters (Johannes, 1980; Bokuniewicz, 1980; Valiela, 1984, Valiela et al., 1992; and Table

TABLE 14-4. Nitrogen Budget for Great Sippewissett Marsh Itemizing the Contribution of Each Mechanism of Input and Output.[a]

	Inputs	Outputs	Net exchange
Precipitation	380		380
Groundwater flow	6,120		6,120
N_2 fixation	3,280		3,280
Tidal exchange	26,200	31,600	−5,350
Denitrification		6,940	−6,940
Sedimentation		1,295	−1,295
Other	9	26	−17
	35,990	39,860	−3,870
Percentage of exchanges via biotic mechanisms	9	18	
Percentage of exchanges via physical mechanisms	91	82	

[a] Units are in kg N yr^{-1} for the entire marsh. Below the table are the calculated percent exchanges due to biological and physiological mechanisms, which emphasize the importance of non-biological processes. Negative signs indicate net export to coastal waters. Adapted from Valiela and Teal (1979) and Valiela (1984).

14-4). Although this topic remains to be studied, it is likely that the transport of nitrogen by groundwater is significant for coastal ecosystems.

The disposal of sewage can also convey large amounts of nitrogen. The New York City metropolitan area contributes so much dissolved nitrogen to the western end of Long Island Sound that ammonia concentrations are as high as 45–100 μM. Away from the city, waters of the Sound carry only 0–5 μM. Nitrate is similarly high near the city, 8–20 μM, while the rest of the Sound has 0.5–8 μM. Near the city the nutrients are high enough not to limit growth of algae, as we saw in the enrichment experiments discussed in Section 2.2.3.3.2.

Transport from land can carry massive amounts of nitrogen to coastal environments. Runoff, rivers, and sewage provide 99.5% of the nitrogen received by Narragansett Bay, and 99.7% of the phosphorus (Table 14-3).

The nitrogen budgets of salt marshes (and many marine systems) are dominated by physical transport processes. In Great Sippewissett Marsh, flow of groundwater and tidal exchange provide 90% of the nitrogen inputs (Table 14-4). Environments where inputs of nitrogen are physically determined have no way to respond to sudden increases in nitrogen loadings. Biologically active processes such as fixation decrease if the supply of DIN increases, while denitrification may increase (cf. Section 14.2.3.4). Such mechanisms may furnish some control over nitrogen inputs. They could be important where fixation and denitrification are large compared to nitrogen inputs.

14.2.3.1.4 Precipitation. The amount of nitrogen conveyed to marine environments by rain, snow, and dry deposition is usually small relative to other inputs (Table 14-4). Near land and especially near areas of industrial and agricultural activity, this source of nitrogen increases (Eriksson, 1952), but since other inputs are also large, so precipitation is less important. For example, precipitation provides 12% of the nitrogen entering the mixed layers of the Baltic Sea (Sen Gupta, 1973), a body of water whose shores are heavily industrialized. Only 0.5% of the nitrogen entering Narragansett Bay (Nixon, 1981), a body of water surrounded by a less urbanized region, is due to precipitation (Table 14-3).

Precipitation may be more important in open, stratified oceanic water, such as the subtropical gyres where there are few other external sources of nitrogen. There is a significant correlation, for example, between the amount of rainfall and the concentration of ammonium in surface water in the Sargasso Sea (Menzel and Spaeth, 1962). Perhaps only in such oligotrophic areas are nutrients borne by rain important in sustaining primary production (Carpenter and Price, 1977).

Industrial contamination of the atmosphere has increased the nitrate content of rainwater, and marine waters downwind of industrial regions must receive increased doses, although this has not yet been clearly documented.* Section 16.4.1 provides more information on these issues.

14.2.3.2 Uptake of Nitrogen

The dynamics of uptake of dissolved nitrogen by phytoplankton and seaweeds has been discussed in Chapter 2, and are reviewed by Collos and Slawyk (1980) and McCarthy (1980). Most marine algae preferentially take up NH_4^+ over NO_3^-, and take up nitrate when ammonium is depleted.** Some producers, especially estuarine and coastal species, may use organic sources of nitrogen, for instance, methylamine by kelp and phytoplankton (Wheeler, 1979, Wheeler and Hellebust, 1981), and methyl ammonium in phytoplankton (Wheeler, 1980). A variety of amino acids and urea are also used by phytoplankton, but this uptake may only amount to a fraction of the nitrogen used by primary producers (McCarthy, 1980).

Particulate nitrogen is consumed by animals, sinks to the sediments, decays, and is regenerated as dissolved inorganic nitrogen, much like particulate carbon.

14.2.3.3 Regeneration of Nitrogen

Uptake of nitrogen by organisms is only a temporary fate. Regeneration and release eventually make organically bound nitrogen newly available. This regeneration is a key process, since inputs of new nitrogen in many marine environments are not sufficient to support the requirements of primary production. As in the case of phosphorus, nitrogen incorporated in particles is recycled by the release from bacteria, zooplankton, and fish, and by the regeneration of dissolved inorganic nitrogen from the benthos (Fig. 14-9). Measurements of these sources of regenerated nitrogen (Table 14-5) show that each is potentially capable of providing substantial amounts of nitrogen relative to the amounts assimilated by producers.

---

*The increased acidity of rainfall associated with industrial regions is due to the sulfuric and nitric acids released into the air. Freshwater is usually not well buffered and its pH can be changed by external addition of CO_2 or other acids; thus acid rain is a serious problem, and many lakes and ponds in North America and Europe have been acidified enough to lose most of their fauna. Acid rain is a far less severe problem in the more strongly buffered seawater.

**The mechanism behind preference for ammonia is not well known. Either actual uptake or reduction by nitrate reductance may be inhibited by high ammonium concentrations. The ultimate reason for the preference for ammonium may be that energy is required to reduce nitrate to the amino form. Uptake of ammonia thus avoids the need to spend energy, since the reduced nitrogen taken up as NH_4^+ can be directly incorporated into proteins.

TABLE 14-5. Sources of Regenerated Nitrogen in Various Marine Environments.[a]

<table>
<tr><th></th><th>Bacteria</th><th>Zooplankton</th><th>Fish</th><th>Benthos</th></tr>
<tr><td>New York Bight, inshore[b]</td><td>13</td><td>6</td><td></td><td>21</td></tr>
<tr><td>New York Bight, shelf break[b]</td><td></td><td>61</td><td></td><td></td></tr>
<tr><td>Southern Calif. Bight[c]</td><td>70</td><td>100</td><td></td><td></td></tr>
<tr><td>Saanich Inlet, B.C.[c]</td><td colspan="2">100</td><td></td><td></td></tr>
<tr><td>Pacific Ocean, off Oregon[d]</td><td></td><td>36</td><td></td><td></td></tr>
<tr><td>Pac. Oc., plume of Columbia R.[d]</td><td></td><td>90</td><td></td><td></td></tr>
<tr><td>N.W. Africa upwelling[e]</td><td></td><td></td><td></td><td></td></tr>
<tr><td>shelf less than 200 m</td><td></td><td>24</td><td>9.5</td><td>27</td></tr>
<tr><td>shelf over 200 m</td><td></td><td>18</td><td>6</td><td></td></tr>
<tr><td>Peru upwelling[e]</td><td></td><td>15</td><td>22</td><td></td></tr>
<tr><td>Long Island Sound[f]</td><td colspan="3">50</td><td>53</td></tr>
<tr><td>Narragansett Bay[g]</td><td></td><td colspan="2">10</td><td>26</td></tr>
<tr><td>Bering Sea, coastal[h]</td><td></td><td>2–16</td><td></td><td></td></tr>
<tr><td>Bering Sea, open sea[h]</td><td></td><td>13</td><td></td><td></td></tr>
</table>

[a] The values are expressed as percentage of the nitrogen required by measured rates of primary production. N.Y. Bight values based only on use of NH_4^+ by producers; others based on DIN use.

[b] Conway and Whitledge (1979).

[c] Harrison (1978).

[d] Jawed (1973).

[e] Whitledge (1978), Rowe et al. (1977): In this case upwelled water provided an additional 117% of the nitrogen required by the phytoplankton production.

[f] Bowman (1977).

[g] Nixon (1981).

[h] Dagg et al. (1982).

14.2.3.3.1 Regeneration in the Water Column. Most organisms in the water column contribute to the regeneration of nitrogen (Table 14-5). The role of bacteria is not settled (Harrison, 1992), but smaller organisms release relatively more nutrients than larger species (Johannes, 1964; Glibert, 1982). In coastal waters off California 90% of the mineralization of NH_4^+ is accomplished by organisms smaller than 35 μm; 40% was done by cells less than 1 μm in diameter (Harrison, 1978). These two size classes were 40 and 10% of the total particulates. The abundance of small and large plankton vary so much, however, that bacteria do not always release more total ammonium than the large zooplankton (Table 14-5). In fact, it has been argued that bacteria are not likely to be effective regenerators of ammonium. Bacteria contain a C/N of about 5, and attain a threshold gross growth efficiency of 50% if the substrate they are metabolizing has a C/N $>$ 6. Nitrogen demands by bacterial growth cannot be met solely by uptake of ambient concentrations of dissolved amino acids; the deficit is met by uptake of ammonium. Bacteria may accumulate rather than release nutrients in seawater. For example, bacteria grown on kelp detritus incorporate 28% of the carbon in the detritus, while nitrogen is taken up at 94% conversion efficiency (Koop et al., 1982). Regeneration of ammonium by bacteria is likely if C/N of readily oxidizable carbon and nitrogen substrates are <10 (Hollibaugh, 1978; Goldman et al., 1987), a requirement that is not often met in sea water. Regeneration by microbes may be more active within organic aggregates, which are known to be far more biologically active than bulk seawater (Alldredge and Cohen, 1987; Gotschalk and Alldredge, 1989).

Protozoans, flagellates, and even larger animals may be important regenerators of nitrogen. Gelatinous zooplankton may be important in regeneration of nitrogen: in the Sargasso Sea just one average-sized siphonophore per cubic meter of water could excrete enough ammonium to satisfy 40–60% of the requirements of phytoplankton (Biggs, 1977). If fish are abundant enough they may also be important releasers of ammonium, as in the Peruvian upwelling ecosystems (Table 14-5).

Nitrogen is regenerated by heterotrophs mainly as ammonium, the form of nitrogen preferentially taken up by phytoplankton. Even when nitrate concentrations are very high in Antarctic waters (17–31 μM) the phytoplankton still use very substantial amounts of ammonium, although the concentrations of ammonium may be only 0.1–2.5 μM (Glibert et al., 1982). Thus, it is obvious that the regeneration of ammonium is important for algal growth.

Regeneraton of nitrogen (and phosphorus) by zooplankton increases as more food is available to the zooplankton (Eppley et al., 1973). *Calanus* excretes 2.6% of the nitrogen in its body per day in winter, when phytoplankton is scarce, but in spring the excretion of nitrogen by *Calanus* reaches 4.6% of body N per day (Butler et al., 1970).

The importance of regenerated versus allochthonously provided "new" nitrogen was assessed by Eppley and Peterson (1979), who roughly estimate these two sources of nitrogen by assuming that uptake of ammonium (and urea) by phytoplankton corresponds to use of regenerated nitrogen, while uptake of nitrate (and N_2) is based on new nitrogen made available by mixing. Using this distinction they calculated that in neritic (open coastal) and inshore water 54 and 70% of the production is based on regenerated nitrogen, while in the open ocean 82 to 87% of production is supported by regeneration. In subtropical gyres 94% of the nitrogen used by producers is regenerated by other organisms; the phytoplankton of these depauperate waters are thus almost completely dependent on recycled nitrogen. In the most productive regions of the oceans, phytoplankton depend on large inputs of new nitrogen, while phytoplankton in oligotrophic waters make do with recycled nitrogen (Dugdale and Wilkerson, 1992).

The issue of "new" vs. "old" primary production has invigorated plankton research because it focuses on certain key concepts. For example, the export of particulate organic matter from the photic zone to deeper waters and to the sea floor has to be, on average, equivalent to new production. This realization prompted much work, including empirical measurements of the ratio of new to total production (called the *f* ratio). Some constraints to the meaning and interpretation of *f* ratios are given in Platt et al. (1992) and Legendre and Gosselin (1989), and are discussed further in Section 14.4.2.2.

14.2.3.3.2 Regeneration in the Benthos. Organic nitrogen is mineralized (converted to an inorganic form, usually NH_4^+) in sediments. The inorganic nitrogen in interstital waters can then be released to the overlying water; this is a major pathway by which regenerated nitrogen is made available to the overlying plankton (Table 14-5). The rate of release of ammonium depends on biological activity (Fig. 14-10, left). The greater the rate of aerobic decay (or oxygen consumption) the more organic nitrogen that is mineralized to ammonium. Since aerobic respiration rates in microbes and animals are strongly temperature dependent, the regeneration of NH_4^+ also depends on temperature (Fig. 14-10, right). Anaerobic decay also regenerates ammonium in large amounts and is temperature dependent.

We have earlier referred to the Redfield atomic ratios of 16N–1P usually found in particulate matter. This ratio is fairly constant over much of the ocean, but is generally lower in coastal areas. A comparison of the nitrogen and phosphorus generated from coastal sediments (Fig. 14-11) shows that there appears to be a consistent depletion of nitrogen compared to phosphorus. This loss may be the result of denitrification. The nitrogen limitation commonly found in coastal waters may be due to remineralization of 25–50% of the organic matter in the sediments, where the regenerated nitrogen may subsequently be lost by denitrifi-

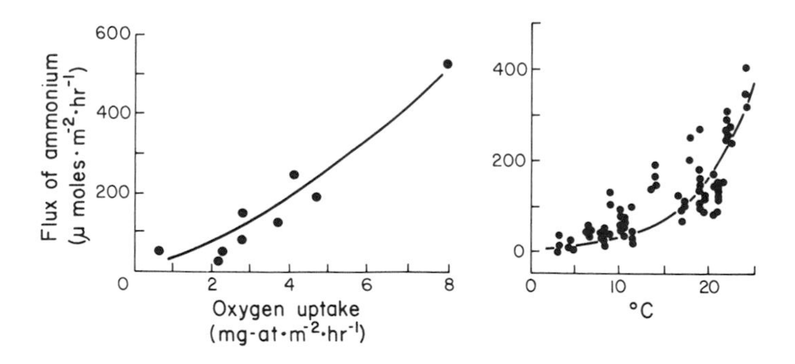

FIGURE 14-10. Effect of biological activity (left) and bottom water temperature (right) on the release of ammonium from sediments in Narragansett Bay, Rhode Island. Adapted from Nixon (1981) and Nixon et al. (1976).

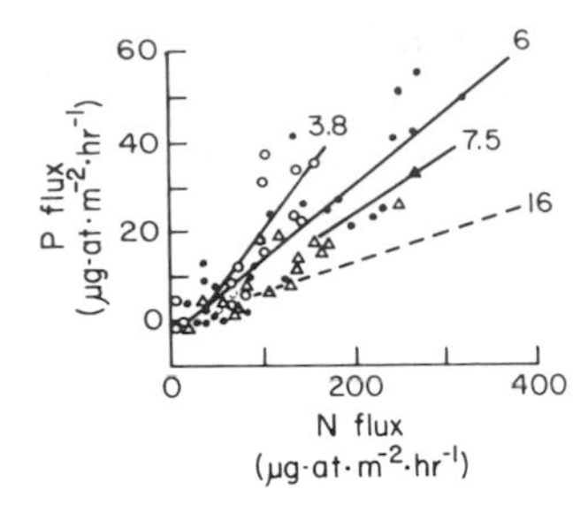

FIGURE 14-11. Relationship of flux of phosphate and dissolved inorganic nitrogen (mainly ammonium) in three stations in Narragansett Bay. The three stations showed fluxes with N/P of 3.8, 6, and 7.5, with correlation coefficients of r^2 = 0.69, 0.76, and 0.83, respectively. The Redfield ratio (N/P = 16) is included for comparison. Adapted from Nixon et al. (1980).

cation. Bioturbation and burrowing by benthic animals mix the organic matter into the surface layers, where mineralization takes place. In the sediments some of the ammonium released from the organic matter is nitrified to nitrate, and denitrification then converts a substantial proportion of the nitrogen to N_2.

The large gradients (100- to 1,000-fold differences) in concentration of nitrogenous compounds between interstitial and overlying water lead to movement of nitrogen out of sediment (Table 14-6). Measured rates of regeneration of inorganic compounds in coastal sediments such as those of Table 14-6 are enough to supply an average of 35% (± SE of 8.7%) of the nitrogen demands of phytoplankton.

14.2.3.4 Losses of Nitrogen

Nitrogen is lost from aquatic systems by burial in sediments, transport by water, or denitrification. Denitrification is the preeminent of these major loss mechanisms. Denitrification depends on the supply of NO_3 furnished by either external sources, such as water overlying sediments, or by local nitrification of NH_4^+ released by decay of organic matter. This latter process has been called "coupled nitrification/

TABLE 14-6. Average Rates of Regeneration (Mean ± Standard Error) of Forms of Inorganic Nitrogen (Nitrate, Nitrite, Ammonium, and Total Dissolved Inorganic Nitrogen) from 10 Coastal Sediments to the Overlying Water.[a]

Nitrogen compounds	Exchange rates (mmole m^{-2} day^{-1})
NO_3^-	0.73 ± 0.42
NO_2^-	0.09 ± 0.06
NH_4^+	2.2 ± 0.6
Total DIN	3 ± 0.9

[a] From Fisher et al. (1982).

denitrification." Denitrification occurs primarily where there is a juxtaposition of oxic and anoxic conditions, because while denitrifiers are anaerobic, nitrification requires oxygen. Higher rates are therefore found in layers of water and sediments between oxic and anoxic conditions.

Seitzinger (1990) reviews studies that show that irrigation of sediments by animals and gas transport and DOC release by marine plants and algae increase delivery of oxygen and nitrate, and hence elevate rates of nitrification and denitrification. Any change in conditions that alters abundance of organisms, such as increased nutrient inputs to seawater, will change the controls of rates of denitrification, so that there are seasonal and secular changes in denitrification (Law et al., 1991), as well as differences in rates among different types of aquatic environments (Table 14-7).

Denitrification makes biologically active DIN unavailable, and because nitrogen availability is important to so many marine processes, this kind of microbial respiration has received much attention. There is, however, a lack of agreement about methods for measurement of this process, but some comparisons are now available (Seitzinger et al., 1993). Desirable methods need high sensitivity to detect small changes in N_2O or N_2, the two gaseous end-products, in air or water where dissolved N_2 is in high concentration (Table 14-2). Methods also need to include denitrification that depends on NO_3^- supplied by external sources (such as overlying water), and denitrification based on NO_3^- produced by nitrification within the sediment sample being studied.

Denitrification accounts for 15–71% of nitrate consumed in various estuaries (Law et al., 1991), so it is not the only sink for NO_3^-. Uptake by producers is often larger, if light is available. For example, over the year, loss of nitrate via denitrification in the Tamar Estuary was only 10% of the nitrate used by phytoplankton. Another pathway of nitrogen

TABLE 14-7. Ranges of Denitrification Rates in Sediments of Different Aquatic Environments.[a]

Location	Rate of denitrification (μmol N $m^{-2} h^{-1}$)
Rivers and streams	0–2,121
Lakes	
Oligotrophic	0.3–56
Moderately eutrophic	10–55
Eutrophic	20–292
Coastal marine	0–888
Shelf	0.5–54
Deep sea	0.03–2.4

[a] From Seitzinger (1990).

loss is fermentative NO_3^- reduction, a process possible in dark, anoxic conditions, and where nitrate supplies are low. In most cases, in particular where there is some degree of bioturbation of sediments, denitrification is more important than reduction.

Denitrification and nitrification produce nitrous oxide, a biogenic gas that is currently increasing in the atmosphere. In the Tamar Estuary, denitrification is responsible for 65% of the N_2O release (Law et al., 1991), and sediments release N_2O to the atmosphere, as long as there is NO_3^- available. If NO_3^- is unavailable, denitrifiers use N_2O as the oxidant source, and N_2O is not released. N_2O release increases in proportion to the availability of NO_3^- (Seitzinger, 1990; Law et al., 1991), so that eutrophication of nearshore waters can be expected to increase the release of this important greenhouse gas to the atmosphere (Law et al., 1992).

Denitrification is quantitatively important in nitrogen budgets for entire ecosystems (Seitzinger, 1988, 1990, and Table 14-4), and rates depend on nitrate, DOC, temperature, and bioturbation. The average

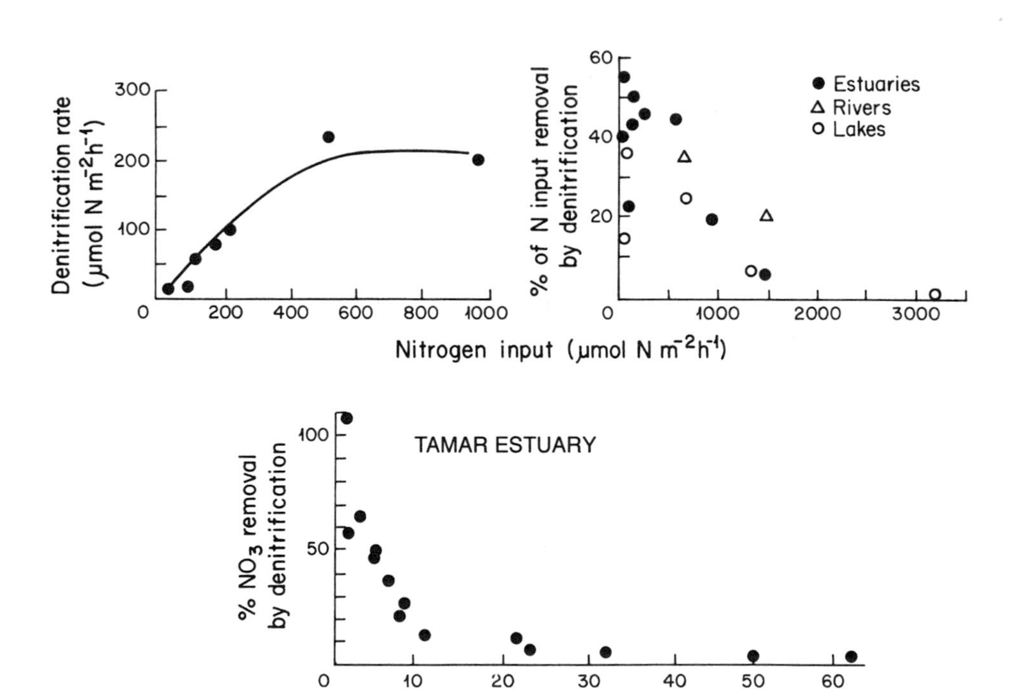

FIGURE 14-12. Rate of denitrification (top left) and % of nitrogen input lost via denitrification (top right) [data of Seitzinger (1990) and Jørgensen and Sørensen (1985)] in several ecosystems and in the Tamar Estuary [bottom, data from Law et al. (1991)].

rate of denitrification in various ecosystems increase as nitrogen supply increases (Table 14-7, Fig. 14-12, top left), but its rate of increase is insufficient to offset external nitrogen loading rates. Average rates reach an asymptote (Fig. 14-12, top left), and so actually decrease as a percent of inputs (Fig. 14-12, top right). Decreases in percent of nitrogen inputs as loading rates increase seem to be the rule in coastal waters (Law et al., 1991; Zimmerman and Benner, 1994). For example, in the Tamar Estuary, different rates of loading took place over different months of the year and % of inputs lost by denitrification are progressively reduced at higher loading rates (Fig. 14-12, bottom). Although losses of anthropogenic nitrogen via denitrification are significant, denitrification by itself is therefore insufficient to prevent nitrogen enrichment of coastal waters.

There is more NH_4, and nitrification of the NH_4 may be more efficient in freshwater sediments than in marine sediments (Seitzinger, 1990). It may therefore be that relatively more NO_3 is made available in freshwaters (cf. Section 2.2.3). Both producers *and* denitrifiers may therefore have a greater supply of NO_3 in freshwaters than in coastal waters. Such contrasts may be in part responsible for the different role of nitrogen as a limiting factor for producers in fresh and coastal waters.

14.2.4 Nitrogen Budgets for Specific Ecosystems

To understand the dynamics of nutrients in specific ecosystems, we need budgets of the inputs and losses, but very few budgets are available. Comparison of the magnitudes of the various sources of nitrogen and of recycling are therefore possible only in a few environments.

In the shallow coastal system of Narragansett Bay (Table 14-3) rivers and sewage are the major sources in the nitrogen and phosphorus budgets. Animals regenerate 70% of the N and 128% of the P that enter the bay. The activity of small animals and microbes, while not measured, probably adds considerably to the recycling of N and P. Regeneration in some environments such as sandy surf zone ecosystems can be even higher, reaching 780% of external nitrogen inputs (Cockroft and McLachlan, 1993). The input to Narragansett Bay have a N:P = 13.3, somewhat lower than the ideal Redfield ratio. perhaps in part creating a nitrogen-limited situation. The loss of nitrogen by denitrification in sediments of Narragansett Bay was equivalent to about half the input of nitrogen from sewage and rivers (Seitzinger et al., 1980).

The Narragansett Bay budget (Table 14-3) shows that in shallow coastal systems, external sources of nutrients are mainly due to horizontal transport, nitrogen recycles slower than phosphorus and is depleted (due to nitrate reduction and denitrification) relative to phosphorus. The Narragansett Bay budget also demonstrates that anthropogenic sources of nutrients can be very large compared to other sources (Table 14-3). Both agricultural fertilizer and waste water contribute to the nitrogen loadings of coastal waters. Since nutrients, as we have seen, are so intimately linked to virtually every process important in marine ecology, very significant changes can be potentially brought upon marine environments by eutrophication due to human activity. The nitrogen in culturally eutrophied waters is of special significance since we have seen that coastal primary production is nitrogen limited.

Another nitrogen budget in which losses as well as inputs are reported is available for a salt marsh (Table 14-4). Most of the inputs and outputs of nitrogen in a salt marsh are conveyed by physical mechanisms (flow of groundwater and tidal exchanges), while biotic mechanisms (N_2 fixation and denitrification) provide smaller inputs and outputs. As we have seen, perhaps in very stable, nutrient-poor oceanic waters fixation may be more important, but most nitrogen transport in marine environments is probably dominated by physical mechanisms.* Considering the variation of each of the entries of Table 14-4, the relatively small deficit of nitrogen is not significant; this ecosystem is roughly in balance with regard to nitrogen.

Another way to make use of nutrient budgets is to tally the inputs, fates, and outputs of each of the major species of nitrogen so as to depict the net transformations that go on which the ecosystem (Fig. 14-13). The exact amounts in each entry are not the important issue, since these are only rough estimates. Instead, notice that the many internal transformations within the marsh ecosystem can be very large, often exceeding annual inputs and outputs. This is invariably the case in many ecosystems. For instance, consider the particulate nitrogen at the bottom of Fig. 14-13. Particulate nitrogen enters the marsh by transport and by fixation. About half of the particulate nitrogen is consumed by animals, of which about two thirds is released as particulate nitrogen in feces. Producers add another substantial amount of particulate nitrogen. The particulate nitrogen released by the marsh to coastal waters exceeds the input; within the marsh major qualitative transformations have taken place, and the nitrogen has been extensively reused while in the ecosystem.

Similar complex patterns of use and reuse can be followed for any of the forms of nitrogen in Fig. 14-13. The point here is that once in an ecosystem, nutrients recirculate in complex ways; knowledge of the

---

*Notice that this discussion deals with inputs and outputs to the ecosystem. Internal transformations due to biological activity—for example, benthic regeneration in Narragansett Bay (Table 14-3)—can be larger than inputs or outputs.

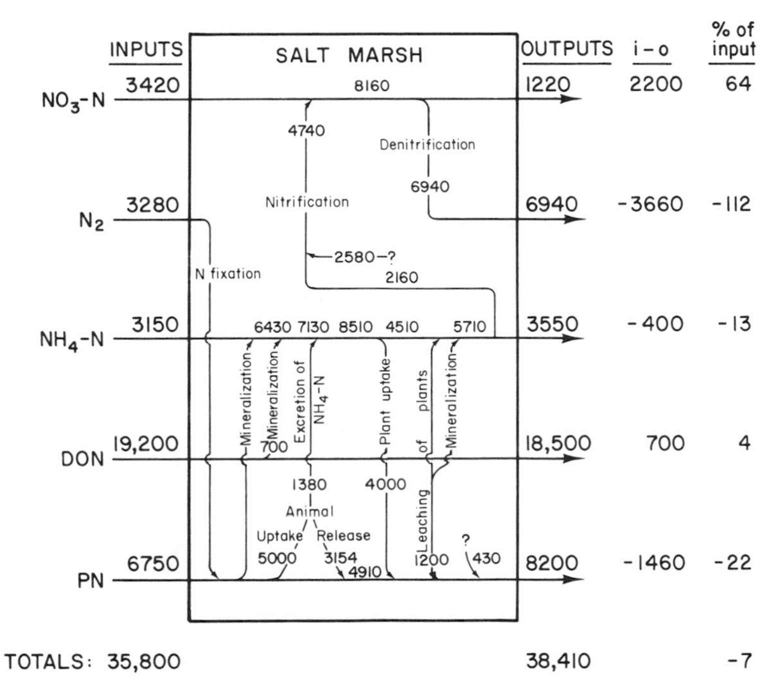

FIGURE 14-13. Nitrogen budget for Great Sippewissett Marsh for a year. Inputs and outputs are indicated along the margins of the box; the two right-most columns show the difference between inputs and outputs and the percentage of the input for each nitrogen species. Inside the box representing the marsh are indicated some of the major transformations. From Valiela (1983), data from Valiela and Teal (1979), Jordan and Valiela (1982), and unpublished data.

inputs and outputs provides only partial understanding of how the salt marsh works or how active processes are.

It is still instructive, however, to compare the chemical state of the inputs vs outputs of nitrogen. The salt marsh essentially converts NO_3 to N_2 gas, ammonium, and particulate nitrogen (Fig. 14–13). The large amounts of DON seem to be flushed back and forth with the tides, and may remain relatively unchanged. Perhaps many of the compounds making up DON are not easily degradable (cf. Section 14.2.2.1). The N_2 exported from the marshes is not useful to coastal organisms but ammonium and particulate nitrogen are energy rich and accessible. Export of these reduced compounds (already mentioned in Sections 7.7.1 and 13.1.2) could provide energy, nitrogen, and carbon to the adjoining coastal waters. Thus, the end products of processes in one environment can be transported and can affect assemblages of organisms in adjoining ecosystems.

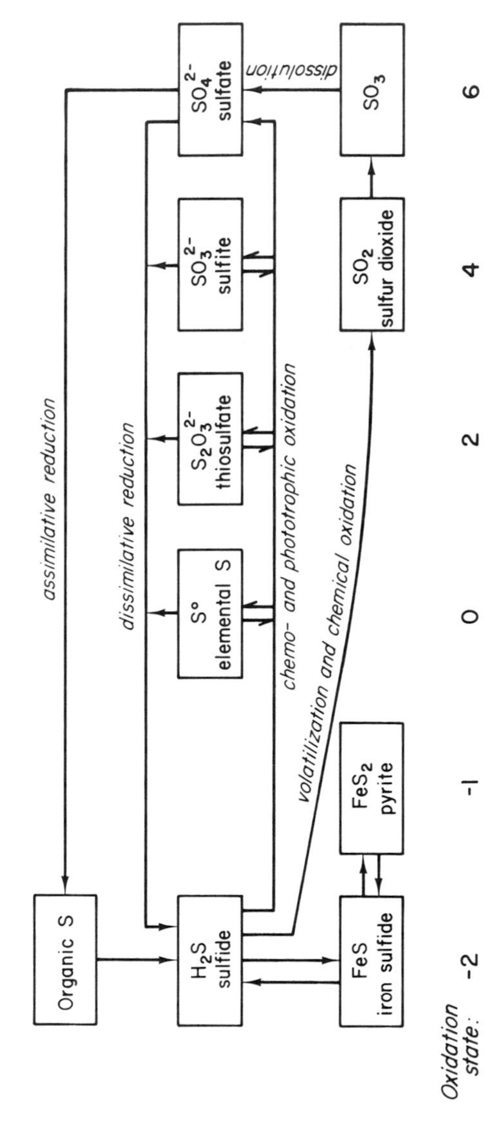

FIGURE 14-14. Scheme of microbial and chemical transformations of sulfur. The oxidation state of S in pyrite is −1 on average; there are probably mixtures of sulfur atoms in 0 and −2 states. Adapted and expanded from Fenchel and Blackburn (1979).

14.3 Sulfur

The dynamics of sulfur include aqueous, sedimentary, and atmospheric phases, as well as a wide range of oxidation states. As a result the sulfur cycle is very complex. We have already covered aspects of sulfur dynamics involved in the decomposition of organic matter. Microbiological reactions of sulfur also involve the nitrogen cycle, as well as many other elements. The cycle of sulfur thus illustrates the couplings among nutrient cycles, production, and decay processes.

14.3.1 Chemical Transformations

In aerobic environments, sulfate (SO_4^{2-}) is by far the most abundant form of sulfur. Although uptake and assimilative reduction of sulfate by bacteria and algae occur (Figs. 14-14, 14-15), these processes do not

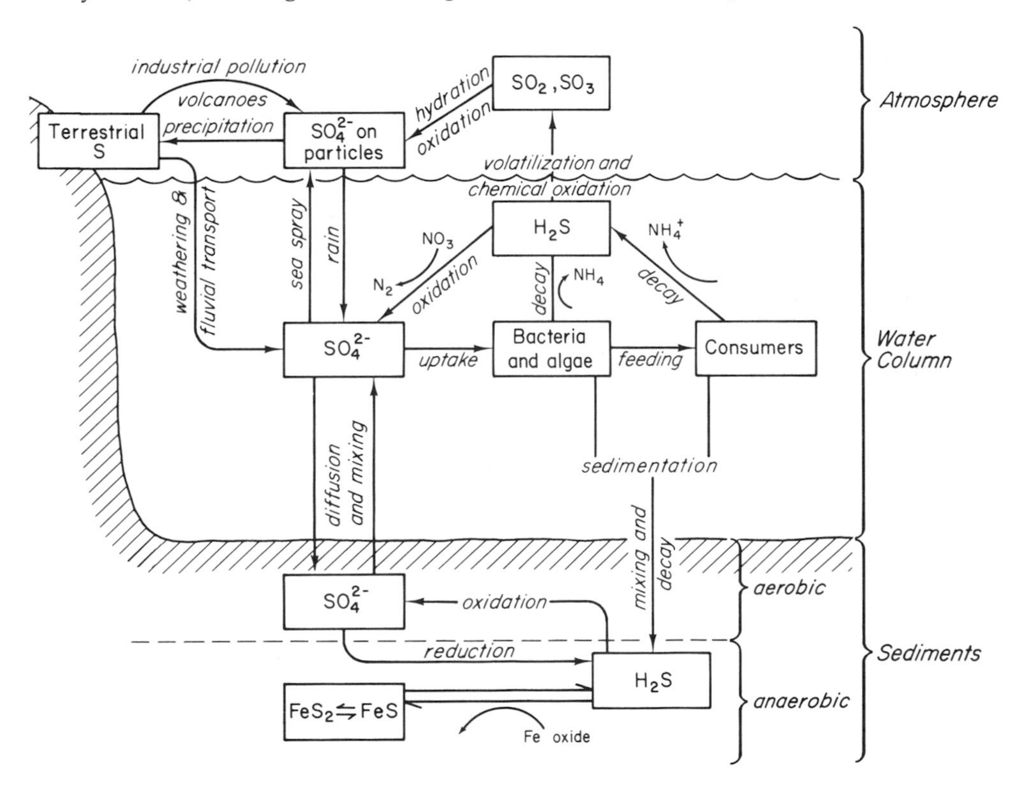

FIGURE 14-15. Simplified scheme of sulfur dynamics. Couplings to nitrogen cycle are indicated on the arrows that represent processes of transfer or transformations. This is representative of a coastal environment; in the deep sea sediments are usually not reduced. The diagram does not include the volatilization of dimethyl sulfide, probably released by producers, and whose flux is probably larger than that of H_2S.

materially change concentration of sulfate because this ion is so abundant in seawater. There is sulfur within cells, including methionine (—S—), cystine (—S—S—), or cysteine (—SH), all reduced organic sulfur compounds. Sulfolipids and sulfate esters are also important. Marine producers contain 0.3–3.3% sulfur, but the decay of this organic sulfur is usually not an important source of sulfide in marine environments, accounting for only a few percent of the sulfide production (Jørgensen, 1977; Deuser, 1979).

Dissimilative reduction of sulfate is the principal source of extracellular sulfide in anaerobic environments (Tables 13-8, 13-9). Sulfate reduction rates range from 0.01 to 10 μmole SO_4^{2-} ml^{-1} day^{-1} (Jørgensen, 1982). The rates depend on the mechanisms that convey sulfate into the anaerobic environment, such as diffusion, mixing by animals, and on the availability of organic matter. Sulfate reducers also depend on fermentation of complex, polymeric organic matter (Table 13-8) by fermenting microorganisms. Such fermentation produces low-molecular-weight substances which in turn are respired by sulfate reducing bacteria (Jørgensen, 1982; Howarth and Teal, 1979). *Desulfovibrio* and *Desulfomatuculum* are the principal taxa involved in sulfate reduction.

Sulfide [actually dimethyl sulfide is the major volatile sulfur compound (Andreae and Raemdouck, 1983)] can diffuse out of a reducing environment* (Fig. 14-15) and be oxidized biologically and chemically into more oxidized sulfur compounds (Fig. 14-14), including elemental sulfur, thiosulfate, sulfite, and sulfate. Oxidation of sulfide occurs where O_2 or nitrate is available, and at the same time sulfide diffuses out of an anaerobic zone; sulfide oxidizers therefore occur in a narrow layer or plate, either in sediments (Fig. 14-16) or in the water column or around anaerobic microzones. The use of nitrate links the nitrogen, sulfur, and carbon cycles, as does the production of ammonium during sulfate reduction (Table 14-1) and aerobic decay (Fig. 14-16).

Some of the sulfide in anaerobic sediments reacts with dissolved iron or oxidized iron minerals to form FeS and FeS_2 (Figs. 14-15, 14-16) (Rickard, 1975; Fenchel and Blackburn, 1979; Luther et al., 1982). Pyrite (FeS_2) can be oxidized by *Thiobacillus* if there is oxygen or nitrate available, and the reaction results in production of acid (this is the process that produces acid mine wastes).

There are reports that sulfide oxidation occurs even in strongly reducing conditions (Fossing and Jørgensen, 1990). The electron acceptors in reduced sediments may be oxidized managanese and iron minerals. Mn(IV) oxides, for instance, may transform sulfides to sulfate.

---

* Anaerobic conditions occur in sediments or in the water column; Fig. 14-16 is representative of a coastal environment, where anaerobic sediments are common. In the Black Sea (cf. Fig. 1-9), fjords, and in certain coastal ponds, water below the mixed layers is anaerobic.

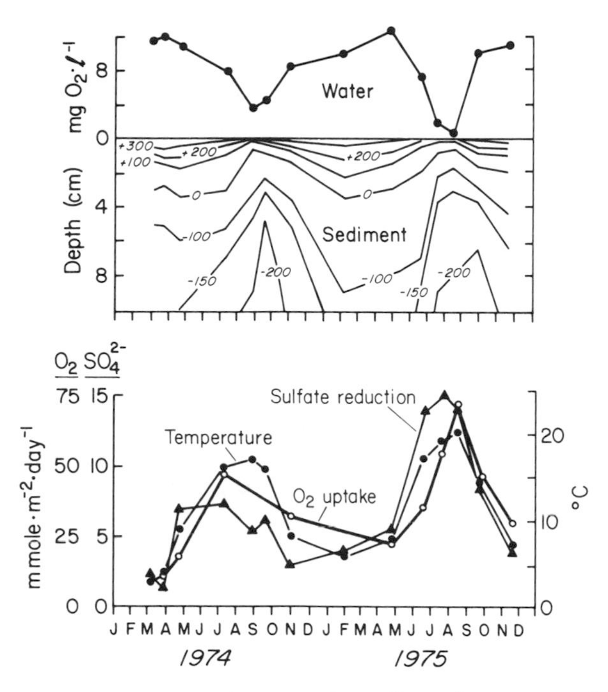

FIGURE 14-16. Top: Vertical and seasonal profiles of oxygen and redox in the water and sediments, respectively, in Limfjorden, Denmark. Positive redox potentials show oxidized conditions, while negative values show reduced conditions. Bottom: Surface temperature and rates of sulfate reduction and dark O_2 uptake in top 10 cm of sediments. Adapted from Jørgensen (1977).

In coastal sediments Mn reduction could account for much of the sulfide produced in anoxic muds (Aller and Rude, 1988), so that the sedimentary sulfur recycles through reduction and oxidation phases.

Bacteria that use sulfate as an electron acceptor are relatively inefficient at extracting energy out of organic compounds (cf. Section 12.3), and can use only about 25% of the energy available in organic matter. The remaining 75% is either buried in sediments as unoxidized organic matter, or is transferred to sulfides (Howarth, 1984; Martens and Klump, 1984; Sampou and Oviatt, 1991) by sulfate reducers.

Dissolved sulfides (HS^-, HS, S^{2-}) produced may be transformed into metal sulfides such as iron monosulfide (FeS), greigite (Fe_3S_4), and ultimately pyrite (FeS_2). These reduced compounds store energy fixed by producers in sediments of salt marshes and estuaries (Howarth, 1984). The formation of reduced sulfur compounds by sulfate reducers thus uncouples on-going carbon metabolism from oxygen and energy use. The sulfides can be reoxidized at some later time (Howes et al., 1984), probably through nitrate, iron, and manganese reduction. In coastal sediments 25–94% of sulfate reduced is eventually reoxidized (Berner, 1989; Chanton and Martens, 1987; Smith et al., 1991).

A number of forms of sulfur of intermediate oxidation state (elemental sulfur and sulfite, Fig. 14-15), and polythionates ($S_{3+n}O_6$, $n = 0 - 3$) are produced by different bacteria or are oxidized chemically. These are usually found in low concentrations or have very fast turnover in seawater.

Thiosulfate is probably the major intermediate in the sulfur cycle (Fossing and Jørgensen, 1990; Jørgensen and Bak, 1991). Thiosulfate is produced by biological sulfide oxidation, or chemical oxidation of dissolved or iron-bound sulfide. Thiosulfate is reduced to sulfides by sulfate reducers, and oxidized to sulfate by thiobacilli. Thiosulfate is also subject to "disproportionation," a type of fermentation in which bacteria obtain energy from the cleavage of the S − S bond in $S_2O_3^{2-}$ to produce sulfates and sulfides. In reduced sediments, more than half of the oxidation to sulfate proceeds through thiosulfate as the intermediary, so this compound may play a key role in the flow of electrons, and the biological activity such flow supports.

14.3.2 Dynamics of Sulfur in Marine Environments

Although the qualitative outlines of the sulfur cycle have long been known, quantitative knowledge of sulfur transfers in natural environments is rare.

Studies in Solar Lake, a hypersaline coastal pond on the edge of the Gulf of Elat, Red Sea, have demonstrated an anaerobic, hypersaline lower layer of water in which there is anoxygenic photosynthesis primarily carried out by cyanobacteria (Cohen et al., 1977, 1977a, 1977b). In Solar Lake, bacterial activity controls the vertical gradients of sulfur and other elements. Sulfide and oxygen, for example, have profiles that look much like those of the Black Sea (Fig. 1-9), although of course the depth of Solar Lake is just a few meters. Oxygen is depleted down the water column, and sulfide increases below the depth where oxygen disappears.

Another comprehensive study of the dynamics of sulfur was done by Jørgensen (1977) on Limfjorden, Denmark. Limfjorden is a shallow brackish water body with an aerobic water column that exhibits a seasonal fluctuation in dissolved oxygen (Fig. 14-16, top). Even though there is seasonal variation in the reduction of the surface sediments, below about 5 cm the sediments are permanently reduced. Near the sediment, surface temperature varies seasonally and seems to drive both the dark uptake of O_2 and the rate of sulfate reduction in the top 10 cm of the sediment (Fig. 14-16, bottom). Sulfate reduction takes place even in winter when the sediments are most oxidized, because of the presence of reduced microzones 50–200 μm in diameter (Jørgensen,

1977a). In such sediments sulfate reducers (*Desulfovibrio*) and sulfide oxidizers (*Beggiatoa*) occur close together, and so the sulfide diffusing out of anaerobic microzones is oxidized; in addition, some of the sulfide may be converted to iron sulfides. The result of both these processes is that little H_2S accumulates in the sediment (Fig. 14-17, left).

The vertical profiles of the various forms of sulfur in sediments are due to various processes. In the upper 10 cm, the sulfate concentrations are like those of the overlying waters (Fig. 14-17, left). This is probably the effect of irrigation caused by the activities of tube-dwelling animals, as in the case of vertical profiles of nitrogen compounds (Section 14.2.3.3.2). Below the zone inhabited by animals, bacterial reduction decreases sulfate (Fig. 14-17, right). Ferric iron in the sediment is reduced to ferrous iron and precipitates with sulfide to form iron monosulfides and pyrite (Luther et al., 1982). Pyrite is not a permanent sink for S. Although there are disagreements as to rate of formation and oxidation of pyrite (Howarth, 1979; King, 1983; Howes et al., 1984), there are seasonal changes in the net oxidation of pyrite.

A considerable portion (170 g organic $C\,m^{-2}\,year^{-1}$) of the primary production of the phytoplankton and eelgrass flats reaches the sediments of Limfjorden and its decay supports the sulfur cycle. Only 12% of this carbon accumulates, and the whole carbon pool in the sediment can be turned over in 3–5 years, by both O_2 respiration and sulfate reduction. In Limfjorden about 53% of the organic carbon is consumed by sulfate reducers. In other coastal environments there are similar high proportions (50% or so) of consumption of organic matter by sulfate reduction (Jørgensen, 1982). In Great Sippewissett Marsh, annual rates of sulfate reduction are high enough to consume over half the annual production of organic carbon by living salt marsh plants (Table 13-10). In such anaerobic environments, therefore, dissimilative

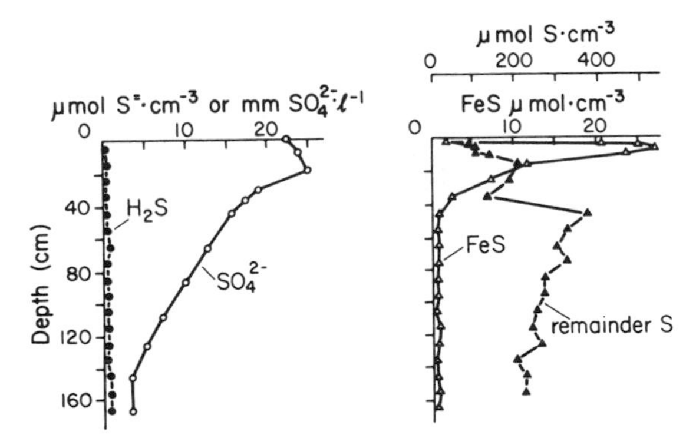

FIGURE 14-17. Vertical profiles of sulfur compounds in sediments of Limfjorden, Denmark. Remainder S is shown as total S minus H_2S, FeS, and SO_4^{2-}, and probably consists of pyrite (FeS_2). Adapted from Jørgensen (1977).

sulfate reduction is a very major pathway by which organic carbon is degraded.

In deeper coastal waters the depth of the aerobic layer of sediment increases. The amount of organic matter reaching the anaerobic layers is lower and the refractory nature of the organic matter increases, so that the organic substrate available for sulfate reducers is less abundant. In deeper waters, then, rates of sulfate reduction are lower and the proportion of organic matter mineralized by sulfate reducers may only amount to 35% of that consumed by aerobic respiration (Jørgensen, 1982).

The rates of organic matter oxidation by sulfate reducers are about the same as the rates of aerobic oxidation, as long as the rate of sediment accretion are high (Fig. 14-18, right side). Under such conditions presumably carbon supply is high, and anoxic conditions are maintained. Metabolism by either oxygen respiration or sulfate reduction become lower where low rates of sedimentation furnish less organic matter (Fig. 14-18, left side). Under low rates of sedimentation, moreover, sulfate reducers are considerably less effective, perhaps because it is more difficult to maintain reducing conditions.

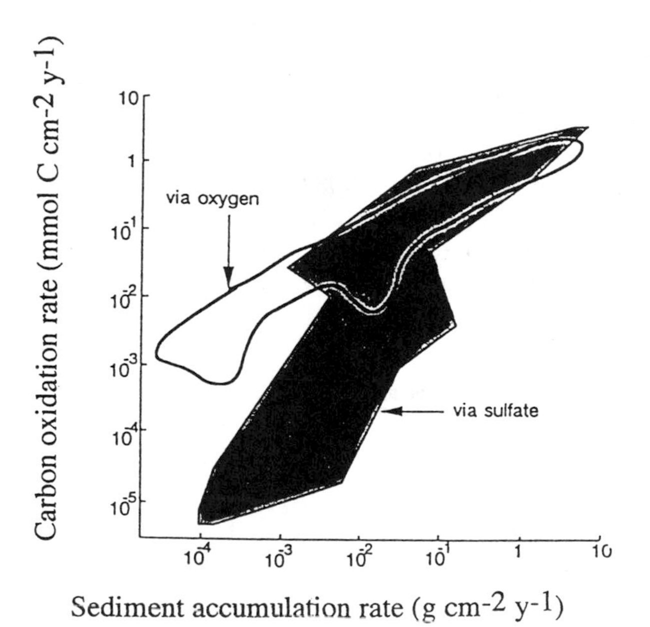

FIGURE 14-18. Envelopes surrounding points that show observations of rates of organic matter oxidation supported by oxygen respiration or by sulfate reduction in sediments exposed to different rates of sedimentation. Redrawn by Blackburn (1991) from data in Canfield (1989).

14.4 Ecosystem Energetics and Stoichiometry

14.4.1 Energy Flow Through Components of Ecosystems

Energy flow budgets for ecosystems were first put together for freshwater systems by C. Juday and R. Lindemann. The first rough energy budgets for marine systems were put together by Clarke (1946) for Georges Bank and Harvey (1950) for the English Channel, followed by Odum and Odum (1955) for a coral reef, Golley et al. (1962) for a mangrove swamp, and Teal (1962) for a salt marsh. More recent research has revised many parts of such budgets (Nixon and Oviatt, 1973; Pomeroy and Wiegert, 1981) and has shown the importance of anaerobic processes in energy transfer (Howarth and Teal, 1980; Howes et al., 1985). We considered the transfer of organic matter of planktonic food webs in Chapter 9.

Most energetics studies focused on aerobic processes. A benthic example is provided by a study of energy budgets in the Ythan estuary ecosystem in Scotland (Baird and Milne, 1981). Only about 2.7% of photosynthetically active radiant energy available to plants is converted into primary production by phytoplankton, benthic microalgae, and macrophytes. Primary production reaches about $630\,g\,C\,m^{-2}\,year^{-1}$, of which macrophytes contributed about 72%.

Secondary production by invertebrates in the Ythan totals about $69\,g\,C\,m^{-2}\,year^{-1}$, of which the zooplankton, meiofauna, and macrofauna contribute about 1, 28, and 71%, respectively. Secondary production by consumers is about one order of magnitude smaller than primary production. Grazers consume a small fraction (7%) of plant production; other major groups of consumers feed on detritus and on other particles. The production of the meiofauna is about 52% of the production of the deposit-feeding macrofauna, even though the biomass of smaller organisms ($2.1\,g\,C\,m^{-2}$) is only 13.6% of the biomass of the macrofauna ($15.1\,g\,C\,m^{-2}$). This reiterates our earlier statements on relatively higher production associated with smaller body size.

There are some other estimates of the relative activity of macro- and meiofauna. In a tidal mudflat in the Wadden Sea, macrofauna biomass is 20 times that of the sum of meiofauna, microfauna, and bacteria (Kuipers et al., 1981). Yet, since the metabolic rate of a nematode, for example, is about 21 times higher than that of an average macrobenthic specimen, 74% of the organic matter is consumed by the small organisms. In other soft-bottom marine habitats respiration of the macrofauna ranges from 2 to 34% of the total respiration of the community (Pamatmat, 1968, 1977; Banse et al., 1971; Smith et al., 1972; Smith, 1973; Davies, 1975).

The meiofauna in the Ythan consume about $69\,g\,C\,m^{-2}\,year^{-1}$, of which 38% is egested as feces and 46% is respired. The major suspension feeder is *Mytilus edulis*. This mussel consumes about 66 g C m^{-2} $year^{-1}$ from the water column, of which 50% is respired, and 25% is egested. From the standpoint of the mussel-predator flow of energy, this is about a 12% ecological efficiency, well in the range of our earlier review.

Macrobenthic deposit-feeders in the Ythan consume about 400 g C m^{-2} $year^{-1}$, respire 29% of this, and defecate 61%. Presumably the large amounts of feces are due to the low quality of food available as detritus. The macrobenthos net production is about 38 g C m^{-2} $year^{-1}$, of which birds, fish, and crabs eat 24.7, 22, and 4%, respectively.

The efficiency of energy transfer within the estuarine ecosystem can be measured by the consumption efficiency, the consumption by trophic link n relative to the net production of trophic link $n - 1$ (recall tht ecological efficiency was $[P_n/C_n] \times 100$). In the Ythan estuary, fish birds, and crabs consume suspension feeders and macrofauna with consumption efficiencies of 71 and 48%, respectively. The prey organisms therefore suffer rather significant annual losses to predators, amounting to about 50–75% of the yearly crop of biomass. Predators thus are very likely to be a major structuring feature of estuarine environments.

The marked loss of fixed energy in the various steps of food chains puts constraints on the amounts of protein harvestable from the sea. Ryther (1969) calculated that about 24×10^7 tons of fish were produced in the oceanic, coastal, and upwelling areas of the sea. His calculation* was based on rates of primary production, estimates of the number of trophic links and assumed ecological efficiencies in each region (10–20%). Oceanic regions, as befits biological deserts, were estimated to produce only about 16×10^5 tons, in spite of making up 90% of the surface area. The upwelling areas, even though only 0.1% of the area, were calculated to yield 12×10^7 tons. The coastal zones produced another 12×10^7 tons of fish, and make up 9.9% of the area of the sea. Clearly there are many uncertainties in each of the elements of the calculation. For example, level and patchiness of primary production, role of gelatinous plankton, ecological efficiency, lack of proper coverage of the Antarctic sea, among other aspects, are all still inadequately known. Ryther's estimate is close to the 200 million tons of fish estimated to be produced annually in the sea by Schaefer (1965).

This 200–240 million tons of wet weight of fish cannot be completely harvested by man, however. A certain proportion of the annual pro-

*Ryther calculated that 20×10^9 tons of carbon were produced per year in the sea by primary producers. This is lower than more recent estimates of 31×10^9 tons of carbon $year^{-1}$ by Platt and Subba Rao (1975). Ryther also did not consider the microbial pathways of energy flow that have received recent attention. It is not clear if these omissions cancel each other; Ryther's estimates should therefore be taken as just that: estimates.

duction has to be left to provide for maintenance of the population, plus other top predators in the sea. Recall, for example, tht predators consumed 50–75% of the production of prey in the Ythan estuary.

It is hard to estimate the sustainable level of harvest of fish from the sea. In 1967, about 60 million tons of fish were harvested worldwide, and this increased to about 100 million tons in the 1990s. Although there is perhaps some room for further increase in fish harvest, the fish resources in the sea are not going to solve the world's need for protein. To make matters worse, much of the current fish harvest is made into fish meal, which is then fed to poultry and other animals. Although poultry are fairly efficient at food conversion, averaging about 20%, the feeding of the fish catch to poultry just adds another trophic link to the food web, which means a substantial loss of energy caught as fish.

14.4.2 Ecosystem Stoichiometry

14.4.2.1 Coastal Waters

Earlier studies on energetics emphasized processing of fixed energy and carbon by animals up a food web. Metabolism was taken to be aerobic respiration, with oxygen being consumed to yield water, carbon dioxide, and energy. Studies in estuaries and salt marshes soon made evident that other respiratory pathways, discussed in Chapter 13, were quantitatively significant.

The relative importance of the different metabolic pathways depends on conditions as well as availability of the different energy sources and electron acceptors. Where there is sulfate in ready supply, sulfate reduction is relatively more important, while where there is more nitrate available, as is often the case where there are large inputs of freshwater, reduction of nitrate (via denitrification and ammonification) may be higher (Table 14-8). The relative importance of sulfate reduction

TABLE 14-8. Relative Importance of Different Pathways in Sediments Located in Seawater- and Freshwater-Dominated Sites of Norsminde Fjord.[a]

	Percent of total electron flow	
	Seawater-dominated site	Freshwater-dominated site
Aerobic respiration	65	44
Denitrification	1.5	4.3
Ammonification	5	33
Sulfate reduction	27	19

[a] Adapted from Jørgensen and Sørensen (1985).

also increases as supply of labile organic matter increases (Sampou and Oviatt, 1991).

The intrincate net nutrient and organic transformations at the level of the ecosystem can be summarized by stoichiometric calculations that relate carbon, oxygen, phosphorus, and nitrogen transformations by equations such as included in Tables 13-8 and 13-9. Is in this context that the concept of total ecosystem metabolism (cf. Section 1.3.1.2) is most relevant. Defining net ecosystem production for such whole-system metabolism studies require considerable understanding of the physical exchange of water, as well as knowledge of gaseous exchanges across the air-water interface (Oviatt et al., 1986; Smith et al., 1991; Sampou and Oviatt, 1991; Smith and Hollibaugh, 1993; Marino and Howarth, 1993).

Smith et al. (1991) provides an example of use of a whole-system stoichiometric approach. They concluded that degradation of allochthonous organic matter in Tomales Bay, California, produced a supply of phosphorus that was exported. In contrast, denitrification within the bay was a sink for nitrogen delivered from outside the bay. In addition, there was a release of CO_2 to the atmosphere as a result of net oxidation of organic matter, so the ecosystem was heterotrophic. The total metabolism of Tomales Bay was supported by inputs of organic matter from its watershed or perhaps from deeper waters.

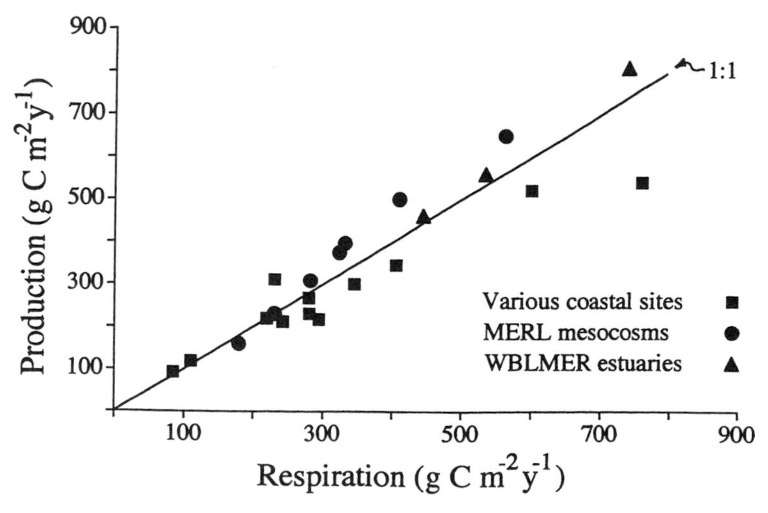

FIGURE 14-19. Relation of net ecosystem production to respiration in various coastal ecosystems. Data on various coastal sites compiled by Smith (1991), points from the Marine Ecology Research Laboratory mesocosms from Oviatt et al. (1986), points from the Waquoit Bay Land Margin Ecosystems Research project from D'Avanzo and Kremer (1995). The straight line shows the position where production and respiration would have a 1:1 relationship.

Ecosystem net production in different coastal ecosystems is related to respiration (Fig. 14-18). The departures of the individual ecosystems from the 1:1 line in Fig. 14-19 may or may not have significance. The estimates of net production and respiration may have an associated variation of perhaps 20–30% or so; if so, the points of Fig. 14-19 fall at some distance from the 1:1 line just by chance. On the other hand, if the deviations away from the line are significant, they would suggest that the individual ecosystems either are heterotrophic, as in the case of Tomales Bay, with some of the organic matter consumed being contributed from outside the ecosystem, or are autotrophic, which implies some organic export to other systems. This topic is discussed further in Section 16.43.

14.4.2.2 Oceanic Waters

A key corollary to the idea of "new" production proposed by Dugdale and Goering (1967) and Eppley and Peterson (1979) was that auto-otrophic activity in a community depended on larger nutrient inports than exports (Platt et al., 1992). Nitrogen has been considered the key nutrient in most discussions of new vs. old nutrients, although in larger geochemical terms phosphorus has to be the ultimate limiting nutrient to primary production (Smith, 1984). Nitrogen fixation can supply nitrogen from the unlimited atmospheric pool, so given appropriate space and time, production would be limited by phosphorus supply (Smith, 1984; Codispoti, 1989). Within time scales most relevant to most ecological processes discussed here (seconds to multiple years), nitrogen supply is likely to limit growth and production of primary producers. Phosphorus is likely to be the limiting factor for producers at longer time scales, perhaps 10^2–10^3 years (Codispoti, 1989).

There are several ways to estimate new production (Harrison et al., 1987; Jenkins and Wallace, 1992). The most direct way is to measure uptake of $^{15}NO_3$ in containers, but there is a question whether such measurements are representative of entire water masses. A second approach is to take the sum of inputs of nitrogen by hydrographic mechanisms, fixation, and atmospheric sources to be a stoichiometric measure of net ecosystem production. In most situations, vertical mixing from below is assumed to be the major source of dissolved nitrogen to surface oceanic waters (Platt et al., 1992). If the surface waters are in steady state, another stoichiometric measure of new production is the particulate exports downward out of the surface layers. Adequate sampling over relevant scales would be demanding, so it is difficult to show that in fact such a steady state holds. A third approach is to measure rate of O_2 consumption in the subsurface layer (where there are no internal O_2 sources). The rate of O_2 consumption is a measure of the organic material sinking from the surface layer, and can be converted stoichiometrically to equivalent C or N use, over appropriately long time periods (Jenkins, 1982). Such calculations have already been discussed in Chapter 2. All these approaches need more evaluation.

New production can be equated to net ecosystem production, the difference between gross primary production and respiration by all the organisms in an ecosystem. New production can be also be thought of as that part of ecosystem production that can be removed without loss of biomass or without having an external supply of organic matter (Smith, 1988). Removal can be by sinking, or by the action of consumers of various kinds. The concept therefore could potentially tell us what proportion of production could go to support food webs in deeper waters of the oceans, or even how much fishing harvest would be sustainable by an ecosystem. We are not yet at the stage where such calculations can be made with confidence, but the possibilities are of interest. The link between new production and net ecosystem production becomes less defined in ecosystems that are subject to larger inputs of allochthonous organic matter, such as estuaries and other coastal waters (Quiñones and Platt, 1991).

The issue of new vs. old production is also relevant to the current controversy regarding whether or not some of the increasing atmospheric CO_2 may be stored in the ocean. Production based on nitrogen in the surface layer fixes C and respiration releases CO_2, but this CO_2 can diffuse back to the atmosphere within fairly short time intervals (Platt et al., 1992). It is the new production, whose equivalent leaves the surface layers, that is relevant to the issue of sequestration of carbon in the sea. This packaging and sinking has been called the "biological pump" that may carry atmospheric CO_2 to the deep layers of the ocean, where waters may be out of contact with the atmosphere for thousands of years.

Chapter 15
Seasonal Changes in Marine Ecosystems

15.1 Introduction

Seasonal cycles take place in all marine ecosystems. In this chapter, we examine some examples to make evident the interplay between biological and chemical processes and meteorological and physical processes. The examples show how some of the detailed mechanisms examined in previous chapters are involved in the control of function in marine ecosystems.

15.2 Water Column Seasonal Cycles

There are several patterns of seasonal cycles in water columns, but the identification of the kind of cycle is not as important as is the insight that the changes may provide as to how planktonic assemblages are governed.

15.2.1 Temperate Seas

The early development of marine research centers took place in the coastal areas of temperate regions, so that naturally enough annual seasonal cycles in plankton were first described for areas such as the English Channel. During winter the nutrients regenerated by mineralization, resuspended from the bottom or transported by mixing of richer deeper water accumulate in surface waters. The accumulated nutrients are not depleted by algal uptake because the absolute amount of light is low and the depth to which surface waters mix exceeds the critical depth, so algae are often too deep for photosynthesis to be very active (Sverdrup, 1953).

We have earlier (Section 2.1.1) mentioned the compensation depth (D_c), where the rate of photosynthesis of a phytoplankton cell (P_c) equals its rate of respiration (R_c). Above D_c, P_c exceeds R_c, and below D_c, P_c is smaller than R_c. Cells will be transported above and below D_c,

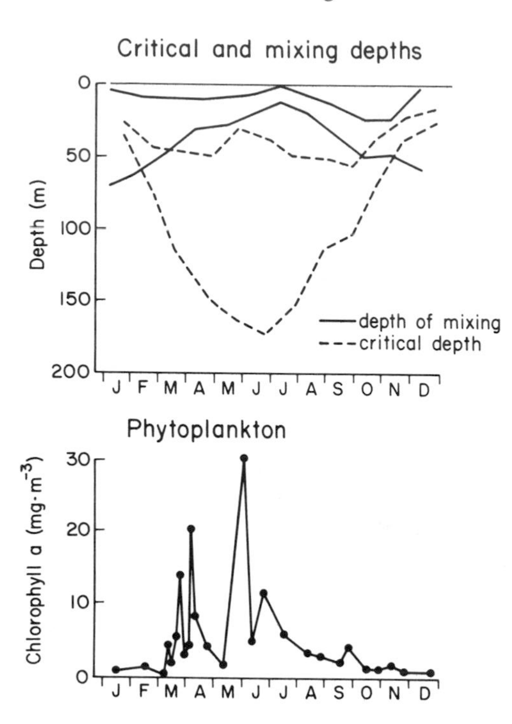

FIGURE 15-1. Top: Maximum and minimum depths (indicated by upper and lower lines) of the mixed layer and of the critical depths in the Strait of Georgia, west of 124°W. Bottom: Seasonal cycle of chlorophyll in nearby Departure Bay. Adapted from Parsons and Le Brasseur (1968).

and they will experience an average compensation light intensity (I_c). The critical depth (D_{cr}) is the depth at which photosynthesis for the water column equals respiration for the water column. When the critical depth is a large value it can be calculated (Parsons et al., 1977) as $D_{cr} = 0.5\ I_o/I_c k$, where k is the extinction coefficient and I_o is solar radiation. The importance of the critical depth is that if it is less than the depth to which surface waters are mixed, no net production in the water column takes place, since P_w is smaller than R_w, where P_w and R_w are the production and respiration in the water column. If the critical depth is greater than the depth of mixing, net production takes place in the water column (P_w exceeds R_w) and phytoplankton growth occurs.

In the early spring the critical depth increases in proportion to the increased light intensity (Fig. 15-1, top and Fig. 15-4, top and 3rd graph). The depth of mixing decreases as wind strength decreases during that same period, and there may be some warming of near-surface water or fresh water input (Legendre, 1990) that increase stratification. The result of all this is that more light is available, and the critical depth falls below the depth of mixing of surface waters (Fig. 15-1, top). The algae are therefore not carried below the critical depth and the spring bloom ensues (Fig. 15-1, bottom). The interaction between critical depth and the mixed layer depth starts off a bloom in temperate seas, and perhaps ends growth of phytoplankton in the fall.

Temperature plays a modest but clear part in the seasonal cycle of phytoplankton production. Even though temperature alters activity of phytoplankton species (cf. Fig. 2-28), in a phytoplankton assemblage there are species adapted for activity at most temperatures; photosynthetic performance decreases significantly only at water temperature below 1°C (Pomeroy and Diebel, 1986). In many temperate coastal areas, for example, the timing of the spring bloom is governed by factors other than temperature, and, fortuituously, takes place while water temperatures are coldest (Fig. 15-2, top).

Cold temperatures seem to affect bacteria more than phytoplankton (Pomeroy and Diebel, 1986; Pomeroy et al., 1991). Bacterial counts, respiration, and growth remain low during the spring phytoplankton bloom off Newfoundland, where temperatures are between −1° and 2°C. Bacterial generation times averaged 30–86 d, and bacteria con-

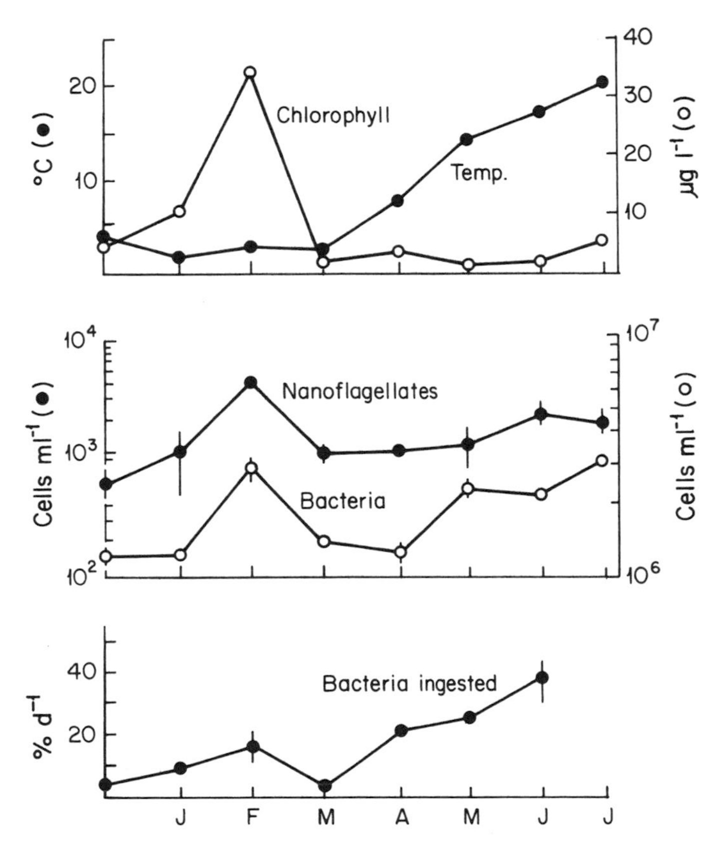

FIGURE 15-2. Seasonal changes in temperature and chlorophyll (top graph), in nanoflagellates and bacteria (middle graph), and in % of bacteria consumed by nanoflagellates (bottom graph) in Vineyard Sound off Massachussetts. Data from Marrasé et al. (1992).

sumed 0.8–8% of net primary production, compared to the 25–50% in warmer waters. It may be that in cold waters the considerable lower bacterial activity allows more of net production to be used by herbivores in the classic food web.

As water warms, bacterial abundance and activity increase (Fig. 15-2, middle). Bacterial abundance increases even though the rate of bacterial consumption by heterotrophic nanoflagellates also increase (Fig. 15-2, bottom). Even though at ambient temperatures nanoflagellates remove up to 41% of the bacteria per day, the increase in bacterial cell counts suggests that bacterial abundance depends more on resources rather than consumers.

The increase in phytoplankton in early spring provides more food for the few grazers in the water column, and the zooplankton respond (Fig. 15-4, top) by increasing consumption and reproduction—the functional and numerical responses to prey density. Microzooplankton increase in abundance between winter and spring (Fig. 15-3, top). The pattern characteristically has larger species, mostly *Calanus* copepods, peaking earlier (Fig. 15-3, middle), and a second later peak of smaller species (Fig. 15-3, bottom). Many species of zooplankton overwinter in deep water in subadult stages. In the spring there are usually too few individuals for their functional response to increased density of algae to prevent further growth of the phytoplankton. The rate of growth of the zooplankton population (the numerical response) involves a time lag, as pointed out in Section 5.3. Growth and reproduction also involve time lags and are temperature dependent, as shown in Chapter 7, and in early spring water temperatures are low. Since phytoplankton populations are light-limited rather than temperature-limited during spring, they grow at very fast rates, perhaps doubling or more each day. Under spring temperatures, it is difficult for the zooplankton to expand their numerical abundance and consumption rate fast enough to limit the increase in phytoplankton. The feeding, reproductive, and migratory rates of zooplankton just are not fast enough to produce enough mouths to restrict growth of a food supply that can double every day.

At some point in spring the algae deplete essential nutrients, principally nitrogen, from the mixed layer (Fig. 15-4, second graph). Diatoms make up the bulk of the spring bloom (or blooms following upwellings or mixing events), with smaller flagellates taking over after nutrients are depleted (Keller and Riebesell, 1989; Legendre, 1990; Bode et al., 1994). Production rates may remain high based on flagellate activity, but eventually the rate of phytoplankton growth slows down. At this time of year, recycled nitrogen is the major source of nitrogen for phytoplankton and the phytoplankton may be deficient in nutrients (Yentsch et al., 1977). Since water temperatures are higher, grazers can be more active, and build up their numbers. Because the rate of growth

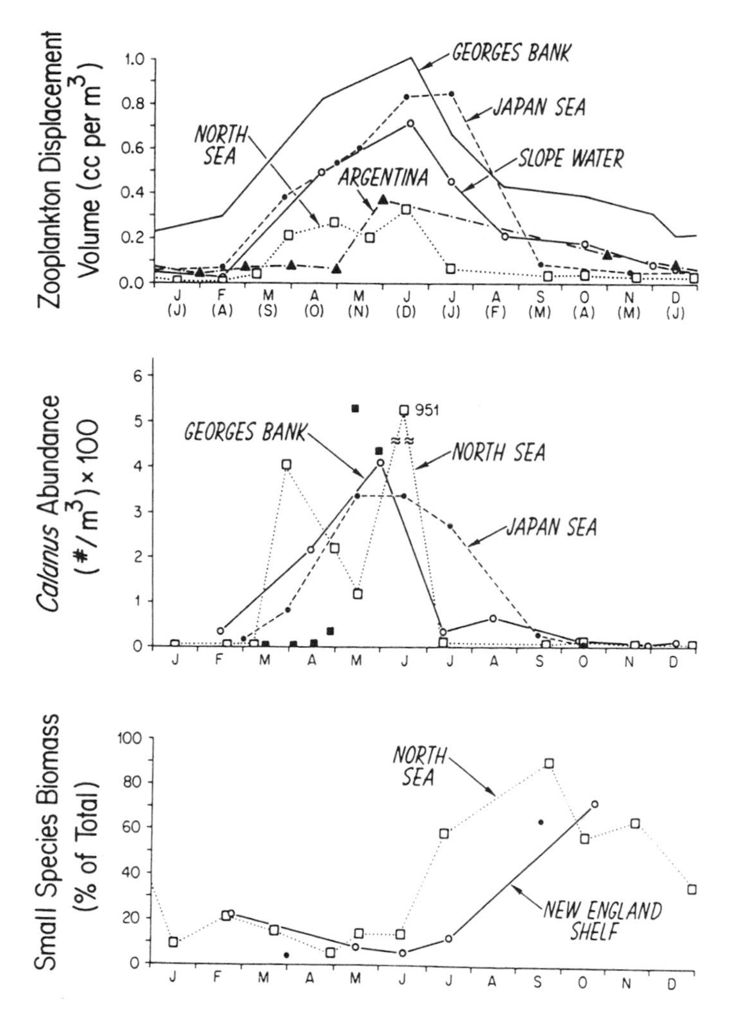

FIGURE 15-3. Seasonal cycles of zooplankton biomass (top), abundance of *Calanus* spp. (middle), and biomass of smaller species (< than 1.2 mm prosome length) (bottom) expressed as % of total zooplankton biomass. From Davis (1987).

of algae is slower than earlier in the season, it becomes possible for grazers to remove phytoplankton cells faster than they are produced (Fig. 15-4, top). The result is a midseason decrease in phytoplankton standing crops. One example of this changed impact of grazing through the year is provided by a study in a Swedish fjord. Bamstedt (1981) found that zooplankton only consumed 3–9% of net primary production early in the growth season, while later on, 60% of the production was consumed daily. Others report that at most 11–34% of the production is consumed by grazers (Keller and Riebesell, 1989) but in any case, warmer temperatures makes it more feasible for grazers to consume significant quantities of phytoplankton.

Depletion of nutrients in surface waters may persist through the summer if the thermally stratified water column that prevents the vertical mixing of deeper, nutrient-rich water with surface, nutrient-poor water persists. There may be intermittent periods of mixing that give rise to short bursts of algal production (Figs. 15-1,15-4) during the summer. These intermittent mixing events could be of meteorological or tidal origin (Legendre, 1990).

The annual plankton cycle of Georges Bank is one of the best studied examples for coastal waters (Fig. 15-5). Populations of large copepods

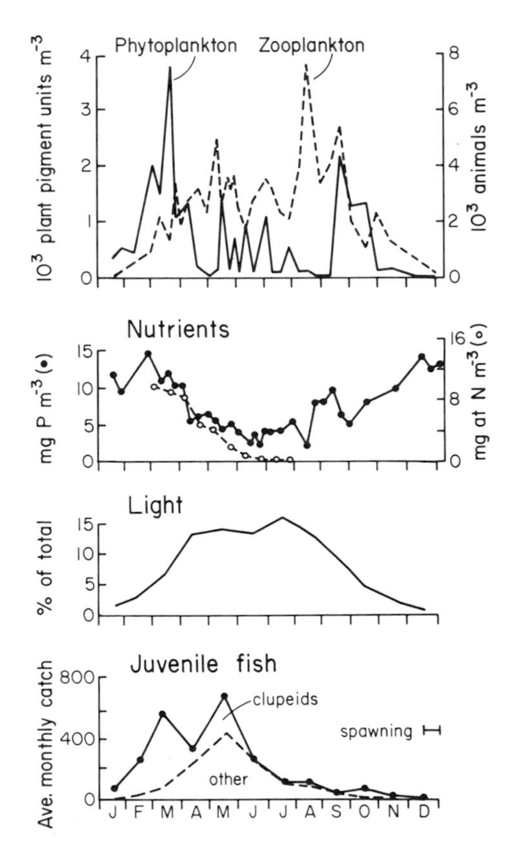

FIGURE 15-4. Seasonal changes in top 0–45 cm of the English Channel. Light is shown as the percentage of the total annual radiation that reached the sea surface in each month. Adapted from Harvey et al. (1935). Nitrate values for North Sea, adapted from Steele and Henderson (1977). Clupeid spawning occurs in December. Juvenile fish are the average number of fish caught on a 2-hr oblique haul with a 2-m ring trawl. Adapted from Russell (1935).

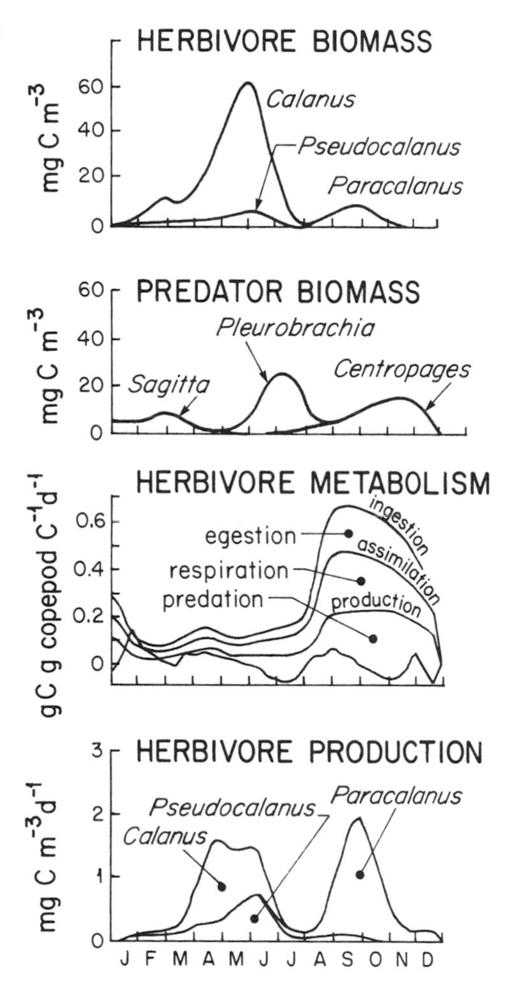

FIGURE 15-5. Seasonal changes of herbivore and predator biomass, and herbivore metabolism and production in Georges Bank. Adapted from Davis (1987).

such as *Calanus* produce only one generation during spring; smaller copepod dominate the fall microplankton and are more productive (Fig. 15-5, third and fourth graphs). Consumption of phytoplankton reflects these differences: ingestion during the autumn is considerably larger, at least in modeling studies. Since phytoplankton production is higher during spring, these simulations suggest that a greater fraction of the production is consumed by the smaller copepods. Loss of phytoplankton during spring may be dominated by sinking rather than grazing, since the larger diatoms would fall faster than the smaller flagellates.

Food use becomes more efficient as a season progresses. The smaller copepods of the fall have higher growth efficiencies than the larger spring forms; *Pseudocalanus*, for example, requires lower phytoplankton concentrations for peak growth than *Calanus*. Moreover, growth efficiency of the larger copepods decreases as the water warms. Small forms may be more efficient at feeding in warmer waters. This may result in higher reproductive success (Runge, 1984) and greater contribution to the total zooplankton production in Georges Bank (Fig. 15-5, bottom). One consequence of all this is that perhaps the larger microzooplankton are food-limited except during the spring bloom. The more trophically efficient smaller species are not food limited over the year, especially in productively constant waters such as Georges Bank (Davis, 1984).

Loss of microzooplankton to predators is largest during autumn (Fig. 15-5, third graph), even though predators are present throughout the season (Fig. 15-5, second graph). It seems unlikely that functional responses may be involved in control of copepod prey. Invertebrates such as the predaceous chaetognaths, ctenophores, and carnivorous copepods (Fig. 15-5, second graph) have rapid numerical responses and may be the more important predators (Davis, 1984a). In Bedford basin off Nova Scotia, medusae were more abundant than ctenophores or chaetognaths, and consumed only 2% of the microzooplankton daily during March, but 25–59% from May through June (Matsakis and Conover, 1991). Such consumption rates could easily decrease zooplankton populations, but there are other gelatinous forms that feed on medusae, so controls by top-down forces become confused.

Juvenile fish are most abundant in the English Channel in spring, near the time of the spring bloom (Fig. 15-4, bottom). The herbivorous young clupeids peak first, followed by other fishes. Juvenile fish grow in size and become more effective predators on zooplankton.* These fish may so decrease biomass of zooplankton as to prevent the zooplankton from entirely removing all algal cells in summer (another "cascade" effect). In the late summer when the juvenile fish either migrate to deeper waters or move elsewhere, zooplankton numbers may increase, and this, combined with depletion of nutrients, may result in the lowest phytoplankton densities of the year.

After September, either because of shortage of food or because of physiological triggers cued by the shorter photoperiod, most grazers

---

*Recall from Section 4.4 that competition for food was evident in young fish so they must limit their food supply. Other predators of copepods such as chaetognaths may also restrict zooplankton grazing. *Noctiluca* are algal feeders but can also effectively eat eggs of copepods such as *Acartia* (Ogawa and Nakahara, 1979). In the Sea of Japan maximum densities of such predators coincide with low peaks of copepod abundance. The abundance of adult fish, as we might expect from Chapter 9, seem unrelated to events lower down in the food web.

begin to die or migrate downward, eventually to become relatively inactive in deeper waters.

With cooling temperatures, or perhaps because of increased meteorological disturbances in early fall, mixing occurs and nutrient concentrations in surface water increase. Even though light intensity is reduced in the early fall, there may still be enough for phytoplankton growth. Again, whatever grazers are still present fail to control the growth of algae and hence a fall bloom of phytoplankton may take place. Soon, however, light levels become low or the mixed layer deepens due to the breakdown of stratification of the water, and the seasonal cycle of production ends, with nutrients accumulating once again in the surface layers.

This seasonal cycle, with some variation, can be seen in many temperate parts of the sea, for example, off Nova Scotia (Platt, 1971) and in the Mediterranean (Steemann Nielsen, 1975). Waters of higher latitudes have somewhat later spring blooms than lower latitudes, and the intensity of the fall versus the spring bloom also varies, depending on local circumstances.

In shallow coastal areas or where the vertical stratification is easily disrupted, nutrients may continue to be delivered to the photic zone after the spring bloom, and growth rates of phytoplankton usually continue at a high level through the summer (Fig. 15-6). Grazers seem unable to catch up to the algae and the phytoplankton may be limited by low light most of the year. Production in such shallow temperate areas seems to follow the seasonal pattern of light intensity, and perhaps temperature, through the year.

The seasonal dynamics of plankton at temperate latitudes seem therefore determined by light, the depth of vertical mixing, nutrient supply, grazing and predation pressure. Local combinations of these variables result in specific seasonal idiosyncracies for each area. How variable these patterns can be is seen in the surveys of continuous plankton recorders carried by commercial vessels in the North Atlantic (Colebrook, 1979). A mixture of the bimodal and unimodal cycles is the rule (Fig. 11-18). In open ocean areas, the warming and stratification of the water column often precede the spring bloom. In these areas, down-mixing of cells to subcritical depths is important. In coastal or shallow areas (B1, C4, D1, D2, E8, E9 in Fig. 11-18) or where haloclines or thermoclines prevent vertical mixing, for example, area B8 (compare to the deeper B7), the blooms often occur before the warming of the surface waters; the shallow depth prevents the loss of cells from the photic zone. Substantial overwintering stocks of copepods can, however, damp out the spring bloom (B1). The marked fall peak in F4 may be due to upwelling of rich water during July–September.

The point that needs emphasis here is not the identity of the type of seasonal pattern but rather that it is the interactions of a relatively few

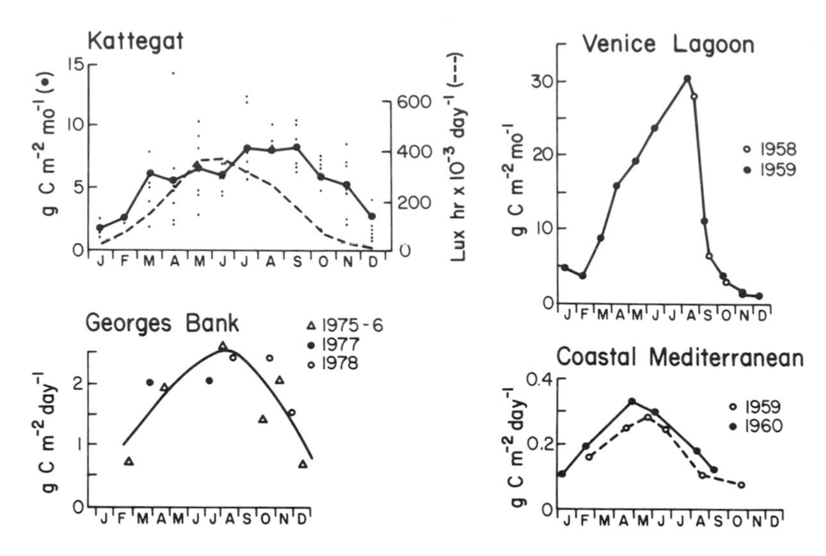

FIGURE 15-6. Seasonal cycle of primary production in shallow temperate waters. Top left: Gross production in Danish coastal water and light intensity in Copenhagen (Steemann Nielsen, 1975). Top right: Net production in the lagoon of Venice (Vatova, 1961). Bottom left: Production in Georges Bank, an open sea shallow area. Average standard deviation of each point is $1\,g\,C\,m^{-2}\,day^{-1}$ (Cohen et al., 1981). Bottom right: Production off the city of Villafranche. From Brouardel and Rinch in Steemann Nielsen (1975).

processes that control the structure of the plankton community over time.

15.2.2 Estuaries

Estuarine seasonal cycles show effects of the mechanisms described in the last section, but the nutrients are borne by seasonally-varying river flow rather than provided by seasonal mixing.

In Narragansett Bay, Rhode Island, the Providence River and sewage treatment plants provide a variable but high-nutrient input (Fig. 15-7), with somewhat enhanced deliveries of ammonium and phosphate in the fall. Nitrogenous nutrients are depleted down the estuary through the year, especially during February–September. Dissolved nitrogen is the primary limiting factor and is in very low concentrations by early March. Although there is a seasonal change in concentrations of phosphate and silicate, these nutrients are not depleted down the estuary.

In Chapter 2 we mentioned the limited role of nitrogen and phosphorus in coastal marine and freshwater, respectively. In estuaries we deal with transitions between fresh and seawater, so it is not surprising to find that there are graded responses by primary producers to N and P along estuaries (Caraco et al., 1987). In Chesapeake Bay, at the extremes of low and high salinity (Fig. 15-8, bottom), ratios of N to P lie above or below Redfield values, and producers are limited by P or N, respectively, throughout the year (Fig. 15-8, top). Phosphorus may be seasonally limiting to estuarine producers at salinities lower than 10–15‰. Enrichment studies and nutrient and production surveys show that P limits producers during spring when river flow dominates nutrient inputs to Chesapeake Bay, and N/P values during spring are much higher than Redfield values (Fig. 15-8, top). During the rest of the year, when wastewater and exchanges with coastal waters are the major sources of nutrients, N/P are lower than 16:1 (Fig. 15-8, top), and nitrogen supply determines activity of producers. Similar results were found in Chinese waters affected by river runoff (Harrison et al., 1990), but in other estuaries inputs of N and P may not vary seasonally in the same way, and N supply may remain as the dominant control over the year (Graneli et al., 1990; Rudek et al., 1991). Silicate may be important at certain times of the

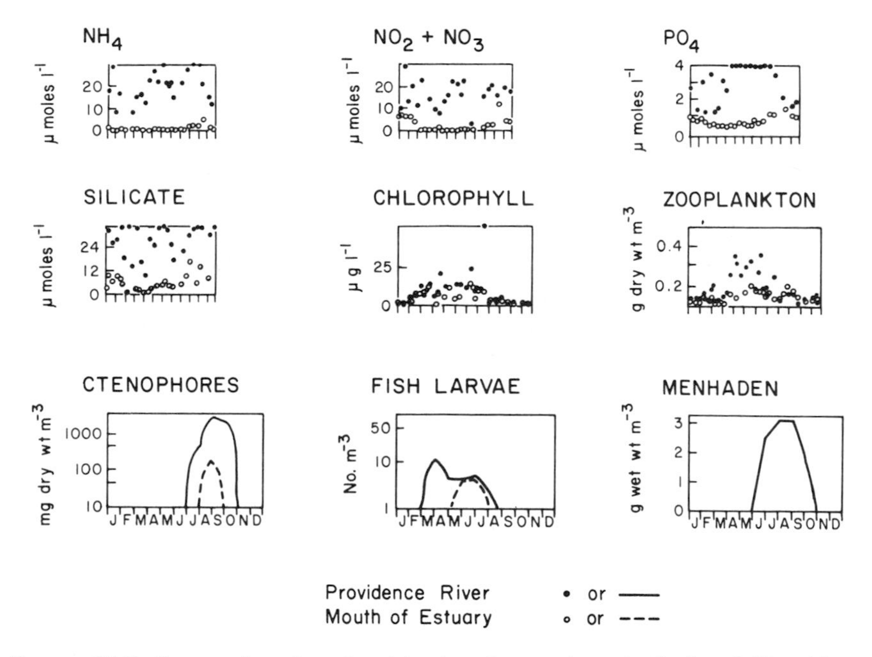

FIGURE 15-7. Seasonal cycles of nutrient and organisms in the head (Providence River) and mouth of the Narragansett Bay estuary. Nutrients and chlorophyll are from the surface. The copepods *Acartia tonsa* and *A. clausii* make up 95% of the zooplankton. *Mnemiopsis leiydi* is the carnivorous ctenophore. The menhaden (*Brevoortia tyrannus*) are found with various other fish species. Adapted from Kremer and Nixon (1978).

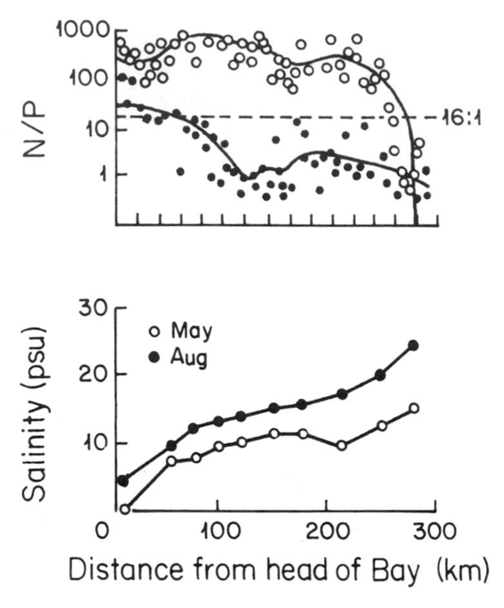

FIGURE 15-8. Differences in ratio of dissolved inorganic nitrogen to phosphorus (top), and in salinity (bottom) along the Chesapeake Bay estuary, from the headwaters to near the mouth of the Bay. Peak river discharge occurs Mar–May. Adapted from Fisher et al. (1992).

year; exhaustion of silicate may be responsible for ending the spring diatom bloom.

The phytoplankton in Narragansett Bay are probably light-limited in winter, as evidenced by the low chlorophyll concentration (Fig. 15-7, middle row and column) throughout the estuary. The phytoplankton (Fig. 15-7 middle row, right column) have more pronounced midseason peaks at the head that at the mouth of the estuary.

The high nutrient concentrations at the head of Narragansett Bay allow growth of phytoplankton through much of the year. There may be greater primary production and a higher zooplankton crop up the estuary during the warm months. The zooplankton may be responsible for the reduction of chlorophyll in the summer (Martin, 1968).

There are no general rules to predict what nutrients limit estuarine producers, and when. Rather, the generality is that estuarine seasonal cycles depend on the temporal occurrence of deliveries of nutrients, the relative magnitudes of the sources of nutrients, and the demands of the specific groups of organisms present. Each estuary may show a different array of these three types of conditions, which result in a seasonal cycle with reasonably well-understood control mechanisms.

There are other estuaries where the river inputs are more strongly seasonal than in Narragansett Bay. One of these is the Nile estuary

(Fig. 15-9) where the river input—shown by the sharply lowered salinity of the seawater near the discharge of the river—comes in late summer at the end of the rainy season in the Nile catchment area.* Nutrients in coastal water increase when the Nile discharges and the phytoplankton blooms quite readily. Numbers of zooplankton increase with only a slight lag behind the algal blooms. The rapidity of the numerical response of the grazers may be related to the warm temperatures. Since there are no data for nitrate, it is unclear whether grazing pressure or exhaustion of nutrients cause the sharp reduction in stodk of algae during autumn, but it seems more than likely that this estuarine system, located in the very nutrient-poor Eastern basin of the Mediterranean, is virtually completely governed by nutrient supply and its exhaustion by algae.

In some estuaries and coastal waters tidal mixing may be a major mechanism providing nutrients. Puget Sound in the Pacific Northwest of the United States is a deep temperate fjord with some estuarine

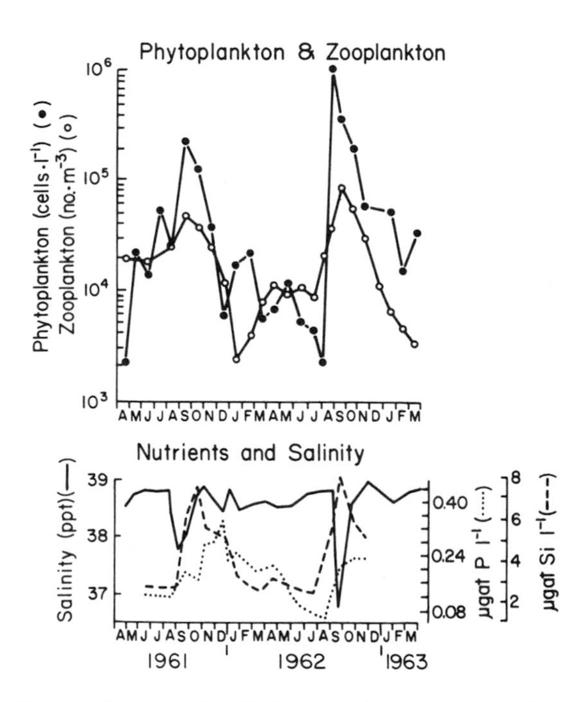

FIGURE 15-9. Seasonal cycle of salinity, nutrients, and production in the Nile estuary. Adapted from Aleem and Dowidar (1967).

*After the construction of the Aswan High Dam in the 1960s, this whole pattern was disrupted, and coastal fisheries reduced because of the lack of nutrient input into the nutrient-poor Eastern Mediterranean (cf. Section 16.8).

properties. Nutrients are regenerated by the benthos and also advected horizontally, and tidally induced vertical advection and turbulence furnish phytoplankton with adequate amounts of nutrients. Production during the winter and spring is determined primarily by light, maintained at a low level by particulates in the water (Winter et al., 1975). The grazers are probably ineffective as a mechanism limiting numbers of algae. The seasonal cycles are quite irregular, determined mainly by hydrological and weather events.

Stratification due to vertical differences in salinity may prevent mixing of phytoplankton in estuaries below the critical depth. This may allow an earlier onset of spring blooms in estuaries than in the open ocean (Parsons and Le Brasseur, 1968).

Internal regeneration of nutrients and associated processes also have seasonal patterns in estuaries, peaking in the warmer months. In Tomales Bay respiration and regeneration increase in late summer as organic matter delivery to the bottom increases (Dollar et al., 1991). In Chesapeake Bay, much of the spring diatom bloom sinks, increases rates of release of regenerated nutrients from sediments (Malone et al., 1986), and depletes oxygen in bottom waters. The low oxygen condition continues through the summer, and makes phosphate more available in the water column. The net effect is to emphasize the externally driven shift from spring P-limitation to N-limitation during summer.

Abundance and activity of estuarine bacteria lag behind the phytoplankton bloom in winter-spring. Bacterial biomass and production are poorly correlated to phytoplankton in estuaries; increased temperatures are much better correlated to increases in bacteria (Ducklow and Shiah, 1993; Hoch and Kirchman, 1993), as discussed in the previous section. The role of bacterial consumers in estuaries is not well known.

The high density of zooplankton at the head of Narragansett Bay supports the growth of large numbers of ctenophores, fish larvae, and adult menhaden (Fig. 15-7, bottom row). These predators feed primarily on zooplankton* and are very abundant, but it is not known whether they reduce the numbers of grazers. Larger fish such as bluefish (*Pomatomus saltatrix*), striped bass (*Roccus saxatilis*), and bottlefish (*Preprilus triacanthus*) feed on menhaden and ctenophores.

Abundance and activity of larger consumers in Chesapeake Bay peak in late summer (Cowan and Houde, 1993). During the peak abundance, ctenophores and medusae have the potential of consuming up to 78% of zooplankton daily (Purcell, 1992). The medusae also feed on the ctenophores, with potential consumption rates reaching over 100%

---

*Menhaden usually feed on phytoplankton, but in summer at the head of Narragansett Bay the dominant algae are small flagellates. These are too small for menhaden and therefore the fish switch feeding methods and feed raptorially on the abundant zooplankton (Kremer and Nixon, 1978).

daily of the ctenophore population. There are, therefore, potential cascade effects that might affect the seasonality of possible top-down controls in estuarine food webs.

The specific mechanisms and timing by which light, nutrients, grazing and predation interact and result in the local phenomena may differ, but the major variables are near-universal. Although at first sight estuaries may have seemed very distinctive environments, the seasonal cycles are determined by the same general kinds of limiting factors that are preeminent elsewhere in the sea modified to an extent by the seasonal input of seawater.

15.2.3 Upwelling Areas

Areas of upwelling have been studied intensively in recent years, but there are few descriptions of seasonal production cycles. In the Gulf of Panama (Fig. 15-10) the offshore winds during January–May create

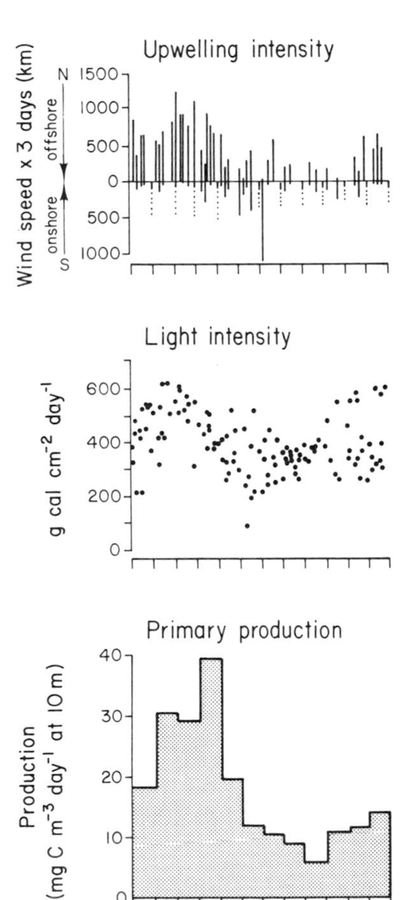

FIGURE 15-10. Seasonal cycle in upwelling in Gulf of Panama. Top: Average intensity of upwelling, measured by wind-speed for three day intervals, onshore and offshore, for 1954–1967. Middle: Light intensity through the year, 1955–1957. Bottom: Rate of primary production. Adapted from Forsbergh (1963) and Smayda (1966).

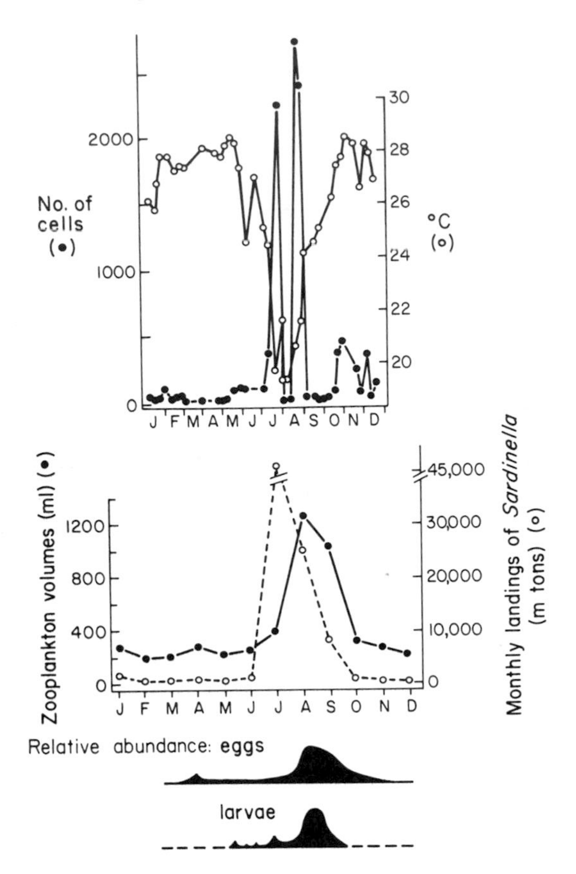

FIGURE 15-11. Seasonal cycle in upwelling of Gulf of Guinea, Ghana. Top: Temperature of the water and phytoplankton abundance (units are number of cells per sample taken). Note the sharp rise in phytoplankton during the time when nutrient-rich cooler water is upwelled. Bottom: Zooplankton (60–80% *Calanoids carinatus*) abundance and landings of the sardine (*Sardinella aurita*). Adapted from Houghton and Mensah (1978).

upwelling of cold deeper waters, new nutrients are brought into the photic zone, and blooms of phytoplankton occur. The high production increases the numbers of particles in the water enough to reduce the depth of the photic zone, even though at the latitude of the Gulf of Panama light intensity is high throughout the year.

In some areas of upwelling such as the Gulf of Panama and off Peru, the advection of cold rich water may last for several months. In the Gulf of Guinea the upwelling season is short-lived and seasonal fluctuations in the production of phytoplankton are more intense (Fig. 15-11, top). Phytoplankton densities are very variable during the upwelling season, probably due to the marked spatial heterogeneity that seems characteristic of upwellings (Fig. 15-12). Small adjoining patches of upwelled water may contain phytoplankton at a different stage of the bloom or succession.

In the Gulf of Guinea the zooplankton are dominated by a herbivorous calanoid copepod and its abundance is low most of the year (Fig. 15-9, middle). As water temperature drops during the brief upwelling season, juvenile copepods migrate or are advected shoreward, feed on the abundant organisms near the surface, molt into adults, and reproduce. This aggregational and numerical response to food density may be capable of reducing phytoplankton abundance to some extent, since there is a small increase in abundance of phytoplankton in October–November, when there is no upwelling (Fig. 15-9, top) but the copepods have migrated offshore. In the upwelling off Peru, relatively large copepods (*Centropages brachiatus, Eucalanus inermis,* and *Calanus chilensis*) consume 0.5–4.7% of the primary production (Dagg et al., 1982). Smaller copepods (*Paracalanus* sp.) are estimated from lab experiments to consume about 33% of the primary production (Paffenhöfer et al., 1982). Other small species may add to the impact of grazers. It thus seems that grazing rates are not high enough to control phytoplankton blooms in upwellings, since the producers more than double per day.

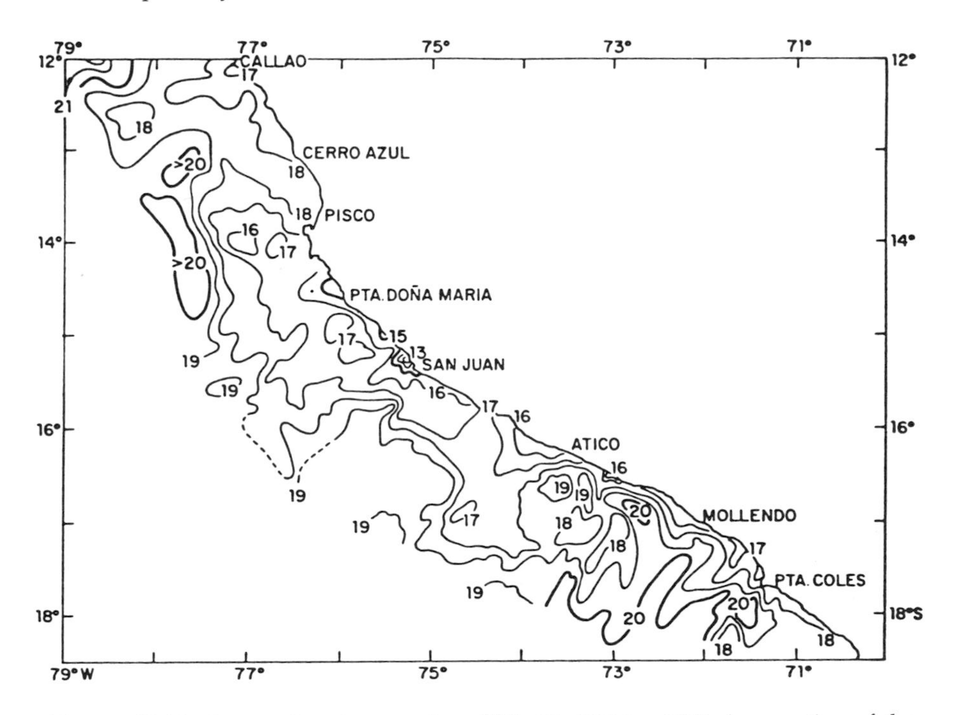

FIGURE 15-12. Sea-surface temperature (°C), 28–30 May 1974, in a section of the Peruvian coastal upwelling. From Zuta et al. (1978).

The clupeid *Sardinella aurita* is the most important economic fish in the Ghanaian fishery in the Gulf of Guinea, making up to 45% of the total landings of marine fish. This sardine feeds on zooplankton, is highly mobile, and shows a marked aggregation response to food density, even over considerable distances (Fig. 15-11, middle). The fish find upwelled water and spawn there (Fig. 15-11, bottom), where the larvae have available supplies of suitably-sized organisms to feed on. Perhaps predation by sardines is responsible for the decrease in zooplankton in August–October. Reduced algal food and migration to deeper, colder water are a more likely explanation for the lowered densities of zooplankton during this period.

The seasonal pattern of upwelling communities, then, seems largely governed by the intensity and timing of nutrient inputs, with the potential for some influence of young fish on the grazers.

15.2.4 Polar Seas

At high latitudes the pattern of seasonal growth of phytoplankton is characteristically unimodal. Light is very low during the winter, and increases in illumination in spring lead to sharp spring blooms both in the Arctic and in the Antarctic (Fig. 15-13). The farther away a station is from the pole, the earlier the peak, showing the effect of earlier onset of increasing light intensity away from the pole (Fig. 15-13, left). The growth of algae at the start of the season can begin even while the ice still covers the water column (Fig. 15-13, right), since growth can occur at quite low light intensities.*

In the Southern Ocean abundant nutrients are provided much of the year by the strong Antarctic Divergence (cf. Section 2.4, Figs. 2-33 and 2-35) near the continent. Farther north the effect of the upwelled water tapers off, especially near the Antarctic Convergence (Balech et al., 1968) and the seasonal peaks are less pronounced.

In the Greenland study there are no nutrient data available to assess whether nutrient depletion takes place in summer (Fig. 15-13, right). There is, however, a surprisingly rapid and early response by the zooplankton. Perhaps in this instance grazers are able to overtake the growth of phytoplankton and cause the late summer decrease in algal stock. The early increase in abundance of zooplankton may not be carried out by a numerical response through reproduction (which would have a longer lag in time) but to an aggregative response, with individual nauplii that over-winter at depth (Digby, 1954) coming to surface water to feed on the summer bloom, and quickly growing and reproducing.

---

*In the North Sea, for example, only $0.03\,\mathrm{g\,cal\,cm^{-2}\,min^{-1}}$ are needed to start the spring bloom (Gieskes and Kraay, 1975).

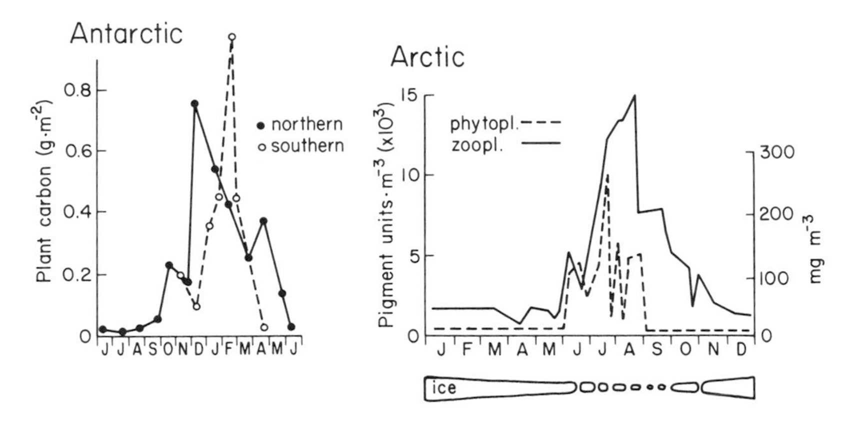

FIGURE 15-13. Seasonal cycles in the plankton of polar waters. Left: Stocks of phytoplankton in Antarctic waters; northern refers to stations nearer the Antarctic convergence while southern refers to stations nearer the continent. Adapted from Currie (1964) and Hart (1942). Right: Seasonal changes in copepods and phytoplankton (both net and membrane-filtered) in the upper 50 m of Scoresby Sound, East Greenland (1950–1951). Adapted from Digby (1953).

15.2.5 Subarctic Pacific

Unlike all other areas we have examined so far, there is no obvious spring bloom in the stock of algae in the Subarctic Pacific (Frost, 1987) (Fig. 15-14, top). If we exclude the coastal zones, there is a remarkably uniform amount of chlorophyll in oceanic water from 40°N to the Bering Sea throughout the year. The values of organic carbon range from 5 to 15 μg liter^{-1}, a very low amount. There is, however, a seasonal peak of production in midsummer that is seldom expressed as an increase in stock of algae (Fig. 15-14, top). Nutrient concentrations are high throughout, even up to 15 μM nitrate-N. Even in September–November there may be 6–7 μM* nitrate-N in surface waters. It seems therefore, that the low and constant algal stock is not due to low nutrient levels or to low light.

The zooplankton are dominated by two species of very large copepods, *Neocalanus plumchrus* (up to 5 mm in length) and *N. cristatus* (up to 10 mm in length). These two species make up 70–80% of the zooplankton in stock in the summer and overwinter as adults at depth (Fulton, 1973). Much of the subarctic Pacific shows a halocline at about

*These are high values compared to other oceanic waters (cf. Section 11.2.2.1, Fig. 2-20, and Table 2-3).

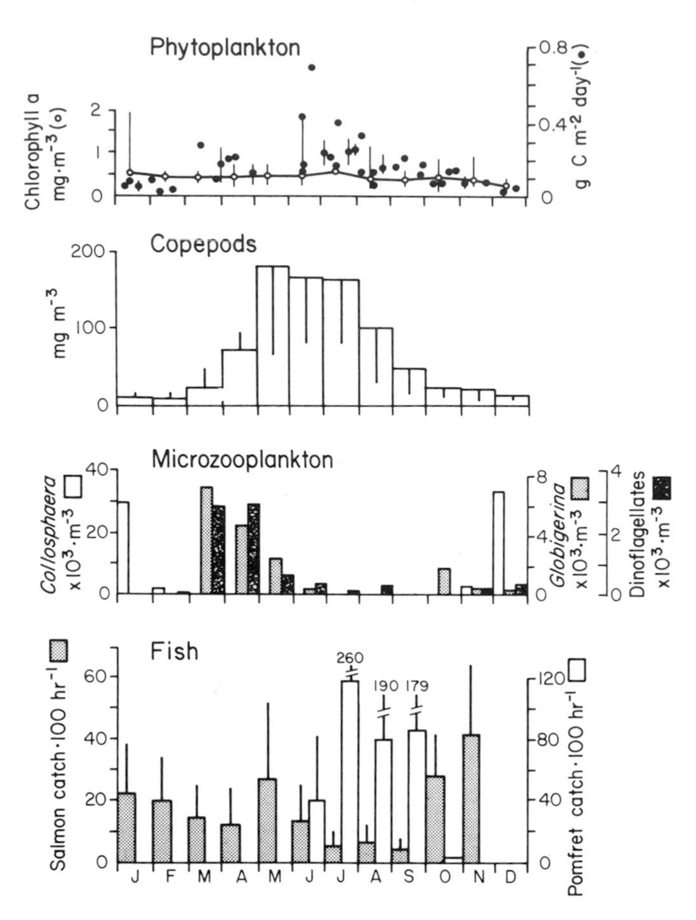

FIGURE 15-14. Seasonal cycle in the Subarctic Pacific, Ocean Station P, 50°N 145°W. Top: Standing crop of chlorophyll and primary production (0–150 m) through the year. Adapted from Le Brasseur (1965), Parsons and Le Brasseur (1968), and from data of G.C. Anderson. Second row: Abundance of large zooplankton in net hauls, 1963–1976. Copepods are the dominant taxon, but weight includes chaetognaths, euphausiids, amphipods, and medusae. Third row: Abundance of protozoans through the year (1966–1967) in the upper 100 m at Ocean Station P. Data from Le Brasseur and Kennedy (1972). Bottom: Number of fish per hour of long line fishing off Station P. Salmon include various species; pomfret is *Brama japonica*. Data from Fulton (1978).

100–120 m (Dodimead et al., 1963) that keeps the upper mixed layer shallow and maintains algae near the surface, where the cells can use the low light intensity during winter to photosynthesize. This halocline is a key feature, because it allows young *N. plumchrus* and *N. cristatus* to find concentrated enough food even in winter. If light is too low during winter and early spring to allow much algal production, the copepods may switch to feed on the numerous microzooplankton (Fig. 15-14, third graph from top) (40–200 μm in length) in the surface water (Le Brasseur and Kennedy, 1972). Eggs are produced by *Neocalanus* from fall to spring and the young copepods come to the surface in very large numbers from early spring to late fall (Fulton, 1973). Thus there are copepods available in the surface waters throughout the year, ready to take advantage of any spurt in algal growth. These young copepods grow fast (developmental response) and since large size is correlated to higher fecundity (cf. Section 4.23), the copepods may also have a rapid numerical response.

N. plumchrus can effectively graze on particles 2–20 μm in length (Frost, 1987), a range that extends to a remarkably small size range for such a large copepod.* The performance of *N. plumchrus* can be contrasted with that of *Calanus pacificus*, a much smaller copepod, which does not feed when food is smaller than 3.9–5 μm. *N. plumchrus* generally filters at low rates, but can increase feeding markedly when exposed to even small increases in abundance of algae (the functional response) (Frost, 1987), another adaptation to make effective use of occasional patches of high-density phytoplankton. Modelling studies show, however, that the grazing rates required to maintain phytoplankton biomass during the first 120 d of the year are far higher than could be managed by *Neocalanus* (Frost, 1987).

Thus, features of hydrography, life history pattern, feeding behavior, and morphology of feeding appendages allow *N. plumchrus* to exploit readily a food resource in low supply and of reasonably small size, and thus to make use of whatever increase in food density occurs. *Neocalanus* may be important as a consumer of larger cells that are not available to the protist microzooplankton and as predators on the smaller microzooplankton. *Neocalanus* may have evolved its life history as a response to a fairly constant food supply, rather than be a cause of the near-constancy of its food supply (Frost, 1987).

---

*We have seen that current views on how copepods feed suggest that passive sieving through appendages is less likely than previously thought (Chapter 6). In spite of this, intersetule distances may still have some role in feeding on very small particles. *N. plumchrus* and *N. cristatus* have relatively large feeding appendages with very closely spaced setules, perhaps an adaptation for a diet of the small cells usually found in surface waters of the subarctic Pacific. *N. plumchrus* has intersetule distances of 1.5–7.2 μm. In spite of the very large size of this copepod, the 1.5 μm is the smallest such distance for the many species surveyed by Heinrich (1963). For comparison, here are ranges of intersetule distances for a few other suspension feeding herbivores: *N. cristatus*: 1.8–9.6; *Undinula darwini*: 3.0–8.4; *Pseudocalanus elongatus*: 1.8–7.2; *Eutideus giesbrechti*: 4.2–6.6. Raptorial species, probably mainly predaceous, include *Eucheta marina*: 16.8–19.2; *Metridia pacifica*: 2.4–21.6; *Labidocera acutifrons*: 12.0–27.6; *Acartia longiremis*: 9.6–19.2.

Simulations also show that grazing by the total microplankton (Fig. 15-11, third graph) may be sufficient to maintain control of chlorophyll concentrations. In the subarctic Pacific, therefore, grazers may be responsible for maintaining the very low levels of phytoplankton standing stock constant through the year (Frost, 1987). Given the right circumstances, especially hydrography, timing, and a level of phytoplankton production that is not extremely high, herbivorous zooplankton *can* limit phytoplankton abundance at certain times of year.

Fish predation may reduce zooplankton biomass in the latter half of the year (Fig. 15-14, second row). Although the fish data are not very substantial (Fig. 15-14, fourth row), there is an apparent increase of some fish synchronously with the decrease in copepods. There are estimates by R.J. Le Brasseur that roughly one fifth of the spring production of herbivores may be consumed annually by salmon maturing at sea. Sanger (1972), however, claims that salmon take less than 4% of the annual zooplankton production. There are other predators on zooplankton, including carnivorous zooplankton, medusae, squid, myctophid fish, juvenile salmon, pomfret, and baleen whales. If all these predators were considered, it seems possible that the zooplankton decline could be due to predation and not just to migration or sinking. At present this is mere speculation.

15.2.6 Subtropical Seas

The Sargasso Sea is a subtropical gyre in the Atlantic, consisting of a lens of 36.5% salinity and 18°C temperature, with a thickness of about 500 m in the center. Below this lens lies a permanent thermocline. There is a seasonal thermocline at about 100 m in depth that may disappear in winter north of Bermuda.

The intensity of light remains high year-round and the amounts of particulate and dissolved organic matter in Sargasso Sea water are extremely low, so that the photic zone extends to about 100 m or more throughout the year.

Nutrient concentrations are low throughout the year in the Sargasso Sea. The supply of nutrients to the euphotic zone is severely limited because of the pronounced vertical stratification. Nutrients, episodically brought to the surface but in low amounts, are used by the phytoplankton as quickly as they are made available. This nutrient regime creates the large area of low primary production in the Atlantic north of the Equator (cf. Fig. 2-30). In the top 100 m, phosphate ranges from 0.02 to 0.16 μM P, nitrite plus nitrate from undetectable to 1.8 μM N, and silicate from 0.3 to 1.8 μM Si. Most of the measurements lie to the low end of these ranges, but because of the fast rate of turnover and regeneration of these nutrients, the concentration of any nutrient gives little idea of rates of use by algae, as discussed in Chapter 2.

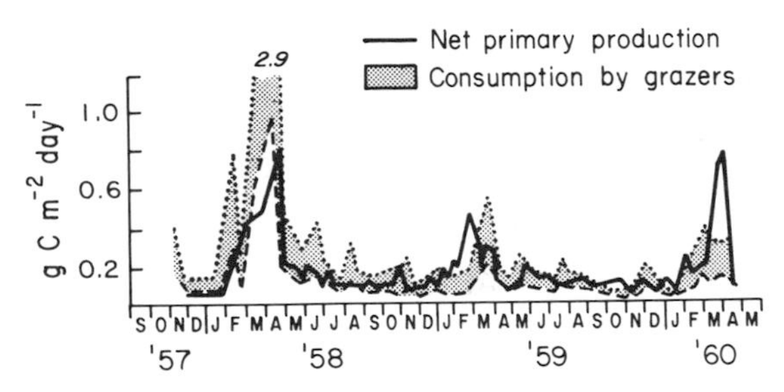

FIGURE 15-15. Seasonal variation in net primary production measured by ^{14}C in the subtropical Sargasso Sea and rough estimates of upper and lower limits of consumption by grazers. The latter are calculated based on standing crops of zooplankton between 0 and 500 m (dashed line) and 0 and 2,000 m (dotted line). The zooplankton were assumed to be herbivorous, and average feeding and respiration rates were used to obtain the consumption rates. Adapted from Menzel and Ryther (1961).

During winter the surface waters of the northern Sargasso may be briefly mixed down to 400 m, and thus nutrients reach their highest values (1–2 μM N, 0.1–0.2 μM P). The "spring" bloom starts in December for January (Fig. 15-15). As stratification is quickly reestablished, the algae are maintained in a shallow surface layer, and the relatively high nutrients and high light prompt a vigorous bloom. Rates of production can be high (up to $2\,g\,C\,m^{-2}\,day^{-1}$) (cf. Section 1.4), but the nutrients are quickly exhausted and the bloom is short-lived.

After the single bloom, net production by phytoplankton is low and irregular for the remainder of the year (Fig. 15-15). The zooplankton also have a single seasonal peak, more or less synchronous with the phytoplankton. There are some zooplankton always present in surface waters so that they can respond quickly and effectively to the increased food supply. Since the nutrients are so quickly used by algae, the periods of high growth rate are short and the algae do not escape control by the grazers. Rough estimates of consumption of algae by the grazers show that virtually throughout the year the demand brackets the production rate by algae (Fig. 15-15). In contrast to coastal environments, almost 100% of the primary production goes to the grazers (values on right of 3rd graph, Fig. 6-9). The potential impact of grazers on phytoplankton is large, to the extent that Sheldon et al. (1972) calculated that a 10% decrease of grazing in the Sargasso Sea for just a few days would allow growth equal to the spring bloom if sufficient nutrients were available. Nutrients are so seldom available that virtually the entire crop of phytoplankton is consumed by grazers, yet consumption by grazers may not be what controls producers. In fact, the dependence of production on nutrients regenerated by the zooplankton (cf. Section 14.233) is such that probably the key role of animals is in regeneration of

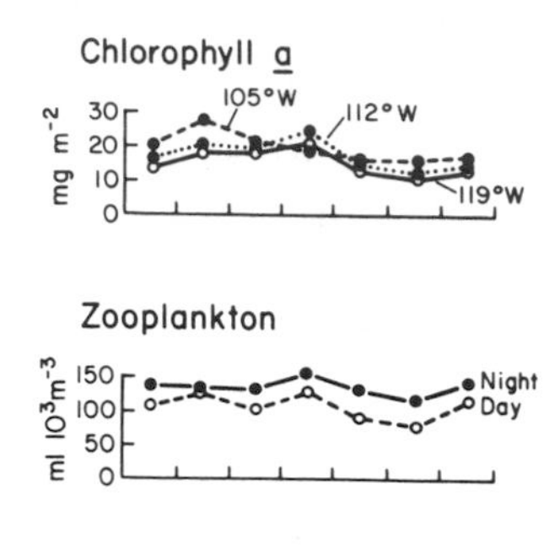

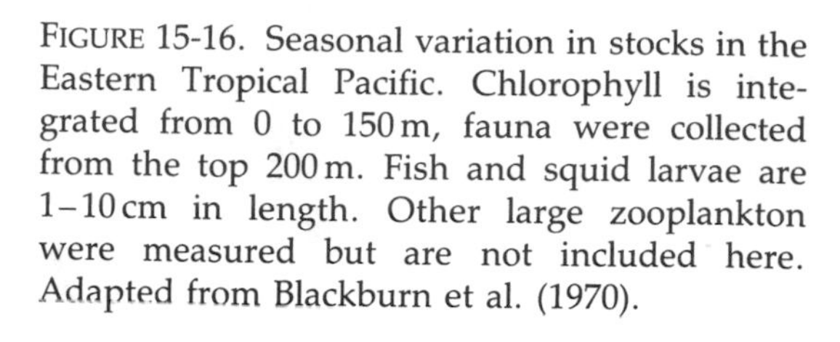
FIGURE 15-16. Seasonal variation in stocks in the Eastern Tropical Pacific. Chlorophyll is integrated from 0 to 150 m, fauna were collected from the top 200 m. Fish and squid larvae are 1–10 cm in length. Other large zooplankton were measured but are not included here. Adapted from Blackburn et al. (1970).

nutrients. There is thus a close relationship between producers and consumers mediated by regeneration of nutrients and grazing by consumers.

In the subtropical North Pacific there is a very extensive nutrient-poor gyre. This region is very homogenous over space (McGowan, 1977; Hayward and McGowan, 1979), and there is very little seasonal change (Fig. 15-16). Both phyto- and zooplankton vary hardly at all over the year. The microplankton (ranging in size from 2 to 200 μm) varied less than twofold through five cruises over 2 years (Beers et al., 1982). The North Pacific gyre resembles the Sargasso Sea, but the spring blooms are even less prominent, and the abundance of other taxa is equally unchanging (Sharp et al., 1980; McGowan and Howard, 1978). This is another pelagic system in which there is a close coupling between phyto- and zooplankton. The very low concentrations of nitrogen limit phytoplankton (cf. Section 2.2333), and zooplankton both graze the algal biomass and through excretion regenerate the nitrogen that largely supports growth of the phytoplankton.

15.2.7 Tropical Marine Waters

In coastal tropical waters of the Great Barrier Reef there are also low and constant nutrients and very low variability in stocks of phytoplankton through the year (Orr, 1933; Marshall, 1933; Russell and Coleman, 1934; Russell, 1934). A fluctuation between 1×10^3 and 4×10^3 cells liter^{-1} is very modest compared to the changes of several orders of magnitude characteristic of cycles in coastal temperate waters. Dinoflagellate densities are virtually constant through the year.

The zooplankton fauna of the Great Barrier Reef is very diverse and there is a modest seasonal pattern, peaking in late summer. The increase in zooplankton comes when algae are very scarce; perhaps the zooplankton feed on the numerous detrital particles or mucilage released by the corals. In any case, the density of zooplankton is very much lower than that in temperate waters and only varies about three-fold, again a very small change compared to fluctuations in temperate marine environments. The corals themselves are of course perennial and are active through the year, primarily as producers through their zooxanthellae, with nitrogen needs provided by predation by the host coral. There are few data on the seasonal changes in coral activity, which are probably slight.

In tropical areas, as in the subtropical gyres, there is a lack of or only slight seasonality of the plankton (Banse and English, 1994). In the open-ocean pelagic communities, there is a remarkable lack of nutrients, primarily due to stratification of the water column and resulting isolation of the photic zone from nutrient-rich water. Whatever blooms take place are soon ended by the reduced quickly of nutrients and the grazers are able to harvest algae almost as quickly as they grow. The herbivores may restrict algal growth and hence may also restrict their own abundance, although the evidence is circumstantial. Carnivorous animals may be rather unimportant in limiting prey populations in tropical waters, at least as far as the production cycle in pelagic communities is concerned, but there is little concrete information on this topic.

15.3 Benthic Seasonal Cycles

In shallow waters the seasonal cycles of water column and benthos are closely linked by sedimentation, resuspension, vertical mixing, and regeneration, processes discussed in earlier chapters. In fact, the benthos in shallow waters may greatly influence the dynamics in the water column, providing regenerated nutrients, as detailed in Chapters 12 and 13. In deeper waters benthic-pelagic coupling are assymetrical: the seasonality in the deep sea, as discussed below, is determined by events from the surface. Seasonal cycles in producers attached to the sea floor in shallow environments have their own characteristics and also merit some mention below.

15.3.1 Seasonality in the Deep Sea Floor

Certain environmental conditions in the deep sea are remarkably constant: temperature remains within 1–2°C, and salinity at about 34.8%. From such data it was easy to conclude that the deep sea floor was among the world's most constant environments. To some extent this is true, but that does not mean that the sea floor is free of seasonal change. Evidence has recently accumulated that shows that there are seasonally varying physical processes, carbon and sediment inputs, and that deep-sea organisms show seasonal rhythms. It has taken more than a century to obtain evidence to support Moseley's (1880) contention that there was a seasonal supply of food to the deep-sea floor " . . . which may give rise to a little annual excitement amongst the inhabitants."

There are strong seasonal shifts, with spring peaks, in the frequency of eddy kinetic energy and currents over the bottom at a depth of 3050 m in the Rockall Trough (Dickson et al., 1988). Such shifts may be common, and could serve as cues for breeding behavior of organisms (Tyler, 1988).

There is a well-defined seasonal fluctuation in the flux of organic matter from surface waters off Bermuda to the deep sea below (Fig. 13-4). Deuser (1986) shows distinct intraannual variation in delivery of organic matter and mineral particles and some interannual differences, over the course of several years. The rates of downward flux amount to only 1–3% of surface production rates, but the fluxes, however small, are closely related to seasonal events in the photic zone (Cole et al., 1987).

The seasonal cues provided by currents and carbon supply affect benthic organisms. Although not all deep-sea species show seasonal reproductive or physiological cycles (Rokop, 1974, 1977), there is mounting evidence that such changes take place in many species. Tyler (1988) reviews data that show seasonally varying growth and reproduction in deep-sea brachiopods, fish, echinoids, ophiuroids, isopods, asteroids, anemones, crabs, and bivalves. In general, reproductive activity follows the peak in organic flux from the surface, suggesting that it is food supply that triggers responses in the benthic fauna, although flux of small planktonic particles may not completely supply the needs of the fauna (Smith, 1987).

15.3.2 Large Attached Algae

The seasonal control of production cycles of macroalgae depends on how hydrography supplies nutrients and on light. Where there are seasonal variations in nutrient abundance, there is a rapid incorporation of nitrogen as nitrate becomes more available in the water column during early winter (Fig. 15-17, top right). The increase in the internal pool of nitrate does not immediately lead to peak growth rates. Maximum growth does not take place until after light intensity increases later in the spring (Fig. 15-17, top left), at which time the rate of photosynthesis increases (Fig. 15-17, bottom left). In kelp and other macroalgae the ability to respond physiologically to increased light intensity is most pronounced in spring (Brinkhuis, 1977, 1977a). Kelp can, however, grow during winter (Hatcher et al., 1977; Chapman and

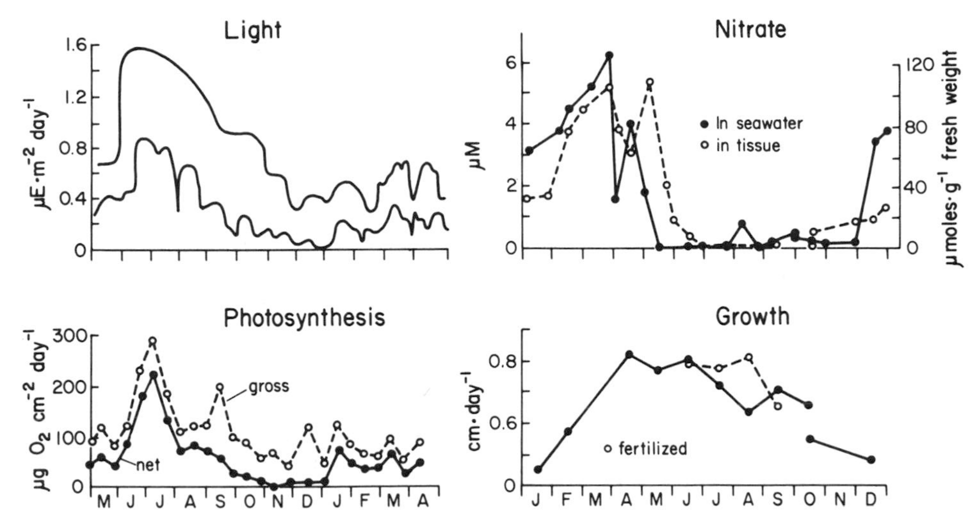

FIGURE 15-17. Seasonal cycle in growth and photosynthesis in the kelp *Laminaria longicruris* on the Nova Scotia coast. Left: Maximum and minimum light intensities (top) and rate of photosynthesis (bottom) during the year. Adapted from Hatcher et al. (1977). Right: Nitrate content in seawater and in tissues of kelp (top) and growth, as cm day^{-1} of blade elongation (bottom) in a kelp forest growing in water 18 m deep. Fertilization experiments done at site 9 m in depth; the growth rate of unfertilized kelp of 9 m was similar to that at 18 m. Adapted from Chapman and Craigie (1977).

Craigie, 1978) by using reserves of carbohydrates stored during the previous growing season.

The decrease in growth after spring is primarily due to the reduction of nitrate in the water and in the tissues (Fig. 15-17, top right). The importance of nitrate is evidenced by the sustained growth of kelp in plots fertilized with nitrate (Fig. 15-17, bottom right). Eventually light is again reduced in early autumn and this decreases both growth and photosynthesis. In sites where upwellings or other mechanisms provide high concentrations of nutrients year-round, light may limit growth of kelp (Gagné et al., 1982). In such sites the role of internal storage of carbon and nitrogen compounds is less important.

Salinity, temperature, and photoperiod may also have some influence on seasonal growth of macroalgae (Hanisak, 1979) but these factors act primarily by interacting with the primary factors of light and nutrients. The ability of macroalgae to store reserves gives them an advantage over phytoplankton, in that growth can take place during early winter and photosynthetic tissues made ready to make early use of the increased light of late winter. The seasonal pattern of growth of dominant macroalgae affects the occurence and production of other macroalgal species over the course of a year through shading.

15.3.3 Vascular Plants

Most communities dominated by vascular plants—salt marshes, mangroves, eelgrass beds—have other associated producers, including epiphytes and benthic microalgae. In salt marshes of New England, growth of the predominant plant, *Spartina alterniflora,* begins in April, and the growth rate in spring is more closely related to temperature rather than light intensity (Figs. 15-18a,b). Similar correlations have been described for eelgrass beds (Harrison and Mann, 1975).

The nutrient supply determines the amount of growth achieved during the year, since more biomass accrues where *S. alterniflora* is experimentally fertilized (Fig. 15-18a). The nitrogen content of above-ground tissues is built up in early spring from stores in the perennial below-ground plant parts,* and decreases through the summer (Fig. 8-6, top). Potential demand for nitrogen by the plants exceeds actual uptake, since even though there is an abundant supply of ammonium nitrogen in the interstitial water of marsh sediment throughout the year (Fig. 15-18c), uptake of nitrogen is inhibited under anoxic conditions (Howes et al., 1981; Mendelssohn et al., 1981). The balance of root and bacterial activity determines the equilibrium between oxidized and reduced forms of sulfur as well as the redox of the marsh sediment (Howes et al., 1981). As temperatures increase, the activity of sulfate-reducing bacteria (cf. Section 14.3) produces sulfide and reduces sediments. The activity of live roots is also temperature dependent, and their release of O_2 or organic oxidants oxidizes the sulfide in the sediment. High temperatures in midsummer increase respiration and since light intensity falls during late summer (Fig. 15-18b), photosynthesis may not be able to compensate for respiratory losses of CO_2. Growth may therefore stop, and tissues may senesce and release dissolved organic carbon. Below the sediments this DOC stimulates activity of fermenters and sulfate-reducing bacteria and the latter release sulfide (cf. Section 14.3). The reduction in below-ground tissues also lowers the oxidative activity of plant roots, so the sediment becomes more anaerobic. This may further reduce uptake of the abundant ammonium and perhaps end growth of the plants for the season. The more nitrogen present in the tissues, the longer senescence of the grass is delayed, as seen by the contrast in fertilized and control grasses (Fig. 15-18a). Grasses richer in nitrogen last longer in the autumn and are eventually killed by freezing temperatures.

Grazers in salt marshes generally consume small amounts of vascular plants, except in certain places where dense flocks of overwintering

---

*Below-ground plant parts store materials over winter and can exceed the weight and production of above-ground parts in marsh plants throughout the year (Valiela et al., 1976; Good et al., 1982; Kistritz et al., 1983).

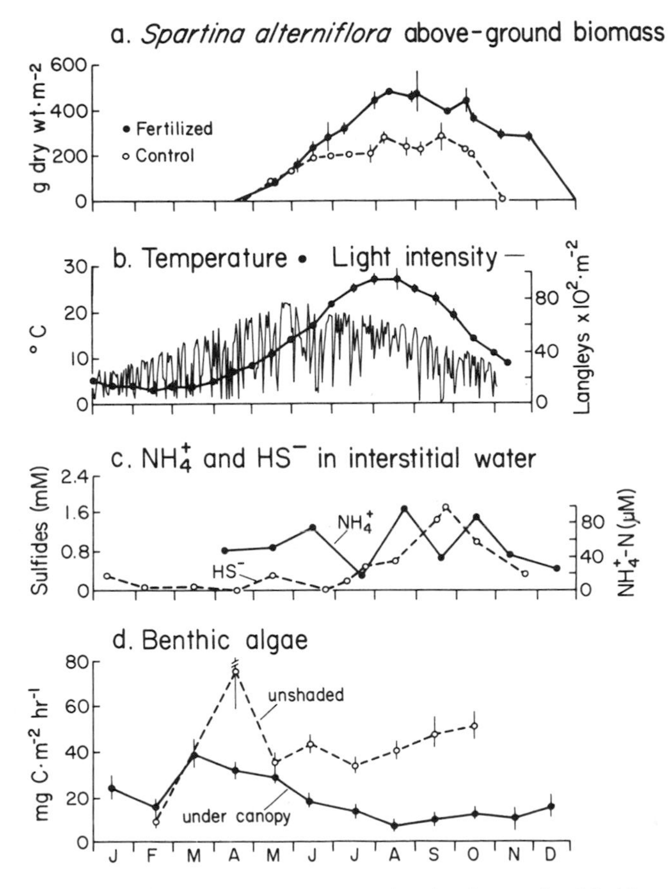

FIGURE 15-18. Seasonal events in a New England salt marsh. (a) Above-ground biomass of *Spartina alterniflora* in fertilized and control plots data of I. From Valiela and J. Teal. (b) Water temperatures and light intensity throughout the year. Changes in temperature in sediments lag a few weeks behind water temperatures. (c) Concentration of ammonium and sulfide in interstitial water of sediments. From Giblin (1982). (d) Production by benthic algae. From Van Raalte et al. (1976).

geese accumulate. In general, herbivores are therefore not very important in the seasonal cycle nor in structuring salt marsh food webs; the latter are primarily detrital.

15.3.4 Benthic Single-Celled Algae

In shallow water there are layers of single-celled algae that grow on the sediment surface. An example of the seasonal cycle of such benthic algae is provided by salt marsh algae. Benthic algae on the surface of salt marsh sediments bloom in spring when light increases (Fig. 15-

18d), but as the grasses form a canopy, the shading reduces production of benthic algae through the entire summer and fall.* This can be demonstrated by the increased production of benthic algae that follows experimental removal of the grass canopy (Fig. 15-18d). Activity of grazers on salt marsh algae increases during the warmer months (May–September), so their impact is delayed until after the early spring peak of benthic algae (March–April). The reduction in algal production after the April peak is likely due to grazers, since during warmer months dense mats of benthic microalgae form where grazers are experimentally excluded. The accumulation of high densities of algal cells on a two-dimensional surface may make the benthic algae more susceptible to grazer control than the widely dispersed phytoplankton.

15.4 Control of Seasonal Cycles

In most cases that we have examined, control of seasonal cycles in the water column is by a sequence of factors. Light usually initiates the cycle as it increases in spring and higher temperature may form faster metabolic rates. The actual start may be mediated by the depth of the mixed layer relative to the photic zone. The rate and peak of primary production are then determined by the supply of available nutrients. In most cases, grazers are not abundant enough to play much of a role until the rate of primary production slows down due to depletion of nutrients within the mixed layers; then grazing rates are more capable of reducing phytoplanktonic populations.

The densities of grazers often are lower in the autumn and winter. Perhaps predation or other mortality factors are responsible, but this is not known. Many pelagic zooplankton move to deeper waters during winter, so the lower abundance may not be due to mortality. The reason for overwintering behavior is not clear. The lack of effect of pelagic fish and other predators on grazers agrees with the conclusion of Chapter 9, where we found that such predators seem unlikely to control prey populations.

Where nutrients are continually renewed, such as in shallow areas, the cycle seems determined primarily by light intensity. Seasonal inputs of nutrients as occur in estuaries and upwellings can prompt sudden and short-lived bursts of production that can be traced all the way up the planktonic food web.

---

*In eelgrass there are also important shading relationships between higher plants and epiphytes, but in this case it is the epiphytes that reduce eelgrass production as they grow over the surface of eelgrass leaves (McRoy and McMillan, 1977).

Where zooplankton can maintain a dense enough population over the less productive part of the year and where phytoplankton growth is not very high—subtropical gyres, subartic Pacific, perhaps the Arctic—the grazers may be able to crop algal blooms fast enough to restrict them.

Where light is plentiful through the year, as in tropical and subtropical seas, blooms may be started by small amounts of nutrients made available by mixing. The amplitude of oscillations of the phytoplankton are damped, and zooplankton grazing rates are comparable to primary production rates. The whole community is closely adjusted, and any increase in resources is quickly consumed.

Seasonal cycles in the deep benthos depend on events in the water column above. In shallow systems, however, as we saw in Chapter 14, regeneration from the bottom may seasonally alter dynamics in the water column. In shallow waters, therefore, benthopelagic coupling is not asymmetrical, as is the case in the deep sea.

The seasonal cycles of production in attached macrophytes is driven principally by changes in light intensity and nutrient supply. In vascular plants, where nutrient uptake is by roots, the redox of the sediment also mediates growth rates.

Consideration of seasonal cycles emphasizes that communities in different marine environments are structured each in its own peculiar fashion. There are generalities, however, in that a few major types of biotic relationships are invariably present, and that the conditions constraining the biological relationships are provided by physically-driven forces.

Chapter 16
Long-Term and Large-Scale Change in Marine Ecosystems

16.1 Introduction

We have so far mainly examined mechanisms and processes that control populational, community, and ecosystem changes over relatively short time periods (up to a year) and relatively small spatial scales. Marine systems are also exposed to changes at rather larger scales and over multiyear (secular) time intervals. Different processes and mechanisms may be involved in changes over different ranges space and time intervals.

In this chapter we first examine long-term ecological changes that occur at large geographical scales, and are related to a mix of "natural" and anthropogenic changes in global atmospheric conditions. We then review ecological effects of other anthropogenic activities that have led to long-term alterations of marine systems. These human activities include depletion of fishery stocks, eutrophication, toxic contamination, spread of exotic species, harmful algal blooms, and interception of freshwater inputs and changes in sediment loads.

16.2 Large-Scale Effects of Long-Term Atmospheric Changes

Large-scale climatic changes are evident throughout the geological history of the earth, and continue, enhanced by human activities (Bolin et al., 1986). Global warming by greenhouse gases, increases of sea level by melting glaciers, and ozone holes created by halogenated carbon compounds, have become part of our common lexicon. Such global-scale atmospheric changes have direct and indirect effects on weather patterns, hydrography, and ecology of marine systems.

16.2.1 Effects of Changes in the Atmosphere

Increases in temperature, carbon dioxide, and UV radiation may have direct ecophysiological effects at populational and community levels.

Temperature increases. It is difficult to find sufficient long-term, comprehensive data with which to estimate global increases in temperature (Wigley et al., 1986; Henderson-Sellers, 1990), but there seems to be agreement that surface temperatures have increased by about 0.4–0.6°C since the turn of the century (Mitchell, 1989; Graham, 1995, and cf. Fig. 16-5), and that the increase in concentration of greenhouse gases in the atmosphere may be responsible for the elevation in temperature (Mitchell, 1989; Ashmore, 1990). Although a mean increase of less than 1°C may seem small, much larger local and regional meteorological changes result from the mean change and have led to major local and regional changes. For example, higher temperatures are one cause of the widespread "bleaching" of corals (a result of loss of zooxanthellae) reported from many tropical areas in recent years and of the 50–97% mortality of corals recorded from Costa Rica to the Galápagos (Glynn, 1991). Higher temperatures may also make corals more susceptible to damage by UV radiation (Glynn, 1988). The disappearance of the more heat-sensitive branching corals has had a series of indirect effects. Massive corals, previously protected from predators by surrounding beds of branching corals, are now subject to attack by such predators as the crown of thorns starfish. The branching corals are faster-growing species and contribute to long-term reef-building; their decrease may retard the process by which reefs are maintained (Glynn, 1991).

Increases in temperature may also have indirect biological effects via alterations to chemistry and physics of water columns. In cooler regions, for example, it is conjectured (Williamson and Holligan, 1990) that warming may enhance stratification, reducing upward mixing of nutrients, and lower downward flux by favoring smaller plankton.

Acutal examples of the effects of globally-driven temperatures regimes on phytoplankton and benthos have recently been described off California. Roemmich and McGowan (1995) found an 80% decrease in abundance of zooplankton biomass in waters off Southern California since 1951. During this period surface waters warmed by as much as 1.5 degrees C, and this enhanced stratification of the water column. The stratification reduced wind-driven upwelling, which reduced nutrient delivery to the upper waters, and hence lowered phytoplankton production; the latter appears to be the cause of the reduction in zooplankton abundance. Barry et al. (1995) found that the temperature increases also seemed to change the animal species assemblage on the rocky shore of California. Northern species tended to decrease in abundance, while southern species, presumably more favored by the increased temperatures, increased in abundance. Large-scale, atmospheric-driven temperature shifts, even though small in magnitude, have significant consequences for the biology of marine systems.

CO_2 increases. Increases in atmospheric CO_2 that are expected to occur by the middle of the next century could affect photosynthesis, respiration, and growth of plants (Morison, 1990). Field experiments in which ambient CO_2 was doubled were done in salt marshes of Chesapeake Bay (Curtis et al., 1989). The results showed that even only one season of treatment increased net primary production, and changed the composition of the sward, specifically favoring C_3 plants, while not affecting C_4 plants.

Increased CO_2 in the atmosphere also has whole-ecosystem and larger-scale effects of high interest to oceanographers: the matter of transfer and storage of atmospheric CO_2 into the oceans, referred to as the "biological pump," is well reviewed by Mann and Lazier (1991). Although there are many disagreements on this topic, most researchers agree that there is transfer of CO_2 across the sea surface, and the rate of transfer depends on the partial pressures of CO_2 in air and surface waters, multiplied by a gas transfer coefficient. Partial pressures of CO_2 in water depend on uptake by cells and on temperature. In certain places, such as equatorial divergences, CO_2 is released to the atmosphere, while in colder waters CO_2 is taken into sea water. Carbon in surface waters may be incorporated into the marine food web, and a part of assimilated carbon sinks as particles, which leads to storage away from the atmosphere. Downward export of carbon from the surface waters can be estimated as equivalent to new production, assuming production in surface waters is at steady state. One of the many difficulties in this field is to calculate new production accurately from intermittent bursts of nitrate brought to surface waters by internal waves, turbulent eddies, and other physical processes. As a whole, the ocean is a large sink for increased atmospheric CO_2, but it is uncertain how large the amount of ocean-stored carbon is relative to carbon storage in land vegetation.

Ozone depletion. The action of chlorofluorocarbon produced industrially has led to significant decreases in ozone in the upper atmosphere, particularly near the poles. We have little evidence as to the direct effects of ozone depletion over polar regions. There is increasingly more energy in the UV range arriving at the surface of the land and sea, and the levels are likely to be biologically damaging. Increases in UV radiation predicted from decreases in ozone could potentially lower primary production and nutrient uptake by producers, lower motility, orientation, and survival of larval forms of zooplankton and fish (Worrest and Häder, 1989).

16.2.2 Consequences of Changes in Weather Patterns

Changes in the physics and chemistry of the atmosphere have major effects on marine systems. There are few data sets of sufficient

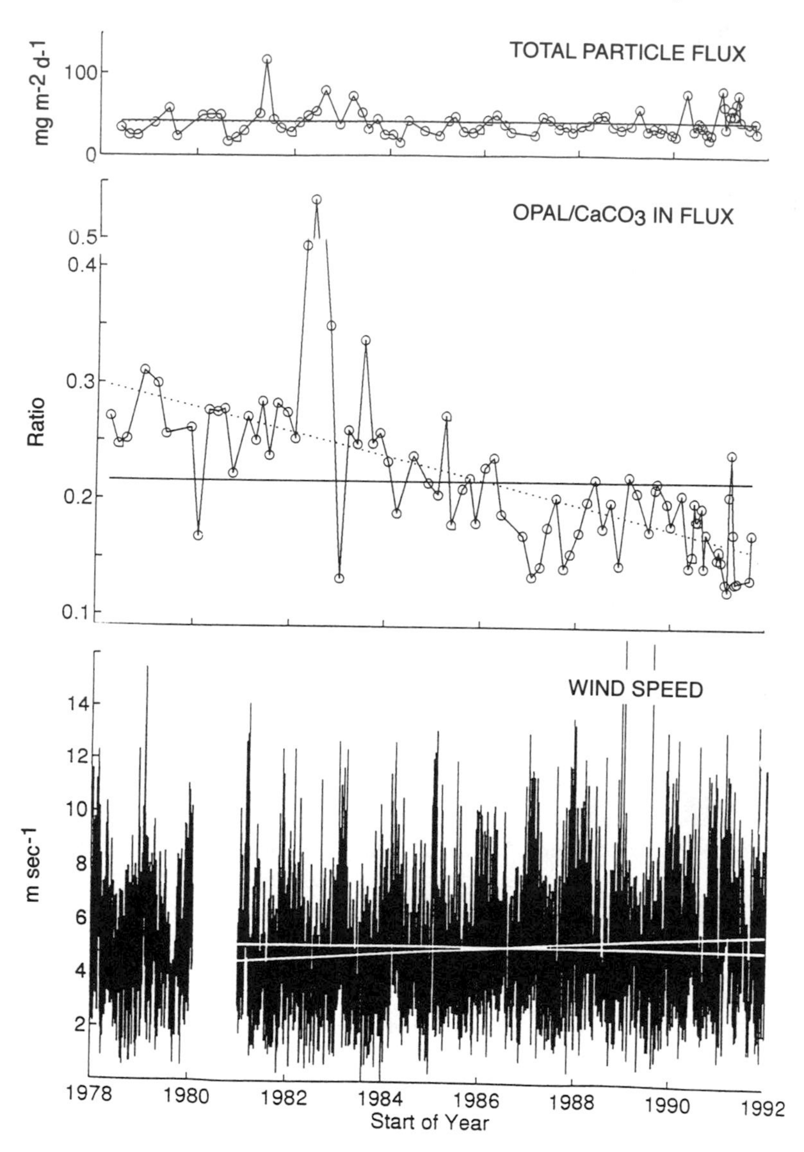

FIGURE 16-1. Total particle flux (top), ratio of silicates (opal) to carbonates (middle) into traps at 3,200 m in the Sargasso Sea off Bermuda, and wind velocity in Bermuda (bottom), all for the period 1978–1992. Adapted from Deuser (in press).

length and quality to discern long-term, large-scale changes in marine ecosystems. Some, such as the California current surveys, have already been mentioned. Another one of the few such studies is the near-continuous record of vertical particle flux in the Sargasso Sea off Bermuda (Fig. 16-1). Particle flux integrates many different aspects of the functioning of the upper layers of the ocean. Over 14 years, there has been no detectable secular trend, even though the seasonal and year to year variations are considerable (Fig. 16-1, top). This is one of the most uniform ecosystems anywhere; comparable data from other ecosystems shows more variation. Nonetheless, there are secular changes even in this data set: the ratio of radiolarians to coccoliths (groups that make up most of the carbonate and silicate flux, respectively) changes over decades (Fig. 16-1, middle). The causes of the shift, and its consequences, are unknown; they may related in some unidentified way to long-term changes in wind regime (Fig. 16-1, bottom), and may affect bottom sediments.

Secular changes in the atmosphere may therefore drive some long-term changes in the ecology of marine systems. We will examine two mechanisms that mediate coupling of atmosphere and oceans: shifts in location of barometric pressure centers, and rate of sea level rise, as examples.

The average location of barometric low and high pressures changes over time, with key consequent weather and hydrographical effects (Fig. 16-2). The shifts in position of barometric centers are driven by events at global scales (Enfield, 1989). Below we review two examples of such changes in weather conditions, one in the North Atlantic, another in the South and Equatorial Pacific.

16.2.2.1 Meteorological/Hydrographical Changes in the North Atlantic and Their Ecological Effects

Multiyear changes in average position of weather-maker centers over the North Atlantic are thought to be behind the major shifts in species assemblages and conditions evident in the longest such records available, those from the English Channel, North Sea, and North Atlantic (Russell, 1973; Colebrook, 1986; Southward et al., 1988; Dickson et al., 1988; Southward, 1991). The trends in year-to-year variation seem related to the dominance of westerly or northerly winds, which depend on the average geographic position of barometric regimes over the North Atlantic (Fig. 16-2).

During intervals of more intense prevailing westerlies, warmer water species predominated in areas previously occupied by colder water taxa (Cushing, 1982). Portuguese man-of-war, goose barnacles, loggerhead turtles, and octopus appear off France and Britain, and swordfish, cod, pollock, and ray appear in more northern waters. Anecdotal, historical, and fishery catch evidence from English waters suggest that since 1660, herring were most abundant during decades of low temperature, and pilchard when temperatures were higher (Russell, 1973; Southward et al., 1988). Shifts in species of chaetognath indicated changing water conditions: *Sagitta elegans* was most common when waters were warmer, *S. setosa* in colder waters. Similar taxonomic

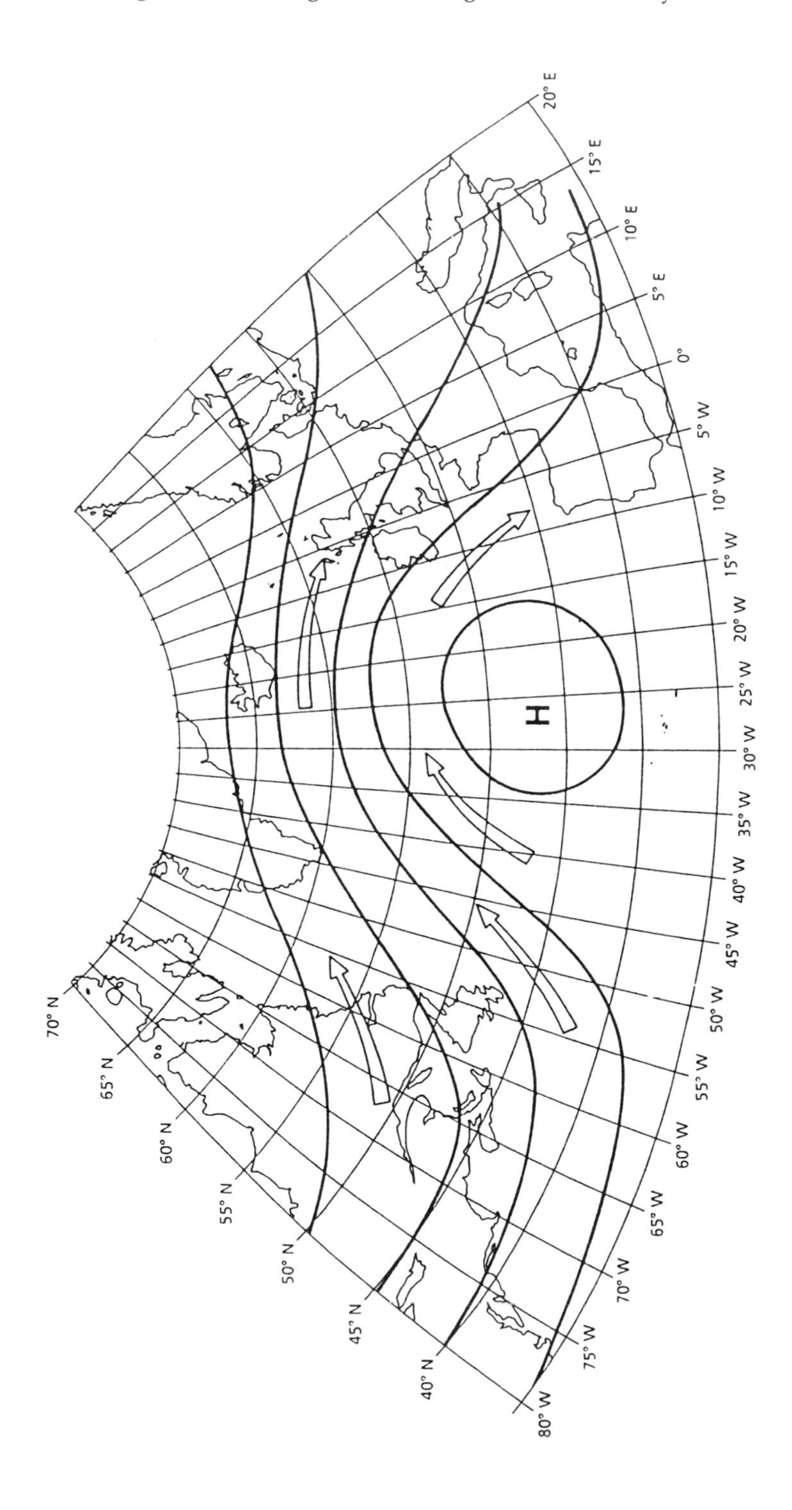
70° N
65° N
60° N
55° N
50° N
45° N
40° N
80° W
75° W
70° W
65° W
60° W
55° W
50° W
45° W
40° W
35° W
30° W
25° W
20° W
15° W
10° W
5° W
0°
5° E
10° E
15° E
20° E
H

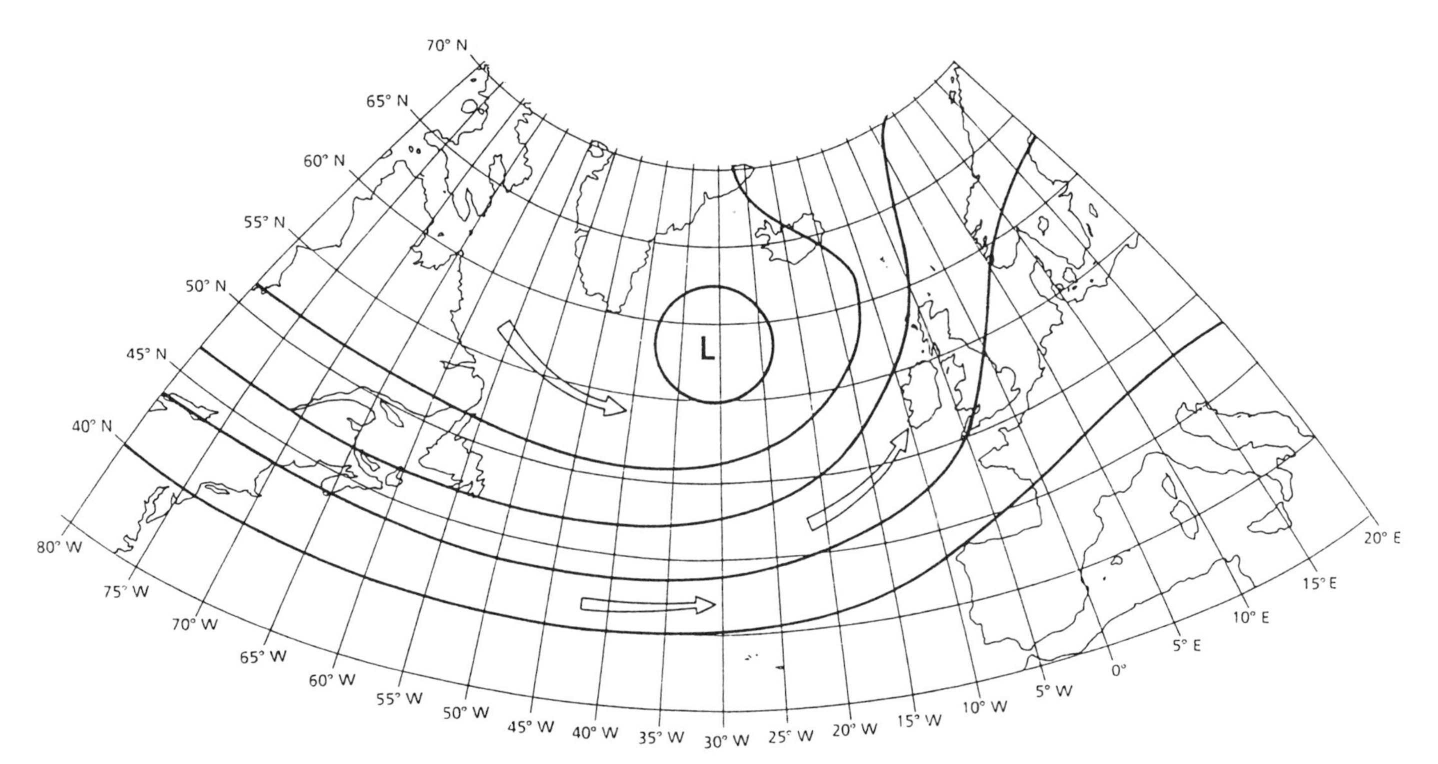

FIGURE 16-2. Different possible locations of high and low atmospheric pressure centers, with consequent shifts in the prevailing winds over the North Atlantic. Adapted from Mann and Lazier (1991).

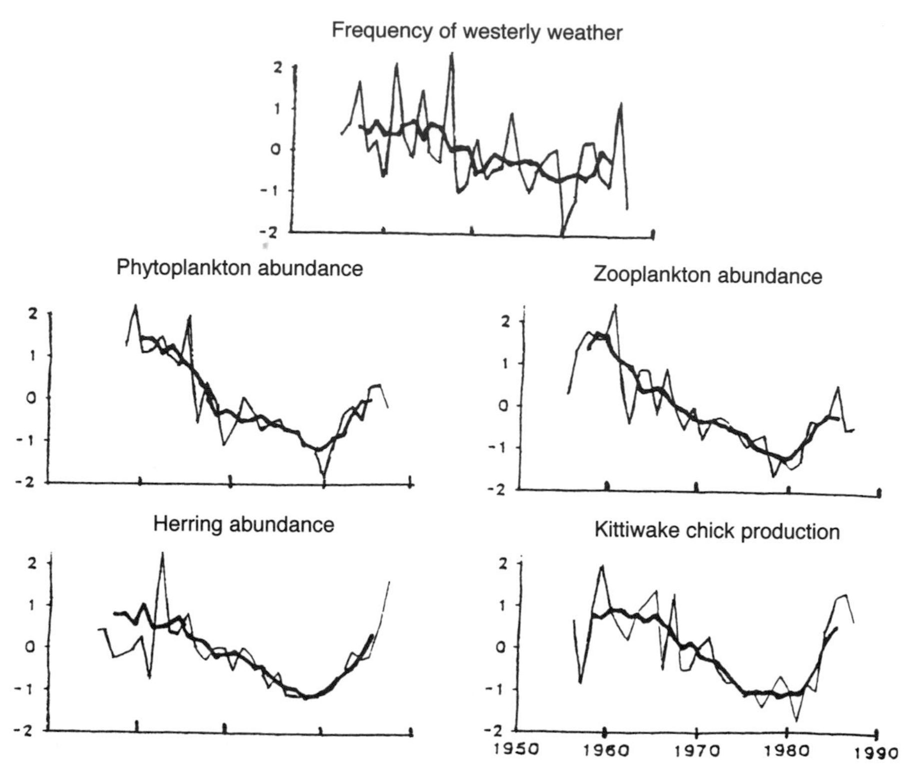

FIGURE 16-3. Time series over several decades of standard deviation relative to overall means (shown on the Y axis) for frequency of westerly weather, abundance of phytoplankton, zooplankton, and herring, and chick production in kittiwakes (oceanic gulls). Thin lines are annual values, thick lines 5-yr running means. Adapted from Aebischer et al. (1990).

changes in barnacles followed the water mass changes (Southward, 1991). Abundance of phytoplankton and microzooplankton in waters surrounding Great Britain decreased markedly as waters became cooler, while abundance of gadoid fish increased during such periods (Cushing, 1982). Secular episodes of colder or warmer waters evidently brought along or favored different assemblages of species.

Multiyear trends in ecological features can be correlated to trends in weather. The predominance of westerly winds over the northwestern North Sea varied over several decades (Fig. 16-3, top), and changes in four different components of the food web changed in parallel fashion (Fig. 16-3, four bottom panels). Note that the year-to-year variations in weather and organisms are poorly correlated; comparisons of data from just 1–3 years show inconsistent relationships. Secular changes such as we are describing only become evident when sufficiently long multiyear data are available. Such data sets are rarely available, but will become increasingly valuable and necessary to assess ecological change in natural environments.

Decadal and year-to-year changes in the food web over Georges Bank in the NW Atlantic were demonstrated by stable isotopic data of scales of the major fish species (Wainwright et al., 1993). The taxonomic assemblage of fish species present did not change, but the fish present evidently changed feeding habits over the last dozen decades. Correlational evidence suggested that there was a connection between the trophic changes and shifts in meteorological conditions.

The available data from the North Atlantic, North Sea, and English Channel make it amply evident that change at regional spatial scales, over decadal intervals and longer, are a major, common, and continuing feature of marine systems. Meteorological factors are certainly involved in prompting the changes, but there is insufficient evidence about specific mechanisms controlling the shifts. Moreover, it is not completely clear whether the changes involve mere geographical displacement of water masses, carrying with it the attendant taxa, or whether, as some suggest (Colebrook, 1986; Wainwright et al., 1993), actual changes in the function and structure of these North Atlantic systems were involved.

16.2.2.2 Atmosphere/Hydrographical Changes in the Pacific and Their Ecological Effects

Shifts in position of atmospheric pressure centers in the Indian Ocean, and in the Equatorial and Subtropical Pacific create hydrographic teleconnections that lead to changes in wind, rainfall, currents, and sea level over much of the Pacific, all the way to the American coast. The shifts tend to have a 4–7 year cycle, and physical mechanisms have been proposed that suggest that the events can be conceived as recurring, prompting the name Southern Oscillation (Rasmusson, 1984; Harrison and Cane, 1984; Enfield, 1989; Graham, 1995). The SO has multiple manifestations over different parts of the world's oceans. One such in the Eastern Pacific is the phenomenon that oceanographers refer to as El Niño, in which large-scale wind-related conditions favor transport of warm, low-salinity, low-nutrient water towards the coasts of Ecuador and Perú, overcoming the effects of local upwelling-producing winds, and raising sea level (Enfield, 1989). El Niño counters the usual upwelling of cold, salty, nutrient-rich water, which is the basis for the exceedingly rich fishery along the west coast of South America. The combined Southern Oscillation plus El Niño is referred to as an ENSO event.

Probably the best-known consequence of ENSO events is the reduction of anchoveta catch off Perú and Ecuador (Fig. 16-4). Anchoveta have furnished the world's largest single fishery (Barber and Chavez,

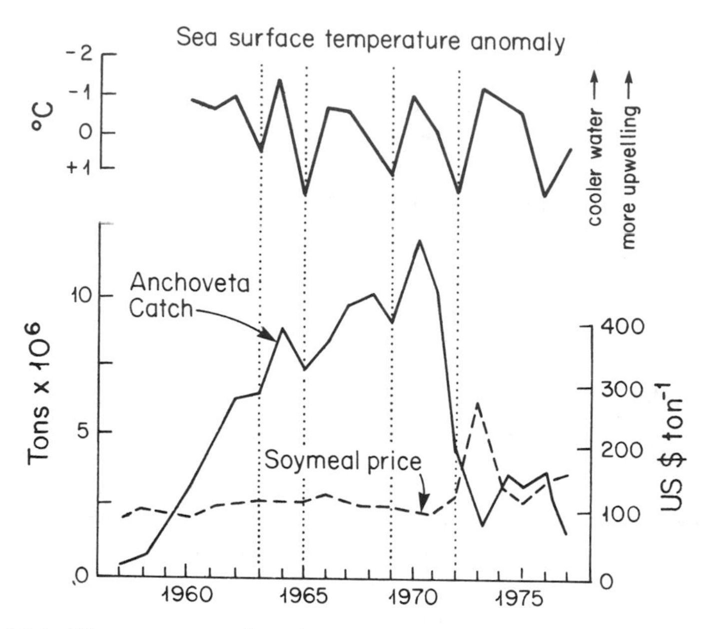

FIGURE 16-4. Time course of anchoveta catch off Perú, annual sea surface temperature anomaly (difference from long-term mean), and world price of soy bean meal, 1955–1978. The changes in catch show combined effects of increased fishing effort and lack of sufficient upwelling. Vertical dotted lines show years of reduced upwelling and low catch; after early 1970s the anchoveta stock was severely depleted by overfishing, and yield did not respond to upwelling intensity. Adapted from Barber and Chavez (1983).

1986), but the annual catch was lowered whenever intensity of upwelling decreased. Not all ENSO events have equally intense effects; the 1972–1973 event was specially destructive to the fishery, and the overexploited stock did not recover readily (cf. Section 9.4.3).

ENSO events have pervasive repercussions into human affairs and fate of other environments. As a result of the collapse of the anchoveta fishery (Fig. 16-4), world supplies of protein meal were reduced, increasing prices of soybean meal in world markets. This motivated large-scale clearing of Amazon rainforest and North American wetlands to grow soybeans. Rainforests and wetlands are involved in the release and storage of greenhouse gases (Mitchell, 1989), and we have yet to learn of the chain of global consequences of the shift in land use initiated by ENSO. Many unexpected far-flung effects of ENSO have been reported (Glantz, 1984). Just a few examples provide an idea of the diversity of purported consequences. In Africa and Australia, dust storms and brush fires increase during ENSO years, while both drought and flooding increase in different regions of South America. In the eastern United States, warm, wet springs breed more mosquitoes, leading to higher incidence of encephalitis. In Montana, drier weather causes mice to seek food at lower elevations, rattlesnakes follow the mice, and the incidence of rattlesnake bites increases. In New Mexico cool, wet springs favor flea-bearing rodents, so bubonic plague cases increase, and warm waters increased shark attacks off Oregon.

ENSO events have manifestations over much of the western coast of the Americas. Probably the best studied (Arntz, 1986; Glynn, 1988; Dayton et al., 1992) ENSO event occurred during 1982–1983, the most intense such event of the 20 that have taken place since 1900. Off California, for example, the 1982–1983 ENSO inhibited the usual upwelling conditions and resulted in warmer surface emperatures and lower nutrients and chlorophyll near shore (McGowan, 1985). The relaxation of winds during ENSO has been linked to reduced phytoplankton production off California (Smith et al., 1982), and diatom abundance off British Columbia (Ware and Thompson, 1991), lowered zooplankton abundance and lower recruitment success in sardines off the West coast of North America (Ware and Thompson, 1991).

The 1982–1983 event produced notable shifts in commercial hauls from California waters. Catches of cold-water species (chinook salmon, squid, crab, and shrimp) were down by more than 70% (McGowan, 1985). The losses were partially offset by increases in catch of warm-water species (albacore, tunas, marlin, and dorado). The northern anchovy, an ecological analog of the anchoveta, showed reduced growth, spawning range, fecundity, and early larval mortality. A strong year class in 1984 returned the anchovy to its earlier abundance (Fiedler, 1984).

The effects of the 1982–1983 ENSO on benthic primary producers were manifold. Increases in storms, wave action, lowered salinity, and warmer temperatures reduced abundance of benthic algae in the eastern Pacific, from Chile to California. The storms resulted in direct mortality of 13–66% of the giant kelp of Califonia (Tegner and Dayton, 1987). An additional 20–67% of the kelp died following the storms. In non-ENSO years kelp growth is light limited, but during the 1983–1983 ENSO the lowered nutrient contents of water limited kelp growth, as well as phytoplankton. Increased temperatures caused bleaching (loss of zooxanthellae) in corals growing near the sea surface. There was 50–98% mortality of corals from Ecuador to Costa Rica.

The effects of ENSO on benthic consumers were as varied as those on producers. Off California, the low nutrients and primary productivity lowered zooplankton abundance during 1983 (McGowan, 1985). Sea urchin recruitment, probably because of poor conditions for larvae, was reduced for several years (Tegner and Dayton, 1987). Little food was available for abalones and urchins, and their growth and gonadal development was curtailed. Warm water favored diseases of starfish, and densities of the batstar *Patiria miniata* diminished by an order of magnitude. In addition, there were many other changes in abundance and distribution of different kelp ecosystem species that became manifest during the recovery from the 1982–1983 event (Dayton et al., 1992). The powerful recruitment capacity and competitive dominance of giant kelp in a few growing seasons led to recreation of kelp beds in many places off the California coast.

The effects of the 1982–1983 event off Perú parallel those off California. Nitrate in surface water went from 3–5 to 0.1 μM, primary production from 16–219 to 3–10 $mg\,C\,m^{-3}\,d^{-1}$ (Barber and Chavez, 1983). Anchoveta suffered enormous losses, to the extent that the 1983 catch was <1% of the catch a decade before (Barber and Chavez, 1986). Long lists of other specific deleterious effects of the 1982–1983 ENSO, similar to those compiled for California, have been recorded for Peruvian and Ecuadorian shores (Arntz, 1986). In addition, increases in sea level of up to 2 m in mangrove areas covered these environments in mud and salt, and caused extensive damage to trees and animals. Marine iguanas in the Galápagos suffered 45% mortality. Most seabird and seal colonies were abandoned during the ENSO.

Much as in the Atlantic, some species benefitted by the changes; scallops and octopus increased—for unknown reasons—after 1982 (Arntz et al., 1988). The scallop catch, for example, increased 60-fold. Mackerel concentrated in the nearshore, where there was a remnant of upwelled cold water containing euphausids, their main food. Warm water-tolerant fish (bonito, yellowfin tuna, and dorado) became more abundant and came inshore to feed on the concentrated mackerel (Barber and Chavez, 1983).

Studies in the Pacific thus compellingly show that marked alterations of the functioning of populations and ecosystems result from large-scale, multiyear, atmospherically-driven changes. The ecological effects are mediated by hydrographical mechanisms. Knowledge of the work of such large-scale mechanisms is key to understanding how marine systems work and to management of the fisheries, because it is essential to predict the array of harvestable species that will be found before or after ENSO events.

16.2.3 Changes in Sea Level Rise

Mean global sea level has changed over geological time, and continues to change (Fig. 16-5), rising today on average about 1–1.5 $mm\,y^{-1}$ (Wyrtki, 1990). Changes in sea level rise result from tectonic and postglacial isostatic adjustments (deformations of ocean basins, subsidence or emergence of land) and effects of atmospheric temperature changes (volume expansion on warmed ocean waters, melting of glaciers).

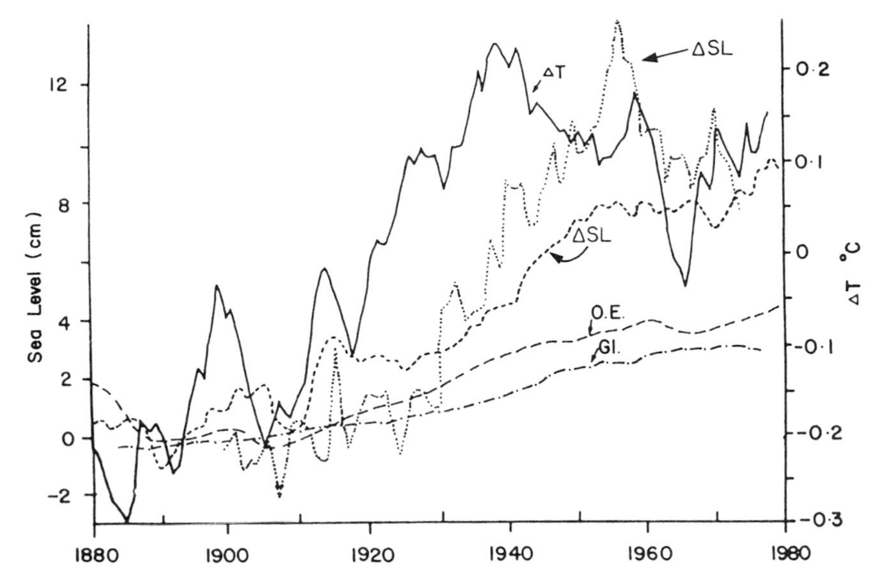

FIGURE 16-5. Changes in estimated mean temperature of the earth (T), global sea level (SL, two different estimates), sea level rise owing to expansion of the oceans above the thermocline (O.E.), and to melting of glaciers (Gl.). From Robin (1986).

Tectonic changes lead to much region-to-region variability in local sea level dynamics. In Japan, for example, the islands are tilting, so the Pacific shores are subsiding at up to 24 mm y^{-1}, while the shores on the Sea of Japan are being uplifted at up to 6 mm y^{-1} (Emory and Aubrey, 1986). Variation from one region to another is so large (Wyrtki, 1990; Robin, 1986) that it is difficult to even calculate global sea level rise (cf. different estimates in Fig. 16-5), let alone generalize about ecological effects of global sea level rise. While global temperatures at the sea surface have increased during the last century, the increase in temperature is not steady, and its time steps are not clearly related to changes in global sea level (Fig. 16-5). Ocean volume expansion and melting of ice to mean global sea level only account for a fraction of sea level rise (Fig. 16-5), so other factors must also be involved.

There is much concern that global warming may be accelerated by the continuing release of greenhouse gases by burning of fossil fuels and of forests, and other sources (Bolin et al., 1986; Mitchell, 1989). Such increased releases could accelerate already visible effects. In low-lying deltas (Mississippi, Bay of Bengal, Nile), rising sea level has led to loss of coastal wetland environments, salinization of fresh water supplies, and loss of habitable and arable land. Depending on the rate of sea level rise, salt marshes and mangroves may be unable to accrete fast enough to keep pace with sea level rise (Stevenson et al., 1988; Day

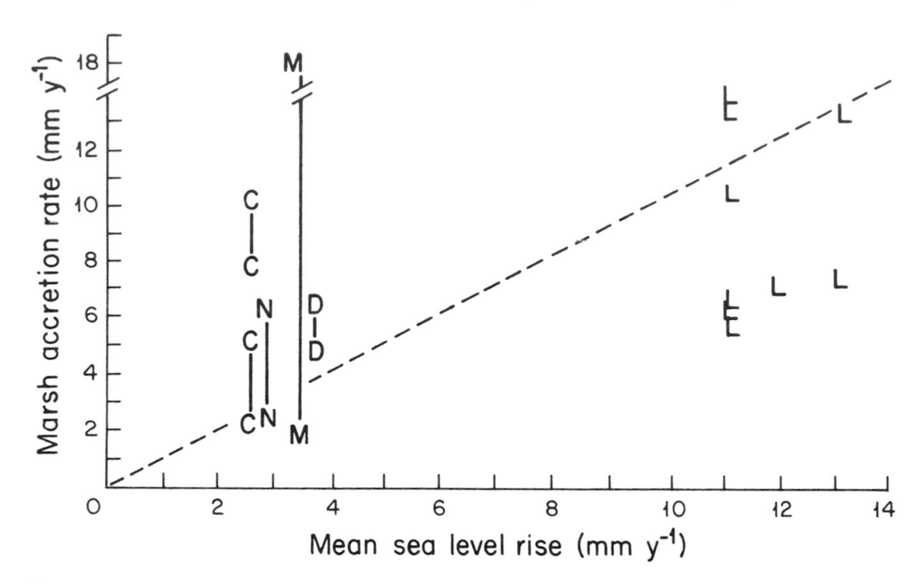

FIGURE 16-6. Accretion rates of salt marshes in relation to rate of sea level rise for marshes in various states of the U.S. (Connecticut, New Jersey, Massachussetts, Delaware, and Louisiana, indicated by their initials). Vertical lines show range in accretion within one site, dashed line shows a 1:1 relation of accretion and sea level rise. Data from Day and Templet (1989).

and Templet, 1989). Salt marsh accretion rates range between 2 to 18 mm y^{-1} (Fig. 16-6). Within any one site accretion is higher on the seaward, growing margin, and less on the more stabilized land margin. At the relatively low rates of sea level rise found in the northeast coast of the United States, salt marshes can easily keep pace. Where sea level is rising faster, such as in Louisiana, the maximum accretion rate of marshes at their growing margins is just sufficient to prevent submergence, but in many sites the salt marshes are being submerged.

Sea level changes can affect aquatic coastal populations and communities. Obviously, increased flooding of shorelines will also affect coastal terrestrial landscapes and many major heavily-populated coastal areas. This topic, which is closely related to human interference with sediment and water transport, is discussed below.

Although sea level changes could be locally important, there is much uncertainty about the facts. If the worse-case scenarios become true, up to 26% of the habitable land on Nile Delta and up to 34% of Bangladesh could be flooded, with additional losses of land from erosion, plus loss of mangroves, and fisheries (Milliman et al., 1990). Such sea level increases would certainly have serious effects over the shorelines of the world. At present, we have limited ability to make predictions about global warming and climate change, and to link these in turn to sea level rise. We are unable to discern whether disastrous scenarios are more likely than less eventful predictions. Whatever the effects of sea level rise, they are likely to only affect environments in near-shore situations.

16.3 Depletion of Fishery Stocks

The world commercial catch of fish has gone from about 30 million metric tons in 1955 to about 100 million metric tons in the early 1990s, according to the Food and Agriculture Organization of the United Nations. This unprecedented increase exceeds expected ceilings for sustainable harvests. The result is that most of the world's marine fish of commercial interest are severely affected by fishing pressure.

The greatest impact of fishing pressure is exerted on coastal and shelf environments. Primary production in coastal and shelf waters may be more than 10 and 3 times, respectively, as large as production in ocean waters (Walsh, 1988). As a result, coastal and shelf waters support more than 90% of the worlds' fish yield, and stocks are generally overfished.

In the New England fishery, for example, 45% of the stocks whose status is known are overfished (NMFS, 1993). Populations of some species have been reduced to less than 10% of the estimated level needed for sustainable catch. In the case of some anadromous species, such as salmon, reduction of suitable spawning habitats contributes to the reduction in stocks.

There are exceptions to the critical state of fish stocks, as in the examples of the stringently controlled Pacific halibut and salmon stocks, and the remarkably successful reestablishment of the striped bass in the coasts of the eastern United States. The recovery of striped bass was the product of a moratorium on catch along the mid-Atlantic states during 1985–1990, argued on the basis of simulation studies of population growth (Goodyear, 1985).

The statistics on groundfish and flounders caught off New England (Fig. 16-7), however, are representative of the dire general situation. The increase in landings before 1965 reflects more effective fishing; the yields could not be sustained because stocks of the 12 species involved have diminished to alarming levels (Anthony, 1993). The New England fishermen catch around 60% of the entire fish population each year, more than twice the estimated sustainable level. This is a dramatic fishing pressure, considering that Iverson (1990) estimated that 25% of the oceans' fish production is harvested annually by marine fisheries, and suggested that was a high rate of harvest.

The annual removal of 100 million metric tons of consumers from the world's oceans is an impressive amount, orders of magnitude larger

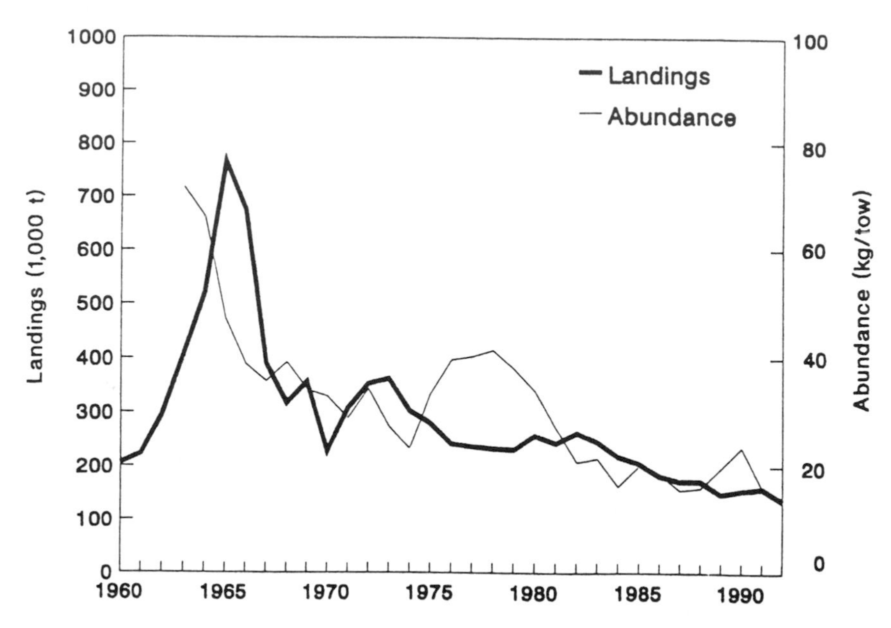

FIGURE 16-7. Total fishery landings and index of abundance for demersal fish off the New England coast, 1960–1992. Index of abundance are mean weight per tow taken in bottom trawled surveys. From Nat. Mar. Fish. Serv., NOAA (1993).

than losses by any toxic pollutant and much more likely to lead to long-term, large-scale changes. Fishing fleets have removed a substantial fraction of biomass from the upper levels of marine food webs in many parts of the oceans. Too little is known as to the potential consequences of this massive alteration. The intensive fisheries at once remove consumers, affecting possible top-down controls, and also reduce the abundance of food items for top predators, thus interferring with bottom-up controls.

The alteration of harvest on top-down controls were evident in examples such as increases in penguin and seal populations after depletion of whales in the Southern Ocean (cf. Chapter 9): abundance and species composition of top predators were thoroughly altered. In addition, removal of fish may be in part responsible for the apparent decoupling of adult fish stocks from their food supply; such removal of consumers must have consequences for the lower links in food webs. Fishery exploitation can also exert bottom-up effects: Springer (1992) reports decreases of 50% in fur seals and 40% in kittiwakes following depletion of pollock, an important item in mammal and bird diets in the Bering Sea. The magnitude of the fishing harvest and the examples of major alterations to marine food webs by predator removal suggest that effects of fishing are ecologically substantial at large spatial scales. The reproductive ability of fish, on the other hand, is such that relatively fast recovery from fishery exploitation is possible, as in the instance of striped bass and other stocks, soon after exploitation is reduced.

16.4 Eutrophication

16.4.1 Anthropogenic Loading to Marine Ecosystems

We have reviewed much material that emphasizes the powerful role played by nutrient supply in limiting primary producers in marine environments. Throughout the world, human beings have carried out activities whose ultimate result is to add nutrients* to receiving waters, most of which end up discharging into the marine waters. In fact, the more people in a watershed, the greater the nutrient loads received by estuaries, for example (Fig. 16-8). For nitrogen, wastewater and fertilizer use account for most of the loadings to coastal waters (Cole et al., 1993; Hinga et al., 1991). Atmospheric deposition may also be

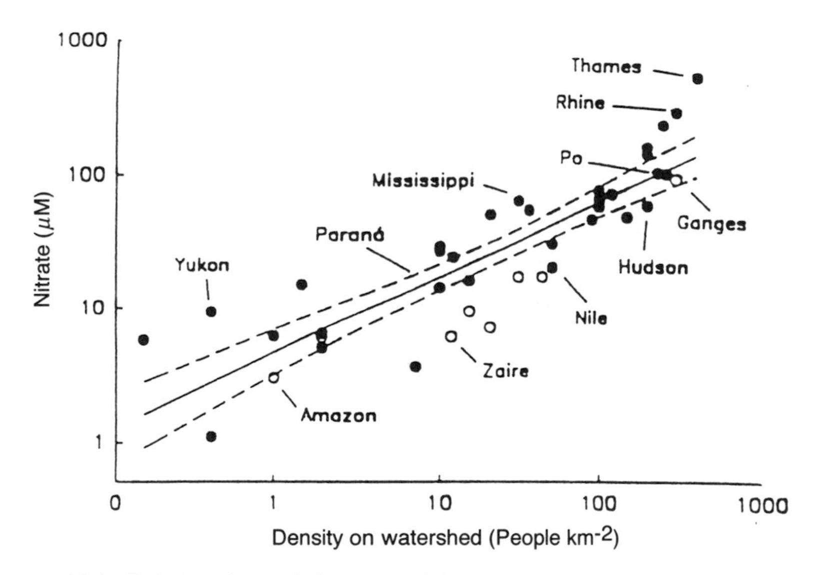

FIGURE 16-8. Relationship of density of human beings on a watershed, and resulting nitrate content in water discharged from the river. From Cole et al. (1993).

---

*Some anthropogenic activities, especially release of wastewater, also release land-derived organic matter into marine environments. This input increases oxygen demand, provides a source of externally furnished energy, alters availability of pollutants, and generally enhances the biogeochemical changes brought about by increased delivery of nutrients to marine environments. At a global scale these inputs are small compared to anthropogenic CO_2 inputs, but organic inputs may be important at regional scales (Spitzy and Ittekot, 1991).

important (Paerl, 1993), but since much of the atmospheric nitrogen is of anthropogenic origin, this third major source is also linked to human activity. The emphasis on human activities here stem from the impressive fact that worldwide, we are producing biologically accessible nitrogen at about 140 Tg y^{-1}, which is of the same magnitude as the 90–130 Tg y^{-1} achieved by all natural processes (Galloway et al., in press).

Estimates of preindustrial and present inputs of anthropogenic nitrogen into marine waters provide an idea of the magnitude of the human influence on nitrogen loadings into the sea (Duce et al., 1991; Galloway et al., 1994). Rivers and atmospheric deposition are thought to have contributed 13 and 12 Tg y^{-1} during preindustrial years. Estimates of current inputs show 49 and 28 Tg y^{-1}, a threefold increase. The inputs of course are not distributed uniformly; coastal and oceanic waters receive different inputs and respond in different ways.

16.4.2 Nitrogen Inputs to the Open Ocean

Nitrogen loading to the open ocean has increased mainly because of atmospheric deposition (16 Tg y^{-1}), although there may be some as yet unquantified horizontal transport from the nearshore across the continental shelf. Since perhaps 0.4 Tg y^{-1} are buried in deep sea sediments (GESAMP, 1990), denitrification must be important (Duce et al., 1991) and has been estimated roughly at 150–180 Tg y^{-1}, much in excess of inputs. Some additional losses have been suggested: for example, some of the N_2O produced by marine nitrifiers may reenter the atmosphere (Kim and Craig, 1993). Part of the deficit may be met by nitrogen fixation, which has not been well quantified. Human activities have added trace metals to atmospheric deposition, which could enhance nitrogen fixation. It may well be, however, that nitrogen content of the deep ocean may be increasing in the long term. Nitrate or N_2 dissolved in deep water could have increased over recent decades, but such increases would be difficult to discern because annual inputs are tiny (0.02% for nitrate) compared to the size of nitrogen pools in sea water (10^5 and 10^7 Tg N for NO_3 and N_2, respectively, Laws, 1983).

It is not at all clear just what the effects of increased delivery of nitrogen to the open ocean have been. Some changes in the nitrogen cycles and on phytoplankton growth patterns and species composition have been suggested (Paerl, 1985; Michaels et al., 1993). There may also be enhancement of carbon sinking to deeper water.

16.4.3 Eutrophication of Coastal Systems

Human populations have especially increased in coastal zones, and the large inputs from land to sea reach coastal waters directly. These features make coastal waters particularly subject to eutrophication. Below we examine eutrophication in large coastal environments, as well as in the smaller, but more numerous, shallow marine coastal bays and lagoons.

16.4.3.1 Eutrophication of Large Coastal Systems

One large-scale example of the long-term process of eutrophication is provided by the Baltic Sea. Loadings of nitrogen to the Baltic have increased about 4-fold since the 1950s (Fig. 16-9). After an initial increase, phosphorus loads were curtailed by the introduction of chemical precipitation methods in sewage treatment plants. Throughout recent decades the loading has resulted in increased nutrient content in different parts of the Baltic (Fig. 16-10). In the northernmost and freshest part of the Baltic, the Bay of Bothnia, phosphorus limits producers (cf. concentrations in Fig. 16-10), while nitrogen, as we would have predicted, is the limiting nutrient in the rest of the Baltic (Granelli et al., 1990).

The increases in nitrogen concentration were accompanied by a chain of indications of increasing eutrophication (Fig. 16-9). Phytoplankton production increased 30–70%, depth of light penetration decreased, filamentous seaweeds proliferated in shallow waters, low oxygen (<2 ml l^{-1}, about 20% saturation) occurs more frequently in many places, and uniformly below depths of 70 m, and sediments are strongly reduced over large areas of the sea floor (gray areas, Fig. 16-10). Benthic production in shallow waters increased as a result of eutrophication, as have certain fish, but there have been mass mortalities of bivalves, and lobster catches have diminished (Elmgren, 1989;

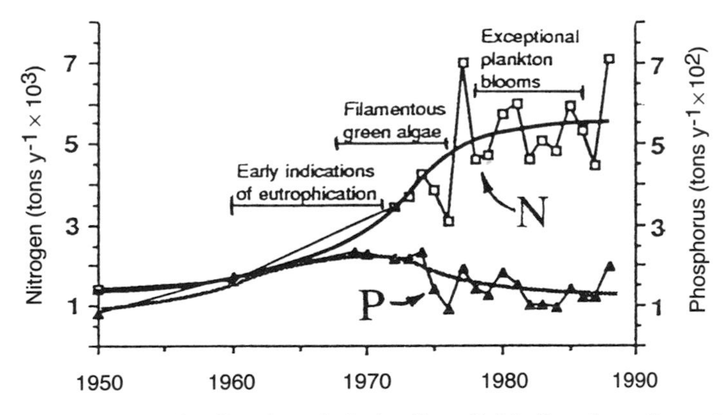

FIGURE 16-9. Nutrient loading into Laholm Bay, Baltic Sea, from its watershed, 1950–1988. Horizontal lines show significant changes in the Bay. Adapted from Rosenberg et al. (1990).

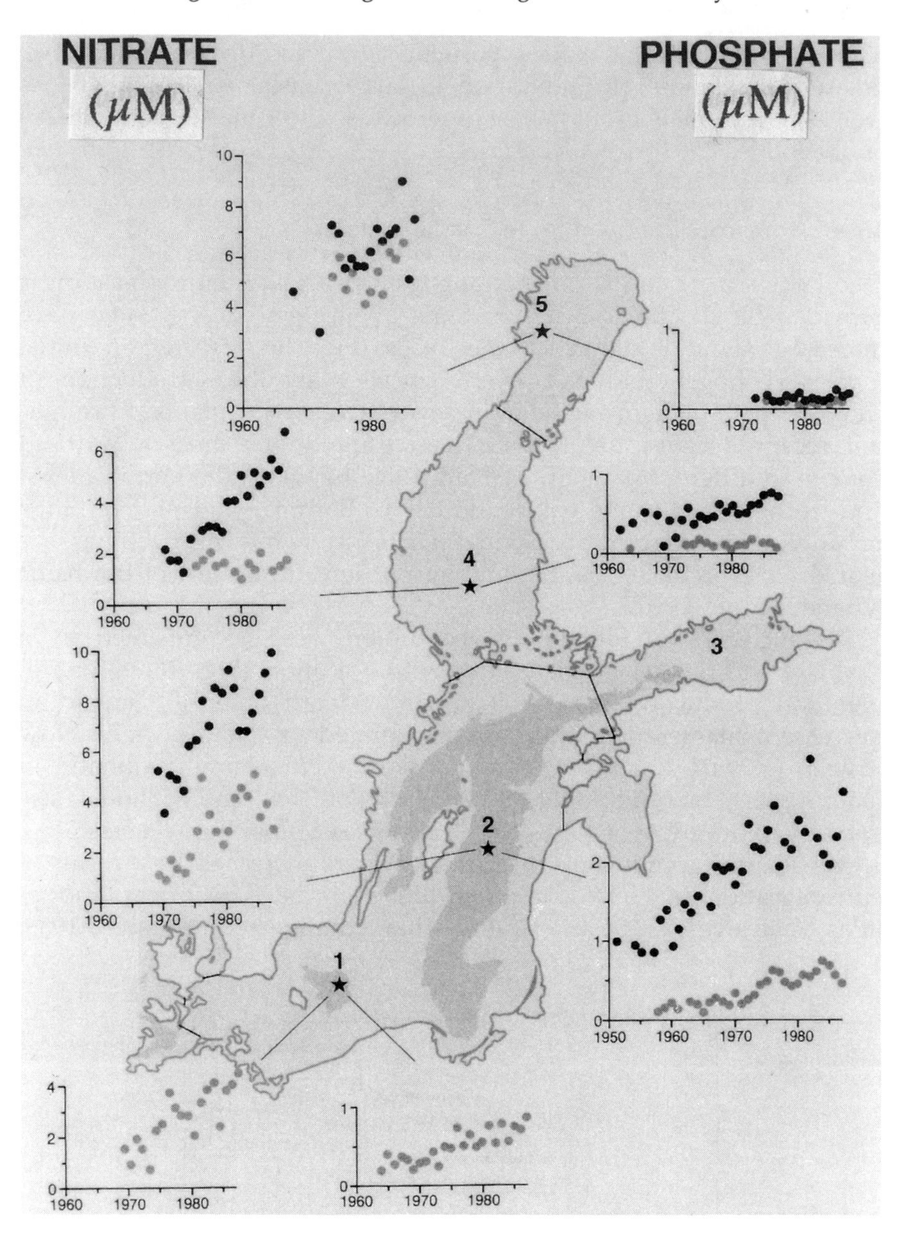

FIGURE 16-10. Concentrations of nitrate (graphs on left) and phosphate (graphs on right) over several decades in different parts of the Baltic Sea. Black circles refer to annual mean concentrations at 100 m, grey circles to concentrations during winter for surface waters. 1: Bornholm Basin, 2: Gotland Sea, 3: Gulf of Finland, 4: Bothnian Sea, 5: Bothnian Bay. The shaded area includes sites where anoxia has reduced benthic fauna. From Elmgren (1989).

Baden et al., 1990). Eutrophication has generally increased production, but there have been deleterious consequences to some components of the Baltic biota, so as we have found elsewhere, the composition of the biota has changed to a new array.

The effects of eutrophication in the Baltic are trammeled by a complex set of other anthropogenic activities. Pollutants may affect consumers, and exploitation of fish and marine mammals has had major effects. For example, the fish catch from the Baltic was 9 times greater in 1980 than in 1900, while marine mammals consumed 30 times less fish in 1980 than in 1900.

Although the example of the Baltic is a compelling example of eutrophication of a large marine system, the effects we have discussed only begin the list of consequences of nutrient enrichment. Many other important alterations and interactive changes have been reported from other systems. As one example, in Chesapeake Bay, nitrogen enrichment not only has the usual effect of increasing primary production, but also allows increased rates of recycling of NH_4 (Kemp et al., 1990). In stratified estuaries, such as Chesapeake Bay, the enrichment leads to lower oxygen in lower layers of the water column. Anoxia reduces areas suitable for this fauna, and also inhibits nitrification. Because nitrification is the prime source of NO_3 for denitrification, denitrification is itself inhibited. The result is that more NH_4 remains unoxidized and is available to producers, which intensifies the level of eutrophication.

16.4.3.2 Eutrophication of Coastal Bays and Lagoons

In shallow water ecosystems, eutrophication creates a complex set of direct and indirect reactions that lead to major changes not only in the producers but in the rest of the ecosystem. The reactions and interactions involved result from the action of many of the specific mechanisms discussed in previous chapters.

Seagrasses often dominate shallow waters exposed to low nutrient inputs. Seagrasses can grow in waters shallow enough so that the bottom is illuminated, and have relatively low nutrient uptake rates (Fig. 16-11), but solve the problem of nutrient acquisition by taking up nutrients via their roots from sediments, where there are usually high concentrations of nutrients. Seagrasses have generally low N to C ratios (Fig. 16-11), but store substantial amounts of nitrogen in their thick leaves, stems, and rhizomes. Growth of seagrasses, as noted in Chapter 2, is generally light- rather than nitrogen-limited, and is slower than those of macroalgae and phytoplankton (Fig. 16-11). In contrast, phytoplankton and fast-growing seaweeds tend to be nutrient- rather than light-limited. While seagrass grow on sediments at depths receiving more than 11% of incident light (Duarte, 1991), macroalgae can grow down to 0.12% (for species with thick fronds) and $<0.003\%$

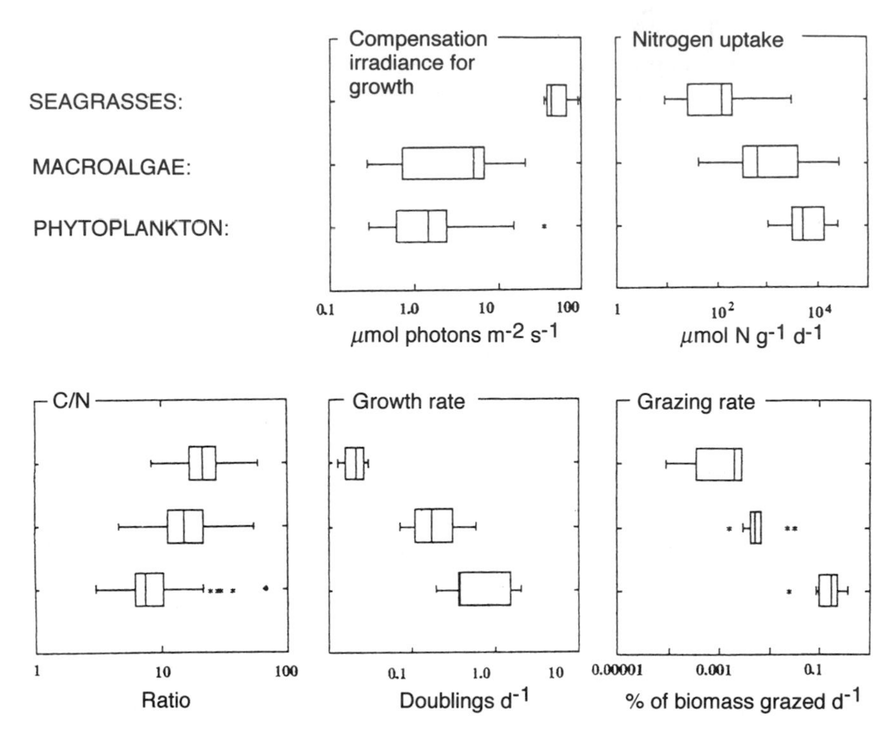

FIGURE 16-11. Compilation of values for selected properties of three major groups of producers (seagrasses, macroalgae, and phytoplankton). Data obtained from many sources, and many species, and shown as box plots. The vertical bounds of the boxes show the 1st and 3rd quartiles of the data for each producer type. The line in center of box shows the median value, the horizontal lines extend to the 95% confidence limits, while asterisks show values lying beyond these limits. Adapted from Duarte (in press).

(for thin macroalgae) of incident light (Markager and Sand-Jensen, 1992).

One of the first signs of increased availability of nutrients is growth of epiphytes on seagrass surfaces and of phytoplankton. We have already seen the phytoplankton response to nitrogen supply on populational or per-unit volume of water basis in Chapter 2; Fig. 16-12 shows that seasonal phytoplankton production in entire coastal ecosystems increases as nitrogen load from the watershed increases (Table 16-1). The increased phytoplankton production resulting from greater nitrogen supply is apparent in spite of the year-to-year variation prompted by other factors.

The shading created by increases in abundance of both phytoplankton and epiphytes is detrimental to the light-limited seagrasses that are unavoidably tied to the bottom (Twilley et al., 1985; Sand-Jensen and Borum, 1991; Duarte, 1994). Thus, the next tier of symptoms

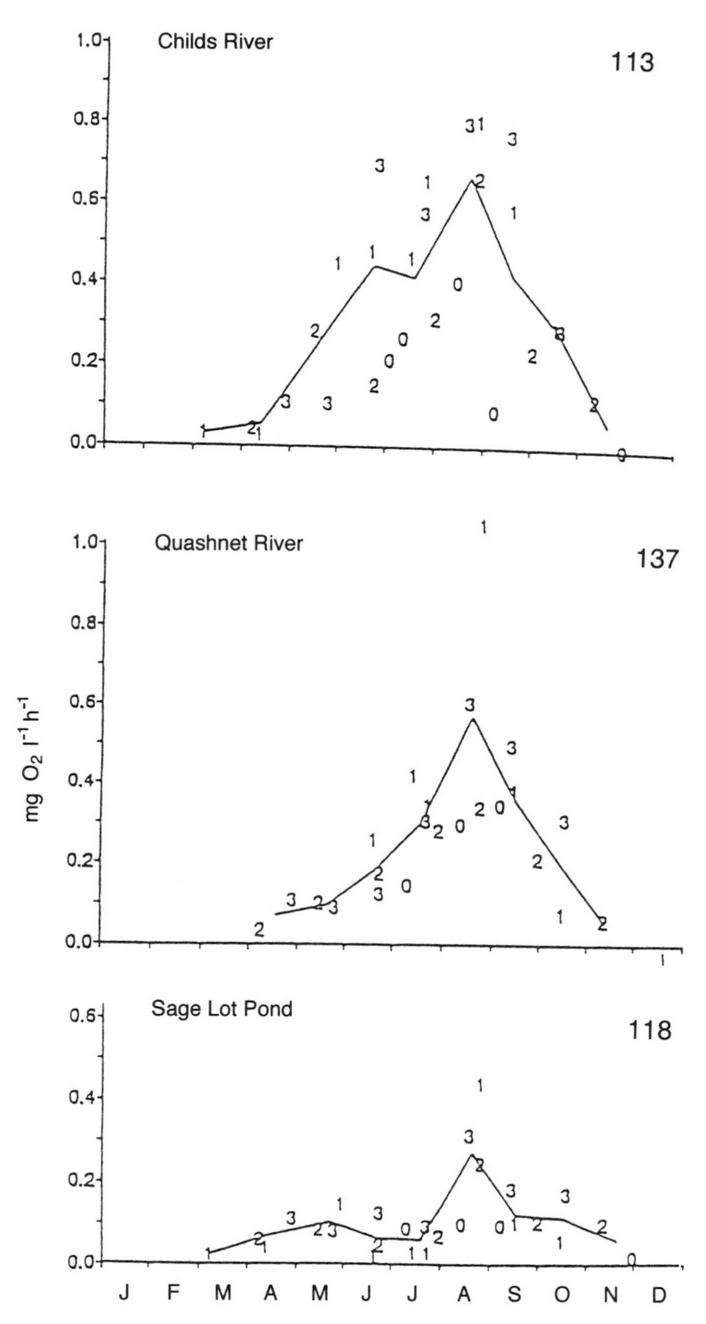

FIGURE 16-12. Seasonal phytoplankton production cycles in three similar shallow estuaries of Waquoit Bay, Massachussetts. The three estuaries receive different rates (CR > QR > SLP) of nitrogen loading from their watersheds; loading rates are shown in Table 16-1. Production data obtained during four years, 1990–1993, shown as 0, 1, 2, or 3. Numbers on the top right of each graph means annual gross production by phytoplankton, in $g\,C\,m^{-2}\,y^{-1}$. Data of K. Foreman.

TABLE 16-1. Mean Biomass of Macroalgae and Eelgrass[a] and Gross and Net Ecosystem Production[b] in Three Estuaries of Waquoit Bay That Receive Different Rates of Nitrogen Loading from Their Watersheds.

	Childs River	Quashnet River	Sage Lot Pond
Average concentration of dissolved inorganic nitrogen in freshwater entering estuaries (μM)	117	78	22
Biomass (g dry weight m^{-2})			
Seaweeds	383	146	100
Seagrass	0	0	100
Ecosystem gross production ($g\ O_2\ m^{-2}\ y^{-1}$)	15,100	5,400	4,600
Ecosystem net production ($g\ O_2\ m^{-2}\ y^{-1}$)	182	67	10

[a] Data of D. Hersh.
[b] Data from C. D'Avanzo and J. Kremer.

of eutrophication in shallow waters involve reduction of seagrass biomass (Table 16-1) and areas of seagrass meadows. Loss of seagrass habitats is important because, among other attributes, these environments support a wide variety of consumers, some of commercial interest (Valiela et al., 1992), and are nurseries for young of many deeper water species. Extensive losses of valuable seagrass habitats, and corresponding losses of fauna, have taken place as a result of eutrophication of many shallow coastal areas (Nienhuis, 1983; Cambridge and McComb, 1984; Giessen et al., 1990; Valiela et al., 1992).

Fast-growing seaweeds do well in enriched waters even if they are turbid, because these macroalgae can take advantage of high-nutrients concentrations, can store nutrients, and can grow at low irradiances (Fig. 16-11). As a result, seafloors of enriched shallow waters are often covered by canopies of seaweeds, as occurs in Waquoit Bay in Massachussetts (Valiela et al., 1992), Venice Lagoon (Sfriso et al., 1992), the Peel-Harvey Estuary in Australia (Lavery et al., 1991), and in many other such environments worldwide. In fact there can be a close link between nitrogen inputs from watersheds and seaweeds. In Waquoit Bay seaweed biomass not only increases in proportion to nitrogen loading rate (Table 16-1), but seaweeds are so tied to nitrogen supply that stable nitrogen isotope ratios in the fronds reflect the isotopic signature of the different nitrogen sources that contributed to the nitrogen load from watersheds. The major sources of nitrogen from watershed to Waquoit Bay are wastewater disposal, atmospheric deposition, and use of fertilizers, in that order. Each of the three sources has reasonable distinct stable isotopic ranges: 10–20‰ for wastes, 2–8‰ for deposition, −3 to 2‰ for fertilizers. Producers in these estuaries of Waquoit Bay that receive more wastewater show elevated stable isotopic ratios for nitrogen (Table 16-2). The elevated ratios also propagate through the food webs via the producers, with fractionation described in Section 8.1. These results demonstrate that it is the nitrogen delivered from each of the watersheds that exerts its effects throughout the estuarine food webs.

If nutrient loading rates increase further, light must become limiting to phytoplankton and macroalgae. At some as yet undetermined point, the ability of macroalgae to survive at low light intensities must be insufficient, and only the phytoplankton may remain as producers.

Eutrophication operates mainly via bottom-up controls, but interacts with top-down controls in shallow-water ecosystems. Higher-nutrient supply makes all producers present more nutritious and fosters species of more palatable producers, so we would expect increased grazing pressure as more nutrient become available. Recall that phytoplankton and some fast-growing seaweeds are less chemically protected than vascular plants and certain seaweeds (Chapter 6). In addition, the hierarchy of nutritive quality is phytoplankton > macroalgae > seagrasses (Fig. 16-11), even without the increased nutritive value conferred by enrichment. Grazing pressure should be therefore lower on seagrasses than on macroalgae, than on phytoplankton (Fig. 16-11). Grazers of phytoplankton and epiphytes therefore may help maintain

TABLE 16-2. Comparison of $\delta\ ^{15}N$ in Plankton Particles, Two Seaweeds, One Seagrass, Three Grazers and One Carnivore in Estuaries of Waquoit Bay Receiving Nitrogen Loads from Watersheds That Include (Childs River) and Not Include (Sage Lot Pond) Significant Amounts of Wastewater Nitrogen, Which Contains Elevated Values of $\delta\ ^{15}N$.[a]

	$\delta\ ^{15}N$ values (‰)		Difference
	Sage Lot Pond	Childs River	(CR-SLP)
Plankton			
Particles	4.5	5.9	1.4
Seaweeds			
Cladophora vagabunda	3.5	5.4	1.9
Gracilaria tikvahiae	6.7	7.9	1.2
Seagrass			
Zostera marina	1.3	—	—
Herbivores			
Sclerodactyla briarias	6.3	9.3	3.0
Podarke obscura	7.2	11.5	4.3
Cyprinodon variegatus	5.6	9.8	4.2
Carnivore			
Menidia menidia	8.8	11.6	2.8

[a] Data of J. McClelland.

dominance of seagrasses in waters exposed to low-nutrient loadings. Grazers can be therefore conceived as a brake on species replacement, but as nutrient loading increases, fast-growing seaweeds overcome controls by grazers, and the ecosystem becomes dominated by macroalgae rather than seagrasses. Probably the same dynamic applies to phytoplankton.

Ecosystem metabolism increases as nitrogen loading increases (Fig. 16-13, top two panels). As ecosystem gross production increases (Table 16-1)—and as seagrasses are replaced by macroalgae, with changes in all the accompanying biota—ecosystem respiration follows suit. The ratio of P to R is surprisingly close to 1 (Fig. 16-13 bottom), regardless of rates of nutrient loading, rate of primary production, and species of producers. The level of productivity and composition of the biota seems immaterial; the close relation between production and respiration in the 3 estuaries of Waquoit Bay (Fig. 16-16) is maintained even though the producers and consumers carrying out the metabolism differ markedly. The near balance between ecosystem production and ecosystem respiration suggests that there is little organic matter left to be exported or buried in sediments, and that allochthonous inputs of organic matter are not needed to maintain ecosystem metabolism.

The apparent balance of respiration and production over the course of months to years should not be taken to mean that shorter-term, intermittent imbalances of respiratory and photosynthetic rates do not occur. These temporary imbalances are sometimes manifested as episodic anoxic events that last hours to days. Eutrophication in Waquoit Bay, as in all such systems, has made anoxic events more frequent (D'Avanzo and Kremer, 1994); a few days of cloudy weather during warm months, while the water column is stratified, are sufficient to permit oxygen consumption by respiration to exceed photosynthetic oxygen production. These episodic events change the structure of communities because they may increase mortality of seagrasses and certain animals, and favor release of nutrients held in sediments into the water column. Thus, eutrophication has short-term consequences (such an anoxic events) whose mechanisms are not the same as those resulting from eutrophication-related long-term events. Their circulating effects significantly affect long-term changes in structure and geochemistry of shallow waters.

Data from some shallow ecosystems exposed to increased enrichment suggest a trend toward net ecosystem autotrophy. Evidence that enrichment leads to autotrophic metabolism is provided by MERL experiments in which increased nitrogen loads led to a shift from P/R = 0.9 to 1.1–1.3 (Oviatt et al., 1986a). Although the Waquoit data suggest an approximate overall balance between production and respiration, most points from the most eutrophic estuary (Childs River) lie slightly above the line showing P/R > 1 (Fig. 16-13, bottom panel,

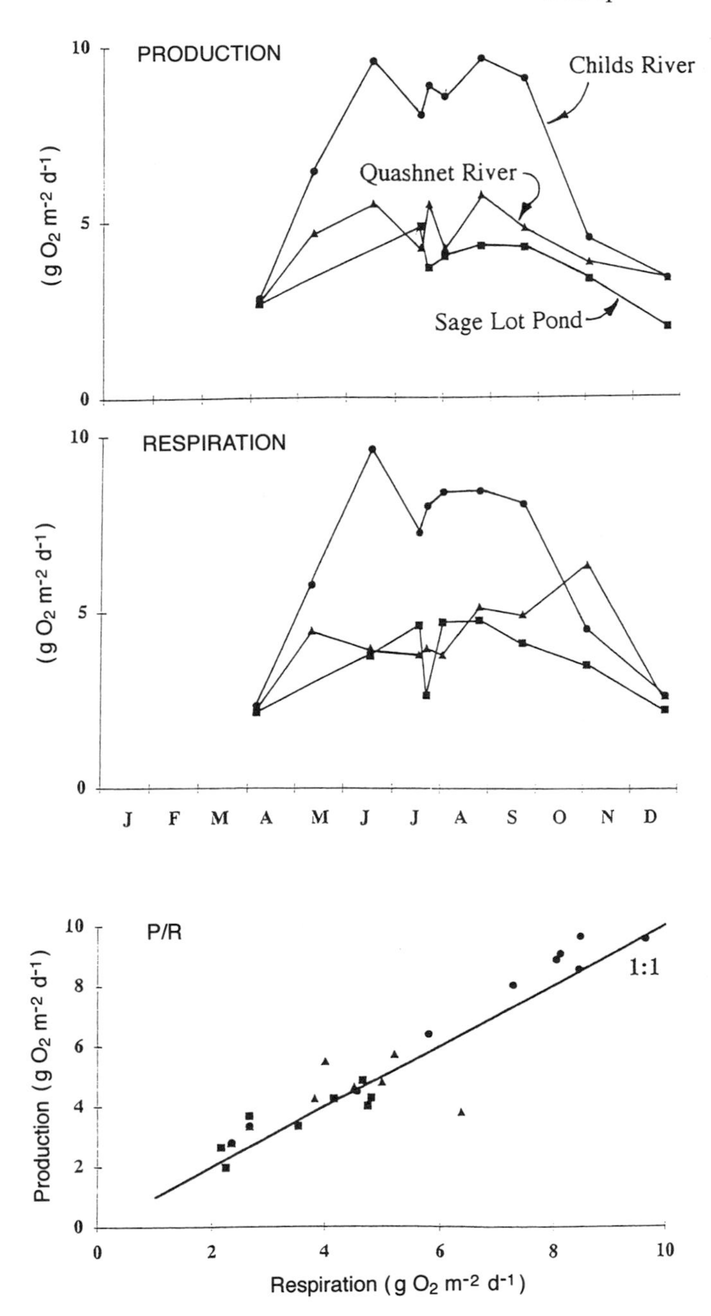

FIGURE 16-13. Gross daytime ecosystem production (top), night ecosystem respiration (middle), and P/R (bottom) in three estuaries of Waquoit Bay that receive different rates of nitrogen loading from their watersheds (cf. Table 16-1). Data obtained using continuous O_2 measurements and diffusion corrections (D'Avanzo and Kremer 1995).

black circles), and net ecosystem metabolism rates are greater than in the other estuaries (Table 16-1). Cross-system comparisons (discussed in Section 14.4.2) suggested that, in contrast, increased nutrient supply promotes ecosystem heterotrophy. Further study is needed to reconcile these different conclusions; the differences may just be a matter of the approach used (experimental enrichment in one kind of system vs. comparisons of many systems), or could be more substantive.

Many of the same features we have just described for temperate bays and lagoons apply to tropical lagoons, where coral reefs are replaced by seaweeds in eutrophic sites (Bell, 1992). In the case of such reefs, the critical nutrient seems to be phosphorus rather than nitrogen. Only small additions of nutrients may have strong effects, because pristine reefs usually have rather low-nutrient concentrations (about 1 μM DIN, 0.1 μM PO_4). Among other effects, increased nutrients improve survival of larvae of crown of thorns seastars, and may therefore indirectly create outbreaks of this consumer of corals.

16.4.4 Interaction of Eutrophication and Global Change

Anthropogenically generated nutrients are delivered to the sea by atmospheric deposition (Paerl, 1985; Duce, 1991) or via flowing freshwater. The atmosphere nutrient load is acquired by industrial or agricultural releases. Flowing freshwaters (rivers or groundwater) are enriched by inputs of wastewater and leached fertilizers. Atmospheric sources tend to be somewhat smaller than the flowing freshwater sources; atmospheric nitrogen contributions of nitrogen to coastal waters, for example, range from 13–50% of total inputs (Duce, 1991).

Nutrient loading by all sources are increasing on a worldwide scale, so that it is not surprising to find that nutrients in many coastal waters have increased over recent decades (Fig. 16-14, top two panels). Even though specific estuaries or regions may show increases or decreases (Radach, 1992), the overall pattern on aggregate tends to show increased nutrient concentrations. The changes in nutrients are paralleled by changes in abundance of organisms, as can be seen by the secular increases in abundance of flagellates, while diatoms increase and decrease, in Fig. 16-14 (bottom two panels). Note that the seasonal variation is larger than the secular trend, and that there is also substantial year-to-year variation. The secular trends, much as in the case of long-term sediment flux data in the Sargasso Sea, are only evident in decadal data records.

Nutrient enrichment is taking place so pervasively that on aggregate eutrophication constitutes one of the most compelling global scale changes in coastal marine environments. It is difficult to find an estuary, for example, that is *not* subject to eutrophication. In addition,

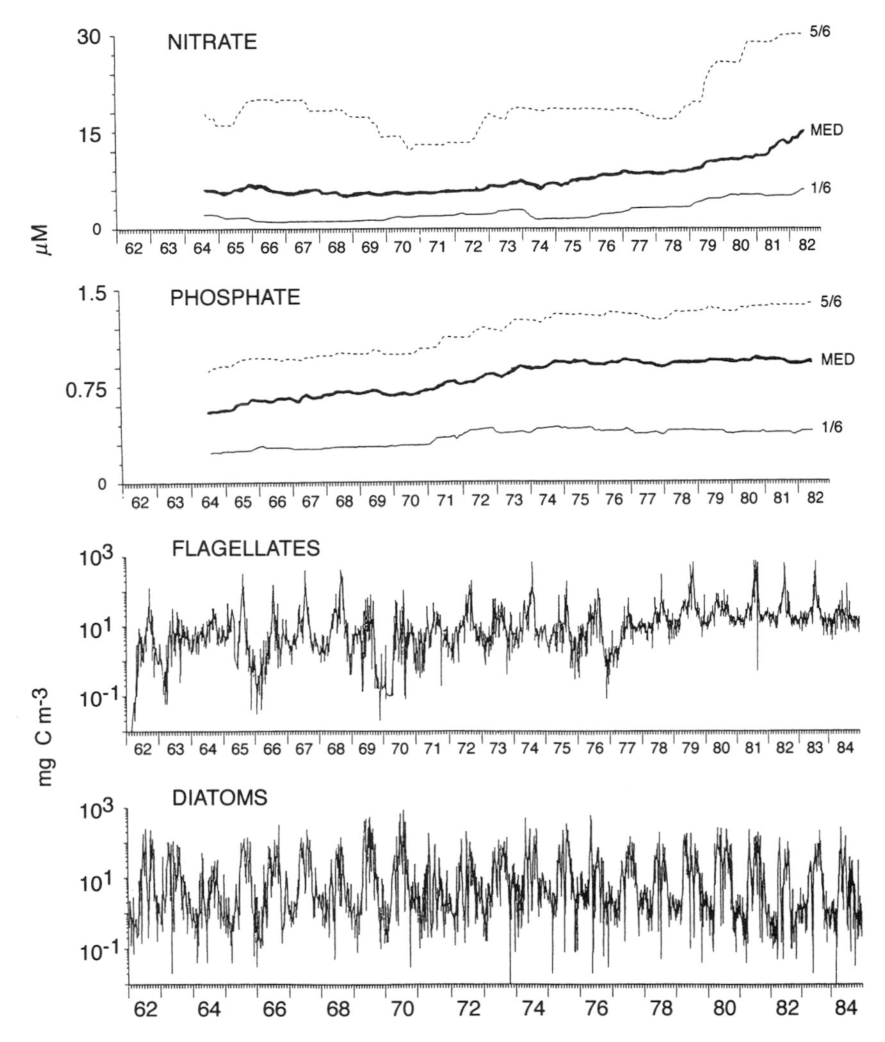

FIGURE 16-14. Time series of nutrients, flagellates, and diatoms, 1962–1984, in the North Sea off Helgoland. The nutrient data are shown as running 5-yr median values (thick lines), bounded by the 1/6th (thin line) and 5/6th (dashed line) quantiles. From Radach (1992).

the changes brought about as a result of eutrophication are linked in manifold ways to other global-scale changes.*

Marine environments are involved in the dynamics of changing composition of atmospheric gases. Coastal waters and sediments are

---

* As less developed nations increase the intensity of agriculture, and use resources, eutrophication of coastal waters will surely increase. We are already witnessing such trends in places such as the Yellow and East China Seas (Galloway, 1994).

sources of nitrous oxide, and coastal wetlands are a source of methane (Ashmore, 1990). Marine producers release significant amounts of DMSP, adding sulfur to an atmosphere already so burdened by anthropogenic sulfur that large areas of the world are exposed to acid rain (Wolfe et al., 1991; Duce et al., 1991; Bates et al., 1992; Spiro et al., 1992). Rates of release of nitrous oxide (Seitzinger, 1990; Law et al., 1992), methane (Sotomayor et al., 1994), and presumably of other gases, increase as nitrogen-enriched coastal systems become more productive. The increasing biomass of coastal seaweeds may store a significant part of the CO_2 being released to the atmosphere (Smith, 1981). We mentioned above the issue of the "biological pump" for carbon in the ocean, as well as the probable increases in nitrogen pools; future studies are likely to reveal important stoichiometric and global links between eutrophication and these features.

16.5 Toxic Contamination

Contamination with diverse classes of toxic chemical substances (petroleum and chlorinated hydrocarbons, heavy metals, radioactive isotopes, and various other agricultural and industrial chemicals) are what most of us understand as pollution.

Although not everyone agrees (Davis, 1993), in almost all cases we find that contamination (much like eutrophication and overfishing) is most pronounced near shore, less so in shelf environments, and that oceanic environments are rather free of impacts (McIntyre, 1992). Anthropogenic contaminants are largely conveyed to the sea via freshwater inputs at the nearshore. Ocean margins are efficient filters of dissolved and particulate trace elements and pollutants in general. In fact, 90–95% of the inputs of trace elements to the sea, for example, accumulates in coastal environments (Martin and Windom, 1991). Although there may be some as yet undefined cross-shelf transport, oceanic waters receive pollutants mainly via atmospheric deposition (Preston, 1992). In addition, the benthic-pelagic coupling and resuspension characteristic of shallow environments may favor recycling of contaminants from sediments, which does not happen in oceanic waters.

There are too many kinds of contaminants to consider them all here, and some pollutants are less likely to create large-scale or long-term change than others (Bayne et al., 1988). Radionuclides, fortunately, are too rare and locally-limited to be of major concern. Similarly, heavy metals are unlikely agents of ecological change. Notable effects of metals have only been seen in a few extraordinary coastal cases [one of the few is mercury pollution in Minamata Bay, Japan (Irukayama,

1966)]. Although metal inputs by rivers and atmospheric deposition into the ocean have increased (Preston, 1992), effects of such inputs in the sea have been hard to demonstrate. Windom (1992) concludes that it "is extremely unlikely that metals in the ocean will pose a threat . . . to the marine environment or to human health."

Rather than examine all classes of toxics, we will focus on two examples that subsume key aspects of toxics as potential sources of long-term large-scale change: petroleum hydrocarbon spills and contamination with industrial chlorinated hydrocarbons.

16.5.1 Petroleum Hydrocarbons

Large amounts of petroleum are released into the sea, but the amounts are decreasing because of more careful handling and disposal. More than 3.2×10^6 metric tons of petroleum were released into marine environments worldwide in 1981 (Nat. Acad. Sci., 1985), but estimates for 1990 are $2.4–3.2 \times 10^6$ metric tons (GESAMP, 1992). Most of the oil is released in small, widely scattered mishaps or as chronic low-level effluents; large tanker accidents are responsible for only 5% of oil spilled. Most of the surface of the oceans are exposed to some level of exposure to oil, and most of the oil sinks to the bottom. Overall, the area of seafloor affected by spills is small, and recovery takes place reasonably soon (GESAMP, 1992).

It is difficult to assess large-scale effects of petroleum contamination. Deleterious effects of polycyclic aromatic hydrocarbons in oil have been documented in local instances where oil contamination was severe (Bayne et al., 1988; Sanders et al., 1980; Teal et al., 1992). Not enough is known about the ecological consequences of the worldwide low-level contamination by hydrocarbons. This is a significant gap in our ability to evaluate the role of toxics. There are many components of ecosystems that seem relatively unaffected by petroleum contamination: denitrifiers, for example, were not affected in a sediment chronically polluted by petroleum products (Bonin et al., 1994).

Most public and research attention has been given to the effects of large catastrophic spills created by accidents such as occurred in Prince William Sound, Alaska, where the supertanker *Exxon Valdez* ran aground in March 1989 (Shaw, 1992). The Exxon Valdez spill, much like other large spills, had severe local effects. Large numbers of birds and mammals were killed; for example, there was a loss of more than 50% of the common murres nesting in the area of the spill, and the colonies failed to reproduce in 1989 and 1990. Most benthic communities were eliminated from the nearshore by the oil, and by attendant attempts to clean beaches. In the short term, the spill was locally devastating, but the duration of the impact of such accidents is relatively short. For example, within a year of the accident, fishery catches in Prince William Sound were as large or larger than before the accident (Windom, 1992).

Experience from other spills, and field observations suggest that recovery from oil spills occurs within months to years. In hard-bottom marine environments recovery is relatively fast, while where oil can persist in sediments, such as in salt marshes (Sanders et al., 1980) and mangroves (Burns et al., 1993), recovery may require 5–25 years.

In contrast to the Prince William Sound spill, some large spills have had minor effects; the largest release of petroleum products occurred during the 1991 war in the Persian Gulf, but overall damage to the marine environment was less than initially expected (Literathy, 1993). Although catastrophic oil spills are therefore hardly of global importance, the potential severe local effects require that efforts be continued at safety and control measures. Perhaps more importantly, particular attention should also be given to low-level, chronic sources of petroleum, which provide most of the worldwide petroleum contamination, and whose effects are much less well-known than those of catastrophic spills.

16.5.2 Chlorinated Hydrocarbons

Agricultural and industrial chlorinated compounds have been dispersed throughout the entire worlds' ocean (Davis, 1993). There is a wide diversity of specific compounds, each with different chemical properties and different toxicities to different species of organisms (Waid, 1986). Certain of these compounds were found to cause decline of specific bird and mammal populations, and since the 1970s the use of such compounds was banned. The result is that inputs of chlorinated hydrocarbons to the sea have diminished, but the compounds are persistent. They continue to be present and affect biota, especially in sediments (Loganathan and Kannan, 1991).

Perhaps the most severely organochlorine-contaminated marine area is the Acushnet estuary in New Bedford, Massachussetts, where industrial wastes released relatively large quantities of polychlorinated biphenyls (PCBs). The sediments remain contaminated with the highest recorded concentrations of PCBs (Brownawell and Farrington, 1986). There are many sublethal effects, probably due to the exposure to PCBs, among the benthic organisms, including high frequencies of neoplasia and reduced gonad development and egg production (Leavitt et al., 1990). Similar sublethal effects have been reported in fish of Puget Sound (Johnson and Lanhdal, 1994), and the North Sea (Watermann and Kranz, 1992). In spite of the sublethal effects, the Acushnet estuary and nearby areas remain highly productive, with high densities of bivalves (D. Leavitt, unpubl. data).

The case histories reviewed show that toxic contaminants can have severe populational effects, but for the most part the toxic effects occur within localized areas. The examples also show that toxic effects are not felt uniformly by all organisms. Many marine organisms have remarkable capacity to survive exposure to contaminants, or reestablish themselves in areas denuded by contaminants, as the contamination diminishes over time. Most toxic compounds become associated with particles, and particles sink to the bottom, where inactivation or burial may take place. Some contaminants retain toxic activity longer than others, but nearly all are degraded in the marine environment over the course of time. Contaminant concentrations and activity diminish by microbial attack on organic molecules, as well as by chemically and physically mediated degradation.

Overall, although we lack sufficient information on the extent of sublethal large-scale and long-term effects, the evidence suggests that contamination by toxic compounds exerts its effects at local rather than larger spatial scales. Toxic effects are species specific, and some taxa in exposed marine ecosystems are affected, often sublethally, and other taxa are unaffected. Contaminants degrade in the marine environment, and recovery of the more susceptible taxa can be expected some time after the contamination event. In cases of chronic contamination, sublethal effects may accumulate, but it is not clear what the cumulative populational and community consequences may be.

16.6 Spread of Exotic Species

The assemblage of species found in most coastal areas are veritable collections of taxa from elsewhere (Carlton, 1989). Hitchhiking on ship hulls, transport in ballast loads, accidental inclusion in shipments of maricultural materials, and purposeful stocking have all contributed to wide dispersal of an enourmous variety of taxa.

There are many examples of the ecological disruptions created by such exotics in freshwater. A classic case is the remarkable series of introductions in the Great Lakes, from the sea lamprey in the 1830s to zebra mussel (Roberts, 1990; Nicholls and Hopkins, 1993) in 1988, with many others in between (Mills et al., 1993). Other examples include the alewife in Connecticut lakes (Brooks and Dotson, 1965), and a predatory cichlid fish from the Amazon into Gatún Lake in Panama (Zaret and Paine, 1973). These are but a few examples, all of which rearranged existing food webs.

One recent instance of invasion by an exotic species that has invaded a marine environment and has had broad repercussions is the case of the Asian clam (*Potamocorbula amurensis*) in San Francisco Bay (Nichols et al., 1990). This suspension-feeding bivalve appeared in 1986 and colonized rapidly, achieving high densities over large areas of the Bay. The Asian clam has displaced a complex of benthic species, which, as it turns out, were themselves exotics. The mean density of asian clams can reach >2,000 ind. m^{-2}, sufficient to significantly change the biology and chemistry of the Bay. As just one indication, feeding by the Asian clam population lowers chlorophyll concentrations in the water column by an order of magnitude compared to preclam conditions (Alpine and Cloern, 1992).

Assemblages of species in coastal areas are therefore highly subject to colonization by species from elsewhere, and these, once established, do markedly reshuffle conditions. In all cases the environments remain highly productive, but with a changed cast of species. While human activities no doubt hasten the vicariant exchange, it may also be that coastal assemblages have always been subject to repeated species invasions.

16.7 Harmful Algal Blooms

Articles in the lay press and in the scientific literature have forcibly raised awareness of the impacts of harmful algal blooms. "Red menace," "new killers," "phantom dinoflagellates" are but a few of the attention-getting headlines that have focused attention on the issue of harmful algal blooms. Such blooms may have become more frequent worldwide in recent decades (Anderson, 1989; Smayda, 1990), although it is hard to say whether the increase may result of more intense scrutiny during recent times (Hallegraeff, 1993).

Blooms of dinoflagellates, chrysophytes, diatoms, bluegreens, and a few other groups have been associated with a variety of harmful effects. Much of the attention has been because of toxic effects for people, including paralytic, neurotoxic, and amnesic shellfish poisoning, and ciguatera fish poisoning (Hallegraeff, 1993). Although human deaths are relatively few, the potential dangers have led to closings of shell-fisheries [perhaps at a cost of several million dollars a year in Georges Bank alone, for example (White et al., 1993)], and other economic disruptions (Smayda, 1992). Here we will concentrate, however, on ecological effects.

Dinoflagellate toxic blooms, usually known as red tides, occur throughout the world's inshore waters (Hallegraeff, 1993) and have recently been found in shallow but open waters such as Georges Bank (White et al., 1993). Toxins produced by dinoflagellates can variously impair or kill zooplankton, fish, birds, or cetaceans (Smayda, 1992; Geraci et al., 1989; Work et al., 1993). There is too little known about the specific target organisms, ecological role of the toxins,* or about the benefits conferred by the toxins to the dinoflagellates. The toxins have a confusing variety and intensity of effects, and the effects depend both on the bloom species and the target organism. Some toxins may reduce grazing by certain specific zooplankton or benthic suspension feeders, but many consumers are unaffected. Some others may even benefit from ingesting dinoflagellates by acquiring a chemical deterrent to being eaten (Kvitek et al., 1991). Further, it is not clear whether toxicity to fish, birds, and mammals has some unknown benefit to dinoflagellates or if it is a serendipitous consequence of substances produced for other, undetermined purposes.

Growth of small (about $2\,\mu m$) chrysophytes [or pelagophytes, according to DeYoe et al. (in press)] *Aureococcus* have led to conditions known as brown tides in a few bays of New York, New England, and Texas (Lonsdale et al., unpublished ms). These phytoplankton make dense (10^6 cells ml^{-1}) blooms and do not respond well to high-nutrient loadings (Keller and Rice, 1989). Brown tide organisms do well at low macronutrient concentrations but have a high iron requirement compared to other phytoplankton. Brown tide organisms appear to be distasteful to larger suspension feeders and probably to protozoans, and shade the already light-limited eelgrass beds over which they characteristically grow. Shellfish suffer higher mortality under brown tides; the New York bay scallop fishery, for example, lost the equivalent of \$2 million y^{-1} during the mid-1980s.

Blooms of certain diatoms may also have harmful consequences (Rinaldi et al., 1991; Taylor et al., 1985). Dense blooms may lead to release of mucilage, increased sinking of cells, increased bacterial activity near the bottom, and the ensuing lower oxygen concentrations reduce populations of shell and finfish.

The causes of harmful algal blooms are not well-established. Frequency of blooms may be related to human activities; red tide outbreaks in Tolo Harbor, for example, are well-correlated to numbers of people in Hong Kong (Lam and Ho, 1989). The presumed increases in bloom frequency may simply be a result of transport and introduction of exotic species in ship ballast and the increasing practice of aquaculture, which provides nutritional substrates for bloom organisms as well as

---

*A curious case is that of a benthic dinoflagellate that digests flecks of sloughed-off tissue from carcasses of fish killed by toxin released by the dinoflagellates, or grows in lesions prompted by the dinoflagellate toxin in still-living fish (Burkholder et al., 1992). The dormant dinoflagellate cysts appear to be activated to leave the sediment and to release toxin, by an as-yet unidentified chemical cue released by the fish. The dinoflagellate is involved in a high proportion of the fish kills recorded in the nearshore of the middle Atlantic states of the United States.

involves transport of exotic bloom species (Hallegraeff, 1993). Nutrient enrichment has most often been suggested as a cause of harmful algal blooms. If that was so, we could have simply added blooms as another item in the already long list of effects of eutrophication. We have treated blooms separately because of the still untested possibility that controls other than nutrient supply may be involved in blooms. Paerl (1988) suggested that algal blooms commonly occur in well-defined water masses that have received increased supplies of nutrients, organic matter, and trace elements. Increased supply of organic matter may be important, both as a nutritive substrate and as well as a chelator of trace elements. Smayda (1990) proposed that blooms are a result of failure of controls by grazers. All these possibilities need evaluation.

It seems likely that changes in meteorological conditions and nutrient enrichment of one kind or another may be the principal direct or indirect mechanism that start blooms. In Laguna Madre, Texas, for example, a brown tide became established following a period of drought and low temperatures (Whitledge, 1993; Stockwell et al., 1993). The cold snap and high salinities (50–60%) killed fin and shellfish, and decay of carcasses increased the ammonium content of water to $>20\,\mu M$. Brown tide organisms use NH_4 rather than NO_3 (DeYoe and Suttle, 1994) and presumably were favored by the unusual circumstances. This brown tide persisted for many months and lowered light penetration enough to reduce seagrass cover and biomass by an order of magnitude (Dunton, 1994). Zooplankton were unable to feed on, let alone control, the brown tide, and their density declined sharply while the brown tide lasted. Survival of fish larvae was reduced during the brown tide event.

The list of ecological consequences of algal blooms is long (Smayda, 1992), and specific effects can be dramatic. For example, one red tide off the Gulf Coast of the United States killed 100 t of fish daily, with a loss of at least 20 million dollars in Florida alone. Similar losses have been reported from other scattered sites (Paerl, 1988). Although many effects have been recorded in different places, they do not all occur at the same time in the same place. In fact, it seems almost characteristic of harmful algal blooms that specific blooms show quite different arrays of effects. This suggests that a suite of different phenomena are included under the term harmful blooms; unraveling the different types of phenomena would help understand the processes and effects. The ecological consequences of such specific combinations of effects have not been sufficiently quantified, nor separated from the effects of other factors in the field.

Overall, the effects of algal blooms vary from weak to intense, blooms tend to occur in restricted, spatially scattered areas, last for relatively short periods of time (days to weeks, occasionally months in brown tide events), and the incidence of blooms may or may not be increasing. Assessment of whether algal blooms are an important agent of large-scale, long-term ecological change therefore has to await future studies. This is not to say that we should ignore the potentially widespread public health and economic dangers posed by harmful algal blooms. Use of existing infomation about conditions favoring blooms can be used for preventive local management, and studies of the ecological controls should be undertaken to provide better understanding of the blooms. Part of the difficulty in understanding this subject must be that there are a variety of different phenomena subsumed under "algal blooms," each with different controls, effects, and mechanisms. Progress in understanding is likely to quickly follow the sorting out of the different types of phenomena involved.

16.8 Interception of Freshwater Inputs and Sediment Loads

Human activities on watersheds have historically thoroughly altered transport of suspended sediments and freshwater to receiving estuaries and coastal water. In some cases, sediment transport has increased as a result of erosion on watersheds. After the "reconquest" of the Iberian peninsula by christian princes, land use became gradually more intense, and eventually much of the live oak and other forests on the watershed of the Ebro River were cut, and much land was devoted to sheep raising (Nelson, 1990). Over the centuries erosion increasingly moved sediment downstream, with the result that the Ebro Delta, initially a minor feature, became well developed (Fig. 16-15). The Ebro Delta is therefore a major coastal feature that owes its existence to human-prompted sediment transport from its watershed. Many major rivers of the world bear much larger sediment loads than they did in the past. For example, the Yellow River carries one order of magnitude more sediment now than before the increases in cultivation of the loess plateau of northern China (Milliman, 1991). Such loads must create significant large-scale ecological change and alter chemical budgets to the worlds' oceans, but there are insufficient data on actual effects. The ecological changes must also be focused on near-shore environments, since most of the river-borne sediment loads are deposited in shallow waters; only 25–30% of land-derived particles make it past the continental shelf (Milliman, 1991).

Human intervention can also diminish sediment loads, particularly after construction of water catchment works and use for irrigation. The Rhone carries perhaps 15% of the loads it did in the 19th century, while the Indus bears only 20% of the loads it carried in the 1940s (Milliman, 1991). The Nile is one of the best studied case histories

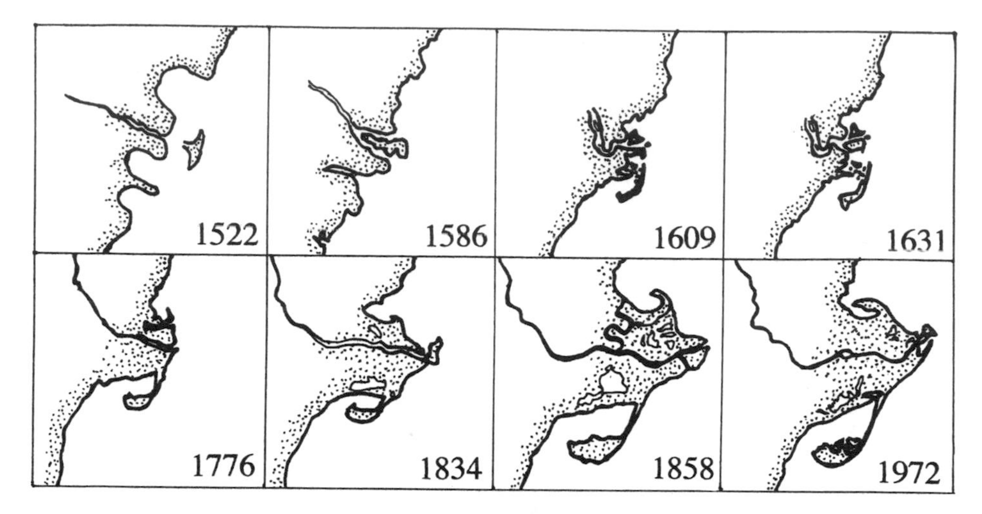

FIGURE 16-15. Development of the Ebro estuary and delta region, recreated from historical maps drawn in the years shown in each panel. Maps are only approximately to the same scale. Adapted from Maldonado (1972).

of the consequences of interception of sediments and water. The couplings between land, river, and sea have been altered by human activities since the beginning of civilization in the case of the Nile, but completion of the High Aswan Dam in 1964–1965 led to a series of profound and accelerated changes (Milliman et al., 1989; Halim, 1991; Stanley and Warne, 1993). The operation of the dam has eliminated the Nile flood cycle north of Aswan, fluvial sediment transport largely ceased, and the coupling between land and sea has been thoroughly altered.

Sediment deposition from the flooding the Nile seasonally fertilized the adjacent valley for milennia and constituted the agricultural basis of ancient and recent Egypt. The waters of the Nile also supplied an annual pulse of adsorbed and dissolved nutrients to the depauperate waters of the eastern Mediterranean. Nile water is now largely used for irrigation or urban uses, and sediments are accumulating in Lake Nasser, the water body formed behind the dam, so that only a small flow of water (largely made up of municipal wastewater and agricultural runoff) and only <2% of the Nile watershed sediment load reach the Nile Delta or the sea (Stanley and Warne, 1993).

The dam provides a much-needed source of power, but it has had many other less-desirable consequences. Human water use near the headwaters now depends on pumping from Lake Nasser or groundwater, and commercial fertilizers have to be applied to crops, all of which increases costs. The reduction of sediment loads has led to erosion of Nile Delta shorelines (Stanley and Warne, 1993), and the reduced freshwater flow, plus pumping of groundwater, have facilitated intrusion of salt into aquifers. Lack of replenishment of sediment plus groundwater pumping have led to subsidence of the Delta, which increases the loss of land. Lack of flooding with freshwater has allowed evaporitic salts to accumulate in soils, lowering agricultural yields. The slower flows in canals within the Delta has made it possible for aquatic plants to choke waterways. The aquatic plants increase evapotranspirative water losses and may foster abundance of snails that are the intermediate host for schistosomiasis. The irrigation systems that were needed to replace the natural delivery of water by floods also breed mosquitoes that carry malaria (Sharaf El Din, 1977).

The effects of the High Aswan Dam quickly became evident at sea also. The year the dam started operation, the standing crop of phytoplankton off the Delta was about 5% of previous amounts, and the abundance of sardines and shrimp decreased. By 1969 the total fisheries yield off the Nile Delta was less than 20% of the 1962–1963 value; the sardine catch decreased by 95%. The coastal lagoons located within the Delta used to provide 50% of Egypt's fish catch (Milliman et al., 1989), but no longer do so now that they are heavily eutrophied by the wastewater that has become their major freshwater source.

Various engineering works designed to reduce flood hazards have also intercepted sediment loads in the Mississippi River. About 400 million tons of suspended sediment were carried annually by the Mississippi during the late 1800s; transport dwindled to 300 million tons by the 1940s–1950s, and to less than 100 million tons by 1970 (Kesel, 1988). Such reductions in sediment supply have many consequences, one of which is that the marshes near the mouth of the river—much like the Nile Delta—are subsiding. Similarly, areas of Mississippi wetlands are being lost at rates of up to $100\,km^2\,y^{-1}$, equivalent to about $0.8\%\,y^{-1}$, because they are not receiving sufficient sediment from upstream to compensate for erosional losses downstream (Day and Templet, 1989). We saw earlier that southern U.S. marshes are just able to keep up with sea level rise (Fig. 16-6). When supplies of sediment do not match losses, these marshes cannot maintain themselves. Marshes of the Southeast United States grow on substrates made up largely of mineral sediments; northern marshes grow on peaty sediments made up largely of undecomposed plant matter. This difference makes southern marshes more subject to changes in rate of supply of mineral sediment.

Interception of freshwater before it reaches estuaries is another and related major challenge to estuarine ecosystems. Discharges by the Nile and Colorado Rivers to the sea, for instance, have been virtually eliminated by construction of dams and use of water for irrigation. Newman and Fairbridge (1986) calculated that discharge of freshwater from continents has diminished by 15% since the 1950s. This is a large amount: the implied lower input of water amounts to about half the annual rate of sea level rise. Such reductions lead to salt water incursions into fresh groundwater, reduce suitable habitats for estuarine biota, and reduce residence times of water in estuaries, which may alter geochemical transformations and exchanges at the land-sea margins. All these changes may alter composition of species assemblages present.

Probably the worst-case example of interception of freshwater sources is that of the Aral Sea. Aggressive plans for agricultural development, particularly for cotton, a crop that demands substantial irrigation, were carried out on the watersheds of the Aral Sea during the 1960s (Micklin, 1988; Precoda, 1991). The immediate result was a dramatic decrease in the river flow into the Aral Sea, from $55\,km^3\,y^{-1}$ in 1960 to near 0 in 1993. The decrease resulted in astounding changes: sea level dropped 14 m, surface area was halved, volume was reduced by 72%, and salinity tripled. In turn, the changes in the Aral itself generated other changes: summers near the Aral became warmer, winters colder, the growing season shorter (in some areas too short for cotton). The cellulose industry based on reeds that grew in 700,000 ha of coastal wetlands cannot find enough reeds, because only 30,000 ha of wetlands remain, and this remnant does survive only owing to intentional managed flooding. Canneries that depended on the Aral fishery to furnish up to 50,000 tons of fish y^{-1} have to import fish from the Atlantic to try to survive. Winds carry salts from the $30{,}000\,km^2$ of now-exposed sea bed and salinize soils in a wide area, making millions of ha of soil unproductive. Aral Sea salts have been identified in air over India and the Pacific and Atlantic Oceans. Groundwater near the Aral Sea is increasingly contaminated because of salts and reuse, and intestinal disorders, hepatitis, typhoid, cholera, and infant mortality are increasing. While there are few data, the litany of effects provides a notion of the level of disturbance that must be exerted on the organisms of the Aral Sea.

16.9 Multiple Factors in Concert: The Case of the Black Sea

In almost all the examples we have examined in this chapter, we have tried to isolate effects of specific agents for heuristic reasons. It is invariably the case, however, that more than one agent of change exert effects simultaneously, and at times, interactively. Climate change combines with sea level rise, eutrophication with algal blooms and toxics, interception of sediments and water go together, and so on, in almost all combinations. In fact, efforts to find specific evidence for the action of specific agents are often frustrated by the pervasive pattern of multiple causes of change. To make this point even more forcefully, we close our list of case histories with one that must surely be the worst case scenario for human impacts in any marine situation, the unfortunate state of the Black Sea (Halim, 1991; Zaitzev, 1992; Mee, 1992). The various agents of change will be familiar; what is new is the complex juxtaposition of factors and the pervasive degree of change.

The facts reviewed below cannot but suggest awe at both the staggering human ability to promote long-term change over huge areas of marine systems, and at the astonishing ability of organisms to establish themselves in the face of profound and intense disturbances. We do not mention changes in sediment transport, or climatic change, simply because the sources (Halim, 1991; Zaitzev, 1992; Mee, 1992) do not. These two agents of change are probably also acting on the Black Sea, but their effects appear to be unnoticed in the maelstrom of other changes.

16.9.1 Fishery Exploitation

The Black Sea was formerly a major source of fish harvest, supporting over 2 million fishermen and their dependants. Recent decades have seen the near-demise of the fisheries: of 26 species of commercially useful species, only 6 remain in exploitable quantities. The catch of bottom fish in the northwest Black Sea dropped from 80,000 $t\,y^{-1}$ before 1949 to 4,000 $t\,y^{-1}$ during 1971–1975. High-value species disappeared, replaced for a time by anchovies, and to a lesser extent by horse mackerel. The fishing fleet, armed with technically advanced gear, particularly along the Turkish coast, turned to the newly abundant species. Anchovies provided most of the yield, but the anchovy population collapsed during 1987–1989. The intense harvest of even the 1–3 year classes suggests that this is another instance where overfishing is involved. In addition, recruitment may have been impaired by the effects of anoxia and of introduced predatory ctenophores on larvae.

16.9.2 Eutrophication and Blooms

Nitrogen and phosphorus loading from the three major rivers discharging onto the northwest Black Sea increased 1.6- and 5-fold, respectively in the 1980s relative to the 1950s. Nitrate concentrations in one major river, the Danube, increased from 36–71 μM in 1970 to 100–200 μM in 1990. The increased nutrient concentrations in the northwest Black Sea led to increased phytoplankton growth, particularly nanoplankton. Dinoflagellate blooms became more important, with peridineans contributing 90% of biomass in recent years, compared to 18% before 1960. Smaller species have replaced larger species, diatoms in particular. Zooplankton has also increased in abundance, especially the smaller species. There have been frequent episodic blooms of *Aurelia aurita* and *Noctiluca miliaris*, both probably exploiting the increased numbers of zooplankton.

The enhanced phytoplankton abundance has lowered water transparency; at a distance of 50 km from the NW coast, secchi disc depths were 15–18 m during the 1960s, but only 2–2.5 m in the 1980s. Benthic seaweeds and seagrasses were therefore shaded, and their cover greatly reduced. Areas of seagrass beds have decreased by an order of magnitude. In the 1950s there was an extensive bed of the red alga *Phyllophora*, covering about $10^4 km^2$, in the shallow northwestern shelf of the Black Sea. By 1990, the area of *Phyllophora* had declined to $50 km^2$. The *Phyllophora* beds helped maintain concentration of oxygen near the shelf bottom, supported their distinctive faunal assemblages, and were economically valuable, since they were a source of agar, dyes, and pharmaceuticals. Today benthic seaweeds are confined to depths less than 5 m, and the species found are those that proliferate under high nutrient regimes, such as *Cladophora vagabunda* and *Enteromorpha intestinalis*.

The benthic fauna found in the coastal Black Sea has been altered by the change in vegetation and by the occurrence of anoxia. The increase in phytoplankton in the water column has resulted in a 15- and 20-fold increase in delivery of organic matter to the bottom. Since shading by phytoplankton left no benthic seaweeds to release oxygen, the organic load leads to anoxic conditions over much of the shelf. Ninety-five percent of the northwest Black Sea shelf and the Sea of Azov are now prone to anoxia near the bottom. In the Romanian coast alone, one anoxic event in 1991 eliminated 50% of the benthic fish. The benthic communities of the central and eastern coasts are also disappearing. Off Crimea and the Caucasian coast the number of macrobenthic species in the 1980s were 1/2 to 1/4 of the number found during the 1960s.

16.9.3 Toxics

Toxics are present but, as we have seen above, it is difficult to discern whether they exert a major impact. All major rivers discharging into the Black Sea are heavily contaminated with heavy metals, chlorinated and petroleum hydrocarbons. The Danube alone discharges annually 1000 t of Cr, 900 t of Cu, 60 t of Hg, 4500 t of Pb, 6000 t of Zn, and 50,000 t of oil. Concentrations of DDE, a derivative of DDT, are high in the eggs of the Dalmatian pelican. There is some teratogenic agent present: up to 25% of the coastal populations of the jellyfish *Aurelia aurita* show anomalous individuals with gonad numbers other than the usual four, while only 2–3% of the offshore individuals are mutants.

Cystoseira barbata, a brown alga known to be sensitive to detergents, has disappeared along the northwest coast. None of these specific effects have been quantified as to their scale or impact.

16.9.4 Introduction of Exotic Species

Many new species have in recent decades been introduced into the Black Sea, most likely via transport in ship ballast water. The introduction that has had the most widespread biological consequences is that of the ctenophore *Mnemiopsis leidyi*. This comb jelly was introduced from the Atlantic coast of North America into the Black Sea sometime in the 1980s. Its population has reached enormous numbers, densities of 100 s of individuals m^{-3}, and has spread to the Sea of Azov, and to the Marmara and Aegean seas. Predation by the new arrival on larval stages has reduced populations of anchovies in the Black Sea. The new ctenophore is said also to be involved in reductions of abundance of *Aurelia*, *Noctiluca*, and various other zooplankton species. In addition, ctenophores are not good fish food. Ctenophores may have replaced food species previously available to fish, and hence may contribute to the precarious position of the fish stock.

Other species have had less consequential effects. For example, a snail, *Rapana thomasiana*, originally from Japan, entered the Black Sea in the 1940s, and exterminated oysters and exploited other bivalve populations along the nearshore. Some introductions have had relatively benign consequences. The bivalve *Mya arenaria*, for example, was brought to the Black Sea from North America during the 1960s. It has become well established in estuaries and shallow waters of the NW Black Sea, where it may have displaced the native *Corbula mediterranea*. *Mya* reaches densities of 1,000 individuals m^{-2}, and serves as a major food item for sturgeon, other fish, and birds. The banks of *Mya* (and other bivalves) have recently been much reduced by anoxic events.

16.9.5 Interception of Freshwater Flow

The agricultural and hydroelectric use of freshwater has increased during recent decades, such that river flow into the Black Sea was reduced by 19–39% during 1971–1975, 27–52% during 1981–1985 and projections are for a further reduction of 43–71% between 1991–2000. This means that there will be increasing flow of salty Mediterranean water in the lower layers through the Bosphorus and increased salinities throughout the Black Sea. Salinity in the Sea of Azov has already increased by more than 2‰. As a result, migratory patterns of anadromous sturgeon have changed, and the fishery has dwindled. In addition, reduced freshwater flow may be responsible for the 30 m rise of the oxic/anoxic boundary reported to have occurred between 1969 and 1988 (Murray et al., 1989). The bottom layers of the Black Sea are anoxic principally because of the lack of water exchange over the shallow threshold of the Bosphorus; the current reduction of flow of new freshwater means that the relative depth of the oxic upper productive layer may be diminishing. This would reinforce the oxygen-reducing effects of eutrophication.

16.10 Implications of Long-Term, Large-Scale Changes

The major, pervasive, and recurring disruptions created by disturbances discussed in this chapter raise several issues. Can assemblages of organisms found in marine environments be thought of as "stable?" It seems evident that external conditions, governed by global-scale processes, can change markedly over periods of time, and that these changing conditions readily and almost routinely alter local assemblages of species, and how they function. "Stability" is therefore as hard to find as it is to define, so it may be more useful to consider how externally driven changes interact with internal control processes discussed in earlier chapters.

Do large-scale external disturbances overwhelm the many internal control mechanisms we have discussed in earlier chapters? They do, depending on the disturbance. If external conditions return to a semblance of the earlier state, recovery readily takes place, as we saw in the case of giant kelp and northern anchovy off California and in the case of oil spills. Disturbances such as ENSO events, however, may leave fairly long-lasting imprints. There are large patches of kelp forest in different places off California that are different in composition because of slight differences in the intensity of initial disturbance or in the histories of recovery from the disturbance.

Do large external perturbations interface with top-down and bottom-up controls? The effects of large-scale atmospherically-driven changes seem mediated primarily through bottom-up controls; changes in wind may change nutrient supply, for example. Changes in temperature may, however, exert effects both through bottom up and top down controls. During recovery there may be cascade effects created by the disruptions but mediated by predators. Anthropogenic perturbations exert effects on all control processes. Among the disturbing realizations that come out of our review is that first, human perturbations are sufficiently extensive that their effects are at a global scale. Second, human beings have managed to massively affect components and processes that are involved in regulation of marine ecosystems. Eutrophication and fish harvest, as we have seen, thoroughly affect both bottom-up and top-down control mechanisms. When we examine marine environments today and in the future we need to realize that human influences on control mechanisms may be as important as "natural" influences.

Are there thresholds beyond which recovery is not possible? Changed conditions may favor a new array of organisms, which may dominate for a few years, as we saw for upwelling communities in between ENSO events, or decades, as we saw in the Atlantic. Obviously, if the changes in global conditions are intense or trend in one direction for sufficiently long periods, there may be no recovery to an earlier state, but rather a shift to another assemblage of organisms, as is so commonly evident in the geological record. Although the idea of a recovery threshold is intuitively simple, in practice it has been most difficult to define and may not be a useful goal of research. In all such changing circumstances the new assemblages of organisms function, and not only is it hard to define recovery, but it is fatuous to try to judge whether the new assemblages are "degraded," less or more "successful."

Do the different agents of ecological change have different impacts over different time and spatial scales? All the agents of change discussed above have significant effects at some level and scale, and all are important at that scale. We can use the information provided earlier in this chapter to compare effects of the various agents at different spatial and temporal scales. We should add that the evaluation below, surely contentious, ignores economic, political, social, or public health aspects.

There is no question that global-scale meteorological changes are major forces that drive alterations to marine ecosystems at all time and spatial scales. There is insufficient knowledge about the specific mechanisms through which meteorological processe create ecological change in many cases. There is only partial information, but much suspicion, about the extent to which these changes are or are not be linked to, and accelerated by, human activities.

In terms of intensity and scale of ecological consequences, eutrophication (GESAMP, 1990; McIntyre, 1992; Windom, 1992) and fishing are demonstrably the most effective agents of change among the clearly anthropogenic challenges to marine ecosystems. Eutrophication and fishing thoroughly restructure structure, function, and the controls of marine ecosystems, and do so at local to global scales and over short to prolonged periods.

Interception of freshwater and sediments are of the utmost local and regional consequence, but the drastic effects have been restricted to a few unfortunate regions, so that these disruptions are less globally widespread and more local than the effects of eutrophication and overfishing. The sum of changes in sediment transport may have global geochemical implications, but data are only fragmentary.

Toxics are widely distributed and not unimportant; after all, unacceptable residues in fish have closed large fisheries (NOAA, 1988), among many such examples. Toxics, however, do not clearly and consistently disrupt ecological function in the pervasive, long-term, large-scale fashion demonstrated by eutrophication and overfishing.

Invasions by exotic species, although widespread along coastal zones, result primarily in partial changes in composition of assemblages of native species. Invaded systems remain biologically active and productive; it is just that the array of species present differs. Species changes are commonplace in marine environments and are likely to occur in any event, given the global changes that continually alter conditions.

Toxic algal blooms can have intensive but local effects, but the dramatic effects tend to be relatively short term. It may be that perhaps their origin is tied to eutrophication in any case.

Global atmospheric changes, eutrophication, and overfishing therefore seem to be the major agents bringing global-scale change to marine ecosystems; interception of freshwater and changes in sediment loads are important regionally. These major agents of change often act simultaneously to change marine ecosystems. Almost inevitably the effects of the major processes are modified locally by the effects of the other processes of change, which exert influence at their own temporal and spatial scales.

Understanding the mechanisms involved, the objective of this chapter, merely points out just what the problems are and sketches their dimensions. It is spurious today to pretend to do "basic" marine science, since many of the key processes we would be studying in "basic" marine ecology are thoroughly affected by "applied" forces. To ignore the changes brought by people to the sea is to ignore the major factors changing marine systems today. These are hugely complex issues, not only from an ecological point of view, but also because their solution has to address the social, political, economic, and public health issues that we have ignored. Moreover, it will be most difficult to find solutions without addressing the ultimate cause of many of the changes, continued growth, and disproportionate use of resources, by the population of the nonmarine species to which we belong.

References

Abele, L.G. 1976. Comparative species richness in fluctuating and constant environments: Coral-associated decapod crustaceans. Science 192:461–463.

Abele, L.G. 1982. Biogeography. Pp. 241–304 in The Biology of Crustacea, Vol. 1. Academic.

Abele, L.G., and K. Walters. 1979. Marine benthic diversity: A critique and alternative explanation. J. Biogeogr. 6:115–126.

Adams, B.B. 1978. The Feeding Strategies of Coastal and Open Ocean Pelagic Copepods. Ph.D. Thesis, Boston University.

Adams, J.A., and J.H. Steele. 1966. Shipboard experiments on the feeding of *Calanus finmarchicus* (Gunnerus). Pp. 19–25 in H. Barnes (ed.), Some Contemporary Studies in Marine Science. George Allen and Unwin.

Aebischer, N.J., J.C. Coulson, and J.M. Colebrook. 1990. Parallel long-term trends across four marine trophic levels and weather. Nature 347:753–755.

Agassiz, L. 1888. Three cruises of the BLAKE. Bull. Mus. Comp. Zool. 14:1–314.

Akre, B.G., and D.M. Johnson. 1979. Switching and sigmoid functional response curves by damselfly naiads with alternative prey available. J. Anim. Ecol. 48:703–720.

Alber, M. 1992. Organic aggregates in detrital food webs: Production from marine macrophyte-derived dissolved organic material, composition, and incorporation by suspension-feeding bivalves. Ph.D. Thesis, Boston Univ. p. 244.

Alber, M., and I. Valiela. 1994. Production of microbial organic aggregates from macrophyte-derived dissolved organic material. Limnol. Oceanogr. 39:37–50.

Alcaraz, M., G.-A. Paffenhöfer, and J.R. Strickler. 1980. Catching the algae: A first account of visual observations of filter feeding calanoid copepods. Amer. Soc. Limnol. Oceanogr. Spec. Symp. 3:241–248.

Aleem, A.A., and N. Dowidar. 1967. Phytoplankton production in relation to nutrients along the Egyptian Mediterranean coast. Pp. 305–327 in Proc. Int. Conf. Trop. Oceanography 1965. Univ. Miami.

Alexander, M. 1977. Introduction to Soil Microbiology. 2nd Ed. Wiley.

Allan, J.D., T.G. Kinsey, and M.C. James. 1976. Abundance and production of copepods in the Rhode River subestuary of Chesapeake Bay. Ches. Sci. 17:86–92.

Alldredge, A.L., and Y. Cohen. 1987. Can microscale chemical patches persist in the sea? Science 235:689–691.

Alldredge, A.L., and W.H. Hamner. 1980. Recurring aggregation of zooplankton by a tidal current. Estuar. Coast. Mar. Sci. 10:31–37.

Alldredge, A.L., and M.W. Silver. 1988. Characteristics, dynamics, and significance of marine snow. Progr. Oceanogr. 20:41–82.

Allee, W.C. 1931. Animal Aggregations: A Study in General Sociology. Univ. of Chicago.

Aller, R.C. 1980. Diagenetic processes near the sediment-water interface of Long Island Sound. I: Decomposition and nutrient element geochemistry (S, N, P). Pp. 237–344 in B. Salzman (ed.), Physics and Chemistry of Estuaries: Studies in Long Island Sound. Advances in Geophysics, Vol. 22. Academic.

Aller, R.C. 1980a. Relationship of tube-dwelling benthos with sediment and overlying water chemistry. Pp. 285–308 in K.R. Tenore and B.C. Coull (eds.), Marine Benthic Dynamics. Univ. South Carolina.

Aller, R.C. 1990. Bioturbation and manganese cycling in hemipelagic sediments. Phil. Trans. Roy. Soc. London A 331:51–68.

Aller, R.C., and P.D. Rude. 1988. Complete oxidation of solid phase sulfides by manganese and bacteria in anoxic marine sediments. Geochim. cosmochim. Acta 52:751–765.

Aller, R.C., J.E. Mackin, and R.T. Cox, Jr. 1986. Diagenesis of Fe and S in Amazon inner shelf muds: Apparent dominance of Fe reduction and implications for the genesis of ironstones. Continent. Shelf Res. 6:263–289.

Alongi, D.M. 1990. Effect of mangrove detrital outwelling on nutrient regeneration and oxygen fluxes in coastal sediments of the central Great Barrier Reef Lagoon. Estuar. Coast. Shelf Sci. 31:581–598.

Alpine, A.E., and J.E. Cloern. 1992. Trophic interactions and direct physical effects control phytoplankton biomass and production in an estuary. Limnol. Oceanogr. 37:946–955.

Altman, P.L., and D.S. Dittmer. 1972. Biology Data Book. 2nd Ed. Vol. 1. Federat. Amer. Soc. Exp. Biol.

Ambler, J.W., and B.W. Frost. 1974. The feeding behavior of a predatory planktonic copepod, *Tortanus discaudatus*. Limnol. Oceanogr. 19:446–451.

Andersen, H.T. (ed.). 1969. The Biology of Marine Mammals. Academic.

Andersen, P., and T. Fenchel. 1985. Bacterivory by microheterotrophic flagellates in seawater samples. Limnol. Oceanogr. 30:198–202.

Anderson, D.M. 1989. Toxic algal blooms and red tides: A global perspective. Pp. 11–16 in T. Okaichi, D.M. Andrerson, and T. Nemoto (eds.), Red Tides: Biology, Environmental Science, and Toxicology. Elsevier.

Anderson, D.M., and F.M. Morel. 1978. Copper sensitivity of *Gonyaulax tamarensis*. Limnol. Oceanogr. 23:283–295.

Anderson, G.C., and R.P. Zeutschel. 1970. Release of dissolved organic matter by marine phytoplankton in coastal and offshore areas of the Northeast Pacific Ocean. Limnol. Oceanogr. 15:402–407.

Anderson, G.C., B.W. Frost, and W.K. Peterson. 1972. On the vertical distribution of zooplankton in relation to chlorophyll concentrations. Pp. 314–345 in A.Y. Takenouti (ed.), Biological Oceanography of the Northern North Pacific Ocean. Idemitsu Shoten.

Anderson, G.R.V., A.H. Ehrlich, P.R. Ehrlich, D.J. Roughgarden, B.C. Russell, and F.H. Talbot. 1981. The community structure of coral reef fishes. Amer. Nat. 117:476–495.

Andersson, A., and Ä. Rudehäll. 1993. Proportion of plankton biomass in particulate organic carbon in the northern Baltic Sea. Mar. Ecol. Progr. Ser. 95:133–139.

Andersson, R.V., E.T. Elliot, J.F. McClellan, et al. 1978. Trophic interactions in soils as they effect energy and nutrient dynamics. Biotic interactions of bacteria, amoebae and nematodes. Microbiol. Ecol. 4:361–371.

Andreae, M.O., and H. Raemdonck. 1983. Dimethyl sulfide in the surface ocean and the marine atmosphere: A global view. Science 221:744–747.

Andrews, P., and P.J.L. Williams. 1971. Heterotrophic utilization of dissolved organic compounds in the sea. III. J. Mar. Biol. Assoc. U.K. 51:111–125.

Anger, K. 1975. On the influence of sewage pollution on inshore benthic communities in the South of Kiel Bay. Part 2. Quantitative studies on community structure. Ilelgol. Wiss. Meeresunters. 27:408–438.

Ankar, S. 1977. The soft bottom ecosystem of the Northern Baltic proper with special reference to the macrofauna. Contr. Asko. Lab. 19:1–62.

Antia, N.J., P.J. Harrison, and L. Oliveira. 1991. The role of dissolved organic nitrogen in phytoplankton nutrition, cell biology and ecology. Phycologia 30:1–89.

Anthony, V.C. 1993. The state of groundfish resources off the northeastern United States. Fisheries 18:12–17.

Arnold, D.E. 1971. Ingestion, assimilation, survival, and reproduction by *Daphnia pulex* fed seven species of blue-green algae. Limnol. Oceanogr. 16:906–920.

Arnold, K.E., and S.N. Murray. 1980. Relationships between irradiance and photosynthesis for marine benthic green algae (Chlorophyta) of differing morphologies. J. Exp. Mar. Biol. Ecol. 43:183–192.

Arntz, W.E. 1986. The two faces of El Niño. Meeresforschung 31:1–46.

Arntz, W.E., E. Valdivia, and J. Zeballos. 1988. Impact of El Niño 1982–83 on the commercially exploited invertebrates (mariscos) of the Peruvian shore. Meeresforschung 32:3–22.

Ashmole, N.P. 1968. Body size, prey size, and ecological segregation in five sympatric tropical terns (Aves: Laridae). Syst. Zool. 17:292–304.

Ashmore, M. 1990. The greenhouse gases. Trends Ecol. Evol., Vol. 5, 296–297.

Atkinson, L.P., and F.A. Richards. 1967. The occurrence and distribution of methane in the marine environment. Deep-Sea Res. 14:673–684.

Avila, C. 1995. Natural products from opisthobranch molluscs: A biological review. Oceanogr. Mar. Biol. Ann. Rev. 33: in press.

Azam, F., and R.E. Hodson. 1977. Dissolved ATP in the sea and its utilization by marine bacteria. Nature 267:696–697.

Azam, F., T. Fenchel, J.G. Field, J.S. Gray, L.A. Meyer-Reil, and T.F. Thingstad. 1983. The ecological role of water-column microbes in the sea. Mar. Ecol. Progr. Ser. 10:257–263.

Aziz, S.A.A., and D.B. Nedwell. 1979. Microbial nitrogen transformations in the salt marsh environment. Pp. 385–398 in R.L. Jefferies and A.J. Davy (eds.), Ecological Processes in Coastal Environments. Blackwell.

Baden, S.P., L.O. Loo, L. Ahl, and R. Rosenberg. 1990. Effects of eutrophication on benthic communities including fish: Swedish west coast. Ambio 19:113–122.

Bailey, K.M., and E.D. Houde. 1989. Predation on eggs and larvae of marine fishes and the recruitment problem. Adv. Mar. Biol. 25:1–85.

Baines, S.R., and M.L. Pace. 1991. The production of dissolved organic matter by phytoplankton and its importance to bacteria: Patterns across marine and freshwater systems. Limnol. Oceanogr. 36:1078–1090.

Baird, D., and H. Milne. 1981. Energy flow in the Ythan estuary, Aberdeenshire, Scotland. Estuar. Coast. Shelf Sci. 13:455–472.

Baker, M.C. 1974. Foraging behavior of blackbellied plovers (*Pluvialis squatarola*). Ecology 55:162–167.

Bakus, G.J. 1981. Chemical defense mechanisms on the Great Barrier Reef, Australia. Science 211:497–499.

Balderston, W.L., and W.J. Payne. 1976. Inhibition of methanogenesis in salt marsh sediments and whole-cell suspensions of methanogenic bacteria by nitrogen oxides. Appl. Environ. Microbiol. 32:264–269.

Balech, E., S.Z. El-Sayed, G. Hasle, M. Neushul, and J.S. Zaneveld. 1968. Primary productivity and benthic marine algae of the Antarctic and Subantarctic. Antarctic Mag. Folio Series 10:1–12.

Bamstedt, U. 1981. Seasonal energy requirements of macrozooplankton from Kosterfjorden, Western Sweden. Kiel. Mecresforsch. 5:140–152.

Bannister, R.C.A. 1978. Changes in plaice stocks and plaice fisheries in the North Sea. Rapp. Proc.-Verb. Reun. Cons. Int. Explor. Mer 172:86–101.

Banse, K. 1974. The nitrogen-to-phosphorus ratio in the photic zone of the sea and the elemental composition of the plankton. Deep-Sea Res. 21:767–771.

Banse, K. 1977. Determining the carbon to chlorophyll ratio of natural phytoplankton. Mar. Biol. 41:199–212.

Banse, K. 1979. On weight dependence of net growth efficiency and specific respiration rates among field populations of invertebrates. Oecologia 38:111–126.

Banse, K. 1980. Microzooplankton interference with ATP estimates of plankton biomass. J. Plankton Res. 2:235–238.

Banse, K. 1991. Rates of phytoplankton cell division in the field and in iron enrichment experiments. Limnol. Oceanogr. 36:1886–1898.

Banse, K. 1993. On the dark bottle in the ^{14}C method for measuring marine phytoplankton production. ICES Mar. Sci. Symp. 197:132–140.

Banse, K., and D.C. English. 1994. Seasonality of coastal zone color scanner phytoplankton pigment in the offshore oceans. J. Geophys. Res. 99:7323–7345.

Banse, K., and S. Mosher. 1980. Adult body mass and annual production biomass relationship of field populations. Ecol. Monogr. 50:355–379.

Banse, K., F.H. Nichols, and D.R. May. 1971. Oxygen consumption by the seabed. III. On the role of the macrofauna at three stations. Vie et Milieu. I (Suppl. 22):31–52.

Barber, R.T. 1966. Interaction of bubbles and bacteria in the formation of organic aggregates in sea water. Nature 211:257–258.

Barber, R.T., and F.P. Chavez. 1983. Biological consequences of El Niño. Science 222:1203–1210.

Barber, R.T., and F.P. Chavez. 1986. Ocean variability in relation to living resources during the 1982–83 El Niño. Nature 319:279–285.

Barber, R.T., and J.H. Ryther. 1969. Organic chelators: Factors affecting primary production in the Cromwell Current upwelling. J. Exp. Mar. Biol. Ecol. 3:191–199.

Barber, R.T., R.C. Dugdale, J.J. MacIsaac, and R.L. Smith. 1971. Variations in phytoplankton growth associated with the source and conditioning of upwelling water. Invest. Pesq. 35:171–193.

Bardach, J.E., and R.M. Santerre. 1980. Climate and fish in the sea. Bioscience 31:206–215.

Barlow, R.G., and R.S. Alberte. 1985. Photosynthetic characteristics of phycoerythrin-containing marine *Synechococcus* spp. I. Responses to growth photon fleix density. Mar. Biol. 86:63–74.

Barnes, H., and S.M. Marshall. 1951. On the variability of replicate plankton samples and some applications of "contagious" series to the statistical distribution of catches over restricted periods. J. Mar. Biol. Assoc. U.K. 30:233–263.

Barnes, R.D. 1980. Invertebrate Zoology. 4th Ed. Saunders.

Barry, J.P., C.H. Baxter, R.D. Sagarin, and S.E. Gilman. 1995. Climate-related, long-term faunal changes in a California rocky intertidal community. Science 267:672–675.

Barsdate, J.J., R.T. Prentki, and T. Fenchel. 1974. Phosphorus cycle of model ecosystems: Significance for decomposer food chains and effect of bacterial grazers. Oikos 25:239–251.

Barsdate, R.J., and V. Alexander. 1975. The nitrogen balance of arctic tundra: Pathways, rates, and environmental implications. J. Env. Qual. 4:111–117.

Bates, T.S. 1992. Sulfur emissions to the atmosphers from natural sources. J. Atm. Chem. 14:315–338.

Bautista, B., V. Rodríguez, and F. Jiménez-Gómez. 1994. Trophic interactions in the microbial food web at a coastal station in the Alboran Sea (Western Mediterranean) in winter. I. Microplankton grazing impact on nanoplankton and free bacteria. Scient. Mar. 58:143–152.

Bayliss, D.E. 1982. Switching by *Lepsiella vinosa* (Gastropoda) in South Australian mangrove. Oecologia 54:212–226.

Bayliss, D.E. 1993. Spatial distribution of Balanus amphitrite and Elminius adelaidae on mangrove pneumatophores. Mar. Biol. 116:(2)251–256.

Baylor, E.R., and W.H. Sutcliffe. 1963. Dissolved organic matter in seawater as a source of particulate food. Limnol. Oceanogr. 8:369–371.

Bayne. B.L. 1973. Physiological changes in *Mytilus edulis* L. induced by temperature and nutritive stress. J. Mar. Biol. Assoc. U.K. 53:39–58.

Bayne, B.L., and C.M. Worral. 1980. Growth and production of mussels *Mytilus edulis* from two populations. Mar. Ecol. Prog. Ser. 3:328–328.

Bayne, B.L., K.R. Clarke, and J.S. Gray. 1988. Biological Effects of Pollutants. Mar. Ecol. Progr. Ser. Special Issue 46.

Baxter, I.G. 1959. Fecundities of winter-spring and summer-autumn herring spawners. J. Cons. Perm. Int. Exp. Mer 25:73–80.

Beddington, J.R. 1975. Mutual interference between parasites or predators and its effect on searching efficiency. J. Anim. Ecol. 44:331–340.

Beddington, J.R., M.P. Hassell, and J.H. Lawton. 1976. The components of arthropod predation. II. The predator rate of increase. J. Anim. Ecol. 45:165–186.

Beers, J.R., and G.L. Stewart. 1971. Microzooplankters in the plankton communities of the upper waters of the eastern tropical Pacific. Deep-Sea Res. 18:861–884.

Beers, J.R., F.M.H. Reid, and G.L. Stewart. 1982. Seasonal abundance of the microplankton populations in the North Pacific central gyre. Deep-Sea Res. 29:227–245.

Beers, J.R., D.M. Steven, and J.B. Lewis. 1968. Primary productivity in the Caribbean Sea off Jamaica and the Tropical North Atlantic off Barbados. Bull. Mar. Sci. 18:86–104.

Berg, H.C., and E.M. Purcell. 1977. Physics of chemoreception. Biophys, J. 20:193–215.

Belehradek, J. 1935. Temperature and Living Matter. Borntraeger, Berlin (Protoplasma—Monogr. 8).

Bell, P.R.F. 1992. Eutrophication and coral reefs: Some examples in the Great Barrier Reef Lagoon. Wat. Res. 26:553–568.

Bell, R.H. 1970. The use of the herb layer by grazing ungulates in the Serengeti. Pp. 111–124 in A. Watson (ed.), Animal Populations in Relation to Their Food Resources. Blackwell.

Bell, S.S. 1980. Meiofauna-macrofauna interactions in a high salt marsh habitat. Ecol. Monogr. 50:487–505.

Bender, M.L., et al. 1987. A comparison of four methods for determining planktonic community production. Limnol. Oceanogr. 32:1085–1098.

Berger, W.H., V.S. Smetacek, and G. Wefer. 1989. Ocean productivity and paleoproductivity: An overview. Pp. 1–34 in Berger, W.H., V.S. Smetacek, and G. Wefer (eds.). Productivity in the Ocean: Present and Past. Wiley & Sons Ltd.

Berkes, F. 1977. Production of the euphausiid crustacean *Thysanoessa raschii* in the Gulf of St. Lawrence. J. Fish. Res. Bd. Canada 34:443–446.

Berner, R.A. 1989. Biogeochemical cycles of carbon and sulfur and their effect on atmospheric oxygen over Phanerozoic time. Paleogeogr. Paleoclim. Paleoecol. 75:97–122.

Bernstein, B.B., B.E. Williams, and K.H. Mann. 1981. The role of behavioral responses to predation in modifying urchins' (*Strongylocentrotus droebachiensis*) destructive grazing and seasonal foraging patterns. Mar. Biol. 63:39–46.

Bertalanffy, L. von. 1957. Quantitative laws in metabolism and growth. Quant. Rev. Biol. 32:217–231.

Beverton, R.J.H. 1962. Long-term dynamics of certain North Sea fish populations. Pp. 242–259 in E.D. LeCren and M.W. Holdgate (eds.), The Exploitation of Natural Animal Populations. Blackwell.

Bidigare, R.R., et al. 1990. Novel chlorophyll-related compounds in marine phytoplankton: Distribution and Geochemical Implications. Energy and Fuels 4:653.

Biggs, D.C. 1977. Respiration and ammonium excretion by open ocean, gelatinous zooplankton. Limnol. Oceanogr. 22:108–117.

Billen, G. 1975. Nitrification in the Scheldt estuary (Belgium and the Netherlands). Estuar. Coast. Mar. Sci. 3:79–89.

Birkeland, C. 1977. The importance of rate of biomass accumulation in early successional stages of benthic communities to the survival of coral recruits. Pp. 15–21 in Proc. 3rd Int. Coral Reef Symp., Vol. 1. Univ. of Miami, Florida.

Birkhead, T.R. 1977. The effect of habitat and density on breeding success in the common guillemot (*Uria aalge*). J. Anim. Ecol. 46:751–764.

Bishop, J.K.B. 1981. Particle sources and sinks from C-FATE (composition flux and transfer experiments) results. Pp. 9–12 in R.F. Anderson and M.P. Bacon (eds.), Sediment Trap Intercomparison Experiment. Woods Hole Oceanographic Institution Tech. Memo No. 1–81.

Bishop, J.K.B., R.W. Collier, D.R. Kettens, and J.M. Edmond. 1980. The chemistry, biology, and vertical flux of particulate matter from the upper 1500 m of the Panama Basin. Deep-Sea Res. 27A:615–640.

Bishop, J.K.B., J.M. Edmond, D.R. Ketten, M.P. Bacon, and W.B. Silker. 1977. The chemistry, biology and vertical flux of particulate matter from the upper 400 m of the equatorial Atlantic Ocean. Deep-Sea Res. 24:511–548.

Bjorndal, K.A. 1980. Nutrition and grazing behavior of the green turtle *Chelonia mydans*. Mar. Biol. 56:147–154.

Bjørnsen, P.K. 1988. Phytoplankton exudation of organic matter: *Why* do healthy cells do it? Limnol. Oceanogr. 33:151–154.

Björnsson, B. 1993. Swimming speed and swimming metabolism of Atlantic cod (*Gadus morhua*) in relation to available food: A laboratory study. Can. J. Fish. Aquat. Sci. 50:2542–2551.

Black, C.C. 1971. Ecological implications of dividing plants into groups with distinct photosynthetic capacities. Adv. Ecol. Res. 7:87–114.

Black, C.C. 1973. Photosynthetic carbon fixation in relation to net CO_2 uptake. Ann. Rev. Plant Physiol. 24:253–286.

Blackburn, M., R.M. Laurs, R.W. Owen, and B. Zeitzschel. 1970. Seasonal and areal changes in standing stocks of phytoplankton, zooplankton, and micronekton in the eastern tropical Pacific. Mar. Biol. 7:14–31.

Blegvad, H. 1925. Continued studies on the quantity of fish food in the sea bottom. Rep. Danish Biol. Stat. 31:25–56.

Blomqvist, S., and L. Håkanson. 1981. A review on sediment traps in aquatic environments. Arch. Hydrobiol. 91:101–132.

Bloom, S.A. 1981. Similarity indices in community studies: Potential pitfalls. Mar. Ecol. Progr. Ser. 5:125–128.

Bode, A., B. Casas, and M. Varela. 1994. Size-fractionated primary productivity and biomass in the Galician shelf (NW Spain): Netplankton versus nanoplankton dominance. Scient. Mar. 58:131–141.

Bohmsack, J.A., and F.H. Talbot. 1980. Species packing by reef fishes on Australian and Caribbean reefs: An experimental approach. Bull. Mar. Sci. 30:710–723.

Boje, R., and M. Tomczak (eds.). 1978. Upwelling Ecosystems. Springer-Verlag.

Bokuniewicz, H. 1980. Groundwater seepage into Great South Bay, New York. Est. Coast. Mar. Sci. 10:437–444.

Bolin, B., B.R. Döös, J. Jäger, and R.A. Warrick (eds.). 1986. The Greenhouse Effect, Climatic Change, and Ecosystems. Wiley & Sons.

Bonin, P., E.R. Ranaivoson, N. Baymond, A. Chalamet, and S.C. Bertrand. 1994. Evidence for denitrification in marine sediment highly contaminated by petroleum products. Mar. Pollut. Bull. 28:89–95.

Bormann, F.H., and G.E. Likens. 1979. Pattern and Process in a Forested Ecosystem. Springer-Verlag.

Bormann, F.H., G.E. Likens, and J.M. Melillo. 1977. Nitrogen budget for an aggrading Northern hardwood ecosystem. Science 196:981–983.

Bosman, A.L., and P.A.R. Hockey. 1986. Seabird guano as a determinant of rocky intertidal community structure. Mar. Ecol. Progr. Ser. 32:247–257.

Bosman, A.L., J.T. Du Toit, P.A.R. Hockey, and G.M. Branch. 1986. A field experiment demonstrating the influence of seabird guano on intertidal primary production. Estuar. Coast. Shelf Sci. 23:283–294.

Bougis, P. 1976. Marine Plankton Ecology. North-Holland.

Bowen, S.H. 1980. Detrital non-protein amino acids are the key to rapid growth of *Tilapia* in Lake Valencia, Venezuela. Science 207:1216–1218.

Bowman, M.J. 1977. Nutrient distribution and transport in Long Island Sound. Estuar. Coast. Mar. Sci. 5:531–548.

Branch, G.M. 1975a. Intraspecific competition in *Patella cochlear* Born. J. Anim. Ecol. 44:263–281.

Branch, G.M. 1975a. Mechanisms reducing intraspecific competition in *Patella* spp.: Migration, differentiation, and territorial behaviour. J. Anim. Ecol. 44:575–600.

Branch, G.M. 1976. Interspecific competition experienced by South African *Patella* species. J. Anim. Ecol. 45:507–529.

Branch, G.M. and C.L. Griffiths. 1988. The Benguela ecosystem. V. The coastal zone. Oceanogr. Mar. Biol. Annu. Rev. 26:395–486.

Brandt, S.B. 1981. Effects of a warm-core eddy on fish distributions in the Tasman Sea off East Australia. Mar. Ecol. Progr. Ser. 6:19–33.

Bratvak, G. 1985. Bacterial biovolume and biomass estimations. Appl. Envir. Microbiol. 49:1488–1493.

Bratvak, G., J.K. Egge, and M. Heldal. 1993. Viral mortality of the marine alga *Emiliania huxleyi* (Haptophyceae) and the termination of algal blooms. Mar. Ecol. Progr. Ser. 93:39–48.

Braudel, F. 1981. The Structures of Everyday Life. Vol. I, The Limits of the Possible. Harper and Row.

Brault, S., and H. Caswell. 1993. Pod-specific demography of killer whales (*Orcinus orca*). Ecology 74:1444–1454.

Brawn, V.M. 1969. Feeding behavior of cod (*Gadus morhua*). J. Fish. Res. Bd. Canada 26:583–596.

Bray, J.R. 1964. Primary consumption in three forest canopies. Ecology 45:165–167.

Bray, J.T., O.P. Bricker, and B.N. Troup. 1973. Phosphate in interstitial waters of anoxic sediments: Oxidation effects during sampling procedure. Science 180:1362–1364.

Bray, R.N., A.C. Miller, and G.G. Geesey. 1981. The fish connection: A trophic link between planktonic and rocky reef communities? Science 214:204–205.

Breen, P.A., and K.H. Mann. 1976. Changing lobster abundance and the destruction of kelp beds by sea urchins. Mar. Biol. 34:137–142.

Brenchley, G.A. 1981. Disturbance and community structure: An experimental study of bioturbation in marine soft-bottom sediments. J. Mar. Res. 39: 767–790.

Brett, J.R., V.E. Shelbourn, and C.T. Shoop. 1969. Growth rate and body composition of fingerling sockeye salmon (*Onchorynchus verka*) in relation to temperature and ration size. J. Fish. Res. Bd. Canada 26:2363–2394.

Briggs, K.B., K.R. Tenore, and R.B. Hanson. 1979. The role of microfauna in detrital utilization by the polychaete. *Nereis succinea* (Frey and Leuckart). J. Exp. Mar. Biol. Ecol. 36:225–234.

Brinkhuis, B.H. 1977. Comparisons of salt marsh fucoid production estimated from three different indices. J. Phycol. 13:328–335.

Brinkhuis, B.H. 1977a. Seasonal variations in salt marsh macroalgae photosynthesis. I. *Ascophyllum nodosum* ecad *scorpioides*. Mar. Biol. 44:165–175.

Brinkhuis, B.H. 1977b. Seasonal variations in salt marsh macroalgae photosynthesis. H. *Fucus vesiculosus* and *Ulva lactuca*. Mar. Biol. 44:177–186.

Brock, T.D. 1971. Microbial growth rates in nature. Bacteriol. Rev. 35:39–58.

Brock, V., and R. Riffenburgh. 1960. Fish schooling: A possible factor in reducing predation. J. Cons. Intern. Explor. Mer 25:307–317.

Brocksen, R.W., G.E. Davis, and C.E. Warren. 1970. Analysis of trophic processes on the basis of density-dependent function. Pp. 468–498 in J.H. Steele (ed.), Marine Food Chains, Univ. California.

Broecker, W.S. 1974. Chemical Oceanography. Harcourt, Brace Jovanovich.

Broenkaw, W.W. 1965. The distribution of nutrients in the Costa Rica dome in the eastern tropical Pacific Ocean. Limnol. Oceanogr. 10:40–52.

Brooks, J.L., and S.I. Dodson. 1965. Predation, body size, and composition of plankton. Science 150:28–35.

Brousseau, D.J. 1978. Spawning cycle, fecundity, and recruitment in a population of soft-shell clam, *Mya arenaria*, from Cape Ann, Massachusetts. Fish. Bull. 76:155–166.

Brousseau, D.J. 1978a. Population dynamics of the soft-shell clam *Mya arenaria*. Mar. Biol. 50:63–71.

Brousseau, D.J. 1979. Analysis of growth rate in *Mya arenaria* using the von Bertalanffy equation. Mar. Biol. 51:221–227.

Brown, C.M., and B. Johnson. 1977. Inorganic nitrogen assimilation in aquatic microorganisms. Adv. Aquatic Microbiol. 1:49–114.

Brown, E.J., R.F. Harris, and J.F. Koonce. 1978. Kinetics of phosphate uptake by aquatic microorganisms: Deviations from a simple Michaelis-Menten equation. Limnol. Oceanogr. 23:26–34.

Brown, W.L., Jr., and E.O. Wilson. 1956. Character displacement. Syst. Zool. 5:49–64.

Brownawell, B.J., and J.W. Farrington. 1986. Biogeochemistry of PCBs in interstitial waters of a coastal marine sediment. Geochim. Cosmochim. Acta 50:157–169.

Bruland, K.W. 1992. Voltammetric determination of trace metal concentrations and organic complexation in seawater. Research Report. Office of Naval Research, Arlington, VA.

Bryan, J.R., J.R. Riley, and P.J. LeB. Williams. 1976. A Winkler procedure for making precise measurements of oxygen concentration for productivity and related studies. J. Exp. Mar. Biol. Ecol. 21:191–197.

Buyant, D.M. 1979. Effects of prey density and site character on estuary usage by overwintering waders (Charadrii). Estuar. Coast. Mar. Sci. 9:369–384.

Brylinsky, M. 1977. Release of dissolved organic matter by some marine macrophytes. Mar. Biol. 39:213–220.

Buchsbaum, R., I. Valiela, and J.M. Teal. 1982. Grazing by Canada geese and related aspects of the chemistry of salt marsh grass. Colonial Waterbirds 4:126–131.

Buchsbaum, R., I. Valiela, T. Swain, M. Dzierzeski, and S. Allen. 1991. Available and refractory nitrogen in detritus of coastal vascular plants and macroalgae. Mar. Ecol. Progr. Ser. 72:131–143.

Buesa, R.J. 1975. Population biomass and metabolic rates of marine angiosperms on the northwestern Cuban shelf. Aquat. Bot. 1:11–23.

Bunt, J.S. 1975. Primary productivity of marine ecosystems. Pp. 169–184 in H. Lieth and R.H. Whittaker (eds.), Primary Productivity of the Biosphere. Springer-Verlag.

Bunt, J.S., K.G. Boto, and G. Boto. 1979. A survey method for estimating potential levels of mangrove forest primary production. Mar. Biol. 52:123–128.

Bunt, J.S., and C.C. Lee. 1970. Seasonal primary production in Antarctic sea ice at McMurdo Sound in 1967. J. Mar. Res. 28:304–320.

Burd, A.C. 1965. Growth and recruitment in the herring of the southern North Sea. Fish. Invest. Ser. 2 23:1–42.

Burd, A.C. 1978. Long term changes in North Sea herring stocks. Rapp. Proc.-Verb. Reun. Cons. Int. Explor. Mer 172:137–153.

Burke, M.V., and K.H. Mann. 1974. Productivity and production: Biomass ratios of bivalve and gastropod populations in an eastern Canadian estuary. J. Fish. Res. Bd. Canada 31:167–177.

Burkholder, J.M., E.J. Noga, C.H. Hobbs, and H.B. Gusgow. 1992. New "phantom" dinoiflagellate is the causative agent of major estuarine fish kills. Nature 358:407–410.

Burns, K.A., S.D. Garrity, and S.C. Levings. 1993. How many years until mangrove ecosystems recover from catastrophic oil spills? Mar. Pollut. Bull. 26:239–248.

Burris, J.E. 1980. Respiration and photorespiration in marine algae. Pp. 411–432 in P.G. Falkowski (ed.), Primary Productivity in the Sea. Plenum.

Buskey, E.J., and D.A. Stockwell. 1993. Effects of a persistent "brown tide" on zooplankton populations in the Laguna Madre of South Texas. Pp. 659–666 in T.J. Smayda and Y. Shimizu (eds.). Toxic Blooms in the Sea. Elsevier.

Buss, L.W. 1979. Bryozoan overgrowth interactions—the interdependence of competition for space and food. Nature 281:475–477.

Butler, E.I., E.D.S. Corner, and S.M. Marshall. 1970. On the nutrition and metabolism of zooplankton. VII. Seasonal survey of nitrogen and phosphorus excretion by *Calanus* in the Clyde Sea area. J. Mar. Biol. Assoc. U.K. 50: 525–560.

Butman, C.A. 1986. Larval settlement of soft-sediment invertebrates: Some predictions based on an analysis of near-bottom velocity profiles. Pp. 487–514 in J.C.J. Nihoul (ed.), Marine Interfaces Ecohydrodynamics. Elsevier.

Butman, C.A. 1987. Larval settlement of soft-sediment invertebrates: The spatial scales of pattern explained by active habitat selection and the emerging role of hydrodynamical processes. Oceanogr. Mar. biol. Ann. Rev. 25:113–165.

Buzas, M.A., and S.J. Culver. 1994. Species pool and dynamics of marine paleocommunities. Science 264-1439-1441.

Calow, P. 1977. Conversion efficiencies in heterotrophic organisms. Biol. Rev. 52:385–409.

Calow, P. 1977a. Ecology, evolution, and energetics: A study in metabolic adaptation. Adv. Ecol. Res. 10:1–60.

Calow, P., and C.R. Fletcher. 1976. A new radiotracer technique involving ^{14}C and ^{61}Cr, for estimating the assimilation efficiencies of aquatic, primary consumers. Oecologia 9:155–170.

Cambridge, M.L., and A.J. McComb. 1984. The loss of seagrasses in Cockburn Sound, Western Australia. I. The time course and magnitude of seagrass decline in relation to industrial development. Aquat. Bot. 20:229–243.

Cammen, L.M. 1980. The significance of microbial carbon in the nutrition of the deposit feeding polychaete *Nereis succinea*. Mar. Biol. 61:9–20.

Cammen, L., P. Rublee, and J. Hobbie. 1978. The significance of microbial carbon in the nutrition of the polychaete *Nereis succinea* and other aquatic deposit feeders. Univ. of North Carolina Sea Grant Publ. UNC-SG-78-12.

Campbell, J.W., and T. Aarup. 1989. Photosynthetically available radiation at high latitudes. Limnol. Oceanogr. 34:1490–1499.

Canfield, D.E. 1989. Sulfate reduction and oxic respiration in marine sediments: Implications for organic carbon preservation in euxinix environments. Deep-Sea Res. 36:121–138.

Canfield, D.E., and D.J. Desmarais. 1991. Aerobic sulfate reduction in microbial mats. Science 251:1471–1473.

Canfield, D.E., B. Thamdrup, and J.W. Hansen. 1993. The anaerobic degradation of organic matter in Danish coastal sediments: Iron reduction, manganese reduction, and sulfate reduction. Geochim. Cosmochim. Acta 57: 3867–3883.

Cannon, H.G. 1928. On the feeding mechanisms of the copepods, *Calanus finmarchicus* and *Diaptomus gracilis*. Brit. J. Exp. Biol. 6:131–144.

Caperon, J., and J. Meyer. 1972. Nitrogen-limited growth of marine phytoplankton II. Uptake kinetics and their role in nutrient limited growth of phytoplankton. Deep-Sea Res. 19:619–632.

Capone, D.G., R.S. Oremland, and B.F. Taylor. 1977. Significance of N_2 fixation to the production of *Thalassia testudinum* communities. In H.B. Stewart, Jr. (ed.), Cooperative Investigations of the Caribbean and Adjacent Regions, Vol. 2. FAO, Rome.

Cappenberg, T.E. 1974. Interrelations between sulfate-reducing and methane-producing bacteria in the bottom deposits of a freshwater lake. II. Inhibition experiments. Ant. v. Leenwenhoek 40:297–306.

Cappenberg, T.E. 1975. A study of mixed continuous cultures of sulfate-reducing and methane-producing bacteria. Microb. Ecol. 2:60–72.

Capriulo, G.M., and E.J. Carpenter. 1980. Grazing by 35 to 202 μm microzooplankton in Long Island Sound. Mar. Biol. 56:319–326.

Capuzzo, J.M., and B.A. Lancaster. 1979. Larval development in the American lobster: Changes in metabolic activity and the O:N ratio. Canad. J. Zool. 57:1845–1848.

Caraco, N., A. Tamse, O. Boutros, and I. Valiela. 1987. Nutrient limitation of phytoplankton growth in brackish coastal ponds. Can. J. Fish. Aquat. Sci. 44:473–476.

Carignan, R., and J. Kalff. 1980. Phosphorus sources for aquatic weeds: Water or sediments? Science 207:987–989.

Carlton, J.T. 1989. Man's role in changing the face of the oceans: Biological conservation of near shore environments. Conserv. Biol. 3:265–273.

Carlton, J.T., and J.B. Geller. 1993. Ecological roulitte: the global transport of nonindigenous maune organisms. Science 261:78–82.

Caron, D.A. 1991. Evolving role of protozoa in aquatic nutrient cycles. Pp. 387–415 in P.C. Reid, C.M. Turley, and P.H. Burkill (eds.), Protozoa and Their Role in Marine Processes, Vol. 25. Springer-Verlag.

Caron, D.A., and J.C. Goldman. 1990. Protozoan nutrient regeneration. Pp. 283–306 in G.M. Capriulo (ed.), Ecology of Marine Protozoa. Oxford Univ. Press.

Caron, D.A., P.G. Davis, L.P. Madin, and J. McN. Sieburth. 1982. Heterotrophic bacteria and bacterivorous protozoa in oceanic aggregates. Science 218:795–797.

Caron, D.A., E.L. Lim, G. Miceli, J.B. Waterbury, and F.W. Valois. 1991. Grazing and utilization of chroococcoid cyanobacteria by protozoa in laboratory cultures and a coastal plankton community. Mar. Ecol. Progr. Ser. 76:205–217.

Carpenter, E.J. 1973. Nitrogen fixation by *Oscillatoria* (*Trichodesmium*) *thiebautii* in the southwestern Sargasso Sea. Deep-Sea Res. 20:285–288.

Carpenter, E.J., and J.L. Culliney. 1975. Nitrogen fixation in marine shipworms. Science 187:551–552.

Carpenter, E.J., and R.R.L. Guillard. 1971. Interspecific differences in nitrate half-saturation constants for three species of marine phytoplankton. Ecology 52:183–185.

Carpenter, E.J., and J.S. Lively. 1980. Review of estimates of algal growth using ^{14}C tracer techniques. Pp. 161–178 in P.G. Folkowski (ed.), Primary Productivity in the Sea. Plenum.

Carpenter, E.J., and J.J. McCarthy. 1975. Nitrogen fixation and uptake of combined nitrogenous nutrients by *Oscillatoria* (*Trichodesmium*) *thiebautii* in the western Sargasso Sea. Limnol. Oceanogr. 20:389–401.

Carpenter, E.J., and C.C. Price IV. 1977. Nitrogen fixation, distribution and production of *Oscillatoria* (*Trichodesmium*) spp. in the Western Sargasso and Caribbean Sea. Limnol. Oceanogr. 22:60–72.

Carpenter, E.J., G.R. Harbison, L.P. Madin, N.R. Swanberg, D.C. Biggs, E.M. Hulburt, V.L. McAlister, and J.J. McCarthy. 1977. *Rhizosolenia* mats. Limnol. Oceanogr. 22:739–741.

Carpenter, E.J., C.D. Van Raalte, and I. Valiela. 1978. Nitrogen fixation by algae in a Massachusetts salt marsh. Limnol. Oceanogr. 23:318–327.

Carpenter, R.C. 1981. Grazing by *Diadema antillarum* (Philippi) and its effects on the benthic algal community. J. Mar. Res. 39:749–765.

Carpenter, S.R., J.F. Kitchell, and J.R. Hodgson. 1985. Cascading trophic interactions and lake productivity. Bioscience 35:634–639.

Cassie, R.M. 1954. Some uses of probability paper in the analysis of size frequency distributions. Austral. J. Mar. Freshw. Res. 5:513–522.

Cassie, R.M. 1962. Frequency distribution models in the ecology of plankton and other organisms. J. Anim. Ecol. 31:65–92.

Cassie, R.M. 1962a. Microdistribution and other error components of ^{14}C primary production estimates. Limnol. Oceanogr. 7:121–130.

Castenholtz, R.W. 1961. The effect of grazing on marine littoral diatom populations. Ecology 42:783–794.

Caswell, H., R.J. Naiman, and R. Morin. 1984. Evaluating the consequences of reproduction in complex salmonid life cycles. Aquaculture 43:123–134.

Caswell, H., F.C. Reed, S.N. Stephenson, and P.A. Werner. 1973. Photosynthetic pathways and selective herbivory: A hypothesis. Amer. Nat. 107: 465–480.

Cates, R.G., and G.H. Orians. 1975. Successional status and the palatability of plants to generalized herbivores. Ecology 56:410–418.

Caughley, G. 1966. Mortality patterns in mammals. Ecology 47:906–918.

Cauwet, G. 1978. Organic chemistry of seawater particulates: Concepts and developments. Oceanol. Acta 1:99–105.

Cavanaugh, C.M., S.L. Gardiner, M.L. Jones, H.W. Jannasch, and J.B. Waterberg. 1981. Procaryotic cells in the hydrothermal vent tubeworm *Riftia pachyptila* Jones: Possible chemoautotrophic symbionts. Science 213:340–342.

Cebrián, J., and C.M. Duarte. 1994. The dependence of herbivory on growth-rate in natural plant communities. Funct. Ecol. 8: in press.

Cederwell, H. 1977. Annual macrofauna production in a soft bottom in the Northern Baltic Proper. Pp. 155–164 in B.F. Keegan, P. O'Leidigh, and P.J. Braden (eds.), Biology of Benthic Organisms. Pergamon.

Chamberlain, W., and J. Shapiro. 1973. Phosphate measurements in natural waters—A critique. Pp. 355–366 in E.J. Griffith, A. Beeton, J.M. Spencer, and D.R. Mitchell (eds.), Environmental Phosphorus Handbook. Wiley.

Chanton, J.P., and C.S. Martens. 1987. Biogeochemical cycling in an organic-rich coastal marine basin. 7. Sulfur mass balance, oxygen uptake, and sulfide retention. Geochim. Cosmochim. Acta 51:1187–1199.

Chapman, A.R.O. 1981. Stability of sea urchin dominated barren grounds following destructive grazing of kelp in St. Margaret's Bay, Southern Canada. Mar. Biol. 62:307–311.

Chapman, A.R.O., and J.S. Craigie. 1977. Seasonal growth in *Laminaria longicruris*: Relations with dissolved inorganic nutrients and internal reserves of nitrogen. Mar. Biol. 40:197–205.

Chapman, A.R.O., and J.S. Craigie. 1978. Seasonal growth in *Laminaria longicruris*: Relation with reserve carbohydrate storage and production. Mar. Biol. 46:208–213.

Chapman, P. 1986. Nutrient cycling in marine ecosystems. J. Limnol. Soc. South Afr. 12:22–24.

Chapman, V.J. 1960. Salt marshes and Salt Deserts of the World. L. Hill Ltd., Interscience.

Chapman, V.J. (ed.). 1977. Wet Coastal Ecosystems. Èlsevier.

Charnov, E.L. 1976. Optimal foraging: Attack strategy of a mantid. Amer. Nat. 110:141–151.

Charnov, E.L., G.H. Orians, and K. Hyatt. 1976. Ecological implications of resource depression. Amer. Nat. 110:247–259.

Cheng, L. (ed.). 1976. Marine Insects. North Holland.

Chervin, M.B. 1978. Assimilation of particulate organic carbon by estuarine and coastal copepods. Mar. Biol. 49:265–275.

Chisholm, S.W. 1992. Phytoplankton size. Pp. 213–237 in P.G. Falkowski P.G. and A.D. Woodhead (eds.), Primary Productivity and Biogeochemical Cycles in the Sea. Plenum Press. 550 pp.

Chisholm, S.W., and F.M.M. Morel (eds.). 1991. What controls phytoplankton production in nutrient-rich areas of the open sea? Limnol. Oceanogr. 36, No. 8. Special issue.

Chisholm, S.W., R.J. Olson, E.R. Zettler, R. Goericke, J.B. Waterbury, and N.A. Welshmeyer. 1988. A novel free-living prochlorophyte abundant in the oceanic euphotic zone. Nature 334:340–343.

Chisholm, S.W., S.L. Frankel, R. Goericke, R.J. Olson, B. Palenik, J.B. Waterbury, L. West-Johnsrud, and E.R. Zettler. 1992. *Prochlorococcus marinus* nov. gen. nov. sp.: An oxyphototrophic marine prokaryote containing divinyl chlorophyll *a* and *b*. Arch. Microbiol. 157:297–300.

Chmyr, V.D. 1967. Radiocarbon method for determining zooplankton production in natural populations (In Russian.). Dokl. Acad. Nauk. SSSR 173:201–203.

Chock, J.S., and A.C. Mathieson. 1976. Ecological studies of the salt marsh ecad *scorpioides* (Hornemann) Hauck of *Ascophyllum nodosum* (L.) Lectolis. J. Exp. Mar. Biol. Ecol. 23:171–190.

Choi, C.I. 1972. Primary production and release of DOC in the Western North Atlantic Ocean. Deep-Sea Res. 19:731–735.

Cifuentes, L.A., J.H. Sharp, and M.L. Fogel. 1988. Stable carbon and nitrogen isotope biogeochemistry in the Delaware estuary. Limnol. Oceanogr. 33: 1102–1115.

Clark, A. 1980. A reappraisal of the concept of metabolic cold adaptation in polar marine invertebrates. Biol. J. Linn. Sioc. 14:77–92.

Clark, L.R., P.W. Geier, R.D. Hughes, and R.F. Morris. 1967. The Ecology of Insect Populations in Theory and Practice. Methuen.

Clark, P.J., and F.C. Evans. 1954. distance to nearest neighbor as a measure of spatial relationships in populations. Ecology 45:445–453.

Clarke, G.L. 1946. Dynamics of production in a marine area. Ecol. Monogr. 16:323–335.

Clarke, T.A. 1970. Territorial behavior and population dynamics of a pomacentrid fish, the garibaldi, *Hypsypops rubicunda*. Ecol. Monogr. 40:189–212.

Claypool, G.E., and I.R. Kaplan. 1974. The origin and distribution of methane in marine sediments. Pp. 99–139 in I.R. Kaplan (ed.), Natural Gases in Marine Sediments. Plenum.

Clayton, R.K. 1971. Light and Living Matter. Vol. 2: The Biological Part. McGraw-Hill.

Clendenning, K.A., T.E. Brown, and H.C. Eyster. 1956. Comparative studies of photosynthesis in *Nostoc muscorum* and *Chlorella pyrenoidosa*. Canad. J. Bot. 34:943–966.

Clutter, R.I. 1969. The microdistribution and social behavior of some pelagic mysid shrimps. J. Exp. Mar. Biol. Ecol. 3:125–155.

Coale, K.H. 1991. Effects of iron, manganese, copper, and zinc enrichments on productivity and biomass in the Subarctic Pacific. Limnol. Oceanogr. 36:1851–1864.

Cochlan, W.P., J. Wilkner, G.F. Steward, D.C. Smith, and F. Azam. 1993. Spatial distribution of viruses, bacteria and chlorophyll *a* in neritic, oceanic and estuarine environments. Mar. Ecol. Progr. Ser. 92:77–87.

Cock, M.J.W. 1978. The assessment of preference. J. Anim. Ecol. 47:805–816.

Cockcroft, A.C., and A. McLachlan. 1993. Nitrogen budget for a high-energy ecosystem. Mar. Ecol. Progr. Ser. 100:287–299.

Codispoti, L.A. 1989. Phosphorus vs. nitrogen limitation of new and export production. Pp. 377–394 in W.H. Berger, V.S. Smetacek, and G. Wefer (eds.), Productivity in the Ocean: Present and Past. Wiley & Sons Ltd.

Cohen, E.B., M.D. Grosslein, M.P. Sissenwine, F. Steimle, and W.R. Wright. 1981. An energy budget of Georges Bank. In M.C. Mercer (ed.), Multispecies Approaches to Fisheries Management Advise. Canad. Spec. Publ. Fish. Aquatic Sci. No. 58.

Cohen, Y., W.E. Krumbein, M. Goldberg, and M. Shilo. 1977. Solar Lake (Sinai) 1. Physical and chemical limnology. Limnol. Oceanogr. 22:597–608.

Cohen, Y., W.E. Krumbein, and M. Shilo. 1977a. Solar Lake, Sinai. 2. Distribution of photosynthetic microorganisms and primary production. Limnol. Oceanogr. 22:609–620.

Cohen, Y., W.E. Krumbein, and M. Shilo. 1977b. Solar Lake, Sinai. 3. Bacterial distribution and production. Limnol. Oceanogr. 22:621–634.

Cole, J.J. 1982. Interactions between bacteria and algae in aquatic ecosystems. Ann. Rev. Ecol. Syst. 13:291–314.

Cole, J.J., S. Findlay, and M.L. Pace. 1988. Bacterial production in fresh and salt water ecosystems: A cross-system overview. Mar. Ecol. Progr. Ser. 43:1–10.

Cole, J.J., S. Honjo, and J. Erez. 1987. Benthic decomposition of organic matter at a deep-water site in the Panama Basin. Nature 327:703.

Cole, J.J., J.M. Lane, R. Marino, and R.W. Howarth. 1993. Molybdenum assimilation by cyanobacteria and phytoplankton in freshwater and salt water. Limnol. Oceanogr. 38:25–35.

Cole, J.J., B.L. Peierls, N.F. Caraco, and M.L. Pace. 1993. Nitrogen loading of rivers as a human-driven process. Pp. 141–157 in M.J. McDonell and S.T.A. Pickett (eds.), Humans as Components of Ecosystems: The Ecology of Subtle Human Effects and Populated Areas. Springer-Verlag.

Cole, L.C. 1946. A study of the cryptozoa of an Illinois woodland. Ecol. Monogr. 16:49–86.

Cole, L.C. 1954. The population consequences of life history phenomena. Quart. Rev. Biol. 29:103–137.

Cole, L.C. 1965. Dynamics of animal population growth. J. Chron. Dis. 18: 1095–1108.

Colebrook, J.M. 1979. Continuous plankton records: Seasonal cycles of phytoplankton and copepods in the North Atlantic Ocean and the North Sea. Mar. Biol. 51:23–32.

Colebrook, J.M. 1986. Environmental influences on long-term variability in marine plankton. Hydrobiologia 142:309–325.

Collos, Y., and G. Slawyk. 1980. Nitrogen uptake and assimilation by marine phytoplakton. Pp. 195–211 in P.G. Falkowski (ed.), Primary Productivity in the Sea. Plenum.

Colwell, R.K., and D.J. Futuyma. 1971. On the measurement of niche breadth and overlap. Ecology 52:567–576.

Comins, H.N., and M.P. Hassell. 1979. The dynamics of optimally foraging predators and parasitoids. J. Anim. Ecol. 48:335–351.

Connell, J.H. 1961. The influence of interspecific competition and other factors on the distribution of the barnacle *Chthamalus stellatus*. Ecology 42: 133–146.

Connell, J.H. 1970. A predator-prey system in the marine intertidal region. I. *Balanus glandula* and several predatory species of *Thais*. Ecol. Monogr. 40:49–78.

Connell, J.H. 1972. Community interactions on marine rocky intertidal shores. Ann. Rev. Ecol. Syst. 3:169–192.

Connell, J.H. 1980. Diversity and the coevolution of competitors, or the ghost of competition past. Oikos 35:131–138.

Connell, J.H. 1985. The consequences of variation in initial settlement vs. post-settlement mortality in rocky intertidal communities. J. Exp. Mar. Biol. Ecol. 93:11–45.

Connell, J.H., and R.E. Slatyer. 1977. Mechanisms of succession in natural communities and their role in community stability and organization. Amer. Nat. 111:1119–1144.

Conover, R.J. 1966. Factors affecting the assimilation of organic matter by zooplankton and the question of superfluous feeding. Limnol. Oceanogr. 11:346–354.

Conover, R.J. 1974. Production in marine planktonic communities. Pp. 119–163 in Proc. 1st Inter. Cong. of Ecology, The Hague 1974. Center for Agric. Publish. and Documentation.

Conover, R.J. 1978. Transformation of organic matter. Pp. 221–499 in O. Kinne (ed.), Marine Ecology, Vol. 4. Wiley.

Conover, R.J., and C.M. Lalli. 1974. Feeding and growth in *Clione limacina* (Phipps), a pteropod mollusc. II. Assimilation, metabolism, and growth efficiency. J. Exp. Mar. Biol. Ecol. 16:131–154.

Conroy, J.W.H. 1975. Recent increases in penguin populations in Antarctica and the subantarctic. Pp. 321–336 in B. Stonehouse (ed.), The Biology of Penguins. Univ. Park.

Conway, H.L., and T.E. Whitledge. 1979. Distribution, fluxes and biological utilization of inorganic nitrogen during a spring bloom in the New York Bight. J. Mar. Res. 37:657–668.

Cook, R.M., and B.J. Cockrell. 1978. Predator ingestion rate and its bearing on feeding time and the theory of optimal diets. J. Anim. Ecol. 47:529–547.

Cooper, J., et al. 1990. Diets and dietary segregation of crested penguins. Pp. 131–156 in L.S. Davis and J.T. Darby (eds.), Penguin Biology. Academic Press.

Cooper, W.E. 1965. Dynamics and predation of a natural population of a freshwater amphipod, *Hyalella azteca*. Ecol. Monogr. 35:377–394.

Copping, A.E., and C.J. Lorenzen. 1980. Carbon budget of a marine phytoplankton-herbivore system with carbon-14 as a tracer. Limnol. Oceanogr. 25:873–882.

Corkett, C.J., and I.A. McLaren. 1969. Egg production and oil storage by the copepod *Pseudocalanus* in the laboratory. J. Exp. Mar. Biol. Ecol. 3:90–105.

Corner, E.D.S., and A.G. Davies. 1971. Plankton as a factor in the nitrogen and phosphorus cycles in the sea. Adv. Mar. Biol. 9:101–204.

Corner, E.D.S., and B.S. Newell. 1967. On the nutrition and metabolism of zooplankton. IV. The forms of nitrogen excreted by *Calanus*. J. Mar. Biol. Assoc. U.K. 47:113–120.

Corner, E.D.S., R.N. Head, and C.C. Kilvington. 1972. On the nutrition and metabolism of zooplankton. VIII. The grazing of *Biddulphia* cells by *Calanus helgolandicus*. J. Mar. Biol. Assoc. U.K. 52:847–861.

Cornwell, J., W.R. Boynton, M. Owens, and J. Cowan. 1991. Phosphorus cycling in Chesapeake Bay sediments: Control by iron and sulfur diagenesis. Estuar. Res. Fed. Proc. Nov. 1991. San Francisco.

Cosgrove, D.J. 1977. Microbial transformations in the phosphorus cycle. Adv. Microb. Ecol. 1:95–134.

Cosper, T.C., and M.R. Reeve. 1975. Digestive efficiency of the chaetognath *Sagitta hispida* Conant. J. Exp. Mar. Biol. Ecol. 17:33–38.

Costopulos, J.J., G.C. Stephens, and S.H. Wright. 1979. Uptake of amino acids by marine polychaetes. Biol. Bull. 157:434–444.

Cottrell, M., and C. Suttlè. 1991. Wide-spread occurrence and clonal variation in viruses which cause lysis of a cosmopolitan, eukaryotic marine phytoplankter, *Micromonas pusilla*. Mar. Ecol. Progr. Ser. 78:1–9.

Cowan, J.H., Jr., and E.D. Houde. 1993. Relative predation potentials of scyphomedusae, ctenophores, and planktivorous fish on ichthyoplankton in chesapeake Bay. Mar. Ecol. Progr. Ser. 95:55–65.

Cowles, T.J. 1979. The feeding response of copepods from the Peru upwelling system: Food size selection. J. Mar. Res. 37:601–622.

Craig, H. 1971. The deep metabolism: Oxygen consumption in abyssal ocean water. J. Geophys. Res. 76:5078–5086.

Craigie, J.S., and J. McLachlan. 1964. Excretion of coloured ultraviolet absorbing substances by marine algae. Canad. J. Bot. 42:23–33.

Crawley, M.J. 1975. The numerical responses of insect predators to changes in prey density. J. Anim. Ecol. 44:877–892.

Creese, R.G. 1980. An analysis of distribution and abundance on populations of the high-shore limpet, *Notoacmea petterdi* (Tenison-Woods). Oecologia 45:252–260.

Crisp, D.J. 1975. Secondary productivity in the sea. Pp. 71–90 in D.E. Reichle, J.F. Franklyn, and D.W. Goodall (eds.), Productivity of World Ecosystems. Natl. Acad. Sci. Washington, D.C.

Crosby, M.P., R.I.E. Newell, and C.J. Langdon. 1990. Bacterial mediation in the utilization of carbon and nitrogen from detrital complexes by *Crassostrea virginica*. Limnol. Oceanogr. 35:625–639.

Crossland, C.J., D.J. Narnes, T. Cox, and M. Devereux. 1980. Compartmentation and turnover of organic carbon in the staghorn coral *Acropora formosa*. Mar. Biol. 59:181–187.

Croxall, J.P., and P.A. Prince. 1979. Antarctic seabird and seal monitoring studies. Polar Rec. 19:573–595.

Cullen, J.J. 1991. Hypotheses to explain high-nutrient conditions in the open sea. Limnol. Oceanogr. 36:1578–1599.

Cullen, J.J., M.R. Lewis, C.O. Davis, and R.T. Barber. 1992. Photosynthetic characteristics and estimated growth rates indicate grazing is the proximate control of primary production in the equatorial Pacific. J. Geophys. Res. 97:639–654.

Cullen, J.J., X. Yang, and H.L. MacIntyre. 1992a. Nutrient limitation of marine photosynthesis. Pp. 69–88 in P.G. Falkowski and A.D. Woodhead (eds.), Primary Productivity and Geochemical Cycles in the Sea. Plenum Press. 550 pp.

Cummins, K.W., and J.C. Wuycheck. 1971. Caloric equivalents for investigations in ecological energetics. Mitt. Int. Verein Limnol. 18:2–158.

Curl, H., Jr., and L.F. Small. 1965. Variations in photosynthetic assimilation ratios in natural, marine phytoplankton communities. Limnol. Oceanogr. 10 (Suppl.):R67–R73.

Currie, R.I. 1964. Environmental features in the ecology of Antarctic Seas. Pp. 89–94 in Biologie Antarctique. Hermann.

Curtis, P.S., B.G. Drake, P.W. Leadley, W.J. Arp, and D.F. Wigham. 1989. Growth and senescence in plant communities exposed to elevated CO_2 concentrations on an estuarine marsh. Oecologia 78:20–26.

Cushing, D.H. 1964. The work of grazing in the sea. Pp. 207–226 in D.J. Crisp (ed.), Grazing in Terrestrial and Marine Environments. Blackwell.

Cushing, D.H. 1968. Grazing by herbivorous copepods in the sea. J. Cons. Int. Explor. Mer 32:70–82.

Cushing, D.H. 1971. Upwelling and the production of fish. Adv. Mar. Biol. 9:255–334.

Cushing, D.H. 1978. Upper trophic levels in upwelling areas. Pp. 101–110 in R. Boje and M. Tomczak (eds.), Upwelling Ecosystems. Springer-Verlag.

Cushing, D.H. 1981. Fisheries Biology: A Study in Population Dynamics. Univ. Wisconsin.

Cushing, D.H. 1982. Climate and Fisheries. Academic Press.

Cyr, H., and M.L. Pace. 1993. Magnitude and patterns of herbivory in aquatic and terrestrial ecosystems. Nature 361:148–150.

Daan, N. 1975. Consumption and production of North Sea Cod, *Gadus morhua*: An assessment of the ecological status of the stock. Neth. J. Sea Res. 9:24–55.

Daan, N. 1978. Changes in cod stocks and cod fisheries in the North Sea. Rapp. Proc.-Verb. Reun. Cons. Int. Explor. Mer 172:39–57.

Dagg, M.J. 1974. Loss of prey body contents during feeding by an aquatic predator. Ecology 55:903–906.

Dagg, M.J. 1977. Some effects of patchy food environments on copepods. Limnol. Oceanogr. 22:99–107.

D'Antonio, C. 1985. Epiphytes on th rocky intertidal red algae *Rhodomela laryx* (Turner) C. Agardh: Negative effects on the host and food for herbivores? J. Exper. Mar. Biol. Ecol. 86:197–218.

D'Avanzo, C., and J. Kremer. 1994. Diel oxygen dynamics and anoxic events in an eutrophic estuary of Waquoit Bay, Mass. Estuaries 17:131–139.

D'Avanzo, C., and J. Kremer. 1995. Ecosystem production and respiration in response to eutrophication in shallow temperate estuaries. Mar. Ecol. Progr. Ser.

D'Avanzo, C., J.W. Kremer, and S.C. Wainright. 1995. Ecosystem production and respiration in response to eutrophication in shallow temperature estuaries. Mar. Ecol. Progr. Ser.

D'Avanzo, C.D., M. Alber, and I. Valiela. 1991. Nitrogen assimilation from amorphous detritus by two coastal consumers. Estuar. Coast. Shelf Sci. 33:203–209.

Davies, J.M. 1975. Energy flow through the benthos in a Scottish Sea Loch. Mar. Biol. 31:353–362.

Davis, C.S. 1982. Processes Controlling Zooplankton Abundance on Georges Bank. Ph.D. Thesis, Boston University.

Davis, C.S. 1984. Interaction of a copepod population with the mean circulation on Georges Bank. J. Mar. Res. 42:573–590.

Davis, C.S. 1984a. Predatory control of copepod seasonal cycles on Georges Bank. Mar. Biol. 82:31–40.

Davis, C.S. 1987. Components of the zooplankton production cycle in the temperate ocean. J. Mar. Res. 45:947–983.

Davis, C.S., and P. Alatalo. 1992. Effects of constant and intermittent food supply on life-history parameters in a marine copepod. Limnol. Oceanogr. 37:1618–1639.

Davis, C.S., S.M. Gallager, and A.R. Solow. 1992. Microaggregations of oceanic plankton observed by towed video microscopy. Science 257:230–232.

Davis, C.S., G.R. Flierl, P.H. Wiebe, and P.J.S. Franks. 1991. Micropatchiness, turbulence and recruitment in plankton. J. Mar Res. 49:109–151.

Davis, H.C., and A. Calabrese. 1964. Combined effects of temperature and salinity on development of eggs and growth of larva of *Mercenaria mercenaria* and *Crassostrea virginica*. Fish. Bull. U.S. Wild. Serv. 63:643–655.

Davis, J.H. 1940. The ecology and geologic role of mangroves in Florida. Pap. Tortuga Lab. 32:1–412.

Davis, W.J. 1993. Contamination of coastal versus open surface waters. Mar. Pollut. Bull. 26:128–134.

Day, J.W., Jr., and P.H. Templet. 1989. Consequences of sea level rise: Implications for the Mississippi Delta. Coast. Manag. 17:2412–257.

Dayton, P.K. 1971. Competition, disturbance, and community organization: the provision and subsequent utilization of space in a rocky intertidal community. Ecol. Monogr. 41:351–389.

Dayton, P.K., M.J. Tegner, P.E. Parnell, and P.B. Edwards. 1992. Temporal and spatial patterns of disturbance and recovery in a kelp forest community. Ecol. Monogr. 62:421–445.

Dawson, E.Y. 1966. Marine Botany—An Introduction. Holt, Rinehart, and Winston.

Dean, T.A., and L.E. Hurd. 1980. Development in an estuarine fouling community: The influence of early colonists on later animals. Oecologia 46: 295–301.

De Angelis, D.L. 1975. Stability and connectance in food web models. Ecology 56:238–243.

Deason, E.E. 1980. Grazing of *Acartia hudsonica* (*A. clausi*) on *Skeletonema costatum* in Narragansett Bay (U.S.A.): Influence of food concentrations and temperature. Mar. Biol. 60:101–113.

de Baar, H.J.W., et al. 1995. Importance of iron for plankton blooms and carbon dioxide drawdown in the Southern Ocean. Nature 373:412–415.

Degens, E.T. 1970. Molecular nature of nitrogenous compounds in sea water and recent marine sediments. Pp. 77–100 in D.W. Hood (ed.), Organic Matter in Natural Waters. Univ. of Alaska.

Degens, E., and K. Mopper. 1976. Factors controlling the distribution and early diagenesis of organic material in marine sediments. Pp. 59–113 in J.P. Riley and R. Chester (eds.), Chemical Oceanography. Vol. 6, 2nd Ed. Academic.

D'Elia, C.F., J.G. Sanders, and W.R. Boynton. 1986. Nutrient enrichment studies in a coastal plain estuary: Phytoplankton growth in large-scale con-inuous cultures. Can. J. Fish. Aquat. Sci. 43:397–406.

Delwiche, C.C., and B.B. Bryan. 1976. Denitrification. Ann. Rev. Microbiol. 30:241–262.

De Manche, J.M., H.C. Curl, Jr., D.W. Lundy, and P.L. Donaghey. 1979. The rapid response of the marine diatom *Skeletonema costatum* to changes in external and internal nutrient concentrations. Mar. Biol. 53:323–333.

DeMott, W.R., and F. Moxter. 1991. Foraging on cyanobacteria by copepods: Responses to chemical defenses and resource abundance. Ecology 75: 1820–1834.

DeMott, W.R., Q.X. Zhang, and W.W. Carmichael. 1991. Effects of toxic cyanobacteria and purified toxins on the survival and feeding of a copepod and three species of *Daphnia*. Limnol. Oceanogr. 36:1346–1357.

Dempsey, M.J. 1981. Marine bacterial fouling: A scanning electron microscope study. Mar. Biol. 61:305–315.

Dennison, W.C., and R.S. Alberte. 1985. Role of daily light period in the depth distribution of *Zostera marina* (eelgrass). Mar. Ecol. Progr. Ser. 25:51–61.

Dennison, W.C., R.C. Aller, R.S. Alberte. 1987. Sediment ammonium availability and *Zostera marina* (eelgrass) growth. Mar. Biol. 94:469–477.

Denslow, J.S. 1980. Pattern of plant species diversity during succession under different disturbance regimes. Oecologia 46:18–21.

Derby, C.D., and J. Atema. 1982. Chemosensitivity of walking legs of the lobster *Homarus americanus*: Neurophysiological response spectrum and thresholds. J. Exp. Biol. 98:303–315.

Deuser, W.G. 1975. Reducing environments. Pp. 1–37 in J.P. Riley and G. Skirrow (eds.), Chemical Oceanography. Vol. 3, 2nd Ed. Academic.

Deuser, W.G. 1986. Seasonal and interannual variations in deep-water particle fluxes in the Saragasso and their relation to surface hydrography. Deep Sea Res. 33:225–246.

Deuser, W.G. 1979. Marine biota, nearshore sediments, and the global carbon balance. Org. Geochem. 1:243–247.

Deuser, W.G. 1980. Carbon-13 in Black Sea water and implications for the origin of hydrogen sulfide. Science 168:1575–1577.

Deuser, W.G. In press. Long-term variability of particle flux in the deep Sargasso Sea.

Deuser, W.G., and E.H. Ross. 1980. Seasonal changes in the flux of organic carbon to the deep Sargasso Sea. Nature 280:364–365.

Deuser, W.G., E.T. Degens, G.R. Harvey, and M. Rubin. 1973. Methane in Lake Kivu: New data bearing on its origin. Science 181:51–54.

Deuser, W.G., E.H. Ross, and R.F. Anderson. 1981. Seasonality in the supply of sediment to the deep Sargasso Sea and implications for the rapid transfer of water to the deep sea. Deep-Sea Res. 28A:495–505.

De Vries, A.L. 1988. The role of antifreeze glycopeptides and peptides in the freezing avoidance of antarctic fishes. Comp. Biochem. Physiol. 90A:611–621.

DeYoe, H.R., and C.T. Suttle. 1994. The inability of the texas "brown tide" alga to use nitrate and the role of nitrogen in the initiation of a persistent bloom of this organism. J. Phycol. 30:800–806.

DeYoe, H.R., A.M. Chan, and C.A. Suttle. In press. Phylogeny of *Aureococcus anophagefferens* (Chrysophyceae) and a morphologically similar bloom-forming alga from Texas as determined by 18S rDNA sequence analysis. J. Phycol.

Dickinson, C.H., and G.J.F. Pugh. 1974. Biology of Plant Litter Decomposition. Vols. 1 and 2. Academic.

Dickson, R.R. 1986. Proc. R. Soc. Edinburgh 88:103–125.

Dickson, R.R., J. Meincke, S.-A. Malmberg, and A.J. Lee. 1988. The "great salinity anomaly" in the northern North Atlantic 1968–1982. Progr. Oceanogr. 20:103–151.

Dietrich, G. 1968. General Oceanography: An Introduction. Interscience.

Digby, P.S.B. 1953. Plankton production in Scoresby Sound, East Greenland. J. Anim. Ecol. 23:289–322.

Digby, P.S.B. 1954. The biology of the marine planktonic copepods of Scoresby Sound, East Greenland. J. Anim. Ecol. 23:298–338.

Di Salvo, L.H., and G.W. Daniels. 1978. Observations on estunarine microfouling using the scanning electron microscope. Microb. Ecol. 2:234–240.

Dittman, D., and C. Robles. 1991. Effect of algal epiphytes on the mussel *Mytilus californianus*. Ecology 72:286–296.

DiTullio, G.R., D.A. Hutchins, and K.W. Bruland. 1993. Interaction of iron and major nutrients controls phytoplankton growth and species composition in the tropical North Pacific Ocean. Limnol. Oceanogr. 38:495–508.

Dixon, D.R. 1976. The energetics of growth and reproduction in a brackish water serpulid *Mercierella enigmatica* (Fauvel). Pp. 197–210 in G. Persoone and E. Jaspers (eds.), Proc. 10th European Symp. Mar. Biol. Ostend, Belgium.

Dodimead, A.J., F. Favorite, and T. Hirano. 1963. Review of the oceanography of the subarctic Pacific region. Int. North Pacific Fish. Comm. Bull. No. 13.

Doherty, P.J. 1983. Tropical territorial damselfish: Is density limited by aggression or recruitment? Ecology 64:176–190.

Doherty, P.J., and D.McB. Williams. 1988. The replenishment of coral reef populations. Oceanogr. Mar. Biol. Annu. Rev. 26:487–551.

Dollar, S.J., S.V. Smith, S.M. Vink, S. Obrebski, and J.T. Hollobaugh. 1991. Annual cycle of benthic nutrient fluxes in Tomales Bay, California, and contribution of the benthos to total ecosystem metabolism. Mar. Ecol. Progr. Ser. 79:115–125.

Donaghay, P.L., and L.F. Small. 1979. Food selection capability of the estuarine copepod *Acartia clausii*. Mar. Biol. 52:137–146.

Douillet, P. 1993. Bacterivory in Pacific oyster *Crassostrea gigas* larvae. Mar. Ecol. Progr. Ser. 98:123–134.

Dowd, J.E., and D.S. Riggs. 1964. A comparison of estimates of Michaelis-Menten kinetic constants for various linear transformations. J. Biol. Chem. 240:863–869.

Doyle, R.W. 1979. Ingestion rate of selective deposit feeders in a complex mixture of particles: Testing the energy-optimization hypothesis. Limnol. Oceanogr. 24:867–874.

Dring, M.J. 1981. Chromatic adaptation of photosynthesis in benthic marine algae: An examination of its ecological significance using a theoretical model. Limnol. Oceanogr. 26:271–284.

Droop, M.R. 1968. Vitamin B_{12} and marine ecology. 4. The kinetics of uptake, growth, and inhibition in *Monochrysis lutherii*. J. Mar. Biol. Assoc. U.K. 48:689–733.

Droop, M.R. 1973. Some thoughts on nutrient limitation in algae. J. Phycol. 9:264–272.

Duarte, C.M. 1991. Seagrass depth limits. Aquat. Bot. 40:363–377.

Duarte, C.M. Submerged aquatic vegetation in relation to different nutrient regimes. Ophelia. In press.

Duarte, C.M., and K. Sand-Jensen. 1990. Seagrass colonization: Patch formation and patch growth in *Cymodocea nodosa*. Mar. Ecol. Progr. Ser. 65:193–200.

Duarte, C.M., and D. Vaqué. 1992. Scale dependence of bacterioplankton patchiness. Mar. Ecol. Progr. Ser. 84:95–100.

Duarte, C.M., M. Masó, and M. Merino. 1992. The relationship between mesoscale phytoplankton heterogeity and hydrographic variability. Deep-Sea Res. 39:45–54.

Duce, R.A. 1991. Chemical exchange at the air-coastal sea interface. Pp. 91–109 in R.F.C. Mantoura, J.-M. Martin, and R. Wollast (eds.), Ocean Margin Processes in Global Change. Wiley-Interscience.

Duce, R.A. 1991a. The atmospheric input of trace species to the world ocean. Global Biogeochem. Cycl. 5:193–259.

Duce, R.A., and E.K. Duursma. 1977. Inputs of organic matter to the sea. Mar. Chem. 5:319–339.

Ducklow, H.W., and C.A. Carlson. 1992. Oceanic bacterial production. Adv. Microb. Ecol. 12:113–181.

Ducklow, H.W., and F.-K. Shiah. 1993. Bacterial production in estuaries. Pp. 261–187 in T.G. Ford (ed.), Aquatic Microbiology: An Ecological Approach. Blackwell Sci. Publ.

Ducklow, H.W., D.A. Purdie, P.J. LeB. Williams, and J.M. Davies. 1986. Bacterioplankton: A sink for carbon ina coastal marine plankton community. Science 232:865–867.

Duffy, J.E. 1990. Amphipods on seaweeds: Partners or pests? Oecologia 83:267–276.

Dugdale, R.C., and J.J. Goering. 1967. Uptake of new and regenerated forms of nitrogen in primary productivity. Limnol. Oceanogr. 12:196–206.

Dugdale, R.C., and F. Wilkerson. 1992. Nutrient limitation of new production in the sea. Pp. 89–122 in P.G. Falkowski and A.D. Woodhead (eds.), Primary Productivity and Biogeochemical Cycles in the Sea. Plenum Press.

Dugdale, R.C., D.W. Menzel, and J.H. Ryther. 1961. Nitrogen fixation in the Sargasso Sea. Deep-Sea Res. 7:298–300.

Duggins, D.O. 1980. Kelp beds and sea otters: An experimental approach. Ecology 61:447–453.

Duggins, D.O., C.A. Simenstad, and J.A. Estes. 1989. Magnification of secondary production by kelp detritus in coastal marine ecosystems. Science 245:170–173.

Dunton, K.H. 1994. Seasonal growth and biomass of the subtropical seagrass *Halodule wrightii* in relation to continuous measurements of underwater irradiance. Mar. Biol. in press.

Durbin, A.G., and E.G. Durbin. 1975. Grazing rates of the Atlantic menhaden *Brevoortia tyrannus* as a function of particle size and concentration. Mar. Biol. 33:265–277.

Durbin, E.G. 1974. Studies on the autecology of the marine diatom *Thalassiosira nordenskioldii* Cleve. I. The influence of daylength, light intensity, and temperature on growth. J. Phycol. 10:220–225.

Dussart, B.M. 1965. Les differentes categories de plancton. Hydrobiologia 26:72–74.

Duxbury, A.D., B.A. Morse, and N. McGary. 1966. The Columbia River effluent and its distribution at sea, 1961–1967. Univ. of Washington, Dept. of Oceanography Tech. Rep. No. 156 (Ref. M66–31).

Ebersole, J.P. 1980. Food density and territory size: An alternative model and a test on the reef fish *Eupomacentrus leucostictus*. Amer. Nat. 115:492–509.

Ebert, T.A. 1977. Estimating growth and mortality rates from size data. Oecologia 11:281–298.

Eckman, J.E. 1979. Small-scale patterns and processes in a soft-substratum, intertidal community. J. Mar. Res. 37:437–457.

Eckman, J.E., W.B. Savidge, and T.F. Gross. 1991. Relationship between duration of cyprid attachment and drag forces associated with detachment of *Balanus amphitrite* cyprids. Mar. Biol. 107:111–118.

Edden, A.C. 1971. A measure of species diversity related to the log-normal distribution of individuals among species. J. Exp. Mar. Biol. Ecol. 6:199–209.

Edmond, J.M. 1970. High precision determination of titration alkalinity and total carbon dioxide content of sea water by potentiometric titration. Deep-Sea Res. 17:737–750.

Edmondson, W.T. 1960. Reproductive rates of rotifers in natural populations. Mem. Ist. Ital. Idrobiol. 12:21–77.

Edwards, G.E., and S.C. Huber. 1981. C_4 pathway. Pp. 238–278 in M.D. Hatch and N.K. Boardman (eds.), The Biochemistry of Plants. Vol. 8, Photosynthesis. Academic.

Edwards, R.R.C., J.H. Steele, and A. Trevallion. 1970. The ecology of the O-group plaice and common dabs in Loch Ewe. III. Prey-predator experiments with plaice. J. Exp. Mar. Biol. Ecol. 4:156–173.

Eggers, D.M. 1976. Theoretical effects of schooling by planktivorous fish predators on rate of prey consumption. J. Fish. Res. Bd. Canada 33:1964–1971.

Ehrlich, P.R., and L.C. Birch. 1967. The "balance of nature" and "population control." Amer. Nat. 101:97–107.

Elliot, E.T., L.G. Castanares, D. Perlmutter, and K.G. Porter. 1983. Trophic-level control of production and nutrient dynamics in an experimental planktonic community. Oikos 41:7–16.

Elliot, J.M., and W. Davison. 1975. Energy equivalents of oxygen consumption in animal energetics. Oecologia 19:195–204.

Elliot, J.M., and L. Persson. 1978. The estimation of daily rates of food consumption for fish. J. Anim. Ecol. 47:977–991.

Ellis, D.V. 1977. Pacific Salmon. Management for People. West Geogr. Series Vol. 13. Univ. of Victoria, Dept. of Geography.

Elmgren, R. 1989. Man's impact on the ecosystem of the Baltic Sea: Energy flows today and at the turn of the century. Ambio 18:326–332.

Elner, R.W., and R.N. Hughes. 1978. Energy maximization in the diet of the shore crab, *Carcinus maenas*. J. Anim. Ecol. 47:107–116.

El-Sayed, S.Z. 1988. Fragile life under the ozone hole. Nat. Hist. 97:72–80.

Elsgaard, L., and B.B. Jørgensen. 1992. Anoxic transformations of radiolabeled hydrogen sulfide in marine and freshwater sediments. Geochim. Cosmochim. Acta. 56:2425–2435.

Elster, H.J. 1954. Uber die Populations dynamik von *Eudiaptomns gracilis* Sars und *Heterocope borealis* Fischer im Bodensee-Obersee. Arch. Hydrobiol. 20 (Suppl.):546–614.

Elton, C. 1927. Animal Ecology. MacMillan.

Elton, C., and R.S. Miller. 1954. The ecological survey of animal communities with a practical system of classifying habitats by structural characters. J. Ecol. 42:460–496.

Elvin, D.W., and J.J. Gossor. 1979. The thermal regime of an intertidal *Mytilus californianus* Conrad population on the central Oregon coast. J. Exp. Mar. Biol. Ecol. 39:265–279.

Elyakova, L.A. 1972. Distribution of cellulases and chitinases in marine invertebrates. Comp. Biochem. Physiol. 43:67–70.

Emerson, S., R. Jahnke, B. Bender, P. Froelich, G. Klinkhammer, C. Bowser, and G. Setlock. 1980. Early diagenesis in sediments from the Eastern Equatorial Pacific. I. Porewater nutrient and carbonate results. Earth Planet. Sci. Lett. 49:57–80.

Emery, K.O., and D.G. Aubrey. 1991. Sea levels, land levels, and tide gauges. Springer-Verlag.

Emlen, J.M. 1968. Optimal choice in animals. Amer. Nat. 102:385–389.

Emlen, J.M., and M.G.R. Emlen. 1975. Optimal choice in diet: Test of a hypothesis. Amer. Nat. 102:385–389.

Enfield, D.B. 1989. El Niño, past and present. Rev. Geophys. 27:159–187.

Engelmann, M.D. 1961. The role of soil arthropods in the energetics of an old field community. Ecol. Monogr. 31:221–238.

Enhalt, D.H. 1976. The atmospheric cycle of methane. Pp. 13–22 in H.G. Schlegel, O. Gottschalk, and N. Pfennig (eds.), Microbial Production and Utilization of Gases. E. Goltze K.G.

Enright, J.T. 1969. Zooplankton grazing rates estimated under field conditions. Ecology 50:1070–1078.

Eppley, R.W. 1972. Temperature and phytoplankton growth in the sea. Fish. Bull. 70:1063–1085.

Eppley, R.W. 1980. Estimating phytoplankton growth rates in the central oligotrophic oceans. Pp. 231–242 in P.G. Falkowski (ed.), Primary Productivity in the Sea. Plenum.

Eppley, R.W. 1989. New production: history, methods, problems. Pp. 85–97 in Berger, W.H., V.S. Smetacek, and G. Wefer (eds.), Productivity in the Ocean: Present and Past. Wiley & Sons Ltd.

Eppley, R.W., and B.J. Peterson. 1979. Particulate organic flux and planktonic new production in the deep ocean. Nature 282:677–680.

Eppley, R.W., and P.R. Sloan. 1965. Carbon balance experiments with marine phytoplankton. J. Fish. Res. Bd. Canada 22:1083–1097.

Eppley, R.W., E.H. Renger, E.L. Venrick, and M.M. Mullin. 1973. A study of plankton dynamics and nutrient cycling in the central gyre of the North Pacific Ocean. Limnol. Oceanogr. 18:534–551.

Eppley, R.W., S.G. Horrigan, J.A. Fuhrman, E.R. Books, C.C. Price, and K. Sellner. 1981. Origins of dissolved organic matter in Southern California coastal waters: Experiments on the role of zooplankton. Mar. Ecol. Prog. Ser. 6:149–159.

Eriksson, E. 1952. Composition of atmospheric precipitation. I. Nitrogen compounds. Tellus 4:215–232.

Es, F.B. van, and L.-A. Meyer-Reil. 1982. Biomass and metabolic activity of heterotrophic marine bacteria. Adv. Microb. Ecol. 6:111–170.

Estep, K.W., P.G. Davis, M.D. Keller, and J.D. Siburth. 1986. How important are oceanic algal nanoflagellates in bactivory. Limnol. Oceanogr. 31:646–650.

Estes, J.A. 1979. Exploitation of marine mammals: r-selection of K-strategists? J. Fish. Res. Bd. Canada 36:1009–1017.

Estes, J.A., and J.F. Palmisano. 1974. Sea otters: Their role in structuring nearshore communities. Science 185:1058–1060.

Estes, J.A., R.J. Jameson, and E.B. Rhode. 1982. Activity and prey selection in the sea otter: Influence of population status on community structure. Amer. Nat. 120:242–258.

Estes, J.A., N.S. Smith, and J.F. Palmisano. 1978. Sea otter predation and community organization in the western Aleutian Islands, Alaska. Ecology 59:822–833.

Estrada, M. 1974. Photosynthetic pigments and productivity in the upwelling region of northwest Africa. Thethys 6:247–260.

Evans, G.T., and F.J.R. Taylor. 1980. Phytoplankton accumulation in Langmuir cells. Limnol. Oceanogr. 25:840–845.

Evans, P.D., and K.H. Mann. 1977. Selection of prey by American lobsters (*Homarus americanus*) when offered a choice between sea urchin and crabs. J. Fish. Res. Bd. Canada 34:2203–2207.

Everson, I. 1977. Antarctic marine secondary production and the phenomenon of cold adaptation. Phil. Trans. Roy. Soc. Lond. B279:55–66.

Farman, J.C., B.G. Gardiner, J.D. Shanklin. 1985. Large losses of total ozone in Antarctica reveal seasonal ClO_x/NO_x interaction. Nature 315:207–210.

Farrell, T.M. 1991. Models and mechanisms of succession: An example from a rocky intertidal community. Ecol. Monogr. 61:95–113.

Farrell, T.M., F. Stagnitti, and J. Roughgarden. 1991. Cross-shelf transport causes recruitment to intertidal populations in central California. Limnol. Oceanogr. 36:279–288.

Farrington, J.W. 1980. An overview of the biogeochemistry of fossil fuel hydrocarbons in the marine environment. Pp. 1–22 in L. Petrakis and F.T. Weiss (eds.), Petroleum in the Marine Environment, Adv. in Chem. Ser. 185, Amer. Chem. Soc.

Farrington, J.W., and B.W. Tripp. 1977. Hydrocarbons in western North Atlantic surface sediments. Geochim. Cosmochim. Acta 41: 1627–1641.

Feeny, R. 1976. Plant apparency and chemical defense. Pp. 1–40 in J. Wallace and R. Mansell (eds.), Rec. Adv. Phytochem. 10. Biochemical Interaction Between Plants and Insects. Plenum.

Feigenbaum, D. 1979. Predation on chaetognaths by typhloscolecid polychaetes: One explanation for headless specimens. J. Mar. Biol. Assoc. U.K. 59:631–633.

Feigenbaum, D., and M.R. Reeve. 1977. Prey detection in the Chaetognatha: response to a vibrating probe and experimental determination of attack distance in large aquaria. Limnol. Oceanogr. 22:1052–1058.

Feller, R.J. 1977. Life history and production of meiobenthic harpacticoid copepods in Puget Sound. Ph.D. Thesis, Univ. of Washington.

Feller, R.J., G.L. Taghon, E.D. Gallagher, G.E. Kenny, and P.A. Jumars. 1979. Immunological methods for food web analysis in a soft bottom benthic community. Mar. Biol. 54:61–74.

Fenchel, T. 1968. The ecology of marine microbenthos. II. The food of marine benthic ciliates. Ophelia 5:73–121.

Fenchel, T. 1969. The ecology of marine microbenthos. IV. Structure and function of the benthic ecosystems, its chemical and physical factors and the microfauna communities with special reference to the ciliated protozoa. Ophelia 6:1–182.

Fenchel, T. 1970. Studies on the decomposition of organic detritus derived from the turtle grass *Thalassia testudinum*. Limnol. Oceanogr. 15:14–20.

Fenchel, T. 1973. Aspects of the decomposition of sea grasses. Pp. 123–145 in C.P. McRoy and C. Helfferich (eds.), Seagrass Ecosystems. Marcel Dekker.

Fenchel, T. 1974. Intrinsic rate of natural increase; the relationship with body size. Oecologia 14:317–326.

Fenchel, T. 1975. Character displacement and coexistence in mud snails (Hydrobiidae). Oecologia 20:19–32.

Fenchel, T. 1982a. Ecology of heterotrophic microflagellates. IV. Quantitative occurrence and importance as bacterial consumers. Mar. Ecol. Prog. Ser. 9:35–42.

Fenchel, T. 1984. Suspended marine bacteria as a food source: a flow of energy and materials in marine ecosystems. Pp. 301–316 in M.J. Fash (ed.), Plenum Press, New York.

Fenchel, T. 1986. The ecology of heterotrophic microflagellates. Pp. 57–97 in Marshall, K.C. (ed.), Advances in Microbial Ecology. Plenum Press.

Fenchel, T. 1987. Excellence in ecology. Ecology-potentials and limitations. Ecology Institute, D-2124 Oldendorf/Luke, FRG. Okinne.

Fenchel, T. 1988. Marine plankton food chains. Ann. Rev. Ecol. Syst. 19:19–38.

Fenchel, T., and T.H. Blackburn. 1979. Bacteria and Mineral Cycling. Academic.

Fenchel, T., and P. Harrison. 1976. The significance of bacterial grazing and mineral cycling for the decomposition of particulate detritus. Pp. 285–299 in J.M. Anderson and A. Macfadyen (eds.), The Role of Terrestrial and Aquatic Organisms in Decomposition Processes. Blackwell.

Fenchel, T., and L.H. Kofoed. 1976. Evidence for exploitative interspecfic competition in mud snails (Hydrobiidae). Oikos 27:367–376.

Fenchel, T., and B.J. Straarup. 1971. Vertical distribution of photosynthetic pigments and the penetration of light in marine sediments. Oikos 22:172–182.

Fenchel, T., T. Perry, and A. Thane. 1977. Anaerobiosis and symbiosis with bacteria in free-living ciliates. J. Protozool. 24:154–163.

Fenical, W. 1982. Natural product chemistry in the marine environment. Science 215:923–928.

Ferguson, R.L., and A.V. Palumbo. 1979. Distribution of suspended bacteria in neritic waters of Long Island during stratified conditions. Limnol. Oceanogr. 24:697–705.

Ferguson, R.L., and P. Rublee. 1976. Contribution of bacteria to standing crop of coastal plankton. Limnol. Oceanog. 21:141–144.

Ferrier, C., and F. Rassoulzadegan. 1991. Density-dependent effects of protozoans on specific growth rates in pico- and nanoplanktonic assemblages. Limnol. Oceanogr. 34(4):657–669.

Fiedler, P.C. 1984. Satellite observations of the 1982–1983 El Niño along the U.S. Pacific Coast. Science 224:1251–1254.

Field, J.G. 1969. The use of the information statistic in the numerical classification of heterogeneous systems. J. Ecol. 57:565–569.

Findlay, S. 1993. Thymidine incorporation into DNA as an estimate of sediment bacterial production. Pp. 505–507 in P.F. Kemp, B.F. Sherr, E.B. Sherr, and J.J. Cole (eds.), Handbook of Methods in Aquatic Microbial Ecology. Lewis Publ.

Findlay, S., M.L. Pace, D. Lints, J.J. Cole, N.F. Caraco, and B. Peierls. 1991. Weak coupling of bacterial and algal production in an heterotrophic ecosystem: The Hudson River estuary. Limnol. Oceanogr. 36:268–278.

Finenko, Z.Z., and V.E. Zaika. 1970. Particulate matter and its role in the productivity of the sea. Pp. 32–45 in J.H. Steele (ed.), Marine Food Chains. Univ. California.

Fischer, A.G. 1961. Latitudinal variation in organic diversity. Amer. Sci. 49:50–74.

Fisher, R.A., A.S. Corbet, and C.B. Williams. 1943. The relation between the number of species and the number of individuals in a random sample of an animal population. J. Anim. Ecol. 12:42–58.

Fisher, T.R., P.R. Carlson, and R.T. Barker. 1982. Sediment nutrient regeneration in three North Carolina estuaries. Estuar. Coast. Shelf Sci. 14:101–116.

Fisher, T.R., E.R. Peele, J.W. Ammerman, and L.W. Harding. 1992. Nutrient limitation of phytoplankton in Chesapeake Bay. Mar. Ecol. Progr. Ser. 82:51–63.

Fitt, W.K., and R.L. Pardy. 1981. Effects of starvation, and light and dark on the energy metabolism of symbiotic and aposymbiotic sea anemones. *Anthopleura elegantissima*. Mar. Biol. 61:199–205.

Fitzwater, S.E., G.A. Knauer, and J.H. Martin. 1982. Metal contamination and its effect on primary production measurements. Limnol. Oceanogr. 27:544–551.

Fleeger, J.W., and M.A. Palmer. 1982. Secondary production of the estuarine, meiobenthic copepod *Microarthridion littorale*. Mar. Ecol. Progr. Ser. 7:157–162.

Focht, D.D., and W. Verstraete. 1977. Biochemical ecology of nitrification and denitrification. Adv. Microb. Ecol. 1:135–214.

Foerster, R.E., and W.E. Ricker. 1941. The effect of reduction of predaceous fish on survival of young sockeye salmon at Cultus Lake. J. Fish. Res. Bd. Canada 5:315–336.

Fogg, G.E., and Than-tun. 1960. Interrelations of photosynthesis and assimilation of elementary nitrogen in a blue-green alga. Proc. Roy. Soc. London B153:111–127.

Fogg, G.E., C. Nalewajko, and W.D. Watt. 1965. Extracellular products of phytoplankton photosynthesis. Proc. Roy. Soc. London B162:517–534.

Fonds, M. 1979. Laboratory observations on the influence of temperature and salinity on development of the eggs and growth of the larvae of *Solea solea*. Mar. Ecol. Prog. Ser. 1:91–99.

Fong, W., and K.H. Mann. 1980. Role of gut flora in the transfer of amino acids through a marine food chain. Canad. J. Fish. Aquat. Sci. 37:88–96.

Foreman, K.H. 1985. Regulation of benthic algal and meiofaunal productivity and standing stock in a salt marsh ecosystem: The relative importance of resources and predation. Pp. 224. Ph.D. Thesis, Boston Univ.

Foreman, R.E. 1977. Benthic community modification and recovery following extensive grazing by *Strongylocentrotus droebachiensis*. Helgol. Wiss. Meeres. 30:468–484.

Forrester, G.E. 1990. Factors influencing the juvenile demography of a coral reef fish. Ecology 71:1666–1681

Forsbergh, E.D. 1963. Some relationships of meteorological, hydrographic and biological variables in the Gulf of Panama. Inter-Amer. Trop. Tuna Comm. Bull. 7:1–109.

Forster, J.R.M., and P.A. Gabbott. 1971. The assimilation of nutrients from compounded diets by the prawns *Palaemon serratus and Pandalus platyceros*. J. Mar. Biol. Assoc. U.K. 51:943–961.

Fossing, H., and B.B. Jørgensen. 1990. Oxidation and reduction of radiolabeled inorganic sulfur compounds in an estuarine sediment, Kysing Fjord, Denmark. Geochim. Cosmochim. Acta 54:2731–2742.

Fournier, R.O. 1971. The transport of organic carbon to organisms living in the deep oceans. Proc. Roy. Soc. Edinburgh 73:203–211.

Fowler, A.J., P.J. Doherty, D. McB. Williams. 1992. Multi-scale analysis of recruitment of a coral reef fish on the Great Barrier Reef. Mar. Ecol. Progr. Ser. 82:131–141.

Fox, L.R., and W.W. Murdoch. 1978. Effects of feeding history on short-term and long-term functional responses in *Notonecta hoffmanni*. J. Anim. Ecol. 47:945–959.

Frank, P.W. 1965. Shell growth in a natural population of the snail *Tegula funebralis*. Growth 29:395–403.

Frankenberg, D., and K.L. Smith, Jr. 1967. Coprophagy in marine animals. Limnol. Oceanogr. 12:443–450.

Free, C.A., J.R. Beddington, and J.H. Lawton. 1977. On the inadequacy of simple models of mutual interference for parasitism and predation. J. Anim. Ecol. 46:543–554.

Frey, B.E., and L.F. Small. 1980. Effects of micronutrients and major nutrients on natural phytoplankton populations. J. Plankt. Res. 2:1–22.

Frey, R.W., and P.B. Basan. 1978. Coastal salt marshes. Pp. 101–170 in R.A. Davis, Jr. (ed.), Coastal Sedimentary Environments. Springer-Verlag.

Friederich, G.E., L.A. Codispoti, and C.M. Sakamoto. 1991. An easy-to-construct automated Winkler titration system. Pp. 14. Tech. Rep. No. 91–96. Monterey Bay Aquarium Res. Inst.

Friedman, M.M., and J.R. Strickler. 1975. Chemoreceptors and feeding in calanoid copepods (Arthropoda:Crustacea). Proc. Natl. Acad. Sci. 72:4185–4188.

Frier, J.O. 1979. Character displacement in *Sphaeroma* spp. (Isopoda: Crustacea). I. Field evidence. Mar. Ecol. Prog. Ser. 1:159–163.

Frier, J.O. 1979a. Character displacement in *Sphaeroma* spp. (Isopoda: Crustacea). II. Competition for space. Mar. Ecol. Prog. Ser. 1:165–168.

Froelich, P.N., G.P. Klinkhammer, M.L. Bender, N.A. Luedke, G.R. Heath, O. Hammond, B. Hartman, and V. Maynard. 1979. Early oxidation of organic matter in pelagic sediments of the eastern equatorial Atlantic: Suboxic diagenesis. Geochim. Cosmochim. Acta 43:1075–1090.

Frost, B.W. 1972. Effects of size and concentration of food particles on the feeding behavior of the marine planktonic copepod *Calanus pacificus*. Limnol. Oceanogr. 17:805–815.

Frost, B.W. 1974. Feeding processes at lower trophic levels in pelagic communities. Pp. 59–77 in C.B. Miller (ed.), The Biology of the Oceanic Pacific. Oregon State.

Frost, B.W. 1975. A threshold feeding behavior in *Calanus pacificus*. Limnol. Oceanog. 20:263–266.

Frost, B.W. 1977. Feeding behavior of *Calanus pacificus* in mixtures of food particles. Limnol. Oceanog. 22:472–491.

Frost, B.W. 1987. Grazing control of phytoplankton stock in the open subarctic Pacific Ocean: A model assessing the role of mesozooplankton, particularly the large calanoid copepods *Neocalanus* spp. Mar. Ecol. Progr. Ser. 39:49–68.

Fry, B. 1988. Food web structure on Georges Bank from stable C, N, and S isotopic compositions. Limnol. Oceanogr. 33:1182–1190.

Fry, B., E.B. Sherr. 1984. δ^{13}C measurements as indicators of carbon flow in marine and freshwater ecosystems. Contrib. Mar. Sci. 27:13–47.

Fuhrman, J.A., and D.G. Capone. 1991. Possible biogeochemical consequences of ocean fertilization. Limnol. Oceanogr. 36:1951–1959.

Fuhrman, J.A., and C.A. Suttle. 1993. Viruses in marine planktonic systems. Oceanography 6:51–63.

Fuhs, G.W., S.D. Demmerle, E. Canelli, and M. Chen. 1972. Characterization of phosphorus-limited plankton algae. Pp. 113–123 in G.E. Likens (ed.), Nutrients and Eutrophication. Spec. Symp. Vol. 1. Amer. Assoc. Limnol. Oceanogr. Allen.

Fuji, A. 1967. Ecological studies on the growth and food consumption of Japanese common littoral sea urchin, *Strongylocentrotus intermedius* (A. Agassiz). Mem. Fac. Fish. Hokkaido Univ. 15:83–160.

Fuji, A., and M. Hashizume. 1974. Energy budget for a Japanese common scallop, *Patinopecten yessoensis* (Jay), in Mutsu Bay. Bull. Fac. Fish. Hokkaido Univ. 25:7–19.

Fulton, J. 1973. Some aspects of the life history of *Calanus plumchrus* in the Strait of Georgia. J. Fish. Res. Bd. Canada 30:811–815.

Fulton, J. 1978. Seasonal and annual variations of net zooplankton at Ocean Station P, 1965–1976. Fish. Mar. Serv. Canada, Data Rep. No. 49.

Fulton, R.S. 1984. Predation, production and the organization of an estuarine copepod community. J. Plank. Ecol. 6:399–415.

Furnas, M., 1990. *In situ* growth rates of marine phytoplankton: approaches to measurement, community and species growth rates. J. Plankton Res. 12: 1117–1151.

Gagné, J.A., K.H. Mann, and A.R.O. Chapman. 1982. Seasonal patterns of growth and storage in *Laminaria longicruris* in relation to differing patterns of availability of nitrogen in the water. Mar. Biol. 69:91–101.

Gagosian, R.B., and D.H. Stuermer. 1977. The cycling of biogenic compounds and their diagenetically transformed products in seawater. Mar. Chem. 5:605–632.

Gaines, S.D., and J. Roughgarden. 1985. Larval settlement rate: a leading determinant of structure in ecological communities of the rocky intertidal zone. Proc. Natl. Acad. Sci. U.S.A. 82:3707–3711.

Gaines, S.D., and J. Roughgarden. 1987. Fish in offshore kelp forests affect recruitment to intertidal barnacle populations. Science 235:479–481.

Gallager, S. 1988. Visual observations of particle manipulation during feeding in larvae of bivalve molluscs. Bull. Mar. Sci. 43:344–365.

Gallagher, E.D., P.A. Jumars, and D.D. Trueblood. 1983. Facilitation of soft-bottom benthic succession by tube builders. Ecology 64:1200–1216.

Gallopin, G.C. 1971. A generalized model of a resource population system. I, II. Oecologia 7:382–413; 414–432.

Galloway, J.N., H. Levy II, and P.S., Kasibhatia. 1994. Year 2020: consequences of population growth and development on deposition of oxized nitrogen. Ambio. 23:120–123.

Gambell, R. 1968. Seasonal cycles and reproduction in sei whales of the southern hemisphere. Disc. Repts. 35:31–134.

Gardner, M.R., and W.R. Ashby. 1970. Connectance of large dynamic (cybernetic) systems: Critical values for stability. Nature 228:784.

Gardner, W.D. 1977. Fluxes, dynamics, and chemistry of particulates in the ocean. Ph.D. Thesis, Massachussetts Inst. of Technology/Woods Hole Ocean-ographic Institution.

Gardner, W.D. 1980. Sediment trap dynamics and calibration: A laboratory evaluation. J. Mar. Res. 38:17–39.

Gardner, W.D. 1980a. Field assessment of sediment traps. J. Mar. Res. 38:41–52.

Gardner, W.S., and D.W. Menzel. 1974. Phenolic aldehydes as indicators of terrestrially derived organic matter in the sea. Geochim. Cosmochim. Acta 38:813–822.

Gargas, E. 1970. Measurement of primary production, dark fixation, and vertical distribution of the microbenthic algae in the Oresund. Ophelia 8: 231–253.

Garrod, D.J., and A.D. Clayden. 1972. Current biological problems in the conservation of deep-sea fishery resources. Symp. Zool. Soc. Lond. 29: 161–184.

Garside, C., and T.C. Malone. 1978. Monthy oxygen and carbon budgets of the New York Bight apex. Estuar. Coast. Mar. Sci. 6:93–104.

Gaskin, D.E. 1978. Form and function in the digestive tract and associated organs in cetacea, with a consideration of metabolic rates and specific energy budgets. Oceanogr. Mar. Biol. Ann. Rev. 16:313–415.

Gasol, J.M., and D. Vaqué. 1993. Lack of coupling between heterotrophic nanoflagellates and bacteria: A general phenomenon across aquatic systems? Limnol. Oceanogr. 38:657–664.

Gaudy, R. 1974. Feeding four species of pelagic copepods under experimental conditions. Mar. Biol. 25:125–141.

Geertz-Hansen. O., K. Sand-Jensen, D.F. Hansen, and A. Christiansen. 1993. Growth and grazing control of abundance of the marine macroalga, *Ulva lactuca* L. in an eutrophic Danish estuary. Aquat. Bot. 46:101–109.

Geiselman, J.A., and O.S. McConnell. 1981. Polyphenols in brown algae *Fucus vesiculosus* and *Ascophyllum nodosum*: Chemical defenses against the marine herbivorous snail, *Littorina littorea*. J. Chem. Ecol. 7:1115–1133.

Geraci, J.R., et al. 1989. Humpback whales (*Megaptera novaeangliae*) fatally poisoned by dinoflagellate toxin. Canad. J. Fish. Aquat. Sci. 46:1895–1898.

Gerard, V.A. 1982. In situ water motion and nutrient uptake by the giant kelp macrocystes pyrifer. Mar. Biol. 69:51–54.

Gerard, V.A. 1986. Photooynthetic characteristics of giant kelp [Macrocystrs Ryufea] determined in-situ. Mar Biol. 90:473–482.

Gerard, V.A., and K.H. Mann. 1979. Growth and production of *Laminaria longicruris* (Phaeophyta) populations exposed to different intensities of water movement. J. Phycol. 15:33–41.

Gerlach, S.A. 1978. Food-chain relationships in substidal silty sand marine sediments and the role of meiofauna in stimulating bacterial productivity. Oecologia 33:55–69.

GESAMP. 1990. The State of the Marine Environment. Joint Group of Experts on the Scientific Aspects of Marine Pollution. Rep. and Stud. 39. United Nations Environmental Programme.

GESAMP. 1992. Impact of oil and related chemicals and wastes on the marine environment. Joint group of experts on the scientific aspects of marine pollution. Pp. 180. Rep. and Stud. 50. United Nations Environmental Programme.

Giblin, A. 1982. Uptake and Remobilization of Heavy Metals in Salt Marshes. Ph.D. Diss. Boston University.

Giere, O. 1981. The gutless marine oligochaete *Phallodrilus leukodermatus*. Structural studies on an aberrant tubificid associated with a bacteria. Mar. Ecol. Progr. Ser. 5:353–357.

Gieskes, W.W.C., and G.W. Kraay. 1975. The phytoplankton spring bloom in Dutch coastal waters of the North Sea. Neth. J. Sea Res. 9:166–196.

Giessen, W.B.J.T., M.M. van Katwijk, C. den Hartog. 1990. Eelgrass condition and turbidity in the Dutch Wadden Sea. Aquat. Bot. 37:71–85.

Gilpin, M.E. 1975. Stability of feasible predator-prey systems. Nature 254:137–139.

Glander, K.E. 1981. Feeding patterns in mantled howling monkeys. Pp. 231–259 in A.C. Kamil and T.D. Sargent (eds.), Foraging Behavior. Garland STPM.

Glantz, M.H. 1984. Floods, fires, and famine: Is El Niño to blame? Oceanus 27:14–19.

Glenn, A.R. 1976. Production of extracellular protein by bacteria. Ann. Rev. Microbiol. 30:41–62.

Glibert, P.M. 1982. Regional studies of daily, seasonal, and size fraction variability in ammonium remineralization. Mar. Biol. 70:209–222.

Glibert, P.M., and J.C. Goldman. 1981. Rapid ammonium uptake by marine phytoplankton. Mar. Biol. Lett. 2:25–31.

Glibert, P.M., F. Lipschultz, J.J. McCarthy, and M.A. Altabet. 1982. Isotope dilution models of uptake and remineralization of ammonium by marine plankton. Limnol. Oceanogr. 27:639–650.

Glombitza, M. 1977. Highly hydroxilated phenols of the phaeophysceae. Pp. 191–204 in D.J. Faulkner and W.M. Fenicol (eds.), Marine Natural Products Chemistry. Plenum.

Glover, H.E. 1980. Assimilation numbers in cultures of marine phytoplankton. J. Plankton Res. 2:69–79.

Glynn, P.W. 1974. The impact of *Acanthaster* on corals and coral reefs in the Eastern Pacific. Environ. Conser. 1:295–304.

Glynn, P.W. 1976. Some physical and biological determinants of coral community structure in the eastern Pacific. Ecol. Monogr. 46:431–456.

Glynn, P.W. 1988. El Niño-Southern Ocean oscillation 1982–1983: Nearshore population, community and ecosystem responses. Ann. Rev. Ecol. Syst. 19:309–345.

Glynn, P.W. 1991. Coral reef bleaching in the 1980s and possible connections with global warming. Trends Ecol. Evol. 6:175–179.

Godshalk, G.L., and R.G. Wetzel. 1987. Decomposition of aquatic angiosperms. I. Dissolved components. Aquat. Bot. 5:281–300.

Godshalk, G.L., and R.G. Wetzel. 1978a. Decomposition of aquatic angiosperms. III. *Zostera marina* L. and a conceptual model of decomposition. Aquat. Bot. 5:329–354.

Goering, J.J., R.C. Dugdale, and D.W. Menzel. 1966. Estimates of *in situ* rates of nitrogen uptake by *Trichodesmium* sp. in the tropical Atlantic Ocean. Limnol. Oceanog. 11:614–620.

Goldman, C.R. 1972. The role of minor nutrients in limiting the productivity of aquatic ecosystems. Pp. 21–33 in G.E. Likens (ed.), Nutrients and Eutrophication: The Limiting Nutrient Controversy. Spec. Symp. Vol. 1 Amer. Soc. Limnol. Oceanogr. Allen.

Goldman, J.C., and E.L. Carpenter. 1974. A kinetic approach to the effect of temperature on algal growth. Limnol. Oceanog. 19:756–766.

Goldman, J.C., J.J. McCarthy, and D.G. Peavey. 1979. Growth rate influence on the chemical composition of phytoplankton in oceanic waters. nature 279:210–215.

Goldman, J.C., D.A. Caron, and M.R. Dennett. 1987. Regulation of gross growth efficiency and ammonium regeneration in bacteria by substrate C:N ratio. Limnol. Oceanogr. 32:1239–1252.

Goldwasser, L., and J. Roughgarden. 1993. Construction and analysis of a large Caribbean food web. Ecology 74:1216–1233.

Goldsborough, W.J., and W.M. Kemp. 1988. Light responses of a submersed macrophyte: Implications for survival in turbid tidal waters. Ecology 69: 1775–1786.

Goldsmith, F.B., and C.M. Harrison. 1976. Description and analysis of vegetation. In S.B. Chapman (ed.), Methods in Plant Ecology. Blackwell.

Golley, F., H.T. Odum, and R.F. Wilson. 1962. The structure and metabolism of a Puerto Rico red mangrove forest in May. Ecology 43:9–19.

Golterman, H.L. 1973. Vertical movement of phosphate in freshwater. Pp. 509–538 in E.L. Griffith, A. Beeton, J.M. Spencer, and D.T. Mitchell (eds.), Environmental Phosphorus Handbook. Wiley.

Gonzalez, J.M., and C.A. Suttle. 1993. Grazing by marine nanoflagellates on viruses and virus-sized particles: Ingestion and digestion. Mar. Ecol. Progr. Ser. 1–10.

Good, R.E., N.F. Good, and B.R. Frasco. 1982. A review of primary production and decomposition dynamics of the belowground marsh component. Pp. 139–158 in V.S. Kennedy (ed.), Estuarine Comparisons. Academic.

Goodman, D. 1982. Optimal life histories, optimal notation, and the value of reproductive value. Amer. Nat. 119:803–823.

Goodyear, C.P. 1985. Toxic materials, fishing, and environmental variation: Simulated effects of on striped bass population trends. Trans. Amer. Fish. Soc. 114:107–115.

Gordon, D.C., Jr. 1971. Distribution of particulate organic carbon and nitrogen at an oceanic station in the central Pacific. Deep-Sea Res. 18:1127–1134.

Gordon, D.C., Jr. 1977. Variability of particulate organic carbon and nitrogen along the Halifax-Bermuda section. Deep-Sea Res. 24:257–270.

Gordon, D.M., P.B. Birch, and A.J. McComb. 1980. The effect of light, temperature, and salinity on photosynthetic rates of an estuarine *Cladophora*. Bot. Mar. 23:749–755.

Goreau, T.F. 1959. The ecology of Jamaican coral reefs. I. Species composition and zonation. Ecology 40:67–90.

Goss-Custard, J.D. 1969. The winter feeding ecology of the redshank *Tringa totanus*. Ibis 111:338–356.

Goss-Custard, J.D. 1970. Factors affecting the diet and feeding rate of the redshank *Tringa totanus*. Pp. 101–110 in A. Watson (ed.), Animal Populations in Relation to Their Food Resources. Blackwell.

Goss-Custard, J.D. 1981. Feeding behavior of redshank, *Tringa totanus*, and optimal foraging theory. Pp. 115–134 in A.C. kamil and T.D. Sargent (eds.), Foraging Behavior. Garland STPM.

Goss-Custard, J.D., A.D. West, and S.E.A. Le V. Dit Durrell. 1993. The availability and quality of the mussel prey (*Mytilus edulis*) of oystercatchers (*haematopus ostralegus*). Neth. J. Sea Res. 31:419–439.

Gotceitas, V., and J.A. Brown. 1993. Risk of predation to fish larvae in the presence of alternative prey: Effect of prey size and number. Mar. Ecol. Progr. Ser. 98:215–222.

Gotschalk, C.C., and A.L. Alldredge. 1989. Enhanced primary production and nutrient regeneration within aggregated marine diatoms. Mar. Biol. 103: 119–129.

Gotto, J.W., and B.F. Taylor. 1976. N_2 fixation associated with decaying leaves of the red mangrove (*Rhizophora mangle*). Appl. Env. Microbiol. 31:781–783.

Govindjee. 1976. Photosynthesis. Wiley.

Govindjee, and P.R. Mohanty. 1972. Photochemical aspects of photosynthesis in blue-green algae. Pp. 171–196 in T.V. Desikachary (ed.), Taxonomy and Biology of Blue-green Algae. Center for Advanced Study in Botany. Madras.

Graham, N.E. 1995. Simulatoin of recent global temperature trends. Science 267:666–671.

Grande, K.D., et al. 1991. Primary production in the North Pacific gyre: A comparison of rates determined by the ^{14}C, O_2 concentration and ^{18}O methods. Deep-Sea Res. 36:1621–1634.

Granelli, E., K. Wallstrom, U. Larsson, W. Granelli, and R. Elmgren. 1990. Nutrient limitation of primary production in the Baltic Sea area. Ambio 19:142–151.

Grant, P.R. 1972. convergent and divergent character displacement. Biol. J. Linn. Soc. 4:39–68.

Grant, W.D., L.F. Boyer, and L.P. Sanford. 1982. The effect of biological processes on the initiation of sediment motion in non-cohesive sediment. J. Mar. Res. 40:659–677.

Grassle, J.F. 1977. Slow recolonization of deep sea sediment. Nature 265:618–619.

Grassle, J.F., and J.P. Grassle. 1974. Opportunistic life histories and genetic systems in marine benthic polychaetes. J. Mar. Res. 32:253–284.

Grassle, J.F., and J.P. Grassle. 1976. Sibling species in the marine pollution indicator *Capitella* (Polychaeta). Science 192:567–569.

Grassle, J.F., and L. Morse-Porteus. 1987. Macrofaunal colonization of disturbed deep-sea environments and the structure of deep-sea benthic communities. Deep-Sea Res. 34:1911–1950.

Grassle, J.P., and C.A. Butman. 1989. Active habitat selection by larvae of the polychaetes, *Capitella* spp I and II, in a laboratory flume. Pp. 107–114, in J.S. Ryland and P.A. Taylor (eds.), Reproduction, Genetics and Distribution of Marine Organisms. 23rd European Marine Biology Symposium. Olsen & Olsen.

Gray, J.S. 1966. The attractive factor of intertidal sand to *protodrilus symbioticus*. J. Mar. Biol. Assoc. U.K. 46:627–645.

Gray, J.S. 1967. Substrate selection by the archiannelid *Protodrilus rubropharyngeus*. Helgol. Wiss. Meeres. 15:252–269.

Green, G. 1977. Ecology of toxicity in marine sponges. Mar. Biol. 40:207–215.

Green, R.H. 1971. A multivariate approach to the Hutchinsonian niche: Bivalve molluscs of central Canada. Ecology 52:543–556.

Greene, R.M., R.J. Geider, and P.G. Falkowski. 1991. Effect of iron limitation on photosynthesis of a marine diatom. Limnol. Oceanogr. 36:1772–1782.

Greze, V.N. 1978. Production in animal populations. Pp. 89–114 in O. Kinne (ed.), Marine Ecology. Vol. 4. Wiley.

Grice, G.D., R.P. Harris, M.R. Reeve, J.F. Heinbokel, and C.O. Davis. 1980. Large scale enclosed water-column ecosystems. An overview of Foodweb I, the final CEPEX experiment. J. Mar. Biol. Assoc. U.K. 60:401–414.

Griffith, D. 1980. Foraging costs and relative prey size. Amer. Nat. 116:743–752.

Griffith, E.J. 1973. Environmental phosphorus—An editorial. Pp. 683–698 in E.J. Griffith, A. Beeton, J.M. Spencer, and D.T. Mitchell (eds.), Environmental Phosphorus Handbook. Wiley.

Griffiths, K.J., and C.S. Holling. 1969. A competition submodel for parasites and predators. Canad. Entom. 101:785–818.

Griffiths, R.J. 1980. Filtration, respiration and assimilation in the black mussel *Choromytilus meridionalis*. Mar. Ecol. Prog. Ser. 3:63–70.

Griffiths, R.J. 1980a. Natural food availability and assimilation in the bivalve *Choromytilus meridionalis*. Mar. Ecol. Prog. Ser. 3:151–156.

Grime, J.P. 1977. Evidence for the existence of three primary strategies in plants and its relevance to ecological and evolutionary theory. Amer. Nat. 111:1169–1194.

Grøntved, J. 1960. On the productivity of microbenthos and phytoplankton in some Danish fjords. Meddr. Danm. Fish.-og Havunders. N.S. 3:55–92.

Gross, M.G. 1977. Oceanography: A View of the Earth. Int. Ed. Prentice-Hall.

Grundmanis, V., and J.W. Murray. 1977. Nitrification and denitrification in marine sediments from Puget Sound. Limnol. Oceanogr. 22:804–813.

Gulbrandsen, R.A., and C.E. Roberson. 1973. Inorganic phosphorus in seawater. Pp. 117–140 in E.J. Griffith, A. Beeton, J.M. Spencer, and D.T. Mitchell (eds.), Environmental Phosphrous Handbook. Wiley.

Gulland, J.A. 1970. Food chain studies and some problems in world fisheries. Pp. 296–315 in J.H. Steele (ed.), Marine Food Chains. Univ. California.

Gulland, J.A. 1971. Ecological aspects of fishery research. Adv. Ecol. Res. 7:115–176.

Haedrich, R.L., G.T. Rowe, and P.T. Polloni. 1980. The megabenthic fauna in the deep sea south of New England. Mar. Biol. 57:165–179.

Hagström, A., U. Larsson, P. Hörstedt, and S. Normark. 1979. Frequency of dividing cells, a new approach to the determination of bacterial growth rates in aquatic environments. Appl. Envir. Microbiol. 37:805–812.

Haines, B.L., and E.L. Dunn. 1976. Growth and resource allocation responses of *Spartina alterniflora* Loisel. to three levels of NH_4-N, Fe, and NaCl in solution culture. Bot. Gazette 137:224–230.

Haines, E., A. Chalmers, R. Hanson, and B. Sherr. 1977. Nitrogen pools and fluxes in a Georgia salt marsh. Pp. 241–254 in M. Wiley (ed.), Estuarine Processes. Vol. II. Academic.

Hairston, N.G. 1959. Species abundance and community organization. Ecology 40:404–416.

Hairston, N.G., F.E. Smith, and L.B. Slobodkin. 1960. Community structure, population control, and competition. Amer. Nat. 94:421–425.

Hairston, N.G., J.D. Allan, R.K. Colwell, D.J. Futuyma, J. Howell, M.D. Lubin, J. Mathias, and J.H. Vandermeer. 1968. The relationships between species diversity and stability: An experimental approach with protozoa and bacteria. Ecology 49:1091–1101.

Halim, Y. 1991. The impact of human alterations of the hydrological cycle on ocean margins. Pp. 301–327 in R.F.C. Mantoura, J.-M. Martin, and R. Wollast (eds.), Ocean Margin Processes in Global Change. Wiley.

Hall, C.A.S. 1988. An assessment of several of the historically most influential theoretical models used in ecology and of the data used in their support. Ecol. Modell. 43:5–31.

Hall, C.A.S., and R. Moll. 1975. Methods of assessing aquatic primary productivity. Pp. 19–53 in J. Lieth and R.H. Whittaker (eds.), Primary Productivity in the Biosphere. Springer-Verlag.

Hall, D.J. 1964. An experimental approach to the dynamics of a natural population of *Daphnia galeata mendotae*. Ecol. 45:94–112.

Hall, D.J., W.E. Cooper, and E.E. Werner. 1970. an experimental approach to the production dynamics and structure of freshwater animal communities. Limnol. Oceanogr. 15:839–928.

Hall, D.J., S.T. Threlkeld, C.W. Burns, and P. Crowley. 1976. The size-efficiency hypothesis and the size structure of zooplankton communities. Ann. Rev. Ecol. Syst. 7:177–208.

Hallegraeff, G.M. 1993. A review of harmful algal blooms and their apparent global increase. Physologia 32:79–99.

Hamilton, W.A. 1979. Microbial energetics and metabolism. Pp. 22–44 in J.M. Lynch and N.J. Poole (eds.), Microbial Ecology. Wiley.

Hamner, P., and W.M. Hamner. 1977. Chemosensory tracking of scent trails by the planktonic shrimp *Acetes sibogae australis*. Science 195:886–888.

Hamner, W.M., L.P. Madin, A.L. Alldredge, R.W. Gilmer, and P.P. Hamner 1975. Underwater observations of gelatinous zooplankton: Sampling problems, feeding biology, and behavior. Limnol. Oceanog. 20:907–917.

Hamre, J. 1978. The effect of recent changes in the North Sea mackerel fishery on stock and yield. Rapp. Proc.-Verb. Reun. Cons. Int. Explor. Mer 172:197–210.

Handa, N. 1977. Land sources of marine organic matter. Mar. Chem. 5:341–359.

Hanisak, M.D. 1979. Growth patterns of *Codium fragile* ssp. *tomentosoides* in response to temperature, irradiance, salinity, and nitrogen source. Mar. Biol. 50:319–332.

Hanlon, R.D.G. 1981. Influence of grazing by Collembola on the activity of senecent fungal colonies grown on media of different nutrient concentration. Oikos 36:362–367.

Hanlon, R.D.G., and J.M. Anderson. 1980. The influence of macroarthropod feeding activity on fungi and bacteria in decomposing oak leaves. Soil Biol. Biochem. 12:255–261.

Hanson, R.B., K.R. Tenore, S. Bishop, C. Chamberlain, M.M. Pamatmat, and J. Tietjen. 1981. Benthic enrichment in the Georgia Bight related to Gulf Stream intrusions and estuarine outwelling. J. Mar. Res. 39:417–441.

Haq, S.M. 1967. Nutritional physiology of *Metridia lucens* and *M. longa* from the Gulf of Maine. Limnol. Oceanogr. 12:40–51.

Harbison, G.R., and R.W. Gilmer. 1976. The feeding rates of the pelagic tunicate *Pegea* and two other salps. Limnol. Oceanogr. 21:517–528.

Harbison, G.R., and V.L. McAlister. 1980. Fact and artifact in copepod feeding experiments. Limnol. Oceanogr. 25:971–981.

Harbison, G.R., L.P. Madin, and N.R. Swanberg. 1978. On the natural history and distribution of oceanic ctenophores. Deep-Sea Res. 25:233–256.

Harding, D., J.H. Nichols, and D. Tungate. 1978. The spawning of plaice (*Pleuronectes platessa* L.) in the southern North Sea and English Channel. Rapp. Proces-Verb. Cons. Intern. Explor. Mer 172:102–113.

Hardy, A.C., and E.R. Gunther. 1935. The plankton of the South Georgia whaling grounds and adjacent waters, 1926–27. Discovery Rep. 2:1–146.

Hargrave, B.T. 1969. Epibenthic algal production and community respiration in the sediments of Marion Lake. J. Fish. Res. Bd. Canada 26:2003–2026.

Hargrave, B.T. 1970. The utilization of benthic microflora by *Hyallela azteca* (Amphipoda). J. Anim. Ecol. 39:427–438.

Hargrave, B.T. 1984. Sinking of particulate matter from the surface water of the ocean. In J.E. Hobbie and P.J. Leb. Williams (eds.), Heterotrophic Activity in the Sea. Plenum.

Hargrave, B.T., and N.M. Burns. 1979. Assessment of sediment trap collection efficiency. Limnol. Oceanogr. 24:1124–1136.

Harris, G.P. 1980. The measurement of photosynthesis in natural populations of phytoplankton. Pp. 129–187 in I. Morris (ed.), The Physiological Ecology of Phytoplankton. Univ. of California.

Harrison, D.E., and M.A. Cane. 1984. Changes in the Pacific during the 1982–83 event. Oceanus 27:21–28.

Harrison, G.W. 1979. Stability under environmental stress: Resistance, resilience, and variability. Amer. Nat. 113:659–669.

Harrison, P.G. 1977. Decomposition of macrophyte detritus in seawater: Effects of grazing by amphipods. Oikos 28:165–169.

Harrison, P.G. 1982. Control of microbial growth and of amphipod grazing by water-soluble compounds from leaves of *Zostera marina*. Mar. Biol. 67:225–230.

Harrison, P.G., and A.T. Chan. 1980. Inhibition of growth of microalgae and bacteria by extracts of eelgrass (*Zostera marina*) leaves. Mar. Biol. 61:21–26.

Harrison, P.G., and K.H. Mann. 1975. Chemical changes during the seasonal cycle of growth and decay in eelgrass (*Zostera marina*) on the Atlantic Coast of Canada. J. Fish. Res. Bd. Canada 32:615–621.

Harrison, P.J., M.H. Hu, Y.P. Yang, and X. Lu. 1990. Phosphate limitation in estuarine and costal waters of China. J. Exp. Mar. Biol. Ecol. 140:79–87.

Harrison, W.G. 1978. Experimental measurements of nitrogen remineralization in coastal waters. Limnol. Oceanog. 23:684–694.

Harrison, W.G. 1992. Regeneration of nutrients. Pp. 385–407 in P.G. Falkowski and A.D. Woodhead (eds.), Primary Productivity and Biogeochemical Cycles in the Sea. Plenum Press.

Harrison, W.G., and G.F. Cota. 1991. Primary production in polar waters: Relation to nutrient availability. Polar Res. 10:87–104.

Harrison, W.G., T. Platt, and M.R. Lewis. 1987. F-ratio and its relationship to ambient nitrate concentration in coastal waters. J. Plankton Res. 9:235–248.

Harrold, C. 1982. Escape responses and prey availability in a kelp forest predatorprey system. Amer. Nat. 119:132–135.

Hart, T.J. 1942. Phytoplankton periodicity in Antarctic surface waters. Discovery Rep. 21:261–356.

Hartley, R.D., and E.C. Jones. 1977. Phenolic components and degradability of cell walls of grass and legume species. Phytochemistry 16:1531–1534.

Hartman, J., H. Caswell, and I. Valiela. 1983. Effects of wrack accumulation on salt marsh vegetation. Oceanol. Acta 1983:99–102.

Hartog, C. den. 1970. The Sea Grasses of the World. North-Holland.

Hartog, C. den. 1977. Structure, function, and classification in seagrass communities. Pp. 89–122 in C.P. McRoy and C. Helfferich (eds.), Seagrass Ecosystems. M. Dekker.

Harvey, G.R., W.G. Steinhauer, and J.M. Teal. 1973. Polychlorobiphenyls in North Atlantic ocean water. Science 180:643–644.

Harvey, H.W. 1945. The Chemistry and Biology of Seawater. Cambridge Univ.

Harvey, H.W. 1950. On the production of living matter in the sea of Plymouth. J. Mar. Biol. Assoc. U.K. 29:97–138.

Harvey, H.W. 1955. The Chemistry and Fertility of Sea Waters. Cambridge Univ.

Harvey, H.W., L.N. Cooper, M.V. Leborn, and F.S. Russell. 1935. Plankton production and its control. J. Mar. Biol. Assoc. U.K. 20:407–441.

Hassell. M.P., and R.M. May. 1974. Aggregation of predators and insect parasites and its effect on stability. J. Anim Ecol. 43:567–594.

Hassell, M.P., and G.C. Varley. 1969. New inductive model for insect parasites and its bearing on biological control. Nature 223:1133–1137.

Hassell, M.P., J.H. Lawton, and J.R. Beddington. 1976. The components of arthropod predation. I. The prey death rate. J. Anim. Ecol. 45:135–164.

Hassell, M.P., J.H. Lawton, and J.R. Beddington. 1977. Sigmoid functional responses by invertebrate predators and parasitoids. J. Anim. Ecol. 46:249–262.

Hatch, M.D., and C.R. Slack. 1970. Photosynthetic CO_2 fixation pathways. Ann. Rev. Plant Physiol. 21:141–162.

Hatcher, B.G., A.R.O. Chapman, and K.H. Mann. 1977. An annual carbon budget for the kelp *Laminaria longicruris*. Mar. Biol. 44:85:–96.

Hauck, R.D., and J.M. Bremner. 1976. Use of tracers for soil and fertilizer research. Adv. Agron. 28:219–266.

Haury, L.R., P.H. Wiebe, and S.H. Boyd. 1976. Longhurst-Hardy plankton recorders: Their design and use to minimize bias. Deep-Sea Res. 23:1217–1229.

Haury, L.R., M.G. Briscoe, and M.H. Orr. 1979. Tidally generated internal wave packets in Massachessetts Bay. Nature 278:312–317.

Haxo, F.T., and L.R. Blinks. 1950. Photosynthetic action spectra of marine algae. J. Gen. Physiol. 33:389–422.

Hay, M.E. 1984. Patterns of fish and urchin grazing on caribbean coral reefs: Are previous results typical? Ecology 65:446–454.

Hay, M.E., and W. Fenical. 1988. Marine plant-herbivore interactions: The ecology of chemical defense. Ann. Rev. Ecol. Syst. 19:111–145.

Hay, M.E., and P.R. Taylor. 1985. Competition between herbivorous fishes and urchins on Caribbean reefs. Oecologia 65:591–598.

Hay, M.E., J.E. Duffy, and W. Fenical. 1990. Host-plant specialization decreases predation on a marine amphipod: An herbivore in plant's clothing. Ecology 71:733–743.

Hay, M.E., J.E. Duffy, V.J. Paul, P.E. Renaud, and W. Fenical. 1990a. Specialist hervivores reduce their susceptibility to predation by feeding on the chemically defended seaweed *Avrainvillea longicaudis*. Limnol. Qceanogr. 35:1734–1743.

Hayes, F.R., and J.R. Phillips. 1958. Lake water and sediment. IV. Radiophosphorus equilibrium with mud, plants, and bacteria under oxidized and reduced conditions. Limnol. Oceanogr. 3:459–475.

Haward, T.L., and J.A. McGowan. 1979. Pattern and structure in an oceanic zooplankton community. Amer. Zool. 1045–1055.

Head, W.D., and E.J. Carpenter. 1975. Nitrogen fixation associated with the marine macroalgae *Codium fragile*. Limnol. Oceanogr. 20:815–823.

Heath, M.R. 1992. Field investigations of the early life stages of marine fish. Adv. Mar. Biol. 28:1–175.

Hedges, J.I., and D.C. Mann. 1979. The lignin geochemistry of marine sediments from the Southern Washington Coast. Geochim. Cosmochim. Acta 43:1809–1818.

Heinle, D.R. 1969. Production of a calanoid copepod, *Acartia tonsa*, in the Patuxent River estuary. Chesap. Sci. 7:59–74.

Heinle, D.R., and D.A. Flemer. 1975. Carbon requirements of a population of the estnarine copepod *Eurytemora affinis*. Mar. Biol. 31:235–247.

Heinrich, A.K. 1962. The life histories of plankton animals and seasonal cycles of plankton communities in the oceans. J. Cons. Int. Explor. Mer 27:15–24.

Heinrich, A.K. 1963. On the filtering ability of copepods in the boreal and tropical region of the Pacific. Trudy Inst. Okeanol. 71:60–71.

Helbling, E.W., V. Villafañe, M. Ferrario, and O. Holm-Hansen. 1992. Impact of natural ultraviolet radiation on rates of photosynthesis and on specific marine phytoplankton species. Mar. Ecol. Progr. Ser. 80:89–100.

Heldal, M., and G. Bratbak. 1991. Production and decay of viruses in aquatic environments. Mar. Ecol. Progr. Ser. 72:205–212.

Helder, W. 1974. The cycle of dissolved inorganic nitrogen compounds in the Dutch Wadden Sea. Neth. J. Sea Res. 8:154–173.

Hellebust, J.A. 1967. Excretion of organic compounds by cultured and natural populations of marine phytoplankton. Pp. 761–766 in G.H. Lauff (ed.), Estuaries. Amer. Assoc. Adv. Sci. Publ. No. 83.

Hellebust, J.A. 1974. Extracellular products. Pp. 838–863 in W.D.P. Stuart (ed.), Algal Physiology and Biochemistry. Bot. Monogr. 10. Univ. Calif.

Hemmingsen, A.M. 1960. Energy metabolism as related to body size and respiratory surfaces, and its evolution. Part II. Rep. Steno Mem. Hosp. Nord. Insulin Lab. 9:1–110.

Hempel, G. 1965. On the importance of larval survival for the population dynamics of marine food fish. Calif. Coop. Oceanic Fish. Inv. Rep. 10:13–23.

Henderson-Sellers, A. 1990. Modelling and monitoring "greenhouse" warming. Trends Ecol. Evol. 5:270–275.

Hendrix, S.D. 1980. An evolutionary and ecological perspective of the insect fauna of ferns. Amer. Nat. 115:171–196.

Henny, C.J., and H.M. Wight. 1969. An endangered osprey population: Estimates of mortality and production. Auk 86:188–198.

Herrera, C.H. 1982. Grasses, grzers, mutualism, and coevolution: A comment. Oikos 38:254–259.

Hessler, R.R., and H.L. Sanders. 1967. Faunal diversity in the deep-sea. Deep-Sea. Res. 14:65–78.

Hewer, H.R. 1964. The determination of age, sexual maturity, longevity and a life-table in the grey seal (*Halichoerus grypus*). Proc. Zool. Soc. Lond. 142:593–623.

Hibbert, C.J. 1977. Energy relations of the bivalve *Mercenaria mercenaria* on an intertidal mudflat. Mar. Biol. 44:77–84.

Hildrew, C.W., and C.R. Townsend. 1977. The influence of substrate on the functional response of *Plectrocnemia conspersa* (Curtis) larvae (Trichoptera: Polycentropodidae). Oecologia 31:21–26.

Hines, M.E., D.A. Bazylinski, J.B. Tugel, and W.B. Lyons. 1991. Anaerobic microbial biogeochemistry in sediments from two basins in the Gulf of Maine: Evidence for iron and manganese reduction. Estuar. Coast. Mar. Sci. 32:313–324.

Hinga, K.R., J. McN. Sieburth, and G. Ross Heath. 1979. The supply and use of organic material at the deep-sea floor. J. Mar. Res. 37:557–579.

Hinga, K.R., A.A. Keller, and C.A. Oviatt. 1991. Atmospheric deposition and nitrogen inputs to coastal waters. Ambio 20:256–260.

Hixon, M.A., and W.N. Brostoff. 1983. Damselfish as keystone species in reverse: intermediate disturbance and diversity of reef algae. Science 220: 511–513.

Hobbie, J.E., and R.T. Wright. 1965. Competition between planktonic bacteria and algae for oceanic solutes. Mem. 1st. Ital. Idrobiol. 18 (Suppl.):175–185.

Hobbie, J.E., O. Holm-Hansen, T.T. Packard, L.R. Pomeroy, R.W. Sheldon, J.P. Thomas, and W.J. Wiebe. 1972. A study of the distribution and activity of microorganisms in ocean water. Limnol. Oceanogr. 17:544–555.

Hobbie, J.E., R.J. Daley, and J. Jasper. 1977. Use of nucleopore filters for counting bacteria by fluorescence microscopy. Appl. Env. Microbiol. 33: 1225–1228.

Hobson, L.A. 1971. Relationships between particulate organic carbon and micro-organisms in upwelling areas off Southwest Africa. Inv. Pesq. 35:195–208.

Hobson, L.A. 1974. Effects of interaction of irradiance, daylength, and temperature on division rates of three species of marine unicellular algae. J. Fish. Res. Bd. Canada 31:391–395.

Hoch, M.P., and D.L. Kirchman. 1993. Seasonal and inter-annual variability in bacterial production and biomass in a temperate estuary. Mar. Ecol. Progr. Ser. 98:283–295.

Hoffmann, E.E., J.M. Klink, and G.-A. Paffenhöfer. 1981. Concentrations and vertical fluxes of zooplankton fed pellets on a continental shelf. Mar. Biol. 61:327–335.

Hogetsu, K., M. Hatanaka, T. Hanaoka, and T. Kawamura (eds.). 1977. Productivity of Biocenoses in Coastal Regions of Japan. JIBP Synthesis 14. Univ. Tokyo.

Holden, M.J. 1973. Are long-term sustainable fisheries for elasmobranchs possible? Rapp. Proc.-Verb. Cons. Intern. Explor. Mer. 164:360–367.

Holden, M.J. 1978. Long-term changes in landings of fish from the North Sea. Rapp. Proc.-Verb. Reun. Cons. Int. Explor. Mer 172:11–26.

Holeton, G.F. 1974. Metabolic adaptations of polar fish: Fact or artifact? Physiol. Zool. 47:137–152.

Hollibaugh, J.T. 1978. Nitrogen regeneration during the degradation of several amino acids by plankton communities collected near Halifax, Nova Scotia, Canada. Mar. Biol. 45:191–201.

Holling, C.S. 1959. Some characteristics of simple types of predation and parasitism. Canad. Entomol. 91:385–398.

Holling, C.S. 1966. The functional response of invertebrate predators to prey density. Mem. Entomol. Soc. Canada 48:1–85.

Holme, N.A., and A.D. Mclntyre. 1971. Methods for the Study of Marine Benthos. IBP Handbook 16. Blackwell.

Holm-Hansen, O. 1970. ATP levels in algal cells as influenced by environmental conditions. Plant Cell Physiol. 11:869–700.

Holm-Hansen, O., and C.R. Booth. 1966. The measurement of adenosine triphosphate in the ocean and its ecological significance. Limnol. Oceanog. 11:510–519.

Holm-Hansen, O., and B. Riemann. 1978. Chlorophyll a determination: Improvements in methodology. Oikos 30:438–447.

Holm-Hansen, O., S.Z. El-Sayed, G.A. Franceschini, and R.L. Cuhel. 1977. Primary production and the factors controlling phytoplankton growth in the Southern Ocean. In G.A. Llano (ed.), Adaptations Within Antarctic Ecosystems. Proc. 3rd SCAR Symp. on Antarct. Biol. Smithsonian Institution.

Honjo, S. 1976. Coccoliths: Production, transportation and sedimentation. Mar. Micropaleont. 1:65–79.

Honjo, S. 1980. Material fluxes and modes of sedimentation in the mesopelagic and bathypelagic zones. J. Mar. Res. 38:53–97.

Honjo, S., and K.O. Emery. 1976. Suspended matter of eastern Asia: Scanning electron microscopy and X-ray peak analysis. Pp. 259–288 in H. Aoki and S. Iizuka (eds.), Volcanoes and Technosphere. Tokai Univ.

Honjo, S., and M.R. Roman. 1978. Marine copepod fecal pellets: Production, preservation, and sedimentation. J. Mar. Res. 36:45–57.

Hooper, F.F. 1973. Origin and fate of organic phosphorus compounds in aquatic systems. Pp. 179–202 in E.J. Griffith, A. Beeton, J.M. Spencer, and D.T. Mitchell (eds.), Environmental Phosphorus Handbook. Wiley.

Hopkinson, C.S. 1985. Shallow-water benthic and pelagic metabolism: Evidence of heterotrophy in the nearshore Georgia Bight. Mar. Biol. 87:19–32.

Hopkinson, C.S., and R.L. Wetzel. 1982. In situ measurements of nutrient and oxgen fluxes in a coastal marine benthic community. Mar. Ecol. Progr. Ser. 10:29–35.

Hopkinson, C.S., J.G. Gosselink, and R.T. Parrondo. 1980. Production of coastal Louisiana marsh plants calculated from phenometric techniques. Ecology 61:1091–1098.

Hoppe, G. 1978. Relations between active bacteria and heterotrophic potential in the sea. Neth. J. Sea Res. 12:78–98.

Hoppe, H.A., T. Levring, and Y. Tanaka. 1979. marine Algae in Pharmaceutical Science. Walter de Gruyter.

Horn, H.S. 1966. Measurement of "overlap" in comparative ecological studies. Amer. Nat. 100:419–424.

Horn, M.H. 1972. The amount of space available for marine and freshwater fishes. Fish. Bull. U.S. Fish. Wildlife Ser. 72:1295–1297.

Horn, M.H. 1989. Biology of marine herbivorous fishes. Oceanogr. Mar. Biol. Ann. Rev. 27:167–272.

Horwood, J.W., and J.D. Goss-Custard. 1977. Predation by the oystercatcher, *Haematopus ostralegus* (L.), in relation to the cockle, *Cerastoderma edule* (L.), fishery in the Burry Inlet, South Wales. J. Appl. Ecol. 14:139–158.

Houghton, R.W., and M.A. Mensah. 1978. Physical aspects and biological consequences of Ghanian coastal upwelling. Pp. 167–180 in R. Boje and M. Tomczak (eds.), Upwelling Ecosystems. Springer-Verlag.

Howard, R.K., and F.T. Short. 1986. Seagrass growth and survivorship under the influence of epiphyte grazers. Aquat. Bot. 24:287–302.

Howarth, R.W. 1979. Pyrite: Its rapid formation in a salt marsh and its importance in ecosystem metabolism. Science 203:49–51.

Howarth, R.W. 1984. The ecological significance of sulfur in the energy dynamics of salt marsh and coastal marine sediments. Biogeochemistry 1:5–27.

Howarth, R.W. 1988. Nutrient limitation of net primary production in marine ecosystems. Ann. Rev. Ecol. Syst. 19:89–110.

Howarth, R.W., and J.M. Teal. 1979. Sulfate reduction in a New England salt marsh. Limnol. Oceanogr. 24:999–1013.

Howarth, R.W., and J.M. Teal. 1980. Energy flow in a salt marsh ecosystem: The role of reduced inorganic sulfur compounds. Amer. Nat. 116:862–872.

Howarth, R.W., A. Giblin, J. Gale, B.J. Peterson, and G.W. Luther. 1982. Reduced sulfur compounds in the pore waters of a New England salt marsh. In R.O. Hallberg (ed.), Environmental Biogeochemistry. Ecol. Bull. 35:135–152.

Howell, A.B. 1930. Aquatic Mammals. Charles Thomas.

Howes, B.L., J.W.H. Dacey, and G.M. King. 1984. Carbon flow through oxygen and sulfate reduction pathways in salt marsh sediments. Limnol. Oceanogr. 29:1037–1051.

Howes, B.L., J.W.H. Dacey, and J.M. Teal. 1985. Annual carbon mineralization and belowground production of *Spartina alterniflora* in a New England salt marsh. Ecology 66:595–605.

Howes, B.W., R.W. Howarth, J.M. Teal, and I. Valiela. 1981. Oxidation-reduction potentials in a salt marsh: Spatial patterns and interactions with primary production. Limnol. Oceanogr. 26:350–360.

Hubbell, S.P. 1973. Population and simple food webs as energy filters. II. Two-species systems. Amer. Nat. 107:122–151.

Hubold, G. 1978. Variations in growth rate and maturity of herring in the Northern North Sea in the years 1955–1973. Rapp. Proc.-Verb. Reun. Cons. Int. Explor. Mer 172:154–163.

Huffaker, C.B. 1958. Experimental studies on predation. II. Dispersion factors and predator-prey oscillations. Hilgardia 27:343–383.

Hughes, R.N. 1970. An energy budget for a tidal flat population of the bivalve *Scrobicularia plana* DaCosta. J. Anim. Ecol. 39:357–381.

Hughes, R.N. 1980. Optimal foraging theory in the marine context. Oceanogr. Mar. Biol. Ann. Rev. 18:423–481.

Hughes, R.N., and R.W. Elner. 1979. Tactics of a predator, *Carcinus maenas* and morphological responses of the prey, *Nucella lapillus*. J. Anim. Ecol. 48:65–78.

Hughes, T.P. 1990. Recruitment limitation, mortality, and population regulation in open systems: A case study. Ecology 71:12–20.

Hulburt, E.M. 1962. Phytoplankton in the Southwestern Sargasso Sea and North Equatorial Current, February, 1961. Limnol. Oceanogr. 7:307–315.

Hulburt, E.M. 1966. The distribution of phytoplankton, and its relationship to hydrography, between southern New England and Venezuela. J. Mar. Res. 24:67–81.

Hulburt, E.M., and R.R.L. Guillard. 1968. The relationship of the distribution of the diatom *Skeletonema tropicum* to temperature. Ecology 49:337–339.

Humphreys, W.F. 1979. Production and respiration in animal populations. J. Anim. Ecol. 48:427–453.

Hunt, G.L., Jr., D. Henneman, and I. Everson. 1992. Distribution and predator-prey interactio ns of macaroni penguins, Antarctic fur seals, and Antarctic krill near Bird Island, South Georgia. Mar. Ecol. Progr. Ser. 86:15–30.

Hunter, J.H., and G.L. Thomas. 1974. Effect of prey distribution and density on the searching and feeding behaviour of larval anchovy *Engraulis mordax* Girard. Pp. 559–574 in J.H.S. Blaxter (ed.), The Early Life History of Fishes. Springer-Verlag.

Hunter, J.R. 1966. Procedure for analysis of schooling behavior. J. Fish. Res. Bd. Canada 23:547–562.

Hunter, J.R., and G.L. Thomas. 1974. Effect of prey distribution and density on the searching and feeding behaviour of larval anchovy *Engraulis mordax*. Pp. 559–574 in J.H.S. Baxter (ed.), The Early Life History of Fish. Springer-Verlag.

Hunter, M.D., and P.W. Price. 1992. Playing chutes and ladders: Heterogeneity and the relative roles of bottom-up and top-down forces in natural communities. Ecology 73:724–732.

Hurlbert, S.H. 1971. The non-concept of species diversity: A critique and alternative parameters. Ecology 52:577–586.

Hutcheson, K. 1970. A test for comparing diversities based on the Shannon formula. J. Theor. Biol. 29:151–154.

Hutchinson, G.E. 1957. Concludng Remarks. Cold Spr. Harbor Symp. Quant. Biol. 22:415–427.

Hutchnison, G.E. 1961. The paradox of the plankton. Amer. Nat. 95:137–145.

Hutchinson, G.E. 1975. A Treatise on Limnology. Vol. III. Limnological Botany. Wiley.

Hutchinson, G.E. 1978. An Introduction to Population Ecology. Yale Univ.

Hylleberg, J. 1975. Selective feeding by *Abarenicola pacifica* with notes on *Abarenicola vagabunda* and a concept of gardening in lugworms. Ophelia 14:113–137.

Hylleberg-Kristensen, J. 1972. Carbohydrates of some marine invertebrates with notes on their food and natural occurrence of the carbohydrates studied. Mar. Biol. 14:130–142.

Ichimura, S. 1967. Environmental gradient and its relation to primary productivity in Tokyo Bay. Records Oceanogr. Works (Japan) 9:115–128.

Ikeda, T. 1970. Relationship between respiration rate and body size in marine phytoplankton animals as a function of the temperature of habitat. Bull. Fac. Fish. Hokkaido. Univ. 21:91–112.

Ikeda, T. 1974. Nutritional ecology of marine zooplankton. Mem. Fac. Fish. Hokkaido. Univ. 22:1–97.

Iles, T.D. 1967. Growth studies on North Sea herring. I. The second year's growth (I-group) of East Anglican herring, 1939–1963. J. Cons. Intern. Explor. Mer 31:56–76.

Iles, T.D. 1968. Growth studies on North Sea herring. II. O-group growth of East Anglican herring. J. Cons. Intern. Explor. Mer 32:98–116.

Imai, I., Y. Ishida, and Y. Hata. 1993. Killing of marine phytoplankton by a gliding bacterium Cytophaga sp., isolated from the coastal sea of Japan. Mar. Biol. 116:527–532.

Inter-American Tropical Tuna Commission 1980. Annual Report of the Inter-American Tropical Tuna Commission. 1979. Pp. 1–227.

Irukayama, K. 1966. The pollution of Minimata Bay and Minimata disease. Adv. Wat. Pollut. Res. 3:153–180.

Isaacs, J.D. 1973. Potential trophic biomasses and trace-substance concentrations in unstructured marine food webs. Mar. Biol. 22:97–104.

Isaacs, J.D., and R.A. Schwartzlose. 1978. Active animals of the deep-sea floor. Sci. Amer. 233:84–91.

Ittekkot, V., U. Brookmann, W. Michaelis, and E.T. Degens. 1981. Dissolved free and combined carbohydrates during a phytoplankton bloom in the Northern North Sea. Mar. Ecol. Progr. Ser. 4:259–305.

Iturriaga, R. 1979. Bacterial activity related to sedimentary particulate matter. Mar. Biol. 55:157–169.

Iturriaga, R., and H.G. Hoppe. 1977. Observations of heterotrophic activity on photoassimilated organic matter. Mar. Biol. 40:101–108.

Iturriaga, R., and B.G. Mitchell. 1986. Chrococcoid cyanobacteria: A significant component in the food web dynamics of the open ocean. Mar. Ecol. Progr. Ser. 28:291–297.

Iverson, R.L. 1990. Control of marine fish production. Limnol. Oceanogr. 35:1593–1604.

Ivlev, V.S. 1961. Experimental Ecology and Feeding of Fishes (D. Scott, translator). Yale Univ.

Ivleva, I.V. 1970. The influence of temperature on the transformation of matter in marine invertebrates. Pp. 96–112 in J.H. Steele (ed.), Marine Food Chains. Univ. California.

Jackson, G.A. 1980. Phytoplankton growth and zooplankton grazing in oligotrophic oceans. Nature 284:439–441.

Jackson, J.B.C. 1977. Habitat area, colonization, and development of epibenthic community structure. Pp. 349–358 in B.F. Keegan, P.O. Leidigh, and P.J.S. Boaden (eds.), Proc. 11th Eur. Mar. Biol. Symp. Pergamon.

Jackson, J.B.C. 1977a. Competition on marine hard substrata: The adaptive significance of solitary and colonial strategies. Amer. Nat. 111:743–767.

Jackson, J.B.C. 1979. Overgrowth competition between encrusting cheilostome ectoprocts in a Jamaican cryptic reef environment. J. Anim. Ecol. 48:805–823.

Jackson, J.B.C., and L. Buss. 1975. Allelopathy and spatial competition among coral reef invertebrates. Proc. Nat. Acad. Sci. U.S.A. 72:5160–5163.

Jacobs, J. 1974. Quantitative measurements of food selection. A modification of the forage ratio and Ivlev's electivity index. Oecologia 14:413–417.

Jannasch, H.W., and C.O. Wirsen. 1973. Deep-sea microorganisms: In situ response to nutrient enrichment. Science 180:641–643.

Jannasch, H.W., and C.O. Wirsen. 1979. Chemosynthetic primary production at East Pacific seafloor spreading centers. Bioscience 29:592–598.

Jannasch, H.W., and C.O. Wirsen. *In press*. Microbial activities in undecomposed and decomposed deep-sea water samples. Appl. Env. Microbiol.

Jannasch, H.W., K. Eimhjellen, C.O. Wirsen, and A. Farmanfarmanian. 1971. Microbial degradation of organic matter in the deep sea. Science 171:672–675.

Jassby, A., and T. Platt. 1976. Mathematical formulation of the relationships between photosythesis and light for phytoplankton. Limnol. Oceanogr. 21: 540–547.

Jawed, M. 1973. Ammonia excretion by zooplankton and its significance to primary productivity during summer. Mar. Biol. 23:115–120.

Jeffrey, S.W. 1980. Algal pigment systems. Pp. 33–58 in P.G. Falkowski (ed.), Primary Productivity in the Sea. Plenum.

Jeffries, C. 1974. Qualitative stability and digraphs in model ecosystems. Ecology 56:238–243.

Jenkins, G.P., J.W. Young, and T.L.O. Davis. 1991. Density dependence of larval growth of a marine fish, the southern bluefin tuna, *Thunnus maccoyii*. Can. J. Fish. and Aquat. Sci. 48:1358–1363.

Jenkins, W.J. 1977. Tritium-helium dating in the Sargasso Sea: A measurement of oxygen utilization. Science 196:291–292.

Jenkins, W.J. 1982. Oxygen utilization rates in North Atlantic subtrizical gyre and primary production in oligotrophic systems. Nature 300:246–248.

Jenkins, W.J., and D.W.R. Wallace. 1992. Tracer-based inferences of new production in the sea. Pp. 299–316 in P.G. Falkowski and A.D. Woodhead (eds.), Primary Productivity and Biogeochemical Cycles in the Sea. Plenum Press.

Jensen, A. 1973. Studies on the phytoplankton ecology of the Trondheim fjord II. Chloroplast pigments in relation to abundance and physiological state of the phytoplankton. J. Exp. Mar. Ecol. 11:137–155.

Jerlov, N.G. 1951. Optical studies of ocean water. Rep. Swed. Deep Sea Exped. 3:1–59.

Jerlov, N.G. 1968. Optical Oceanography. Elsevier.

Johannes, R.E. 1964. Phosphorus excretion and body size in marine animals: Microzooplankton and nutrient regeneration. Science 146:923–924.

Johannes, R.E. 1965. Influence of marine protozoa on nutrient regeneration. Limnol. Oceanogr. 10:434–442.

Johannes, R.E. 1980. The ecological significance of the submarine discharge of groundwater. Mar. Ecol. Prog. Ser. 3:365–373.

Johnson, C.R., D.G. Muir, and A.L. Reysenbach. 1991. Characteristic bacteria associated with surfaces of coralline algae: a hypothesis for bacterial induction of marine invertebrate larvae. Mar. Ecol. Progr. Ser. 74:281–294.

Johnson, K.M., C.M. Burney, and J. McN. Sieburth. 1981. Enigmatic marine ecosystem metabolism measured by direct diel CO_2 and O_2 flux in conjunction with DOC release and uptake. Mar. Biol. 65:49–60.

Johnson, L.L., and J.T. Landahl. 1994. Chemical contaminants, liver disease, and mortality rates in english sole (*Pleuronectes vetulus*). Ecol. Appl. 4:59–68.

Johnson, P.W., and J. McN. Sieburth. 1979. Chroococcoid cyanobacteria in the sea: A ubiquitous and diverse phototrophic biomass. Limnol. Oceanogr. 24:928–934.

Johnson, R.G. 1976. Conceptual models of benthic marine communities. Pp. 149–159 in T.J. Schopf (ed.), Models in Paleobiology. Freeman and Cooper.

Johnston, R. 1963a. Antimetabolites as an aid to the study of phytoplankton nutrition. J. Mar. Biol. Assoc. U.K. 43:409–425.

Johnston, R. 1963b. Seawater, the natural medium of phytoplankton. I. General features. J. Mar. Biol. Assoc. U.K. 43:427–456.

Joint, I.R., N.J.P. Owens, and A.J. Pomroy. 1986. Seasonal production of photosynthetic picoplankton and nanoplankton in the Celtic Sea. Mar. Ecol. Progr. Ser. 28:251–258.

Jolley, E.T., and A.K. Jones. 1977. The interaction between *Navicula muralis* and an associated species of *Flavobacterium*. Br. Phycol. J. 12:315–328.

Jones, B.C., and G.H. Geen. 1977. Food and feeding of spiny dogfish (*Squalus acanthias*) in British Columbia waters. J. Fish. Res. Bd. Canada 34:2067–2078.

Jones, G.P. 1990. The importance of recruitment to the dynamics of a coral reef fish population. Ecology 71:1691–1698.

Jones, R. 1964. A review of methods of estimating population size from marking experiments. Rapp. Proc.-Verb. Cons. Int. Explor. Mer 155:202–209.

Jones, R. 1973. Density dependent regulation of the numbers of cod and haddock. Rapp. Proc.-Verb. Reun. Cons. Int. Explor. Mer 164:156–173.

Jones, R. 1978. Competition and coexistence with particular reference to gadoid fish species. Rapp. Proc.-Verb. Reun. Cons. Int. Explor. Mer 172:292–300.

Jones, R., and J.R.G. Hislop. 1978. Changes in North Sea haddock and whiting. Rapp. Proc.-Verb. Reun. Cons. Int. Explor. Mer 172:58–71.

Jones, R.C. 1973a. The stock and recruitment relation as applied to the North Sea haddock. Rapp. Proc.-Verb. Reun. Cons. Int. Explor. Mer 164:156–173.

Jordan, T.E., and I. Valiela. 1982. A nitrogen budget of the ribbed mussel, *Geukensia demissa*, and its significance in nitrogen flow in a New England salt marsh. Limnol. Oceanogr. 27:75–90.

Jørgensen, B.B. 1977. The sulfur cycle of a coastal marine sediment (Limfjorden, Denmark). Limnol. Oceanogr. 22:814–832.

Jørgensen, B.B. 1977a. Bacterial sulfate reduction within reduced microniches of oxidized marine sediments. Mar. Biol. 41:7–17.

Jørgensen, B.B. 1977b. The distribution of colorless sulfurbacteria (*Beggiatoa* spp.) in a coastal marine sediment. Mar. Biol. 41:19–28.

Jørgensen, B.B. 1980. Mineralization and the bacterial cycling of carbon, nitrogen, and sulfur in marine sediments. Pp. 239–251 in D.C. Ellwood, J.N. Hedges, M.J. Leatham. J.M. Lynch, and J.H. Slater (eds.), Contemporary Microbial Ecology. Academic.

Jørgensen, B.B. 1982. Mineralization of organic matter in seabed: The role of sulfate reduction. Nature 296:643–645.

Jørgensen, B.B., and F. Bak. 1991. Pathways and microbiology of thiosulfate transformations and sulfate reduction in marine sediment (Kattegat, Denmark). Appl. Environ. Microbiol. 57:847–856.

Jørgensen, B.B., and J. Sørensen. 1985. Seasonal cycles of O_2, NO_3^- and SO_4^{2-} reduction in estuarine sediments: The significance of an NO_3^- reduction maximum in spring. Mar. Ecol. Progr. Ser. 24:65–74.

Jørgensen, C.B. 1962. The food of filter-feeding organisms. Rapp. Proc.-Verb. Reun. Cons. Int. Explor. Mer 153:99–107.

Jørgensen, C.B. 1966. Biology of Suspension Feeding. Pergamon.

Jørgensen, C.B. 1981. Mortality, growth, and grazing import of a cohort of bivalve larvae, *Mytilus edulis*. L. Ophelia 20:185–192.

Jørgensen, N.O.G., and E. Kristensen. 1980. Uptake of amino acids by three species of *Nereis* (Annelida: Polychaeta). I. Transport kinetics and net uptake from natural concentrations. Mar. Ecol. Prog. Ser. 3:329–340.

Jørgensen, N.O.G., and E. Kristensen. 1980a. Uptake of amino acids by three species of *Nereis* (Annelida: Polychaeta). II. Effects of anaerobiosis. Mar. Ecol. Prog. Ser. 3:341–346.

Jørgensen, S.E. (ed.). 1979. Handbook of Environmental Data and Ecological Parameters. Pergamon.

Josefson, A.B. 1982. Regulation of population size, growth, and production of a deposit feeding bivalve: A long-term field study of three deep-water populations off the Swedish west coast. J. Exp. Mar. Biol. Ecol. 59:125–150.

Jumars, P.A. 1975. Methods for measurement of community structure in deepsea macrobenthos. Mar. Biol. 30:245–252.

Jumars, P.A. 1976. Deep-sea species diversity: Does it have a characteristic scale? J. Mar. Res. 34:217–246.

Jumars, P.A. 1978. Spatial autocorrelation with RUM (Remote Underwater Manipulator): Vertical and horizontal structure of a bathyal benthic community. Deep-Sea Res. 25:589–604.

Jumars, P.A., A.R.M. Nowell, and R.L.F. Self. 1981. A simple model of flow-sediment-organism interaction. Mar. Geol. 42:155–172.

Jurasz, C.M. and V.P. Jurasz. 1979. Feeding modes of the humpback whale, *Megaptera novacangliae* in Southeast Alaska. Sci. Rep. Whales Res. Inst. 31: 69–83.

Kabanova, Y.G. 1969. Primary production of the Northern part of the Indian Ocean. Okeanologiya 8:270–278.

Kaestner, A. 1970. Invertebrate Zoology. Vols. I-III. Translated by H.W. Levi. Interscience.

Kalle, K. 1966. The problem of the gelbstoff in the sea. Mar. Biol. Ann. Rev. 4:91–104.

Kalmijn, A.J. 1978. Electric and magnetic sensory world of sharks, skates, and rays. Pp. 507–528 in E.S. Hodgson and R.F. Mathewson (eds.), Sensory Biology of Sharks, Skates, and Rays. Office of Naval Research, Dept. of the Navy.

Kamamura, A. 1974. Food and feeding ecology in the Southern sei whale. Sci. Rep. Whales Res. Inst. 26:25–144.

Kanwisher, J.W. 1966. Photosynthesis and respiration in some seaweeds. Pp. 407–420 in H. Barnes (ed.), Some Contemporary Studies in Marine Science. Allen and Unwin.

Kanwisher, J.W., K.D. Lawson, and L.R. McCloskey. 1974. An improved, self-contained polarographic dissolved oxygen probe. Limnol. Oceanogr. 19: 700–704.

Kaplan, W., J.M. Teal, and I. Valiela. 1977. Denitrification in salt marsh sediments: Evidence for seasonal temperature selection among populations of denitrifiers. Microb. Ecol. 3:193–204.

Kaplan, W., I. Valiela, and J.M. Teal. 1979. Denitrification in a salt marsh ecosystem. Limnol. Oceanogr. 24:726–734.

Karl, D.M. 1986. Determination of in situ microbial biomass, viability, metabolism, and growth. Pp. 85–176 in J.S. Poindexter and E.R. Leadbetter (eds.), Bacteria in Nature. Plenum Press.

Karl, D.M., C.O. Wirsen, and H.W. Jannasch. 1980. Deep sea primary production at the Galapagos hydrothermal vents. Science 207:1345–1347.

Karner, M., D. Fuks, and G.J. Herndl. 1992. Bacterial activity along a trophic gradient. Microb. Ecol. 24:243–257.

Kasuya, T. 1972. Growth and reproduction of *Stenella caeruleoalba* based on the age determination by means of dentinal growth layers. Sci. Rep. Whales Res. Inst. 24:57–79.

Kasuya, T. 1991. Density dependent growth in North Pacific sperm whales. Mar. Mamm. Sci. 7:230–257.

Kay, D.G., and A.E. Brafield. 1973. The energy relations of the polychaete Neanthes (*Nereis*) *virens* (Sars). J. Anim. Ecol. 42:673–692.

Keating, K.I. 1978. Blue-green algal inhibition of diatom growth: Transitions from mesotrophic to eutrophic community structure. Science 199:971–973.

Keeling, C.D., R.B. Bacastow, A.E. Bainbridge, C.A. Ekdahl, Jr., P.R. Guenther, and L.S. Waterman. 1976. Atmospheric carbon dioxide variations at Mauna Loa Observatory, Hawaii. Tellus 28:537–551.

Keenleyside, M.H.A. 1955. Some aspects of the schooling behavior of fish. Behavior 8:183–248.

Keller, A.A., and R.L. Rice. 1989. Effects of nutrient enrichment on natural populations of the brown tide phytoplankton *Aureococcus anophagefferens* (Chrysophyceae). J. Phycol. 25:636–646.

Keller, A.A., and U. Riebesell. 1989. Phytoplankton carbon dynamics during a winter-spring diatom bloom in an enclosed marine ecosystem: Primary production, biomass and loss rates. Mar. Biol. 103:131–142.

Kelly, G.J. 1989. A comparison of marine photosynthesis with terrestrial photosynthesis: A biochemical perspective. Oceanogr. Mar. Biol. Ann. Rev. 27:11–44.

Kemp, P.F., B.F. Sherr, E.B. Sherr, and J.J. Cole. 1993. Handbook of Methods of Aquatic Microbial Ecology. Lewis Publishers.

Kemp, W.M., P. Sampow, J. Caffey, M. Maper, K. Henkunsion, and W.R. Boynton. 1990, 1991. Ammonium recycling versus denitrification in Chesapeake Bay sediments. Limnol. Oceanogr. 35:1545–1563.

Kenyon, K.W. 1969. The Sea Otter in the Eastern Pacific Ocean. Govt. Printing Office.

Kerr, S.R. 1971. Analysis of laboratory experiments on growth efficiency of fishes. J. Fish. Res. Bd. Canada 28:801–808.

Kerr, S.R. 1971a. Prediction of fish growth efficiency in nature. J. Fish. Res. Bd. Canada 28:809–814.

Kesel, R.H. 1988. The decline in suspended load of the Lower Mississippi River and its influence on adjacent wetlands. Environ. Geol. Water Sci. 11:271–281.

Ketchum, B.H., and N. Corwin. 1965. The cycle of phosphorus in a plankton bloom in the Gulf of Maine. Limnol. Oceanogr. 10:R148–R161.

Kierstead, H., and L.B. Slobotkin. 1953. The size of water masses containing plankton blooms. J. Mar. Res. 12:141–147.

Kiey, J. 1973. Primary production in the Indian Ocean I. Pp. 115–126 in B. Zeitzschel and S.A. Gerlach (eds.), The Biology of the Indian Ocean. Springer-Verlag.

Kim, K.R., and H. Craig. 1993. Nitrogen-15 and oxygen-18 characterization of nitrous oxide: A global prespective. Science 262:1855–1857.

Kimmerer, W.J. 1987. The theory of secondary production calculations for continuously reproducing populations. Limnol. Oceanogr. 32:1–13.

King, G.M. 1983. Sulfate reduction in Georgia salt marsh soils: An evaluation of pyrite formation by use of ^{35}S and ^{55}Fe tracers. Limnol. Oceanogr. 28: 987–995.

King, G.M., and T. Berman. 1984. Potential effects of isotopic dilution on apparent respiration in ^{14}C heterotrophy experiments. Mar. Ecol. Progr. Ser. 19:175–180.

King, G.M., and W.J. Wiebe. 1978. Methane release from soils of a Georgia salt marsh. Geochim. Cosmochim. Acta 42:343–348.

King, W.B. 1970. The trade wind zone oceanography pilot study. Part VII. Observation of sea birds March 1964 to June 1965. Spec. Sci. Rep. Fish. U.S. Fish Wild. Serv. 586:1–136.

Kinne, O. (ed.). 1970. Marine Ecology, Vol. 1. Environmental Factors. Part I. Wiley Interscience.

Kinne, O., and G.-A. Paffenhöfer. 1965. Hydraulic structure and digestion rate as a fuction of temperature and salinity in *Clava multicornis* (Cnidaria, Hydrozoa). Helgol. Wiss. Meeres. 12:329–341.

Kinsey, D.W. 1981. Is there outwelling in Georgia? Estuaries 4:277–278.

Kiørboe, T. 1993. Turbulence, phytoplankton cell size, and the structure of pelagic food webs. Adv. Mar. Biol. 29:1–72.

Kiørboe, T., F. Møhlenberg, and O. Nøhr. 1981. Effect of suspended bottom material on growth and energetics in *Mytilus edulis*. Mar. Biol. 61:283–288.

Kirchman, D.L. 1990. Limitation of bacterial growth by dissolved organic matter in the subarctic Pacific. Mar. Ecol. Progr. Ser. 62:47–54.

Kirchman, D.L., S. Graham, D. Reish, and R. Mitchell. 1982. Bacteria induce settlement and metamorphosis of *Janua (Dexiospira) brasiliensis* Grube (Polychaeta: Spirorbidae). J. Exp. Mar. Biol. Ecol. 56:153–163.

Kirchman, D.L., S. Graham, D. Reish, and R. Mitchell. 1982a. Lectins may mediate in the settlement and metamorphosis of *Janua (Dexiospira) brasiliensis* Grube (Polychaeta: Spirorbidae). Mar. Biol. Letters 3:131–142.

Kirk, J.T.O. 1992. The nature and measurement of the light environment in the ocean. Pp. 9–30 in P.G. Falkowski and A.D. Woodhead. Primary Productivity and Biogeochemical Cycles in the Sea. Plenum Press.

Kislaliogu, M., and R.N. Gibson. 1976. Prey handling time and its importance in food selection by the 15-spined stickleback, *Spinachia spinachia* (L.). J. Exp. Mar. Biol. Ecol. 25:115–158.

Kistritz, R.U., K.J. Hall, and I. Yesaki. 1983. Productivity, detritus flux, and nutrient cycling in a *Carex lyngbyei* tidal marsh. Estuaries 6:227–236.

Klump, J.V., and C.S. Martens. 1981. Biogeochemical cycling in an organic rich coastal marine basin. II. Nutrient sediment-waters exchange processes. Geochim. Cosmochim. Acta 45:101–121.

Knauer, G.A., and J.H. Martin. 1981. Primary production and carbon-nitrogen fluxes in the upper 1,500 m of the northeast Pacific. Limnol. Oceanogr. 26:181–186.

Knauer, G.A., J.A. Martin, and K.W. Bruland. 1979. Fluxes of particulate carbon, nitrogen, and phosphorus in the upper water column of the northeast Pacific. Deep-Sea Res. 26A:97–108.

Kneib, R.T. 1981. Size-specific effects of density on the growth, fecundity, and mortality of the fish *Fundulus heteroclitus* in an intertidal salt marsh. Mar. Ecol. Progr. Ser. 6:203–212.

Koch, M.S., I.A. Mendelssohn, and K.L. McKee. 1990. Mechanism for the hydrogen sulfide-induced growth limitation in wetland macrophytes. Limnol. Oceanogr. 35:399–408.

Koehl, M.A.R. 1986. Seaweeds in moving water: form and mechanical function. Pp. 603–604 in T. Givnish (ed.), On the Economy of Plant Form and Function. Cambridge Univ. Press.

Koehl, M.A.R., and R.S. Alberte. 1988. Flow, flapping, and photosynthesis of Nereocystis leutkeana: A functional comparison of undulate and flat blade morphologies. Mar. Biol. 99:435–444.

Koehl, M.A.R., and J.R. Strickler. 1981. Copepod feeding currents: Food capture at low Reynolds number. Limnol. Oceanogr. 26:1062–1073.

Kofoed, L.H. 1975. The feeding biology of *Hydrobia ventrosa* (Montague). I. The assimilation of different components of food. J. Exp. Mar. Biol. Ecol. 19:1–9.

Kofoed, L.H. 1975a. The feeding biology of *Hydrobia ventrosa*. 2. Allocation of the components of the carbon-budget and the significance of the reactions of dissolved organic material. J. Exp. Mar. Biol. Ecol. 19:243–256.

Kohl, D.H., G.D. Shearer, and B. Commoner. 1971. Fertilizer nitrogen: Contribution to nitrate in surface waters in a corn belt watershed. Science 174:1331–1334.

Kohn, A.J. 1971. Diversity, utilization of resources, and adaptive radiation in shallow water marine invertebrates of tropical islands. Limnol. Oceanogr. 16:332–348.

Kolber, Z.S., et al. 1994. Iron limitation of phytoplankton photosynthesis in the equatorial Pacific Ocean. Nature 371:145–149.

Komaki, Y. 1967. On the surface swarming of euphausiid crustaceans. Pac. Sci. 21:433–448.

Kooyman, G.L., R.W. Davis, J.P. Croxall, and D.P. Costa. 1982. Diving depth and energy requirements of King penguins. Science 217:726–727.

Koslow, J.A. 1992. Fecundity and the stock-recruitment relationship. Can. J. Fish. Aquat. Sci. 49:210–217.

Krebs, C.T. 1976. The Effects of Sewage Sludge on the Marsh Fiddler Crab, *Uca pugnax*. Ph.D. Diss., Boston Univ.

Krebs, J.R., A.I. Houston, and E.L. Charnov. 1981. Some recent developments in optimal foraging. Pp. 3–18 in A.C. Kamil and T.D. Sargent (eds.), Foraging Behavior. Garland STDM.

Kremer, J.M., and S.W. Nixon. 1978. A Coastal Marine Ecosystem. Springer-Verlag.

Kriss, A.E. 1963. Marine Microbiology (Deep Sea). Oliver and Boyd.

Kristensen, E., and T.H. Blackburn. 1987. The fate of organic carbon and nitrogen in experimental marine sediment systems. Influence of bioturbulation and anoxia. J. Mar. Res. 45(1):231–257.

Kristensen, E., F.Ø. Andersen, and T.H. Blackburn. 1992. Effects of benthic macrofauna and temperature on degradation of macroalgal detritus: The fate of organic carbon. Limnol. Oceanogr. 37:1404–1419.

Krom, M.D., and R.A. Berner. 1980. Adsorption of phosphate in anoxic marine sediments. Limnol. Oceanogr. 25:797–806.

Krom, M.D., and R.A. Berner. 1980a. The diffusion coefficients of sulfate, ammonium, and phosphate ions in anoxic marine sediments. Limnol. Oceanogr. 25:327–337.

Krom, M.D., N. Kress, S. Brenner, and L.I. Gordon. 1991. Phosphorus limitation of primary productivity in the eastern Mediterranean Sea. Limnol. Oceanogr. 36:424–432.

Kuenzler, E.J. 1961. Phosphorus budget of a mussel population. Limnol. Oceanogr. 6:400–415.

Kuhn, T.S. 1970. The structure of scientific revolutions. 2nd ed. Univ. of Chicago.

Kuipers, B.R., P.A.W.J. de Wilde, and F. Creutzberg. 1981. Energy in a tidal flat ecosystem. Mar. Ecol. Progr. Ser. 5:215–221.

Kuparinen, J., and P.K. Bjørnsen. 1992. Bottom-up and top-down controls of the microbial food web in the Southern Ocean: Experiments with manipulated microcosms. Polar Biol. 12:189–195.

Kuparinen, J., and H. Kuosa. 1993. Autotrophic and heterotrophic picoplankton in the Baltic Sea. Adv. Mar. Biol. 29:73–128.

Kurten, B. 1953. On the variation and population dynamics of fossil and recent mammal populations. Acta Zool. Fenn. 76:1–122.

Lack, D. 1947. Darwin's Finches. Cambridge Univ.

Ladle, M. 1972. Larval Simuliidae as detritus feeders in chalk streams. Mem. Ist. Ital. Idrobiol. 29 (Suppl.):429–439.

Lai, D.V., and P.L. Richardson. 1977. Distribution and movement of Gulf Stream rings. J. Phys. Oceanogr. 7:670–683.

Lam, C.W.Y., and K.C. Ho. 1989. Red tides in Tolo Harbour, Hong Kong. Pp. 49–52 in T. Okaichi, D.M. Anderson, and T. Nemoto (eds.), Red Tides: Biology, Environmental Science, and Toxicology. Elsevier.

Lam, R.K., and B.R. Frost. 1976. Model of copepod filtering responses to changes in size and concentration of food. Limnol. Oceanogr. 21:490–500.

Lampert, W. 1970. Release of dissolved organic carbon by grazing zooplankton. Limnol. Oceanogr. 23:831–834.

Lampitt, R.S. 1978. Carnivorous feeding by a small marine copepod. Limnol. Oceanogr. 23:1228–1230.

Lancelot, C. 1979. Gross excretion rates of natural marine phytoplankton and heterotrophic uptake of excreted products in the Southern North Sea, as determined by short-term kinetics. Mar. Ecol. Progr. Ser. 1:179–186.

Land, L.S., J.C. Lang, and B.N. Smith. 1975. Preliminary observations on the carbon isotopic composition of some reef coral tissues and symbiotic zooxanthellae. Limnol. Oceanogr. 20:283–287.

Landenberger. D.E. 1968. Studies on selective feeding in the Pacific starfish *Pisaster* in Southern California. Ecology 49:1002–1075.

Landry, M.R. 1977. A review of important concepts in the trophic organization of pelagic ecosystems. Helgol. Wiss. Meeres. 30:8–17.

Landry, M.R. 1978. Predatory feeding behavior of a marine copepod, *Labidocera trispinosa*. Limnol. Oceanogr. 23:1103–1113.

Landry, M.R. 1980. Detection of prey by *Calanus pacificus*: Implications of the first antenna. Limnol. Oceanogr. 25:545–549.

Landry, M.R. 1981. Switching between herbivory and carnivory by the planktonic marine copepod *Calanus pacificus*. Mar. Biol. 65:77–82.

Langdon, C.J., and R.I.E. Newell. 1990. Utilization of detritus and bacteria as food sources by two bivalve suspension-feeders, the oyster *Crassostrea virginica* and the mussel *Geukensia demissa*. Mar. Ecol. Progr. Ser. 58:299–310.

Langmuir, I. 1938. Surface motion of water induced by the wind. Science 87:119–123.

Lapointe, B.E., M.M. Littler, and D.S. Littler. 1992. Nutrient availability to marine macroalgae in silicoclastic versus carbonate-rich coastal waters. Estuaries 15:75–82.

Larkin, P.A. 1977. An epitaph for the concept of maximum yield. Trans. Amer. Fish. Soc. 106:1–11.

Larsson, U., and A. Hagström. 1979. Phytoplankton exudate release as an energy source for the growth of pelagic bacteria. Mar. Biol. 52:199–206.

Lasker, R. 1966. Feeding, growth, respiration and carbon utilization of an euphausiid crustacean. J. Fish. Res. Bd. Canada 23:1291–1317.

Lasker, R. 1970. Utilization of zooplankton energy by a Pacific sardine population in the California Current. Pp. 265–284 in J.H. Steele (ed.), Marine Food Chains. Univ. California.

Lasker, R., J.B.J. Wells, and A.D. McIntyre. 1970. Growth, reproduction, respiration and carbon utilization of the sand-dwelling harpacticoid copepod, *Asellopsis intermedia*. J. Mar. Biol. Assoc. U.K. 50:147–160.

Lavery, P.S., R.J. Lukatelich, and A.J. McComb. 1991. Changes in the biomass and species composition of macroalgae in an eutrophic estuary. Estuar. Coast. Shelf Sci. 33:1–22.

Law, C.S., A.P. Rees, and N.J.P. Owens. 1991. Temporal variability of denitrification in estuarine sediments. Estuar. Coast. Shelf Sci. 33:37–56.

Law, C.S., A.P. Rees, and N.J.P. Owens. 1992. Nitrous oxide: Estuarine sources and atmospheric flux. Estuar. Coast. Shelf Sci. 35:301–314.

Lawlor, L.L. 1980. Structure and stability in natural and randomly constructed competitive communities. Amer. Nat. 116:394–408.

Lawrence, J.M. 1975. On the relationship between marine plants and sea urchins. Oceanogr. Mar. Biol. Ann. Rev. 13:213–286.

Laws, E.A., and D.G. Redalje. 1979. Effect of sewage enrichment on the phytoplankton population of a subtropical estuary. Pac. Sci. 33:129–144.

Laws, R.M. 1977a. Seals and whales of the Southern Ocean. Phil. Trans. Roy. Soc. Lond. B279:81–96.

Laws, R.M. 1977b. The significance of vertebrates in the Antarctic Marine Ecosystem. Pp. 411–438 in G.A. Llano (ed.), Adaptations Within Antarctic Ecosystems. Proc. 3rd SCAR Symp. Antarct. Biol. Smithsonian Inst.

Laws, R.M. 1983. Ecological dynamics of antartic seals and whales. Proceeding of the Biomass Colloquim (1982). T. Nemoto and T. Mash (eds.), Reserach. Tokoyo, Japan. No. 27. p. 247.

Laws, R.M. 1985. The ecology of the Southern Ocean. Amer. Sci. 73:26–40.

Lawton, P., and R.N. Hughes. 1985. Foraging behavior of the crab *Cancer pagurus* feeding on the gastropods *Nucella* lapillus and *Littorina littorea*: Comparisons with optimal foraging theory. Mar. Ecol. Prog. Ser. 27:143–154.

Leach, J.H. 1970. Epibenthic algal production in an intertidal mudflat. Limnol. Oceanog. 15:514–521.

Leatham, G.F., V. King, and M.A. Stahmann. 1980. In vitro protein polymerization by quinones or free radicals generated by plant or fungal oxidative enzymes. Phytopathology 70:1134–1140.

Leavitt, D.F., J. McDowell Capuzzo, D. Miosky, R. Smolowitz, B.A. Lancaster, and C.C. Reinisch. 1994. Incidence of haematopoetic neoplasms in *Mya arenaria*: Monthly monitoring of prevalence and indeces of physiological condition. Mar. Biol. 105:313–321.

Le Brasseur, R.J. 1965. Seasonal and annual variation of net zooplankton at Ocean Station P, 1956–1964. Fish Res. Bd. Canada, MS Rep. Ser. No. 202.

Le Brasseur, R.J., and O.D. Kennedy. 1972. Microzooplankton in coastal and oceanic areas of the Pacific Subarctic water mass: A preliminary report. Pp. 355–365 in A.Y. Takenouti (ed.), Biological Oceanography of the Northern North Pacific Ocean. Idemitsu Shoten.

LeCren, E.D. 1965. Some factors regulating the size of populations of freshwater fish. Mitt. Intern. Verein. Limnol. 13:88–105.

Lee, C., and C. Cronin. 1982. The vertical flux of particulate organic nitrogen in the sea: Decomposition of amino acids in the Peru upwelling area and the equatorial Atlantic. J. Mar. Res. 40:227–251.

Lee, C., J.W. Farrington, and R.B. Gagosian. 1979. Sterol geochemistry of sediments in the western North Atlantic Ocean and adjacent coastal areas. Geochim. Cosmochim. Acta 43:35–46.

Legendre, L., and M. Gosselin. 1989. New production and export of organic matter to the deep ocean: Consequences of some recent discoveries. Limnol. Oceanogr. 34:1374–1380.

Lehman, J.T. 1976. The filter feeder as an optimal forager, and the predicted shapes of feeding curves. Limnol. Oceanogr. 21:501–516.

Lehman, J.T. 1988. Hypolimnetic metabolism in Lake Washington: Relative effects of nutrient load and food web structure on lake productivity. Limnol. Oceanogr. 33(part 1):1334–1347.

Lehman, J.T., and D. Scavia. 1982. Microscale patchiness of nutrients in plankton communities. Science 216:729–730.

Leighton, D.L. 1966. Studies of food preference in algivorous invertebrates of Southern California kelp beds. Pac. Sci. 20:104–113.

Lenz, J. 1977. On detritus as a food source for pelagic filter-feeders. Mar. Biol. 41:39–48.

Leong, R.J.H., and C.P. O'Connell. 1969. A laboratory study of particulate and filter feeding on the Northern Anchovy (*Engraulis mordax*). J. Fish. Res. Bd. Canada 26:557–582.

Leslie, P.H. 1966. The intrinsic rate of increase and overlap of successive generations in a population of guillemots (*Uria aalge* Pont.). J. Anim. Ecol. 35:291–301.

Lessard, E.S., and E. Swift. 1985. Species specific grazing rates of heterotrophic dinoflagellates in oceanic waters, measured with a stual-lakel radioisotope technique. 81(3):289–296.

Letelier, R.M., R.R. Bidigare, D.V. Hebel, M. Ondrusek, C.D. Winn, and D.M. Karl. 1993. Temporal variability of phytoplankton community structure base on pigment analysis. Limnol. Oceanogr. 38:1420–1437.

Levin, S.A., and L.A. Segel. 1976. Hypothesis for origin of planktonic patchiness. Nature 259–659.

Levins, R. 1966. The strategy of model building in population biology. Amer. Sci. 54:421–431.

Levinton, J.S. 1979. Deposit feeders, their resources, and the study of resource limitation. Pp. 117–141 in R.J. Livingston (ed.), Ecological Processes in Coastal and Marine Systems. Plenum.

Levinton, J.S., and G.R. Lopez. 1977. A model of renewable resources and limitation of deposit-feeding benthic populations. Oecologia 31:177–190.

Lewin, R.A. (ed.). 1962. Physiology and Biochemistry of Algae. Academic.

Lewis, J.B. 1977. Organic Production of Coral Reefs. Biol. Rev. 52:305–347.

Lewis, J.B. 1992. Heterotrophy in corals: zooplankton predation by the hydrocoral *Millepora complanata*. Mar. Ecol. Progr. Ser. 90:251–256.

Lewis, J.R. 1964. The Ecology of Rocky Shores. Hodder & Stoughton (formerly The English Universities Ltd).

Li, W.K.W., D.V. Subba Rao, W.G. Harrison, J.C. Smith, J.J. Cullen, B. Irwin, and T. Platt. 1983. Autotrophic picoplankton in the tropical ocean. Science 219:292–295.

Li, W.K.W., B.D. Irwin, and P.M. Dickie. 1993. Dark fixation of ^{14}C: variations related to biomass and productivity of phytoplankton and bacteria. Limnol. Oceanogr. 38:483–494.

Lieth, H. 1975. Historical survey of primary productivity research. Pp. 7–16 in H. Lieth and R.H. Whittaker (eds.), Primary Productivity of the Biosphere. Springer-Verlag.

Lindquist, A. 1978. A century of observations on sprat in the Skagerrak and the Kattegat. Rapp. Proc.-Verb. Reun. Cons. Int. Explor. Mer 172:187–196.

Linley, E.A.S., R.C. Newell, and S.A. Bosma. 1981. Heterotrophic utilization of microalgae released during fragmentation of kelp (*Eklonia maxima* and *Laminaria pallida*). I. Development of microbial communities associated with the degredation of kelp mucilage. Mar. Ecol. Progr. Ser. 4:31–41.

Linthurst, R.A., and R.J. Reimold. 1978. An evaluation of method for estimating the net aerial primary productivity of estuarine angiosperms. J. Appl. Ecol. 15:919–931.

Literathy, P. 1993. Considerations for the assessment of environmental consequences of the 1991 Gulf War. Mar. Pollut. Bull. 27:349–356.

Livingstone, D.A. 1963. Chemical composition of rivers and lakes. Chapter G in M. Fleischer (ed.), Data of Geochemistry, 6th Ed. Geol. Survey Prof. Pap. 440-G.

Lloyd, M. 1967. Mean crowding. J. Anim. Ecol. 36:1–30.

Lockwood, S.J. 1980. Density-dependent mortality in O-group plaice (*Pleuronectes platessa* L.) populations. J. Cons. Int. Explor. Mer 39:148–153.

Lockyer, C. 1976. Body weights of some species of large whales. J. Cons. Int. Expl. Mer 36:259–273.

Loder, T.C., W.B. Lyons, S. Murray, and H.D. McGuinness. 1978. Silicate in anoxic pore waters and oxidation effects during sampling. Nature 273: 373–374.

Loganathan, B.G., and K. Kannan. 1991. Time perspectives of organochlorine contamination in the global environment. Mar. Pollut. Bull. 22:582–584.

Longhurst, A., M. Colebrook, J. Gulland, R. Le Brasseur, C. Lorenzen, and P. Smith. 1972. The instability of ocean populations. New Scientist 1 June 1972:2–4.

Lonsdale, P. 1977. Clustering of suspension-feeding macrobenthos near abyssal hydrothermal vents at oceanic spreading centers. Deep-Sea Res. 24:857–863.

Lopez, G.R., J.S. Levinton, and L.B. Slobodkin. 1977. The effect of grazing by the detritivore *Orchestia grillus* on *Spartina* litter and its associated microbial community. Oecologia 30:111–127.

Lorenzen, C.J. 1976. Primary production in the sea. Pp. 173–185 in D.H. Cushing and J.J. Walsh (eds.), The Ecology of the Seas. Saunders, CBS College Publishing.

Lorenzen, C.J. 1979. Ultraviolet radiation and phytoplankton photosynthesis. Limnol. Oceanogr. 24:1117–1120.

Lorenzen, C.J., F.R. Shuman, and J.T. Bennett. 1981. In situ calibration of a sediment trap. Limnol. Oceanogr. 26:580–585.

Lotka, A.J. 1925. Elements of Physical Biology. Williams and Wilkins. Reprinted by Dover 1956.

Lotka, A.J. 1956. Elements of Mathematical Biology. Dover.
Loucks, O.L. 1970. Evolution of diversity efficiency, and community stability. Amer. Zool. 10:17–25.
Lovley, D.R., and E.J.P. Phillips. 1987. Competitive mechanisms for inhibition of sulfate reduction and methane production in the zone of ferric iron reduction in sediments. Appl. Environ. Microbiol. 53:2636–2641.
Lovley, D.R., and E.J.P. Phillips. 1988. Novel mode of microbial metabolism: Organic carbon metabolism coupled to dissimilatory reduction of iron and manganese. Appl. Environ. Microbiol. 54:1472–1480.
Lubchenco, J. 1978. Plant species diversity in a marine intertidal community: Importance of herbivore food preference and algal competitive ability. Amer. Nat. 112:23–39.
Lubchenco, J. 1983. *Littorina* and *Fucus*: Effects of herbivores, substratum heterogeneity, and plant escapes during succession. Ecology 64:1116–1123.
Lubchenco, J., and J. Cubit. 1980. Heteromorphic life history of certain marine algae as adaptations to variations in herbivory. Ecol. 6:676–687.
Lubchenco, J., and B.A. Menge. 1978. Community development and persistence in a low rocky intertidal zone. Ecol. Monogr. 48:67–94.
Lubin, D., and J.E. Frederick. 1991. The ultraviolet radiation environment of the antarctic peninsula: the roles of ozone and cloud cover. J. Appl. Meteorol. 30:478–493.
Luecke, C., and W.J. O'Brien. 1981. Prey location volume of a planktivorous fish: A new measure of prey vulnerability. Canad. J. Fish. Aquat. Sci. 38:1264–1270.
Lugo, A.E., G. Evink, M.M. Brinsom, A. Bruce, and S.C. Snedaker. 1975. Diurnal rates of photosynthesis, respiration, and transpiration in mangrove forests of South Florida. Pp. 335–352 in F.B. Golley and E. Medina (eds.), Tropical Ecological Systems. Springer-Verlag.
Lund, J.W.G. 1966. Summation. Pp. 227–249 in C.H. Oppenheimer (ed.), Marine Biology, Vol. 2. New York Academy of Science.
Luther, G.W., III, A. Giblin, R.W. Howarth, and R.A. Ryans. 1982. Pyrite and oxidized iron mineral phases formed from pyrite oxidation in salt marsh and estuarine sediments. Geochim. Cosmochim. Acta. 46:2665–2669.
MacArthur, R.H. 1965. Pattern of species diversity. Biol. Rev. 40:510–533.
MacArthur, R.H., and E.O. Wilson. 1967. The Theory of Island Biogeography. Princeton Univ.
MacCall, A.D. 1979. Population estimates for the waning years of the Pacific sardine fishery. Calif. Coop. Fish. Invest. Rep. 20:72–82.
MacDonald, D.D.M., and T.J. Pitcher. 1979. Age-groups from size-frequency data: A versatile and efficient method of analyzing distribution mixtures. J. Fish. Res. Bd. Canada 36:987–1001.
MacGregor, J.S. 1957. Relation between fish conditions and population size in the sardine (*Sardinops caerulea*). Fish. Bull. U.S. Fish. Wildl. Serv. 60:215–230.
MacIntosh, N.A. 1970. Whales and krill in the twentieth century. Pp. 195–212 in M.W. Holdgate (ed.), Antarctic Ecology. Academic.
MacIsaac, J.J., and R.C. Dugdale. 1969. The kinetics of nitrate and ammonia uptake by natural populations of marine phytoplankton. Deep-Sea Res. 16:45–57.
Mackas, D.L., K.L. Denman, and M.R. Abbott. 1985. Plankton patchiness: Biology in the physical vernacular. Bull. Mar. Sci. 37:652–674.

MacLeod, P., and I. Valiela. 1975. The effect of density and mutual interference by a predator: A laboratory study of predation by the nudibranch *Coryphella rufibranchialis* on the hydroid *Tubularia larynx*. Hydrobiologia 47:339–346.

Madin, L.P., and J.E. Purcell. 1992. Feeding, metabolism, and growth of Cyclosalpa bakeri in the subarctic Pacific. Limnol. Oceanogr. 37:1236–1251.

Maillet, G.L., and D.M. Checkley, Jr. 1991. Storm-related variation in the growth rate of otoliths of larval menhaden *Brevoortia tyrannus*: A time series analysis of biological and physical variables and implications for larva growth and mortality. Mar. Ecol. Progr. Ser. 79:1–16.

Majak, W., J.S. Craigie, and J. McLachlan. 1966. Photosynthesis in the Rhodophyceae. Canad. J. Bot. 44:541–549.

Maldonado, A. 1972. El Delta del Ebro. Bolet. Estratigr. (Univ. Barcelona) 1:1–486.

Malone, T.C., W.M. Kemp, H.W. Duckow, W.R. Boynton, J.H. Tuttle, and R.B. Jonas. 1986. Lateral variation in the production and fate of phytoplankton in a partially stratified estuary. Mar. Ecol. Progr. Ser. 32:149–160.

Manahan, D.T., S.H. Wright, G.C. Stephens, and M.A. Rice. 1982. Transport of dissolved amino acids by the mussel, *Mytilus edulis*: Demonstration of net uptake from natural seawater. Science 215:1253–1255.

Mann, K.H. 1973. Seaweeds: Their productivity and strategy for growth. Science 182:975–981.

Mann, K.H. 1978. Production on the bottom of the sea. Pp. 225–250 in D.H. Cushing and J.J. Walsh (eds.), The Ecology of the Seas. Blackwell.

Mann, K.H. 1982. Ecology of Coastal Waters. Univ. California.

Mann, K.H., and J.R.N. Lazier. 1991. Dynamics of Marine Ecosystems. Blackwell Sci. Publ.

Marais, J.F.K. 1980. Aspects of food intake, food selection, and alimentary canal morphology of *Mugil cephalus* (Linneaus, 1958), *Liza tricuspidens* (Smith, 1935), *L. richardsoni* (Smith, 1846) and *L. dummerili* (Steindauher, 1869). J. Exp. Mar. Biol. Ecol. 44:193–209.

Margalef, R. 1951. Diversidad de especies en las comunidades naturales. Publ. Inst. Biol. Aplic. Barcelona 9:5–27.

Margalef, R. 1957. La teorźa de la información en ecología. Mem. Real Acad. Cienc. Artes. (Barcelona) 33:373–449.

Margalef, R. 1967. The food web in the pelagic environment. Helgol. Wiss. Meeres. 18:548–559.

Margalef, R. 1968. Perspectives in Ecological Theory. Univ. of Chicago.

Margalef, R. 1974. Ecología. Omega.

Margalef, R. 1978. Phytoplankton communities in upwelling areas. The example of NW Africa. Oecol. Aquat. 3:97–132.

Margalef, R. 1978a. General concepts of population dynamics and food links. Pp. 617–704 in O. Kinne (ed.), Marine Ecology, Vol. 4. J. Wiley & Sons, Ltd.

Margulis, L. 1981. Symbiosis in Cell Evolution: Life and Its Environment in the Early Earth. W.H. Freeman.

Marino, R., and R.W. Howarth. 1993. Atmospheric oxygen exchange in the Hudson River: Dome measurements and comparison with other natural waters. Estuaries 16:433–445.

Marinucci, A.C., J.E. Hobbie, and J.V.K. Helfrich. 1983. Effect of litter nitrogen on decomposition and microbial biomass in *Spartina alterniflora*. Microb. Ecol. 9:23–40.

Marr, J.C. 1960. The causes of major variations in the catch of the Pacific sardine *Sardinops cerulea*. Proc. World Sci. Meet. Biol. Sardines and Related Species 3:667–791.

Marr, J.W.S. 1962. The natural history and geography of the Antarctic Krill (*Euphausia superba* Dana). Discovery Rep. 32:33–464.

Marrasé, C., J.H. Costello, T. Granata, and J.R. Strickler. 1990. Grazing in a turbulent environment: II. Energy dissipation, encounter rates, and efficacy of feeding currents in *Centropages hamatus*. Proc. Natl. Acad. Sci. 87:1653–1657.

Marrasé, C., E.L. Lim, and D.A. Caron. 1992. Seasonal and daily changes in bacterivory in a coastal plankton community. Mar. Ecol Progr. Ser. 82: 281–289.

Marsh, A.G., A. Grémare, and K.R. Tenore. 1989. Effect of food type and ration on growth of juvenile *Capitella* sp. I (Annelida: Polychaeta): macro- and micronutrients. Mar. Biol. 102:519–527.

Marsh, J.A., Jr., and S.V. Smith 1978. Productivity measurements of coral reefs in flowing water. Pp. 361–378 in D.R. Stoddart and R.E. Johannes (eds.), Coral Reefs: Research Methods. UNESCO.

Marshall, S.M. 1933. The production of microplankton in the Great Barrier Reef Region. Great Barrier Reef Exp. 1928–29. Sci. Repts. 2:111–157.

Marshall, S.M., and A.P. Orr. 1955. The Biology of a Marine Copepod, *Calanus finmarchicus* (Gunnerus). Oliver and Boyd.

Marshall, S.M., and A.P. Orr. 1964. Grazing by copepods in the sea. Pp. 227–238 in D.J. Crisp (ed.), Grazing in Terrestrial and Marine Environments. Blackwell.

Marshall, S.M., and A.P. Orr. 1966. Respiration and feeding in some small copepods. J. Mar. Biol. Assoc. U.K. 46:513–530.

Marsho, T.V., R.P. Burchard, and R. Fleming. 1975. Nitrogen fixation in the Rhode River estuary of Chesapeake Bay. Canad. J. Microbiol. 21:1348–1356.

Martens, C.S. 1976. Control of methane sediment-water bubble tranport by macrofaunal nitrification in Cape Lookout Bight, North Carolina. Science 192:998–1000.

Martens, C.S., and R.A. Berner. 1974. Methane production in the interstitial waters of sulfate depleted estuarine sediments. Science 185:1167–1169.

Martens, C.S., and R.A. Berner. 1979. Interstitial water chemistry of anoxic Long Island Sound sediments. I. Dissolved gases. Limnol. Oceanogr. 22:10–25.

Martens, C.S., and J.V. Klump. 1984. Biogeochemical cycling in an organic-rich coastal marine basin. 4. An organic carbon budget for sediments dominated by sulfate reduction and methanogenesis. Geochim. Cosmochim. Acta 48: 1987–2004.

Martin, J.H. 1968. Phytoplankton-zooplankton relationships in Narragansett Bay. Seasonal changes in zooplankton excretion rates relative to phytoplankton abundance. Limnol. Oceanogr. 13:63–71.

Martin, J.H. 1970. Phytoplankton-zooplankton relationships in Narragansett Bay. IV. The seasonal importance of grazing. Limnol. Oceanogr. 15:413–418.

Martin, J.H., R.M. Gordon, and S.E. Fitzwater. 1991. The case for iron. Limnol. Oceanogr. 36:1793–1802.

Martin, J.H., et al. 1994. Testing the iron hypothesis in ecosystems of the equatorial Pacific Ocean. Nature 371:123–129.

Martin, J.-M., and H.L. Window. 1991. Present and future roles of ocean margins in regulating marine biogeochemical cycles of trace elements. Pp. 45–67 in R.F.C. Mantoura, J.-M. Martin, and R. Wollast (eds.), Ocean Margin Processes in Global Change. Wiley-Interscience.

Matsakis, S., and R.J. Conover. 1991. Abundance and feeding of medusae and their potential impact as predators on other zooplankton in Bedford Basin (Nova Scotia, Canada) during spring. Can. J. Fish. Aquat. Sci. 48:1419–1430.

Mattson, W.J., Jr. 1980. Herbivory in relation to plant nitrogen content. Ann. Rev. Ecol. Syst. 11:17–25.

Mattson, W.J., and N.D. Addy 1975. Phytophagous insects as regulators of forest primary production. Science 190:515–522.

May, R.M. 1973. Stability and Complexity in Model Ecosystems. Princeton Univ.

May, R.M. 1975. Patterns of species abundance and diversity. Pp. 81–120 in M.L. Cody and J.M. Diamond (eds.), Ecology and Evolution of Communities. Harvard Univ.

Mayzaud, P. 1973. Respiration and nitrogen excretion of zooplankton. II. Studies of the metabolic characteristics of starved animals. Mar. Biol. 21: 19–28.

Mayzaud, P., and S. Dallot. 1973. Respiration et excretion azotée du zooplancton. I. Etude des niveaux métaboliques de quelques espèces de Mediterranée occidentale. Mar. Biol. 19:307–314.

Mayzaud, P., and S.A. Poulet. 1978. The importance of the time factor in the response of zooplankton to varying concentrations of naturally occurring particulate matter. Limnol. Oceanogr. 23:1144–1154.

McAllister, C.D. 1969. Aspects of estimating zooplankton production from phytoplankton production. J. Fish. Res. Bd. Canada 26:199–220.

McAllister, C.D. 1970. Zooplankton rations, phytoplankton mortality, and the estimation of marine production. Pp. 419–457, in J.H. Steele (ed.), Marine Food Chains. Univ. California.

McCall, P.L. 1977. Community patterns and adaptive strategies of the infaunal benthos of Long Island Sound. J. Mar. Res. 35:221–266.

McCarthy, J.J. 1972. The uptake of urea by natural populations of marine phytoplankton. Limnol. Oceanogr. 17:738–748.

McCarthy, J.J. 1980. Nitrogen. Pp. 191–234 in I. Morris (ed.), The Physiological Ecology of Phytoplankton. Univ. California.

McCarthy, J.J., and E.J. Carpenter. 1979. *Oscillatoria* (*Trichodesmium*) *thiebautii* (Cyanophyta) in the central north Atlantic Ocean. J. Phycol. 15:75–82.

McCarthy, J.J., and E.J. Carpenter. 1984. Nitrogen cycling in near-surface waters of the open ocean. Pp. 487–512 in E.J. Carpenter and D.G. Capone (eds.), Nitrogen in the Marine Environment. Academic.

McCarthy, J.J., and J.C. Goldman. 1979. Nitrogenous nutrition of marine phytoplankton in nutrient-depleted waters. Science 203:670–672.

McCarthy, J.J., W.R. Taylor, and M.E. Loftus. 1974. Significance of nanoplankton in the Chesapeake Bay Estuary and problems associated with the measurement of nanoplankton productivity. Mar. Biol. 24:7–16.

McDonough, R.J., R.W. Sanders, K.G. Porter, and D.L. Kirchman. 1986. Depth distribution of bacterial production in a stratified lake with anoxic hypolimnion. Appl. Envir. Microbiol. 52:992–1000.

McGowan, J.A. 1977. What regulates pelagic community structure in the Pacific? Pp. 423–444 in N.R. Anderson and B.J. Zahuranec (eds.), Oceanic Sound Scattering Prediction. Plenum.

McGowan, J.A. 1984. The California El Niño, 1983. Oceanus 27:48–57.

McGowan, J.A. 1985. El Niño 1983 in the southern California Bight. Pp. 166–184 in W.S. Wooster and D.H. Fluharty (eds.), El Niño North, Niño Effects in the Eastern Subarctic Pacific Ocean. Washington Sea Grant. Univ. of Washington.

McGowan, J.A., and T.L. Hayward. 1978. Mixing and oceanic productivity. Deep-Sea Res. 25:771–793.

McGregor, J.S. 1957. Fecundity of the Pacific Sardine (*Sardinops caerulea*). Fish. Bull. U.S. Wildl. Serv. 17:427–449.

McFarland, W.N., and J. Prescott. 1959. Standing crop, chlorophyll content and *in situ* metabolism of a giant kelp community in Southern California. Contr. Mar. Sci. 6:110–132.

McIntyre, A.D. 1992. The current state of the oceans. Mar. Pollut. Bull. 25:28–32.

McLaren, I.A. 1965. Some relationships between temperature and egg size, body size, development rate, and fecundity of the copepod *Pseudocalanus*. Limnol. Oceanogr. 10:528–538.

McManus, G.B., and J.A. Fuhrman. 1986. Bacterivory in seawater seawater studied with the use of inert flourescent particles. Limnol. Oceanogr. 31: 420–426.

McNaughton, S.J. 1977. Diversity and stability of ecological communities: A comment on the role of empiricism in ecology. Amer. Nat. 111:515–525.

McNeill, S. 1973. The dynamics of a population of *Leptopterna dolabrata* (Heteroptera: Miridae) in relation to its food resources. J. Anim. Ecol. 42: 495–508.

McNeill, S., and T.R.E. Southwood. 1978. The role of nitrogen in the development of insect/plant relationships. Pp. 77–98 in J.B. Harborne (ed.), Biochemical Aspects of Plant and Animal Coevolution. Academic.

McQueen, D.J., J.R. Post, and E.L. Mills. 1986. Trophic relationships in freshwater pelagic ecosystems. Can. J. Fish. Aquat. Sci. 43:1571–1581.

McRoy, C.P. 1974. Seagrass productivity: Carbon uptake experiments in eelgrass, *Zostera marina*. Aquaculture 4:131–137.

McRoy, C.P., and C. McMillan. 1977. Production ecology and physiology of sea grasses. Pp. 53–87 in C.P. McRoy and C. Helfferich (eds.), Seagrass Ecosystems. M. Dekker.

Mee, L.D. 1992. The Black Sea in crisis: A need for concerted international action. Ambio 21:278–286.

Mendelssohn, I.A., K.L. McKee, and W.H. Patrick. 1981. Oxygen deficiency in *Spartina alterniflora* roots: Metabolic adaptation to anoxia. Science 214:439–441.

Menge, B.A. 1991. Relative importance of recruitment and other causes of variation in rocky intertidal community structure. J. Exp. Mar. Biol. Ecol. 146:69–100.

Menge, B.A. and T.M. Farrell. 1989. Community structure and interaction webs in shallow marine hard-bottom communities: Tests of and environmental stress model. Adv. Ecol. Res. 19:198–262.

Menzel, D. 1974. Primary production, dissolved and particulate organic matter. Pp. 659–678 in E.D. Goldberg (ed.), The Sea. Wiley.

Menzel, D.W., and J.J. Goering. 1966. The distribution of organic detritus in the ocean. Limnol. Oceanogr. 11:333–337.

Menzel, D.W., and J.H. Ryther. 1960. The annual cycle of primary production in the Sargasso Sea off Bermuda. Deep-Sea Res. 6:351–367.

Menzel, D.W., and J.H. Ryther. 1961. Zooplankton in the Sargasso Sea off Bermuda and its relation to organic production. J. Cons. Perm. Inter. Explor. Mer 26:250–258.

Menzel, D.W., and J.H. Ryther. 1970. Distribution and cycling of organic matter in the oceans. Pp. 31–54 in D.W. Hood (ed.), Organic Matter in Natural Waters. Inst. Mar. Sci. Univ. Alaska.

Menzel, D.W., and J.P. Spaeth. 1962. Occurrence of ammonia in Sargasso Sea waters and in rainwater at Bermuda. Limnol. Oceanogr. 7:159–162.

Menzel, D.W., E.M. Hulburt, and J.R. Ryther. 1963. The effects of enriching Sargasso Sea water on the production and species composition of the phytoplankton. Deep-Sea Res. 10:209–219.

Mertz, D.B. 1970. Notes on methods used in life history studies. Pp. 4–17 in J.H. Connell, D.N. Mertz, and W.W. Murdoch (eds.), Readings in Ecology and Ecological Genetics. Harper and Row.

Meyer, J.L., E.T. Schultz, and G.S. Helfman. 1983. Fish schools: An asset to corals. Science 220:1047–1049.

Meyer-Reil, L.-A. In press. Bacterial biomass and heterotrophic activity in sediments and overlying waters. In J. Hobbie and P.J. Leb. Williams (eds.), Heterotrophic Activity in the Sea. Plenum.

Meyer-Reil, L.-A., and A. Faubel. 1980. Uptake of organic matter by meiofauna organisms and interrelationships with bacteria. Marine Ecol. Progr. Ser. 3:251–256.

Michaels, A.F., D.A. Siegel, R.J. Johnson, A.H. Knap, and J.N. Galloway. 1993. Episodic inputs of atmospheric nitrogen to the Saragasso Sea: Contributions to new production and phytoplankton blooms. Global Biogeochemical Cycles GBCYEP. 7:339–351.

Micklin, P.P. 1988. Dessication of the Aral Sea: A water management disaster in the Soviet Union. Science 241:1170–1175.

Miller, C.B., and B.W. Frost. 1982. Subarctic pacific ecosystem research (SUPER). EOS Trans. 63:62.

Milliman, J.D. 1991. Flux and fate of fluvial sediment and water in coastal seas. Pp. 69–89 in R. F. C. Mantoura, J.-M. Martin, and R. Wollast (eds.), Ocean Margin Processes in Global Change. Wiley.

Milliman, J.D., J.M. Broadus, and F. Gable. 1990. Environmental and economic implications of rising sea level and subsiding deltas: The Nile and Bengal examples. Ambio 18:340–345.

Mills, E.L. 1980. The structure and dynamics of shelf and slope ecosystems off the North East Coast of North America. Pp. 25–47 in K.R. Tenore and B.C. Coull (eds.), Marine Benthic Dynamics. Univ. South Carolina.

Mills, E.L., J.H. Leach, J.T. Carlton, and C.L. Secor. 1993. Exotic species in the Great Lakes: A history of biotic crises and anthropogenic introductions. J. Great Lakes Res. 19:1–54.

Minas, H.J., M. Minas, and T.T. Packard. 1986. Productivity in upwelling areas deduced from hydrographic and chemical fields. Limnol. Oceanogr. 31: 1182–1206.

Mitchell, J.F.B. 1989. The "greenhouse" effect and climate change. Rev. Geophys. 27:115–139.

Mitchell, R. 1978. Mechanism of biofilm formation in seawater. Pp. 45–50 in R.H. Gray (ed.), Proc. Ocean Thermal Energy Conversion (OTEC) Biofouling and Corrosion Symp. Oct. 10–12, 1977, Seattle, Wash. U.S. Dept. of Energy and Pacific Northwest Lab.

Miyazaki, N. 1977. Growth and reproduction of *Stenella coeruleoalba* off the Pacific coast of Japan. Sci. Rep. Whales Res. Inst. 19:21–48.

Moe, R.L., and P.C. Silva. 1977. Antarctic marine flora: Uniquely devoid of kelps. Science 196:1206–1208.

Moloney, C.L., and J.G. Field. 1989. General allometric equations for rates of nutrient uptake, ingestion and respiration in plankton organism. Limnol. Oceanogr. 34:1290–1299.

Moment, G.B. 1962. Reflo'xive selection: A possible answer to an old puzzle. Science 136:262–263.

Mootz, C.A., and C.E. Epifanio. 1974. An energy budget for *Menippe mercenaria* larvae fed *Artemia nauplii*. Biol. Bull. 146:44–55.

Moran, M.A., and R.E. Hodson. 1994. Dissolved humic substances of vascular plant origin in a crantal manine environment. Limnol. Oceanogr. 39:762–771.

Moran, M.A., R. Benner, and R.E. Hodson. 1989. Kinetics of microbial degradation of vascular plant material in two wetland ecosystems. Oecologia 79:158–167.

Moran, M.A., L.R. Pomeroy, E.S. Sheppard, L.A. Atkinson, and R.E. Hodson. 1991. Distribution of terrestrially derived dissolved organic matter on the southeastern U.S. continental shelf. Limnol. Oceanogr. 36:1134–1149.

Morel, A. 1974. Optical properties of pure water and pure seawater. Pp. 1–24, in N.G. Jerlov and E. Steemann Nielsen (eds.), Optical Aspects of Oceanography. Academic.

Morel, F.M.M., and J. Morgan. 1972. A numerical method for computing equilibria in aqueous chemical systems. Env. Sci. Tech. 6:58–67.

Moriarty, D.J.W. 1977. Improved method using muramic acid to estimate biomass of bacteria in sediments. Oecologia 26:317–323.

Morisita, M. 1959. Measuring the dispersion of individuals and analysis of the distributional patterns. Mem. Fac. Sci. Kyushu Univ. Ser. E (Biol.) 2:215–235.

Morison, J.I.L. 1990. Plant and ecosystem responses to increasing atmospheric CO_2. Trends Ecol. Evol. 5:69–70.

Morita, R.Y. 1980. Microbial life in the deep sea. Canad. J. Microbiol. 26: 1375–1385.

Morris, I. 1980. Paths of carbon assimilated in marine phytoplankton. Pp. 139–159 in P.O. Falkowski (ed.), Primary Productivity in the Sea. Plenum.

Morris, I., and H. Glover. 1981. Physiology of photosynthesis by marine coccoid cyanobacteria—Some ecological implications. Limnol. Oceanogr. 26:957–961.

Morrison, S.J., and D.C. White. 1980. Effects of grazing by estuarine gammaridean amphipods on the microbiota of allochthonous detritus. Appl. Env. Microbiol. 40:659–671.

Mortrain-Bertrand, A., C. Descolas-Gros, and H. Jupin. 1988. Pathway of dark inorganic carbon fixation in two species: influence of light regime and regulated facts on diet variations. J. Plankt. Res. 10:199–217.

Moseley, H.N. 1880. Dee sea dredging and life in the deep sea. Nature 21: 591–593.

Muller, P.J., and E. Suess. 1979. Productivity, sedimentation rate, and sedimentary organic matter in the oceans. I. Organic carbon preservation. Deep-Sea Res. 26A:1347–1362.

Mullin, M.M. 1969. Production of zooplankton in the ocean: The present status and problems. Oceanogr. Mar. Biol. Ann. Rev. 7:293–314.

Mullin, M.M., E.F. Stewart, and F.J. Fuglister. 1975. Ingestion by planktonic grazers as a function of concentration of food. Limnol. Oceanogr. 20:259–262.

Mullineaux, L.S., and C.A. Butman. 1991. Initial contact, exploration and attachment of barnacle (*Balanus amphitrite*) cyprids settling in flow. Mar. Biol. 110:93–103.

Munk, P. 1992. Foraging behaviour and prey size spectra of larval herring *Clupea harengus*. Mar. Ecol. Progr. Ser. 80:149–158.

Munk, P., and T. Kiørboe. 1985. Feeding behaviour and swimming activity of larval herring (*Clupea harengus*) in relation to density of copepod nauplii. Mar. Ecol. Progr. Ser. 24:15–21.

Murdoch, W.W. 1966. "Community structure, population control, and competition"—a critique. Amer. Nat. 100:219–226.

Murdoch, W.W. 1969. Switching in general predators: Experiments on predator specificity and stability of prey populations. Ecol. Monogr. 39: 335–354.

Murdoch, W.W. 1971. The development response of predators to changes in prey density. Ecology 52:132–137.

Murdoch, W.W., and A. Oaten. 1975. Predation and population stability. Adv. Ecol. Res. 9:1–132.

Murphy, G.I. 1967. Vital statistics of the Pacific sardine (*Sardinops caerulea*) and the population consequences. Ecology 48:731–736.

Murphy, G.I. 1968. Pattern in life history and the environment. Amer. Nat. 102:390–404.

Murray, J.W., H.W. Jannaoch, S. Honipo, R.F. Andexon, W.S. Reeburgh, Z. Top, G.E. Frsedeuch, L.A. Codiepoti, and E. Izdar. 1989. Unexpected changes in the oxic/anoxic interface in the Black Sea. Nature 338:411–413.

Muscatine, L. 1990. The role of symbiotic algae in carbon and energy flux in reef corals. Pp. 75–87 in Z. Dubinski (ed.). Ecosystems of the World. Vol. 25. Coral Reefs. Elsevier Publ. Amsterdam.

Muscatine, L., and R.E. Marian. 1982. Dissolved inorganic introgen flux in symbiotic and nonsymbiotic medusae. Limnol. Oceanogr. 27:910–917.

Myers, J.P., P.G. Connors, and F.A. Pitelka 1979. Territory size in wintering sanderlings: The effects of prey abundance and intruder density. Auk 96: 535–561.

Naito, Y., and M. Nishiwaki. 1972. The growth of two species of the harbour seal in the adjacent waters of Hokkaido. Sci. Rep. Whales Res. Inst. 24: 127–144.

National Academy of Sciences. 1985. Oil in the Sea. Nat. Acad. Press.

National Marine Fisheries Service. 1993. Our Living Oceans. NOAA Tech. Memo. NMFS-F/SPO-15.

National Oceanic and Atmospheric Administration. 1988. PCB and Chlorinated Pesticide Contamination in U.S. Fish and Shellfish: A Historical Assessment Report. NOAA Tech. Mamo. NOS OMA 39.

Navarro, J.M., and J.E. Winter. 1982. Ingestion rate, assimilation efficiency, and energy balance in *Mytilus chilensis* in relation to body size and different algal concentrations. Mar. Biol. 67:255–266.

Nelson, C.H. 1990. Estimated post-Messinian sediment supply and sedimentation rates on the Ebro continental margin, Spain. Mar. Geol. 95:395–418.

Nelson-Smith, A. 1968. In J.D. Carthy and D.R. Arthur (eds.), The Biological Effects of Oil Pollution on Littoral Communities. Vol. 2, Suppl. Field Studies Council, London.

Neundorfer, J.V., and W.M. Kemp. 1993. Nitrogen versus phosphorus enrichment of brackish waters: Responses of the submersed plant *Potamogeton perfoliatus* and its associated algal community. Mar. Ecol. Progr. Ser. 94: 71–82.

Neushul, M. 1971. Submarine illumination in *Macrocystis* beds. Pp. 241–254 in W.J. North (ed.), The Biology of Giant Kelp Beds (*Macrocystis*) in California. J. Cramer.

Newell, R.C., M.I. Lucas, B. Velimisov, and L.J. Seiderer. 1980. Quantitative significance of dissolved organic losses following fragmentation of kelp (*Ecklonia maxima* and *Laminaria pallida*). Mar. Ecol. Progr. Ser. 2:45–59.

Newell, R.C., M.I. Lucas, and E.A.S. Linley. 1981. Rate of degradation and efficiency of conversion of phytoplankton debris by maine microorganisms. Mar. Ecol. Progr. Ser. 6:123–136.

Newman, P., R. Stolarski, M. Schoeberl, and A. Kruger. 1991. The 1990 Antarctic ozone hole as observed by TOMS. Geophys. Res. Lett. 18:661–664.

Newman, W.S., and R.W. Fairbridge. 1986. The management of sea-level rise. Nature 320:319–321.

Nicholls, K.H., and G.J. Hopkins. 1993. Recent changes in Lake Erie (North Shore) phytoplankton: Cumulative impacts of phosphorus loading reductions and the zebra mussel introduction. J. Great Lakes Res. 637–647.

Nichols, F.H., J.K. Thompson, and L.E. Schemel. 1990. Remarkable invasion of San Francisco Bay (Californis, USA) by Asian clam *Potamocorbula amurensis*. 2. Displacement of a former community. Mar. Ecol. Progr. Ser. 66:95–101.

Nicholson, A.J. 1954. An outline of the dynamics of animal populations. Austral. J. Zool. 2:9–65.

Nicotri, M.E. 1980. Factors involved in herbivore food preference. J. Exp. Mar. Biol. Ecol. 42:13–26.

Nienhuis, P.H. 1983. Temporal and spatial patterns of eelgrass (*Zostera marina* L.) in a former estuary in the Neterlands, dominated by human activities. Mar. Technol. Soc. J. 17:69–77.

Nienhuis, P.H., and E.T. van Ierland. 1978. Consumption of eelgrass, *Zostera marina*, by birds and inertebrates during the growing season in Lake Grevelinge (SW Netherlands). Neth. J. Sea Res. 12:180–194.

Nilson, C.S., and G.R. Cresswell. 1981. The formation and evolution of East Australian Current Eddies. Progr. Oceanogr. 9:133–183.

Nissenbaum, A., and I.R. Kaplan. 1972. Chemical and isotopic evidence for the *in situ* origin of marine humic substances. Limnol. Oceanogr. 17:570–582.

Nixon, S.W. 1980. Between coastal marshes and coastal waters—A review of twenty years of speculation and research on the role of salt marshes in estuarine productivity and water diverstiy. Pp. 437–525 in R. Hamilton and K.B. MacDonald (eds.), Estuarine and Wetland Processes. Plenum.

Nixon, S.W. 1981. Remineralization and nutrient cycling in coastal marine ecosystems. Pp. 111–138 in B.J. Neilson and L.E. Cronin (eds.), Estuaries and Nutrients. Humana.

Nixon, S.W. 1986. Nutrients and the productivity of estuarine and coastal marine ecosystems. J. Limnol. Soc. South Africa 12:43–71.

Nixon, S.W. 1988. Physical energy inputs and the comparative ecology of lake and marine ecosystems. Limnol. Oceanogr. 33 (part 2):1005–1025.

Nixon, S.W. 1992. Quantifying the relationship between nitrogen input and the productivity of marine ecosystems. Pro. Adv. Mar. Sci. Conf. 5:57–83.

Nixon, S.W., and C.A. Oviatt. 1973. Ecology of a New England salt marsh. Ecol. Monogr. 43:463–498.

Nixon, S.W., C.A. Oviatt, and S.S. Hale. 1976. Nitrogen regeneration and the metabolism of coastal marine bottom communities. Pp. 269–283 in J.M. Anderson and A. MacFadyen (eds.), The Role of Terrestrial and Aquatic Organisms in Decomposition Processes. Blackwell.

Nixon, S.W., J.R. Kelly, B.N. Furnas, C.A. Oviatt, and S.S. Hale. 1980. Phosphorus regeneration and the metabolism of coastal marine bottom communities. Pp. 219–242 in K.R. Tenore and B.C. Coull (eds.), Marine Benthic Dynamics. Univ. South Carolina.

Nygaard, K., and A. Tobiesen. 1993. Bacterivory in algae: a survival strategy during nutrient limitation. Limnol. Oceanogr. 38:273–279.

O'Brien, W.J., N.A. Slade, and G.L. Vinyard. 1976. Apparent size as the determinant of prey selection by bluegill sunfish. Ecology 57:1304–1310.

O'Connors, H.B., Jr., C.B. Wurster, C.D. Powers, D.C. Biggs, and R.G. Rowland. 1978. Polychlorinated biphenyls may alter marine trophic pathways by reducing phytoplankton size and production. Science 201:737–739.

Odum, E.P. 1969. The strategy of ecosystem development. Science 164:267–270.

Odum, E.P. 1971. Fundamentals of Ecology. 3rd Ed. Saunders.

Odum, E.P., C.E. Connell, and L.B. Davenport. 1962. Population energy flow of three primary consumer components of old-field ecosystems. Ecology 43:88–96.

Odum, H.T., and E.P. Odum. 1955. Trophic structure and productivity of a windward coral reef community on Eniwetok Atoll. Ecol. Monogr. 25: 291–320.

Odum, W.E., P.W. Kirk, and J.C. Zieman. 1979. Non-protein nitrogen compounds associated with particles of vascular plant detritus. Oikos 32:363–367.

Ogawa, Y., and T. Nakahara. 1979. Interrelationships between pelagic fishes and plankton in the coastal fishing ground of the Southwestern Japan Sea. Mar. Ecol. Progr. Ser. 1:115–122.

Ogden, J.C., R.A. Brown, and N. Salesby. 1973. Grazing by the echinoid *Diadema antillarum* Philippi: Formation of halos around West Indian patch reefs. Science 182:715–717.

Ogura, N. 1972. Rate and extent of decomposition of dissolved organic matter in surface seawater. Mar. Biol. 13:89–93.

Ogura, N. 1975. Further studies on decomposition of dissolved organic matter in coastal seawater. Mar. Biol. 31:101–111.

Ohman, M.D. 1990. The demographic benefits of diel vertical migration by zooplankton. Ecol. Monogr. 60:257–281.

Olesiuk, P.F., M.A. Biggs, and G.M. Ellis. 1990. Life history and population dynamics of resident killer whales (*Orcinus orca*) in the coastal waters of British Columbia and Washington State. Rep. Int. Whal. Comm. Special issue 12:209–243.

Olson, R.J., E.R. Zettler, J. Dusenberry, and S.W. Chisholm. 1991. Advances in oceanography through flow cytometry. Pp. 351–399 in S. Demers (ed.), Particle analysis in oceanography. NATO ASI Ser. Vol. G27. Springer-Verlag.

Omori, M. 1978. Zooplankton fisheries of the world: A review. Mar. Biol. 48:199–205.

Onuf, C.P., J.M. Teal, and I. Valiela. 1977. Interactions of nutrients, plant growth, and herbivory in a mangrove ecosystem. Ecology 58:514–526.

Oremland, R.S. 1975. Methane production in shallow water tropical marine sediments. Appl. Microbiol. 30:602–608.

Oremland, R.S., and B.F. Taylor. 1977. Diurnal fluctuation of O_2, N_2 and CH_4 in the rhizosphere of *Thalassia testudinum*. Limnol. Oceanogr. 22:566–570.

Oremland, R.S., and B.F. Taylor. 1978. Sulfate reduction and methanogenesis in marine sediments. Geochim. Cosmochim. Acta 42:209–214.

Orians, G.H. 1962. Natural selection and ecological theory. Amer. Nat. 96: 257–263.

Orians, G.H. 1975. Diversity, stability and maturity in natural ecosystems. Pp. 139–150 in W.H. van Dobben and R.H. Lowe-McConnell (eds.), Unifying Concepts in Ecology. A.W. Junk.

Oritsland, T. 1977. Food Consumption of seals in the Antarctic pack ice. Pp 749–768 in G.A. Llano (ed.), Adaptations Within Antarctic Ecosystems. Proc. 3rd SCOR Symp. on Antarctic Biol. Smithsonian Inst.

Orr, A.P. 1933. Physical and chemical conditions in the sea in the neighbourhood of the Great Barrier Reef. Great Barrier Reef Exped. 1928–29, Sci. Reps. 2:37–86.

Orth, R.J. 1977. Effect of nutrient enrichment on growth of the eelgrass Zostera marina in the Chesapeake Bay, Virginia, USA. Mar. Biol. 44:187–194.

Osman, R.W., and R.B. Whitlach. 1978. Patterns of species diversity: Fact or artifact? Paleobiology 4:41–54.

Ostfeld, R.S. 1982. Foraging strategies and prey switching in the California sea otter. Oecologia 53:170–178.

Otsuki, A., and T. Hanya. 1972. Production of dissolved organic matter from dead green algal cells. I. Aerobic microbial decomposition. Limnol. Oceanogr. 17:248–257.

Otsuki, A., and T. Hanya. 1972a. Production of dissolved organic matter from dead green algal cells. II. Anaerobic microbial decomposition. Limnol. Oceanogr. 17:258–264.

Oviatt, C.A., A.A. Keller, P.A. Sampou, and L.L. Beatty. 1986a. Patterns of productivity during eutrophication: A mesocosm experiment. Mar. Ecol. Progr. Ser. 28:69–80.

Oviatt, C.A., D.T. Rudnick, A.A. Keller, P.A. Sampou, and G.T. Almquist. 1986. A comparison of system (O_2 and CO_2) and C-14 measurements of metabolism in estuarine ecosystems. Mar. ecol. Progr. Ser. 28:57–67.

Owen, D.F., and R.G. Wiegert. 1976. Do consumers maximize plant fitness? Oikos 27:488–492.

Owen, D.F., and R.G. Wiegert. 1981. Mutualism between grasses and grazers: An evolutionary hypothesis. Oikos 36:376–378.

Owen, R.W. 1989. Microscale and finescale variations of small plankton in coastal and pelagic environments. J. Mar. Res. 47:197–240.

Paasche, E. 1966. Action spectrum of coccolith formation. Physiol. Plant. 19: 770–779.

Paasche, E. 1973. Silicon and the ecology of marine plankton diatoms. I. *Thalassisira pseudonana* (*Cyclotella nana*) grows in a chemostat with silicate as the limiting nutrient. Mar. Biol. 19:117–126.

Passche, E. 1980. Silicon. Pp. 259–284 in I. Morris (ed.), The Physiological Ecology of Phytoplankton. Univ. California.

Pacala, S.W., and J. Roughgarden. 1982. An experimental investigation of the relationship between resource partitioning and interspecific competition in two two-species insular *Anolis* lizard communities. Science 217:444–446.

Pace, M.L., and E. Funke. 1991. Regulation of planktonic microbial communities by nutrients and herbivores. Ecology 72:904–914.

Pace, M.L., S. Shimmel, and W.M. Darley. 1979. The effect of grazing by a gastropod, *Nassarius obsoletus*, on the benthic microbial community of a salt marsh mudflat. Estuar. Coast. Mar. Sci. 9:121–134.

Packard, T.T., D. Blasco, J.J. MacIsaac, and R.C. Dugdale. 1971. Variations in nitrate reductase activity in marine phytoplankton. Inv. Pesq. 35:209–220.

Paerl, H.W. 1985. Enhancement of marine primary production by nitrogen enriched acid rain. Nature 315:747–749.

Paerl, H.W. 1988. Nuisance phytoplankton blooms in coastal, estuarine, and inland waters. Limnol. Oceanogr. 33:823–847.

Paerl, H.W. 1993. Emerging role of atmospheric nitrogen deposition in coastal eutrophication: Biogeochemical and trophic perspectives. Can. J. Fish Aquat. Sci. 50:2254–2269.

Paffenhöfer, G.-A. 1968. Nahrungsaufnahme, Stoffumsatz und Energiehaushalt desmanrinen Hydroidenpolypen *Clava multicornis*. Helgol. Wiss. Meeresunters. 18:1–44.

Paffenhöfer, G.-A. 1971. Grazing and ingestion rates of nauplii, copepodids, and adults of the marine planktonic copepod *Calanus helgolandicus*. Mar. Biol. 11:286–298.

Paffenhöfer, G.-A., and S.C. Knowles. 1979. Ecological implications of fecal pellet size, production and consumption by copepods. J. Mar. Res. 37: 35–49.

Paffenhöfer, G.-A., J.R. Strickler, and M. Alcaraz. 1982. Suspension-feeding by herbivorous calanoid copepods: A cinematographic study. Mar. Biol. 67: 193–199.

Paine, R.T. 1966. Food web complexity and species diversity. Amer. Nat. 100:65–75.

Paine, R.T. 1969. A note on trophic complexity and community stability. Amer. Nat. 103:91–93.

Paine, R.T. 1971. The measurement and application of the calorie to ecological problems. Ann. Rev. Ecol. Syst. 2:145–164.

Paine, R.T. 1980. Food webs: linkage, interaction strength, and community infrastructure. J. Anim. Ecol. 49:667–685.

Paine, R.T. 1988. Food webs: Roadmaps of interactions or grist for theoretical development. Ecology 69:1648–1654.

Paine, R.T. 1988a. Habitat suitability and local population persistence of the sea palm *Postelsia palmaeformis*. Ecology 69:1787–1794.

Paine, R.T., and S.A. Levin. 1981. Intertidal landscapes: Disturbance and the dynamics of pattern. Ecol. Monogr. 51:145–178.

Paine, R.T., and R.L. Vadas. 1969. The effects of grazing by sea urchins, *Strongylocentrotus* spp. on benthic, algae populations. Limnol. Oceanogr. 14:710–719.

Painter, H.W. 1970. A review of the literature on inorganic nitrogen metabolism in microorganisms. Water Res. 4:393–450.

Paloheimo, J.E. 1974. Calculation of instantaneous birth rate. Limnol. Oceanogr. 19:692–694.

Palumbi, S.R. 1992. Marine speciation on a small planet. Trends Ecol. Evol. 7:114–117.

Pamatmat, M.M. 1968. Ecology and metabolism of a benthic community on an intertidal sandflat. Int. Rev. Ges. Hydrobiol. 53:211–298.

Pamatmat, M.M. 1977. Benthic community metabolism: A review and amendment of present status and outlook. Pp. 89–111 in B.C. Coull (ed.), Ecology of Marine Benthos. Univ. South Carolina.

Pandian, T.J. 1975. Mechanisms of heterotrophy. Pp. 61–249 in O. Kinne (ed.), Marine Ecology. Vol. II, Pt. 1. Wiley.

Paranjape, M.A. 1967. Moulting and respiration of euphausiids. J. Fish. Res. Bd. Canada 24:1229–1240.

Parker, M. 1975. Similarities between the uptake of nutrients and the ingestion of prey. Verb. Internat. Verein. Limnol. 19:56–59.

Parker, P.L., E.W. Behrens, J.A. Calder, and D. Shultz. 1972. Stable carbon isotope ratio variations in the organic carbon from Gulf of Mexico sediments. Contr. Mar. Sci. 16:139–147.

Parsons, T.R. 1963. Suspended organic matter in seawater. Progr. Oceanogr. 1:205–239.

Parsons, T.R., and R.J. Le Brasseur. 1968. A discussion of some critical indices of primary and secondary production for large scale ocean surveys. Calif. Mar. Res. Comm. Calif. Coop. Oceanic Fish. Invest. Rep. 12:54–63.

Parsons, T.R., and R.J. Le Brasseur. 1970. The availability of food to different trophic levels in the marine food chain. Pp. 325–343 in J.H. Steele (ed.), Marine Food Chains. Univ. California.

Parsons, T.R., and H. Seki. 1970. Importance and general implications of organic matter in aquatic environments. Pp. 1–27 in D.W. Hood (ed.), Organic Matter in Natural Waters. Univ. Of Alaska.

Parsons, T.R., and J.D.H. Strickland. 1962. On the production of particulate organic carbon by heterotrophic processes in sea water. Deep Sea Res. 8:211–222.

Parsons, T.R., and M. Takahashi. 1974. A rebuttal to the comment by Hecky and Kilham. Limnol. Oceanogr. 19:366–368.

Parsons. T.R., R.J. Le Brasseur, and J.D. Fulton. 1967. Some observations on the dependence of zooplankton grazing on the cell size and concentration of phytoplankton blooms. J. Oceanogr. Soc. Japan 23:10–17.

Parsons, T.R., Y. Maita, and C.M. Lalli. 1984. A Manual of Chemical and Biological Methods for Seawater Analysis. Pergamon Press.

Parsons, T.R., M. Takahashi, and B. Hargrave. 1977. Biological Oceanographic Processes. 2nd Ed. Pergamon.

Patrick, R., M.H. Hohn, and J.H. Wallace. 1954. A new method for determining the pattern of the diatom flora. Not. Naturae 259:1–12.

Patterson, M.R., K.P. Sebens, and R.R. Olson. 1991. In situ measurements of flow effects on primary production and dark respiration in reef corals. Limnol. Oceanogr. 36:936–948.

Paul, E.A., and R.L. Johnson. 1977. Microscopic counting and adenosine 5'-triphosphate measurements in detormining microbial growth in soils. Appl. Environ. Microbiol. 34:263–269.

Paul, V.J., and M.E. Hay. 1986. Seaweed suaceptibility to herbivory: Chemical and morphological correlates. Mar. Ecol. Progr. Ser. 33:255–264.

Paulsen, D.M., H.W. Paerl, and P.E. Bishop. 1991. Evidence that molybdenum-dependent nitrogen fixation is not limited by high sulfate concentrations in marine environments. Limnol. Oceanogr. 36:1325–1334.

Pawlik, J.R., and C.A. Butman. 1993. Settlement of a marine worm as a function of current velocity: Interacting effects of hydrodynamlcs and behavior. Limnol. Oceanogr. 38:1730–1740.

Payne, M.R. 1977. Growth of a fur seal population. Phil. Trans. Roy. Soc. Lond. B279:67–79.

Pearcy, W.G. 1962. Ecology of young winter flounder in an estuary. Bull. Bingh. Oceanogr. Coll. 18:1–78.

Pearre, S., Jr. 1980. Feeding by chaetognatha: The relation of prey size to predator size in several species. Marine Ecology Progr. Ser. 3:125–134.

Pearse, J.S., and A.H. Hines. 1979. Expansion of a central California kelp forest following the mass mortality of sea urchins. Mar. Biol. 51:83–91.

Pearson, T.H., and R. Rosenberg. 1976. A comparative study of the effects on the marine environment of wastes from the coastal industries in Scotland and Sweden. Ambio 5:77–79.

Pearson, T.H., and R. Rosenberg. 1978. Macrobenthic succession in relation to organic enrichment and pollution of the marine environment. Oceanogr. Mar. Biol. Ann. Rev. 16:229–311.

Peckol, P., B. DeMeo-Anderson, J. Rivers, I. Valiela, M. Maldonado, and J. Yates. 1994. Growth, nutrient uptake capacities and tissue constituents of the macroalgae *Cladophora vagabunda* and *gracilaria tikvahiae* related to site specific nitrogen loading rates. Mar. Biol. 121:175–185.

Pedrós-Aliós, C. 1993. Diversity of bacterioplankton. Trends Ecol. Evol. 8: 86–90.

Pedrós-Aliós, C., J. Garcia-Cantizano, and J.I. Calderón. 1993. Bacterial production in aerobic water columns. Pp. 519–530 in P.E. Kemp., B.F. Sherr, E.B. Sherr, and J.J. Cole (eds.), Handbook of Methods in Aquatic Microbial Ecology. Lewis Publ.

Peduzzi, P., and G.J. Herndl. 1992. Zooplankton activity fueling the microbial loop: Differential growth response of bacteria from oligotrophic and eutrophic waters. Limnol. Oceanogr. 37:1087–1092.

Peer, D.L. 1970. Relation between biomass, productivity, and loss to predators in a population of a marine benthic polychaete, *Pectinaria hyperborea*. J. Fish. Res. Bd. Canada 27:2143–2153.

Peet, R.K. 1974. The measurement of species diversity. Ann. Rev. Ecol. Syst. 5:285–307.

Peña, M.A., M.R. Lewis, and W.G. Harrison. 1990. Primary productivity and size structure of phytoplankton biomass on a transect bb the equator at 135°W in the Pacific Ocean. Deep-Sea Res. 37:295–315.

Peng, T.-H., and W.S. Broeker. 1991. Factors limiting the reduction of atmospheric CO_2 by iron fertilization. Limnol. Oceanogr. 36:1919–1927.

Penhale, P.A., and W.O. Smith, Jr. 1977. Excretion of dissolved organic carbon by eelgrass (*Zostera marina*) and its epiphytes. Limnol. Oceanogr. 22:400–407.

Pennings, S.C., and V.J. Paul. 1992. Effects of plant toughness, calcification, and chemistry on herbivory by *Dolabella* Ecology 73:1606–1619.

Perry, M.J., and R.W. Eppley. 1981. Phosphate uptake by phytoplankton in the central North Pacific Ocean. Deep-Sea Res. 28:39–49.

Perry, M.J., M.C. Talbot, and R.S. Alberte. 1981. Photoadaption in marine phytoplankton: Response of the photosynthetic unit. Mar. Biol. 62:91–101.

Persson, L.-E. 1981. Were macrobenthic changes induced by thinning out of flatfish stocks in the Baltic proper? Ophelia 20:137–152.

Peterman, R.M. 1980. Testing for density-dependent marine survival in Pacific salmonids. Pp. 1–23 in W.J. McNeil and D.C. Himsworth (eds.), Salmonid Ecosystems of the North Pacific. Oregon State Univ.

Peterman, R.M., and M. Gatto. 1978. Estimation of functional responses of predators of juvenile salmon. J. Fish. Res. Bd. Canada 35:797–808.

Peters, R.H. 1976. Tautology in evolution and ecology. Amer. Nat. 110:1–12.

Peters, R.H. 1988. Some general problems for ecology illustrated by food web theory. Ecology 69:1673–1676.

Peters, R.H. 1991. A Critique for Ecology. Cambridge Univ. Press.

Peterson, B.J. 1980. Aquatic primary productivity and the ^{14}C-CO_2 method: A history of th productivity problem. Ann. Rev. Ecol. Syst. 11:359–385.

Peterson, B.J., and B. Fry. 1987. Stable isotopes in ecosystem studies. Ann. Rev. Ecol. Syst. 18:293–320.

Peterson, B.J., J.E. Hobbie, and J.F. Haney. 1978. *Daphnia* grazing on natural bacteria. Limnol. Oceanogr. 23:1039–1044.

Peterson, B.J., R.W. Howarth, and R.H. Garritt. 1985. Multiple stable isotopes used to trace the flow of organic matter in estuarine food webs. Science 227:1361–1363.

Peterson, C.H. 1979. Predation, competitive exclusion, and diversity in the softsediment benthic communities of estuaries and lagoons. Pp. 233–264 in R.J. Livingston (ed.), Ecological Processes in Coastal and Marine Systems. Plenum.

Peterson, C.H. 1992. Competition for food and its community-level implications. Benthos Res. 42:1–11.

Petipa, T.S., E.V. Pavlova, and G.N. Mironov. 1970. The food web structure, utilization and transport of energy by trophic levels in the planktonic communities. Pp. 142–167 in J.H. Steele (ed.), Marine Food Chains. Univ. California.

Pfeiffer, W.J., and R.G. Wiegert. 1981. Grazers on *Spartina* and their predators. Pp. 87–112 in L.R. Pomeroy and R.G. Wiegert (eds.), The Cology of a Salt Marsh. Springer-Verlag.

Pianka, E.R. 1966. Latitude gradients in species diversity: A review of concepts. Amer. Nat. 100:33–46.

Pianka, E.R. 1973. The structure of lizard communities. Ann. Rev. Ecol. Syst. 4:53–74.

Pickard, G.L. 1964. Descriptive Physical Oceanography: An Introduction. Pergamon.

Pielou, E.C. 1966. The measurement of diversity in different types of biological collection. J. Theor. Biol. 13:131–144.

Pielou, E.C. 1969. An Introduction to Mathematical Ecology. Wiley.

Pierce, G.G., and J.G. Ollason. 1987. Eight reasons why foraging theory is a complete waste of time. Oikos 49:111–118.

Pielou, E.C. 1977. Mathematical Ecology. Wiley.

Platt, H.M., and R.M. Warwick. 1980. The significance of freeliving nematodes to the littoral ecosystem. Pp. 729–759 in Price, J.H., D.E.G. Irvine, and W.F. Farnham (eds.), The Shore Environment. Vol. 2: Ecosystems. Academic Press.

Platt, J.R. 1964. Strong inference. Science 146:347–353.

Platt, T. 1971. The annual production by phytoplankton in St. Margaret's Bay, Nova Scotia. J. Cons. Perm. Int. Explor. Mer 33:324–333.

Platt, T. 1972. Local phytoplankton abundance and turbulence. Deep-Sea Res. 19:183–197.

Platt, T., and K.L. Denman. 1975. Spectral analysis in ecology. Ann. Rev. Ecol. Syst. 6:189–210.

Platt, T., and C.L. Gallegos. 1980. Modeling primary production. Pp. 339–362 in P. Falkowski (ed.), Primary Productivity in the Sea. Plenum.

Platt, T., and D.V. Subba Rao. 1975. Primary production of marine microphytes. Pp. 249–280 in J.P. Cooper (ed.), Photosynthesis and Productivity in Different Environments. Cambridge Univ.

Platt, T., L.M. Dickie, and R.W. Trites. 1970. Spatial heterogeneity of phytoplankton in a near-shore environment. J. Fish. Res. Bd. Canada 27:1453–1473.

Platt, T., P. Jauhari, and S. Sathyendranath. 1992. The importance and measurement of new production. Pp. 273–284 in P.G. Flakowski and A.D. Woodhead (eds.), Primary Productivity and Biogeochemical Cycles in the Sea. Plenum Press.

Poindexter, J.S. 1981. Oligotrophy: Fast and famine existence. Adv. Microbial Ecol. 5:63–89.

Polachek, T., D. Mountain, D. McMillan, W. Smith, and P. Berrien. 1992. Recruitment of the 1987 year class of Georges Bank haddock (*Melanogrammus aeglefinus*): The influence of unusual larval transport. Can. J. Fish. Aquat. Sci. 49:484–496.

Pollard, R.T. 1977. Observations and theories of Langmùir circulations and their role in near surface mixing. Pp. 235–251 in M. Angel (ed.), A Voyage of Discovery. Pergamon.

Pomeroy, L.R. 1959. Algal productivity in salt marshes of Georgia. Limnol. Oceanogr. 4:386–397.

Pomeroy, L.R. 1960. Residence times of dissolved phosphate in matural waters. Science 131:1731–1732.

Pomeroy, L.R. 1974. The ocean's food web, a changing paradigm. Bioscience 24:499–504.

Pomeroy, L.R., and D. Diebel. 1986. Temperature regulation of bacterial activity during the spring bloom in Newfoundland coastal waters. Science 233: 359–361.

Pomeroy, L.R., and R.E. Johannes. 1968. Occurrence and respiration of ultraplankton in the upper 500 m of the ocean. Deep-Sea Res. 15:381–391.

Pomeroy, L.R., and W.J. Wiebe. 1988. Energetics of microbial food webs. Hydrobiologia 159:7–18.

Pomeroy, L.R., and R.G. Wiegert (eds.). 1981. The Ecology of a Salt Marsh. Springer-Verlag.

Pomeroy, L.R., H.M. Mathews, and Hong Saik Min. 1963. Excretion of phosphate and soluble organic phosphoric compounds by zooplankton. Limnol. Oceanogr. 8:50–55.

Pomeroy, L.R., W.J. Wiebe, D. Deibel, R.J. Thompson, G.T. Rowe, and J.D. Pakulski. 1991. Bacterial responses to temperature and substrate concentration during the Newfoundland spring bloom. Mar. Ecol. Progr. Ser. 75: 143–159.

Poole, R.W. 1974. An Introduction to Quantitative Ecology. McGraw-Hill.

Porter, J.W. 1972. Predation by *Acanthaster* and its effect on coral species diversity. Amer. Nat. 106:487–492.

Porter, J.W. 1974. Community structure of coral reefs on opposite sides of the isthmus of Panama. Science 186:543–545.

Porter, J.W. 1976. Autotrophy, heterotrophy, and resource partitioning in Caribbean reef-building corals. Amer. Nat. 110:731–742.

Porter, J.W., and K.G. Porter. 1977. Quantitative sampling of demersal plankton migrating from different coral reef substrates. Limnol. Oceanogr. 22:553–556.

Porter, K.G. 1973. Selective grazing and differential digestion of algae by zooplankton. Nature 244:179–180.

Porter, K.G. 1976. Enhancement of algal growth and productivity by grazing zooplankton. Science 192:1332–1334.

Porter, K.G., and Y.S. Feig. 1980. The use of DAPI for identifying and counting aquatic microflora. Limnol. Oceanogr. 25:943–947.

Postma, H., and J.W. Rommets. 1979. Dissolved and particulate organic carbon in the North Equatorial Current of the Atlantic Ocean. Neth. J. Sea Res. 13:85–98.

Potts, M., and B.A. Whitton. 1977. Nitrogen fixation by blue-green algal communities in the intertidal zone of the lagoon of Aldabra Atoll. Oecologia 27:275–283.

Poulet, S.AA. 1978. Comparison between five coexisting species of marine copepods feeding on naturally occurring particulate matter. Limnol. Oceanogr. 23:1126–1143.

Poulet, S.A., and P. Marsot. 1980. Chemosensory feeding and food gathering by omnivorous marine copepods. Pp. 198–218 in W.C. Kerfoot (ed.), Evolution and Ecology of Zooplankton Communities. Spec. Symp. Vol. 3. Amer. Soc. Limnol. Oceanogr. Univ. Press of New England.

Power, M.E. 1992. Top-down and bottom-up forces in food webs: Do plants have primacy? Ecology 73:733–746.

Prakash, A., M.A. Rashid, A. Jensen, and D.V. Subba Rao. 1973. Influence of humic substances on the growth of marine phytoplankton: Diatoms. Limnol. Oceanogr. 18:516–524.

Prakash, A., A. Jensen, and M.A. Rashid. 1975. Humic substances and aquatic productivity. Pp. 219–268 in D. Povoledo and H.L. Golterman (eds.), Proc. Int. Meet. Humic Substances, Nieuwersluis, 1972. PUDOC.

Prakash, A., R.W. Sheldon, and W.H. Sutcliffe. 1991. Geographic variation of oceanic ^{14}C dark uptake. Limnol. Oceanogr. 36:30–39.

Pratt, D.M. 1966. Competition between *Skeletonema costatum* and *Olisthodiscus luteus* in Narragansett Bay and in culture. Limnol. Oceanogr. 11:447–455.

Precoda, N. 1991. Requiem for the Aral Sea. Ambio 20:109–114.

Preston, F.W. 1948. The commonness and rarity of species. Ecology 29:254–283.

Preston, F.W. 1962. The canonical distribution of commonness and rarity. Ecology 39:185–215.

Preston, M.R. 1992. The interchange of pollutants between the atmosphere and the ocean. Mar. Pollut. Bull. 24:477–483.

Prezelin, B., and B.M. Sweeney. 1979. Photoadaptation of photosynthesis in two bloom-forming dinoflagellates. Pp. 101–106 in D.L. Taylor and H.H. Seliger (eds.), Toxic Dinoflagellates Blooms. Elsevier North Holland.

Price, P.W., C.E. Bouton, P. Gross, B.A. McPheron, J.N. Thompson, and A.E. Weis. 1980. Interactions among three trophic levels: Influence of plants on interactions between insect herbivores and natural enemies. Ann. Rev. Ecol. Syst. 11:41–65.

Prince, J.S. 1974. Nutrient assimilation and growth of some seaweeds in mixtures of seawater and secondary sewage treatment effluents. Aquaculture 4:69–80.

Prinslow, T., I. Valiela, and J.M. Teal. 1974. The effect of detritus and ration size on the growth of *Fundulus heteroclitus* L., a salt marsh killifish. J. Exp. Mar. Biol. Ecol. 16:1–10.

Proctor, L.M., and J.A. Fuhrman. 1990. Viral mortality of marine bacteria and cyanobacteria. Nature 343:60–62.

Pulliam, R.H. 1974. On the theory of optimal diets. Amer. Nat. 108:59–74.

Purcell, J.E. 1980. Influence of siphonophore behavior upon their natural diets: Evidence for aggressive mimicry. Science 209:1045–1047.

Purcell, J.E. 1992. Effects of predation by the scyphomedusan *Chrysaora quinquecirrha* on zooplankton populations in Chesapeake Bay, USA. Mar. Ecol. Progr. Ser. 87:65–76.

Qasim, S.Z. 1970. Some problems related to the food chain in a tropical estuary. Pp. 45–51 in J.H. Steele (ed.), Marine Food Chains. Oliver and Boyd.

Qasim, S.Z. 1979. Primary production in some tropical environments. Pp. 31–69 in M.J. Dunbar (ed.), Marine Production Mechanisms. Cambridge Univ.

Quiñones, R.A., and T. Platt. 1991. The relationship between the *f*-ratio and the P:R ratio in the pelagic ecosystem. Limnol. Oceanogr. 36:211–213.

Radach, G. 1992. Ecosystem functioning in the German Bight under continental nutrient inputs by rivers. Estuaries 15:477–496.

Radovich, J., and A.D. McCall. 1979. A management model for the central stock of the Northern Anchovy, *Engraulis mordax*. Calif. Coop. Fish. Invest. Rep. 20:83–88.

Raimbault, P., M. Rodier, and I. Taupier-Letage. 1988. Size fraction of phytoplankton in the Ligurian Sea and the Algerian Basin: Size distribution versus total concentration. Mar. Microb. Food Webs 3:1–7.

Raitt, D.F.S. 1968. The population dynamics of the Norway pout in the North Sea. Dept. Agric. Fish. Scotland, Mar. Res. 24 pp.

Ramus, J. 1978. Seaweed anatomy and photosynthetic performance: The ecological significance of light guides, heterogeneous absorption and multiple scatter. J. Phycol. 14:352–362.

Rand, A.L. 1954. Social feeding behavior of birds. Fieldiana: Zoology 36:1–71.

Rasmusson, E.M. 1984. El Niño: The ocean/atmosphere connection. Oceanus 27:4–12.

Rau, G.H., T.L. Hopkins, and J.J. Torres. 1991. $^{15}N/^{14}N$ and $^{13}C/^{12}C$ in Weddell Sea invertebrates: Implications for feeding diversity. Mar. Ecol. Prog. Ser. 77:1–6.

Rau, G.H., D.G. Ainley, J.L. Bengtson, J.J. Torres, and T.L. Hopkins. 1992. $^{15}N/^{14}N$ and ^{12}C in Weddell Sea buds, seals, and fish: Implications for diet and trophic structure. Mar. Ecol. Prog. Ser. 84(1):1–8.

Rauck, G., and J.J. Zijlstra. 1978. On the nursery aspects of the Wadden sea for some commercial fish species and possible long-term changes. Rapp. Proc.-Verb. Renn. Cons. Int. Explor. Mer 172:266–275.

Raven, J.A., and K. Richardson. 1986. Marine environments. Pp. 337–396 in N.R. Baker and S.P. Long (eds.), Photosynthesis in Contrasting Environments. Elsevier Sci. Publ.

Real, L.A. 1979. Ecological determinants of functional response. Ecology 60: 481–485.

Reddy, K.R., and W.H. Patrick. 1976. Effect of frequent changes in aerobic and anaerobic conditions on redox potential and nitrogen loss in a flooded soil. Soil Biol. Biochem. 8:491–495.

Redfield, A.C. 1934. On the proportion of organic derivatives in sea water and their relation to the composition of plankton. Pp. 176–192 James Johnston Memorial Volume. Univ. Press of Liverpool.

Redfield, A.C., B.H. Ketchum, and F.A. Richards. 1963. The influence of organisms on the composition of sea-water. Pp. 26–77 in M.N. Hill (ed.), The Sea, Vol. 2. Interscience.

Reeburgh, W.S. 1976. Methane consumption in Cariaco Trench waters and sediments. Earth Planet. Sci. Lett. 28:337–344.

Reeburgh, W.S., and D.T. Heggie. 1977. Microbial methane consumption reaction and their effect on methane distributions in freshwater and marine sediments. Limnol. Oceanogr. 22:1–9.

Rees, C.P. 1975. Competitive interactions and substratum preferences of two intertidal isopods. Mar. Biol. 30:21–25.

Reeve, M.R. 1970. The biology of Chaetognatha. I. Quantitative aspects of growth and egg production in *Sagitta hispida*. Pp. 168–189 in J.H. Steele (ed.), Marine Food Chains. Univ. California.

Reeve, M.R., and M.A. Walter. 1972. Conditions of culture, food-size selection and the effects of temperature and salinity on the rate and generation drive in *Sagitta hispida* Conant. J. Exp. Mar. Biol. Ecol. 9:191–200.

Reeve, M.R., T.C. Cooper, and M.A. Walter. 1975. Visual observations on the program of digestion and the production of fecal pellet in the chaetognath *Sagitta hispida* Conant. J. Exp. Mar. Biol. Ecol. 17:39–46.

Reeve, M.R., M.A. Walter, and T. Ikeda. 1978. Laboratory studies of ingestion and food utilization in lobate and tentaculate ctenophores. Limnol. Oceanogr. 23:740–751.

Reichle, D.E., R.A. Goldstein, R.I. Van Hook, Jr., and G.J. Dodson. 1973. Analysis of insect composition in a forest canopy. Ecology 54:1077–1084.

Reiswig, H.M. 1974. Water transport, respiration, and energetics of three tropical marine sponges. J. Exp. Mar. Biol. Ecol. 14:231–249.

Rex, M.A. 1981. Community structure in the deep-sea benthos. Ann. Rev. Ecol. Syst. 12:331–353.

Rhee, G.-Y. 1972. Competition between an alga and an aquatic bacterium for phosphate. Limnol. Oceanogr. 17:505–514.

Rhee, G.-Y. 1978. Effects of N:P atonic ratio and nitrate limitation on algal growth, cell composition, and nitrate uptake. Limnol. Oceanogr. 23:10–25.

Rhoads, D.C., and L.F. Boyer. 1983. The effects of marine benthos on physical properties of sediments: A successional perspective. In M. Tevesz and P. McCall (eds.), The Biogenic Alterations of Sediment. Plenum Geobiology Series, Vol. 2. Plenum.

Rhoads, D.C., and R.A. Lutz (eds.). 1980. Skeletal Growth of Aquatic Organisms. Plenum.

Rhoads, D.C., P.L. McCall, and J.Y. Yingst. 1978. Disturbance and production on the estuarine sea floor. Amer. Sci. 66:577–586.

Rhoads, D.C., and D.K. Young. 1971. Animal-sediment relations in Cape Cod Bay, Massachusetts. II. Reworking by *Molpadia oolitica* (Holothuroidea). Mar. Biol. 11:255–261.

Rhode, K. 1992. Latitudinal gradients in species diversity: The search for the primary cause. Oikos 65:514–527.

Rice, D.L. 1982. The detritus nitrogen problem. New observations and perspectives from organic geochemistry. Mar. Ecol. Progr. Ser. 9:153–162.

Rice, E.L., and S.K. Pancholy. 1972. Inhibition of nitrification by climax ecosystems. Amer. J. Bot. 53:1033–1040.

Richards, F.A. 1968. Chemical and biological factors in the marine environment. Pp. 259–303 in J.F. Brahtz (ed.), Ocean Engineering. Wiley.

Richards, F.A. 1970. The enhanced preservation of organic matter in anoxic marine environments. Pp. 399–412 in D.W. Hood (ed.), Organic Matter in Natural Waters. Inst. Mar. Sci. Alaska Occ. Publ. 1.

Richards, F.A., and W.W. Broenkow. 1971. Chemical changes, including nitrate reduction in Darwin Bay, Galapagos Archipelago, over a 2 month period, 1969. Limnol. Oceanogr. 16:758–765.

Richards, S.W., D. Merriman, and L.H. Calhoun. 1963. Studies on the marine resources of southern New England. IX. The biology of the little skate, *Raja erinacea* Mitchell. Bull Bingham Oceanogr. Coll. 18:4–67.

Richardson, P. 1976. Gulf stream rings. Oceanus 19:65–68.

Richerson, P., R. Armstrong, and C.R. Goldman. 1970. Contemporaneous disequilibrium, a new hypothesis to explain the "paradox of the plankton." Proc. Nat. Acad. Sci. 67:1710–1714.

Richman, S., D.R. Heinle, and R. Huff. 1977. Grazing by adult estuarine calanoid copepods of the Chesapeake Bay. Mar. Biol. 42:69–84.

Rickard, D.T. 1975. Kinetics and mechanisms of pyrite formation at low temperature. Amer. J. Sci. 275:636–652.

Ricker, W.E. 1975. Composition and interpretation of biological statistics of fish populations. Bull. Fish. Res. Bd. Canada 191:1–382.

Ricklefs, R.E. 1979. Ecology. 2nd Ed. Chiron.

Rieper, M. 1979. Microautoradiographic studies on North Sea sediment bacteria. Mar. Ecol. Prog. Ser. 1:337–345.

Rijnsdorp, A.D., and P.I. van Leeuwen. 1992. Density-dependent and independent changes in somatic growth of female North Sea plaice between 1930 and 1985 as revealed by back-calculation of otoliths. Mar. Ecol. Progr. Ser. 88:19–32.

Riley, G.A. 1947. A theoretical analysis of the zooplankton populations of Georges Bank. J. Mar. Res. 6:104–113.

Riley, G.A. 1956. Oceanography of Long Island Sound, 1952–1954. IX. Production and utilization of organic matter. Bull. Bingham Oceanogr. Coll. 18:324–341.

Riley, G.A. 1956a. Review of the oceanography of Long Island Sound. Deep-Sea Res. 3 (Suppl.):224–238.

Riley, G.A. 1963. Theory of food chain relations in the ocean. Pp. 438–463 in M. N. Hill (ed.), The Sea, Vol. 2. Interscience.

Riley, J.P., and R. Chester. 1971. Introduction to Marine Chemistry. Academic.

Rinaldi, A., et al. 1991. Mucilage aggregates during 1991 in the Adriatic and Tyrrhenian seas. Red Tide Newsl. 4:2–3.

Roberts, L. 1990. Zebra mussel invasion threatens U.S. Waters. Science 249: 1370–1372.

Robin, G. de Q. 1986. Changing the sea level. Pp. 323–359 in B. Bolin, B.R. Döös, J. Jäger, and R.A. Warrick (eds.), The Greenhouse Effect, Climatic Change, and Ecosystems. Wiley & Sons.

Robinson, J.D., K.H. Mann, and J.A. Novitsky. 1982. Conversion of the particulate fraction of seaweed detritus to bacterial biomass. Limnol. Oceanogr. 27:1072–1079.

Rodríguez, J. 1994. Some comments on the size-based structural analysis of the pelagic ecosystem. Scient. Mar. 58:1–10.

Roemmich, D., and J. McGowan. 1995. Climatic warming and the decline of zooplankton in the California Current. Science 267:1324–1326.

Roesler, C.S., M.J. Perry, and K.L. Carder. 1989. Modeling in situ phytoplankton absorption from total absorption spectra in productive inland waters. Limnol. Oceanogr. 34:1510–1523.

Rogers, D.E. 1980. Density-dependent growth of Bristol Bay sockeye salmon. Pp. 257–283 in W.J. McNeil and D.C. Himsworth (eds.), Salmonid Ecosystems of the North Pacific. Oregon State Univ.

Rogers, D.J., and M.P. Hassell. 1974. General models for insect parasite and predator searching behavior: Interference. J. Anim. Ecol. 43:239–253.

Rokop, F.J. 1974. Reproductive patterns in the deep-sea benthos. Science 186: 743–745.
Rokop, F.J. 1977. Seasonal reproduction in the deep sea. Mar. Biol. 43:237–246.
Roman, M.R., and P.A. Rublee. 1980. Containment effects in copepod grazing experiments: A plea to end the black box approach. Limnol. Oceanogr. 25:982–990.
Rosenberg, R., et al. 1990. Marine eutrophication case studies in Sweden. Ambio 19:102–108.
Rosenfeld, J.K. 1979. Ammonium adsorption in near shore anoxic sediments. Limnol. Oceanogr. 24:356–364.
Rosenthal, G.A., and D.H. Janzen. 1979. Herbivores: Their interaction with secondary plant metabolites. Academic.
Roughgarden, J. 1974. Species packing and the competitive function with illustrations from coral reef fish. Theor. Pop. Biol. 5:163–186.
Roughgarden, J., S. Gaines, and H. Possingham. 1988. Recruitment dynamics in complex life cycles. Science. Wash. 241:1460–1466.
Rowe, G.T., C.H. Clifford, and K.L. Smith. 1977. Nutrient regeneration in sediments off Cap Blanc, Spanish Sahara. Deep-Sea Res. 24:57–63.
Rowe, G.T., and W.D. Gardner. 1979. Sedimentation rates in the slope water of the Northwest Atlantic Ocean measured directly with sediment traps. J. Mar. Res. 37:581–600.
Rubenstein, D.I., and M.A.R. Koehl. 1977. The mechanisms of filter feeding. Some theoretical considerations. Amer. Nat. 111:981–994.
Rudd, J.W.M., and C.D. Taylor. 1980. Methane cycling in aquatic environments. Adv. Aquat. Microbiol. 2:77–150.
Rudek, J., H.W. Pearl, M.A. Mallin, and P.W. Bates. 1991. Seasonal and hydrological control of phytoplankton nutrient limitation in the lower Neuse River Estuary, North Carolina. Mar. Ecol. Prog. Ser. 75:133–142.
Runge, J.A. 1980. Effects of hunger and season on the feeding of *Calanus pacificus*. Limnol. Oceanogr. 25:134–145.
Russell, F.S. 1934. The zooplankton. III. A comparison of the abundance of zooplankton in the barrier reef lagoon with that of some regions in northern European waters. Great Barrier Reef Exp. 1928–29. Sci. Rep. 2:176–201.
Russell, F.S. 1935. The seasonal abundance and distribution of the pelagic young of teleostean fishes caught in the righ-trawl in offshore waters in the Plymouth area. Part II. J. Mar. Biol. Assoc. U.K. 20:147–179.
Russell, F.S. 1973. A summary of the observations of the occurrence of planktonic stages of fish off Plymouth 1924–1972. J. Mar. Biol. Assoc. U.K. 53:347–355.
Russell, F.S., and J.S. Colman. 1934. The zooplankton. II. The composition of the zooplankton of the Barrier Reef Lagoon. Great Barrier Reef Exp. 1928–29. Sci. Rep. 2:159–201.
Russell-Hunter, W.D. 1970. Aquatic Productivity. McMillan.
Ryther, J.H. 1969. Photosynthesis and fish production in the sea. Science 166:72–76.
Ryther, J.H., and W.M. Dunstan. 1971. Nitrogen, phosphorus and eutrophication in the coastal marine environment. Science 171:1008–1013.
Ryther, J.H., and C.S. Yentsch. 1957. The estimation of phytoplankton production in the ocean from chlorophyll and light data. Limnol. Oceanogr. 2:281–286.

Safina, C. 1990. Bluefish mediation of foraging competition between roseate and common terms. Ecology 71:1804–1809.

Saiz, E., P. Tiselius, P.R. Jonsson, P. Verity, and G.-A. Paffenhöffer. 1993. Experimental records of the effects of food patchiness and predation on egg production of *Acartia tonsa*. Limnol. Oceanogr. 38:280–289.

Saks, N.M., and E.G. Kahn. 1979. Substrate competition between a salt marsh diatom and a bacterial population. J. Phycol. 15:17–21.

Sale, P.F. 1977. Maintenance of high diversity in coral reef fish communities. Amer. Nat. 111:337–359.

Sale, P.F. 1978. Reef fishes and other vertebrates: A comparison of social structure. Pp. 313–346 in E.S. Reese and F.J. Lighter (eds.), Contrasts in Behavior. Wiley-Interscience.

Sameoto, D.D. 1971. Life history, ecological production, and an empirical mathematical model of the population of *Sagitta elegans* in St. Margaret's Bay, Nova Scotia. J. Fish. Res. Bd. Canada 28:971–985.

Sameoto, D.P. 1973. Annual life cycle and production of the chaetognath *Sagitta elegans* in Bedford Basin, Nova Scotia. J. Fish. Res. Bd. Canada 30:333–344.

Samina, L.V. 1976. Rate and intensity of filtration in some Caspian sea bivalve molluscs. Oceanology 15:496–498.

Sammarco, P.W. 1980. *Diadema* and its relationship to coral spat mortality, grazing, competition, and biological disturbance. J. Exp. Mar. Biol. Ecol. 45:245–272.

Sampou, P., and C.A. Oviatt. 1991. A carbon budget for a eutrophic marine ecosystem and the role of sulfur metabolism in sedimentary carbon, oxygen, and energy dynamics. J. Mar. Res. 49:825–844.

Sampou, P., and C.A. Oviatt. 1991a. Seasonal patterns of sedimentary carbon and anaerobic respiration along a simulated eutrophication gradient. Mar. Ecol. Progr. Ser. 72:271–282.

Sanders, H.L. 1956. The biology of marine bottom communities. Bull Bingh. Oceanogr. Coll. Yale Univ. 15:344–414.

Sanders, H.L. 1968. Marine benthic diversity: A comparative study. Amer. Nat. 102:243–282.

Sanders, H.L. 1969. Benthic marine diversity and the stability-time hypothesis. Pp. 71–81 in G.M. Woodwell and H.H. Smith (eds.), Diversity and Stability in Ecological Systems. Brookhaven Symp. Biol. 22. Brookhaven Natl. Lab.

Sanders, H.L., J.F. Grassle, G.R. Hampson, L.S. Morse, S. Garner-Price, and C.J. Jones. 1980. Anatomy of an oil spill: Long term effects from the grounding of the barge *Florida* off West Falmouth, Mass. J. Mar. Res. 38: 265–380.

Sanders, R.W., D.A. Caron, and U.-G. Berninger. 1992. Relationships between bacteria and heterotrophic nanoplankton in marine and fresh waters: An inter-ecosystem comparison. Mar. Ecol. Progr. Ser. 86:1–14.

Sand-Jensen, K. 1975. Biomass, net production and growth dynamics in an eelgrass (*Zostera marina* L.) population in Vellerup Vig, Denmark. Ophelia 14:185–201.

Sanger, G.A. 1972. Fisheries potentials and estimated biologicla productivity of the Subarctic Pacific region. Pp. 561–574 in A.Y. Takenouti (ed.), Biological Oceanography of The Northern North Pacific Ocean. Idemitsu Shoten.

Sansone, F.J., and C.J. Martens. 1978. Methane oxidation in Cape Lookout Bight, North Carolina. Limnol. Oceanogr. 23:349–355.

Sardá, R., K. Foreman, and I. Valiela. 1995a. Macroinfauna of a Southern New England salt marsh: Seasonal dynamics and production. Mar. Biol. 121: 431–445.

Sardá, R., I. Valiela, and K. Foreman. 1995b. Life cycle, demography, and production of *Marenzelleria viridis* (Verrill, 1873) in a salt marsh of southern New England. J. Mar. Biol. Assoc. UK.

Satchell, E.R., and T.M. Farrell. 1993. Effects of settlement density on spatial arragement in four intertidal barnacles. Mar. Biol. 116(2):241–245.

Saunders, R.L. 1963. Respiration in the Atlantic Cod. J. Fish. Res. Bd. Canada 20:373–386.

Saville, A. 1978. The growth of herring in the northwestern North Sea. Rapp. Proc.-Verb. Reun. Cons. Int. Explor. Mer 172:164–117.

Schaefer, M.B. 1957. A study of the dynamics of the fishery for yellowfin tuna in the eastern tropical Pacific Ocean. Bull. Interamer. Trop. Tuna Comm. 2:245–285.

Schaefer, M.B. 1965. The potential harvest of the sea. Trans. Amer. Fish. Soc. 94:123–128.

Schaefer, M.B. 1970. Men, birds, and anchovies in the Peru Current—dynamic interactions. Trans. Amer. Fish. Soc. 99:461–467.

Schaeffer, W.M. 1974. Optimal reproductive effort in fluctuating environments. Amer. Nat. 108:783–790.

Scheer, G. 1978. Application of phytosociologic methods. Pp. 175–196 in D.R. Stoddart and R.E. Johannes (eds.), Coral Reefs: Research Methods. UNESCO.

Scheltema, R. 1967. Relationship of temperature to the larval development of *Nassarius obsoleta* (Gastropoda). Biol. Bull 132:253–255.

Schindler, D.W. 1974. Whole-lake eutrophication experiments with phosphorus, nitrogen and carbon. Int. Ver. Theor. Angew. Limnol. Verh. 19:3221–3231.

Schindler, D.W. 1977. Evolution of phosphorus limitation in lakes. Science 195:260–267.

Schlichter, D. 1980. Adaptations of enidarians for integumentary absorption of dissolved organic material. Rev. Can. Biol. 39:259–283.

Schmitt, R.J., and S.J. Holbrook. 1990. Population responses of surfperch released from competition. Ecology 71:1653–1665.

Schoenter, A., and T.W. Schoener. 1981. The dynamics of the species-area relation in marine fouling systems: 1. Biological correlates of changes in the species-area slope. Amer. Nat. 118:339–360.

Schoener, T.W. 1971. Theory of feeding strategies. Ann. Rev. Ecol. Syst. 2:369–404.

Schoener, T.W. 1974. Resource partitioning in ecological communication. Science 185:27–39.

Schoener, T.W. 1983. Field experiments on interspecific competition. Amer. Nat. 122:240–285.

Scholander, P.F., W. Flagg, W. Walters, and L. Irving. 1953. Climatic adaptation in arctic and tropical poikilotherms. Physiol. Zool. 26:67–92.

Schottler, U. 1979. On the anaerobic metabolism of three species of *Nereis* (Annelida). Mar. Ecol. Progr. Ser. 1:249–254.

Schultz, J.C., and I.T. Baldwih. 1982. Oak leaf quality declines in response to defoliation by gypsy moth larvae. Science 217:149–151.
Schwinghamer, P. 1981. Characteristic size distributions of integral benthic communities. Canad. J. Fish. Aquat. Sci. 38:1255–1263.
Scott, J.A., N.R. French, and J.W. Leatham. 1979. Patterns of consumption in grasslands. Pp. 89–105 in N. French (ed.), Perspectives in Grasslands Ecology. Ecol. Stud. 32. Springer-Verlag.
Scranton, M.I., and P.G. Brewer. 1977. Occurrence of methane in the near-surface waters of the Western subtropical North-Atlantic. Deep-Sea Res. 24:127–138.
Scranton, M.I., and J.W. Farrington. 1977. Methane production in waters off Walvis Bay. J. Geophys. Res. 82:4947–4953.
Sebens, K.P., and K. De Riemer. 1977. Diel cycles of expansion and contraction in coral reef anthozoans. Mar. Biol. 43:247–256.
Sebens, K.P., and A.S. Johnson. 1991. Effects of water movement on prey capture and distribution in reef corals. Hydobiologia 226:91–101.
Seitzinger, S.P. 1988. Denitrification in freshwater and coastal marine ecosystems: Ecological and geochemical significance. Limnol. Oceanogr. 33:702–724.
Seitzinger, S.P. 1990. Denitrification in aquatic sediments. Pp. 301–322 in N.P. Revsbech and J. Sørensen (eds.), Denitrification in Soil and Sediment. Plenum Press.
Seitzinger, S.P., and S.W. Nixon. 1985. Eutrophication and the rate of denitrification and N_2O production in coastal marine sediments. Limnol. Oceanogr. 30:1332–1339.
Seitzinger, S.P., S.W. Nixon, M. Pilson, and S. Burke. 1980. Denitrification and N_2O production in near-shore marine sediments. Geochim. Cosmochim. Acta 44:1053–1860.
Seitzinger, S.P., L.P. Nielsen, J. Caffrey, and P.B. Christensen. 1993. Denitrification measurements in aquatic sediments: A comparison of three methods. Biogeochemistry.
Seki, H. 1968. Relation between production and mineralization in Aburatsubo Inlet, Japan. J. Fish. Res. Bd. Canada 25:625–637.
Sekiguchi, H., and T. Kato. 1976. Influence of *Noctiluca* predation on the *Acartia* populations in Isle Bay, coastal Japan. J. Oceanogr. Soc. Japan 32:195–198.
Self, R.F.L., and P.A. Jumars. 1978. New resource axes for deposit feeders? J. Mar. Res. 36:627–641.
Sellner, K.G. 1981. Primary productivity and the flux of dissolved organic matter in several marine environments. Mar. Biol. 65:101–112.
Sen Gupta, R. 1973. Nitrogen and phosphorus budgets in the Baltic Sea. Mar. Chem. 1:267–280.
Sergeant, D.E. 1973. Environment and reproduction in seals. J. Reprod. Fertil. Suppl. 19:555–561.
Sfriso, A., B. Paveni, A. Marcomini, and A.A. Orio. 1992. Macroalgae, nutrient cycles, and pollutants in the Lagoon of Venice. Estuaries 15:517–528.
Shafir, A., and J.G. Field. 1980. Importance of a small carnivorous isopod in energy transfer. Mar. Ecol. Progr. Ser. 3:203–215.
Shanks, A.L., and J.D. Trent. 1980. Marine snow: Sinking rates and potential role in vertical flux. Deep-Sea Res. 27A:137–143.

Shapiro, J. 1973. Blue-green algae: Why they become dominant. Science 179: 382–384.
Sharaf El Din, S.H. 1977. Effect of the Aswan High Dam on the Nile flood and coastal circulation pattern along the Mediterranean Egyptian coast. Limmol. Oceanogr. 22:194–207.
Sharp, J.H. 1973. Size classes of organic carbon in seawater. Limnol. Oceanogr. 18:441–447.
Sharp, J.H. In press. Inputs into microbial food chains. In J. Hobbie and P.L. LeB. Williams (eds.), Heterotrophic Activity in the Sea. Plenum.
Sharp, J.H., M.J. Perry, E.H. Renger, and R.W. Eppley. 1980. Phytoplankton rate processes in the oligotrophic waters of the central North Pacific Ocean. J. Plankton Res. 2:335–353.
Shaw, D.G. 1992. The Exxon *Valdez* oil-spill: Ecological and social consequences. Envir. Cons. 19:253–258.
Shelbourn, J.E. 1962. A predator-prey size relationship for plaice larvae feeding on *Oikopleura*. J. Mar. Biol. Assoc. U.K. 42:243–252.
Shelbourn, J.E., J.R. Brett, and S. Shirata. 1973. Effect of temperature and feeding regime on the specific growth rate of sockeye salmon fry (*Oncorhynchus nerka*), with a consideration of size effect. J. Fish. Res. Bd. Canada 30:1191–1194.
Sheldon, R.W., and S.R. Kerr. 1976. The population density of monsters in Loch Ness. Limnol. Oceanogr. 21:796–979.
Sheldon, R.W., A. Prakash, and W.H. Sutcliffe, Jr. 1972. The size distribution of particles in the ocean. Limnol. Oceanogr. 17:327–340.
Sheldon, R.W., W.H. Sutcliffe, Jr., and A. Prakash. 1973. The production of particles in the surface waters of the ocean with particular reference to the Sargasso Sea. Limnol. Oceanogr. 18:719–733.
Sherr, E., and B. Sherr. 1988. Role of microbes in pelagic food webs: A revised concept. Limnol. Oceanogr. 33:1225–1227.
Short, F.T. 1987. Effects of sediment nutrients on seagrasses: Literature review and mesocosm experiment. Aquat. Bot. 27:41–57.
Shukla, S.S., J.K. Syers, J.D.H. Williams, D.E. Armstrong, and R.F. Harris. 1971. Sorption of inorganic phosphate by lake sediments. Soil Sci. Soc. Amer. Proc. 35:244–249.
Shulenberger, E., and J.L. Reid. 1981. The Pacific shallow oxygen maximum, deep chlorophyll maximum, and primary productivity, reconsidered. Deep-Sea Res. 28A:901–919.
Siebers, D. 1979. Transintegumentary uptake of dissolved amino acids in the sea star *Asterias rubens*. A reassessment of its nutritional role with special reference to the significance of heterotrophic bacteria. Mar. Ecol. Progr. Ser. 1:169–177.
Sieburth, J.McN. 1968. The influence of algal antibiosis on the ecology of marine microorganisms. Pp. 63–94 in M.R. Droop and E.J.F. Wood (eds.), Advances in Microbiology of the Sea. Academic.
Sieburth, J.McN. 1969. Studies on algal substances in the sea. III. The production of extracellular organic matter by littoral marine algae. J. Exp. Mar. Biol. Ecol. 3:290–309.
Sieburth, J.McN., and J.T. Conover. 1965. *Sargassum* tannin, an antibiotic that retards fouling. Nature 208:52–53.

Sieburth, J.McN., and A. Jensen. 1969. Studies on algal substances in the sea II. The formation of gelbstoff (humic material) by exudates of Phaeophyta. J. Exp. Mar. Biol. Ecol. 3:275–289.

Sieburth, J.McN., V. Smetacek, and J. Lenz. 1978. Pelagic ecosystem structure: Heterotrophic compartments of the plankton and their relationship to plankton size fractions. Limnol. Oceanogr. 23:1256–1263.

Siegismund, H.R. 1982. Life cycle and production of *Hydrobia ventrosa* and *H. neglecta* (Mollusca: Prosobranchia). Mar. Ecol. Progr. Ser. 7:75–82.

Silliman, R.P. 1968. Interaction of food level and exploitation in experimental fish populations. Fish. Bull. U.S. Fish Wildl. Serv. 66:425–430.

Silver, M.W., and A.L. Alldredge. 1981. Bathypelagic marine snow: Vertical transport system and deep-sea algal and detrital community. J. Mar. Res. 39:510–530.

Silver, M.W., and K.W. Bruland. 1981. Differential feeding and fecal pellet organization of salps and pteropods and the possible origin of the deep-water flux and olive-green "cells." Mar. Biol. 62:263–273.

Silverman, M.P., and H.L. Ehrlich. 1964. Microbial formation and degradation of minerals. Adv. Appl. Microbiol. 6:153–206.

Silvertown, J.W. 1982. No evolved mutualisms between grasses and grazers. Oikos 38:253–254.

Simenstad, C.A., and G.M. Caillet. 1986. Contemporary Studies in Fish Feeding. Junk Publ. 354 pp.

Simenstad, C.A., J.A. Estes, and K.W. Kenyon. 1978. Aleuts, sea otters, and alternate stable state communities. Science 200:403–411.

Simenstad, C.A., K.L. Fresh, and E.O. Salo. 1982. The role of Puget Sound and Washington coastal estuaries in the life history of Pacific salmon: An unappreciated function. Pp. 343–364 in V.S. Kennedy (ed.), Estuarine Comparisons. Academic.

Simenstad, C.A., D.O. Duggins, and P.D. Quay. 1993. High turnover of inorganic carbon in kelp habitats as a cause of $\delta^{13}C$ variability in marine food webs. Mar. Biol. 116:147–160.

Simon, M., B.C. Cho, and F. Azam. 1992. Significance of bacterial biomass in lakes and ocean: Comparison to phytoplankton biomass and biogeochemical implications. Mar. Ecol. Progr. Ser. 86:103–110.

Simpson, E.H. 1949. Measurement of diversity. Nature 163:688.

Sinclair, A.R.E. 1975. The resource limitation of trophic levels in tropical grassland ecosystems. J. Anim. Ecol. 44:497–520.

Skauen. D.M., N. Marshall, and R.J. Frajala. 1971. A liquid scintillation method for assaying ^{14}C labeled benthic microflora. J. Fish. Res. Bd. Canada 28:769–770.

Skirrow, G. 1975. The dissolved gases—carbon dioxide. Pp. 1–192 in J.P. Wiley and G. Skirrow (eds.), Chemical Oceanography, Vol. 2. Academic.

Skopintsev, B.A. 1972. A discussion of some views of the sizes, distribution, and composition of organic matter in deep ocean waters. Oceanology 12: 471–474.

Slater, J.H. 1979. Microbial population and community dynamics. Pp. 45–66 in J.M. Lynch and N.J. Poole (eds.), Microbial Ecology. Wiley.

Slichter, D. 1980. Adaptations of enidarians for integumentary absorption of dissolved organic material. Rev. Can. Biol. 39:259–282.

Slobodkin, L.B., and L. Fishelson. 1974. The effect of the cleaner-fish *Labroides dimidiatus* on the point diversity of fishes on the reef front at Eilat. Amer. Nat. 108:369–376.

Slobodkin, L.B., F.E. Smith, and N.G. Hairston. 1967. Regulation in terrestrial ecosystems, and the implied balance of nature. Amer. Nat. 101:109–124.

Small, L.F. and J.F. Hebard. 1967. Respiration of a vertically migrating marine crustacean *Euphausia pacifica* Hansen. Limnol. Oceanogr. 12:272–280.

Small, L.F., S.W. Fowles, and M.Y. Unlu. 1979. Sinking rates of natural copepod fecal pellets. Mar. Biol. 51:233–241.

Smayda, T.J. 1970. The suspension and sinking of phytoplankton in the sea. Oceanogr. Mar. Biol. Ann. Rev. 8:357–414.

Smayda, T.J. 1980. Phytoplankton species succession. Pp. 493–570 in I. Morris (ed.), The Physiological Ecology of Phytoplankton. Univ. California Press.

Smayda, T.J. 1990. Novel and nuisance phytoplankton blooms in the sea: Evidence for a global epidemic. Pp. 29–40 in E. Granéli, B. Sundström, L. Edler, and D.M. Anderson (eds.), Toxic Marine Phytoplankton. Elsevier.

Smayda, T.M. 1992. Global epidemic of noxious phytoplankton blooms and food chain consequences in large ecosystems. Pp. 275–307 in K. Sherman, L.M. Alexander, and B.D. Gold (eds.), Food Chains, Yields, Models, and Management of Large Marine Ecosystems. Westview Press.

Smedes, G.W., and L.E. Hurd. 1981. An empirical test of community stability: Resistance of a fouling community to a biological patch-forming disturbance. Ecology 62:1561–1572.

Smith, F.E. 1972. Spatial heterogeneity, stability, and diversity in ecosystems. Trans. Conn. Acad. Arts. Sci. 44:309–335.

Smith, K.L. 1973. Respiration of a sublittoral community. Ecology 54:1065–1075.

Smith, K.L., K.A. Burns, and J.M. Teal. 1972. *In situ* respiration of benthic communities in Castle Harbor, Bermuda. Mar. Biol. 12:196–199.

Smith, K.L., Jr. 1987. Food energy supply and demand: A discrepancy between particulate orgamic carbon flux and sediment community oxygen consumption in the deep ocean. Limmit. Oceanogr. 32:201–220.

Smith, K.L., Jr., and J.M. Teal. 1973a. Deep-sea benthic community respiration: An *in-situ* study at 1850 m. Science 179:282–283.

Smith, K.L., Jr., and J.M. Teal. 1973b. Temperature and pressure effects on thecosomatous pteropods. Deep-Sea Res. 20:853–858.

Smith, K.L., Jr., G.A. White, and M.B. Laver. 1979. Oxygen uptake and nutrient exchange of sediments measured *in situ* using a free vehicle grab respirometer. Deep-Sea Res. 26A:337–346.

Smith K.L., Jr. 1992. Benthic boundary layer communities and carbon cycling at abyssal depths in the central North Pacific. Limmit. Oceanogr. 37:1034–1056.

Smith, R.C., R.W. Eppley, and K.S. Baker. 1982. Correlation of primary production as measured aboard ship in southern California coastal waters and as estimated from satellite chlorophyll images. Mar. Biol. 66:281–288.

Smith, R.C., K.S. Baker, O. Holm-Hansen, and R.J. Olson. 1980. Photoinhibition of photosynthesis in natural waters. Photochem. Photobiol. 31: 585–592.

Smith, S. 1991. Stoichiometry of C:N:P fluxes in shallow-water marine ecosystems. Pp. 259–286 in J. Cole, G. Lovett, and S. Findlay (eds.), Comparative analysis of ecosystems. Ecosystems. Springer-Verlag.

Smith, S.H. 1968. Species succession and fishery exploitation in the Great Lakes. J. Fish. Res. Bd. Canada 25:667–693.

Smith, S.V. 1984. Phosphorus versus nitrogen limitation in the marine environment. Limnol. Oceanogr. 29:1149–1160.

Smith, S.V. 1988. Mass balance in coral reef-dominated areas. Pp. 209–226 in Coastal-offshore ecosystem interactions. Lecture Notes Coastal Estuarine Stud. v. 22. Springer-Verlag.

Smith, S.V. 1991. Stoichiometry of C:N:P fluxes in shallow-water marine ecosystems. Pp. 259–286 in J. Cole, G. Lovett, and S. Findlay (eds.), Analyses of Ecosystems: Patterns, Mechanisms and Theories. Springer-Verlag.

Smith, S.V., and J.T. Hollibaugh. 1993. Coastal metabolism and the oceanic organic carbon balance. Rev. Geophys. 31:75–89.

Smith, S.V., W.J. Kimmerer, E.A. Laws, R.E. Brock, and T.W. Walsh. 1981. Kaneohe Bay sewage diversion experiment: Perspectives on ecosystem respnses to nutritional perturbations. Pac. Sci. 35:279–395.

Smith, S.V., J.T. Hollibaugh, S.J. Dollar, and S. Vink. 1991. Tomales Bay metabolism: C-N-P stoichiometry and ecosystem heterotrohy at the land-sea interface. Estuar. Coast. Shelf Sci. 33:223–257.

Smith, T.J. III., and W.E. Odum. 1981. The effects of grazing by snow geese on coastal salt marshes. Ecology 62:98–106.

Smith, V.H. 1983. Low nitrogen to phosphorus ratios favor dominance by blue-green algae in lake phytoplankton. Science 221:669–671.

Smith, V.H. 1991. Competition between consumers. Limnol. Oceanogr. 36: 820–823.

Smith, W.O., R.T. Barber, and S.A. Huntsman. 1977. Primary production off the Coast of Northwest Africa: Excretion of dissolved organic matter and its heterotrophic uptake. Deep-Sea Res. 24:35–47.

Snelgrove, P.V.R., J.F. Grassle, and R.F. Petrecca. 1992. The role of food patches in maintaining high deep-sea diversity: Field experiments with hydrodynamically unbiased colonization trays. Limnol. Oceanogr. 37:1543–1550.

Solomon, M.E. 1949. The natural control of animal populations. J. Anim. Ecol. 18:1–35.

Somero, G.N. 1991. Biochemical mechanisms of cold adaptation and stenothermality in Antarctic fish. Pp. 232–247 in G. di Presco, B. Maresca, and B. Tota (eds.), Biology of Antarctic Fish. Springer-Verlag.

Somero, G.N., and P.W. Hochachka. 1976. Biochemical adaptation to temperature. Pp. 125–190 in R.C. Newell (ed.), Adaptation to Environment. Butterworths.

Sommer, U. 1988. Some size relationships in phytoplankton mobility. Hydrobiologia 161:125–131.

Sørensen, J., and B.B. Jørgensen. 1987. Early diagenesis in sediments from Danish coastal waters: Microbial activity and Mn-Fe-S geochemistry. Geochim. Cosmochim Acta 51:1583–1590.

Sorokin, Y.I. 1964. On the primary production and the bacterial activities in the Black Sea. J. Cons. Int. Explor. Mer 29:41–60.

Sotomayor, D., J.E. Corredor, and J.M. Morell. 1994. Methane flux from mangrove sediments along the southwestern coast of Puerto Rico. Estuaries 17:140–147.

Sournia, A. 1969. Cycle annual du phytoplankton, et de la production primaire dans les mers tropicales. Mar. Biol. 3:287–303.

Sournia, A. 1977. Analyse et bilan de la production primaire dans les récifs coralliens. Ann. Inst. Océanogr. 53:47–74.

Sousa, W.P. 1979. Experimental investigations of disturbance and ecological succession in a rocky intertidal algal community. Ecol. Monogr. 49:227–254.

Sousa, W.P. 1979a. Disturbance in marine intertidal boulder fields: The cross-equilibrium maintenance of species diversity. Ecology 60:1225–1239.

Sousa, W.P. 1980. The responses of a community to disturbance: The importance of successional age and species life histories. Oecologia 45:72–81.

Soutar, A., and J.D. Isaacs. 1969. History of fish populations inferred from fish scales in anaerobic sediments off California. Coop. Fish. Inv. Rep. 13:63–70.

Southward, A.J. 1991. Forty years of changes in species composition and population density of barnacles on a rocky shore near Plymouth. J. Mar. Biol. Assoc. U.K. 71:495–513.

Southward, A.J., and E.C. Southward. 1968. Uptake and incorporation of labelled glycine by pogonophores. Nature 218:875–876.

Southward, A.J., G.T. Boalch, and L. Maddock. 1988. Fluctuations in the herring and pilchard fisheries of Devon and Cornwall linked to change in climate since the 16th century. J. Mar. Biol. Assoc. U.K. 68:423–445.

Southward, G.M. 1967. Growth of Pacific halibut. Rep. Int. Pac. Halibut Comm. 43:1–40.

Southwood, T.R.E. 1978. Ecological Methods. 2nd Ed. Chapman and Hall/Wiley.

Spencer, D.W. 1981. The sediment trap intercomparison experiment: Some preliminary data. Pp. 57–104 in R.F. Anderson and M.P. Bacon (eds.), Sediment Trap Intercomparison Experiment. Woods Hole Oceanographic Institutions Techn. Memo. WHOI-1–81.

Spiro, P.A., D.J. Jacob, and J.A. Logan. 1992. Global inventory of sulfur emmissions with a $1^0 \times 1^0$ resolution. J. Geophys. Res. 97:6023–6036.

Spitzy, A., and V. Ittekot. 1991. Dissolved and particulate organic matter in rivers. Pp. 5–17 in R.F.C. Mantoura, J.-M. Martin, and R. Wollast (eds.), Ocean Margin Processes in Global Change. Wiley-Interscience.

Springer, A. 1992. Walleye pollock: How much difference do they really make? Fish. Oceanogr. 1:80–96.

Stanley, D.J., and A.G. Warne. 1993. Nile Delta: Recent geological evolution and human impact. Science 260:628–634.

Staresinic, N., G.T. Rowe, D. Shaughnessey, and A.J. Williams. 1978. Measurement of the vertical flux of particulate matter with a free drifting sediment trap. Limnol. Oceanogr. 23:559–563.

Stavn, R.H. 1971. The horizontal-vertical distribution hypothesis: Langmuir circulations and *Daphnia* distributions. Limnol. Oceanogr. 16:453–466.

Stearns, S.C. 1976. Life-history tactics: A review of the ideas. Quart. Rev. Biol. 5:3–47.

Steele, J.H. 1974a. The Structure of Marine Ecosystems. Harvard Univ.

Steele, J.H. 1974b. Spatial heterogeneity and populations stability. Nature 248:83.

Steele, J.H., and B.W. Frost. 1977. The structure of plankton communities. Phil. Trans. Roy. Soc. London 280:485–534.

Steele, J.H., and E.W. Henderson. 1977. Plankton patches in the Northern North Sea. Pp. 1–19 in J.H. Steele (ed.), Fisheries Mathematics. Academic.

Steemann Nielsen, E. 1975. Marine Photosynthesis with Special Emphasis on the Ecological Aspects. Amsterdam: Elsevier Oceanography Series 13.

Steemann Nielsen, E., and A. Jensen. 1957. Primary ocean production, the autotrophic production of organic matter in the oceans. Galathea Rep. 1:49.

Steemann Nielsen, E., and E.G. Jorgensen. 1968. The adaptation of plankton algae. III. With special consideration of the importance in nature. Physiol. Plant. 21:647–654.

Steemann Nielsen, E., and S. Wium-Anderson. 1970. Copper ions as poison in the sea and freshwater. Mar. Biol. 6:93–97.

Stevenson, J.C., L.G. Ward, and M.S. Kearney. 1988. Sediment trapping an marsh systems: Implications of tidal flux studies. Mar. Geol. 80:37–59.

Stewart, M.G. 1979. Absorption of dissolved organic nutrients by marine invertebrates. Oceanogr. Mar. Biol. Ann. Rev. 17:163–192.

Stewart, M.G. 1981. Kinetics of dipeptide uptake by the mussel *Mytilus edulis*. Comp. Biochem. Physiol. 69A:311–315.

Stewart, M.G., and R.C. Dean. 1980. Uptake and utilization of amino acids by the shipworm *Bankia gouldi*. Comp. Biochem. Physiol. 66B:443–450.

Stimson, J. 1970. Territorial behavior of the owl limpet, *Lottia gigantea*. Ecology 51:113–118.

Stimson, J. 1973. The role of territory in the ecology of the intertidal limpet *Lottia gigantea* (Gray). Ecology 54:1020–1030.

Stockner, J.G., and N.J. Antia. 1986. Algal picoplankton from marine and freshwater ecosystems: a multidisciplinary perspective. Can. J. Fish. Aquat. Sci. 43:2472–2503.

Stockwell, D.A., E.J. Buskey, and T.E. Whitledge. 1993. Studies on conditions conducive to the development and maintenance of a persistent "brown tide" in Laguna Madre, Texas. Pp. 693–698 in T.J. Smayda and Y. Shimizu (eds.), Toxic Blooms in the Sea. Elsevier.

Stoecker, D. 1980. Relationships between chemical defense and ecology in benthic ascidians. Mar. Ecol. Prog. Ser. 3:257–265.

Stoecker, D., R.R.L. Guidllard, and R.M. Kavee. 1981. Selective predation by *Favella ehrenbergii* (Tintinnia) on and among dinoflagellates. Biol. Bull. 160: 136–145.

Stoecker, D.K., and J. McDowell Capuzzo. 1990. Predation on protozoa; Its importance to zooplankton. J. Plankton Res. 12:891–908.

Stolarski, R.S., P. Bloomfield, R.D. McPeters, and J.R. Harman. 1991. Total ozone trends deduced from Nimbus 7 TOMS data. Geophys. Res. Lett. 18:1015–1018.

Stonehouse, B. (ed.). 1975. The Biology of Penguins. Univ. Park.

Stoner, A.W., and M. Ray. 1993. Aggregation dynamics in juvenile queen conch (*Strombus gigas*): Population structure, mortality, growth, and migration. Mar. Biol. 116:571–582.

Stoner, D.S. 1990. Recruitment of a tropical colonial ascidian: Relative importance of pre-settlement vs. post-settlement processes. Ecology 71:1682–1690.

Stoul, J.D., K.R. Tate, and L.F. Molloy. 1976. Decomposition processes in New Zealand soils with particular respect to rates and pathways of plant degredation. Pp. 97–144 in J.M. Anderson and A. Macfadyen (eds.), The Role of Terrestrial and Aquatic Organisms in Decomposition Processes. Blackwell.

Strickland, J.D.H. 1965. Production of organic matter in the primary stages of the oceanic food chains. Pp. 478–610 in J.B. Riley and G. Skirrow (eds.), Chemical Oceanography. Academic.

Strickland, J.D.H., and K.H. Austin. 1960. On the forms, balance, and cycle of phosphorus observed in the coastal and oceanic waters of the Northeastern Pacific. J. Fish. Res. Bd. Canada 17:337–345.

Strickland, J.D.H., and T.R. Parsons. 1968. A practical handbook of seawater analysis. Fish. Res. Bd. Canada Bull. 167:1–311.

Strickler, J.R. 1975. Intra and interspecific information flow among planktonic copepods: Receptors. Int. Ver. Theor. Angew. Limmol. Verh. 19:2951–2958.

Strickler, J.R., and A.K. Skal 1973. Setae of the first antennae of the copepod *Cyclops scutifer* (Sars.): Their structure and importance. Proc. Nat. Acad. Sci. 70:2656–2659.

Stuart, V., M.I. Lucas, and R.C. Newell. 1981. Heterotrophic utilization of particulate matter from the kelp *Laminaria pallida*. Mar. Ecol. Progr. Ser. 4:357–348.

Stuermer, D.H., and J.R. Payne. 1976. Investigation of seawater and terrestrial humic substances with carbon-13 and proton nuclear magnetic resonance. Geochim. Cosmochim. Acta 40:1109–1114.

Stumm, W., and J.J. Morgan. 1981. Aquatic Chemistry. 2nd Ed. Wiley.

Suchanek, T.H., S.L. Williams, J.C. Ogden, D.K. Hubbard, and I.P. Gill. 1985. Utilization of shallow-water seagrass detritus by Caribbean deep-sea macrofauna: $\delta^{13}C$ evidence. Deep-Sea Res. 32:201–214.

Suess, E. 1980. Particulate organic carbon flux in the oceans—surface productivity and oxygen utilization. Nature 288:260–263.

Sullivan, B.K., P.H. Doering, C.A. Oviatt, A.A. Keller, and J.B. Frithsen. 1991. Interactions with thew benthos alter pelagic food web structure in coastal waters. Can. J. Fish. Aquat. Sci. 48:2276–2284.

Sunda, W., and R.R.L. Guillard. 1976. The relationship between Cu ion activity and the toxicity of Cu to phytoplankton. J. Mar. Res. 34:511–527.

Sunda, W.G., and J.A.M. Lewis. 1978. Effect of complexation by natural organic ligands on the toxicity of copper to a unicellular algae, *Monochrysis lutheri*. Limnol. Oceanogr. 23:870–876.

Sunda, W.G., R.T. Barber, and S.A. Huntsman. 1981. Phytoplankton growth in nutrient rich seawater: Importance of copper-manganese cellular interactions. J. Mar. Res. 39:567–586.

Sundby, S., and P. Fossum. 1990. Feeding conditions of Arcto-Norwegian cod larvae compared to the Rothschild-Osborn theory on small-scale turbulence and plankton contact rates. J. Plankt. Res. 12:1153–1162.

Sushchenya, L.M. 1968. Elements of energy balance in the amphipod *Orchestia bottae* Mil.-Edw. (Amphipoda-Talitroidea). Biol. Morya 15:52–70. In Russian. Fish. Res. Bd. Canada Transl. 1225.

Sushchenya, I.M. 1970. Food rations, metabolism, and growth of crustaceans. Pp. 127–141 in J.H. Steele (ed.), Marine Food Chains. Univ. California.

Sutcliffe, W.H. 1960. On the diversity of the copepod population in the Sargasso Sea of Bermuda. Ecology 41:585–587.

Sutherland, J.P. 1974. Multiple staple points in natural communities. Amer. Nat. 108:859–873.

Sutherland, J.P. 1978. Functional roles of *Schizoporella* and *Styela* in the fouling community at Beaufort, N.C. Ecology 59:257–264.

Sutherland, J.P. 1980. Dynamics of the epibenthic community on roots of the mangrove *Rhizophora mangle,* at Bahia de Buche, Venezuela. Mar. Biol. 58: 75–84.

Sutherland, J.P., and R.H. Karlson. 1977. Development and stability of the fouling community at Beaufort, N.C. Ecol. Monogr. 47:425–446.

Suttle, C.A., and F. Chen. 1992. Mechanisms and rates of decay of marine viruses in seawater. Appl. Environ. Microbiol. 58:3721–3729.

Suttle, C.A., A.M. Chan, and M.T. Cottrell. 1990. Infection of phytoplankton by viruses and reduction of primary productivity. Nature 347:467–469.

Sverdrup, H.U. 1953. On conditions for the vernal blooming of phytolankton. J. Cons. Int. Explor. Mer 18:287–295.

Swain, T. 1979. Tannins and lignins. Pp. 657–682 in G.E. Rosenthal and D.H. Janzen (eds.), Herbivores: Their Interaction with Secondary Plant Metabolites. Academic.

Swift, M.J., O.W. Heal, and J.M. Anderson (eds.). 1979. Decomposition in Terrestrial Ecosystems. Stud. in Ecol. Vol. 5. Univ. California.

Szyper, J.P., J. Hirota, J. Caperon, and D.A. Ziemann. 1976. Nutrient regeneration by the larger net zooplankton in the southern basins of Kaneshe Bay. Pac. Sci. 30:363–372.

Taghon, G.L., R.F. Self, and P.A. Jumars. 1978. Predicting particulate selection by deposit feeders: A model and its implications. Limnol. Oceanogr. 23: 752–759.

Taguchi, S. 1976. Relationship between photosynthesis and cell size of marine diatoms. J. Phycol. 12:185–189.

Takahashi, M., and S. Ichimura. 1968. Vertical distribution and organic matter production of photosynthetic sulfur bacteria in Japanese lakes. Limnol. Oceanogr. 13:644–655.

Talbot, F.H., B.C. Russell, and G.R.V. Anderson. 1978. Coral reef fish communities: Unstable high diversity systems? Ecol. Monogr. 48:425–440.

Talling, D.F. 1960. Comparative laboratory and field studies of photosynthesis by a marine planktonic diatom. Limnol. Oceanogr. 5:62–77.

Taylor, F.J.R. 1978. Problems in the development of an explicit hypothetical phylogeny of the larval eukaryotes. Biosystems 10:67–89.

Taylor, F.R.J., N.J. Taylor, and J.R. Welsby. 1985. A bloom of the planktonic diatom, *Ceratulina pelagica,* off the coast of northeastern New Zealand in 1983, and its contribution to an associated mortality of fish and benthic fauna. Int. Rev. Ges. Hydrobiol. 70:773–795.

Taylor, L.R., R.A. Kempton, and I.P. Woiwod. 1976. Diversity statistics and the log-series model. J. Anim. Ecol. 45:255–272.

Teal, J.M. 1962. Energy flow in the salt marsh ecosystem of Georgia. Ecollogy 43:614–624.

Teal, J.M., and F.G. Carey. 1967. Effects of pressure and temperature on the respiration of euphausiids. Deep-Sea Res. 14:725–733.

Teal, J.M., and J.W. Kanwisher. 1961. Gas exchange in a Georgia salt marsh. Limnol. Oceanogr. 6:388–399.

Teal, J.M., and J.W. Kanwisher. 1966. The use of pCO_2 for the calculation of biological production, with examples from waters off Massachusetts. J. Mar. Res. 24:4–14.

Teal, J.M., I. Valiela, and D. Berlo. 1979. Nitrogen fixation by rhizosphere and freeliving bacteria in salt marsh sediments. Limnol. Oceanogr. 24:126–132.

Teal, J.M., I. Valiela, and D. Berlo. 1979. Nitrogen fixation by rhizosphere and freeliving bacteria in salt marsh sediments. Limnol. Oceanogr. 24:126–132.

Teal, J.M., J.W. Farungton, K.A. Burno, J.J. Stegerran, B.W. Jupp, B. Woodin, and C. Prininey. 1992. The West Falmouth oil spill after 20 years: Fate of fuel oil compounds and effects on animals. Mar. Pollut. Bull. 24:607–614.

Tegner, M.J., and P.K. Dayton. 1987. El Niño effects on Southern California kelp forest communities. Adv. Ecol. Res. 17:243–279.

Tempel, D., and W. Westenheide. 1980. Uptake and incorporation of dissolved amino acids by interstitial turbellaria and polychaeta and their dependence on temperature and salinity. Mar. Ecol. Progr. Ser. 3:41–50.

Tenore, K.R. 1977. Growth of *Capitella capitata* cultured on various levels of detritus derived from different sources. Limnol. Oceanogr. 22:936–941.

Tenore, K.R. 1983. What controls the availability of detritus derived from vascular plants: Organic nitrogen enrichment or caloric availability? Mar. Ecol. Progr. Ser. 10:307–309.

Tenore, K.R., and W.M. Dunstan. 1973. Comparison of feeding and biodeposition of three bivalves at different feed levels. Mar. Biol. 21:190–195.

Tenore, K.R., and D.L. Rice. 1980. A review of trophic factors affecting secondary production of deposit feeders. Pp. 325–340 in K.R. Tenore and B.C. Coull (eds.), Marine Benthic Dynamics. Univ. of South Carolina.

Tenore, K.R., J.C. Goldman, and J.P. Clarner. 1973. The food chain dynamics of the oyster, clam, and mussel in an aquaculture food chain. J. Exp. Mar. Biol. Ecol. 12:157–165.

Tenore, K.R., J.H. Tietjen, and J.J. Lee. 1977. Effects of meiofauna on incorporation of aged eelgrass detritus by the polychaete *Nepthys incisa*. J. Fish. Res. Bd. Canada 34:563–567.

Tenore, K.R., R.B. Hanson, B.E. Donseif, and C.N. Wiederhold. 1979. The effect of organic nitrogen supplement on the utilization of different sources of detritus. Limnol. Oceanogr. 84:350–355.

Tenore, K.R., L. Cammen, S.E.G. Findlay, and N. Philips. 1982. Perspectives of research on detritus: Do factors controlling the availability of detritus to macroconsumers depend on its source? J. Mar. Res. 40:473–480.

Thiel, H., D. Pfannkuche, G. Schrieker, K. Lochts, A.J. Gooday, C.H. Hemelan, R.F.G. Mante, C.M. Turley, and F. Riemann. 1988–1989. Phytodetritus on the deep-sea floor in a central oceanic region of the Northeast Atlantic. Biol. Oceanogr. 6:203–239.

Thistle, D. 1981. Natural physical disturbances and communities of marine soft bottoms. Mar. Ecol. Progr. Ser. 6:223–228.

Thomas, W.H. 1969. Phytoplankton nutrients enrichment experiments off Baja California and in the eastern equatorial Pacific Ocean. J. Fish. Res. Bd. Canada 26:1133–1145.

Thomas, W.H. 1979. Anomalous nutrient–chlorophyll interrelationships in the offshore eastern tropical Pacific Ocean. J. Mar. Res. 37:327–335.
Thomas, W.H., and A.N. Dodson. 1972. On nitrogen deficiency in tropical Pacific phytoplankton. II. Photosynthetic and cellular characteristics of a chemostat-grown diatoms. Limnol. Oceanogr. 17:515–523.
Thomas, W.H., E.H. Renger, and A.N. Dodson. 1971. Near surface organic nitrogen in the eastern tropical Pacific Ocean. Deep-Sea Res. 18:65–71.
Thompson, D.J. 1978. Towards a realistic predator-prey model: The effect of temperature on the functional response and life history of larvae of the damselfly, *Ischnura elegans*. J. Anim. Ecol. 47:757–767.
Thorson, G. 1975. Bottom communities (sublittoral or shallow shelf). Pp. 401–534 in J. Hedgepeth, Treatise on Marine Ecology and Paleoecology. Geol. Soc. Amer. Mem. 67.
Thresher, R.E., P.D. Nichols, J.S. Gunn, B.D. Bruce, and D.M. Furlani. 1992. Seagrass detritus as the basis of a coastal planktonic food chain. Limnol. Oceanogr. 37:1754–1758.
Tietjen, J.H., and J.J. Lee. 1977. Feeding behavior of marine nematodes. Pp. 21–35 in B.C. Coull (ed.), Ecology of Marine Benthos. Univ. South Carolina.
Tiews, K. 1978. On the disappearance of bluefin tuna in the North Sea and its ecological implications for herring and mackerel. Rapp. Proc.-Verb. Cons. Intern. Explor. Mer 172:301–309.
Tijssen, S.B. 1979. Diurnal oxygen rhythms and primary production in the mixed layer of the Atlantic Ocean at 20°N. Neth. J. Sea Res. 13:79–84.
Tinbergen, L. 1960. The natural control of insects in pine woods. I. Factors influencing the intensity of predation by songbirds. Arch. Neerl. Zool. 13: 265–343.
Todd, C.D. 1981. The ecology of nudibranch molluscs. Oceanogr. Mar. Biol. Ann. Rev. 19:141–234.
Tolbert, N.E. 1974. Photorespiration. Pp. 474–504 in W.D. Stewart (ed.), Algal Physiology and Biochemistry. Univ. California.
Tomosada, A. 1978. A large warm eddy detached from the Kuroshio east of Japan. Bull. Tokai. Reg. Fish. Res. Lab. 94:59–103.
Topinka, J.A., and J.V. Robbins. 1976. Effect of nitrate and ammonium enrichment on growth and nitrogen physiology in *Fucus spiralis*. Limnol. Oceanogr. 21:659–664.
Tranter, D.J., and B.S. Newell. 1963. Enrichment experiments in the Indian Ocean. Deep-Sea Res. 10:1–9.
Trathan, P.N., J. Priddle, J.L. Watkins, D.G.M. Miller, and A.W.A. Murray. 1993. Spatial variability of Antarctic krill in relation to mesoscale hydrography. Mar. Ecol. Progr. Ser. 98:61–71.
Trench, R.K. 1974. Nutritional potentials in *Zoanthus sociatus* (Coelenterata, Anthozoa). Helgol. Wiss. Meeresunt. 26:174–216.
Trench, R.K. 1979. The cell biology of plant-animal symbiosis. Ann. Rev. Pl. Physiol. 30:485–592.
Trevallion, A. 1971. Studies on *Tellina tenuis* Da Costa. III. Aspects of general biology and energy flow. J. Exp. Mar. Biol. Ecol. 7:95–122.
Trevallion, A., R.R.C. Edwards, and J.H. Steele. 1970. Dynamics of a benthic bivalve. Pp. 285–295 in J.H. Steele (ed.), Marine Food Chains, Univ. California.

Trowbridge, C.D. 1991. Diet specialization limits herbivorous sea slug's capacity to switch among food species. Ecology 72:1880–1888.
Tseytlin, B.V. 1976. Gigantism in deep-water plankton-consuming organisms. Oceanology 15:499–503.
Turner, R.E. 1978. Community plankton respiration in a salt marsh estuary and the importance of macrophytic leachates. Limnol. Oceanogr. 23:422–451.
Turner, R.E., W.W. Woo, and H.R. Jitts. 1979. Estuarine influences on a continental shelf plankton community. Science 206:218–220.
Turpin, D.H., and P.J. Harrison. 1979. Limiting mutrient patchiness and its role in phytoplankton ecology. J. Exp. Mar. Biol. Ecol. 39:151–166.
Tyler, A.V. 1970. Rates of gastric emptying in young cod. J. Fish. Res. Bd. Canada 27:1177–1189.
Tyler, P.A. 1988. Seasonality in the deep sea. Oceanogr. Mar. Biol. Ann. Rev. 26:227–258.
Ursin, E. 1973. On the prey size preferences of cod and dab. Meed. Danm. Fiskog Havunders. N.S. 7:85–98.
Valiela, I. 1983. Nitrogen in salt marsh ecosystems. Pp. 649–678 in E.J. Carpenter and D.G. Capone (eds.), Nitrogen in the Marine Environment. Academic.
Valiela, I. 1984. Marine ecological processes. Pp. 546. Springer-Verlag.
Valiela, I., and R. Buchsbaum. 1989. The role of phenolics in marine organisms and ecosystems. Pp. 23–28 in M.-F. Thompson, R. Sarojini, and R. Nagabhushanam (eds.), Bioactive Compounds From Marine Organisms. Oxford & IBH Publ. Co.
Valiela, I., and C.S. Rietsma. 1984. Nitrogen, phenolic acids, and other feeding cues for salt marsh detritivores. Oecologia 63:350–356.
Valiela, I., and J.M. Teal. 1979. The nitrogen budget of a salt marsh ecosystem. Nature 280:652–656.
Valiela, I., J.M. Teal, and N.Y. Persson. 1976. Production and dynamics of experimentally-enriched salt marsh vegetation: Below-ground biomass. Limnol. Oceanogr. 21:245–252.
Valiela, I., J.E. Wright, J.M. Teal, and S.B. Volkmann. 1977. Growth, production and energy transformations in the salt-marsh killfish *Fundulus* heteroclitus. Mar. Biol. 40:135–144.
Valiela, I., L. Koumjian, T. Swain, J.M. Teal, and J.E. Hobbie. 1979. Cinnamic acid inhibition of detritus feeding. Nature 280:55–57.
Valiela, I., B. Howes, R. Howarth, A. Giblin, K. Foreman, J.M. Teal, and J.E. Hobbie. 1982. Regulation of primary production and decomposition in a salt marsh ecosystem. Pp. 151–168 in B. Gopal, R.E. Turner, R.G. Wetzel, and D.F. Whigham (eds.), Wetlands, Ecology, and Management. Nat. Inst. of Ecol. India.
Valiela, I., J.M. Teal, C. Cogswell, J. Hartman, S. Allen, R. Van Etten, and D. Goehringer. 1985. Some long-term consequences of sewage contamination in salt marsh ecosystems. Pp. 301–316 in P.J. Godfrey, E.R. Kaynor, S. Pelczarski, and J. Benforado (eds.), Ecological Considerations in Wetlands Treatment of Municipal Waste waters. Van Nostrand Reinhold Co.
Valiela, I., J.M. Teal, S.D. Allan, R. Van Etten, D. Goehungel, and S. Volkman. 1985. Decomposition in salt marsh ecosystems: The phases and major factors affecting disappearance of above-ground organic matter. J. Exp. Mar. Biol. Ecol. 89:29–54.

Valiela, I., J. Costa, K. Foreman, J.M. Teal, B. Howes, and D. Aubrey. 1990. Transport of groundwater-borne nutrients from watersheds and their effects on coastal waters. Biogeochemistry 10:177–197.
Valiela, I., K. Foreman, M. LaMontagne, D. Hersh, J. Costa, P. Peckol, B. DeMeo-Anderson, C. D'Avanzo, M. Babione, C.-H. Sham, J. Brawley, and K. Lajtha. 1992. Couplings of watersheds and coastal waters: Sources and consequences of nutrient enrichment of Waquoit Bay, Massachussetts. Estuaries 15:433–457.
Valley, S.L. (ed.). 1965. Handbook of Geophysics and Space Environments. McGraw Hill.
Van Blaricom, G.R. 1982. Experimental analysis of structural regulation in a marine sand community exposed to oceanic swell. Ecol. Monogr. 52:283–305.
Van der Assam, J. 1967. Territory in the three-spined stickleback *Gasterosteus aculeatus* L. Behavior Suppl. 16:1–164.
Van Raalte, C.D. 1975. Epibenthic salt marsh algae: Light and nutrient limitation. Ph.D. Thesis, Boston Univ.
Van Raalte, C.D., I. Valiela, E.J. Carpenter, and J.M. Teal. 1974. Nitrogen fixation: Presence in salt marshes and inhibition by addition of combined nitrogen. Estuar. Coast. Mar. Sci. 2:301–305.
Van Raalte, C.D., I. Valiela, and J.M. Teal. 1976. Productivity of benthic algae in experimentally fertilized salt marsh plots. Limnol. Oceanogr. 21:862–872.
Van Raalte, C.D., I. Valiela, and J.M. Teal. 1976a. The effect of fertilization on the species of salt marsh diatoms. Water Res. 10:1–4.
Van Sickle, J. 1977. Mortality rates from size distribution. Oecologia 27:311–318.
Van Someren, V.G.L., and T.H.E. Jackson. 1959. Some comments on protective resemblance amongst African Lepidoptera (Rhopalocera). J. Lepidopterists Soc. 13:121–150.
Van Tamelen, P.G. 1987. Early successional mechanisms in the rocky interitdal: The rôle of direct and indirect interactions. J. Exp. Mar. Biol. Ecol. 112:39–48.
Van Valkenberg, J., K. Jones, and D.R. Heinle. 1978. A comparison by size class and volume of detritus versus phytoplankton in Chesapeake Bay. Estuar. Coast. Mar. Sci. 6:569–582.
Van Wazer, J.R. 1958. Phosphorus and Its Compounds. Vol. I. Interscience.
Van Wazer, J.R. 1973. The compounds of phosphorus. Pp. 169–178. in E.J. Griffith, A. Beeton, J.M. Spencer, and D.T. Mitchell (eds.), Environmental Phosphorus Handbook. Wiley.
Vaqué, D., and M.L. Pace. 1992. Grazing on bacteria by flagellates and cladocerans in lakes of contrasting food web structure. J. Plankt. Res. 14:307–321.
Vatova, A. 1961. Primary production in the High Venice Lagoon. J. Cons. Perm. Inter. Explor. Mer 26:148–155.
Veen, J.F. de. 1978. Changes in North Sea sole stocks (*Solea solea* (L.)). Rapp. Proc.-Verb. Cons. Int. Explor. Mer 172:124–136.
Velimirov, B., J.A. Ott, and R. Novak. 1981. Microorganisms on macrophyte debris: Biodegradation and its implication for the food web. Kieler Meeresforsch. 5:333–344.
Venrick, E.L., J.R. Beers, and J.F. Heinbokel. 1977. Possible consequences of containing microplankton for physiological rate measurements. J. Exp. Mar. Biol. Ecol. 26:55–76.

Verity, P.G. 1991. Measurement and simulation of prey uptake by marine planktonic cliates fed plastidic and aplastidic nanoplankton. Limnol. Oceanogr. 36:729–749.

Verity, P.G., and D. Stoecker. 1982. The effects of *Olisthodiscus luteus* Carter on the growth and abundance of tintinnids. Mar. Biol. 72:79–87.

Vermeij, G.J., and A.D. Covich. 1978. Coevolution of freshwater gastropods and their predators. Amer. Nat. 112:833–843.

Vernberg, W.B., and B.C. Coull. 1974. Respirations of an interstitial ciliate and benthic energy relationships. Oecologia 16:259–264.

Victor, B.C. 1983. Recruitment and population dynamics of a coral reef fish. Science 219:419–420.

Vidal, J. 1980. Physioecology of zooplankton. I. Effects of phytoplankton concentration, temperature, and body size on the growth rate of *Calanus pacificus* and *Pseudocalanus* sp. Mar. Biol. 56:111–134.

Vince, S., and I. Valiela. 1973. The effects of ammonium and phosphate enrichment on chlorophyll *a*, pigment ratio, and species composition of phytoplankton of Vineyard Sound. Mar. Biol. 19:69–73.

Vince, S., I. Valiela, N. Backus, and J.M. Teal. 1976. Predation by the salt marsh killifish *Fundulus heteroclitus* (L.) in relation to prey size and habitat structure: Consequences for prey distribution and abundance. J. Exp. Mar. Biol. Ecol. 23:255–266.

Vince, S.W., I. Valiela, and J.M. Teal. 1981. An experimental study of the structure of herbivorous insect communities in a salt marsh. Ecology 62: 1662–1678.

Vine, I. 1971. Risk of visual detection and pursuit by a predator and the selective advantage of flocking behavior. J. Theor. Biol. 30:405–428.

Vine, P.J. 1974. Effects of algal grazing and aggressive behavior of the fishes *Pomacentrus lividus* and *Acanthurus sohal* on coral-reef ecology. Mar. Biol. 24:131–136.

Vinogradov, A.P. 1953. The elementary chemical composition of marine organisms. Sears Found. Mar. Res. Mem. 2:1–647.

Vinogradov, M.E. 1970. Vertical Distribution of the Oceanic Zooplankton. Israel Program for Scientific Translations.

Vinogradov, M.E. 1972. Vertical stratification of zooplankton in the Kurile-Kamchatka trench. Pp. 333–340 in A.Y. Takenouti (ed.), Biological Oceanography of the Northern North Pacific Ocean. Idemitsu Shoten.

Virnstein, R.W. 1977. The importance of predation by crabs and fishes on benthic infauna in the Chesapeake Bay. Ecology 58:1199–1217.

Vitousek, P.M., J.R. Gosz, C.C. Grier, J.M. Melillo, and W.A. Reiners. 1982. A comparative analysis of potential nitrification and nitrate mobility in forest ecosystems. Ecol. Monogr. 52:155–177.

Volkmann-Rocco, B. and G. Fava. 1969. Two sibling species of *Tisbe* (Copepoda, harpacticoidea): *Tisbe reluctans* and *T. persimilis* n. sp. Mar. Biol. 3:159–164.

Volterra, V. 1926. Variazioni e fluttuazione del numero d'individui in specie animali conviventi. Mem. Acad. Lincei Roma 2:31–113.

Von Arx, W.S. 1962. An Introduction to Physical Oceanography. Addison-Wesley.

Vooys, C.G.N. de. 1979. Primary production in aquatic environments. Pp. 259–252 in B. Bolin, E.T. Degens, S. Kempe, and P. Ketner (eds.), the Global Carbon Cycle. Wiley.

Voytek, M.A. 1989. Ominous future under the ozone hole. Environmental Defense Fund. Washington, D.C.

Waid, J.S. (ed.). 1986. PCBs and the Environment. Vols. I–III. CRC Press.

Wainwright, S.C., M.J. Fogarty, R.C. Greenfield, and B. Fry. 1993. Long-term changes in the Georges Bank food web: Trends in stable isotopic compositions of fish scales. Mar. Biol. 115:481–493.

Wakeham, S.G., and J.W. Farrington. 1980. Hydrocarbons in contemporary aquatic sediments. Pp. 3–32 in R.A. Baker (ed.), Contaminants and Sediments. Ann Arbor. Sci. Publ.

Walsh, J.J. 1975. A spatial simulation model of the Peru upwelling ecosystem. Deep-Sea Res. 22:201–246.

Walsh, J.J. 1976. Models of the sea. Pp. 388–446 in D.H. Cushing and J.J. Walsh (eds.), The Ecology of the Seas. Blackwell.

Walsh, J.J. 1988. On the Nature of Continental Shelves. Academic Press.

Walsh, J.J., J.C. Kelley, T.E. Whitledge, J.J. Maclsaac, and S.A. Huntsman. 1974. Spin-up of the Baja California upwelling ecosystem. Limnol. Oceanogr. 19:553–572.

Wangerski, P.J., and R.R.L. Guillard. 1960. Low molecular weight organic base from the dinoflagellate *Amphidinium carteri*. Nature 185:689–690.

Ware, D.M., and R.E. Thompson. 1991. Link between long-term variability in upwelling and fish production in the Northeast Pacific Ocean. Can. J. Fish. Aquat. Sci. 48:2296–2306.

Warren, C.E., and G.E. Davis. 1967. Laboratory studies on the feeding, bioenergetics, and growth of fish. Pp. 175–214 in S.D. Gerking (ed.), The Biological Basis of Freshwater Fish Production. Blackwell.

Warwick, R.M. 1980. Population dynamics and secondary production of benthos. Pp. 1–24 in K.R. Tenore and B.C. Coull (eds.), Marine Benthic Dynamics. Univ. of South Carolina.

Warwick, R.M., and R. Price. 1975. Macrofauna production in an estuarine mudflat. J. Mar. Biol. Assoc. U.K. 55:1–18.

Warwick, R.M., and R. Price. 1979. Ecological and metabolic studies on freeliving nematodes from an estuarine mudflat. Estuar. Coast. Mar. Sci. 9:257–271.

Wassman, P. 1993. Regulation of vertical export of particulate organic matter from the euphotic zone by planktonic heterotrophs in eutrophicated aquatic environments. Mar. Pollut. Bull. 26:636–643.

Waterbury, J.B., S.W. Watson, and F. Valois. 1980. Preliminary assessment of the importance of *Synechococcus* spp. as oceanic primary producers. Pp. 516–517 in P.G. Falkowski (ed.), Primary Productivity in the Sea. Plenum.

Watermann, B., and H. Kranz. 1992. Pollution and fish diseases in the North Sea. Mar. Pollut. Bull. 24:131–138.

Watt, W.D., and F.R. Hayes. 1963. Tracer study of the phosphorus cycle in seawater. Limnol. Oceanogr. 8:276–285.

Weber, L.H., S.Z. El-Sayed, and I. Hampton. 1986. The variance spectra of phytoplankton, krill and water temperature in the Antarctic Ocean south of South Africa. Deep-Sea Res. 33:1327–1343.

Weisberg, S.B., and V.A. Lotrich. 1986. Food limitatin of a Delaware salt marsh population of the mummichog, *Fundulus heteroclitus* (L.). Oecologia 68: 168–173.

Weisse, T. 1991. The annual cycle of heterotrophic freshwater nanoflagellates: Role of bottom-up versus top-down control. J. Plankton Res. 13:167–185.

Welch, H.E. 1968. Relationships between assimilation efficiencies and growth efficiencies for aquatic consumers. Ecology 49:755–759.

Wellington, G.M. 1982a. An experimental analysis of the effects of light and zooplankton on coral formation. Oecologia 52:311–320.

Wellington, G.M. 1982b. Depth zonation of corals in the Gulf of Panama: control and facilitation by resident reef fishes. Ecol. Monogr. 52:223–241.

Werme, C. 1981. Resource Partitioning in a Salt Marsh Fish Community. Ph.D. Diss., Boston Univ.

Werner, E.E. 1977. Species packing and niche complementarity in three sunfishes. Amer. Nat. 111:553–578.

Werner, E.E., and D.J. Hall. 1976. Niche shifts in sunfishes: Experimental evidence and significance. Science 191:404–406.

Werner, I., and J.T. Hollibaugh. 1993. *Potamocurbula amurensis*: Comparisons of clearance rates and assimilation efficiencies for phytoplankton and bacterioplankton. Limnol. Oceanogr. 38:949–964.

Wetzel, R.G. 1975. Limnology. Saunders.

Wharton, W.G., and K.H. Mann. 1981. Relationship between destructive grazing by the sea urchin, *Strongylocentrotus droebachiensis*, and the abundance of American lobster, *Homarus americanus*, on the Atlantic Coast of Nova Scotia. Can. J. Fish. Aquat. Sci. 38:1339–1349.

Wheeler, P.A. 1979. Uptake of methylamine (an ammonium analogue) by *Macrocystis pyrifera* (Phaeophyta). J. Phycol. 15:12–17.

Wheeler, P.A. 1980. Use of methylammonium as an ammonium analogue in nitrogen transport and assimilation studies with *Cyclotella cryptica* (Bacillariophyceae). J. Phycol. 16:328–334.

Wheeler, P.A., and J.A. Hellebust. 1981. Uptake and concentration of alkylamines by a marine diatom. Plant Physiol. 67:367–372.

White, A.W. 1981. Sensitivity of marine fishes to toxins from red-tide dinoflagellate *Gonyaulax excavata* and implication for fish kills. Mar. Biol. 65: 255–260.

White, A.W., J. Nassif, S.E. Shumway, and D.K. Whittaker. 1993. Recent occurence of paralytic shellfish toxins in offshore shellfish in the northeastern United States. Pp. 435–440 in T.J. Smayda and Y. Shimizu (eds.), Toxic Phytoplankton Blooms in the Sea. Elsevier.

White, P.A., J. Kalff, J.B. Rasmussen, and J.M. Gasol. 1991. The effect of temperature and algal biomass on bacterial production and specific growth rate in freshwater and marine habitats. Microb. Ecol. 21:99–118.

Whitledge, T.E. 1978. Regeneration of nitrogen by zooplankton and fish in the Northwest Africa and Peru upwelling ecosystems. Pp. 90–110 in R. Boje and M. Tomczak (eds.), Upwelling Ecosystems. Springer-Verlag.

Whitledge, T.E. 1993. The nutrient and hydrographic conditions, prevailing in Laguna Madre, Texas before and during a brown tide bloom. Pp. 711–716 in T.J. Smayda and Y. Shimizu (eds.), Toxic Blooms in the Sea. Elsevier.

Whitney, D.M., A.G. Chalmers, E.B. Haines, R.B. Hanson, L.R. Pomeroy, and B. Sherr. 1981. The cycles of nitrogen and phosphorus. Pp. 163–182 in L.R. Pomeroy and R.G. Wiegert (eds.), The Ecology of a Salt Marsh. Springer-Verlag.

Whittaker, R.H. 1972. Evolution and measurement of species diversity. Taxon 21:213–251.

Whittaker, R.H. (ed.). 1973. Ordination and Classification of Communities. Junk. b.v.-Publishers.

Whittaker, R.H., and D. Goodman. 1979. Classifying species according to their demographic strategy. I. Population fluctuations and environmental heterogeneity. Amer. Nat. 113:185–200.

Whittaker, R.H., and G.E. Likens 1975. The biosphere and man. Pp. 305–328 in H. Lieth and R.H. Whittaker (eds.), Primary Productivity of the Biosphere. Springer-Verlag.

Whittle, K.J. 1977. Marine organisms and their contribution to organic matter in the oceans. Mar. Chem. 5:381–411.

Wickett, W.P. 1967. Ekman transport and zooplankton concentrations in the North Pacific Ocean. J. Fish. Res. Bd. Canada 24:581–594.

Wiebe, P.H. 1970. Small-scale spatial distribution in oceanic zooplankton. Limnol. Oceanogr. 15:205–217.

Wiebe, P.H. 1982. Rings of the Gulf Stream. Scientific Amer. 246:60–70.

Wiebe, P.H., S.H. Boyd, and C. Winget. 1976. Particulate matter sinking to the deep sea floor at 2,000 m in the Tongue of the Ocean, Bahamas, with a description of a new sedimentation trap. J. Mar. Res. 34:341–354.

Wiebe, P.H., E.M. Hulbert, E.J. Carpenter, A.E. Jahn, G.P. Knapp, S.H. Boyd, P.B. Ortner, and J.L. Cox. 1976a. Gulf stream cold core rings: Large scale interaction sites for open ocean plankton communities. Deep-Sea Res. 23:695–710.

Wiebe, W.T., and D.F. Smith. 1977. Direct measurement of dissolved organic carbon release by phytoplankton and incorporation by microheterotrophs. Mar. Biol. 42:213–223.

Wiebe, W.J., W.M. Sheldon, Jr., and L.R. Pomeroy. 1992 Bacterial growth in the cold: evidence for an enhanced substrate requirement. Appl. Envir. Microbiol. 58:359–364.

Wiegert, R.G., and F.C. Evans. 1964. Primary production and disappearance of dead vegetation on an old field in southeastern Michigan. Ecology 45: 49–63.

Wiegert, R.G., and F.C. Evans. 1967. Investigations of secondary productivity in grasslands. Pp. 499–518 in K. Petrusewicz (ed.), Secondary Productivity of Terrestrial Ecosystems. Polish Acad. Sci.

Wiegert, R.G., and D.F. Owen. 1971. Trophic structure, available resources, and population density in terrestrial vs aquatic ecosystems. J. Theoret. Biol. 30:69–81.

Wiens, J.A. 1976. Population responses to patchy environments. Ann. Rev. Ecol. Syst. 7:81–120.

Wigley, T.M.L., P.D. Jones, and P.M. Kelly. 1986. Empirical climate studies. Pp. 271–322 in B. Bolin, B.R. Döös, J. Jäger, and R.A. Warrick (eds.), The Greenhouse Effect, Climatic Change, and Ecosystems. Wiley & Sons.

Williams, D.McB. 1980. Dynamics of the ponacentrid community on small patch reefs in One Tree Lagoon (Great Barrier Reef). Bull. Mar. Sci. 30: 159–170.

Williams, P.J. Le B. 1975. Biological and chemical dissolved organic material in sea water. Pp. 301–364 in G. Riley and O. Skirrow (eds.), Chemical Oceanography, Vol. 2. 2nd Ed.

Williams, P.J. Le B. 1981. Incorporation of microheterotrophic processes into the classical paradigm of the planktonic food web. Kieler Meeresforsch. Suppl. 5:1–28.

Williams, R.B., and M.B. Murdoch. 1972. Compartmental analysis of the production of *Juncus roemerianus* in a North Carolina salt marsh. Chesap. Sci. 13:69–79.

Williams, R.B., M.N. Murdoch, and L.K. Thomas. 1968. Standing crop and importance of zooplankton in a system of shallow estuaries. Chesap. Sci. 9:42–51.

Williams, S.J., and M.H. Ruckelshaus. 1993. Effects of nitrogen availability and herbivory on eelgrass (*Zostera marina*) and epiphytes. Ecology 74:904–918.

Williams, W.T., G.N. Lance, L.J. Webb, J.G. Tracey, and M.B. Dale. 1969. Studies in the numerical analysis of complex rain-forest communities. III. The analysis of successional data. J. Ecol. 57:515–535.

Williamson, P., and P.A. Holligan. 1990. Ocean productivity and climate change. Trends Ecol. Evol. 5:299–303.

Wilson, E.O., and W.H. Bossert 1971. A Primer of Population Biology. Sinauer.

Wilson, W.H. 1980. A laboratory investigation of the effects of a terebellid polychaete on the survivorship of nereid polychaete larvae. J. Exp. Mar. Biol. Ecol. 46:73–80.

Wiltse, W.I., K.H. Foreman, J.M. Teal, and I. Valiela. 1984. Importance of predators and food resources to the macrobenthos of salt marsh creeks. J. Mar. Res. 42:923–942.

Windom, H.L. 1992. Contamination of the marine environment from land-based sources. Mar. Pollut. Bull. 25:32–36.

Winter, D.F., K. Banse, and G.C. Anderson. 1975. The dynamics of Phytoplankton blooms in Puget Sound, a fjord in the Northwestern United States. Mar. Biol. 29:139–176.

Winter, J.E. 1978. A review of the knowledge of suspension-feeding in lamellibranchiate bivalves, with special reference to artificial aquaculture systems. Aquaculture 13:1–33.

Winters, G.H. 1976. Recruitment mechanisms of Southern Gulf of St. Lawrence Atlanctic herring (*Clupea harengus harengus*). J. Fish. Res. Bd. Canada 33: 1751–1763.

Wishner, K.F. 1980. the biomass of the deep-sea benthopelagic plankton. Deep-Sea Res. 27A:203–216.

Witman, J.D. 1987. Subtidal coexistence: Storms, grazing, mutualism, and the zonation of kelps and mussels. Ecol. Monogr. 57:167–187.

Witman, J.D., and K.P. Sebens. 1992. Regional variation in fish predation intensity: A historical perspecitive in the Gulf of Maine. Oecologia 90:305–315.

Wohlschlag, D.E. 1964. Respiratory metabolism and ecological characteristics of some fishes in McMurdo Sound, Antarctica. Antarctic Res. Ser. 1:33–62.

Wolfe, G.V., T.S. Bates, and R.J. Charlson. 1991. Climatic and environmental implications of biogas exchanges at the sea surface: Modeling DMS and the marine biologic sulfur cycle. Pp. 383–400 in R.F.C. Mantoura, J.-M. Martin, and R. Wollast (eds.), Ocean Margin Processes in Global Change. Wiley-Interscience.

Wolff, T. 1976. Utilization of seagrass in the deep sea. Aquat. Bot. 2:161–174.

Womersley, H.B.S., and R.E. Norris. 1959. A free-floating marine red algae. Nature 184:828–829.

Wood, E.J.F., W.E. Odum, and J.C. Zieman. 1967. Influence of seagrasses on the productivity of coastal lagoons. Pp. 495–502 in Lagunas Costeras, un Symposio. Mem. Simp. Intern. Lagunas Costeras. UNAM-UNESCO, Mexico D.F.

Woodin, S.A. 1977. Algal "gardening" behavior by nereid polychaetes: Effects on soft-bottom community structure. Mar. Biol. 44:39–42.

Woddin, S.A. 1978. Refuges, disturbance, and community structure: A marine soft bottom example. Ecology 59:274–284.

Wooton, J.T. 1991. Direct and indirect effects of nutrients on intertidal community structure: Variable consequences of seabird guano. J. Exp. Mar. Biol. Ecol. 151:139–153.

Work, T.M., et al. 1993. Domoic acid intoxication of brown pelicans and cormorants in Santa Cruz, California. Pp. 643–649 in T.J. Smayda and Y. Shimizu (eds.), Toxic Phytoplankton Blooms in the Sea. Elsevier.

Worrest, R.C., and D.-P. Häder. 1989. Effects of stratospheric ozone depletion on marine organisms. Envir. Conserv. 16:261–263.

Wright, R.T. In press. Dynamics of pools of dissolved organic carbon. In J.E. Hobbie and P.J. Le B. Williams (eds.), Heterotrophic Activity in the Sea. Plenum.

Wright, R.T., and J.E. Hobbie. 1965. The uptake of organic solutes in lake water. Limnol. Oceanogr. 10:22–28.

Wroblewski, J.S. 1984. Formulation of growth and mortality of larval northern anchovy in a tubulent feeding environment. Mar. Ecol. Progr. Ser. 20:13–2.

Wroblewski, J.S., and J.J. O'Brien. 1976. A spatial model of phytoplankton patchiness. Mar. Biol. 35:161–175.

Wroblewski, J.S., J.J. O'Brien, and T. Platt. 1975. On the physical and biological scales of phytoplankton patchiness in the ocean. Mem. Soc. R. Sci. Liege 7:43–57.

Wroblewski, J.S., J.G. Richman, and G.L. Mellor. 1989. Optimal wind conditions for the survival of larval northern anchovy, *Engraulis mordax*: a modelng investigation. Fish. Bull. U.S. 87:387–395.

Wyatt, T. 1973. The biology of *Oikopleura dioica* and *Fritillaria borealis* in the Southern Bight. Mar. Biol. 22:137–158.

Wylie, J.L., and D.J. Currie. 1991. The relative importance of bacteria and algae as food sources for crustacean zooplankton. Limnol. Oceanogr. 36: 708–728.

Wyman, M., R.P.F. Gregory, and N.G. Carr. 1985. Novel role for phycoerythrin in a marine cyanobacterium, *Synechococcus* strain DC2. Science 230: 818–820.

Wyrtki, K. 1975. El Niño: The dynamic response of the Equatorial Pacific Ocean to atmospheric forcing. J. Phys. Oceanogr. 5:572–584.

Wyrtki, K. 1990. Sea level rise: The facts and the future. Pac. Sci. 44:1–16.

Yamaguchi, M. 1975. Estimating growth parameters from growth rate dates. Oecologia 20:321–332.

Ydenberg, R.C., and A.I. Houston. 1986. Optimal tradeoffs between competing behavioral demands in the Great Tit. Anim. Behav. 34:1041–1050.

Yellin, M.B., C.R. Agegian, and J.S. Pearse. 1977. Ecological benchmarks in the Santa Cruz County kelp forests before the re-establishment of sea otters. Center for Coastal Marine Studjes. Univ. Calif., Santa Cruz.

Yentsch, C.S. 1967. The measurement of chloroplastic pigments—thirty year progress? Pp. 255–270 in H.L. Golterman and R.J. Clymo (eds.), Chemical Environment in the Aquatic Habitat. N.V. Noord-Hollandsche Uitgevers Maatschappij.

Yentshc, C.S. 1980. Light attenuation and phytoplankton photosynthesis. Pp. 95–127 in I. Morris (ed.), The Physiological Ecology of Phytoplankton. Univ. California.

Yentsch, C.M., C.S. Yentsch, and L.R. Strube. 1977. Variations in ammonium enhancement, an indication of nitrogen deficiency in New England coastal phytoplankton populations. J. Mar. Res. 35:537–555.

Yingst, J.Y. 1976. The utilization of organic matter in shallow marine sediments by an epibentic deposit-feeding holothurian. J. Exp. Mar. Biol. Ecol. 23:55–69.

Yingst, J.Y., and D.C. Rhoads. 1980. The role of bioturbation in the enhancement of bacterial growth rates in marine sediments. Pp. 407–421 in K.R. Tenore and B.C. Coull (eds.), Marine Benthic Dynamics. Univ. South Carolina.

Yong, D.K. 1971. Effects of infauna on the sediment and seston of a subtidal environment. Vie Mil. Suppl. 22:557–571.

Young, D.K., and M.W. Young. 1978. Regulation of species densities of seagrass-associated macrobenthos: evidence from field experiments in the Indian River estuary, Florida. J. Mar. Res. 36:591–604.

Zach, R., and J.N.M. Smith. 1981. Optimal foraging in wild birds? Pp. 95–110 in A.C. Kamil and T.D. Sargent (eds.), Foraging Behavior. Garland STPM.

Zaika, V.E., and N.P. Makarova. 1979. Specific production of free-living marine nematodes. Mar. Ecol. Progr. Sci. 1:153–158.

Zaitzev, Yu.P. 1992. Recent changes in the trophic structure of the Black Sea. Fish. Oceanogr. 1:180–189.

Zaret, T.M., and T.R. Paine. 1973. Species introduction in a tropical lake. Science 182:449–455.

Zeitzschel, B. (ed.). 1973. The Biology of the Indian Ocean. Springer-Verlag.

Zeitzschel, B. 1980. Sediment-water interaction in nutrient dynamics. Pp. 195–18 in K.R. Tenore and B.C. Coull (eds.), Marine Benthic Dynamics. Univ. South Carolina.

Zenkevitch, L. 1963. Biology of the Seas of the USSR. Interscience.

Zimmerman, A.R., and R. Benner. 1994. Denitrification, nutrient regeneration, and carbon mineralization in sediments of Galveston Bay. Texas. Mar. Ecol. Progr. Ser. 114:275–301.

Zimmerman, R.C., and J.N. Kremer. 1984. Episodic nutrient supply to a kelp forest ecosystem in Southern California. J. Mar. Res. 42:591–604.

Zimmerman, R.C., R.D. Smith, and R.S. Alberte. 1987. Is growth of eelgrass nitrogen limited? A numerical simulation of the effects of light and nitrogen on the growth dynamics of *Zostera marina*. Mar. Ecol. Progr. Ser. 41:167–176.

Zuberer, D.A., and W.S. Silver. 1978. Biological dinitrogen fixation (acetylena reduction) associated with Florida mangroves. Appl. Env. Microbiol. 35: 567–575.

Zuta, S., J. Rivera, and A. Bustamante. 1978. Hydrologic aspects of the main upwelling areas off Peru. Pp. 235–257 in R. Boje and M. Tomczak (eds.), Upwelling Ecosystems. Springer-Verlag.

Zwarts, L., and J.H. Wanink. 1993. How the food supply harvestable by waders in the Wadden Sea depends on the variation in energy density, body weight, biomass, burying depth and behaviour of tidal-flat invertebrates. Neth. J. Sea Res. 31:441–476.

Index

Q

R

S

Z